Statistics in Engineering
With Examples in MATLAB® and R
Second Edition

CHAPMAN & HALL/CRC

Texts in Statistical Science Series

Joseph K. Blitzstein, *Harvard University, USA*
Julian J. Faraway, *University of Bath, UK*
Martin Tanner, *Northwestern University, USA*
Jim Zidek, *University of British Columbia, Canada*

Recently Published Titles

Extending the Linear Model with R
Generalized Linear, Mixed Effects and Nonparametric Regression Models, Second Edition
J.J. Faraway

Modeling and Analysis of Stochastic Systems, Third Edition
V.G. Kulkarni

Pragmatics of Uncertainty
J.B. Kadane

Stochastic Processes
From Applications to Theory
P.D Moral and S. Penev

Modern Data Science with R
B.S. Baumer, D.T Kaplan, and N.J. Horton

Generalized Additive Models
An Introduction with R, Second Edition
S. Wood

Design of Experiments
An Introduction Based on Linear Models
Max Morris

Introduction to Statistical Methods for Financial Models
T. A. Severini

Statistical Regression and Classification
From Linear Models to Machine Learning
Norman Matloff

Introduction to Functional Data Analysis
Piotr Kokoszka and Matthew Reimherr

Stochastic Processes
An Introduction, Third Edition
P.W. Jones and P. Smith

Theory of Stochastic Objects
Probability, Stochastic Processes and Inference
Athanasios Christou Micheas

Linear Models and the Relevant Distributions and Matrix Algebra
David A. Harville

An Introduction to Generalized Linear Models, Fourth Edition
Annette J. Dobson and Adrian G. Barnett

Graphics for Statistics and Data Analysis with R
Kevin J. Keen

Statistics in Engineering
With Examples in MATLAB and R, Second Edition
Andrew V. Metcalfe, David A. Green, Tony Greenfield, Mahayaudin M. Mansor, Andrew Smith, and Jonathan Tuke

For more information about this series, please visit: https://www.crcpress.com/go/textsseries

Statistics in Engineering

With Examples in MATLAB® and R

Second Edition

Andrew Metcalfe
David Green
Tony Greenfield
Mahayaudin Mansor
Andrew Smith
Jonathan Tuke

CRC Press
Taylor & Francis Group
Boca Raton London New York

CRC Press is an imprint of the
Taylor & Francis Group, an **informa** business

A CHAPMAN & HALL BOOK

CRC Press
Taylor & Francis Group
6000 Broken Sound Parkway NW, Suite 300
Boca Raton, FL 33487-2742

First issued in paperback 2020

© 2019 by Taylor & Francis Group, LLC
CRC Press is an imprint of Taylor & Francis Group, an Informa business

No claim to original U.S. Government works

Version Date: 20191213

ISBN 13: 978-0-367-57062-0 (pbk)
ISBN 13: 978-1-4398-9547-4 (hbk)

Visit the Taylor & Francis Web site at
http://www.taylorandfrancis.com

and the CRC Press Web site at
http://www.crcpress.com

Contents

Preface

Engineering is a wide ranging discipline with a common theme of mathematical modeling. Probability and statistics is the field of mathematics that deals with uncertainty and variation, features that are part of every engineering project. Engineering applications have provided inspiration for the development of mathematics, and just a few examples from the last century alone are Shannon's theory of communication, Shewhart's focus on the improvement of industrial processes, Wiener's contribution to signal processing and robotics, and Gumbel's research into extreme values in hydrology.

We aim to motivate students of engineering by demonstrating that probability and statistics are an essential part of engineering design, enabling engineers to assess performance and to quantify risk. Uncertainties include variation in raw materials, variation in manufacturing processes, and the volatile natural environment within which engineered structures operate. Our emphasis is on modeling and simulation. Engineering students generally have a good mathematical aptitude and we present the mathematical basis for statistical methods in a succinct manner, that in places assumes a knowledge of elementary calculus. More mathematical detail is given in the appendices. We rely on a large number of data sets that have either been provided by companies or are available from data archives maintained by various organizations. We appreciate the permission to use these data and thank those involved. All of these data sets are available on the book website.

A feature of the book is the emphasis on stochastic simulation, enabled by the generation of pseudo-random numbers. The principal reason for this emphasis is that engineers will generally perform stochastic simulation studies as part of their design work, but it has other advantages. Stochastic simulation provides enlightening demonstrations of the concept of sampling distributions, which is central to statistical analysis but unfamiliar to students beginning the subject. Stochastic simulation is also part of modern statistical analysis, enabling bootstrap methods and Bayesian analyses.

The first eight chapters are designed to be read in sequence, although chapters two and three could be covered in reverse order. The following six chapters cover: multiple regression; the design of experiments; statistical quality control; probability models; and sampling strategies. We use the multiple regression model to introduce the important topics of time series analysis and the design of experiments.

The emphasis on stochastic simulation does not diminish the importance of physical experiments and investigations. We include an appendix of ten experiments that we have used with classes of students, alternated with computer based practical classes. These simple experiments, for example descent times of paper helicopters with paperclip payloads and cycles to failure when bending paperclips, offer scope for discussing principles of good experimentation and observing how well mathematical models work in practice. They also provide an opportunity for students to work together in small groups, which have suggested intriguing designs for paper helicopters.

The choice of the mathematical software MATLAB® and R rather than a statistical package provides a good combination of an interactive environment for numerical computation, statistical analysis, tools for visualization, and facility for programming and simulation. In particular, MATLAB is very widely used in the engineering community for design and simulation work and has a wide range of inbuilt statistical functions. The R software has similar capabilities, and has the potential advantage of being open source. It too has a wide range of inbuilt statistical functions augmented with hundreds of specialist packages that are available on the CRAN website. Many advances in statistical methodology are accompanied by new packages written in R.

The exercises at the end of each chapter are an essential part of the text, and are organized by targeting each section within the chapter and followed by more general exercises. The exercises fall into three categories: routine practice of the ideas presented; additions to the explanatory material in the chapter including details of derivations, special cases and counter examples; and extensions of the material in the chapter. Additional exercises, and solutions including code are given to odd numbered exercises on the website.

We thank John Kimmel for his generous support and encouragement. We are also grateful to several anonymous reviewers for their helpful comments and constructive advice.

Andrew Metcalfe
David Green
Tony Greenfield
Mahayaudin Mansor
Andrew Smith
Jonathan Tuke

MATLAB® is a registered trademark of The MathWorks, Inc. For product information please contact:
The MathWorks, Inc.
3 Apple Hill Drive
Natick, MA, 01760-2098 USA
Tel: 508-647-7000
Fax: 508-647-7001
E-mail: info@mathworks.com
Web: www.mathworks.com

1

Why understand statistics?

Engineers need to take account of the uncertainty in the environment and to assess how engineered products will perform under extreme conditions. They have to contend with errors in measurement and signals that are corrupted by noise, and to allow for variation in raw materials and components from suppliers. Probability and statistics enable engineers to model and to quantify uncertainty and to make appropriate allowances for it.

1.1 Introduction

The Voyager 1 and Voyager 2 spacecraft were launched from Cape Canaveral in 1977, taking advantage of a favorable alignment of the outer planets in the solar system. Thirty five years later Voyager 1 entered interstellar space traveling "further than anyone or anything in history" [The Times, 2017]. The trajectory of Voyager 2 included flybys of Jupiter, Saturn, Uranus and Neptune and the spacecraft is now in the heliosheath where the solar wind is compressed and turbulent. The robotic spacecraft have control systems that keep their high gain antennas pointing towards the earth. They have the potential to transmit scientific data until around 2020 when the radioisotope thermoelectric generators will no longer provide sufficient power.

The work of renowned engineers such as Rudolf Kalman and Norbert Weiner in electrical engineering, in particular control theory and robotics, Claude Shannon in communication theory, and Waloddi Weibull in reliability theory is directly applicable to the space program. Moreover, statistics is an essential part of all engineering disciplines. A glance at the titles of journals published by American Society of Civil Engineers (ASCE), American Society of Mechanical Engineers (ASME), and Institute of Electrical and Electronics Engineers (IEEE) give an indication of the wide range of applications. These applications have also led to advances in statistical theory, as seen, for example, in the work of: Emil Gumbel in hydrology and Walter Shewhart [Shewhart, 1939] in manufacturing. In this book we consider examples from many engineering disciplines including: hydrology; water quality; strengths of materials; mining engineering; ship building; chemical processes; electrical and mechanical engineering; and management.

Engineers have always had to deal with uncertainty, but they are now expected to do so in more accountable ways. Probability theory provides a mathematical description of random variation and enables us to make realistic risk assessments. Statistics is the analysis of data and the subsequent fitting of probability models.

1.2   Using the book

The first eight chapters are best read in sequence, although Chapter 3 could be read before most of Chapter 2. The exercises include: routine examples; further detail for some of the explanations given in chapters; extensions of the theory presented in chapters; and a few challenges. Numerical answers to odd numbered exercises are given on the book website, together with sets of additional exercises.

The following chapters cover more advanced topics, although the mathematical detail is kept at a similar level to that in the preceding chapters. Chapter 9 on multiple regression is a pre-requisite for the two chapters on the design of experiments. The other three chapters, on statistical quality control, probability models, and sampling strategies, can be read in any order. Of these, probability models relies only on the material on probability in Chapter 2, whereas the other two chapters assume some familiarity with most of the material in the first eight chapters.

Appendix D.11 contains experiments which have been designed to complement computer based investigations. The experiments are simple, such as descent time of paper helicopters with paper-clip payloads and cycles to failure for bending paperclips, but the rationale is that doing something is more convincing and more memorable than reading about it. The website includes answers to odd numbered exercises, additional exercises, and informal derivations of key results. These include the Central Limit Theorem, Gumbel's extreme value distribution, and more detail on the multiple regression model. The proofs of these results rely only on standard calculus and matrix algebra and show something of the diverse applications of fundamental mathematics. These proofs are intriguing, but are not needed to follow the main text.

1.3   Software

We have chosen to implement the analyses in two software environments: R and MATLAB®. The flexibility of command line programming is offset against the convenience of menu driven statistical packages such as Minitab. Appendices D and G are tutorials to get you started in R and MATLAB respectively. Both R and MATLAB have built in functions for a wide range of modern statistical analyses. To see what a particular command such as `plot` does in R, type `help(plot)` and in MATLAB type `help plot`. R has the great advantage of being open source, and hence free for anyone to use, and many research statisticians chose to develop the subject by providing new packages for R. However, MATLAB is very well suited to engineering applications and is widely used in industry and universities. Moving between the two is partially facilitated by Hiebeler's MATLAB/R Reference [Hiebeler, 2010] and complemented where possible in a direct function comparison at the end of each chapter. Other useful computing resources are Short's R Reference card [Short, 2004], and many of the posts on the World Wide Web (www) once one has learnt the basic principles of a computing language. In general there are several ways of programming a statistical routine and also packages and toolboxes that automate the entire process. We have aimed to use R and MATLAB to demonstrate calculations that follow the statistical development in the book, and to show the use of the standard statistical functions.

2

Probability and making decisions

Three approaches to defining probability are introduced. We explain the fundamental rules of probability and use these to solve a variety of problems. Expected monetary value is defined and applied in conjunction with decision trees. Permutations and combinations are defined and we make a link with the equally likely definitions of probability. We discuss the concept of random digits and their use for drawing simple random samples from a population. See relevant examples in Appendix E:

Appendix E.1 *How good is your probability assessment?*
Appendix E.2 *Buffon's needle*

2.1 Introduction

The Australian Bureau of Meteorology Adelaide Forecast gives the chance of any rain tomorrow, a summer day in South Australia, as 5%. We will see that it is natural to express the chance of an uncertain event, such as rain tomorrow, occurring as a probability on a scale from 0 to 1. If an event is as likely to occur as it is not to occur, then it has a probability of occurrence of 0.5. An impossible event has a probability of 0 and a certain event has a probability of 1. Formally, the Bureau's chance of 5% is a probability of 0.05, and as this is considerably closer to 0 than 1 we think it is unlikely to rain tomorrow. There are several ways of giving a more precise interpretation. One is to imagine that similar weather patterns to today's have been observed in Adelaide on many occasions during the Australian summer, and that on 5% of these occasions it has rained on the next day. Another interpretation is based on the notion of a fair bet (formally defined in Section 2.3.3). The weather forecaster thinks that the possibility of paying out \$95 if it rains is offset by the more likely outcome of receiving \$5 if it is dry. Many engineering decisions are based on such expert assessments of probability. For example, after drilling a well an oil company must decide either to prepare it for oil production or to plug and abandon it. Before drilling at a specific location, the company will seek a geologist's opinion about the probability of finding sufficient oil to justify production. There are various strategies, including the notion of a fair bet, for making assessments of probabilities. One, the quadratic rule [Lindley, 1985], which has been used in the U.S. in the training of weather forecasters, is covered in Experiment E.1. Others are discussed later in the chapter.

There is, however, a basic approach to defining probability, which is applicable in special cases when we can define the outcomes of some experiment so that they are **equally likely** to occur. In this context, an **experiment** is any action that has an uncertain outcome. Typical experiments that are supposed to have equally likely possible outcomes are games of chance played with carefully constructed apparatus such as dice, cards, and roulette wheels. The claim that outcomes are equally likely is based on the symmetry of the apparatus. For example, all the cards in a deck should have the same physical dimensions and all slots and frets on the roulette wheel should have the same physical dimensions[1]. The equally likely definition of probability was developed in the context of gambling games by Gerolamo Cardano (1501-1576) and other mathematicians including Galileo, Pascal and Fermat in the sixteenth century [David, 1955]. Cards and dice may seem unconnected to engineering, but the generation of digits from 0 up to 9, such that each digit is equally likely to appear as the next in sequence, is the basis of stochastic simulation. Simulation studies have a wide range of applications including engineering design and analysis.

2.2   Random digits

2.2.1   Concepts and uses

You can use the R function `sample()` to obtain a sequence in which each digit from 0 up to 9 appears equally likely to occur at each turn. To obtain such a sequence of length 20:

```
> sample(0:9, size=20, replace=TRUE)
 [1] 6 2 4 2 8 3 0 8 8 1 3 5 1 6 5 2 8 6 7 0
```

although your sequence will be different. The R function `sample(x,n, replace=TRUE)` takes a sample of size n from the object x as follows. Imagine each element of x has an associated ticket that is put into a hat. Then n tickets are drawn from the hat, sequentially with the selected ticket returned to the hat before the next ticket is drawn, in an idealized manner such that at every draw each ticket in the hat is equally likely to be drawn. The third argument in the function call, `replace=TRUE`, gives sampling with replacement, which means that the ticket is returned to the hat after each draw.

If a ticket is not replaced after being drawn the sampling is without replacement and this, more common application, is the default for the `sample()` function. So, when sampling without replacement the function call `sample(x,n,replace=FALSE)` can be shortened to `sample(x,n)`. One use of such sequences is the selection of a random sample from a finite population.

Example 2.1: Island ferry safety [random selection]

An island is served from a city on the mainland by a fleet of ferries, which each carry 12 inflatable life rafts in hard shelled canisters. The port authority requires that the crew of each ferry demonstrates the launch of one life raft each year. The life raft will be chosen at random from the 12, in such a way that each life raft has the same chance of selection, so that all concerned can agree that there is no possibility of performing the demonstration with a specially prepared life raft.

---

[1]This understanding of probability can be traced back over thousands of years. The oldest known dice were excavated as part of a 5 000-year-old backgammon set, at Shahr-i Sokhta in Iran. The concept of equally likely outcomes is the basis for the *kleroterion* allotment machine, that was used by the Athenians in the third century BC for selecting jurors and other representatives. Substantial remnants of a kleroterion can be seen in the Agora Museum, in Athens.

A sequence of random digits can be used to make the selection. One way to do this is as follows: number the life rafts from 01 to 12; pair consecutive random digits; and take the life raft corresponding to the first pair in the range from 01 up to 12. With the sequence 6 2 4 2 8 3 0 8 .., the pairing gives: 62, 42, 83, 08, .., and life raft 8 would be selected. If we were asked to sample more than one lifeboat, identifying the 12 lifeboats by the pairs from 01 to 12 only, and ignoring all pairs between 13 and 99, might require a long sequence of digits. You are asked to devise a more efficient identification in Exercise 2.5. Direct use of the software functions is more convenient, as in Exercise 2.6.

Apart from sampling, sequences of random digits are used for computer simulations of random processes. One example is the simulation of random wave forces and the calculation of their effects on off-shore structures such as oil rigs and wave energy converters. Other uses include Monte-Carlo strategies for computing probabilities.

2.2.2 Generating random digits

How can we produce sequences of random digits between 0 and 9, without relying on software? In principle we might: flip a fair coin, taking a head as 0 and a tail as 1, to obtain a sequence of random binary digits; take consecutive sets of four digits, and convert from binary to base 10; accept the digit if the base 10 integers is between 0 and 9, and reject base 10 integers between 10 and 15. For most purposes this would be far too slow. A variant on a roulette wheel, with an equal number of each digit from 0 to 9, would be a rather more convenient generator of random digits but it would still be far too slow for simulations. There is also the impracticality of making a roulette wheel, or any other mechanical apparatus, perfectly fair [2].

Given the limitations of mechanical devices for producing random digits, it is natural to consider electronic alternatives. In 1926, John Johnson, a physicist at Bell Labs, noticed random fluctuations in the voltage across the terminals of a resistor that he attributed to thermal agitation of electrons. This noise can be amplified, sampled, and digitized into a sequence of 0s and 1s by taking sampled voltages below the nominal voltage as 0 and sampled voltages above the nominal voltage as 1. Such an electronic device needs to be very stable and, in particular, the nominal voltage has to be maintained so that a sampled value is equally likely to be a 0 as a 1. Provided the sampling rate is slow relative to typical fluctuations in voltage, the next binary digit is equally likely to be a 0 as a 1, regardless of previous values. Similar principles are exploited in solid state devices called hardware random number generators (HRNG), such as ComScire's R2000KU, which produces 2 megabits per second and is guaranteed, by the manufacturer, to pass any test for randomness. A renowned HRNG is ERNIE (Electronic Random Number Indicator Equipment), first used in 1957 to draw winning numbers in the Premium Bond investment scheme run by the Office of National Savings in the UK. ERNIE has been through several incarnations since 1957, but continues to be used for the draw.

---

[2]In 1875, an engineer Joseph Jagger took advantage of a bias that he detected in one of six roulette wheels at the Beaux-Arts Casino in Monte-Carlo to win over two million francs. A rather riskier strategy employed by the adventurer Charles Wells, who is the most likely inspiration for the well known song "The Man Who Broke the Bank at Monte-Carlo", is the Martingale betting system (Exercise 2.59).

2.2.3 Pseudo random digits

Despite the availability of HRNG hardware, the sequence of 20 random digits produced by `sample()` in R, and most other software, is not random but generated according to some algorithm. To obtain the same sequence shown in Section 2.2.1, type

```
> set.seed(16)
> sample(0:9,20,replace=T)
```

The seed, here 16, is the initial value for the algorithm. If no seed is specified, some coding of the computer clock time will be used. Once begun, the algorithm is completely deterministic but the resulting sequence appears to be practically equivalent to a record of an experiment in which each of the digits from 0 up to 9 was equally likely to occur at the next turn. Such algorithms are called **pseudo random number generators** (PRNGs). A relatively simple example of a PRNG is given in Exercise 2.7, and [Kroese et al., 2011] is a detailed reference. John von Neumann, a renowned mathematician, quipped that

> *"Any one who considers arithmetical methods of producing random digits is, of course in a state of sin".*

So, why do we use them? Reasons for using a PRNG are: they do not involve additional hardware; hardware does not need to be checked for stability; and if a seed is specified, the sequences are reproducible. Reproducible sequences can be used to document how a random selection was made. Also, using a reproducible sequence in a computer simulation allows us to investigate any anomalous results.

Donald E Knuth discusses why PRNGs work as well as they do in his multi-volume monograph *The Art of Computer Programming* [Knuth, 1968]. As a consequence of its construction, a PRNG must repeat after a given number of digits, known as its cycle length (or period), but this number can be very large[3].

Example 2.2: Linear Congruential Generator [simple PRNG]

An early algorithm used to generate random numbers was the Linear Congruential Generator (LCG), specified by the recursive formula using modular integer arithmetic[4]

$$Z_i \;=\; a\,Z_{i-1} + c \;(\mathrm{mod}\; m),$$

where m is called the *modulus*
    a is called the *multiplier*
    c is called the *increment* and
    Z_0 is the *seed* or starting value.

The variable Z_i will be an integer between 0 and $m-1$. To get a number U in the range $[0,1)$, set $U_i = Z_i/m$. This can be transformed to an integer between 0 and 9 (inclusive) by multiplying by 10 and removing the decimal fraction (using the `floor` command).

Choosing the 4 parameters m, a, c and Z_0 in a good way, provides sequences of numbers Z_i and U_i that appear for most purposes to be random.

---

[3]The default in R and MATLAB is the Mersenne Twister [Matsumoto and Nishimura, 1998] with a period length of $2^{19\,937} - 1$.

[4]The expression $x = y \;(\mathrm{mod}\; m)$ for non-negative integers x and y and positive integer m means that x is the remainder of the integer division y/m, so that $(y - x) = km$ for some non-negative integer k.

We will assume we have some long sequence that is less than the cycle length. We consider briefly how we might check that sequences from PRNGs appear random. The most basic requirement is that the digits $0, \ldots, 9$ appear in the same proportions. Apart from this requirement, there should be no easily recognizable patterns in the sequence of generated numbers, such as there is in the following example.

Example 2.3: RANDU [limitations of PRNG]

An example of an LCG with poorly chosen parameters was RANDU, specified by $m = 2^{31}$, $a = 2^{16} + 3 = 65\,539$, $c = 0$, with Z_0 odd, which was distributed in the 1960s, and has bad statistical properties and a period of only 2^{29}.

> *"... its very name RANDU is enough to bring dismay into the eyes and stomachs of many computer scientists!"*
>
> –Donald Knuth

In fact, RANDU can be re-written as

$$Z_{i+2} \;=\; 6Z_{i+1} - 9Z_i \,(\text{mod } 2^{31}),$$

which implies that its resultant sequences have a marked pattern that can be readily visualized (see Exercise 2.9).

More demanding tests require the correct proportions of double digits, the correct proportions of digits following a specific digit, and so on. A set of routines to evaluate randomness called the Diehard tests[5] were developed by George Marsaglia, an American mathematician and computer scientist, who established the lattice structure of linear congruential generators in his paper memorably titled *Random numbers fall mainly in the planes*.

A good PRNG is far better than some ad hoc strategy based on long lists of numbers. A strategy such as using the first digits from long lists of numbers that range over orders of magnitude, such as the magnitudes of $248\,915$ globally distributed earthquakes, photon fluxes for $1\,452$ bright objects identified by the Fermi space telescope [Sambridge et al., 2011], and populations of cities, is not suitable. The reason is that, apart from the lack of 0s, measurements with a lower first digit $(1, 2, \ldots)$ occur more frequently than those with a higher first digit $(\ldots, 8, 9)$. This latter result is commonly known as Benford's Law (see Exercise 2.61).

2.3 Defining probabilities

We discuss three approaches to defining probabilities: equally likely outcomes; relative frequencies; and subjective assessments. But, no matter which approach is suitable for a particular application, the rules for manipulating probabilities are the same.

Definition 2.1: Sample space

A set of possible outcomes of an experiment, defined so that exactly one of them must occur, is known as a **sample space**.

---

[5]The Diehard tests and the TestU01 library provide standards for checking the output from PRNGs.

Definition 2.2: Mutually exclusive (disjoint)

Outcomes that comprise a sample space are said to be **mutually exclusive** because no two of these outcomes can occur simultaneously. A commonly used synonym for mutually exclusive is **disjoint**.

Definition 2.3: An event occurs

An **event** A is some subset of the sample space, and is said to **occur** if one of its elements is the outcome of the experiment.

Definition 2.4: Collectively exhaustive

A set of events is **collectively exhaustive** if it covers the sample space[6]. In particular, the set of all possible outcomes is collectively exhaustive.

The sample space is not necessarily unique and we set up a sample space that will enable us to answer the questions that we pose in a convenient manner.

2.3.1   Defining probabilities – Equally likely outcomes

In some experiments we can claim that all outcomes are **equally likely** (EL) because of symmetry, such as that which a gambling apparatus is designed to possess. How can we deduce the proportion of occasions on which a particular event A will occur in the long run? In the case of EL outcomes, the deduced proportion is referred to as a probability and is given by the following definition.

Definition 2.5: Probability for equally likely events

If all outcomes of an experiment are equally likely, then for any event A, the probability of the event A occurring is defined as

$$P(A) \;\; = \;\; \frac{\text{number of EL outcomes in } A}{\text{total number of EL outcomes}}.$$

This probability measure ranges between 0, when an event is impossible, and 1 when an event is certain. It also follows that

$$P(\text{not } A) \;\; = \;\; \frac{\text{total number of EL outcomes - number of EL outcomes in } A}{\text{total number of EL outcomes}}$$
$$= \;\; 1 - P(A).$$

Definition 2.6: Complement

The event "not A" is called the **complement** of A, denoted by $\overline{A}$.

---

[6] A more precise statement is that the union equals the sample space. See Section 2.3.

Example 2.4: Decagon spinner [equally likely outcomes]

A regular decagon laminar with a pin through its center (Figure 2.1) is spun on the pin. The laminar comes to rest with one edge resting on the table and the corresponding number is the outcome of the experiment. The sample space is the set $\{0, 1, 2, 3, 4, 5, 6, 7, 8, 9\}$. If the device is carefully constructed so as to preserve the symmetry, and the initial torques differ, the outcomes can be considered as equally likely.

The event "spinning a 7" is the set $\{7\}$ and so the probability of this event, $P(7)$, is $1/10$. The event "spinning an odd number" is the set $\{1, 3, 5, 7, 9\}$.

It follows that the probability of obtaining an odd number is

$$P(\text{odd number}) \quad = \quad \frac{\text{number of elements in } \{1, 3, 5, 7, 9\}}{\text{number of elements in } \{1, \ldots, 10\}} \quad = \quad \frac{5}{10}.$$

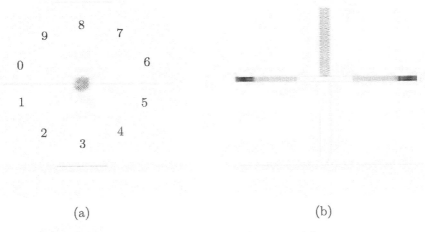

(a)   (b)

FIGURE 2.1: Decagon spinner (a) at rest (b) spinning.

Example 2.5: Two decagon spinners [equally likely outcomes]

Two decagons are spun, or equivalently the same decagon is spun twice. Either way, a sample space of 100 equally likely outcomes is shown as a set of points in Figure 2.2.

(a) Find P(total of 3). There are four points that give a total of 3: $(3, 0), (2, 1), (1, 2)$ and $(0, 3)$. Hence, P(total of 3) $= 4/100$.

Similarly we can count points in the sample space to obtain:

(b) P(total of 4) $= 5/100$.

(c) P(doubles) $= 10/100$.

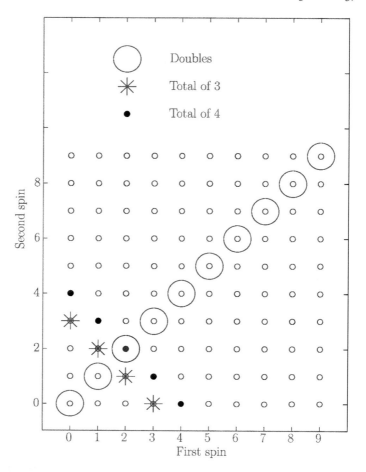

FIGURE 2.2: Sample space for two spins of a decagon spinner.

Now consider P(total of 3 or doubles)[7] .

(d) Calculate P(total of 3 or doubles). We can count points in the sample space to obtain 14/100.

However, we can express it in terms of the probabilities of the constituent events "total of 3" and "doubles". If we look at the sample points corresponding to a "total of 3 or doubles" we see that they are those corresponding to "total of 3" together with those corresponding to "doubles". Hence

$$P\big((\text{total of 3}) \text{ or } (\text{doubles})\big) = P(\text{total of 3}) + P(\text{doubles})$$
$$= 4/100 + 10/100 = 14/100.$$

---

[7]Doubles is the same digit on both spins.

That was easy, but the next part of this example requires a precise definition of "or". If a tutor asks you whether you wish to take a course in calculus or statistics there is some ambiguity. Can you take both courses, or are you restricted to just one of the two courses? The **inclusive or** includes both. The mathematical convention is that "or" in "*A* or *B*" is the inclusive or and therefore includes: *A* and not *B*, *B* and not *A*, and both *A* and *B*. If we want the **exclusive or**, one or the other but not both, we have to specify "*A* or *B* but not both" or some equivalent statement such as "precisely one of *A* or *B*".

(e) Calculate P((total of 4) or (doubles)). To begin with, we note that the event "total of 4 or doubles" includes the point $(2, 2)$, which is both a double and has a total of 4, because the mathematical convention is that "or" includes both. The probability can be found by counting the points in the sample space and equals 14/100. This is less than the sum of the probabilities of the constituent events. The explanation is that the sample point $(2, 2)$ is both a double and has a total of 4. If the probabilities of the constituent events are added then the sample point $(2, 2)$ is counted twice. To compensate for this double counting, its probability is removed once. So,

$$
\begin{aligned}
\text{P}\big((\text{total of 4}) \text{ or } (\text{doubles})\big) &= \text{P}(\text{total of 4}) + \text{P}(\text{doubles}) \\
&\quad - \text{P}\big((\text{total of 4}) \text{ and } (\text{doubles})\big) \\
&= 5/100 + 10/100 - 1/100 \\
&= 14/100.
\end{aligned}
$$

This calculation is an example of the **addition rule of probability**. It is quite general and applicable to $P\big((\text{total of 3}) \text{ or } (\text{doubles})\big)$, although in this case $P\big((\text{total of 3}) \text{ and } (\text{doubles})\big) = 0$ because it is impossible to have double digits that add to 3.

Returning to the context of checking sequences of digits from PRNGs, we would expect 10% of consecutive pairs to be doubles (the same digit repeated), and 14% of consecutive pairs to give a total of 4 or doubles.

The concept of equally likely outcomes is of fundamental importance because it can provide a standard for measuring probability. Imagine a Grecian urn, which is opaque, containing 100 balls that are identical apart from their color. A specified number b of the balls are black and the rest are white, and the balls are thoroughly mixed. The weather forecaster, who gave the chance of rain tomorrow as 5%, considers that if b equals 5 the probability of drawing a black ball from the urn is the same as rain tomorrow.

2.3.2 Defining probabilities – Relative frequencies

In most engineering situations the outcomes of experiments are not equally likely, and we rely on a more versatile definition of the probability of an event A occurring. This definition is based on the notion of a long sequence of N identical experiments, in each of which A either occurs or does not occur.

Definition 2.7: Probability in terms of relative frequencies

If an experiment is performed a large number of times N, then the probability of an event A occurring is

$$P(A) \approx \frac{\text{number of times } A \text{ occurs}}{N}.$$

This ratio is actually the **relative frequency** of occurrence of A, which is the sample proportion of occurrences of A, and becomes closer to $P(A)$ as N increases, a result known as the **law of large numbers**. Insurance companies make risk calculations for commonly occurring accidents on the basis of relative frequencies, often pooling their claims records to do so. The concept of relative frequency does not, however, give a sound basis for assessing the probabilities of outcomes in non-repeatable experiments such as the launching of communications satellites. The experiment is non-repeatable because the technology is rapidly changing.

Example 2.6: Auto warranty [Venn diagram]

From extensive records, a marketing manager in a motor vehicle manufacturing company knows that 5% of all autos sold will have bodywork faults (B) and 3% will have mechanical faults (M), which need correcting under warranty agreements. These figures include 1% of autos with both types of fault. The manager wishes to know the probability that a randomly selected customer will be sold an auto that has to be returned under warranty, that is $P(B \text{ or } M)$.

We can show the sample space and the logical relationships between the events B and M using a **Venn diagram**[8] (Figure 2.3).

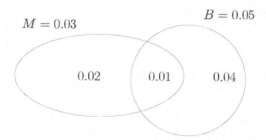

FIGURE 2.3: Venn diagram for sample space for bodywork (B) and mechanical (M) faults in cars.

It is also convenient to use set notation and nomenclature when working with probabilities. Table 2.1 is a summary of the equivalences. In the Venn diagram the rectangular frame represents the sample space, which is the **universal set** Ω. In this case, the required universal set is all autos produced by the vehicle manufacturing company. The **set** of outcomes that give the event M is represented by the ellipse and the set of outcomes that give the event B is represented by the circle. The overlap, known as the **intersection**, is the set of outcomes that give both B and M. We have divided the sample space into four mutually exclusive and collectively exhaustive events: B and M; B and not M; M and not B; neither B nor M.

---

[8]Venn diagrams are typically not to scale.

TABLE 2.1: Probability notation and nomenclature.

Probability description	Set language	Set notation
sample space	universal set	Ω
event A	set A	$A \quad (A \subset \Omega)$
or	union	$\cup$
and	intersection	$\cap$
not A	complement of A	$\overline{A}$
impossible event	empty set	$\emptyset$

The **intersection** of B and M, which is the set of outcomes that are in both B and M, is written

$$B \cap M.$$

If B and M are mutually exclusive, $B \cap M$ is the **empty set** $\emptyset$, and $P(\emptyset) = 0$. The **union** of B and M is the set of outcomes that are in B or M or both, and it is written as

$$B \cup M.$$

Example 2.6: (Continued) Auto warranty

We now return to the Venn diagram and explain how the probabilities associated with disjoint events are obtained. We are given that the probability $P(B \cap M) = 0.01$ and so we write 0.01 in the intersection. We are given that $P(B) = 0.05$ and so $P(B \cap \overline{M}) = 0.05 - 0.01 = 0.04$, which we write in the remaining part of the circle representing those outcomes in B and not in M. Similarly we have that $P(M \cap \overline{B}) = 0.03 - 0.01 = 0.02$. Adding the three probabilities for the disjoint events, we obtain

$$
\begin{aligned}
P(B \cup M) &= P(M \cap \overline{B}) + P(B \cap M) + P(B \cap \overline{M}) \\
&= 0.02 + 0.01 + 0.04 = 0.07.
\end{aligned}
$$

Alternatively, we could use the addition rule of probability, which follows from the Venn diagram:

$$
\begin{aligned}
P(B \cup M) &= P(B) + P(M) - P(B \cap M) \\
&= 0.05 + 0.03 - 0.01 = 0.07.
\end{aligned}
$$

Notice that we subtract the probability of B and M, to avoid counting it twice. Finally, notice that the probability of no fault needing correction under warranty, which is represented by the region outside both the ellipse and circle, is $1 - 0.07 = 0.93$.

2.3.3 Defining probabilities – Subjective probability and expected monetary value

In some applications the notion of an imagined sequence of identical experiments is rather artificial and it is more plausible to model expert opinion in terms of a **fair bet**. Before defining a fair bet, we need to define the concept of **expected monetary value**.

Definition 2.8: Expected Monetary Value (EMV)

Suppose a decision is associated with a set of outcomes that are mutually exclusive and collectively exhaustive. Each outcome has a probability of occurrence, p_i, and an associated payoff, M_i, which can be positive or negative, the latter representing a loss. The expected monetary value (EMV) of the decision is defined as the sum of the products of the probability of each outcome with its payoff.

$$\text{EMV} \;=\; \sum p_i M_i.$$

In most cases the decision maker will not receive the actual EMV as a consequence of an individual decision, but if many decisions are being made, the total return will be close to the sum of the EMVs provided none of the decisions involves a much larger payoff than the others. Maximizing EMV is often used as a criterion for engineering decisions, although other criteria that penalize uncertainty, such as conditional value at risk (Exercise 2.41), are sometimes preferred.

Definition 2.9: A Fair Bet

A **fair bet** is a bet with an EMV of 0.

As an example of a **subjective probability**, an engineer considers it a fair bet to receive 100 monetary units if a planetary exploration vehicle is successfully landed, and to pay 300 monetary units if it is not. If the bet is fair the engineer's expected return will be 0. Let the probability of success be p. Then:

$$
\begin{aligned}
100 \times p + (-300) \times (1 - p) &= 0 \\
\Rightarrow p &= 0.75 .
\end{aligned}
$$

So, the engineer implies that the probability of success is 0.75.

In practice, decisions are not necessarily based on the notion of a fair bet. We often insure against big losses despite the bet being in favor of the insurance company, if it is to stay in business, and we sometimes choose to gamble despite the casino's intention to make a profit. Such actions can be modeled by introducing the concept of **utility** (Exercise 2.42).

We do not have to resort to the concept of a fair bet to define a subjective probability. Another option is to imagine an urn containing balls of two colors. The engineer considers that a successful landing is as likely as drawing a black ball from an urn containing four balls of which three are black and one is white, provided each one of the four balls is equally likely to be drawn.

Example 2.7: Oil production [EMV for decisions]

An expert geologist assesses the probability of finding sufficient oil to justify production at a location in the Bass Strait as 40% (0.4). The cost of a 1 year concession is 20% of revenue. Drilling costs would be $500 million and, if production is set, the oil revenue for the year would be $2 000 million. Is the concession worthwhile?

If oil is found, the profit will be 80% of the oil revenue less the drilling cost, ($0.8 \times 2\,000 - 500$). If no oil is found, the loss will be the drilling cost, a profit of (-500).

The EMV is:

$$(0.8 \times 2\,000 - 500) \times 0.4 + (-500) \times (1 - 0.4) \quad = \quad 140.$$

The concession is worthwhile, according to an EMV criterion, because the EMV is greater than 0.

2.4   Axioms of probability

All the rules of probability can be deduced from just three axioms, together with an extension[9] of the second axiom.

Definition 2.10: Axioms of probability

If events E_1 and E_2 are mutually exclusive and the sample space is denoted by Ω, the three axioms of probability are as follows.

Axiom 1: non-negativity	$P(E_1) \geq 0$
Axiom 2: additivity	If $P(E_1 \cap E_2) = 0$, then $P(E_1 \cup E_2) = P(E_1) + P(E_2)$
Axiom 3: normalization	$P(\Omega) = 1$

The addition rule follows from the argument in Exercise 2.13.

2.5   The addition rule of probability

We have considered three interpretations of probabilities. Whichever is adopted for a particular case, the rules of probability are identical. In this section we reiterate the addition rule and show that it is consistent with Definition 2.6 of complement.

Definition 2.11: Addition rule of probability

For two events A and B

$$P(A \cup B) \quad = \quad P(A) + P(B) - P(A \cap B).$$

where $\cup$ and $\cap$ are "or" and "and", with "or" including the possibility that both occur.

---

[9]The extension of the second axiom, which we do not need for this chapter, is a generalization to a countably infinite union of mutually exclusive events. If $E_i \cap E_j = \emptyset$ for $i \neq j$, then $P\left(\bigcup_{i=1}^{\infty} E_i\right) = \sum_{i=1}^{\infty} P(E_i)$.

Definition 2.12: Mutually exclusive

If A and B cannot both occur, then the events are said to be **mutually exclusive** or **disjoint**, and $P(A \cap B) = 0$.

The addition rule extends to any countable number of events, by mathematical induction (Exercise 2.14). For three events A, B and C

$$
\begin{aligned}
P(A \cup B \cup C) &= P(A) + P(B) + P(C) \\
&\quad - P(A \cap B) - P(A \cap C) - P(B \cap C) + P(A \cap B \cap C).
\end{aligned}
$$

Example 2.8: Construction awards [Probability of union of 3 events]

A construction company has recently completed a skyscraper in a capital city. The building has been entered for design awards under three categories: appearance (A), energy efficiency (E), and work environment (W). An independent architect made a first assessment of the probability that it would win the various awards as 0.40, 0.30 and 0.30 respectively, and added that it would not win more than one award. On reflection, she decided that it might win any two awards with probability 0.08 and all three awards with probability 0.01. Why did she modify her original assessment, and what is the modified probability that the building wins at least one award?

If it is assumed that the building cannot win more than one award the events are mutually exclusive. Then

$$
\begin{aligned}
P(\text{wins an award}) &= P(A \cup E \cup W) \\
&= P(A) + P(E) + P(W) \\
&= 0.4 + 0.3 + 0.3 \;=\; 1.00
\end{aligned}
$$

The architect revised her assessments because she did not think the building was certain to win an award. With the revised assessments the probability that it wins an award becomes:

$$
\begin{aligned}
P(A \cup E \cup W) &= P(A) + P(E) + P(W) \\
&\quad - P(A \cap E) - P(A \cap W) - P(E \cap W) + P(A \cap E \cap W) \\
&= 0.4 + 0.3 + 0.3 - 0.08 - 0.08 - 0.08 + 0.01 \;=\; 0.77
\end{aligned}
$$

If you prefer to use a Venn diagram (Figure 2.4) you should start with the intersection of all three areas. Then be aware that the event of winning two awards, A and E for example, includes the event of wining A and E and W. Whenever we assign subjective probabilities, we need to ensure that they are all consistent with each other.

2.5.1 Complement

From the definition of "not", $(A \cup \overline{A})$ is a certain event and $(A \cap \overline{A})$ is an impossible event. Using the addition rule

$$
P(A \cup \overline{A}) \;=\; P(A) + P(\overline{A}) - P(A \cap \overline{A}).
$$

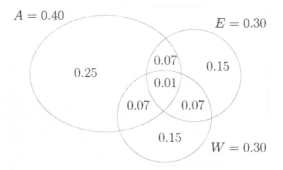

FIGURE 2.4: Venn diagram for sample space for design awards: appearance (A), energy efficiency (E), and work environment (W).

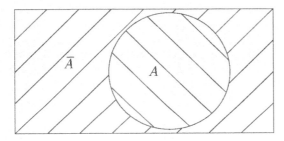

FIGURE 2.5: The set A, its complement $\bar{A}$, where $A \cap \bar{A} = \emptyset$ and $A \cup \bar{A} = \Omega$.

Now $(A \cup \bar{A})$ is certain and A and $\bar{A}$ are mutually exclusive (Figure 2.5), symbolically:

$$
\begin{aligned}
\mathrm{P}(A \cup \bar{A}) &= \mathrm{P}(\Omega) &= 1, \\
\mathrm{P}(A \cap \bar{A}) &= \mathrm{P}(\emptyset) &= 0.
\end{aligned}
$$

It follows that

$$
\begin{aligned}
1 &= \mathrm{P}(A) + \mathrm{P}(\bar{A}) \\
\Rightarrow \mathrm{P}(\bar{A}) &= 1 - \mathrm{P}(A).
\end{aligned}
$$

We have already used this result as it follows directly from the definitions of probability.

Example 2.9: Wave energy [complement]

An engineer considers that her design for a heaving-buoy wave-energy device will receive funding for trials off the coast of Oregon and off the coast of Scotland, with probability 0.15 and 0.20 respectively. After further consideration the engineer thinks that the probability it will receive funding for both trials is 0.10, because if it receives funding for one trial it is more likely to receive funding for the other. What is the probability that the device does not receive funding for a trial?

Not receiving funding is the complement of receiving funding for at least one trial. If we write O and S for receiving funding for the trials off Oregon and Scotland respectively, "at least one of O" and S is equivalent to "O or S". Hence, using the addition rule:

$$
\mathrm{P}(\text{no trial}) = 1 - \mathrm{P}(O \cup S) = 1 - (0.15 + 0.20 - 0.10) = 0.75.
$$

You can check this answer by sketching a Venn diagram.

2.6   Conditional probability

2.6.1   Conditioning on information

Concrete hardens and gains strength as it hydrates. The hydration process is rapid at first but continues slowly over many months. Waiting several years to measure the ultimate strength is impractical, so standardization bodies such as the American Concrete Institute (ACI), and the International Standards Institute (ISI) specify that strengths of test cylinders (or cubes) be measured after 28 days. Concrete is usually mixed in batches, and as part of the quality assurance process, four test cylinders are made from each batch of concrete. One cylinder will be chosen randomly from the four test cylinders and tested after 7 days. The other three will be tested after 28 days, and the gain in strength between 7 days and 28 days is typically around 25%.

Under ACI rules the batch is deemed satisfactory (S) if the average compressive strengths of the three cylinders exceeds the required strength, and none has a strength less than 90% of that required (ACI). The 7-day test provides an early warning that the batch may not be satisfactory.

A design engineer has specified high strength concrete for a structure. The engineer responsible for quality assurance sets a required compressive strength of 85 MPa (12 328 psi), and stipulates that a batch will be designated as questionable (Q) if the 7-day test result is less than 70 MPa.

From records, 20% of batches are Q, 90% of batches are S, and 8% of batches are both Q and S. Without any 7-day test result, the probability that a batch will be S is 0.9. If the 7-day test is a useful warning, the probability that a batch will be S depends on whether or not it was Q. This is an example of a **conditional probability**, and the probability of S, conditional on Q is written as $P(S|Q)$.

This conditional probability can be deduced from the information we have been given. The conditioning on Q restricts attention to the 20% of batches that are Q, which includes the 8% that are both Q and S. The proportion of Q that is S is $8/20 = 0.4$, and this is $P(S|Q)$. We see that the 7-day test does provide a useful warning because $P(S|Q)$, which equals 0.4, is substantially less than $P(S) = 0.9$. The information we are given is shown in the Venn diagram in Figure 2.6(a), the restriction to Q, and the interpretation of $P(S|Q)$ as the proportion of Q that is also S, is shown in Figure 2.6(b).

Also note that $P(Q|S)$ is well defined even though S occurs after Q. $P(Q|S)$ represents the proportion of satisfactory batches (S) that were designated questionable (Q). In this case $P(Q|S)$ equals $0.08/0.90 = 0.089$. The foregoing discussion justifies the following definition of conditional probability.

2.6.2   Conditional probability and the multiplicative rule

For two events A and B, the conditional probability of A given B is written as $P(A|B)$, where the vertical line | in the probability argument is read as "given that" or "conditional on".

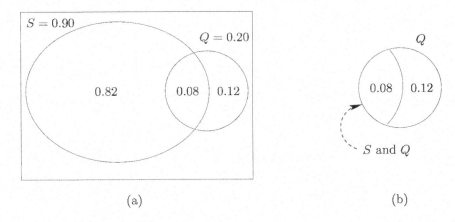

(a)   (b)

FIGURE 2.6: Venn diagram (a) for sample space of concrete cubes, (b) for sample space given the conditioning on Q.

Definition 2.13: Conditional Probability

$$P(A|B) = \frac{P(A \cap B)}{P(B)}, \quad \text{provided} \quad P(B) > 0.$$

The definition of conditional probability is consistent with the axioms. The conditioning on event B reduces the sample space to B. The division by $P(B)$ is the normalization, so $P(B \mid B) = 1$. The definition of conditional probability can be rearranged to give the **multiplicative rule** of probability.

Definition 2.14: Mutilplicative Rule of Probability

$$\begin{aligned} P(A \cap B) &= P(A)\,P(B|A) \\ &= P(B)\,P(A|B). \end{aligned}$$

The fact that $P(A \cap B)$ can be expressed in two ways, which follows formally from the letters A and B being exchangeable, turns out to be very useful in applications. The multiplicative rule extends inductively. For example, for three events

$$P(A \cap B \cap C) = P(C|A \cap B)\,P(B|A)\,P(A).$$

Example 2.10: Free brake check [conditional probabilities]

An auto-repair shop offers a free check of brakes and lights, including head light alignment. Over the years, the mechanic has found amongst autos inspected under this offer: 30% have a brake defect (B), 65% have a lighting defect (L), and 26% have both defects. Suppose that an auto is randomly selected from the population of autos taking up the offer.

(a) Calculate the probability that the auto has a brake defect given that it has a lighting defect.

Using the notation in the definition above, $P(B|L) = P(B \cap L)/P(L) = 0.26/0.65 = 0.40$.

(b) Calculate the probability that the auto has a lighting defect given that it has a brake defect.

This conditional probability is $P(L|B) = P(L \cap B)/P(B) = 0.26/0.3 = 0.87$.

Notice that the conditional probability of a brake defect given a lighting defect, 0.40, is greater than the unconditional probability of a brake defect, 0.30, and the conditional probability of a lighting defect given a brake defect, 0.87, is greater than the unconditional probability of a lighting defect, 0.65. A plausible explanation is that both defects tend to be associated with a lack of maintenance.

2.6.3   Independence

The special case for which the occurrence of one event is independent of whether or not the other event occurred is particularly important.

Definition 2.15: Independence of two events

The events A and B are **independent** if and only if

$$P(A \cap B) = P(A)\,P(B).$$

Equivalently $P(A|B) = P(A)$.

Do not confuse A and B being independent with A and B being mutually exclusive. If A and B are mutually exclusive, $P(A \cap B) = 0$ and $P(A|B) = 0$.

Example 2.11: Sports cars [independence of two events]

The marketing manager in a motor vehicle manufacturing company has found that 10% of its sports cars are returned under warranty for electrical repairs, and 2% are returned for problems, with the electronic navigation and engine management system. Assume that electrical repairs are independent of electronic problems and calculate the probability that a randomly selected sports car is returned for both reasons.

If we write L and E to represent return for electrical repair and electronic repair respectively, and assume returns are independent:

$$P(L \cap E) = P(L)\,P(E) = 0.1 \times 0.02 = 0.002.$$

Example 2.12: Ice loading [a sequence of n independent events]

It may be easier to find the probability that an event does not occur than the probability that it does occur. In the heavy ice loading districts of the U.S., power supply line designs are based on cables having a layer of ice of 13 mm thickness.

Suppose this thickness of ice is exceeded, on average, in one year in twenty. If such years occur randomly and independently, find the probability of at least one such year in the next ten years.

Let C represent the event that ice exceeds a thickness of 13 mm in any one year.

$$P(C) = \frac{1}{20}$$

$$P(\overline{C}) = 1 - \frac{1}{20}$$

$$P(\text{at least one exceedance in 10 years}) = 1 - P(\text{no exceedance in 10 years})$$

$$= 1 - \left(1 - \frac{1}{20}\right)^{10} = 0.4013.$$

Example 2.13: Annual maxima [average recurrence interval]

The **average recurrence interval** (ARI) of annual maxima, such as floods or ice loading exceeding some specific level (H) is the reciprocal of the probability of exceeding H in any one year. For example, if $P(H < \text{annual maximum}) = 0.01$, then ARI $= 1/0.01 = 100$ years[10].

Suppose H is such that the ARI is T years. Assume annual maxima are independent. Find an expression for the probability that H will be exceeded at least once in n years, in terms of n and T. Calculate the probabilities for $T = 100$ and $n = 10, 20, 50, 100$.

The probability that the annual maximum exceeds H in any one year is $1/T$. Hence the probability that H is not exceeded in any year is

$$1 - \frac{1}{T}.$$

Assuming annual maxima are independent,

$$P(H \text{ not exceeded in } n \text{ years}) = \left(1 - \frac{1}{T}\right)^n,$$

and the general formula is

$$P(H \text{exceeded at least once in } n \text{ years}) = 1 - \left(1 - \frac{1}{T}\right)^n.$$

With $T = 100$ the probabilities are 0.096, 0.182, 0.395, 0.634 for 10, 20, 50 and 100 years respectively.

Example 2.14: Safety circuits [independent events]

A launch rocket for a space exploration vehicle has a sensing system that can stop the launch if the pressure in the fuel manifold falls below a critical level. The system consists of four pressure switches (Sw_i for $i = 1, \ldots, 4$), which should open if the pressure falls below the critical level. They are linked into an electrical connection between terminals A and B as shown in Figure 2.7. If current cannot flow from A to B the launch is

---

[10]Imagine 1 000 000 years. If the ARI is 100 years we expect arouns 10^4 years in which the flood exceeds H. The probability of exceeding H is $10^4/10^6 = 0.01$.

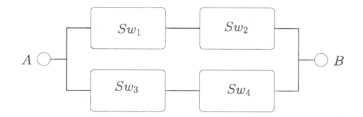

FIGURE 2.7: Circuit diagram for fuel manifold safety system.

stopped. Assume that each switch fails to open when it should with probability q and that failures of switches are independent. Find the probability that the launch will not be stopped if the pressure drops below the critical level in terms of q.

The safety system fails to stop the launch if current flows from A to B. Current will flow if either Sw_1 and Sw_2 or Sw_3 and Sw_4 fail to open. So the probability that the launch will not be stopped is

$$P\bigg(\big(Sw_1 \cap Sw_2 \text{ fail to open}\big) \cup \big(Sw_3 \cap Sw_4 \text{ fail to open}\big)\bigg)$$

$$= P\big(Sw_1 \cap Sw_2 \text{ fail to open}\big) + P\big(Sw_3 \cap Sw_4 \text{ fail to open}\big)$$

$$-P\bigg(\big(Sw_1 \cap Sw_2 \text{ fail to open}\big) \cap \big(Sw_3 \cap Sw_4 \text{ fail to open}\big)\bigg)$$

$$= q^2 + q^2 - q^2 q^2 = 2q^2 - q^4 \,.$$

The safety system is prone to another failure mode in which the current is cut off when the pressure is above the critical level. You are asked to investigate this in Exercise 2.29.

Independence extends to collections of any number n of events.

Definition 2.16: Independence for more than two events

A collection of events $A_1, \ldots, A_n$ is independent if, for every collection of k events $A_{i_1}, \ldots, A_{i_k}$, for $k \leq n$, we have that

$$P\left(\bigcap_{j=1}^{k} A_{i_j}\right) = P(A_{i_1} \cap A_{i_2} \cap \ldots \cap A_{i_k})$$

$$= P(A_{i_1})\,P(A_{i_2})\ldots P(A_{i_k})\,.$$

In particular, for the case of $n = 3$ events A, B and C, this condition requires that

1. $P(A \cap B) = P(A)\,P(B)\,,$

2. $P(A \cap C) = P(A)\,P(C)\,,$

3. $P(B \cap C) = P(B)\,P(C)\,,$ and also that

4. $P(A \cap B \cap C) = P(A)\,P(B)\,P(C)\,.$

It is essential to check all the requirements, as we see in the next example.

Example 2.15: Annual inspection of cars [three events]

An auto repair shop has kept records of the proportions of occasions on which it has needed to: fix hand brake adjustment (H); make a head lamp adjustment (L); and replace a spare tire with insufficient depth of tread (T), before the annual government inspection. The following probabilities are estimated from the records: $P(H) = 0.1$, $P(L) = 0.2$, $P(T) = 0.3$, $P(H \cap L) = 0.02$, $P(H \cap T) = 0.03$, $P(T \cap L) = 0.06$, and $P(H \cap L \cap T) = 0.015$. If you draw a Venn diagram, similar to Figure 2.4, and start from the triple intersection you will see that these probabilities are consistent. The first three requirements for the independence of three events are satisfied, but the fourth is not since $P(H) \times P(L) \times P(T)$ is 0.006 which is different from 0.015. The events are not independent because the probability of the occurrence of all three safety issues is higher than it would be if the three issues were independent. A possible explanation is that all three safety issues occurring is associated with bad maintenance, but one or two safety issues could occur by oversight.

2.6.4 Tree diagrams

Tree diagrams are a useful means of representing the sample space for questions that involve conditional probabilities. The diagram begins from the left hand side of the paper. An uncertain event is represented by a node, and any possible events that could follow are represented by lines leaving that node. Events leaving a node are mutually exclusive and mutually exhaustive. Labels representing these events are placed at the end of lines and can become nodes for subsequent uncertain events. The probability associated with an event is placed above the line joining it to the preceding node. This probability is conditional on the occurrence of the event represented by the previous node. So the sum of the conditional probabilities on lines leaving any node must be 1.

Example 2.16: Computer bugs [tree diagrams]

In a batch of 90 computers, 10 have a hardware bug. Use a tree diagram to find the probability of exactly one computer with a bug in a random sample of three, drawn without replacement. Let B represent a computer with the hardware bug and G represent a computer without the bug. The tree diagram shown in Figure 2.8 is a sample space. Notice that the sum of probabilities on the lines leaving each node is 1.

The eight sequences shown in the tree diagram are mutually exclusive and collectively exhaustive events. The three sequences GGB, GBG and BGG correspond to exactly one computer with a bug. We now consider the probability of the sequence GGB, and assume that at each stage of the sampling, the remaining computers have the same chance of selection.

The multiplicative rule gives

$$P(GGB) = P(1^{st}G) \times P\left(2^{nd}G \mid 1^{st}G\right) \times P\left(3^{rd}B \mid 2^{nd}G, 1^{st}G\right).$$

For the first draw $P(1^{st}G) = \dfrac{80}{90}$.

Then there are 89 computers remaining of which $80 - 1 = 79$ are good and so

$$P(2^{nd}G \mid 1^{st}G) = \dfrac{79}{89}.$$

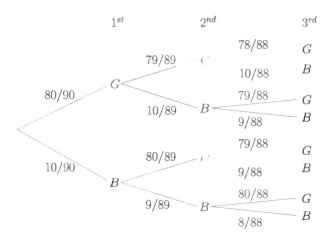

FIGURE 2.8: Tree diagram for sample space for drawing three computers from a batch of 90.

Finally at the third draw, there are 88 computers remaining of which 10 have a bug and $80 - 2 = 78$ are good and hence

$$P\left(3^{rd}B \mid 2^{nd}G, 1^{st}G\right) \;=\; \frac{10}{88}.$$

Thus

$$P(GGB) \;=\; \frac{80}{90} \times \frac{79}{89} \times \frac{10}{88} \;=\; 0.08966.$$

The probabilities $P(GBG)$ and $P(BGG)$ are identical to $P(GGB)$ as only the order of terms in the numerator of the product differs. Since GGB, GBG and BGG are mutually exclusive, the probability of exactly one computer with a bug is

$$3 \times 0.08966 \;=\; 0.269.$$

Example 2.17: Satellite launch [tree diagram]

A telecommunications company plans to launch a satellite. The probability of a successful launch is 0.90. If the launch is successful the probability of correct positioning is 0.8. If the position is wrong there is a 0.5 chance of correcting it. Even if the satellite is correctly positioned there is a probability of 0.3 that the solar powered batteries fail within a year. Let L, P, R and S represent a successful launch, successful initial positioning, successful correction to position, and solar batteries lasting longer than one year respectively. Use a tree diagram to find the probability the satellite will be handling messages in a year's time (Figure 2.9).

The sample space consists of six mutually exclusive and exhaustive sequences, each representing an event. Labeling events from left to right, they are:

$$LPS; \;\; LP\overline{S}; \;\; L\overline{P}RS; \;\; L\overline{P}R\overline{S}; \;\; L\overline{P}\,\overline{R}; \;\; \text{and} \;\; \overline{L}.$$

Two of these events, LPS and $L\overline{P}RS$, correspond to the satellite handling messages in

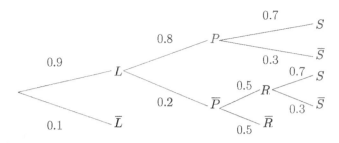

FIGURE 2.9: Tree diagram for sample space for satellite launch.

a year's time. The probabilities on the branches leaving any node in a tree diagram are conditioned on events to the left of the node, and you should check that the probabilities leaving a node sum to 1. It follows that the probability of any sequence is given by the product of the probabilities on the branches. In this case

$$
\begin{aligned}
\mathrm{P}(LPS) &= 0.9 \times 0.8 \times 0.7 = 0.504 \\
\mathrm{P}(L\overline{P}RS) &= 0.9 \times 0.2 \times 0.5 \times 0.7 = 0.063
\end{aligned}
$$

and the probability that the satellite will be handling messages in a years time is equal to $0.504 + 0.063 = 0.567$. We now write an R-function that takes the probabilities as the argument. This facilitates changing the estimated probabilities for a sensitivity analysis.

We define

$$
\begin{aligned}
p_1 &= \mathrm{P}(L) \\
p_2 &= \mathrm{P}(P|L) \\
p_3 &= \mathrm{P}(R|L\overline{P}) \\
p_4 &= \mathrm{P}(S|L \cap (P \cap \overline{P}R))
\end{aligned}
$$

The required probability is

$$
\begin{aligned}
P(LPS \cup L\overline{P}RS) &= \mathrm{P}(LPS) + P(L\overline{P}RS) \\
&= p_1\,p_2\,p_4 + p_1(1 - p_2)p_3\,p_4.
\end{aligned}
$$

A function, written in R, to perform the arithmetic follows.

```
> satprob <- function(p1,p2,p3,p4)
+ {
+     p1*p2*p4+p1*(1-p2)*p3*p4
+ }
> satprob(0.9,0.8,0.5,0.7)
[1] 0.567
```

2.7 Bayes' theorem

Thomas Bayes, who was born in London in 1702 and died in Tunbridge Wells, England, in 1761, studied logic and theology at the University of Edinburgh before he was ordained as a

Nonconformist minister. He also had a keen interest in mathematics. His famous posthumous paper [Bayes, 1763], has far reaching consequences.

2.7.1 Law of total probability

Sometimes the probability of an event E cannot be determined directly, but can be expressed through conditional probabilities. The conditioning events $\{A_i\}$, for i from 1 up to n, must be mutually exclusive and collectively exhaustive so that they form a **partition** of, and hence constitute, the sample space.

$$
\begin{aligned}
P(E) &= P(E \cap \Omega) \\
&= P\big(E \cap (A_1 \cup A_2 \cup \ldots \cup A_n)\big) \\
&= P\big((E \cap A_1) \cup \ldots \cup (E \cap A_n)\big) \\
&= P(E \cap A_1) + \cdots + P(E \cap A_n).
\end{aligned}
$$

This relationship is illustrated for $n = 5$ in Figure 2.10. The area of E equals the sum of

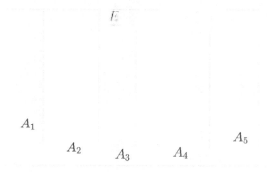

FIGURE 2.10: The vertical lines partition the rectangle representing the sample space into five events: $A_1, A_2, \ldots, A_5$. An event E is $(E \cap A_1) \cup (E \cap A_2) \cup \cdots \cup (E \cap A_5)$, where $A_1 \cup A_2 \cup \cdots \cup A_5 = \Omega$.

the areas E and A_1, E and A_2, E and A_3, E and A_4, and E and A_5. The **law of total probability** is obtained by expanding each of the $P(E \cap A_i)$ with the multiplicative rule.

Theorem 2.1 The Law of Total Probability

If the events $A_1, A_2, \ldots, A_n$ form a partition of the sample space, then for any other event E, we may write

$$
P(E) = P(A_1) P(E|A_1) + P(A_2) P(E|A_2) + \cdots + P(A_n) P(E|A_n).
$$

Example 2.18: Snowmelt flooding [law of total probability]

Spring snow-melt is a major contribution to flooding in river basins in the north of the U.S. Let F represent the event of a spring flood and E_1, E_2 and E_3 represent the snow accumulation, none/light, normal, and heavy respectively, during the preceding winter. An engineer has proposed the management of wash-lands as a flood defense strategy.

From records, the engineer estimates $P(E_1) = 0.30$, $P(E_2) = 0.45$, and $P(E_3) = 0.25$. A hydrologist estimates the conditional probabilities of flooding, given the proposed management of wash-lands and a snow accumulation scenario, as: $P(F|E_1) = 0.05$; $P(F|E_2) = 0.10$; $P(F|E_3) = 0.20$. Then the probability of flooding in some future year, for which we don't know the snow accumulation, is:

$$P(F) \quad = \quad 0.05 \times 0.30 + 0.10 \times 0.45 + 0.20 \times 0.25 \quad = \quad 0.11 \ .$$

2.7.2 Bayes' theorem for two events

Example 2.19: Screening test [Bayes' theorem]

A **screening test** is a relatively cheap and easily applied test for some underlying condition that, if present, will require detailed investigation or treatment. For example, recent research has demonstrated that the application of a pressure transient to a water main can detect leaks before pipe failure. The pressure transient can be created by opening and then rapidly closing a valve. The pressure is monitored with a sensor and this signal is analyzed. The result of the test and analysis is either that a leak is detected or that no leak is detected (see Figure 2.11). If a leak is detected the main will be inspected with closed circuit television (CCTV), a far more expensive procedure than the pressure transient test (PTT).

The pressure transient is the screening test and the detailed investigation is inspection by CCTV. In common with most screening tests, the PTT is fallible but we assume that CCTV is infallible. There is a probability that the PTT detects a leak when there is no leak, and a probability that PTT does not detect a leak when there is a leak. The probabilities of these two errors are assessed from laboratory tests as a and b respectively. Write L for the event of a leak in the main and D for the event that PTT detects a leak in the main. Then

$$P(D|\overline{L}) \quad = \quad a \quad \text{and} \quad P(\overline{D}|L) = b.$$

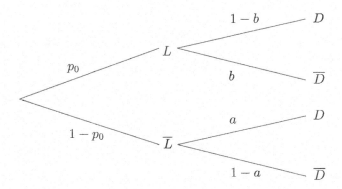

FIGURE 2.11: Screening test for water mains.

Now suppose that an engineer's assessment that a water main has a leak before PTT is p_0. This assessment would be based on: the material the main is constructed from; the age of the main; the soil type; and the record of bursts in neighboring pipes. Bayes' theorem uses the definition of conditional probability to give an updated probability of a leak given the PTT result.

$$P(L|D) = \frac{P(L \cap D)}{P(D)} = \frac{P(L) P(D|L)}{P(L) P(D|L) + P(\overline{L}) P(D|\overline{L})}$$

$$= \frac{p_0(1-b)}{p_0(1-b) + (1-p_0)a}.$$

For example, suppose $p_0 = 0.1$, $a = 0.02$, $b = 0.05$, and PTT detects a leak. The revised probability of a leak is

$$\frac{0.1 \times 0.95}{0.1 \times 0.95 + 0.9 \times 0.02} = 0.84.$$

In general, there is a trade-off between making a screening test very **sensitive**, in which case there is a higher chance of detecting a condition when it does not exist and making it highly **specific** to the condition, which tends to increase the chance of failing to detect a potential condition when it does exist.

2.7.3 Bayes' theorem for any number of events

Bayes' theorem is usually given for a set of underlying events $\{A_i\}$ that are mutually exclusive and exhaustive.

Theorem 2.2 Bayes' theorem

If the events $A_1, A_2, \ldots, A_n$ form a partition of the sample space, then for any other event E, we may write

$$P(A_i|E) = \frac{P(E|A_i) P(A_i)}{\sum_i P(E|A_i) P(A_i)}.$$

Note that the law of total probability has been used to expand $P(E)$ in the denominator. A past[11] online BBC News Magazine headline was "Meet the bots that edit Wikipedia". ClueBot NG is one of several hundred autonomous computer programs that help keep the encyclopedia running. Bot is short for robot, and amongst other tasks they rapidly detect and erase wiki-vandalism. The data in the following table (Table 2.2) is fictitious but we use it to demonstrate Bayes' law. We can now calculate the probability that a change was in each of the four categories, given that it has been erased by a Bot. First we use the theorem of total probability to calculate the probability that a change is erased by a Bot.

P(erased by a Bot) $= 0.81 \times 0.02 + 0.05 \times 0.40 + 0.10 \times 0.93 + 0.04 \times 0.99 = 0.1688$

---

[11] 25th July 2012.

TABLE 2.2: Changes to Wikipedia and robot erases of changes (fictitious).

reason for change	percentage %	P(Bot erases)
legitimate	81	0.02
advertising	5	0.40
mischievous	10	0.93
malicious	4	0.99

Now

$$
\begin{aligned}
\text{P(legitimate} \mid \text{erased by a Bot)} &= \frac{\text{P(legitimate} \cap \text{erased by a Bot)}}{\text{P(erased by a Bot)}} \\[2mm]
&= \frac{\text{P(legitimate) P(erased by a Bot} \mid \text{legitimate)}}{\text{P(erased by a Bot)}} \\[2mm]
&= \frac{0.81 \times 0.02}{0.1688} = 0.0960.
\end{aligned}
$$

Similarly

$$
\begin{aligned}
\text{P(advertising} \mid \text{erased by a Bot)} &= 0.1185 \\
\text{P(mischievous} \mid \text{erased by a Bot)} &= 0.5509 \\
\text{P(malicious} \mid \text{erased by a Bot)} &= 0.2350
\end{aligned}
$$

We check that these possibilities, all of which are conditional on the change being erased by a Bot, add to 1.

2.8 Decision trees

Tree diagrams can be extended by associating monetary values with uncertain events, and adding nodes to represent decisions.

Example 2.20: Macaw Machining (adapted from [Moore, 1972]) [decision tree]

Mia Mallet, the engineering manager of Macaw Machining, has been told by the sales manager that Crow Cybernetics needs 1000 control valves designed and manufactured to meet a specification that Crow will supply. However, Crow will only place an order with Macaw if Macaw produces a satisfactory prototype. Crow is prepared to pay $800 per valve.

Mia estimates the cost of producing a prototype is $48 000, and this cost must be met by Macaw whether or not it is given the order.

Next, Mia talks to the production engineer Hank Hammer. Hank says the cost of tooling will be \$80 000 and the marginal cost of production will be \$560 per valve using machined parts. Hank adds that the marginal cost per valve could be reduced to \$480 if they use some molded parts. The cost of the die for the molded parts would be \$40 000 but the remaining tooling cost would be reduced by \$16 000. This sounds good, until Hank points out that the molded parts might not meet the specification. In this case they will have to revert to machined parts and incur the full cost of tooling as well as the cost of the redundant die.

Before she decides whether or not it is worth submitting a prototype to Crow, Mia needs some estimates of probabilities. Mia estimates the probability that the prototype meets Crow's specification and Macaw receives the order as 0.4. Hank estimates the probability that the molded parts meet the specification as 0.5.

Do you think it is worthwhile producing a prototype? If so, would you risk buying the die? We will answer these questions in terms of maximizing expected monetary value (EMV), using a decision tree.

The decision tree is similar to a tree diagram, with the refinement that some of the nodes are decision points, shown as filled squares, rather than uncertain events shown as open circles (Figure 2.12).

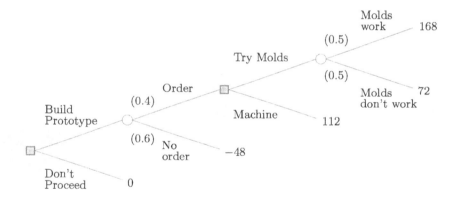

FIGURE 2.12: Initial decision tree for production of prototype and manufacture of valves.

We will work backwards from the right hand side, replacing uncertain events and the associated profits by the EMV and selecting the option with the highest EMV at decision points. But first we need to calculate the profit associated with each possible scenario. For example, suppose we produce a prototype, get the order, try molded parts and the molded parts are satisfactory. The profit, in unit of \$1 000, is: minus the cost of producing the prototype; plus the revenue from selling 1000 valves to Crow, minus the tooling cost; minus the die cost; plus the reduction in tooling cost; minus the marginal cost of producing 1000 valves with molded parts.

$$-48 + 0.8 \times 1\,000 - 80 - 40 + 16 - 0.48 \times 1\,000 \quad = \quad 168.$$

Similar arithmetic gives the following table. Now we roll back to the decision about purchasing a die to make molded parts. If we try molds, then there is a 0.50 probability of ending up with 168 and a 0.5 probability of ending up with 72. The EMV is

$$168 \times 0.5 + 72 \times 0.5 \quad = \quad 120.$$

Scenario	Profit ($1 000)
Do not produce prototype	0
Produce prototype but no order	−48
Prototype, order, use machine parts	112
Prototype, order, try molded parts, mold satisfactory	168
Prototype, order, try mold but not satisfactory	72

This is greater than the profit of 112 if we use machined parts without trying molded parts, so we decide to try molded parts (Figure 2.13).

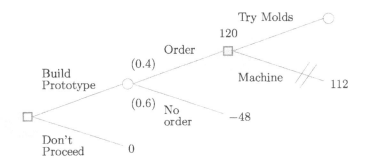

FIGURE 2.13: Rolling back to the decision about purchase of die for molded parts.

Rolling back to the original decision, if we decide to produce a prototype there is a 0.6 probability of −48 and a 0.4 probability of an EMV of 120. The EMV for producing the prototype is therefore

$$120 \times 0.4 + (-48) \times 0.6 \quad = \quad 19.2$$

This is greater than 0, so using an EMV criterion we advise Mia to produce a prototype and purchase a die to try molded parts.

The EMV criterion is risk neutral, rather than risk averse, which is more reasonable for relatively small contracts than it would be for large ones. We have ignored the delay between Macaw's outlay and receiving money from Crow - we ask you to consider discounting in Exercise 2.40. The decision is sensitive to assumed probabilities, and writing an R script would enable us to carry out a sensitivity analysis. Decision tree analyses are particularly useful for comparing business opportunities if you have only enough resources to follow up a few of them.

2.9 Permutations and combinations

A permutation is an arrangement of a given number of objects. A combination is a choice of a given number of objects from a collection of a greater than, or equal, number of objects. Permutations and combinations are commonly used in probability calculations when outcomes are equally likely, and in other mathematical applications.

The number of **permutations** (arrangements) of r distinguishable objects is

$$r \times (r-1) \times \ldots \times 1 \;=\; r!$$

because the first in line is any of the r objects, the second is any of the $(r-1)$ remaining objects, and so on until only one is left for the rth in line.

Example 2.21: Service improvements [permutations of r objects]

An electricity supply company lists seven improvements to service and offers a prize of $1 000 if you list them in the same order of importance as a panel of experts, and agree to take out a direct debit payment scheme. What is the cash value of this inducement if customers and experts are equally likely to select any one of the orderings?

There are 7! arrangements of the seven improvements to service

$$7! \;=\; 7 \times 6 \times \cdots \times 1 \;=\; 5\,040.$$

If the winning order is equally likely to be any one of these, the expected value of the prize is

$$1\,000 \times \frac{1}{5\,040} + 0 \times \left(1 - \frac{1}{5\,040}\right) \;=\; 0.1984,$$

which is slightly less than 20 cents.

The number of **permutations of r objects from n** is

$$P_r^n \;=\; n \times (n-1) \times \ldots \times (n-r+1),$$

because the first object is any one of n objects, the second object is any one of the remaining $(n-1)$ objects, and the rth object is any one of the remaining $(n-r+1)$ objects.

Example 2.22: Committee positions [permutations of r from n objects]

How many ways are there of assigning 3 from 5 people on the committee of a local chapter of an engineering institution to the posts of chair, treasurer and secretary?

Any 1 of 5 people for the chair, with any 1 of 4 people for treasurer with any 1 of 3 people for secretary.
That is

$$5 \times 4 \times 3 \;=\; 60$$

arrangements. This is illustrated in Figure 2.14.

The number of **combinations (choices) of r objects from n** is:

$$\binom{n}{r} \;=\; \frac{P_r^n}{r!} \;=\; \frac{n!}{(n-r)!r!} \,,$$

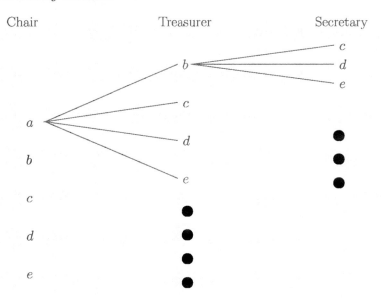

FIGURE 2.14: Sixty ways of arranging five people (a, b, c, d, e) to three offices.

because each choice can be arranged in $r!$ ways.
That is,

$$\binom{n}{r} \times r! \;=\; P^n_r.$$

Example 2.23: Computer packages [combinations]

A company offers its latest computer, at list price, with a choice of any 5 from a list of 8 software packages added as a special offer. How many choices can be made?

$$\binom{8}{5} \;=\; \frac{8!}{(8-5)!5!} \;=\; \frac{8 \times 7 \times 6}{3 \times 2 \times 1} \;=\; 56.$$

Therefore, there are 56 possible choices.

2.10 Simple random sample

Probability is used to indicate how likely it is that certain events will occur given assumptions about the underlying population and random process. In contrast, the aim of statistics is to infer[12] something about a population from a sample. We would like our sample to be representative, but we cannot be sure about this without knowing about the population. Usually, the best we can do is to draw a sample according to some random sampling scheme.

---

[12]In statistics, an **inference** is an uncertain statement about a characteristic of a population based on data. In contrast, a **deduction** is a logical consequence of the premises on which the detection is based. Given a model and the laws of probability we can deduce the probabilities of specified events.

This has the advantages of being seen as fair and of providing a basis for quantifying sampling error. The most straight-forward random sampling scheme is the **simple random sample**.

Suppose a population has N members. Imagine each member of the population has an associated ticket that is put into a hat. Then n tickets are drawn from the hat, in an idealized manner such that at every draw each ticket in the hat is equally likely to be drawn. This is simple random sampling and gives a simple random sample (SRS). If a ticket is not replaced after being drawn, the sampling is without replacement. Formally, a SRS of n from N is a sample drawn in such a way that every possible sample of size n has the same probability of selection. It is usual to sample without replacement and SRS, without further specification, will mean without replacement.

With SRS, the probability of selecting a particular member of the population is n/N. This result follows from the equally likely outcomes definition of probability. In the case of no replacement there are $\binom{N}{n}$ possible samples of size n. The number of samples of size n that contain a particular member of the population, X say, is $\binom{N-1}{n-1}$ because we are taking $n-1$ items from the remaining $N-1$ items to go with X. Since all samples of size n are equally likely,

$$\text{P(sample contains } X) \;=\; \frac{\binom{N-1}{n-1}}{\binom{N}{n}} \;=\; \frac{n}{N}.$$

You are asked to consider the case of SRS with replacement in Exercise 2.57.

Example 2.24: Disk drives [SRS]

A manufacturer of computers has been offered a batch of 6 000 disk drives from a company that has gone into liquidation. The manufacturer usually buys disk drives from a known reliable source and on average, they last for 130 hours under accelerated test conditions. The manufacturer is interested in the batch of disk drives for a new, cheaper model, and would require an average lifetime, under the accelerated (artificially extreme) test conditions, of at least 90 hours.

The disk drives are packed in cartons of 50, and it is agreed that the manufacturer may test a random sample of 25 drives before making any offer for the whole consignment. The following strategy would result in a simple random sample of the disk drives. Number the cartons from 1 up to 120. Define a rule for numbering drives within a carton. Then, carton 1 contains drives $0001,\ldots,0050$, carton 2 contains drives $0051,\ldots,0100$ and so on until carton 120 contains drives $5\,951,\ldots,6\,000$. The R command `sample(1:6000, 25)` will give a simple random sample (SRS) of 25 from 6 000 disk drives[13].

In a SRS scheme every item in the population has the same probability of selection. However, a sampling scheme in which every item in the population has the same chance of selection is not necessarily a SRS scheme. In the last example, Example 2.24, the disc drives are packed in 120 cartons of 50. Suppose now that the manufacturer is now taking a sample of 120 for a non-destructive visual assessment. If we take a SRS of 1 from each carton, every disc has

---

[13] Alternatively use a table of random numbers and work down columns of four consecutive digits. The four digit number 0001 corresponds to drive $1,\ldots,6\,000$ corresponds to drive 6 000. Four digit numbers in the range 6 001 up to 9 999 and 0000 are ignored. Continue until you have selected 25 different drives.

the same probability, 1/50, of being in the sample of size 120. This is not a simple random sampling scheme from the population of 6 000 discs, because it distributes the sample across the cartons. In a simple random sampling scheme a particular SRS could come from just a few cartons. Distributing the sample across the cartons is preferable to a simple random sampling scheme, although it relies on the notion of a simple random sampling scheme within cartons. It is an example of a stratified sampling scheme (Chapter 14).

It is not always practical, or even possible, to use simple random sampling schemes in an engineering context. In the next chapter you will see that we often resort to assuming that data are an SRS from some population. In general, we will use "random sample" to indicate that a sample is assumed to be an SRS from the corresponding population, and reserve SRS for cases where a simple random sampling scheme has been implemented. If a simple random sampling scheme is impractical, we try to make the assumption realistic.

Example 2.25: Fuel economy [approximation to SRS]

An inspector from the United States Environmental Protection Agency (EPA) intends estimating the fuel economy and emission rates of atmospheric pollutants of a particular compact car, under typical urban driving conditions. The age of the car has been specified as within two years of manufacture. The inspector has resources to test three cars, each for one day.

Implementing a simple random sampling scheme of all such vehicles in the U.S. is not feasible, partly because the cost of tracing the owners would be excessive. Moreover, implementing a simple random sampling scheme of all possible drivers and all possible scenarios to emulate typical driving conditions is not possible. A reasonable approximation would be to take an SRS of three local car hire agencies that include such compacts in their fleet, and to hire one compact from each agency. The inspector would specify three different daily drives designed to emulate typical urban driving conditions and provide guidelines for the drivers.

2.11   Summary

2.11.1   Notation

$\emptyset$	empty set
Ω	sample space/universal set
$\overline{A}$	complement event of A (not A)
$\mathrm{P}(A)$	probability of event A
$\mathrm{P}(A \mid B)$	probability of event A conditional on event B
$\cup$	union/OR
$\cap$	intersection/AND
$\binom{n}{r}$	number of combinations of r objects from n objects
P_r^n	number of permutations of r objects from n objects

2.11.2   Summary of main results

For an empty set $\emptyset$ and sample space Ω, we have

$$\mathrm{P}(\emptyset) = 0 \quad \text{and} \quad \mathrm{P}(\Omega) = 1.$$

For events A, B and C we have

$0 \leq \mathrm{P}(A) \leq 1$

$\mathrm{P}(A \cup B) = \mathrm{P}(A) + \mathrm{P}(B) - \mathrm{P}(A \cap B)$ the addition rule

$\mathrm{P}(A \mid B) = \dfrac{\mathrm{P}(A \cap B)}{\mathrm{P}(B)}$ conditional probability

$\mathrm{P}(A \cap B) = \mathrm{P}(A)\,\mathrm{P}(B \mid A) \; = \; \mathrm{P}(B)\,\mathrm{P}(A \mid B)$ the multiplicative rule

$\mathrm{P}(A \cap B) = 0$ mutually exclusive/disjoint

$\mathrm{P}(A \cap B) = \mathrm{P}(A)\,\mathrm{P}(B)$ independence between A and B

$\mathrm{P}(A \cap B) = \mathrm{P}(A)\,\mathrm{P}(B),$

$\mathrm{P}(A \cap C) = \mathrm{P}(A)\,\mathrm{P}(C),$

$\mathrm{P}(B \cap C) = \mathrm{P}(B)\,\mathrm{P}(C),$ independence between A, B and C

$\mathrm{P}(A \cap B \cap C) = \mathrm{P}(A)\,\mathrm{P}(B)\,\mathrm{P}(C)$

For arranging and choosing r objects out of n possible objects we have:

$P_r^n = n(n-1)\dots(n-r+1) = \frac{n!}{(n-r)!}$ the number of arrangements (permutations)

$\binom{n}{r} = \frac{P_r^n}{r!} = \frac{n!}{(n-r)!r!}$ the number of combinations (choices)

A SRS of size n from a population of size N is a sample drawn in such a way that every possible choice of n items from N has the same probability of being chosen.

2.11.3   MATLAB® and R commands

In the following x and y are nonzero integers and p1 to p4 are any real numbers. For more information on any built in function, type **help(function)** in R or **help function** in MATLAB.

R command	MATLAB command
`set.seed(x)`	`rng(x)`
`sample(0:x, size=y, replace=TRUE)`	`datasample(0:x, y,'Replace',true)`
`sample(0:x, size=y, replace=FALSE)`	`datasample(0:x, y,'Replace',false)`
`satprob <- function(p1,p2,p3,p4){`	`function phm = satprob(p1,p2,p3,p4)`
`p=p1*p2*p4+p1*(1-p2)*p3*p4`	`phm(1) = p1*p2*p4+p1*(1-p2)*p3*p4`
`phm = rep(0,2)`	`phm(2) = 1 - phm(1)`
`phm[1] = p`	`end`
`phm[2] = 1 - p`	
`return(phm)}`	

2.12 Exercises

Section 2.2 Random digits

Exercise 2.1: Random digits

Generate 10 sequences of 20 pseudo-random digits.

(a) For each sequence of 20 digits record:
  (i) the number of cases of two consecutive digits being the same digit repeated (for example 00 or 33);
  (ii) the longest run of digits in natural order including 9, 0 as a natural order (for example 2 3 4 5 and 8 9 0 1 are runs of length 4);
  (iii) any instance of three consecutive digits being the same digit repeated three times;
  (iv) the number of instances of two consecutive digits being the same digit repeated if you treat each sequence of 20 digits as 10 distinct pairs.

  For example, with the following sequence obtained using R
  ```
  > set.seed(1)
  > sample(0:9,size=20,replace=TRUE)
  [1] 2 3 5 9 2 8 9 6 6 0 2 1 6 3 7 4 7 9 3 7
  ```
  A. is 1 (66)
  B. is 2 (23)
  C. is 0
  D. is 0 (23 59 28 96 60 21 63 74 79 37)

(b) Referring to (a),
  (i) What proportion of two consecutive digits are the same digit?
  (ii) What is the longest run of digits in natural order? Does this increase if you concatenate the 10 sequences?
  (iii) How many instances of three consecutive digits have you considered and for what proportion of these is the same digit repeated three times?
  (iv) What proportion of the 100 distinct pairs of digits are the same digit repeated?

Exercise 2.2: Pseudo-random binary process

Use a software function to generate a pseudo random binary sequence (PRBS) in which consecutive digits are equally likely to be 0 or 1. How might you generate a short random binary sequence using an everyday object, and why is this not practical for engineering applications?

Exercise 2.3: PRBS and change of base (radix)

The following sequences are excerpts from a long PRBS record format in blocks of four.

(a) ...0001 1100 0111 0010 1001 ...
  Transform each block of four binary digits into a base 10 integer, ignoring any integers greater than 9. What number do you obtain?

(b) ...1110 0100 1010 0010 ...

Transform each block of four binary digits into hexadecimal
(0, 1, ..., A, B, C, D, E, F). What hexadecimal number do you obtain?

Exercise 2.4: Random selection of one unit

A public health inspector intends taking a sample of water from the kitchen tap at one
house in a street of houses numbered from 1 through to 186.

(a) Which number house would be chosen using the following sequence of random
digits if consecutive blocks of three digits represent a house number,

```
> sample(0:9,size=20,replace=TRUE)
[1] 5 0 2 2 8 2 7 9 9 0 7 2 1 9 4 4 9 5 9 7
```

and

(i) 000 and numbers above 186 are ignored;

(ii) Numbers in the ranges: 001 through to 186; 201 through to 386; 401 through
to 586; 601 through to 786; and 801 through to 986 with 0, 200, 400, 600, 800
respectively subtracted represent a house number between 1 and 186;

(iii) each consecutive sequence of three digits is reduced to the range 001 through
to 186 by taking the remainder after division by 186 (arithmetic modulo 186)
and adding 1.
For example in R, 873 would become:

```
> (873 %% 186) + 1
[1] 130)
```

(b) For each of (i) (ii) (iii), does every house have an equal chance of selection?

(c) What values of i, j, k in the R function sample(i:j,k) would give a random
selection of one house such that each house has the same probability of selection?
How could this pseudo random selection be linked to today's date, expressed in
the form yyyymmdd, so that it is reproducible?

Exercise 2.5: Life rafts 1

Refer to Example 2.1. You have been asked to select three lifeboats from 12, in such a
way that all 12 lifeboats are equally likely to be selected.

(a) Associate the integers 01 and 51 with lifeboat 1, 02 and 52 with lifeboat 2 and so
on up to 12 and 62 with lifeboat 12. Are all lifeboats equally likely to be selected?
Use the sequence of 20 digits given in Section 2.2.1 to select three lifeboats.

(b) Associate five two digit integers with each lifeboat. Use the sequence of 20 digits
given in Section 2.2.1 to select three lifeboats.

(c) Consider all one hundred two digit integers from 00 to 99. Take the remainder
after division by 12 and associate the non-zero remainder with the same numbered
lifeboat. Associate a remainder of 0 with lifeboat 12. Would all the lifeboats be
equally likely to be selected?

Exercise 2.6: Life rafts 2

You have been asked to select 3 from 12 life rafts such that each is equally likely to be
selected.

(a) Suggest a way of doing so using the sequence:

$$62428308813516528670$$

obtained from `sample(0:9,20,replace=T)` in R.

(b) Give a simple function call that will produce such an SRS of 3 from 12.

Exercise 2.7: PRNG

Investigate the performance of the following algorithm as a pseudo-random generator of digits D_j, where

$$
\begin{aligned}
I_j &= 106\, I_{j-1} + 1\,283 (\text{modulo } 6\,075), \quad \text{with } I_0 = 1\,234. \\
D_j &= \left\lfloor \frac{10\, I_j}{6\,075} \right\rfloor,
\end{aligned}
$$

where $\lfloor x \rfloor$ is the integer part of x.

Exercise 2.8: Linear Congruential Generator

A linear congruential generator (LCG) has the form

$$x_{t+1} = ax_t + c\,(mod\,m).$$

Take $a = 134775813$, $c = 1$, $m = 2^{32}$ and with $x_0 = 123456789$,

(a) Calculate $x_1, x_2, \ldots, x_5$ using R.   (Note that in R, $x\%\%y$ is $x(mod\,y)$).

(b) Now divide $x_1, x_2, \ldots, x_5$ by m, multiply by 10, and truncate to the greatest integer less than the product, writing down the sequence of 5 digits that you obtain.

Exercise 2.9: RANDU

Generate 30 000 numbers U_i using RANDU given in Example 2.3, setting $U_i = Z_i/2^{31}$ and create a set of 10 000 triples $(x, y, z) = (U_i, U_{i+1}, U_{i+2})$ from the 30 000 data you have generated.

(a) Plot the points (x, y, z) as a 3D scatter plot (see for example the command `scatter3` in MATLAB), rotating the view until you see why Knuth and his colleagues were so dismayed.

(b) Show that RANDU can be re-written as

$$Z_{i+2} = 6Z_{i+1} - 9Z_i \pmod{2^{31}}$$

and give an explanation as to why this indicates how this LCG fails the three dimensional criteria above.

Section 2.3 Defining probabilities

Exercise 2.10: Decagonal spinners

Two perfectly balanced regular decagonal spinners are spun. Spins are independent and each spinner is equally likely to show any one of the digits $0, \cdots, 9$.

Find probabilities for the following events.

(a) The total is a multiple of 3.

(b) The total is a multiple of 4.

(c) The total is a multiple of either 3 or 4.

(d) The total is a multiple of one of 3 or 4 but not both.

Exercise 2.11: Pumps

Five faulty pumps are mixed up with twelve good ones.

Find the probabilities of the following events if selection is at random:

(a) one selected pump is good;

(b) two selected pumps are good;

(c) two selected pumps are bad;

(d) out of two selected pumps one is good and one is bad;

(e) at least one out of two selected pumps is good.

Exercise 2.12: Hexadecimals

Hexadecimal is a positional numeral system with a radix (or base) of 16. It uses 16 distinct symbols: $0, \ldots, 9, A, B, C, D, E, F$. A pseudo random number generator appears to generate these symbols independently and with equal probabilities. Write down the probabilities of the following events as fractions.

(a) The next symbol is 8.

(b) The next two symbols are 53.

(c) The next two symbols are identical (i.e. one of $00, 11, \ldots, FF$).

(d) The next three symbols are ACE (in that order).

(e) The next four symbols are $7, A, C, E$ in any order.

Section 2.4 Axioms of probability

Exercise 2.13: Addition rule from the axioms of probability

(a) Draw a Venn diagram.

(i) Using Axiom 3, show that

$$\mathrm{P}(A) \;=\; \mathrm{P}\big(A \cap \overline{B}\big) + \mathrm{P}(A \cap B),$$

(ii) and

$$P(A \cup B) = P(A \cap \overline{B}) + P(A \cap B) + P(\overline{A} \cap B).$$

(b) Hence deduce the addition rule of probability for two events by applying Axiom 2.

Section 2.5 The addition rule of probability

Exercise 2.14: Addition rule

Show that for any three events A, B and C

$$P(A \cup B \cup C) = P(A) + P(B) + P(C)$$
$$-P(A \cap B) - P(A \cap C) - P(B \cap C) + P(A \cap B \cap C),$$

by using the addition rule for any two events A and B,

$$P(A \cup B) = P(A) + P(B) - P(A \cap B),$$

and substituting $(B \cup C)$ for B.

Exercise 2.15: Auto adjustments

A garage owner estimates, from extensive past records, that 70% of cars submitted for a Ministry of Transport Test need lamp adjustments, 60% need brake adjustments, and 50% of the cars need both adjustments.

(a) What is the probability that a car selected at random needs at least one adjustment?

(b) What is the conditional probability of a car requiring a lamp adjustment given that a brake adjustment was necessary?

(c) What is the conditional probability of a car requiring a brake adjustment given that a lamp adjustment was necessary?

Exercise 2.16: Roadside check

A roadside check of cars as they enter a tunnel found that: 38% had incorrect headlamp alignment (H); 23% had excessively worn tires (W); and 10% had both defects.

(a) Represent the sample space on a Venn diagram. Show the probabilities given in the question on your diagram.

(b) What is the probability that a randomly selected car will have either incorrect headlamp alignment or excessively worn tires or both defects.

(c) What is the probability that a randomly selected car has neither defect?

(d) What is the probability that a randomly selected car has incorrect headlamp alignment given that it has excessively worn tires?

(e) What is the probability that a randomly selected car has excessively worn tires given that it has incorrect headlamp alignment?

(f) Are the events H and W independent or not? Give a reason for your answer.

Exercise 2.17: Aluminum propellers

A manufacturer of aluminum propellers finds that 20% of propellers have high porosity (H) and have to be recycled. Fifteen percent of propellers are outside the dimensional specification (D) and have to be recycled. Of the propellers that have high porosity, 40% are outside the dimensional specification.

(a) What is the probability that a randomly chosen propeller has high porosity and is also outside the dimensional specification?

(b) Represent the sample space on a Venn diagram, and show the probabilities of the events $H \cap D$, $H \cap \overline{D}$, and $D \cap \overline{H}$ on your diagram.

(c) What is the probability that a randomly selected propeller will have either high porosity or be outside the dimensional specification or have both defects?

(d) What is the probability that a propeller has neither defect?

(e) What is the probability that a propeller has high porosity given that it is outside the dimensional specification?

Exercise 2.18: Car design awards

A new design of electric car has been entered for design awards. A journalist assesses the probability that it wins an award for appearance (A) as 0.4, efficiency (E) as 0.3, and comfort (C) as 0.2. Moreover, the probability that it wins any two awards, but not all three awards, is assessed as 0.09, and the probability that it wins all three awards is assessed as 0.01.

(a) Represent the sample space by a Venn diagram.

(b) What is the probability that the car wins at least one award?

(c) What is the probability that the car does not win an award?

(d) What is the probability that the car wins all three awards given that it wins both A and E?

(e) What is the probability that the car wins all three awards given that it wins at least two awards?

Exercise 2.19: Building design awards

A new university engineering building has been entered for design awards. An architect assesses the probability that it wins an award for aesthetics (A) as 0.5, efficiency (E) as 0.5, and ergonomics (G) as 0.5. Moreover, the probability that it wins any two awards, including the possibility of all three, is 0.25. However, the architect thinks it unlikely that it will win all three awards and assesses the probability of this event as 0.01.

(a) Represent the sample space by a Venn diagram.

(b) What is the probability that it does not win an award?

(c) Explain why for any two events from the collection of three events $\{A, E, G\}$, the two events are independent.

(d) Is the collection of three events $\{A, E, G\}$ independent?

Exercise 2.20: Boat design awards

A yacht has been entered for design awards. A naval architect assesses the probability that it wins awards for: aesthetics (A) as 0.6; handling (H) as 0.4; speed (S) as 0.2; A and H as 0.3; A and S as 0.1; and S and H as 0.1. When asked for the probability that it wins all three awards the naval architect replies that it will be the product of $P(A)$, $P(B)$ and $P(C)$, which is 0.048.

(a) Represent the sample space by a Venn diagram.

(b) What is the probability that it wins an award?

(c) What is the probability that it does not win an award?

(d) Does $P(A \cap H \cap S) = P(A)P(H)P(S)$ imply that the collection of three events $\{A, H, S\}$ is independent?

Section 2.6 Conditional probability

Exercise 2.21: Binary console

(a) A machine, on the operating console of a dam, displays numbers in binary form by the use of lights. The probability of an incorrect digit is 0.01 and errors in digits occur independently of one another. What is the probability of

    (i) a 2 digit number being incorrect?

    (ii) a 3 digit number being incorrect?

    (iii) a n-digit number being incorrect?

(b) What is the probability of guessing a 12 digit PIN number on a Say G'day card (a prepaid phone-card in Australia)?

Exercise 2.22: Intersection

Cars approaching a cross roads on Hutt Road turn left with probability 0.2, go straight with probability 0.7 and turn right with probability 0.1. Assume the directions that cars take are independent.

(a) What is the probability that the next 3 cars go in different directions?

(b) What is the probability that the next 3 cars go in the same direction?

(c) What is the probability that the next 3 cars all turn left?

Exercise 2.23: Component failure

The failure of A or B, and C in Figure 2.15 will stop the system functioning. If the

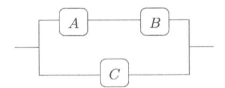

FIGURE 2.15: Component failure.

components A, B and C have independent probabilities of failure a, b and c what is the probability the system functions (in terms of a, b, and c)?

Exercise 2.24: Aircraft engine failure

Suppose that the probability of failure of an aircraft engine in flight is q and that an aircraft is considered to make a successful flight if at least half of its engines do not fail. If we assume that engine failures are independent, for what values of q is a two-engined aircraft to be preferred over a four engined one?

Exercise 2.25: Annual floods

The probability that the annual maximum flood level exceeds a height mark on a building in a flood plain area is p. Assume annual maximum floods are independent (water years are typically defined to make this plausible). The annual recurrence interval (ARI) for exceeding the mark is defined as

$$T = \frac{1}{p}.$$

(a) Justify this definition.

(b) Express the probability of at least one exceedance of the mark in an n year period.

   (i) Calculate this probability if $T = 100$ for $n = 50, 100$, and 200.

   (ii) For what value of n is the probability about 0.5?

Exercise 2.26: Twin engined aircraft

In a twin-engined plane the unconditional probability an engine fails at some time during a flight is q. However engine failures are not independent:

$$P(\text{right engine fails some time during the flight} \mid \text{left}$$
$$\text{engine fails some time during the flight}) = a.$$

Similarly for the left engine failing at some time during the flight given that the right fails at some time during the flight. The probability that one engine fails when the other does not:

$$P(\text{right engine fails} \mid \text{left engine does not fail}) \quad = \quad b,$$

and similarly for left engine fails at some time during the flight given right engine does not fail.

(a) Find the probabilities of $0, 1$ and 2 engines failing in terms of a, b and q.

(b) Now eliminate b, and express the probabilities in terms of a and q only.

Exercise 2.27: Core sample

The probability that a core sample will strike oil is 0.25 and the probability it will strike gas is 0.08. The probability of striking both oil and gas is 0.05. What are the probabilities of:

(a) striking oil or gas,

(b) striking oil or gas but not both,

(c) striking oil conditional on striking gas,

(d) striking gas conditional on striking oil.

Exercise 2.28: Bounds for probability

Show that for any events A, B and C

$$P(A) + P(B) + P(C) - P(A \cap B) - P(A \cap C) - P(B \cap C) \leq P(A \cup B \cup C)$$
$$\leq P(A) + P(B) + P(C).$$

Exercise 2.29: Pressure switches

Refer to Example 2.14 and Figure 2.7. The safety system malfunctions if it fails to stop the launch if the pressure is below the critical level or if it stops the launch when the pressure is above the critical level. Suppose that the each of the switches opens when there is not a critical pressure drop with probability p and that such failures are independent.

(a) Find the probability that the launch will be stopped when the pressure is above the critical level, in terms of p.

(b) If the probability of a critical drop is θ, find the expression for the probability that the safety system malfunctions in terms of θ, p and q.

(c) Suppose that failing to stop the launch if there is a critical pressure drop entails a loss of 100 monetary units, whereas stopping a launch when there is no critical pressure drop entails a loss of 1 monetary unit. If θ equals 0.02 and $pq = 0.1$ (with $0 < p, q < 1$), find the optimal values of p and q.

Exercise 2.30: Coordinates in space

Consider a random experiment for which the sample space consists of four equally likely outcomes

$$\mathcal{S} \;=\; \{(1,0,0),\ (0,1,0),\ (0,0,1),\ (1,1,1)\},$$

and so $P((1,0,0)) = \frac{1}{4}$, $P((0,1,0)) = \frac{1}{4}$, $P((0,0,1)) = \frac{1}{4}$ and $P((1,1,1)) = \frac{1}{4}$.
Let the events:

- E be the event that the first coordinate is 1,
- F be the event that the second coordinate is 1,
- G be the event that the third coordinate is 1.

Are E, F, and G independent events ?

Exercise 2.31: Favorable events

An event B is said to favor an event A if $P(A \mid B) > P(A)$. If B favors A and C favors B does it follow that C favors A? Give a proof or counter example.

Section 2.7 Bayes' Theorem

Exercise 2.32: Offshore drilling

An oil company manufactures its own offshore drilling platforms. An ultrasonic device will detect a crack in a weld with a probability of 0.95. It will also wrongly signal cracks in 10% of good welds. From past experience about 2% of new welds contain cracks. What is the probability of a crack if the ultrasonic device signals one?

Exercise 2.33: Flood mitigation

A flood mitigation scheme using washlands to protect an historic city has been assessed, under three rainfall scenarios. An engineer has assessed conditional probabilities of flood damage to the city (flood) in any year as:

$$\begin{aligned}
P(\,flood \mid dryyear) &= 0.2,\\
P(\,flood \mid typicalyear) &= 0.4,\\
P(\,flood \mid wetyear) &= 0.5
\end{aligned}$$

A meteorologist has specified probabilities for the three scenarios by:
$P(dry) = 0.3, P(typical) = 0.6$. Find the probability of flooding in a year.

Exercise 2.34: Main bearings

The chief engineer on a cargo ship applies an acoustic test of the condition of the engine main bearing while the ship is in port. There is a 5% chance the test indicates a defective bearing when it is good, and a 10% chance the test indicates a good bearing when it is defective. Before applying the test, the chief engineer assesses the probability of a defective main bearing as 0.01. If the acoustic test indicates a defective bearing, what should the probability of a defective bearing be revised to?

Exercise 2.35: Excavator

If an excavator has an engine problem, a warning light on the excavator indicates an engine problem exists with probability 0.96. However, the warning light will also indicate a problem, when a problem does not exist, with probability 0.02. At the start of a shift, before seeing the excavator, the driver considers that the probability of an engine problem is 0.03. Draw a tree diagram to represent the sample space. If the warning light comes on, what probability should the driver assign to the engine problem?

Exercise 2.36: Email

An email system sends incoming mail to either the In-Folder (I) or the Trash Folder (T). You classify incoming mail as Useful (U), in which case you want it sent to I or as a Nuisance (N) in which case you would like it sent to T. If incoming mail is U, the system unhelpfully sends it to T with probability 0.1. If the incoming mail is N, the system unhelpfully sends it to I with probability 0.05. Suppose a proportion 0.35 of your incoming mail is N.

(a) What is the probability that an incoming mail is sent to T?

(b) What is the probability that an incoming mail is U given that it is sent to T?

Exercise 2.37: Mine safety

At the beginning of each shift the safety of the walls and roof in a mine gallery are assessed by a laser system. The laser system provides either a warning (W) of instability or no warning ($\overline{W}$). If a warning is issued an engineer will perform a detailed check and declare the gallery as safe (S) or unsafe ($\overline{S}$). The probability that the laser system issues a warning if the gallery is safe is 0.02, and the probability it fails to issue a warning if the gallery is unsafe is 0.01. Assume that at the beginning of a shift, before the laser system is used, the probability that the mine is unsafe is 0.004.

(a) Represent the sample space by drawing a tree diagram. Show the probabilities given in the question on your diagram.

(b) At the beginning of the shift, what is the probability the system will issue a warning?

(c) Given that the laser system issues a warning at the beginning of the shift, what is the probability that the engineer finds the gallery to be unsafe?

Section 2.8 Decision trees

Exercise 2.38: Decision tree

Use an expected monetary value (EMV) criterion, EMV being the product of an amount of money with the probability that it accrues, to quantify benefits. Represent the problem as a tree, going from left to right, using squares for decision points and circles for contingencies. Add the corresponding probabilities to the lines leaving circles. Then work backwards from the right hand side, calculating the EMV at decision points. EMV and a the decision tree were covered in Example 2.20.

A design engineer in Poseidon Pumps has an idea for a novel electronically controlled pump which would have many industrial applications. It would cost 15 monetary units (MU) to develop a prototype pump which would have a 0.3 probability of being successful. If the prototype is successful, the tooling for full scale production would cost an additional 90 MU. Without advertising, the probabilities of high, medium and low sales are estimated as $0.6, 0.3$ and 0.1. With 5 MU of advertising the probabilities of high, medium and low sales are estimated as $0.7, 0.2$ and 0.1 respectively. The revenue from high medium and low sales, having allowed for the cost of materials and labour during manufacture, is estimated as $200, 100$, and 50 at current prices.

(a) Draw a tree diagram using squares for decisions and circles for contingencies (chance events). Show the probabilities for the different outcomes following circles.

(b) Calculate the net value of each route through the tree and put these net values at the ends.

(c) Using an EMV criterion, should Poseidon Pumps build the prototype, and if it is successful should they advertise?

[Hint: consider the decision about advertising first.]

Exercise 2.39: Airport construction

You have to decide whether or not to bid for an airport construction project, and if you do bid whether or not to carry out a survey of ground conditions before preparing the tender.

(a) Cost of preparing tender is 20.

(b) Profit depends on ground conditions which can be good (G) with profit 1000, fair (F) with profit 500, poor (P) with profit 200 and bad (B) with profit -500 (a loss). Profit estimates have not allowed for the tender cost or the cost of a possible survey.

(c) Cost of survey would be 30. The survey will establish the ground conditions.

(d) Probability you win the contract is 0.10.

(e) Probability of G, F, P, B ground conditions are: $0.1, 0.5, 0.2, 0.2$ respectively.

    (i) Draw a decision tree, marking decisions with squares and contingencies with circles.

    (ii) What decisions would you make if you use an EMV criterion?

Exercise 2.40: Discounting

Discounting is the converse of compound interest. Suppose interest rates are α per annum. Then, $1 now will be worth $$1(1 + \alpha)$ at the end of one year. Conversely, a promise of $1 in a year's time is worth $$1(1+\alpha)^{-1}$ now, and the latter is the **discounted value**.

(a) (i) If $1 000 is invested at 4% p.a. over a three year period what will the amount in $ be at the end of the three years?

(ii) What is the current worth, that is discounted value, of $1 000 to be paid at the end of three years?

(b) Consider the decision tree for Macaw Manufacturing in Example 2.20. Now suppose that even if it wins the order, Macaw will not receive payment from Crow before the end of 12 months. In the meantime Macaw has to find the money for development and manufacture. Discount the payment from Crow by 10% per annum and assume that Macaw incurs development and manufacturing costs now. Use an EMV criterion with the discounted payment from Crow and advise Macaw.

Exercise 2.41: CVaR

A company has options on two construction contracts, A and B, but only has the resources to complete one of them. An engineer has assessed the options and has produced tables of possible profit to the company with associated probabilities.

Contract A

Profit	-3	-2	-1	0	1	2	3	4	5	6	7
probability	0.05	0.05	0.1	0.1	0.1	0.1	0.1	0.1	0.1	0.1	0.1

Contract B

Profit	-4	-3	-2	-1	0	1	2	3	4	5	6	7
probability	0.05	0.05	0.05	0.05	0.05	0.1	0.1	0.1	0.1	0.1	0.1	0.15

(a) Calculate the expected monetary value (EMV) for the two options.

(b) The conditional value at risk at the 20% cut-off, CVaR(0.20), is the expected profit conditional on the profit being amongst the least favorable outcomes which have a combined probability of 0.20.

(i) For option A the CVaR(0.20) is the expected profit given that the profit is negative (-3 or -2 or -1). What is the numerical value of CVaR(0.20) for option A?

(ii) For option B the CVaR(0.20) is the expected profit given that the profit is negative (-4 or -3 or -2 or -1). What is the numerical value of CVaR(0.20) for option B?

(c) Which option would you recommend if the company wishes to use a EMV criterion? Which option would you recommend if the company wishes to use a CVaR(0.20) criterion?

Exercise 2.42: Utility 1

A company has a large order for electricity generation and distribution plant from an overseas customer. The profit to the company will be 10 monetary units (mu) but there is a probability (0.1) of the customer defaulting. If this occurs the profit to the company will be -6. Unfortunately the company would be in jeopardy if it makes such a large loss. There is an asymmetry between large losses and large profits. The benefits of a profit of 10 mu to the company are less in magnitude than the detriment of a 6 mu loss. The notion of utility can be used to model this asymmetry. The CEO draws up a relationship between profit and utility to the company, which is shown in the following table. The utility is equal to the profit when the profit is positive, but larger

Profit (mu)	-6	-5	-4	-3	-2	-1	0	...	10
Utility	-110	-80	-50	-20	-6	-2	0	...	10

in absolute magnitude than losses. The company can insure against the default. The insurance premium is 2 mu to cover a loss of 6 mu, so if the company takes out insurance and the customer defaults the profit to the company is -2 because the company will not recover the cost of the premium.

(a) Would the company accept the order without insurance if it uses an EMV criterion?

(b) Would it be worth accepting the order with insurance using an EMV criterion?

(c) Would the company accept the order without insurance if it uses an expected utility criterion (similar to EMV except utility replaces mu)?

(d) Would the company accept the order with insurance if it uses an expected utility criterion?

Exercise 2.43: Utility 2

A company has a risk averse utility function for money (x) given by $U(x) = 1 - e^{-0.05x}$. The probability of a fire destroying the premises in any one year is 0.001. The cost of rebuilding would be 80 monetary units (mu). The cost of insurance would be 0.5 mu.

(a) Show that insurance is worthwhile if expected utility is considered.

(b) The risk premium is the difference between the premium and the EMV of the loss. What is the risk premium in this case?

Section 2.9 Permutations and combinations

Exercise 2.44: Radios

A manufacturer makes digital radios with three wavebands. In how many ways can three from seven non-overlapping wavebands be chosen?

Exercise 2.45: Counting 1

Use a combinatorial argument to explain the following results, stating any restrictions on the values of the integer r.

(a) $\binom{n}{r} = \binom{n}{n-r}.$

(b) $\binom{n}{0} = 1.$

(c) $\binom{n}{r} = \binom{n-1}{r-1} + \binom{n-1}{r}.$

(d) $\sum_{k=0}^{n} \binom{n}{k}\binom{m-n}{n-k} = \binom{m}{n}.$

(e) Prove part (c) above, which is known as *Pascal's Formula*, using the algebraic definition of

$$\binom{n}{r} = \frac{n!}{r!(n-r)!}.$$

Show that this corresponds to obtaining each subsequent row of Pascal's triangle by adding the two entries diagonally above.

Exercise 2.46: Counting 2

Consider arranging n objects which can be grouped into sets of identical types.

(a) How many anagrams are there of the inner-city Sydney suburb WOOLLOOMOOLOO?

(b) How many arrangements are there of four 0s and three 1s?

(c) How many arrangements that do not begin with a 0 are there of four 0s and three 1s?

(d) Suppose there are n objects where n_1 of them are identical of type 1, n_2 of them are identical of type 2, ... and n_k of them are identical of type k, so that

$$\sum_{i=1}^{k} n_i = n_1 + n_2 + \ldots + n_k = n.$$

Explain why there are

$$\frac{n!}{n_1! n_2! \cdots n_k!}$$

different arrangements. The usual notation for $\dfrac{n!}{n_1! n_2! \ldots n_k!}$ is $\binom{n}{n_1, n_2, \ldots, n_k}.$

(e) What is the coefficient of $xy^3 z^2$ in the expansion of $(x + y + z)^6$?

(f) How many arrangements are there of n_1 objects of type 1 and n_2 objects of type 2?

(g) Explain why the number of arrangements of n_1 objects of type 1 and n_2 objects of type 2 is identical to the number of choices of n_1 objects from $n = n_1 + n_2$ distinct objects.

Exercise 2.47: Counting 3

Consider arrangements in which repetitions are allowed. Explain why the number of arrangements of r from n distinct objects if repetitions are allowed is

$$n^r$$

and hence deduce that

$$\ln(n!) \quad < \quad n \ln(n).$$

Exercise 2.48: Counting 4

Consider the number of ways in which r objects can be chosen from a set of n distinct objects without regard to the order of the selection, but in which we allow repetition.

(a) How many ways are there of choosing 2 tickets for 3 baseball games if you can choose two tickets for the same game?

(b) Explain why the number of ways to choose r elements from n distinct elements, if elements can be repeated, is equivalent to finding the number of non-negative integer solutions to the equation:

$$x_1 + x_2 + \ldots + x_n \quad = \quad r.$$

(c) Explain why the number of ways to choose r elements from n distinct elements, if elements can be repeated, is

$$\binom{n+r-1}{r}.$$

Exercise 2.49: Car registration plates

(a) How many different identity plates can be issued if a plate has four digits (from $0, 1, \cdots, 9$) and repetitions are allowed?.

(b) How many different identity plates can be issued if a plate has three letters (from $A, B, \cdots, Z$) and repetitions are allowed, followed by four digits as in part (a)?

(c) How many different identity plates can be issued if a plate has two letters, two digits and two letters in sequence?

Exercise 2.50: Yacht crew

A crew of 4 for a yacht is to be selected from 12 sailors, 7 from the U.S. and 5 from Canada. How many different crews can be formed

(a) if there is no restriction on the selection of the crew;

(b) if the crew must consist of 2 from the U.S. and 2 from Canada;

(c) if the crew must include at least one from Canada?

Exercise 2.51: Binomial expansion

The binomial expansion for positive integer n is given by

$$(a+b)^n = \underbrace{(a+b)(a+b)\ldots(a+b)}_{n}$$

$$= \binom{n}{0}a^n + \binom{n}{1}a^{n-1}b + \binom{n}{2}a^{n-2}b^2 + \ldots + \binom{n}{n-1}ab^{n-1} + \binom{n}{n}b^n$$

$$= a^n + na^{n-1}b + \frac{n(n-1)}{2}a^{n-2}b^2 + \ldots + nab^{n-1} + b^n.$$

(a) Expand $(a+b)^3$, $(1+x)^4$, $(1+x)^5$.

(b) Justify the binomial expansion by a combinatorial argument.

(c) Use the binomial expansion to prove that

$$\sum_{k=0}^{n} \binom{n}{k} = 2^n.$$

(d) Prove the binomial theorem by mathematical induction.

Exercise 2.52: Gamma function

The gamma function is defined by

$$\Gamma(\alpha) = \int_0^{\infty} t^{\alpha-1}e^{-t}dt, \alpha < 0.$$

Use integration by parts to show that

$$\Gamma(\alpha) = (\alpha - 1)!$$

Hence deduce a value for 0!

Exercise 2.53: Morse code

Morse code is made up of short (S) and long (L) pulses.

(a) How many arrangements are there of 3 Ss and 2 Ls?

(b) Explain why this is the same as the number of ways of choosing 3 from 5.

(c) How many arrangements are there of r Ss and $(n-r)$ Ls?

Exercise 2.54: Injection moulding

A small injection moulding company has three machines A, B and C, which produce proportions of out of specification items $0.01, 0.02$ and 0.05 respectively. Seventy percent (70%) of production is from A, 20% from B, and the remaining 10% is from C.

(a) What proportion of the total production is out of specification?

(b) An inspector finds an item that is out of specification. What is the probability that it was produced on machine C?

Exercise 2.55: Alloy engine blocks

Aluminum engine blocks for a sports car are inspected for porosity using X-ray tomography and classed as perfect, good or poor. The proportions of blocks in the three categories are 0.10, 0.75, and 0.15 respectively, and poor blocks are recycled. An engineer suggests weighing engine blocks to estimate porosity instead of accurate X-ray tomography. A critical weight (c) is set such that proportions 0.05, 0.20 and 0.90 of blocks in the three categories, respectively, fall below c.

(a) What is the probability that a block is poor if it is above (c).

(b) How many owners in one million will be affected if the probability that a poor engine block causes noticeable mechanical problems is 0.02?

Exercise 2.56: Arrangement of books

A small library consists of 100 different books. An imaginary super robot can rearrange the books into any other order in Planck time ($5.39 \times 10^{-44}s$). The current age of the universe is estimated as 13.8 billion years by NASA's Wilkinson Microwave Anisotropy Probe (WMAP) project.

(a) How many arrangements of 100 books are there?

(b) How long would it take the robot to go through them all? Give your answer as a multiple of the estimated age of the universe.

Section 2.10 Simple random sample

Exercise 2.57: SRS with replacement I

A sample of size n is drawn from a population of size N using SRS with replacement.

(a) Is the probability of a particular member being selected greater or less or the same as using SRS without replacement?

(b) What is the probability of a particular member being selected using SRS with replacement?

Exercise 2.58: SRS: Automobile manufacturer

An automobile manufacturer receives pistons for a particular engine from three suppliers A, B and C.

(a) A batch of 100 pistons has just been received from each of the three suppliers. An engineer will take an SRS of 6 from the 300 pistons to check that the dimensions are within specification.

    (i) What is the probability that all 6 pistons in the SRS come from A?

    (ii) What is the probability that there is no piston from C in the SRS?

(b) Batches of 50, 80, and 100 have been received from A, B and C. An engineer will take an SRS of 6 from the 230 pistons.

    (i) What is the probability that all 6 pistons in a SRS of 6 pistons come from C?

    (ii) What is the probability that no piston in a SRS of 6 pistons is from A?

(c) Suggest a better sampling scheme than an SRS of 6 from the combined batches.

Miscellaneous problems

Exercise 2.59: The Martingale betting system

Suppose we have a perfectly fair coin. You can bet x units, up to a maximum stake $64, that the coin will land head up if it is flipped. The coin is flipped. If it lands head up you receive x, but if it lands tail up you forfeit x. The martingale strategy is to start with $1 and double your bet after every loss. Then, at the first win you recover all previous losses and receive a $1 profit.

(a) Show that your expected profit is 0.

(b) There have been four consecutive tails since you first bet so you have lost $1 + 2 + 4 + 8 = 15$ dollars . Therefore you present value is -15, and your next stake is 16 dollars. What is your expected value after one more flip of the coin?

(c) Does the expected profit depend on the maximum stake allowed?

Exercise 2.60: PRNG

Is a sequence from the decimal representation of π a reasonable source of random digits between 0 and 9 such that each digit is equally likely to appear as the next in sequence? If so, would this be a PRNG with an infinite cycle length?

Exercise 2.61: Benford's Law

Benford's Law describes the fact that the distribution of leading (or leftmost) digits of the vast collection of data sets follows a well-defined logarithmic trend, rather than an intuitive uniformity. That is,

$$P(i) \quad = \quad \log_{10}\left(\frac{i+1}{i}\right),$$

which in practice means that the most common leading digit is a 1, with probability of approximately 0.301, and the least common leading digit is 9, with an approximate probability of 0.046.

Verify this fact by checking the larger data sets on the book website. There are several theoretical arguments to justify Benford's Law, including [Lee, 2012].

3

Graphical displays of data and descriptive statistics

We consider how to take a sample so that it is likely to be a fair representation of the population from which it is drawn. You will learn how to present data using diagrams and numerical summary measures. In some applications the time order of the observations is particularly relevant, and we refer to the data as a time series. We consider a descriptive approach that describes a time series as a combination of a trend, seasonal effects and an random component.

3.1   Types of variables

Consider a set of items, and define a **variable** as a feature of an item which can differ from one item to the next. Variables that are measured on a numerical scale can conveniently be classified as **discrete** or **continuous**. Discrete variables are usually a count of the number of occurrences of some event and are therefore non-negative integers $\{0, 1, 2, ...\}$. Continuous variables such as mass, temperature and pressure are measured on some underlying continuous scale.

Other variables provide a verbal rather than a numerical description, and are referred to as **categorical** variables. If the categories of a categorical variable can be placed in some relevant order it is referred to as an **ordinal** variable.

We illustrate the use of these terms in the following six examples.

Example 3.1: Integrated circuit chips [discrete variable]

A typical very-large-scale integrated circuit chip has thousands of contact windows, which are holes of 3.5 microns diameter etched through an oxide layer by photo-lithography. A window is defective (closed) if the hole does not pass through the oxide layer. A factory produces wafers that contain 400 chips, arranged in four sectors that are referred to as north, east, south and west. There is a test pattern of 20 holes systematically located within each sector. A robot records the number of closed windows in each test pattern. The number of closed windows is a discrete variable that can take any integer value between 0 and 20. A sector is scrapped if any closed windows are found in the test pattern.

Example 3.2: Linear energy transfer [continuous vs discrete variables]

Characterizing the vulnerability of space-borne electronic devices to single-event upsets, which cause a change in logic state, is critical for the success of space missions. The upset-rate depends on the linear energy transfer (LET) of the incident particles, the exposed cross-section, and the fluence.

The results of tests on a programmable gate array device are given in Table 3.1. The count of the number of upsets is a discrete variable which can take integer values from 0 upwards, with no clearly defined upper limit. The LET is a continuous variable. The fluence is the average number of particles that intersect a unit area per second. The average is not restricted to integer values and the fluence is considered as a continuous variable.

TABLE 3.1: Numbers of " power on reset single-event functional interrupt" upsets for the XQR4VLX200 field programmable gate array device under test conditions (courtesy of Xilinx).

LET MeV cm^2 mg^{-1}	Fluence $\times 10^3$	Count
2.0	100 000	6
4.0	100 000	11
6.8	50 000	11
16.9	19 400	10
22.0	10 000	6
30.0	13 400	19
90.3	9 180	28

Example 3.3: Flame retardant [continuous variable]

Antimony trioxide is a flame-retardant used in the manufacture of polymers and paints and as an opacifying agent for glass, ceramics and enamels. A company sells its top grade, for use when color is critical, in bags with declared contents of 25 kg. The bags are filled and weighed automatically and the tare (average mass of the bag) is subtracted to give the mass of antimony trioxide. The masses are is automatically recorded to the nearest 1 gm. Although a mass is recorded to the nearest integer, it is considered as a continuous variable because, for example, 25 018 gm represents a mass between 25 017.50 and 25 018.50 gm.

Example 3.4: Paint color [categorical variable]

Metal roofing is manufactured as sheet steel coated with aluminum zinc alloy and an optional color finish. There are four colors: white, green, red, and gray. The color is an example of a categorical variable. The manufacturer monitors sales of the different colors.

Example 3.5: Profitability [ordinal variable]

[Pan and Chi, 1999] investigated the effect of entry timing, mode of market entry, market focus, and location on the profitability of multinational companies in China.

The level of profitability (net profit as a percentage of sales) was categorized on a 7-level ordinal scale: heavy loss; slight loss; no profitability; profitability less than 3%; profitability between 3% and 8%; profitability between 8% and 15%; and profitability greater than 15%.

They used the ordinal variable profitability in their analysis, rather than introduce some numerical scale. A disadvantage of introducing a numerical scale is that, for example, you have to relate the benefit of moving from slight loss to no profitability with the benefit of moving from no profitability to small profitability in a quantitative way. Moreover, it may be hard to obtain more precise information about companies' level of profitability.

Example 3.6: Rivers and streams [ordinal vs categorical variables]

A table from the U.S. Environmental Protection Agency, reproduced as Figure 3.1, shows the percentage of rivers and streams in three classes of water quality, an ordinal variable with categories 'good', 'threatened' and 'impaired', by user group, which is a categorical variable.

National Summary
Designated Use Support in Assessed Rivers and Streams*

* Waters assessed for more than one designated use are included in multiple designated use groups below.

Description of this table

Designated Use Group	Miles Assessed	Percent Good	Percent Threatened	Percent Impaired	% Good / % Threatened / % Impaired
Fish, Shellfish, And Wildlife Protection And Propagation	761,058	55.8	.4	43.9	
Recreation	388,034	56.9	1.0	42.1	
Agricultural	352,881	95.3	.2	4.5	
Aquatic Life Harvesting	252,847	35.7	.1	64.2	
Industrial	196,867	97.8	.0	2.2	
Public Water Supply	196,798	77.5	.1	22.5	
Other	85,231	97.9	.0	2.1	
Aesthetic Value	26,213	88.9	.0	11.1	
Exceptional Recreational Or Ecological Significance	2,478	72.6	.0	27.4	

FIGURE 3.1: Water quality in rivers categorized by use.

3.2   Samples and populations

A **population** is a set of items which can be defined as finite or infinite. For example, it is common to consider all the production from a process that continues with its present settings, for an infinite period, as an imaginary infinite population. It is usually impractical, or impossible, to measure the value of a variable for all items in a population so we consider a finite subset of the population, known as a **sample**, instead.

We measure values of the variable for items in the sample. These values are known as **data**, and we need data to make estimates of probabilities that are used to model random variation within a population. The aim of modeling random variation is to predict the consequences, and when appropriate investigate strategies to reduce the random variation or construct systems that can accommodate it. There are five essential steps in any data collection exercise.

1. State the objective of the investigation.

2. Specify the variables that need to be measured.

3. Define the population.

4. Decide on a sampling strategy, draw the sample, and record the values of the variables.

5. Analyze the results and draw conclusions.

The definition of the population will depend on our point of view. Once we have defined our population we hope to obtain a representative sample. But why do we sample, and so raise the possibility that our sample is not representative of the population?

There are several compelling reasons for drawing a sample rather than investigating the entire population.

• If testing is destructive, then we must sample.

• In many cases, we define a hypothetical infinite population. For example, all future production if our process continues on its present settings. Then it follows from the definition of the population that the data we have now is a sample.

• If the population is large, then investigating every member of the population may not be practically possible.

• Sampling saves time and money.

• One hundred percent inspection of mass produced product by human operators does not generally find all the defective items, because the task is so repetitive and uninteresting. Even if such inspection does remove all defective items, obtaining high quality by removing defects is inefficient and wasteful. A far better strategy is to take samples from the process and use the results to identify how we can run the process so that it does not produce defective items.

A **probability based sampling scheme** has the property that every item in the population has a known non-zero probability of being in the sample. In a simple random sample (SRS) every item has the same probability of selection, but other sampling strategies can also have this property.

A **haphazard sample** is chosen without any specific rule, but it does not satisfy the definition of a probability based sampling scheme because the probability of selection is not known for each item in the population. Nevertheless, haphazard samples are often used as approximation to SRSs.

Example 3.7: Airline check-in [probability based sample]

An airline employee asks every 10th passenger leaving the check-in queue a few questions about the service. If the employee begins by asking the kth next passenger, where k is a random digit from $\{1, 2, \cdots, 9, 0\}$ and 0 corresponds to the 10^{th}h next passenger, then this is a probability based sample. Every passenger has the same probability of selection (1/10). It is not a SRS because, for example, a passenger cannot be in the same sample as the preceding or following passenger.

Example 3.8: Airline locker luggage [probability based sample]

A manager in an airline wants information about the masses of luggage stored in overhead lockers. She decides to randomly select two flights within each of these three categories:

- domestic flight of less than one hour,
- domestic flight of greater than one hour,
- international flight.

She arranges for the contents of a random sample of ten lockers from each flight to be weighed.

This is a probability based sampling scheme. The probability of a locker being selected is known because it follows from the number of flights in each category and the number of lockers in each aircraft which are known. But it is not an SRS because the population is subdivided into three categories. If an SRS of six flights was to be selected, then they could all be international or all domestic of less than one hour. The subdivision of the population ensures the sample contains a known proportion of lockers from each category. These ideas are followed up in Chapter 14.

We cannot be sure our sample is representative of the population, but if we use a probability based sampling scheme we can quantify how likely it is to be representative. Also, if we know something about the population we can ensure our sample is representative of the population in this respect. Providing we use a probability based sampling scheme, the larger the sample the more likely it is to be representative of the population. However, a small sample may suffice if items in a population do not vary much.

Example 3.9: Ranger Robots [steps in data collection]

Sam Spade is the design engineer in a small company Ranger Robots (RR), a spin-off from a university school of mechanical engineering, that designs and assembles specialist robots. Orders from customers are often for a single robot and rarely exceed ten, so the robots are assembled by hand. Each design requires different miniature electric motors and Sam has to rely on different suppliers. Sam's latest job is to design and supply a vertical climbing robot and he has ordered 240 direct-drive DC motors from Elysium Electronics (EE). EE is itself a small company and this is the first time Sam has dealt with it.

Sam's immediate objective is to ascertain whether the batch of 240 motors that has just been delivered will operate effectively and reliably in the robots. From Sam's perspective the batch of 240 motors is the population.

He can't measure the effectiveness and reliability of motors directly but he can measure electrical characteristics, such as peak torque at stall and stalling current and electrical time constant, and mechanical characteristics, such as the diameters of the armature and permanent magnet, without damaging the motor. He could subject motors to a highly accelerated lifetime test (HALT) but this will destroy the motors that are tested, or at least leave them unusable in the robots.

Sam now considers the sampling issue. Measuring the electrical and mechanical characteristics of every motor is possible in principle, but would be a lengthy and monotonous process, and a HALT test of every motor would destroy the entire batch! Instead he will test a sample, but how can he arrange that the sample is likely to be reasonably representative of the population? The motors all look the same so one possibility is to take a haphazard sample of, let's say, 6 motors.

Generally, there are potential pitfalls with taking a haphazard sample and a random sampling scheme should be used if it is practical to do so. For example, a haphazard sample will usually be chosen in a convenient manner, and if EE is cutting corners in its manufacturing processes it may place the better motors at the top of the container in the hope that they will be sampled.

Another scenario is that the haphazard sample values of torque are slightly below the specification but RR considers that they will be adequate for the application and expects a concession on the price. EE may suspect that all the motors have been tested and that the sample of 6 consists of the 6 motors with the lowest torques. Such situations can be avoided if RR and EE can agree that taking a SRS is a fair procedure. It is still possible that all six motors in the sample are below the specification for torque, when most of the batch meets the specification, but this is unlikely with a SRS. For example, if 10% of the batch is below specification, then the probability that all six motors in a SRS of six motors are below specification is approximately one in a million (0.1^6). Moreover, random sampling schemes help justify the assumptions underlying statistical analyses.

Sam finds that the motors are packed in two cartons of 120. He takes a SRS of 3 motors from each carton. If there is some substantial difference between the motors in each carton, and motors within a carton are relatively similar, then his sample should detect it.

After a few months, Sam was impressed with the good quality of EE motors and now regularly uses its products. His sampling of incoming batches is limited to a single motor. The main purpose of this check is to ascertain that the specification has been understood correctly and that no mistake has been made when shipping the order.

Example 3.10: Elysium Electronics [haphazard sample]

Elysium Electronics specializes in small high-tech components and its continuing success depends on its ability to meet specifications.

From EE's perspective the batch of motors is a sample in time of the hypothetical infinite population of all motors EE will produce if its processes remain unchanged. The production manager, Frances Fretsaw, samples 2 motors from each batch before they are packed and checks that they are within specification. She is familiar with the manufacturing processes at EE and is willing to rely on a haphazard sample of the motors from each batch, rather than draw a SRS.

3.3 Displaying data

Newspapers, scientific periodicals and the magazines published by professional engineering societies need to present data in a manner that will attract readers' attention, and graphics are often used. The aim is to make graphics entertaining and informative.

3.3.1 Stem-and-leaf plot

A **stem-and-leaf plot** provides a useful display, and check, of data that flags extreme observation and possibly erroneous observations. We illustrate a stem-and-leaf with gold grade from a mine in Canada.

Example 3.11: Gold grades [stem-and-leaf plot]

An engineer in a Canadian mining company was asked to estimate the gold grade of a sub-vertical vein-gold deposit in a mine. The engineer had identified the face area to be sampled and drilled 160 cores, perpendicular to the face, at the center points of a lattice superimposed on a plan of the face. The cores are evenly distributed over the area, but there is a possible snag. There might be a spatial periodicity in the gold grade which matches the spacing of the points in the lattice, so the gold grade measurements could be systematically higher or lower than the average level. However, spatial periodicity is likely to have a longer wavelength than the spacing between cores and such a frequency matching is unlikely. Nevertheless, in case there was some underlying pattern the engineer drilled a further 21 cores at randomly selected points on the face. So, the total sample size is 181 cores.

Each drill core was 1 m in length from a 105 mm diameter percussion drill-hole and weighed around 25 kg, but the assay of its gold content was based on a small sub-sample. The aim is that the small sub-sample will be representative of the core. In an attempt to achieve this goal, the core is fed into a rotating cone splitter so that most of the cuttings are less than 1 mm in length. The cuttings are placed on a table and a sample of 5 kg is taken. A table is used rather than a hopper because smaller particles will tend to gravitate to the base of the hopper. The 5 kg sub-sample is crushed to pass through a 100 mesh sieve (150 microns) and a 500 gm sub-sample is taken. This 500 gm sub-sample is pulverized and a 30 gm sub-sample is sent for assay. The results in parts per million of gold (ppm) are given on the website. In the following R commands we: read the data into a data frame; check that the data frame does contain the data we intend to analyze, using stem() and tail() to display the first 6 and last 6 observations; and use stem() to obtain a stem-and-leaf plot.

```
> gold <- read.table("gold_grade_C.txt",header=T)
```

```
> head(gold);tail(gold)
> stem(gold$gold)
```

The stem-and-leaf plot provides a first impression of the data, but wouldn't be suitable for a report. The leading digits on the left of the | are the stems, and each digit to the right of | is a leaf and represents a single datum. For this data set, the function defaults to stems that are even (increasing by 2), and leaves that go from 0 up to 9, repeated for the implied odd stem. The print-out tells us the location of the decimal point. So the second row starts with 0.20 and end with 0.39, and the third row starts with 0.42 and ends with 0.59. Some of the rows are ambiguous and, for example, we can't tell whether the minimum is 0.04 or 0.14, or whether the maximum is 3.87 or 3.97.

```
The decimal point is 1 digit(s) to the left of the |
 0 | 44445567
 2 | 0222334566677880011224556666778889
 4 | 22223345566778889900111233333344889999
 6 | 00012334455566899900000112345556889
 8 | 0111122333444788899124445567
10 | 80466799
12 | 001335679
14 | 000089189
16 | 5778
18 | 6
20 | 06
22 | 72
24 |
26 |
28 |
30 | 2
32 | 4
34 |
36 | 8
38 | 7
```

If we want to know the actual minimum or maximum, we can use **min()** and **max()**.

```
> min(gold$gold)
[1] 0.14
> max(gold$gold)
[1] 3.87
```

3.3.2   Time series plot

Observations are often taken in time order, and their variation over time may be of particular interest. For example, are sales increasing or decreasing, do they vary with time of year, can we identify the consequences of marketing a new product? A plot of a variable against time is known as a **time series plot**, and points are usually joined with line segments. Examples of time series plots are a plot of annual peak streamflow against year, and a plot showing a reduction in traffic fatalities over 1990 to 2012.

Example 3.12: Animas River [time series plot]

Flood mitigation strategies rely on estimates of probabilities that peak flows in rivers will exceed particular values. These probabilities have to be estimated from records of flows, which can be thought of as a sample in time. As an example, we give the annual peak flows, measured in cubic feet per second (cfs), and gage height, measured in feet (ft) for the Animas River at Durango in Colorado from 1924 until 2012 (see website).We need to consider how representative environmental records will be of the future, particularly given the evidence of climatic changes.

The first step is to plot the data as a time series. The year is shown on the horizontal axis and the variable is plotted against the vertical axis. In a time series plot, points for each year are joined by line segments[1].

```
> peak.ts <- ts(animas[,1],start=1924)
> gage.ts <- ts(animas[,2],start=1924)
> plot(cbind(peak.ts,gage.ts),main="")
```

The resultant plots are given in Figure 3.2. There is no obvious trend, but climatic change can be subtle and we consider sensitive methods of analysis in Section 9.8.

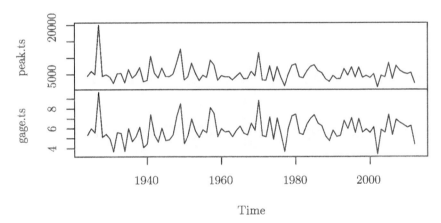

FIGURE 3.2: Annual peak streamflow (cubic feet per second) and Gage Height (feet) for the Animas River from 1924 to 2012.

Example 3.13: Road safety [time series plot]

The New York City Pedestrian Safety Study & Action Plan, August 2010 gives annual traffic fatalities from 1990 until 2009. The data are reproduced in Table 3.2.

The number of deaths in 2010, 2011 and 2012 are 269, 245, and 274 respectively. An article in *The New York Times* of March 13, 2013 had the headline "Traffic Fatalities in City Increased in 2012, but Officials Point to Larger Picture". A time series plot of the fatalities from 1990 to 2012 is shown in Figure 3.3.

---

[1]The function `ts()` makes a time series object that `plot()` will plot as a time series. The start time is an argument `start =` for `ts()`. The function `cbind()` binds vectors together into an object with several columns.

TABLE 3.2: Traffic fatalities in New York City 1990–2009.

Year	Fatalities	Year	Fatalities
1990	701	2000	376
1991	630	2001	395
1992	592	2002	389
1993	539	2003	358
1994	485	2004	294
1995	482	2005	326
1996	424	2006	330
1997	495	2007	280
1998	362	2008	293
1999	423	2009	256

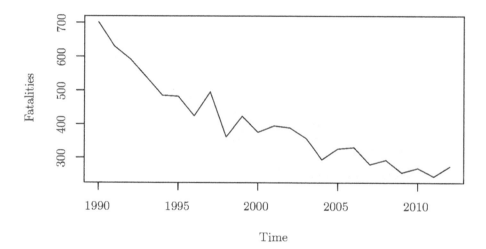

FIGURE 3.3: Annual traffic fatalities in New York City between 1990 and 2012.

```
> fatalities <- c(701, 630, 592, 539, 485, 482, 424, 495, 362, 423, 376,
395, 389, 358, 294, 326, 330, 280, 293, 256, 269, 245, 274)
> fatalities <- ts(fatalities,start=1990)
> plot(fatalities,main="")
```

There is clear evidence that the number of traffic fatalities has decreased over this period, but there is some random variation about the trend and the higher figure in 2012 does not imply that the annual fatalities are on the increase. We can consider the random variation about the trend as a sample in time from a hypothetical population of all possible random variates. Considering the trend, it looks like an exponential decrease over the period up to 2012. However we should not expect this trend to extrapolate into the future, although further reduction may be achieved if the city continues to introduce successful road safety measures.

3.3.3 Pictogram

Pictograms can be used to provide a visual summary of the main features of a data set. They are often amusing and can make an immediate impression, but they can, intentionally or not, be misleading. A common device is to scale the linear dimension of some icon in proportion to the change in some variable, so that the area is scaled by the square of the change. To illustrate the difference we use a quotation from the Global Gas Flaring Reduction Partnership brochure, October 2011: "Data on global gas flaring show that efforts to reduce gas flaring are paying off".

Example 3.14: Gas flares [pictogram]

"From 2005 to 2010, the global estimate for gas flaring decreased from 172 billion cubic meters (bcm) to 134 bcm (22%)". We compare the two scalings in Figure 3.4[2].

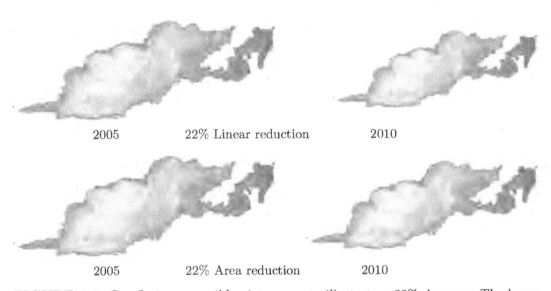

2005	22% Linear reduction	2010

2005	22% Area reduction	2010

FIGURE 3.4: Gas flaring   possible pictograms to illustrate a 22% decrease. The lower pictogram is a fairer representation.

---

[2]No pictogram was included in the brochure.

Example 3.15: Press advertisement [pictogram]

A variation on the theme appeared in a press advertisement in England during the "Westland affair" of 1986. It is reproduced in Figure 3.5.

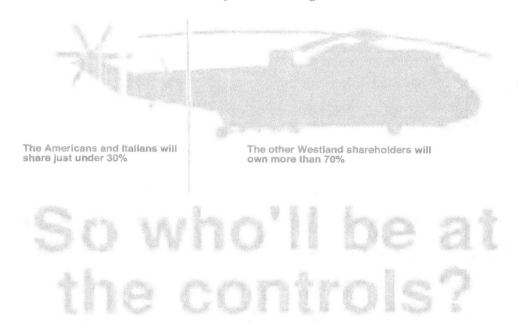

FIGURE 3.5: Press advertisement in England during the "Westland affair" of 1986.

Example 3.16: Gender gaps [pictogram]

An nice example of a pictogram with correct scaling, by Jen Christiansen, was used in an article "Gender Gaps" in the May 2013 edition of the *Scientific American*, Figure 3.6. The clever use of area encodes a considerable amount of information.

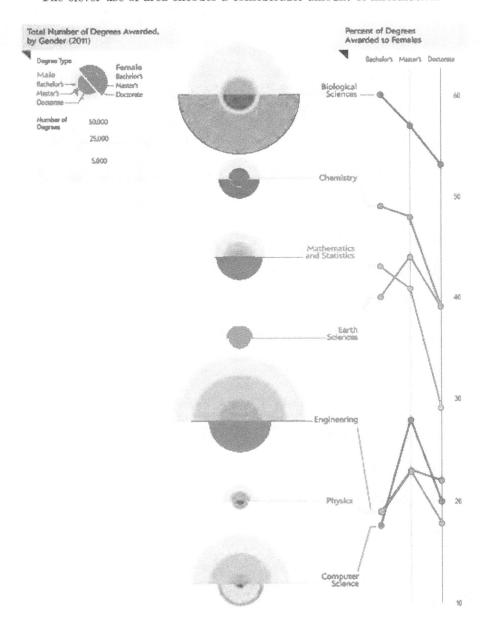

FIGURE 3.6: Diagram after a color figure by Jen Christiansen published in the *Scientific American*.

68 | Statistics in Engineering, Second Edition

3.3.4 Pie chart

A **pie chart** shows how the total quantity of some variable can be attributed to a set of mutually exclusive and exhaustive categories. The circle is divided into sectors with areas proportional to the quantity of the variable attributable to each category. Pie charts, with the same scaling, can be used to show how the total quantity and the division into categories has changed over time. Pie charts are commonly used in annual reports of corporations to compare this year with last year.

Example 3.17: Energy consumption [pie chart]

Pie charts for U.S. Energy consumption by user group (quadrillion Btu) in 1960, where the total was 45, compared to 2010, when the total was 98, are shown in Figure 3.7.

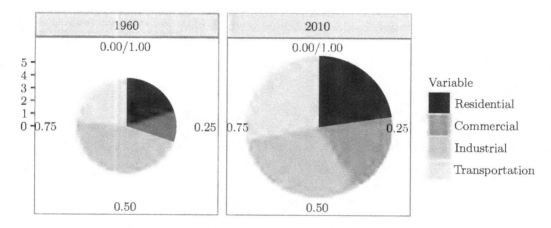

FIGURE 3.7: Energy usage in U.S. 1960 and 2010 [Energy Information Administration (EIA)].

3.3.5 Bar chart

An alternative to a pie chart is a **bar chart**. The heights of the bars represent the totals and the areas of the bars are divided into rectangles with areas proportional to the quantity attributable to each category. Bar charts are useful when we have data for several years.

Example 3.18: Energy generation [bar chart]

The Energy Information Administration (EIA) provides data for U.S. energy generation by source. Figures (thousands Megawatt hours) for ten years from 2003 with some merging of categories are given in Table 3.3. A bar chart for the data is shown in Figure 3.8.

TABLE 3.3: Based on U.S. Energy generation by source 2003-2012 [EIA].

Year	Coal	Petroleum	Gas	Nuclear	Hydro	Other	Total
2003	1 973 737	119 406	665 508	763 733	275 806	93 532	3 893 725
2004	1 978 301	121 145	725 352	788 528	268 417	97 299	3 981 046
2005	2 012 873	122 225	774 424	781 986	270 321	100 150	4 063 984
2006	1 990 511	64 166	830 618	787 219	289 246	109 499	4 073 265
2007	2 016 456	65 739	910 043	806 425	247 510	117 469	4 165 649
2008	1 985 801	46 242	894 688	806 208	254 831	137 905	4 127 683
2009	1 755 904	38 936	931 611	798 855	273 445	156 207	3 956 967
2010	1 847 290	37 061	999 010	806 968	260 203	180 028	4 132 570
2011	1 733 430	30 182	1 025 255	790 204	319 355	208 135	4 108 572
2012	1 517 203	22 900	1 241 920	769 331	276 535	231 253	4 061 154

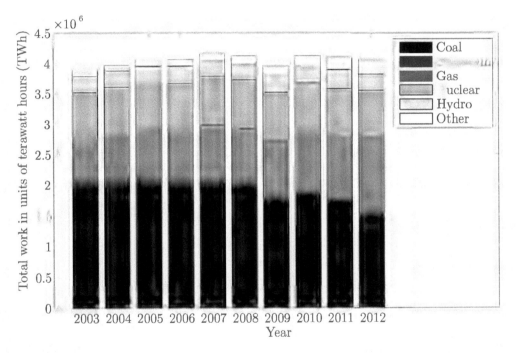

FIGURE 3.8: Based on U.S. Energy generation by source 2003-2012 [EIA].

Example 3.19: Cash earnings [misleading bar chart?]

Although it is not necessary to start graphical scales at 0, they should be clearly labeled. The bar chart reproduced in Figure 3.9 does not have the baseline labeled and the increments between the first half of 2012 and the second half of 2012, and the second half of 2012 and the first half of 2013, are not on the same scale. The consequence is that the increase of cash earnings is exaggerated.

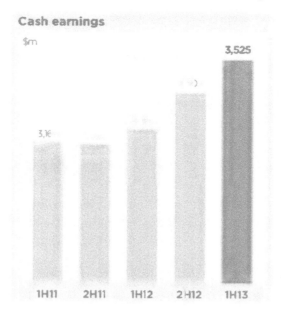

FIGURE 3.9: From a Shareholder Newsletter sent out by an Australian Bank in August 2013.

3.3.6 Rose plot

Rose plots are also known as radar, star, and spider plots. A rose plot is used for displaying data that are distributed around a circle. Examples are wind directions (where the wind is coming from), and wave headings (where the waves are heading to).

Headings of waves during hurricanes with winds exceeding 80 knots in the North Sea are given in Table 3.4, with a rose plot given in Figure 3.10.

```
> hurricane <- read.csv("data/hurricane.csv")
> ggplot(hurricane,aes(x=Heading,y=Number.of.hurricanes)) +
> geom_bar(stat='identity',fill='white',col='black') +
> coord_polar(start=-pi/16)
```

3.3.7 Line chart for discrete variables

A workshop that was used for the manufacture of brake linings now houses several computer numerical control (CNC) milling machines. The air quality in the workshop has been monitored over one week by taking 143 one-liter samples of air and counting the number of asbestos-type fibers in the liter sample. Samples were taken using gas sampling bags, at a height of 1.5 m at different times of day, and at locations corresponding to the mid-points of a lattice drawn on a plan of the workshop. The data are given in the Table 3.5. The **frequency** associated with a particular number of fibers is the number of one-liter samples of air containing exactly that number of fibers. For example, 46 one-liter samples of air contained exactly 1 fiber. The relative frequency of m fibers is:

$$\frac{\text{number of samples of air containing exactly } m \text{ fibers}}{\text{number of samples of air}}.$$

TABLE 3.4: Headings of waves during hurricanes with winds exceeding 80 knots in the North Sea.

Heading	Number of hurricanes	Heading	Number of hurricanes
North (N)	398	South (S)	978
North-northeast (NNE)	403	South-southwest (SSW)	253
Northeast (NE)	508	Southwest (SW)	194
East-northeast (ENE)	362	West-southwest (WSW)	144
East (E)	395	West (W)	194
East-southeast (ESE)	367	West-northwest (WNW)	168
Southeast (SE)	413	Northwest (NW)	342
South-southeast (SSE)	421	North-northwest (NNW)	337

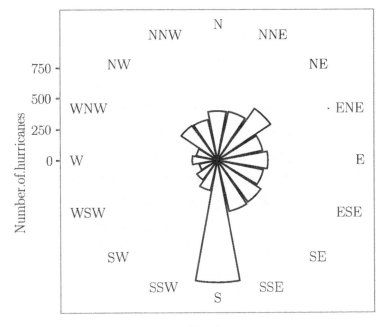

FIGURE 3.10: Headings of waves during hurricanes with winds exceeding 80 knots in the North Sea.

For example, the relative frequency of 1 fiber is $46/143 = 0.322$. The six **relative frequencies** are shown on the **line chart** in Figure 3.11, and they add to 1 (with a rounding error of 0.001). We use lines to emphasize that the number of fibers, m, is a discrete variable[3].

---

[3]In the U.S. the Occupational Safety & Health Administration permissible exposure limit (PEL) is 0.1 fiber per cubic centimeter (100 fiber per liter) of air (Standard Number 1910.1001). A flow rate of 1 liter/minute over eight hours, with particles remaining on a membrane filter, is equivalent to 480 gas sampling bags. However, the analysis of the sampling bags has shown that the fiber per liter of air in the workshop is well below the PEL.

TABLE 3.5: Relative frequency of number of asbestos-type fibers in one-liter samples of air.

Observed number	Frequency	Relative frequency
0	34	0.238
1	46	0.322
2	38	0.266
3	19	0.133
4	4	0.028
5	2	0.014

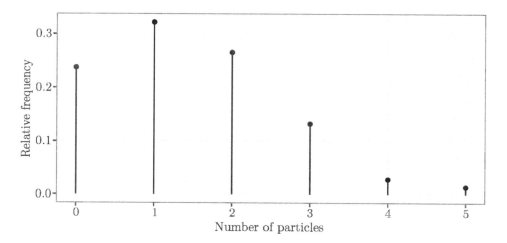

FIGURE 3.11: Relative frequencies of numbers of asbestos-type fibers in a workshop.

```
> x <- 0:5
> y <- c(34,46,38,19,4,2)
> f <- y / sum(y)
> df <- data.frame(x,y,f)
> library(ggplot2)
> theme_set(theme_bw())
> qplot(x=x,y=f,data=df) + geom_segment(aes(x=x,xend=x,y=f,yend=0))+
  labs(x="number of particles",y="relative frequency")
> dev.off()
> colnames(df) <- c("Observed number", "Frequency",
                "Relative frequency of particles")
> df
> library(xtable)
> tab <- xtable(df,caption="Relative frequency of particles in air.",
  label='tab:milling',digits=3)
> print(tab,include.rownames=FALSE)
```

3.3.8 Histogram and cumulative frequency polygon for continuous variables

The American Concrete Institute defines high strength concrete as concrete with a compressive strength greater than 6 000 psi (41 MPa). The data in **cubes_NuT.txt** are the compressive strengths of 180 concrete cubes, tested after 28 days curing[4]. The cubes were made as part of a research project into admixtures for high strength concrete. See Figure 3.12. We will draw a **histogram** to display the data. The first step is to find the smallest and largest datum.

```
> cubes <- scan("data/cubes_NuT.txt")
Read 180 items
> min(cubes)
[1] 43.8
> max(cubes)
[1] 68.8
```

We now divide this range into contiguous cells[5], and count the number of data in each cell (**frequency**). The number of cells should be few enough for there to be some data in most cells, and it is convenient to take cells of equal width with breaks that require only two significant figures. For these data 14 cells of width 2, starting at 42 and ending at 70 is a reasonable choice. Sturges' formula for the number of cells, $log_2 n + 1$, where n is the number of data is sometimes used as a guide to the minimum number of cells that should be used. In this case

```
> n <- length(cubes)
> #Sturges
> log(n,2) + 1
[1] 8.491853
```

which is somewhat less than the 14 cells we suggest.

It happens that the same 14 cells are the default if we use the **hist()** function without specifying **breaks** as an argument. The 14 lower break points for the cells are given in the first column, the fourteen upper break points are given in the second column, and the 14 frequencies are given in the third column. So, for example, there is one datum between 42 and 44, and 43 data between 60 and 62. Data can coincide with a break point. The default in R is that the cell $(a, b]$ includes a datum x_i if

$$a < x_i \leq b,$$

with the exception that the first interval includes the lower break point. So the frequency of 43 for cell $(60, 62]$ includes the four 62.0s. The fourth column has the **cumulative sums** of the counts, known as the **cumulative frequencies**. In general, cumulative sums can be calculated by the function **cumsum()**.

```
> h1 <- hist(cubes)
> tab <- data.frame(lower = h1$breaks[-15],
+                   upper = h1$breaks[-1],
+                   frequency = h1$counts,
+                   cumulative.frequency = cumsum(h1$counts))
```

---

[4]Department of Civil Engineering at the University of Newcastle upon Tyne.
[5]Cells are also commonly called "bins"or "class intervals", but R uses "cells".

```
> tab
   lower upper frequency cumulative.frequency
1     42    44         1                    1
2     44    46         0                    1
3     46    48         0                    1
4     48    50         2                    3
5     50    52         0                    3
6     52    54         2                    5
7     54    56         8                   13
8     56    58        28                   41
9     58    60        26                   67
10    60    62        43                  110
11    62    64        26                  136
12    64    66        25                  161
13    66    68        12                  173
14    68    70         7                  180
```

`> hist(cubes,freq=FALSE,main="",xlab="strength")`

The histogram is made up of contiguous rectangles superimposed on the cells, with area equal to the relative frequency of data in that cell.

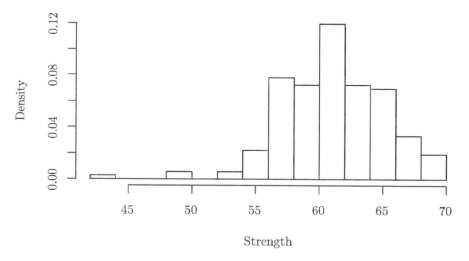

FIGURE 3.12: Histogram of compressive strengths (MPa) of 180 high strength concrete cubes [Department of Civil Engineering, University of Newcastle upon Tyne].

It follows that the height of a rectangle is the relative frequency divided by the width of the cell. This quotient is defined as the **relative frequency density**, which is commonly abbreviated to density. Since the cells have equal width of 2, the highest rectangle is that for the cell $(60, 62]$. The frequency is 43, the relative frequency is $43/180 = 0.239$, and the relative frequency density is $0.239/2 = 0.119$.

The total area of the histogram is 1 because the sum of the relative frequencies equals 1. Also, as the sample size becomes larger the area of the rectangle above each cell becomes closer to the probability that the value of the variable for a randomly selected item falls within that cell.

Provided the cells have equal width, we would obtain the same shape if the height of the rectangles is set equal to the frequency but we lose the relationship between area and probability. If the cells have different widths, then we must make the heights of rectangles equal to relative frequency density. The `hist( )` command in R sets up equal width cells and plots frequency as the default, if you do not specify break points for the cells. To obtain a histogram with total area of 1, plot density by adding the argument `freq=FALSE`.

If we allow cells to be narrower towards the middle of the histogram and wider in the tails, then we can show more detail where we have sufficient data to do so and avoid gaps in the tails. We demonstrate this with a set of measurements of ship hull roughness[6] in Figure 3.13.

Example 3.20: Hull roughness [histogram]

The 550 data in `shiphullrough.txt` were made on a ship in dry dock using a hull roughness analyzer. A hull roughness analyzer consists of a hand held carriage with an optical sensor that measures changes in height, attached to a microprocessor. The microprocessor records the height, in microns, from the highest peak to the lowest trough over a 50 mm transect (see Figure 3.14).

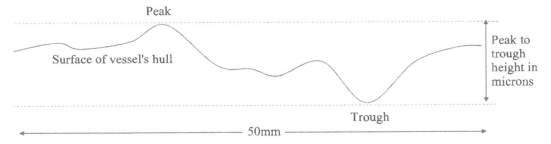

FIGURE 3.13: Measurements of ship hull roughness.

A plan of the wetted area of the hull was divided into 110 equal sub-areas, and 5 transects were made within each sub-area. The breaks for the histogram command were chosen after seeing the default histogram. We also use R to construct a table with: the lower break point and upper break point of each cell, the cell width, the frequency for the cell, the relative frequency, and the relative frequency density.

```
> ships <- scan("data/shiphullroughness.txt")
Read 550 items
> head(ships)
[1] 240.9  80.0  98.6  86.9  87.8  66.3
> breaks <- c(45,55,65,75,80,85,90,95,105,115,125,135,205,280)
> h1 <- hist(ships,
+        breaks=breaks,
+        freq=FALSE,
+        main="",
+        xlab="highest peak to lowest trough height (microns)")
> n <- length(ships)
> width <- breaks[-1]-breaks[-14]
```

---

[6]A large proportion of the total resistance to motion of a slow moving merchant ship is due to friction between sea water and the hull. Even a moderate degree of hull roughness below the water line will lead to increases of around 20% in fuel costs. Self polishing co-polymer paints have been developed to help maintain a smooth paint finish on ship's hulls.

```
> lower <- breaks[-14]
> upper <- breaks[-1]
> frequency <- h1$counts
> rel_freq <- round((frequency/n),3)
> density <- round(rel_freq/cellwidth,4)

> tab <- data.frame(lower,upper,width,frequency,rel_freq,density)
> tab
   lower upper width frequency rel_freq density
1     45    55    10         7    0.013  0.0013
2     55    65    10        24    0.044  0.0044
3     65    75    10        66    0.120  0.0120
4     75    80     5        41    0.075  0.0150
5     80    85     5        55    0.100  0.0200
6     85    90     5        86    0.156  0.0312
7     90    95     5        77    0.140  0.0280
8     95   105    10        70    0.127  0.0127
9    105   115    10        46    0.084  0.0084
10   115   125    10        26    0.047  0.0047
11   125   135    10        14    0.025  0.0025
12   135   205    70        25    0.045  0.0006
13   205   280    75        13    0.024  0.0003
```

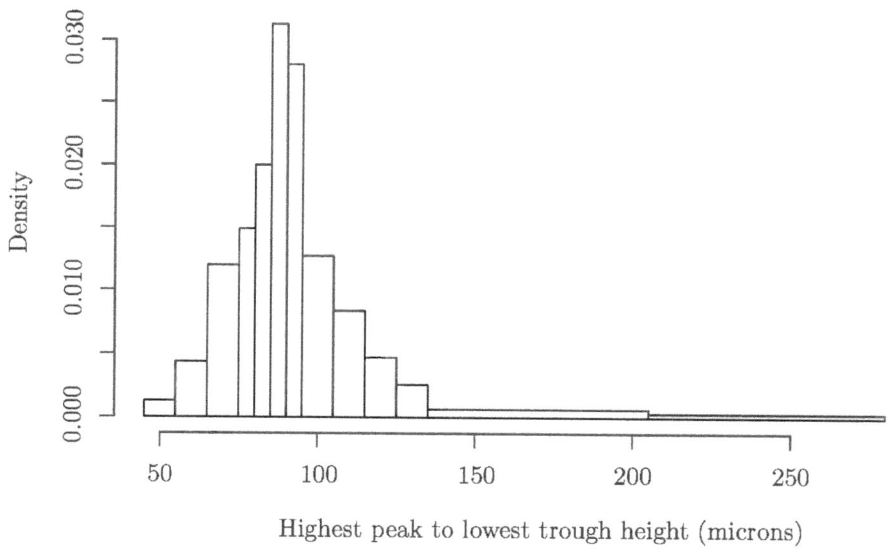
Density
0.000
0.010
0.020
0.030
50
100
150
200
250
Highest peak to lowest trough height (microns)

FIGURE 3.14: Histogram for ship hull roughness.

The **cumulative frequency polygon** (cfp) is an alternative display of data that have been grouped into cells. It is a plot of the proportion of data less than or equal to a given value against that value. The proportion of data less than the upper break point of each cell is the cumulative frequency divided by the number of data. For the concrete cubes the cumulative frequency for the cell $(60, 62]$ is 110. Therefore a proportion $110/180 = 0.611$ of the cubes have compressive strengths less than or equal to 62. The following R commands draw the cumulative frequency polygon and display it below the histogram. The plots are shown in Figure 3.15. The proportion of data less than or equal to some given value is the area under the histogram from its left hand end to that value.

```
> cubes <- scan("data/cubes_NuT.txt")
Read 180 items
> n <- length(cubes)
> h1 <- hist(cubes)
> cf <- cumsum(h1$counts)
> cp <- cf/n
> par(mfrow=c(2,1))
> hist(cubes,freq=F,main="")
> plot(h1$breaks,c(0,cp),
+    type="l",
+    xlab="strength (x)",
+    ylab=expression("cumulative proportion" <= "x"))
```

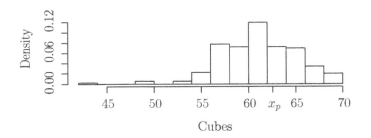

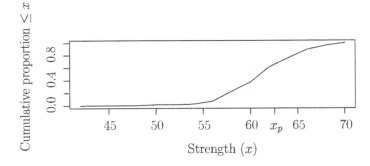

FIGURE 3.15: Histogram (upper frame) and cumulative frequency polygon (lower frame) for the strength of concrete cubes. The height of the cumulative frequency polygon, at some point x_p, gives the area under the histogram from the left end up to x_p.

3.3.9 Pareto chart

The Pareto principle is named after the economist Vilfredo Pareto who observed in his 1906 *Manuale di Economia Politica* that 80% of the land in Italy was owned by about 20% of the population. He found similar results for other countries. The business management consultant Joseph M. Juran suggested that similar results could be found in business: 80% of business from about 20% of customers; 80% of defective items caused by around 20% of possible faults. Rooney (2002) quotes Microsoft's CEO as saying: "About 20 percent of the bugs causes 80 percent of all errors, and – this is stunning to me – 1 percent of bugs caused half of all errors." The rule extends in a self-similar fashion so, 80% of the remaining 20% of errors would be prevented by fixing the most commonly reported 20% of the remaining 80% reported bugs. These percentages are just based on empirical observations and there is no compelling reason for such self-similarity in economic or industrial processes. Nevertheless, if all faults require similar resources to fix, it is sensible to attend to faults that cause the majority of defects first.

A **Pareto chart** is a similar to a bar chart for frequencies attributable to categories, but the bars are arranged in decreasing height from left to right or top to bottom. The Royal Automobile Association of South Australia Inc (RAA) provides roadside assistance to members. The number of call-outs a year exceeds $600,000$ and the most common reasons for call-outs are given in the Table 3.6. The Other category includes faulty: locks, wipers, brakes, steering, suspension, windscreen, and seat-belts. If a call-out was for more than one reason it is recorded under each reason.

TABLE 3.6: Reason for call-out of RAA roadside assistance.

Reason	%
Battery failure	46
Lock outs	13
Electrical faults	11
Other	9
Wheel and tyre problems	6
Fuel-related	6
Ignition trouble	5
Cooling faults	4

The R code for reading in the .xlsx file containing the RAA data into R and plotting a Pareto chart (Figure 3.16) is given below.

```
> library(gdata)
> raa <- read.xls("data/RAA.xlsx")
colnames(raa) <- c("Reason","Percent")
raa$Reason <- factor(raa$Reason,levels=raa$Reason[order(raa$Percent)])
library(ggplot2)
theme_set(theme_bw(12))
qplot(raa,aes(x=Reason,y=Percent,label=Percent,hjust=2)) +
  geom_bar(stat='identity',fill='white',col='black') +
  coord_flip()
```

Battery failure is the reason for nearly half the call-outs, and the RAA patrol vehicles carry a range of new batteries that they offer to supply and fit if the old battery needs replacing.

The U.S. Environmental Protection Agency gives Pareto charts for the water quality of rivers, lakes and estuaries. The chart for assessed rivers and streams is reproduced in Figure 3.17.

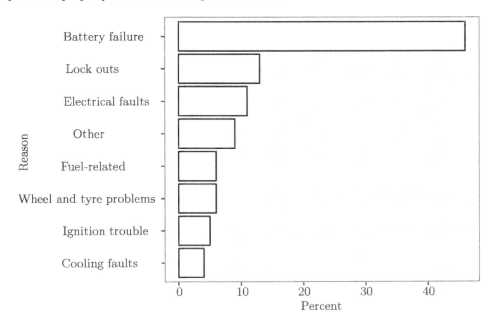

FIGURE 3.16: Pareto chart for the RAA call-out data.

3.4   Numerical summaries of data

3.4.1   Population and sample

We consider our data as a sample from some population. In some cases, such as sampling from a batch of electric motors when we define the batch as the population (Example 3.9), we can set up a list of the population and draw a SRS from the population. In other cases, such as the drill cores from the vein-gold deposit (Example 3.11), the population is finite but it is not practical or necessary to list it because we drill cores at the center points of a grid drawn on a map of the mine. Then there are the many cases for which the population is hypothetical and infinite. This holds for environmental variables such as stream-flow (Example 3.12) and industrial processes when the population is defined as all future production if the process continues on its present setting.

No matter how the population is defined, we can describe it in mathematical terms as

$$\{x_j\}, \quad j = 1, .., N,$$

where N is the population size which can be finite or infinite, j denotes an item in the population, and x_j is the value of the variable x for item j. We do not know the x_j, so we take a sample of n from the population and measure x for the items in the sample. The sample is denoted by

$$\{x_i\}, \quad i = 1, \ldots, n.$$

Generally, we require a succinct numerical summary of the population which provides an indication of a typical value of the variable, its location, and some measure of the spread of the values. We estimate such summary values by calculating corresponding quantities from the sample. Numerical summaries of the population are known as **parameters** of the

Cause of Impairment Group	Miles Threatened or Impaired
Pathogens	157,964
Sediment	125,045
Nutrients	98,585
Organic Enrichment/Oxygen Depletion	84,770
Polychlorinated Biphenyls (PCBs)	78,339
Metals (other than Mercury)	75,770
Temperature	67,266
Mercury	63,837
Habitat Alterations	62,373
Flow Alteration(s)	40,691
Cause Unknown	39,187
Cause Unknown - Impaired Biota	38,997
Salinity/Total Dissolved Solids/Chlorides/Sulfates	34,239
Turbidity	29,308
pH/Acidity/Caustic Conditions	28,566
Pesticides	16,819
Other Cause	11,950
Ammonia	11,768
Fish Consumption Advisory	9,948
Total Toxics	8,999
Toxic Inorganics	7,879
Algal Growth	5,945
Toxic Organics	4,474
Dioxins	4,321
Oil and Grease	2,874
Biotoxins	2,150
Nuisance Exotic Species	1,406
Cause Unknown - Fish Kills	1,371
Taste, Color and Odor	871
Trash	751
Radiation	680
Chlorine	618
Noxious Aquatic Plants	295
Nuisance Native Species	127

http://iaspub.epa.gov/waters10/attains_nation_cy.control

FIGURE 3.17: National summary of causes of impairment in assessed rivers and streams.

population and these are unknown. The corresponding values calculated from the sample are known as **statistics**. So, statistics are estimates of the corresponding population parameters.

3.4.2 Measures of location

In statistical analysis the average, in the sense of the sum of a set of values divided by their number, is known as the **arithmetic mean**. Other means are defined (see exercises), but when we use **mean** without any qualification it is an arithmetic mean.

Definition 3.1: Population mean

The population mean is denoted by μ, defined as

$$\mu = \frac{1}{N} \sum_{j=1}^{N} x_j,$$

where

$$\sum_{j=1}^{N} x_j = x_1 + \cdots + x_N.$$

The subscript and superscript on the sum symbol Σ can be omitted if they are clear from the context. The case of infinite N is covered in more detail in Chapter 5, but imagining that N becomes arbitrarily large will do here. Although the equation for μ defines the population mean we cannot compute μ because we do not have the data for the entire population.

Definition 3.2: Sample mean

The sample mean, $\bar{x}$, is an estimate of the population mean, μ, and is computed from

$$\bar{x} = \frac{1}{n} \sum_{i=1}^{n} x_i.$$

Example 3.21: Fault creep [calculation of mean]

The following data are 6 measurements of small baseline subset (SBAS) fault creep (mm relative to 0 before 1996 event) at a site on the South Hayward Fault made during the first half of 2010 (Berkeley Seismological Laboratory).

$$84 \quad 95 \quad 96 \quad 91 \quad 89 \quad 83$$

The sum is 538 and the mean, to the nearest decimal place, is $538/6 = 89.7$ In R

```
> fault <- c(84, 95, 96, 91, 89, 83)
> sum(fault)
[1] 538
> mean(fault)
[1] 89.66667
```

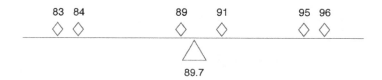

FIGURE 3.18: The mean value is the balance point for the data set expressed as weights.

The mean may also be represented as the balance point for the data set as in Figure 3.18.

The differences between the individual observations and the mean are known as *deviations from the mean*. The sum of the deviations from the mean is identically zero, as proved below.

$$\sum_{i=1}^{n}(x_i - \bar{x}) \;=\; \sum_{i=1}^{n}x_i - \sum_{i=1}^{n}\bar{x} \;=\; \sum_{i=1}^{n}x_i - n\bar{x} \;=\; \sum_{i=1}^{n}x_i - \frac{n}{n}\sum_{i=1}^{n}x_i \;=\; 0.$$

We often refer back to this fact. Any difference between the sum of calculated deviations from the mean and 0 is due to rounding error.

Example 3.21: (Continued) Fault creep [rounding error]

```
> fault - mean(fault)
[1] -5.6666667  5.3333333  6.3333333  1.3333333 -0.6666667 -6.6666667
> sum(fault - mean(fault))
[1] -2.842171e-14
```

Essentially zero, but we see that the rounding error is of the order 10^{-14}.

Definition 3.3: Median

The median value of a variable x in a population, which will be denoted by M, is the value such that half the members of the population have x_j below, or equal to, M.

Definition 3.4: Sample median

The sample median is the value such that half the data are less than or equal to it. We will use $\widehat{M}$ to represent the sample median.

An approximate value can be found from the cumulative frequency polygon by reading across from 0.5 to the polygon and down to the horizontal axis to read the approximate median value of about 91 (Figure 3.19)[7].

---

[7]We have used a common notation for representing a sample estimate of a population parameter, when it is not convenient to use corresponding Roman and Greek letters. The sample estimate is written as the population parameter with a hat over it. In the case of the sample median we write $\widehat{M}$.

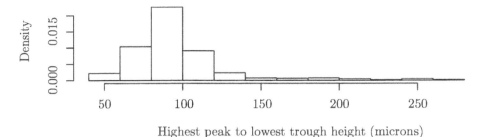

Highest peak to lowest trough height (microns)

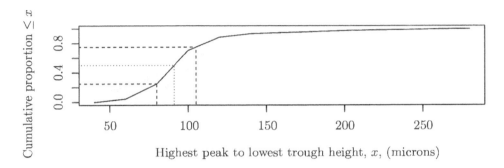

Highest peak to lowest trough height, x, (microns)

FIGURE 3.19: Histogram and cumulative frequency polygon for the ship hull roughness data. Approximate constructions for the median (dotted line), lower quantile (dashed line), and upper quantile (dashed line) are shown.

The median is calculated from the original data by placing them into ascending order and taking the middle value, if the number of data is odd, or the mean of the two values in the middle if the number of data is even.

Definition 3.5: Order statistics

When the data are placed in ascending order they are known as **order statistics**, and the notation $x_{i:n}$ is used for the ith smallest in a sample of size n. So the order statistics are

$$x_{1:n} \leq x_{2:n} \leq \ldots \leq x_{n:n}.$$

The sample median is given by

$$\widehat{M} = x_{\left((n+1)/2\right):n}.$$

If n is even we obtain $\widehat{M}$ from linear interpolation between $x_{n/2:n}$ and $x_{n/2+1:n}$, which is the mean of the two values.

The median is less affected by a few extremely high values, or low values, than the mean, and a substantial difference between the mean and median indicates outlying values. If the median is substantially less than the mean there are some high outlying values and we say the data are positively skewed. If the mean is substantially less than the median, there are some low outlying values and the data are said to be negatively skewed.

Example 3.22: Engineering salaries [mean and median]

Macaw Engineering has three engineers currently working for them, two junior engineers with annual salaries of 60 monetary units (mu) and Mike Mallet, the senior engineer who is paid 180 mu.

The mean salary for engineers at Macaw Engineering is 100 mu, so it looks like a great place for graduate engineers to start their working career. However, the median salary of 60 mu is considerably less, and provides more realistic summary for new entrants.

Example 3.23: Wind speeds [mean and median]

The number of days in a year with wind-speeds above 70 mph at Boulder between 1969 and 2002 (NOAA Earth System Research Laboratory) are: $6, 3, 0, 10, 1, 0, 10, 3, 9, 5, 1, 0, 3, 16, 11, 5, 3, 11, 6, 9, 5, 9, 2, 3, 6, 2, 3, 13, 1, 1, 16, 9, 3$ and 3.

```
> ndays <- c(6, 3, 0, 10, 1, 0, 10, 3, 9, 5, 1, 0, 3, 16,
11, 5, 3, 11, 6, 9, 5, 9, 2, 3, 6, 2, 3, 13,
1, 1, 16, 9, 3, 3)
> sort(ndays)
 [1]  0  0  0  1  1  1  1  2  2  3  3  3  3  3  3  3  3  5
[19]  5  5  6  6  6  9  9  9  9 10 10 11 11 13 16 16
> median(ndays)
[1] 4
> mean(ndays)
[1] 5.529412
```

The number of days cannot be negative and there are three years with 0 days with winds above 70 mph. In contrast there is no clear upper limit (except the quite unrealistic 365) and there are a few years with more than 10 such days. The mean is noticeably higher than the median, and the data have a longer tail towards the right. Such data are said to be positively skewed.

If data are skewed, then the median and mean will differ, but this does not imply that we should use the median in place of the mean. We should present both statistics as they provide somewhat different information. The gold grades shown in Example 3.11 are positively skewed and the mean of 0.7922 is substantially greater than the median value of 0.65. It is the mean value that is more relevant for the mining company, because it gives a direct assessment of the monetary value of the gold in the field.

Example 3.24: Flood prevention [positively skewed data]

The data in `propflood.txt` are the costs of 41 schemes to prevent flooding undertaken by the erstwhile Northumbrian Water Ltd. The costs are positively skewed. The mean is substantially greater than the median. The sample mean is appropriate for predicting the mean cost of future schemes.

```
> flood <- read.xls("~/Dropbox/SiE/data/propflood.xls")
> head(flood)
  X.props    cost
1      10  567703
2       8 1807692
3       7   21333
4       6  152672
5       5  706645
6       5  207082
> median(flood$cost)
[1] 152672
> mean(flood$cost)
[1] 412084.4
```

Another measure of centrality is the **mode** of the distribution.

Definition 3.6: Mode for discrete data

For discrete data, the mode is the most commonly occurring value, and this definition applies in both the population and the sample.

For example, the mode of the distribution of asbestos type fibers (Figure 3.11) is 1.

Example 3.25: LaGuardia Airport [mode for a discrete variable]

A wind rose plot for LaGuardia Airport, New York, available at

http://en.wikipedia.org/wiki/File:Wind_rose_plot.jpg

is reproduced in Figure 3.20. The concentric circles are labelled with relative frequencies, expressed as percentage. Nearly 12% of the wind directions were recorded from the south and this is the modal direction. The second most common direction is from the north-west with 10% of the records. The smallest percentage of the records is from the east south-east direction. The plot also shows wind speeds (m/s) by using different colored bands within spokes. The scale on the right shows that 22.6% of the wind speeds, cumulated over all directions, were between 1.54 and 3.09 m/s, and that speeds below 1.54 m/s are recorded as calm (0.00 m/s). The given percentages add to 96.5% so 3.5% of wind speeds were recorded as higher than 15.5 m/s. Pilots prefer to take off and land into the wind, and engineers try to allow for this when designing airports. The two runways at LaGuardia are at 32 deg and 122 degrees from true north respectively. Most airport runways are bidirectional.

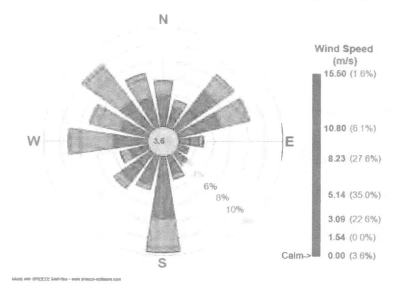

FIGURE 3.20: A wind rose plot for LaGuardia Airport, New York.

If a variable is continuous, a sample mode can be defined within the cell for which the histogram has a maximum relative frequency density.

Example 3.26: Paver strength [mode for a continuous variable]

Table 3.7 presents grouped data on the compressive strength of concrete paving blocks (pavers). The construction shown in Figure 3.21 identifies a modal value on the histogram.

The mode is the point at which a line (dashed in Figure 3.21) through the intersection of the two solid cross lines, parallel to the vertical y-axis, meets the horizontal axis. The value of the mode is 61.7 MPa. The construction relies on the cells containing and adjacent to the mode having the same width.

TABLE 3.7: Compressive strengths in Mega-Pascals (MPa), of a random sample of 200 pavers.

Compressive strength interval	Frequency	Relative frequency	Relative frequency density	Cumulative relative frequency	(%)
40 - 50	20	0.10	0.010	0.10	10
50 - 55	30	0.15	0.030	0.25	25
55 - 60	40	0.20	0.040	0.45	45
60 - 65	50	0.25	0.050	0.70	70
65 - 70	30	0.15	0.030	0.85	85
70 - 80	30	0.15	0.015	1.00	100

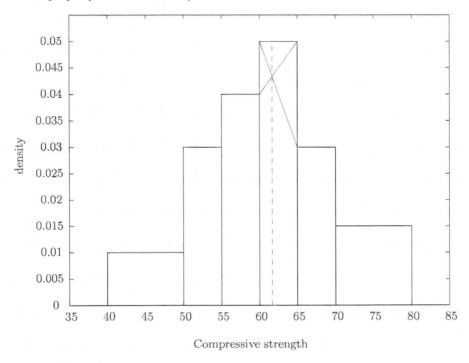

FIGURE 3.21: Compressive strengths in Mega-Pascals (MPa), of a random sample of 200 pavers, showing construction for finding the mode.

For positively skewed distributions the mean is greater than the median, and the median is usually greater than the mode. For negatively skewed distributions the mean is less than the median, and the median is usually less than the mode. In some cases the histogram can have more than one clearly defined maximum value. If there are two clear maxima the distribution is described as bi-modal.

Example 3.27: Old Faithful [scatter plot]

The histograms of the times between eruptions and the durations of eruptions of Old Faithful geyser in Yellowstone National Park, both appear bi-modal (Figure 3.22). The lower left hand panel is a **scatter plot** (Section 3.9.1) of the waiting time until the next eruption against the duration of the last eruption. It seems that long waiting times tend to follow long durations. The lower right panel shows the sequence of the first 30 waiting times, suggesting that long and short waiting times tend to alternate.

An alternative to the histogram construction for the mode is to fit a smooth curve rather than a histogram to the data, and look for the maximum. The construction of this **kernel smoother** will be explained in Example 5.16, but is easy to implement using the R function `density()` as follows, the results of which are illustrated in Figure 3.23.

```
> hist(OF$waiting,main="",freq=FALSE,xlab='Waiting')
> lines(density(OF$waiting))
```

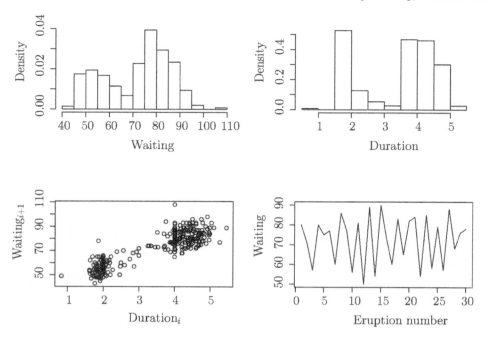

FIGURE 3.22: Upper panels: histograms for waiting time and duration of eruptions of the Old Faithful geyser in Yellowstone National Park. Lower left: waiting time until eruption $(i+1)$ against duration i. Lower right: time series plot of waiting time for eruption i against eruption number i.

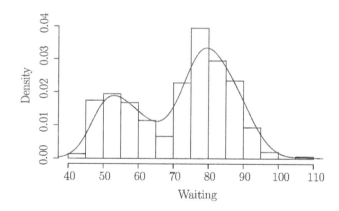

FIGURE 3.23: Histogram and density of waiting times for Old Faithful.

The modes for the distribution of waiting times are about 53 and 79 minutes.

A variation on the arithmetic mean is the weighted mean.

Definition 3.7: Weighted mean

The value x_i is given a weight w_i and the weighted mean is defined by

$$\bar{x}_w = \frac{\sum x_i w_i}{\sum w_i}.$$

The arithmetic mean corresponds to all the weights being 1. A weighted mean is used when the items being averaged represent different proportions of the whole. If a public transport system takes an average of 100, 000 passengers per hour during four peak hours and an average of 10, 000 passengers an hour during 14 off-peak hours then the overall average passengers per hour is

$$\frac{100,000 \times 4 + 10,000 \times 14}{4 + 14} = 30,000$$

Example 3.28: Water supplies [weighted mean]

A water company operates in three divisions A, B, and C. An engineer has been asked to estimate the overall proportion of properties with lead communication pipes between the water main and the boundary of the property. The engineer has taken simple random samples from each division, and the results of the survey are given in Table 3.8. The estimate of the total number of properties with lead communication pipes is

TABLE 3.8: Proportion of properties with lead communication pipes.

Division	Number of properties (1 000s)	Sample size	Number with lead communication pipes	Proportion
A	358	300	47	0.1567
B	214	250	58	0.2320
C	107	200	73	0.3650

$$0.1567 \times 358\,000 + 0.2320 \times 214\,000 + 0.3650 \times 107\,000 = 144\,790.$$

The total number of properties is $358\,000 + 214\,000 + 107\,000 = 679\,000$. So, the estimate of the overall proportion is $144\,790/679\,000 = 0.213$. This is the weighted mean of the three proportions with weights proportional to the number of properties in each division.

$$\frac{0.1567 \times 358 + 0.2320 \times 214 + 0.365 \times 107}{358 + 214 + 107} = 0.213$$

3.4.3   Measures of spread

Definition 3.8: Range

A natural measure of the spread of a set of data is the **range**, defined as the largest less the smallest value.

$$range \;=\; x_{n:n} - x_{1:n}.$$

The limitation of this measure is that it is highly dependent on the sample size. We expect the range to be wider if we take a larger sample, and if we draw items for a sample sequentially the range can only increase or remain the same when the next item is drawn. Also, the range may not be defined in an infinite population. Nevertheless, the range is commonly used in industrial quality control when samples of the same size are routinely collected on a daily or weekly basis.

A quantity that is well defined in the population as well as in the sample is the **interquartile range** (given in Definition 3.13). The lower quartile (LQ) in the population is the value of the variable such that 25%, one quarter, of the population have values less than or equal to it.

Definition 3.9: Sample Lower Quartile

We will use the following definition of the sample lower quartile ($\widehat{LQ}$)

$$\widehat{LQ} \;=\; x_{0.25\times(n+1):n}.$$

If $(n + 1)$ is not a multiple of 4 we interpolate linearly between $x_{z:n}$ and $x_{z+1:n}$, where $z = \texttt{floor}\big((n + 1)/4\big)$. The upper quartile ($UQ$) in the population is the value of the variable such that 25%, one quarter, of the population items have values greater than it.

Definition 3.10: Sample Upper Quartile

We will use the following definition of the sample upper quartile ($\widehat{UQ}$)

$$\widehat{UQ} \;=\; x_{0.75\times(n+1):n}.$$

The use of $(n+1)$ gives a symmetry to the definitions of $\widehat{LQ}$ and $\widehat{UQ}$. For example, suppose n is 99. Then $\widehat{LQ}$ is $x_{25:99}$ which is the 25th smallest (25 values are less than or equal to it) and $\widehat{UQ}$ is $x_{75:99}$, which is the 25th largest[8] (25 values are greater than or equal to it). This is the reason for using $(n + 1)$ when calculating the median, which is the 50% **quantile**. The median and quantiles are special cases of quantiles.

---

[8]Moreover, $\widehat{UQ}$ is the 75th smallest (75 values are less than or equal to it).

Definition 3.11: Population quantile

A quantile corresponding to a proportion (or probability) p, written as $q(p)$ is the value such that a proportion p of the population is less than or equal to $q(p)$.

Definition 3.12: Sample quantile

If a random sample is taken from the population, then the estimate of $q(p)$ is

$$\widehat{q}(p) = x_{p(n+1):n},$$

with interpolation when needed.

If p is close to 0 or 1, estimates of corresponding quantiles will only be useful if we have very large samples[9]. The graphical constructions from the cfp are not generally precisely the same as the values calculated from the order statistics. The difference will only be slight, but the order statistics should be used if the original data are available and the graphical estimates are approximations to these.

Definition 3.13: Interquartile range

For a continuous variable, half the data will lie between LQ and UQ and the interquartile range is

$$IQR = UQ - LQ.$$

The sample estimate of this quantity is

$$\widehat{IQR} = \widehat{UQ} - \widehat{LQ}.$$

For a discrete variable the precise proportion will depend on the number of data equal to the quartiles, but the IQR is generally used only for continuous variables.

Example 3.29: Hardap dam [Adamson et al., 1999] interquartile range

We will calculate the median and quartiles for the peak inflows to the Hardap Dam in Namibia over 25 years (Table 3.9). We first sort the data into ascending order.

```
> sort(peak)
 [1]   30   44   83  125  131  146  197  230  236  347  364  408
[13]  412  457  477  554  635  765  782  911 1506 1508 1864 3259
[25] 6100
```

---

[9]This is one of the reasons why we fit probability distributions to model histograms as the sample size increases (Chapter 5). We can estimate $q(p)$ by extrapolation into the tails of a fitted probability distribution.

TABLE 3.9: Hardap dam inflows.

Year	Peak	Year	Peak	Year	Peak
1962-3	1864	1971-2	6100	1980-1	125
1963-4	44	1972-3	197	1981-2	131
1964-5	146	1973-4	3259	1982-3	30
1965-6	364	1974-5	554	1983-4	765
1966-7	911	1975-6	1506	1984-5	408
1967-8	83	1976-7	1508	1985-6	347
1968-9	477	1977-8	236	1986-7	412
1969-0	457	1978-9	635		
1970-1	782	1979-0	230		

The median is the $0.5 \times (25 + 1) = 13^{th}$ smallest datum which is 412. The lower quartile is the $0.25 \times (25 + 1) = 6.5^{th}$ smallest datum which is

$$146 + 0.5 \times (197 - 146) \quad = \quad 171.5$$

The upper quartile is the $0.75 \times (25 + 1) = 19.5^{th}$ smallest datum which is

$$782 + 0.5 \times (911 - 782) \quad = \quad 846.5$$

The inter-quartile range is $846.5 - 171.5 = 675$. The R function **summary()** can be used

```
> summary(peak)
  Min. 1st Qu.  Median mean 3rd Qu.    Max.
  30.0   197.0   412.0  862.8  782.0  6100.0
```

The R function **summary()** uses a slightly different definition (Exercise 3.10) of the sample quartiles and the IQR is $782 - 197 = 585$. The difference is noticeable in this case, partly because the sample is small, but either would suffice for a descriptive statistic.

We have seen that the sum of deviations from the mean is 0, so the mean deviation from the mean is also 0 and provides no information about variability. We need to lose the signs attached to the deviations if we are to use them in a measure of variability, and one way of doing this is to square them.

Definition 3.14: Population variance

The population variance (σ^2) of a variable x is the mean of the squared differences between the individual values of x_j and the population mean. defined by

$$\sigma^2 \quad = \quad \frac{\sum (x_j - \mu)^2}{N}.$$

Definition 3.15: Population standard deviation

The population standard deviation σ is the positive square root of the variance.

Definition 3.16: Sample variance

The sample variance (s^2) is defined by

$$s^2 = \frac{1}{n-1} \sum_{i=1}^{n} (x_i - \bar{x})^2.$$

Notice that the division is by $n-1$, rather than n, and there are reasons for preferring this definition (Exercise 3.3). We say that s^2 is an estimate of σ^2 on $n-1$ degrees of freedom, because given arbitrary numbers for $n-1$ of the deviations from $\bar{x}$, the nth deviation is determined. This is because the sum of all n deviations is constrained to equal 0.

Definition 3.17: Sample standard deviation

The sample standard deviation is the positive square root of the sample variance.

The unit of measurement of the standard deviation is the same as that of the variable. The variance has a unit of measurement which is the square of the unit of measurement of the variable.

Example 3.30: Metal film resistor [calculation of standard deviation]

The product information for a metal film resistor states that the resistance is 220 kilohms $(k\Omega)$ with a tolerance of $\pm1\%$ at 20 degrees Celsius. The manufacturer does not provide a definition of tolerance, but with modern manufacturing processes for electronic components it is likely that only a few parts per million (ppm) are outside the manufacturer's tolerance. An engineer takes a sample of 5 from a recent delivery. The resistances are $219.7, 219.8, 220.0, 220.3, 219.7$. The mean, deviations from the mean, squared deviations from the mean, sum of squared deviations divided by $(5-1)$, variance, and standard deviation are calculated in R as follows

```
> x
  [1] 219.7 219.8 220.0 220.3 219.7
> n=length(x)
> n
  [1] 5
> mean(x)
  [1] 219.9
> x-mean(x)
  [1] -0.2 -0.1  0.1  0.4 -0.2
> (x-mean(x))^2
  [1] 0.04 0.01 0.01 0.16 0.04
> sum((x-mean(x))^2)/(5-1)
```

```
    [1] 0.065
> var(x)
    [1] 0.065
> sd(x)
    [1] 0.254951
```

Notice that R uses the denominator $(n-1)$ when calculating the variance and the standard deviation. For distributions with a central mode, approximately 2/3 of the data lie within one standard deviation of the mean. If the histogram of a data set is bell-shaped we expect about 95% of the values to lie within two standard deviations of the mean, and practically all the values to be within six standard deviations of the mean.

Example 3.31: Hearing protection [practical interpretation of sd]

The brochure for E.A.R. band semi-aural hearing protectors gives the attenuation performance shown in Table 3.10. Although variation in the hearing protectors is negligible, the attenuation provided will vary from one person to the next because of variations in human ears. The assumed protection, as stated in the product brochure, is set at one standard deviation below the mean. Therefore, about one out of six users will have a noise attenuation less than the assumed protection.

TABLE 3.10: Attenuation performance for E.A.R. band semi-aural hearing protectors.

Frequency (Hz)	63	125	250	500	1 000	2 000	4 000	8 000
mean attenuation (dB)	20.5	19.4	16.0	16.5	20.9	31.4	35.3	36.0
standard deviation (dB)	4.2	5.4	4.1	4.2	2.5	4.3	3.6	4.0
assumed protection (dB)	16.3	14.0	11.9	12.3	18.4	27.1	31.7	32.0

If a variable is restricted to non-negative values, it may be appropriate to express the standard deviation as a fraction or percentage of the mean.

Definition 3.18: Coefficient of variation

The **coefficient of variation** (CV) is defined for non-negative variables as the ratio of the standard deviation to the mean.

The CV is a dimensionless quantity and is often used in the electronics industry.

Example 3.32: Hardap and Resistance CVs [comparison of CVs]

The estimated CV $(\widehat{CV})$ of the Hardap Dam inflows (Example 3.29) is $1310.4/862.8 = 1.52$. In contrast the $\widehat{CV}$ of the resistances of resistors (Example 3.30) is $0.2550/219.9 = 0.0012$.

Another way of discarding the signs associated with deviations from the mean is to take absolute value. The mean absolute deviation from the mean is a reasonable descriptive measure of the variability of a data set, but it is rarely used because there are theoretical reasons for preferring the variance and standard deviation.

Definition 3.19: Median absolute deviation from the median (MAD)

The median absolute deviation from the median (MAD) is sometimes used as a measure of variability that is insensitive to outlying values. In a sample

$$\widehat{MAD} \quad = \quad y_{(n+1)/2:n} \quad \text{where } y_i = \left| x_i - \widehat{M} \right|.$$

Example 3.33: Heat of sublimation of platinum [MAD]

Hampson and Walker (1960) published the following data for the heat of sublimation of platinum (kcal/mole):

136.2	136.6	135.8	135.4	134.7
135.0	134.1	143.3	147.8	148.8
134.8	135.2	134.9	146.5	141.2
135.4	134.8	135.8	135.0	133.7
134.2	134.9	134.8	134.5	134.3
135.2				

```
> sublimation=c(136.2,136.6,135.8,135.4,134.7,135.0,134.1,143.3,147.8,
+ 148.8,134.8,135.2,134.9,146.5,141.2,135.4,134.8,135.8,135.0,133.7,
+ 134.2,134.9,134.8,134.5,134.3,135.2)
> par(mfrow=c(1,2))
> plot(sublimation,xlab="time order",ylab="heat of sublimation")
> boxplot(sublimation,ylab="heat of sublimation")
> devmed <- sublimation-median(sublimation)
> sort(devmed)
 [1] -1.4 -1.0 -0.9 -0.8 -0.6 -0.4 -0.3 -0.3 -0.3 -0.2 -0.2 -0.1 -0.1
[14]  0.1  0.1  0.3  0.3  0.7  0.7  1.1  1.5  6.1  8.2 11.4 12.7 13.7
> sum(devmed)
[1] 50.3
> median(abs(devmed))
[1] 0.65
> sd(sublimation)
[1] 4.454296
```

The data are plotted in Figure 3.24. Notice that the sum of deviations from the median is not generally 0. The MAD is 0.65 whereas the standard deviation is 4.45. The standard deviation is greatly influenced by the outlying values.

3.5   Box-plots

A box plot is a useful graphical display for a small data set, and a box plot for the Hardap dam peak inflows is shown in the left hand panel of Figure 3.25.

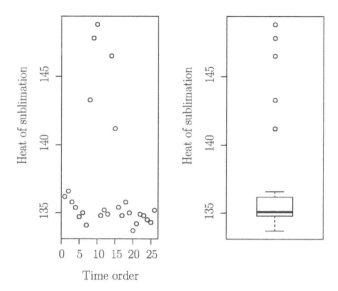

FIGURE   3.24: Twenty   six   estimates   of   heat   sublimation   of   platinum,
[Hampson and Walker, 1960].

```
> par(mfrow=c(1,2))
> boxplot(peak,ylab="Annual maximum inflow")
> plot(as.ts(peak),ylab="Annual maximum inflow")
```

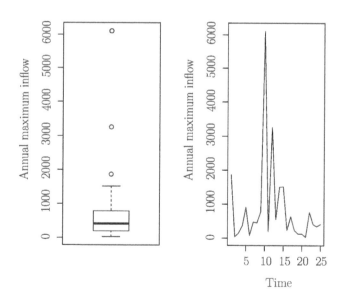

FIGURE 3.25: Box-plot and time series plot of the Hardap dam inflows.

The rectangle extends from $\widehat{LQ}$ to $\widehat{UQ}$ and the median is shown by the horizontal line
within the rectangle. Outlying values are typically defined as values that are more than 1.5
$\widehat{IQR}$ above the $\widehat{UQ}$ or below the $\widehat{LQ}$. Any outlying values are shown individually. The lines
extend from the $\widehat{UQ}$ to the greatest value that is not an outlier, and from $\widehat{LQ}$ to the least

value which is not an outlier. In the case of Hardap Dam there are no low outliers. The right hand panel is a time series plot for the peak inflows. There is no apparent trend over the period.

Box plots are particularly useful when we wish to compare several samples.

Example 3.34: Cable corrosion [boxplots for comparing samples]

[Stahl and Gagnon, 1995] compare the load that causes failure for samples of new cable and corroded cable from the George Washington Bridge. There were 23 pieces of new cable tested in 1933 and 18 pieces of corroded cable tested in 1962. Box plots are shown in Figure 3.26. There is a clear reduction in median strength and an increase in variability of the corroded cable.

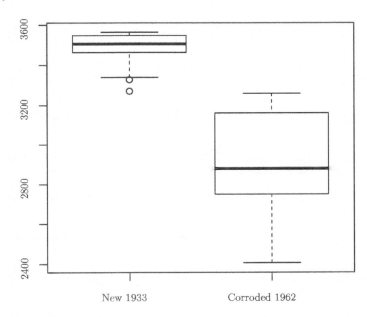

FIGURE 3.26: Side-by-side box-plots of the load that causes failure (kN) for new versus corroded cables from the George Washington Bridge.

3.6   Outlying values and robust statistics

3.6.1   Outlying values

Box plots, and other graphics for data, are very useful for detecting outliers and are used routinely at the beginning of any analysis. In some cases, outliers can be deleted as erroneous. For example: a decimal point has been omitted or misplaced; measuring equipment has been set up incorrectly; a Pitot tube has become blocked; a sensor has failed; a tipping bucket rain gage has ceased to tip due to torrential rain.

In some other cases, outlying data may be unreliable: during flash floods measurement

of water level at a gaging station may not be possible and assessment of peak flow may
have to be made from peoples' recollection of water levels, and rough calculations of the
relationship between such water levels and flow rather than a calibrated rating curve; measuring equipment may have been set up by inexperienced operators; rainfall attributed to
Sunday may have accumulated over the entire weekend. In these cases it would be unwise
to discard the data but it is worth investigating how much they affect the analysis.

However, in many cases there will be no reason to question the veracity of outlying
values and they represent the nature of the population. For example: rogue waves have
now been well documented and new mathematical models are being developed to explain
the physics; advances in particle physics and astronomy are attributable to observations of
outlying values; the out-turn cost of some civil engineering projects may be far higher than
anticipated because of unexpected underground streams; Alexander Fleming's discovery of
penicillin was a consequence of his noticing an outlying culture of staphylococci.

3.6.2   Robust statistics

The median and MAD are examples of robust statistics because they are relatively insensitive to outlying values. We have already seen that the mean is sensitive to outlying values,
and that the difference between the mean and median is informative. The variance, and
standard deviation, are also sensitive to outlying values. Provided we are confident that the
outlying values are reliable records, this sensitivity is appropriate. We consider these issues
in the following example.

Example 3.35: Data transmission [robust statistics]

[Hampson Jr and Walker, 1961] set aside the seven largest and the seven smallest observations of heat of sublimation of platinum (Example 3.33) before calculating the
mean of the remainder which equals 134.9^{10}. They had performed the experiments
themselves, so they were aware of potential inaccuracies in the procedure, but a less
subjective analysis is to use the median value as the estimate of the physical constant

```
> median(sublimation)
[1] 135.1
```

For comparison, the mean of all 26 observations is

```
> mean(sublimation)
[1] 137.0346
```

Another robust measure of location is the $100\,p\%$ trimmed mean, which is calculated
by sorting the data into ascending order and discarding the lowest $100\,p\%$ and highest
$100\,p\%$. The 20% trimmed mean for the platinum data is

```
> mean(sublimation,trim=0.2)
[1] 135.2812
```

You should report all the data collected, as did Hampson and Walker, even if you decide
to summarize them with a trimmed mean.

---

[10]Very close to the accepted value.

In some industrial processes we may rely on personnel transmitting data to a computer which generates weekly reports. Examples are records of fuel used and payload on domestic flights. If some returned data are likely to be erroneous it may be better to monitor variability from week to week using MAD, or the IQR, rather than s. For data that have a histogram that approximates a bell shape

$$s \approx \frac{MAD}{0.65} \approx \frac{IQR}{1.35}$$

3.7 Grouped data

We have already seen that data are often grouped. In the case of discrete data this is because several items take the same values of the variable. In the case of data for a continuous variable this is because they are recorded as observations within specific cells, for constructing a histogram.

In this section we modify the formulae for mean and variance to allow for the grouping. Although data for a continuous variable are likely to be held in a computer, so that grouping before calculating the mean and variance is unhelpful, the formulae for grouped data have an important theoretical interpretation in the context of the underlying population.

3.7.1 Calculation of the mean and standard deviation for discrete data

Suppose we have a discrete set of K values $\{x_k\}$ for $k = 1, ..., K$, with frequencies $\{f_k\}$. The sample size n is

$$n = \sum_{k=1}^{K} f_k.$$

Then the mean of the data is

$$\bar{x} = \frac{\sum_{k=1}^{K} x_k f_k}{n},$$

the variance is

$$s^2 = \frac{\sum_{k=1}^{K} (x_k - \bar{x})^2 f_k}{n-1}$$

and the standard deviation is the square root (s) of the variance. These results follow directly from the definitions of sample mean and variance with repeated additions replaced by multiplication.

Example 3.36: Car pool [mean and sd of grouped data]

A college has decided to implement a car sharing scheme which will include publicity and some parking incentives for participants.

The student union will monitor the effect of the scheme, and one measure will be average car occupancy. A survey of vehicles arriving at campus car parks between 7:30 a.m and 9:00 a.m. will be performed before the launch of the scheme, at three monthly intervals for the year following the launch, and annually thereafter. Data from the before the launch survey are given in Table 3.11.

TABLE 3.11: The number of occupants in cars arriving at the campus before car sharing.

Number of occupants including driver x_k	Number of cars f_k
1	734
2	102
3	25
4	9
5	1
Total (n)	871

The mean occupancy is

$$\bar{x} \;=\; \frac{1 \times 734 + 2 \times 102 + 3 \times 23 + 4 \times 9 + 5 \times 1}{871} \;=\; 1.21$$

The variance s^2 of occupancy is

$$\frac{(1 - 1.21)^2 \times 734 + (2 - 1.21)^2 \times 102}{(871 - 1)}$$

$$+ \; \frac{(3 - 1.21)^2 \times 23 + (4 - 1.21)^2 \times 9 + (5 - 1.21)^2 \times 1}{(871 - 1)}$$

$$= \; 0.2995.$$

The standard deviation of occupancy is $s \;=\; \sqrt{0.2995} \;=\; 0.55$.

3.7.2 Grouped continuous data [Mean and standard deviation for grouped continuous data]

Sometimes data are only available as a table with cell boundaries and frequencies, or equivalently as a histogram. We can calculate an approximate mean and standard deviation by assuming all the data in a cell are at the mid-point of that cell. Then we have a discrete set of values and frequencies and we can use the formulas of Section 3.7.1. We consider this in detail, here and in Section 3.7.3, because the argument is used to justify the definition of the mean and standard deviation of an infinite population in Chapter 5.

Example 3.37: North Sea waves [statistics from grouped data]

Wave headings and heights recorded over one year at a location in the Irish Sea are given in Table 3.12. Referring to the R code below, the total number of waves, added over headings, in the 13 cells for height are given in f and the cell mid points are given in x. The following R commands calculate the mean, variance, standard deviation and $\widehat{CV}$ from the grouped data.

TABLE 3.12: Wave heights (for all waves during automated recording period) and headings at a location in the Irish Sea.

Wave height (m)	N	NE	E	SE	S	SW	W	NW	Total
0 − 1	453 036	327 419	403 390	571 102	947 294	1 074 776	804 533	439 132	5 020 682
1 − 2	178 960	109 887	113 367	111 698	174 120	291 679	268 823	179 909	1 428 443
2 − 3	45 305	23 131	25 478	17 797	22 798	42 112	57 887	51 699	286 207
3 − 4	12 718	5 851	6 193	3 105	3 024	5 636	12 837	15 288	64 652
4 − 5	3 939	1 798	1 543	535	416	804	3 056	4 772	16 863
5 − 6	1 319	593	384	91	63	123	749	1 556	4 878
6 − 7	473	193	93	17	12	19	183	520	1 510
7 − 8	175	60	21	3	2	3	45	173	482
8 − 9	65	17	5	0	0	0	10	58	155
9 − 10	24	5	1	0	0	0	4	18	52
10 − 11	9	1	0	0	0	0	0	6	16
11 − 12	3	0	0	0	0	0	0	2	5
12 − 13	1	0	0	0	0	0	0	0	1

```
> f=c(5020682,1428443,286207,64652,16863,4878,1510,482,155,52,16,5,1)
> x=0:12+0.5 ; print(x)
 [1]  0.5  1.5  2.5  3.5  4.5  5.5  6.5  7.5  8.5  9.5 10.5 11.5 12.5


> print(f)
 [1] 5020682 1428443  286207   64652 16863 4878 1510  482  155
[10]      52      16       5       1
> n=sum(f) ; print(n)
[1] 6823946
> m=sum(x*f)/n ; print(m)
[1] 0.8371984
> v=sum((x-m)^2*f)/(n-1) ; print(v)
[1] 0.4199117
> s=sqrt(v) ; print(s)
[1] 0.6480059
> CV=s/m ; print(CV)
[1] 0.7740171
```

The mean wave height is 0.88 m, the variance of the wave height is 0.42 m^2 and the standard deviation is 0.65 m. The coefficient of variation is 0.77.

3.7.3   Mean as center of gravity

The formula for the mean of grouped data provides a mechanical explanation for the difference between the mean and median. It makes the explanation easier to follow if we consider a non-negative variable, but the result is quite general. Suppose the histogram is a uniform lamina with density equal to 1 per unit area, so its total mass is 1. A histogram rectangle with mid-point x_k has a mass f_k/n. The moment of this rectangle about the origin is therefore:

$$x_k \times \frac{f_k}{n}$$

The sum of all these moments is, by its definition, $\bar{x}$. The x-coordinate of the center of gravity, denoted by G_x, is the point such that the moment about the origin of the mass of the lamina acting at that point equals the sum of the moments of the rectangles that make up the lamina. That is

$$G_x \times \text{mass of lamina} \;=\; G_x \times 1 \;=\; \text{sum of the moments of the rectangles} \;=\; \bar{x}$$

So, $\bar{x}$ is the x-coordinate of the center of gravity of the histogram and the histogram will balance at this point. The median is the value such that half the area, and hence half the mass, of the histogram lies to either side. If the histogram has a long tail to the right, then the point at which it balances will be to the right of the median and it is positively skewed.

Example 3.38: Discs and wheels [mean and median for grouped data]

Manufactured discs and wheels should lie in a plane. If a bicycle wheel is rotated with a clock-gage held against the rim, the deflection of the gage during a revolution is a measure of **run-out**. Run-out is a non-negative variable and the ideal value is 0. The data in buckle.txt are run-outs (mm) of 100 mass produced steel rimmed bicycle wheels as they come off an automated machine that tightens the spokes. The histogram is shown in Figure 3.27. A vertical line through the median (H), which equals 0.725, bisects the

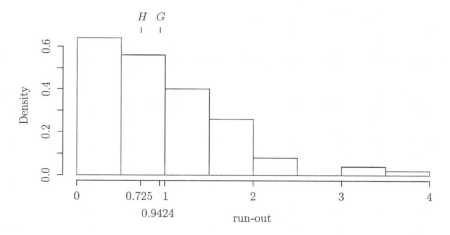

FIGURE 3.27: Run-outs (mm) of 100 mass produced steel rimmed bicycle wheels: median (H) and mean (G).

area. It will not balance at H because the long tail to the right, indicating positive skewness, will cause it to rotate in a clockwise direction. The histogram balances at the mean (G) which equals 0.942.

```
> x=scan("buckle.txt")
Read 100 items
> H=median(x)
> G=mean(x)
> hist(x,xlab="run-out",main="",freq=FALSE)
```

```
> axis(1,at=G,padj=2)
> axis(1,at=H)
> axis(3,at=H,lab=expression(H))
> axis(3,at=G,lab=expression(G))
> print(c(H,G))
[1] 0.7250 0.9424
```

3.7.4 Case study of wave stress on offshore structure.

A finite element model for a static analysis of an offshore structure design resulted in stress against wave height relationships for each element at each joint. Separate relationships were given for four pairs of wave headings N-S, NE-SW, E-W, and SE-NW. We use a simple approximation for fatigue damage from stress cycling, known as **Miner's rule**. Miner's rule states that the damage is proportional to the cube of the stress amplitude and that the damage from stress cycles is additive. The columns of wave data in Table 3.12 are paired to correspond to the pairs of wave headings. Let f_ℓ represent the number of waves in a cell which has a mid-point wave height and heading corresponding to a stress y_ℓ. Then the total damage D is calculated as

$$D \;=\; \sum_{52 \text{ cells}} \frac{y_\ell^3 f_\ell}{A},$$

where A is a constant which depends on the geometry. The joint is assumed to fail when D equals 1. The stress against wave height and heading relationship for a particular element and joint is given in Table 3.13. The value of A is 8.75×10^{12}. The total damage is

$$D \;=\; \left((26.3)^3 \times (453\,036 + 947\,294) + \cdots + (699.1)^3 \times (1 + 0) \right) / (8.75 \times 10^{12}),$$

which equals 0.139 per year and the corresponding average lifetime is 7.2 years. Although this estimate is based on a simple approximation and a single year of wave data, it is useful for design purposes. The calculations are repeated for all the other elements at joints, to assess the safety of the structure and locate its weakest points.

3.8 Shape of distributions

While the location and spread of a set of data are the most important features, two other aspects which may be seen in histograms of large data sets are asymmetry and a lack of data in the flanks because of a high peak or heavy tails. This detail becomes particularly important when we attempt to predict the extreme values taken by variables. Applications include river flood levels, high tides, and tensile failure of materials. Another consequence is that some general results for approximately bell-shaped distributions, such as the fact that 95% of the data fall within two standard deviations of the mean, are no longer reliable.

3.8.1 Skewness

A useful measure of asymmetry is based on the average cubed deviation from the mean. Values near zero indicate symmetry, while large positive values indicate a long tail to the

TABLE 3.13: Stress corresponding to wave height for a particular element at a joint on an offshore structure.

Wave height (m)	Stress wave heading N–S (Pa)	Stress wave heading NE–SW (Pa)	Stress wave heading E–W (Pa)	Stress wave heading SE–NW (Pa)
0.5	26.3	13.4	14.8	25.2
1.5	78.8	40.1	44.4	75.5
2.5	121.2	67.6	75.7	128.4
3.5	167.1	134.4	100.0	178.3
4.5	211.9	174.9	124.8	226.4
5.5	255.6	188.9	150.2	272.5
6.5	305.9	211.6	175.7	319.0
7.5	362.9	243.0	201.5	365.9
8.5	423.9	294.4	239.3	416.7
9.5	489.1	365.9	289.2	471.6
10.5	554.3	437.4	339.1	526.5
11.5	624.3	515.4	392.9	589.2
12.5	699.1	599.8	450.5	659.5

right and large negative values indicate a long tail to the left (see Exercise 3.29). However, the average cubed deviation will have dimensions of the measurement cubed and it is difficult to assess what is 'large', so a non-dimensional measure of asymmetry, called sample **skewness** denoted $\widehat{\gamma}$, is constructed by dividing the average sum of cubes by the cube of the standard deviation.

$$\widehat{\gamma} = \frac{\sum(x_i - \bar{x})^3/(n-1)}{s^3}$$

Absolute values of $\widehat{\gamma}$ in excess of about 0.5 correspond to noticeable asymmetry in an histogram and absolute values in excess of 2 are unusual. The skewness of the strengths of the 180 concrete cubes (Figure 3.12) is −0.60.

3.8.2   Kurtosis

The extent of the tails of a histogram, relative to the standard deviation, is measured by a non-dimensional quantity, based on the average fourth power of deviations from the mean, known as **kurtosis** (denoted $\widehat{\kappa}$):

$$\widehat{\kappa} = \frac{\sum(x_i - \widehat{x})^4/(n-1)}{s^4}$$

A bell-shaped histogram distribution has a kurtosis around 3. Values of kurtosis that are substantially greater than 3 usually indicate relatively extensive tails and a high peak compared with a typical bell-shaped histogram. A flat histogram would have a kurtosis nearer 2. The value of $\widehat{\kappa}$ for the strength of the concrete cubes is 4.5.

Hydrologists calculate skewness and kurtosis for flood records at many sites to help decide on appropriate theoretical distributions for predicting floods (Chapter 5). In a manufacturing context, high kurtosis may indicate some contaminating distribution which is more variable than the predominant distribution. This may warrant investigation. Furthermore, range charts used in statistical quality control (Chapter 10) are sensitive to deviation in kurtosis from 3.

The sample skewness and kurtosis are highly variable in small samples and alternative measures of asymmetry and weight in the tails, based on order statistics, are often preferred, particularly in hydrological applications. These measures are known as **L−moments** because they are linear functions of order statistics. You are asked to investigate these in Exercise 3.11.

3.8.3   Some contrasting histograms

Histograms for the six data sets discussed in the following examples are shown in Figure 3.28, and the summary statistics are given in Table 3.14.

Example 3.39: Precision components [bell-shaped]

Bank Bottom Engineering Services manufactured precision components. The company has provided data from a batch of 206 robot arms. Fourteen measurements have been recorded for each arm. The first measurement is the difference between the length of the arm and the nominal value (mm). A histogram for these 206 differences has a bell shape. The mean and median differ by less than one tenth of the standard deviation. The skewness and kurtosis are close to 0.0 and 3.0 respectively. The CV is not applicable because the differences can be negative.

Example 3.40: Aerotrain [rectangular shape]

The Aerotrain in Kuala Lumpur International Airport runs between the main terminal and the satellite terminal. It departs from the main terminal every three minutes, so the maximum waiting time for a passenger is 3 minutes. We'd expect about one third of passengers to wait less than a minute, one third to wait between 1 and 2 minutes, and about one third to wait longer than 2 minutes. A computer simulation of passenger movement through the terminal was implemented. A histogram for the waiting times (seconds) of 1 000 passengers is shown. As expected it is approximately flat between 0 and 3, roughly symmetric about the mean, and the skewness is close to 0. The histogram is bounded between 0 and 3, there are no tails and the kurtosis is less than 2.

Example 3.41: Dublin City Council [negatively skewed]

Dublin City Council and Sonitus have set up a website that provides statistics of noise levels over 5 minute intervals, from continuous monitoring at various sites in Dublin. A histogram of 72 night time noise levels, level exceeded 10% of the time ($L10, dB$) outside Ballymun Library between midnight and 6 a.m. on a summer night is shown. There is a clear tail to the left which is characteristic of negative skewness. The mean is less than the median, the difference is about one quarter of the standard deviation, and the skewness is -0.8. The kurtosis is 2.90, as a consequence of the lack of a tail on the right hand side. You are asked to plot these data as a time series in Exercise 3.36.

Example 3.42: LAN data [high positive skewness]

The local area network (LAN) data are the number of packet arrivals in 4000 consecutive 10 milli-second intervals seen on an Ethernet at the Bellcore Morristown Research and Engineering Facility. The histogram has an extraordinarily long tail to the right. The mean is about one third of a standard deviation to the right of the median and the skewness is remarkably high at 2.98. The kurtosis is 11.28 which is high relative to 3 but kurtosis can take high values. In contrast kurtosis values less than 2.0 are unusual. The *CV* is very high at 1.88. For some purposes it would be convenient to work with the logarithm of the number of packets with one added, as 0 is a possible observation. You are asked to plot the data as a time series in Exercise 3.37.

Example 3.43: Earthquake [high kurtosis]

Ground acceleration (g) during the San Fernando earthquake on February 9, 1971 at 06:00 hours PST measured at Pacoima Dam. The sampling interval is 0.02s and there are 2 087 data over a 41.74 second period. The data are available from the Canadian Association for Earthquake Engineering website hosted by the department of Civil Engineering at the University of Ottawa. The histogram has very long tails and the kurtosis is nearly 22. Although the difference between the median and mean is negligible relative to the standard deviation the skewness is −1.0. The time series plot (Exercise 3.38) provides an explanation.

Example 3.44: Ferry fuel usage [verified outlying values]

The volume of fuel used (m^3) by a ferry for 141 passages between the same two ports. The extraordinarily high skewness and kurtosis of the fuel consumption volumes is due to two outlying observations. We checked the ship's log, and found there had been gale force headwinds on these passages. If these outlying data had been ignored when calculating the mean, fuel costs would probably be underestimated. Outliers are even more important when it comes to deciding minimum fuel requirements.

The following R function, `moments()`, was written to calculate the statistics, given in Table 3.14.

```
> moments<- function(x) {
  n <- length(x)
  dm <- (x-mean(x))
  CV <- sd(x)/mean(x)
  cubdm <- dm^3
  qudm <- dm^4
  scub <- sum(cubdm)
  squd <- sum(qudm)
  sk <- (scub/(n-1))/sd(x)^3
  kurt <- (squd/(n-1))/sd(x)^4
  return(round(c(mean(x),median(x),sd(x),CV,sk,kurt),2))
}
```

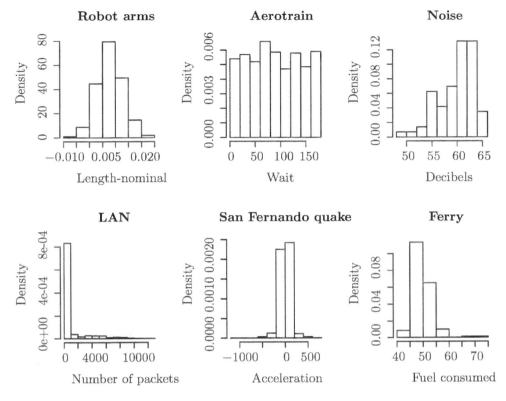

FIGURE 3.28: Contrasting histograms.

Used with, for example, the Ballymun Library L10 data, we get

```
> moments(noise$L10)
[1] 59.75 60.59  3.53  0.06 -0.79  2.90
```

TABLE 3.14: Comparative statistics for six data sets.

data	mean	median	sd	CV	skewness	kurtosis
Robot arm	0.0084	0.008	0.0050	N/A	0.00	2.98
Aerotrain wait	89.95	87.00	51.92	0.58	0.04	1.81
Dublin L10 noise	59.75	60.59	3.53	0.06	-0.79	2.90
LAN	980.01	336.00	1838.48	1.88	2.89	11.28
Earthquake	0.05	0.31	107.50	N/A	-1.02	21.93
Ferry fuel	49.80	49.47	3.51	0.07	2.36	15.93

3.9 Multivariate data

You may have noticed that for several of the data sets there was more than one variable measured for each item. For example, the Ballymun noise items are 5-minute intervals. For each interval, there are two measurements *Leq* and *L*10. The robot arm data had 14 measurements on each arm. If there are two variables measured for each item the data are referred to as **bivariate**. In general, if there are several variables measured for each item the data are referred to as **multivariate**. Bivariate data can be displayed with **scatter plots** and **bivariate histograms**. Multivariate data can be displayed with **parallel coordinate plots**.

3.9.1 Scatter plot

The data are taken from the earthquake catalog of the Institute of Geophysics at the University of Iran for the period 2006-2012 [Nemati, 2014] and are given in the Iran earthquake table on the website. There are 22 earthquakes and each datum consists of 3 numbers: the magnitude on the M_N scale ([Nuttli, 1973]); number of aftershocks; and depth (km). If we consider the magnitude (x) and the common logarithm (y) of the number of aftershocks we have 22 data pairs which we denote by (x_i, y_i) for $i = 1, \ldots, n$ where $n = 22$. These can be displayed on a **scatter plot**. A symbol for each datum (a circle is the R default) is shown centered at the point with coordinates (x_i, y_i). By default, R uses x and y scales that extend only over the range of the data, so they are generally different and do not include 0. The scatter plot is shown in Figure 3.29.

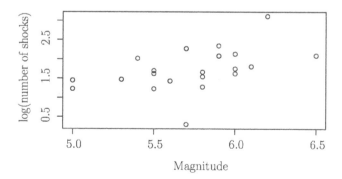

FIGURE 3.29: Scatter plot of the magnitude (x) and the common logarithm (y) of the number of aftershocks.

```
> quake.dat=read.table("Iran_quake.txt",header=T)
> print(head(quake.dat))
    M numashock depth
1 5.8        19  18.0
2 6.0        56  21.7
3 5.9       219  18.0
4 5.5        17  16.7
5 5.9       121  14.0
6 5.7       185  20.5
> attach(quake.dat)
> lognshock=log10(numashock)
> x=M
> y=lognshock
> plot(x,y,xlab="Magnitude",ylab="log(number shocks)")
```

There is a tendency for the logarithm of the number of aftershocks to increase with magnitude, but there is considerable scatter of the points.

The R package plot3d provides a 3D scatter plot which can be used to display data comprised of 3 variables. Figure 3.30 is a 3D scatter plot of aftershocks and depth.

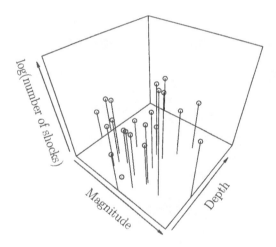

FIGURE 3.30: 3D scatter plot of the magnitude x, the common logarithm (y) of the number of aftershocks and depth.

```
> scatter3D(M,depth,lognshock,xlab="Magnitude",
+ zlab="log(number shocks)",ylab="depth",colvar=NULL,col=NULL,type="h")
```

There is no clear evidence of a relationship between depth and magnitude or the logarithm of aftershocks from the plot[11].

---

[11] The order of the 3 variables is x, y and z. The colvar= and col= suppress the color options, and type=''h'' adds the vertical lines from the points to the xy-plane.

3.9.2   Histogram for bivariate data

The principle is similar to constructing a histogram for a single variable (the **univariate** case). However, the cells are now rectangles rather than line segments. We construct blocks (formally rectangular prisms) over the cells. The heights of the blocks indicate the relative frequency densities, which are the relative frequencies divided by the cell areas. Then the total volume of the histogram equals 1.

Example 3.45: Sea states [bivariate histogram]

Data for sea states measured from the Forties Platform, Central North Sea are given in Table 3.15. Sea states are typically based on 10–20 minutes of observations of significant wave heights and mean zero crossing period, taken every 3 hours. The significant wave height was originally defined as the mean trough to crest height of the highest third of the waves. It is now usually defined as four times the standard deviation of the surface elevation. A histogram of the data is given in Figure 3.31. The mean zero crossing period tends to increase as the wave height increases.

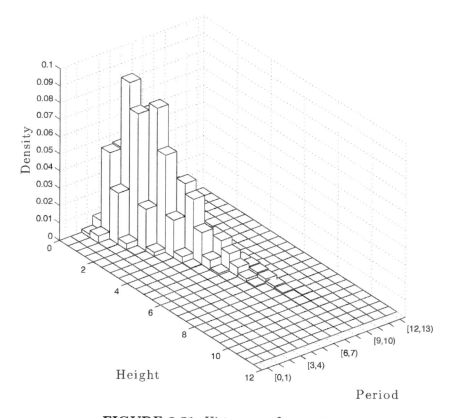

FIGURE 3.31: Histogram of sea states.

TABLE 3.15: Sea states measured from the Forties Platform, Central North Sea, between June 1974 and August 2001, classed into cells by significant wave height (m) and mean zero crossing period (s). [Health and Safety Executive, UK].

height ≥	<												
11.0	11.5	0	0	0	0	0	0	0	0	0	0	1	0	0
10.5	11.0	0	0	0	0	0	0	0	0	0	0	3	1	0
10.0	10.5	0	0	0	0	0	0	0	0	1	2	4	4	0
9.5	10.0	0	0	0	0	0	0	0	0	1	6	6	4	0
9.0	9.5	0	0	0	0	0	0	0	0	4	13	5	4	0
8.5	9.0	0	0	0	0	0	0	0	0	16	26	9	3	0
8.0	8.5	0	0	0	0	0	0	0	2	21	47	6	2	0
7.5	8.0	0	0	0	0	0	0	0	2	67	58	3	0	1
7.0	7.5	0	0	0	0	0	0	0	6	128	50	5	2	1
6.5	7.0	0	0	0	0	0	0	1	51	219	60	5	0	0
6.0	6.5	0	0	0	0	0	0	3	165	270	42	5	1	0
5.5	6.0	0	0	0	0	0	1	9	462	309	44	1	0	0
5.0	5.5	0	0	0	0	0	3	92	887	299	29	5	0	0
4.5	5.0	0	0	0	0	0	3	477	1 273	203	19	2	0	0
4.0	4.5	0	0	0	0	2	24	1 613	1 255	201	26	3	0	1
3.5	4.0	0	0	0	0	2	318	3 029	996	148	23	4	0	0
3.0	3.5	0	0	0	0	10	1 889	3 546	776	133	30	4	0	0
2.5	3.0	0	0	0	2	205	4 940	3 151	675	153	22	1	0	0
2.0	2.5	0	0	0	17	2 102	7 079	2 624	612	131	28	10	0	0
1.5	2.0	0	0	0	302	6 717	6 545	1 808	394	79	22	11	0	2
1.0	1.5	0	0	11	2 592	8 091	4 765	1 334	247	66	31	12	2	3
0.5	1.0	0	0	356	4 390	4 470	2 031	565	178	70	30	10	1	0
0.0	0.5	2	0	153	635	450	227	105	54	24	17	12	0	0
≥		0	1	2	3	4	5	6	7	8	9	10	11	12
period <		1	2	3	4	5	6	7	8	9	10	11	12	13

3.9.3   Parallel coordinates plot

Parallel coordinate plots are useful for multivariate data. There is an ordinate for each variable, and each item is plotted on each coordinate. The points for each item are joined by lines.

Example 3.46: Robot arms [parallel coordinates plot]

A parallel coordinates plot for four robot arms (numbers 10 to 13) is shown in Figure 3.32. It can be generated using the following R command.

```
> library(MASS)
> parcoord(ROB[10:13,],lty=c(1:4),var.label=TRUE)
```

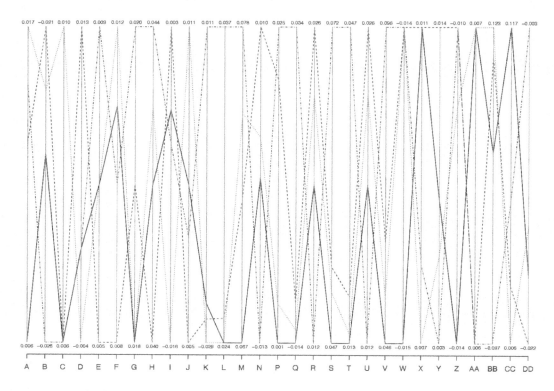

FIGURE 3.32: Sample parallel coordinates plot for robot arms 10–13.

The `parcoord()` function sets the vertical axis for each variable from the minimum value in the sample being plotted to the maximum value in the sample being plotted [12]. Arm number 10, represented by the full line, has a value of 0.006 for variable A, about -0.023 on variable B, 0.006 on variable C and around 0.001 on variable D, and so on. It appears somewhat lower than the other three robot arms on most of the first 20 variables, but there are no clearly outlying arms on a parallel coordinates plot of all 206 arms (Exercise 3.32).

Example 3.47: Sea ports [parallel coordinates plot]

The parallel coordinate plot can be useful for comparing groups of multivariate observations. [Frankel, 1998] summarized China coastal port capability in 1995. In Table 9.6, we give the region of China, the total number of ocean berths (tob), the number of general cargo berths (gcb), and the annual throughput during 1995 in million twenty foot equivalent units (TEU). We calculate the number of specialist berths (sp), which were dedicated to RoRo, container, bulk, tanker, and multi-purpose, as the difference between tob and gcb. The `as.numeric()` codes N,E and S using alphabetical order as 2, 1, 3. We then draw a parallel coordinates plot in which the line type is 2 (dashed) for north, 1 (full) for east, and 3 (dotted) for south (Figure 3.33).

---

[12]The function `ggplot.parcoords()` allows more choice about the axes.

```
> CP=read.table("Chinaports.txt",header=T)
> attach(CP)
> reg=as.numeric(region)
> sp=tob-gcb
> Ports=cbind(tob,sp,TEU)
> parcoord(Ports,lty=reg,var.label=TRUE)
```

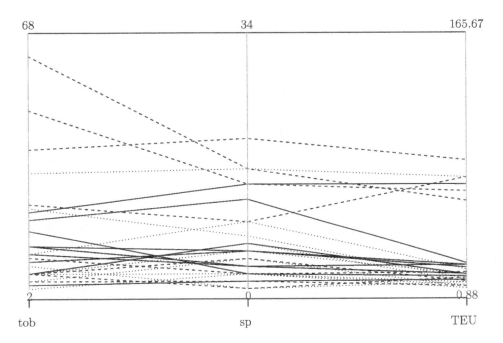

FIGURE 3.33: Parallel coordinates plot for ChinaPorts data.

We can see that the three largest ports are all in the north, and that ports appear to do better in terms of throughput if more of their berths are specialist berths rather than classed as general cargo berths. The benefits of installing the more modern specialist berths will be quantified using statistical analysis in Chapter 9. Figure 3.30 illustrates parallel coordinates for China ports.

3.10 Descriptive time series

We described a time series plot in Section 3.3.2 and remarked that trends and seasonal variation can often be seen in such plots. We aim to quantify these features to provide insight into possible generating physical causes, and to make short term forecasts.

3.10.1 Definition of time series

A **time series** is a record of observations of some variable over time. The observations constitute a discrete sequence, and it is convenient if the time between them, the sampling interval, is constant. The observations can be instantaneous values, such as a voltage measured by a digital storage oscilloscope, or be aggregated over the sampling interval, as with daily rainfall or weekly sales. The sampling interval can be very short, nano-seconds for digital oscilloscopes, or relatively long, as in 50 years for the methane concentrations from the Vostok ice core.

3.10.2   Missing values in time series

As for any data set, we should check outlying values. In addition, many time series analyses rely on data being evenly spaced and this raises questions about missing values: are there missing values and, if so, what should we do about them?

Economic and social time series published by government agencies are usually complete but there may be changes in definitions and methods of calculation that we should be aware of. In contrast, rainfall time series usually have periods of missing data because of the failure of remote instrumentation. Rainfall time series are used by civil engineers to calibrate stochastic rainfall models that are used, for example, for the design of urban drainage systems.

A first step is to check data files for missing value codes and then for missing values. Occasional missing values can be replaced by interpolation. If the data file is relatively small the interpolation can be manual but for large data files a computer algorithm will need to be developed. If there are substantial periods of missing data, sub-series without missing values can be analyzed and the results merged. Alternatively, an analysis that allows for unequal time steps can be used.

3.10.3   Decomposition of time series

Here we concentrate on monthly time series as might be used for planning decisions made by engineering companies. We take monthly electricity use in the U.S. from 1973 to 2013 as an example. It is useful to think of such time series in terms of an underlying trend, a seasonal component, and a random component. Following the methods offered by R we consider an additive model

$$Y_t \;=\; T_t + S_t + G_t$$

and a multiplicative model

$$Y_t \;=\; T_t \times S_t \times G_t$$

where Y_t is the variable, T_t is the **trend**, S_t is the **seasonal component**, and G_t is the **random component**. For monthly data the seasonal component has a period of twelve time steps.

3.10.3.1   Trend - Centered moving average

For either model, we can estimate the trend by a **centered moving average** with 13 terms. The idea is that the seasonal component will be removed by averaging over a calendar year and that the random component will be reduced by this averaging. If the data are $\{y_t\}$ this centered moving average is:

$$T_t \;=\; \frac{0.5 \times y_{t-6} + y_{t-5} + \cdots + y_t + \cdots + 0.5 \times y_{t+6}}{12}, \qquad \text{for } t = 1, \cdots, n.$$

This formula applies the same weight $1/12$ to each calendar month. The first and last observations are averaged (factor 0.5) because they correspond to the same calendar month.

For example, if t corresponds to July of Year 1, then $t-6$ corresponds to January of Year 1 and $t+6$ to January of Year 2. The reason for splitting the January contributions is that the moving average does then align with July. If we took the mean of January up to December in the same year the mean would correspond to half way between June and July[13]. The centered moving average will start after the first six months and end before the final six months of the time series.

3.10.3.2　Seasonal component - Additive monthly model

For the additive model, we subtract the centered moving average from the monthly observations:

$$y_t - T_t, \qquad 7 \le t \le n - 6.$$

We then take the mean of these difference for each calendar month[14] as an initial estimate of a seasonal difference.

$$\widetilde{S}_m \;=\; \sum_{t \text{ for month } m} (y_t - T_t)/(n/12 - 1), \qquad m = 1, \cdots, 12$$

The mean of these 12 monthly differences, $\widetilde{S}_m$, will be close to 0. However, we adjust the $\widetilde{S}_m$ by subtracting their mean value so that the adjusted monthly differences S_m have a mean of precisely 0.

$$S_m \;=\; \widetilde{S}_m - \sum_{k=1}^{12} \widetilde{S}_k/12, \qquad m = 1, \cdots, 12 \text{ with } k = 1, \cdots, 12.$$

The reason for making this adjustment is that the seasonal differences, S_m, do not affect yearly mean values. Finally, the random component is found as

$$y_t - T_t - S_{m(t)}, \qquad 7 \le t \le n - 6,$$

where $m(t)$ is the calendar month corresponding to time t.

3.10.3.3　Seasonal component - Multiplicative monthly model

Similar principles are applied to fit the multiplicative model. We begin by calculating the ratios of observation to trend

$$y_t/T_t, \qquad 7 \le t \le n - 6,$$

and then we take the mean of these ratios for each calendar month ($m = 1, ..., 12$)

$$\widetilde{S}_m \;=\; \sum_{t \text{ for month } m} (y_t/T_t)/(n/12 - 1).$$

The mean of these 12 indices should be close to 1, but they are adjusted by dividing by their mean so that their mean is precisely 1.

$$S_m \;=\; \widetilde{S}_m / \sum_{k=1}^{12} \widetilde{S}_k/12,$$

---

[13]The centered moving average for July in a particular year can also be expressed as the mean of the mean of January to December in that year and the mean of February in that year up to January of the next year.

[14]The formula assumes that n runs over complete years ($n = 12 \times$ number of years).

Finally, the irregular component is found as

$$y_t/(T_t \times S_{mt}), \qquad 7 \le t \le n - 6,$$

where $m(t)$ is the calendar month correspond to time t.

3.10.3.4   Seasonal adjustment

With the additive model, a **seasonally adjusted** (also known as a **deseasonalized**) time series is obtained by subtracting the estimated seasonal effects from the original time series:

$$y_t - S_t.$$

In the case of the multiplicative model, the seasonally adjusted time series is

$$y_t/S_t.$$

Example 3.48: Electricity usage [trend and seasonal correction]

The data can be found on the website and represent monthly electricity usage in the U.S. from January 1973 to June 2013.

The R function `decompose()`[15] fits seasonal effects and plots: the time series (Observed); trend (Trend); seasonal contributions (Seasonal); and the seasonally adjusted series with the estimated trend subtracted (Random). The figures are given in Figure 3.34 and Figure 3.35 for additive and multiplicative models respectively.

```
> elec <- read.table("data/elecuse.txt",header=T)
> el.ts <- ts(elec$elecuse,frequency=12,start=c(1971,1))
> plot(decompose(el.ts,type="additive"))
> plot(decompose(el.ts,type="multiplicative"))
```

Over the period 1970 until 2010 electricity usage in the U.S. has approximately doubled. During the same period the population has increased from 203.2 million to 308.7 million[16] (U.S. Census Bureau), but seems to have leveled since around 2005. Season variation is around plus or minus 10% of the trend, probably because of heavier use of air conditioning in the summer, and heating in the winter, than in spring or autumn. The random component has a range of around 7% of the trend which is somewhat smaller than season variation. The range of the random variation is more nearly constant with the multiplicative model, which would be preferred in this case.

3.10.3.5   Forecasting

The objectives of time series analysis are to understand how some variable has changed in the past, and and to use this information to help make realistic assessments of will happen in the future.

In general, seasonal effects will be less prone to change than a trend. To make a short

---

[15]The function `ts()` makes a time series object and `frequency=` the period of the seasonal component, 12 for months in the year and `start=` gives start year and month. There is no restriction to an integer number of years.

[16]The population (in millions) in 1980, 1990, and 2000 is given as 226.5, 248.7, and 281.4 respectively.

Decomposition of additive time series

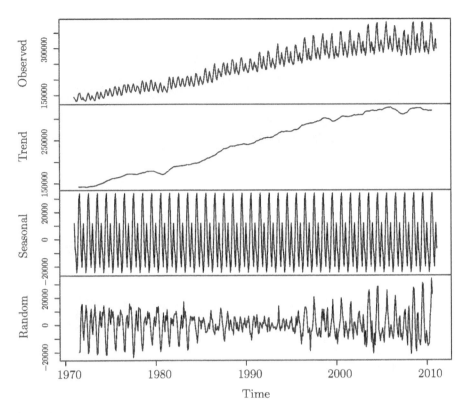

FIGURE 3.34: Decomposition of electricity use data with an additive model.

term forecast up to a few time steps ahead, it may be reasonable to extrapolate a linear trend, fitted to the more recent part of the time series, and then apply the estimated seasonal effects. We need to be aware that trends change and to anticipate this by monitoring the overall situation.

Example 3.49: Manufacturing solar panels [short term forecast]

The production engineer in a company that manufactures solar panels needs to forecast sales for the next three months. She knows that the demand is considerably lower during winter months because people prefer to install solar panels when the weather is better. She has analyzed monthly sales figures for the past five years and has estimated seasonal indices, and a linear trend over the period.

The overall trend is a monthly increase of around 500 units per month. The monthly increase was substantially higher immediately following government financial incentives for households to install solar panels, and noticeably lower after a competitor set up business, but has been reasonably consistent over the past few months. She is not aware of any imminent changes to the market. To make the forecasts, she projects the linear trend for three months and then multiplies by the appropriate monthly indices.

The best strategy for making longer term forecasts is to find a *leading variable* which will strongly influence future values of the variable to be forecast.

Decomposition of multiplicative time series

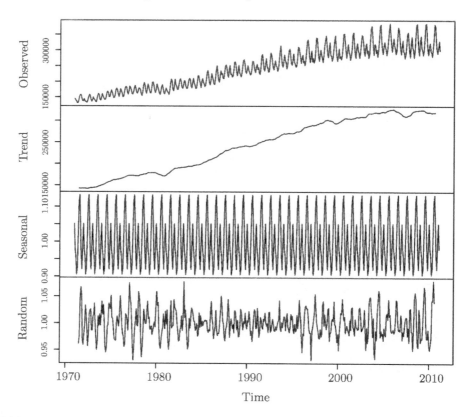

FIGURE 3.35: Decomposition of electricity use data with a multiplicative model.

Example 3.50: Marine paints [leading variable]

A company that manufactures marine paint can use statistics available in the public domain, *World Shipyard Monitor* for example, to forecast the numbers of ships by size and type to be built over the next three years. The company can then forecast demand for its paints under an assumption that it maintains its current market share.

Alternatively, we can look for a variable, or variables, that help account for the trend and which are easier to forecast.

Example 3.51: Electricity use [explanatory variables]

Factors which affect recorded electricity use include: demographics such as the population size and typical household composition; power generation at the property such as deployment of solar panels; and increased use of electric vehicles. The implications for power generation also need to be considered. Electric vehicles will usually be charged overnight which will assist with load balancing. The uptake of electric vehicles might be modeled with a Bass curve ([Bass, 1969], and Exercise 9.9).

It may be more useful to consider electricity use relative to the size of the population as this ratio may be easier to explain and predict, and population change itself can be predicted with reasonable accuracy for several years ahead. See Figure 3.36.

The R `aggregate()` function sums the time series variable over the period defined by the frequency = argument.

```
> elec_year=aggregate(el.ts)
> pop=read.table("uspop.txt",header=T)
> elecupp=elec_year/pop$populn
> plot(elecupp,xlab="Year",ylab="electricity use per person")
```

The power supply industry also needs short term forecasts for day to day operations. In particular, the gas supply industry relies on hourly forecasts up to a day ahead. The seasonal variation now corresponds to hours within the day and the trend depends on weather forecasts. Spot prices in the electricity supply market are quoted for 30 minute, or shorter, intervals.

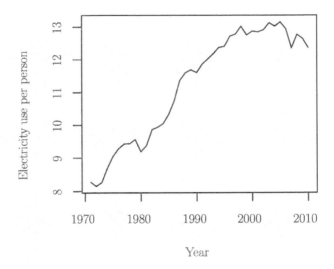

FIGURE 3.36: Average electricity use per person for the years 1971 to 2010.

3.10.4 Index numbers

In the U.S. the Bureau of Labor Statistics publishes Producer Price Indices (PPI). These measure price increases in a range of market sectors, the Nonresidential Building Construction Sector for example, and they are used to adjust prices in contracts that extend over many months

A price index is the ratio of the cost of a basket of goods now to its cost in some base year. In the Laspeyre formula the basket is based on typical purchases in the base year. The basket of goods in the Paasche formula is based on typical purchases in the current year. Data for calculating an index of motoring cost is given in Table 3.16. The "clutch" represents all mechanical parts, and the quantity allows for this.

TABLE 3.16: Motoring costs.

item	Base year		Base year $+ t$	
	quantity	unit price	quantity	unit price
(i)	(q_{i0})	(p_{i0})	(q_{it})	(p_{it})
car	0.33	18 000	0.5	20 000
petrol (*liter*)	2 000	0.80	1 500	1.60
servicing (h)	40	40	20	60
tyre	3	80	2	120
clutch	2	200	1	360

The *Laspeyre Price Index* at time t relative to base year 0 is:

$$LI_t = \frac{\sum q_{i0}p_{it}}{\sum q_{i0}p_{i0}}.$$

The *Paasche Price Index* at time t relative to base year 0 is:

$$PI_t = \frac{\sum q_{it}p_{it}}{\sum q_{it}p_{i0}}.$$

The following calculations in R give the two indices.

```
> p0=c(18000,0.8,40,80,200)
> q0=c(0.33,2000,40,3,2)
> pt=c(20000,1.60,60,120,360)
> qt=c(0.5,1500,20,2,1)
> L=sum(q0*pt)/sum(q0*p0)
> print(c("Laspeyre",round(L,2)))
[1] "Laspeyre" "1.36"
> P=sum(qt*pt)/sum(qt*p0)
> print(c("Paasche",round(P,2)))
[1] "Paasche" "1.25"
```

The Laspeyre Price Index, 1.36, is higher than the Paasche Price Index, 1.25, and this is typical. Can you explain why?

It is a consequence of the definitions that the index for year t relative to year 0 is the product of the indices for: year 1 relative to year 0, year 2 relative to year 1, and so on up to year t relative to year $(t-1)$.

3.11 Summary

3.11.1 Notation

$\bar{x}/\mu$	mean of sample/population	s/σ	standard deviation of sample/population
$\widehat{M}/M$	median of sample/population	$\widehat{\gamma}/\gamma$	skewness of sample/population
$\widehat{q}(p)/q(p)$	quantile of sample/population	$\widehat{\kappa}/\kappa$	kurtosis of sample/population
s^2/σ^2	variance of sample/population		

3.11.2 Summary of main results

Graphical:

- Always plot the data.

- Provide: a clear description of the variable; explain how the sample was obtained and how it relates to the population; label axes; and provide the unit of the measurements.

- For data relating to a discrete variable, a line plot is a plot of relative frequency against the discrete values of the variable. This relates to the probability mass function in Chapter 4.

- For data relating to a continuous variable, classed into contiguous cells (bins), the relative frequency is the number of data in each cell divided by the total number of data, The histogram is a plot of relative frequency divided by cell width (relative frequency density) against the variable. This relates to the probability density function in Chapter 5.

- The cumulative frequency polygon is a plot of the proportion of data less than the right-hand endpoint of intervals against right-hand endpoints. This relates to the cumulative distribution function in Chapter 5.

A box-plot consists of a box extending from the lower quartile to the upper quartile and lines from the ends of the box to the least value and greatest value, that are not classed as outliers, respectively. Outliers are plotted as individual points.
Numerical: (see Table 3.17)

- The main measures of centrality are sample mean (average) and median (middle value of sorted data).

- The most commonly used measure of spread is the variance together with its square root, the standard deviation.

- Approximately 2/3 of a data set will be within one standard deviation of the mean.

- If a variable is restricted to non-negative values, then the CV is the ratio of the standard deviation to the mean.

- The order statistics are the data sorted into ascending order. The sample quantile corresponding to a proportion p is the $p(n+1)^{th}$ smallest datum.

- The lower quartile is estimated by the $0.25(n+1)^{th}$ smallest datum in the sample, and an approximation to this estimate can be found as the value of the variable corresponding to a cumulative proportion of 0.25 on the cumulative frequency polygon. The upper quartile is similarly estimated as the $0.75(n + 1)^{th}$ smallest datum in the sample.

- The inter-quartile range is the difference between the upper and lower quartiles. It is a measure of spread and the central half of the data lies between the quartiles.

TABLE 3.17: Population parameters and sample statistics.

Quantity	Sample (n) statistic	Finite population (N) parameter
mean	$\bar{x} = \sum x_i/n$	$\mu = \sum x_j/N$
Median	$\widehat{M} = x_{(n+1)/2:n}$	M such that $P(x_j \leq M) = 0.5$
Quantile	$\widehat{q}(p) = x_{p(n+1):n}$	$q(p)$ such that $P\big(x_j \leq q(p)\big) = p$
Variance	$s^2 = \sum(x_i - \bar{x})^2/(n-1)$	$\sigma^2 = \sum(x_j - \mu)^2/N$
Standard deviation	$s = \sqrt{s^2}$	$\sigma = \sqrt{\sigma^2}$
Skewness	$\widehat{\gamma} = \big(\sum(x_i - \bar{x})^3/(n-1)\big)/s^3$	$\gamma = \big(\sum(x_i - \mu)^3/N\big)/\sigma^3$
Kurtosis	$\widehat{\kappa} = \big(\sum(x_i - \bar{x})^4/(n-1)\big)/s^4$	$\kappa = \big(\sum(x_i - \mu)^4/N\big)/\sigma^4$

3.11.3 MATLAB and R commands

In the following `filename` is a variable containing the name of .txt file (for example "gold_grade_C.txt"), and x is a vector containing any real values. For more information on any built in function, type `help(function)` in R or `help function` in MATLAB.

R command	MATLAB command
gold <- read.table(filename,header=T)	
	gold = importdata(filename,'', 1)
min(x)	min(x)
max(x)	max(x)
mean(x)	mean(x)
median (x)	median (x)
summary(x)	summary(dataset(x))
var(x)	var(x)
sd(x)	std(x)
sum(x)	sum(x)
sqrt(x)	sqrt(x)
hist(x)	hist(x)
boxplot(x)	boxplot(x)
median (abs(x- median (x)))	mad(x,1)
mydata <- c(701, 630, 592)	mydata = [701, 630, 592]

3.12 Exercises

Section 3.1 Types of variables

Exercise 3.1: Electronics

A company manufactures electronic components and also produces mobile phones. Classify the following variables as discrete, continuous, or categorical. If categorical, then state whether the categories can be ordered.

(a) The color of phones which can be white, red, silver, or gold.

(b) The numbers of phones ordered each day.

(c) Total manufacturing costs each day.

(d) Resistances of resistors.

(e) Temperature at 12 noon in the main factory.

(f) Safety assessment of factory practices by government inspector classified as: dangerous (shut down), poor (need immediate attention), satisfactory, and excellent.

(g) Number of employees on leave each day.

(h) Number of employees, including drivers, in cars arriving at car park.

(i) Gain of amplifiers.

(j) Number of different models of phone produced.

Exercise 3.2: Hydrology

The following variables arise in hydrology. Classify them as discrete, continuous, or categorical. In the case of continuous variables suggest a unit of measurement. In the case of a categorical variable can it be ordered?

(a) River flow at midday.

(b) Number of exceedances of a depth of 1 m at a particular location each year.

(c) Stream water velocity.

(d) Height of water above the crest of a weir.

(e) Flood risk level classed as yellow, orange or red.

(f) Phosphorus content of water.

Section 3.2 Samples and populations

Exercise 3.3: Population variance

If we knew the population mean μ, then the natural estimate of the population variance would be

$$\widehat{\sigma^2} \;=\; \frac{\sum (x_i - \mu)^2}{n}$$

(a) Find the value of a that minimizes the function

$$\psi(a) \;=\; \sum (x_i - a)^2$$

(b) Hence explain why

$$s^2 \frac{n-1}{n} \;\leq\; \widehat{\sigma^2}$$

and suggest a rationale for using the denominator $n-1$ when calculating s^2.

Exercise 3.4: Population sampling

A population of size $2\,000$ consists of $1\,000$ items for which the variable x_i takes the value -1 and $1\,000$ items for which x_i is $+1$.

(a) What is the variance and standard deviation of the population?

(b) If you take a sample of $n = 2$, with replacement, what values can s^2 take, and what are the probabilities that it takes these values? The expected value of s^2 is the sum of the products of the values it can take with the probability that it takes these values. What is the expected value of s^2?

(c) If you take a sample of $n = 2$, with replacement, what values can s take, and what are the probabilities that it takes these values? The expected value of s is the sum of the products of the values it can take with the probability that it takes these values. What is the expected value of s?

(d) If you take a sample of $n = 1$, what is the value of s? Why is this appropriate if we do not know the population mean?

Section 3.3 Displaying data

Exercise 3.5: Cable lengths

The yield for 120 sample lengths of a given cable, measured in $\mathrm{Nmm^{-2}}$ to the nearest integer, can be grouped as follows:

Yieldpoint	50-79	80-89	90-99	100-109	110-119
Frequency	3	6	13	25	24

Yieldpoint	120-129	130-149	150-179	180-239
Frequency	2	18	7	3

(a) Draw a histogram ensuring that the total area is 1.

(b) Draw a cumulative frequency polygon of the data.

(c) Calculate the approximate median, lower quartile, upper quartile, and inter-quartile range.

Exercise 3.6: PAH concentrations

The following data are Polycyclic Aromatic Hydrocarbon (PAH) concentrations (ng/l) in 200 jars of water filled from taps on a domestic supply network.

PAH concentration	0-100	100-200	200-300	300-500
number of jars	30	50	70	50

(a) Consider the construction of a histogram that is to be scaled to a total area of one. What is the area and height of the lowest of the four rectangles in such a histogram?

(b) Calculate an approximate mean PAH concentration.

(c) Calculate an approximate variance and standard deviation of PAH concentrations.

(d) Calculate approximate lower and upper quartiles (0.25 and 0.75 quantiles) of PAH concentrations, and the inter-quartile range.

Exercise 3.7: Sea level 1

The following data are annual maximum sea water levels (mAHD) at a location near Port Adelaide over 50 years.

Annual maximum	0.5-2.5	2.5-3.5	3.5-4.0	4.0-5.5
Number of years	4	6	18	22

(a) Calculate the relative frequencies and the relative frequency densities for each bin (class interval).

(b) Sketch a histogram for the data ensuring that the total area is one, and labelling both axes correctly.

(c) Explain, with a sketch, why placing rectangles with heights equal to frequencies (number of years) gives a misleading impression.

Exercise 3.8: Sea level 2

Refer to the data in Exercise 3.7, with the bin from 4.0-5.5 has been subdivided into two bins

Annual maximum	4.0-4.5	4.5-5.5
Number of years	17	5

(a) Calculate an approximate mean of annual maximum sea water level.

(b) Calculate an approximate variance and standard deviation of annual maximum sea water level.

(c) Calculate an approximate upper quartile (0.75 quantile) of annual maximum sea water level by interpolation within the appropriate bin.

Section 3.4 Numerical summaries of data

Exercise 3.9: Quartiles

The following data are annual maximum flows in $m^3 s^{-1}$ for the Fish River at Dirichas in Namibia for 12 years from the 1977-1978 water year: 160.0, 87.9, 178.4, 0.1, 61.7, 97.8, 193.4, 114.4, 209.1, 230.3, 77.4 and 189.0. Calculate the following statistics:

(a) mean

(b) median

(c) lower quartile

(d) upper quartile

(e) draw a box-plot for the data.

Exercise 3.10: Quartiles in R

Read the information given by the R command `help(quantile)`.

(a) Repeat Exercise 3.9 using the R default.

(b) Which quantile type corresponds to Definition 3.12?

Exercise 3.11: L-moments

L-moments are linear functions of order statistics.

(a) For a random variable X the r-th population L-moment is

$$\lambda_r cr^{-1} \sum_{k=0}^{r-1} (-1)^k \binom{r-1}{k} E[X_{r-k:r}].$$

The first L-moment is

$$\lambda_1 \;=\; E[X]$$

and the fourth L-moment is

$$\lambda_4 \;=\; (E[X_{4:4}] - 3E[X_{3:4}] + 3E[X_{2:4}] - E[X_{1:4}])/4.$$

Write down expressions for λ_2 and λ_3.

(b) Estimators for the first four sample L-moments from a random sample of n observations are

$$\ell_1 \;=\; \binom{n}{1}^{-1} \sum_{i=1}^{n} x_{i:n}$$

$$\ell_2 \;=\; \tfrac{1}{2}\binom{n}{2}^{-1} \sum_{i=1}^{n} \left\{ \binom{i-1}{1} - \binom{n-i}{1} \right\} x_{i:n}$$

$$\ell_3 \;=\; \tfrac{1}{3}\binom{n}{3}^{-1} \sum_{i=1}^{n} \left\{ \binom{i-1}{2} - 2\binom{i-1}{1}\binom{n-i}{1} + \binom{n-i}{2} \right\} x_{i:n}$$

$$\ell_4 \;=\; \tfrac{1}{4}\binom{n}{4}^{-1} \sum_{i=1}^{n} \left\{ \binom{i-1}{3} - 3\binom{i-1}{2}\binom{n-i}{1} + 3\binom{i-1}{1}\binom{n-i}{2} - \binom{n-i}{3} \right\} x_{i:n}.$$

The L-skewness and L-kurtosis are defined by ℓ_3/ℓ_2 and ℓ_4/ℓ_2 respectively.

Calculate the first four sample L-moments and the L-skewness and L-kurtosis for annual flood $(m^3 s^{-1})$ series for the following rivers in Namibia.

River Kuiseb at Schlesein Weir for 28 years from 1963

841.395	0.520	308.973	168.825	456.171	20.583	205.704
122.120	98.683	218.828	396.783	133.934	133.934	267.432
71.488	73.035	8.867	73.035	1.801	0.488	58.138
123.619	151.333	29.079	138.782	175.974	13.230	16.247

River Omatako at Ousema for 27 years from 1962

7.045	434.807	13.660	182.210	113.326	43.529	46.925
31.227	59.177	77.246	111.592	45.779	216.573	8.754
169.276	30.294	160.911	74.309	38.153	20.976	28.472
20.215	111.592	30.294	106.270	91.176	64.479	

River Fish at Seeheim for 21 years from 1962

860.696	356.158	585.374	1205.186	418.487	400.286	8300.113
356.158	6125.126	58.438	2476.661	340.309	78.601	368.169
173.868	0.000	38.444	593.790	178.715	101.400	204.815

River Ugab at Petersburg for 21 years from 1962

22.263	38.935	3.054	87.027	9.249	31.529	28.203
27.621	42.655	2.745	15.736	4.205	21.718	14.996
86.053	36.495	29.879	119.781	301.795	67.064	14.718

Exercise 3.12: Descriptive statistics

Given a set of data $\{x_i\}$, for $i = 1, \ldots, n$ show that

$$\sum_{i=1}^{n} \sum_{j=1}^{n} \frac{(x_i - x_j)^2}{2n(n-1)} = s^2.$$

Hint: Express $x_i - x_j = (x_i - \bar{x}) - (x_j - \bar{x})$.

Exercise 3.13: Sums

If a is some constant, $\sum_{i=1}^{n} a = na$ follows from the definition. Provide another justification for this result by considering

$$\Sigma(x_i - a) = \Sigma x_i - \Sigma a.$$

Exercise 3.14: Geometric mean

A corporation deposited \$1M with a bank that specializes in loans to start up businesses for a fixed term of three years. In the first year the corporation received interest of 9%, in the second year the interest was only 4%, and in the third year the interest was 8%. Assume that interest is paid yearly and compounded.

(a) How much interest does the corporation receive?

(b) What fixed interest rate compounded yearly would give the same amount at the end of the three years?

(c) What is the arithmetic mean of 9, 4 and 8?

(d) Does your answer to (b) depend on the order in which the different yearly rates are applied?

(e) The geometric mean of a set of n numbers $\{x_i\}$ is the nth root of their product.

$$(\Pi_{i=1}^{n} x_i)^{1/n},$$

where

$$\Pi_{i=1}^{n} x_i \;\; = \;\; x_1 \times x_2 ... \times x_n$$

(i) Let $y_i = ln(x_i)$ and show that the geometric mean of x is exponential of the mean of y.

(ii) Prove that the geometric mean of two numbers is less than their arithmetic mean (The result is true for any n).

Exercise 3.15: Average speed

A freight train runs up an incline over 10 miles at 30 mph and descends over another 10 miles at 60 mph. What constant speed would allow it to cover the 20 miles in the same time?

Exercise 3.16: Harmonic mean

If you travel 60 miles at 30 mph, 60 miles at 60 mph and 60 miles at 90 mph, what is your 'average speed'?

Exercise 3.17: Interest rates

You invest \$1 000 in a risky venture for four years with interest rates of 10%, 20%, 28% and 15%, compounded yearly, applied for the first, second, third and fourth year respectively.

(a) What single rate compounded over 4 years would leave you with the same amount of money?

(b) Would your answer change if the three interest rates had been 20%, 15%, 28% and 10% during the first, second, third and fourth year respectively?

Exercise 3.18: Discounting

An engineer makes an interest free loan of \$1000 to a charity that provides water wells for villagers in an overseas country. The loan will be repaid in five years. Assume the bank base interest rate over the five years remains constant at 4% per annum.

(a) How much interest would she receive at the end of five years if she deposited the money in a bank at the base rate of 4% per annum.

(b) What amount deposited with the bank now, at 4% per annum, would yield \$1000 in five years time?

(c) What is the current value a certain \$1000 in five years time?

(d) What is the current value of her donation to the charity (for tax purposes)?

Exercise 3.19: Hand calculations

Consider 20 sets of ten numbers defined by

$$a \;=\; c(1:10) \quad \text{and} \quad b \;=\; a + 10^i,$$

where i runs from 1 up to 20.

(a) Write a script in R, and in MATLAB, that will calculate the standard deviation using the formula, sometimes known as the hand calculation formula (hcf),

$$\left(\sum x_i^2 - \left(\sum x_i\right)^2 / n\right)/(n-1)$$

and print i and the result of the formula.

(b) Repeat (a) using the built in *sd()* function in R in place of the hcf. If you have a hand held calculator try this exercise on it as well.

(c) Repeat (a) except that you subtract the smallest number in the sample from all the sample values before using the hcf.

Exercise 3.20: Iterative scheme

Suppose you have a sample of size n, $\{x_1, \ldots, x_n\}$. Write a script to implement the following iteration where := represents the assignment of a current value.

$$n = 1; x = x_1; S = 0$$

and then for $i = 2$ up to n

$$
\begin{aligned}
n &:= n + 1 \\
d &:= (x_i - \bar{x})/n \\
\bar{x} &:= \bar{x} + d \\
S &:= S + n * (n-1) * d^2.
\end{aligned}
$$

Now print out $\bar{x}$ and s where $s = S/(n-1)$.

(a) Check that the iterative scheme works with an arbitrary set of numbers.

(b) Check that the iterative scheme works with the data from Exercise 3.19.

(c) Prove that $\bar{x}$ is $\bar{x} - \sum_{i=1}^{n} x_i$

(d) Prove that S is $\sum_{i=1}^{n}(x_i - \bar{x})^2$

Exercise 3.21: Carbon content of coal

The following data are carbon contents (%) of coal. Calculate the mean, variances with divisors n and $n-1$, standard deviations with divisors n and $n-1$, median and range.

$$
\begin{array}{cccccccccc}
87 & 86 & 85 & 87 & 86 & 87 & 86 & 81 & 77 & 85 \\
86 & 84 & 83 & 83 & 82 & 84 & 83 & 79 & 82 & 73
\end{array}
$$

Exercise 3.22: Furniture

A company manufactures bedroom furniture and the length of the door for a particular unit is specified as 2 m. Measurements for the lengths of 100 doors are made in made in m $\{x_i\}$. The mean and standard deviation of these measurements are 1.997 and 0.002 m respectively. Now suppose that the deviations from 2 m were measured in mm $\{y_i\}$.

Write down the mean, variance, and the standard deviation of the $\{y_i\}$ in mm.

Exercise 3.23: Speed of light

Given $\{x_i\}$ for $i = 1, \ldots, n$ prove that

$$\sum (x_i - \bar{x})^2 \;=\; \sum x_i^2 - \left(\sum x_i \right)^2 / n$$

This formula might be useful for hand calculations but it is prone to rounding errors and should never be programmed on a computer.

(a) Use both the left hand side and the right hand side of this formula to calculate the standard deviation of the following five estimates of the speed of light (ms^{-1}).

$$299\,792\,458.351$$
$$299\,792\,458.021$$
$$299\,792\,458.138$$
$$299\,792\,458.251$$
$$299\,792\,458.283$$

(b) Repeat (a) after subtracting $299\,792\,458$ from each datum.

(c) Newcomb's 3rd set of measurements of the passage time of light are available in the MASS package of R as newcomer. Repeat (a) for the data:

$$newt \;=\; (newcomb/1\,000) + 24,$$

which are the times in millionths of a second for light to travel $9, 902.145$ m between Fort Myer and the United States Naval Observatory, then situated on the Potomac River.

(d) Michelson's measurements of the speed of light are available in the MASS package of R as Michelson. Repeat (a) for the data:

$$misl \;=\; michelson + 299\,000,$$

which are estimates of the speed of light in kms^{-1}

Exercise 3.24: Proportions

Denote a set of data by $\{x_i\}$ for $i = 1 \ldots n$. Define $\hat{\sigma}$ by

$$\hat{\sigma}^2 \;=\; \sum (x_i - \bar{x})^2 / n$$

(a) Express s in terms of $\hat{\sigma}$ and n. Comment on this relationship.

(b) Explain why the proportion (p) of $\{x_i\}$ that is more than $k\hat{\sigma}$ from $\bar{x}$ is given by

$$p = \sum_I \frac{1}{n},$$

where I is the set of all i such that $|x_i - \bar{x}| > k\hat{\sigma}$

(c) Explain why

$$p \le \sum_{i=1}^{n} \frac{1}{n}\left(\frac{x_i - \bar{x}}{k\hat{\sigma}}\right)^2$$

(d) Explain why

$$p \le \frac{1}{k^2}$$

(e) For any set of data, what is the minimum proportion of the data within plus or minus $2\hat{\sigma}$ of $\bar{x}$?

Section 3.5 Box plots

Exercise 3.25: Floods

Annual flood for River Fish at Gras in Namibia. Fifteen water years from 1974-75. The data in cubic meters per second are:

29, 123, 119, 19, 27, 74, 0, 0, 36, 35, 105, 1, 89, 269, 66

(a) Calculate the median.
(b) Calculate the lower quartile.
(c) Calculate the upper quartile.
(d) Draw the box plot.

Exercise 3.26: River flows 1

The following series is the minimum monthly flows (m^3s^{-1}) in each of the 20 years 1957 to 1976 at Bywell on the River Tyne.

21, 36, 4, 16, 21, 21, 23, 11, 46, 10, 25, 12, 9, 16, 10, 6, 11, 12, 17, 3

(a) Draw a box plot of the data.
(b) Calculate the mean, standard deviation (s), and skewness.
(c) Take the logarithms of the data, repeat (i) and (ii) and compare the results.
(d) Does exponential of the mean of the logarithms equal the mean of the original data?

Section 3.6 Outlying values and robust statistics

Exercise 3.27: Erfenis dam

The January inflows 10^6 m^3 to Erfenis Dam in South Africa 1960-1984 are:

12, 4, 8, 20, 5, 19, 91, 165, 5, 3, 8, 25, 24, 1, 103, 53, 78, 3, 23, 9, 1, 33, 6, 0, 36

[Adamson PT, Robust and exploratory data analysis in arid and semi arid hydrology. Department of Water Affaire, Republic of South Africa 1989]

(a) Plot the data as time series.

(b) plot the data as a time series.

(c) Calculate the mean and standard deviation, and the coefficient of variation.

(d) Calculate the median and the median of the absolute deviations from the median.

(e) Calculate the quartiles and the IQR.

Section 3.7 Grouped data

Exercise 3.28: Grouped data

You are given four sets of grouped data, (A), (B), (C) and (D) below. In each case assume that measurements were made to sufficient precision for none to lie on a boundary between grouping intervals (bins).

(a) Construct a table with 5 columns (the first two are given in the question): bin; frequency; relative frequency; relative frequency density; and cumulative relative frequency as a percentage.

(b) Draw a properly scaled histogram (area of one) and a cumulative frequency polygon (join points with straight lines) on the same side of one sheet of graph paper, the histogram to be above the polygon. Use linear scales for the axes and label them correctly.

(c) Make graphical estimates of the median (M) upper quartile (UQ) and lower quartile (LQ) from your cumulative frequency polygon, showing your construction. Hence calculate the inter-quartile range, ($UQ - LQ$).

(d) Calculate the approximate sample mean and approximate sample standard deviation from the grouped data.

(e) We will define the sample third moment measure of skewness by

$$\widehat{\gamma} \;=\; \frac{\sum (x_i - \bar{x})^3 /(n-1)}{s^3}.$$

For grouped data, this becomes

$$\widehat{\gamma} \;=\; \frac{\sum \left((x_k - \bar{x})^3 f_k\right)/(n-1)}{s^3}.$$

Use the formula for grouped data to calculate $\widehat{\gamma}$ for your data.

(f) Another measure of skewness is the Pearson mode skewness defined as: (mean-mode)/standard deviation. Make a graphical estimate of the mode and so calculate the Pearson mode skewness.

(g) Another measure of skewness is the Pearson median skewness defined as: 3(mean-median)/standard deviation. Calculate the Pearson median skewness.

(h) Comment on the qualitative differences between these three measures of skewness.

The four sets of grouped data:

(A) Cycles until failure for 23 deep-groove ball bearings in endurance tests [Lieblein and Zelen, 1956]. Measurements sufficiently precise for none to lie on a boundary.

Cycles until failure (millions)	0-20	20-40	40-60	60-80	80-100	100-140
Number of units (frequency)	1	2	8	4	3	5

(B) The following 74 polychlorinated biphenyl (PCB) readings were taken at a drainage outflow at the edge of a site that was being developed for residential accommodation. Intermittent high readings appeared to be associated with high rainfall [Stewardson and Coleman, 2001].

PCB mass in sample (ng)	0.0-0.25	0.25-0.5	0.5-1.0	1.0-2.0	2.0-5.0	5.0-10.0
Number of samples (frequency)	22	14	18	11	5	4

(C) Annual peak flows of the Mekong at Vientiane for 79 years: 1913–1991.

Peak flow $(Mm^3 s^{-1})$	10-12	12-14	14-16	16-18	18-20	20-22	22-26
Number of years (frequency)	3	7	20	19	16	8	6

(D) Percent increase in operating current after 4 000 hours for 15 GaAs lasers tested at 80°C, page 642 of [Meeker and Escobar, 2014].

Percentage increase in operating current	6-7	7-8	8-10	10-13
Number of lasers (frequency)	5	5	2	3

Section 3.8 Shape of distributions

Exercise 3.29: Skewness and kurtosis

The following data are annual maximum flows in $m^3 s^{-1}$ for the Omaruru at Etemba in Namibia for 11 years from the 1969-1970 water year:

63.5, 404.3, 433.1, 66.8, 760.2, 1.2, 380.7, 2.1, 172.8, 180.7, 136.0.

The annual maximum flows for the following 10 years are:

17.0, 153.3, 550.0, 591.0, 702.0, 172.1, 202.5, 76.6, 242.7, 17.4.

(a) Calculate the following statistics for the 11 years from 1969-1970, for the following 10 years, and then for the entire sequence of 21 years:

(i) mean.

(ii) standard deviation.

(iii) median.

(iv) mean description mean absolute deviation from the mean.

(v) mean absolute deviation from the median.

(b) Calculate the trimmed mean after removing the 1st, 2nd, 20th, and 21st order statistics.

(c) Show how the mean and standard deviation of the 21 flows can be obtained from the means and standard deviations of the sets of 11 and 10 flows.

Exercise 3.30: River flows 2

(a) The following data are annual maximum flows in $m^3 s^{-1}$ for the Loewen at Geduld in Namibia for 12 years from the 1977-1978 water year:

86.3, 23.8, 62.4, 66.3, 70.4, 6.0, 134.0, 56.1, 70.4, 255.4, 103.8, 38.8.

Calculate the mean, standard deviation, and skewness.

(b) Calculate the mean, standard deviation, skewness and kurtosis of the annual maximum peak discharges ($10^3 m^3 s^{-1}$ and of the associated flood volumes at Concordia on the River Uruguay between 1898 and 1993 (can be found on the website)

Exercise 3.31: Cans of paint

A manufacturer supplies paint in cans with declared contents of 1 liter. The following data are deviations from 1 liter in units of 5 ml for a sample of 31 cans.

Deviation (5 ml)	−1	0	1	2	3
Number of cans	10	15	3	2	1

(a) Draw a line chart for these data.

(b) Calculate the mean, and standard deviation.

(c) The skewness $\hat{\gamma}$ is a measure of asymmetry, and is defined as:

$$\hat{\gamma} = \frac{\sum_{i=1}^{n}(x_i - \bar{x}/(n-1)}{s^3}$$

Positive/negative values indicate data tail out to the right/left. Calculate $\hat{\gamma}$ for these data.

Section 3.9 Multivariate data

Exercise 3.32: Robot arms

Draw a parallel coordinates plot for the 206 robot arms (Example 3.46). Are there any noticeable features?

Section 3.10 Descriptive time series

Exercise 3.33: Time series

Plot the following data as time series, and comment on whether or not there appears to be a trend or periodic repeating pattern. (the data are from NOAA Earth System Research Laboratory, and are pre-produced on the book website.)

(a) Average monthly measurements of atmospheric carbon dioxide, made at Mauna Lao, from August 1969 until December 2017 (MaunaLoa_CO2.txt).

(b) Hourly measurements of ozone taken at Arrival Heights, Antarctica, during August 2018 (ArrivalHeights_Ozone.txt).

Exercise 3.34: Laspeyre

Calculate the *Laspeyre Price Index* (LI_t) for the current year relative to the base year from the data in the table below.

Item	Base year		Current year	
	quantity	unit price	quantity	unit price
(i)	(q_{i0})	(p_{i0})	(q_{it})	(p_{it})
Concrete (m^3)	100	50	600	60
Bricks (m^3)	500	40	90	80
Timber (m^3)	200	60	50	100
Steel (kg)	50	10	50	14
Labor (h)	300	12	240	17

Exercise 3.35: Paasche

Calculate the *Paasche Price Index* PI_t for the data in Exercise 3.34

(a) Explain why the PI_t is usually lower than the LI_t.

(b) Calculate the *Irving-Fisher price index* as the geometric mean of LI_t and PI_t. (The geometric mean of a sample of n items is the nth root of their product.)

Exercise 3.36: Dublin City Council

The 72 night noise levels in Example 3.41 were recorded over consecutive nights.

(a) Plot the data as a time series.

(b) Are there any noticeable features?

Exercise 3.37: LAN data

(a) Plot the numbers of packets arriving in 10 milli-second intervals (Example 3.42) as a time series.

(b) Are there any noticeable features?

Exercise 3.38: Earthquake data

Plot the ground acceleration data in Example 3.43 as a time series and comment on your plot.

4

Discrete probability distributions

We define a discrete random variable and its probability distribution. Expectation is introduced as averaging in the population. A Bernoulli trial is defined and leads to binomial, negative binomial and Poisson distributions. The hypergeometric distribution is considered as a modification of a binomial distribution for a finite population.

4.1 Discrete random variables

A discrete random variable is a rule that typically assigns a non-negative integer to the outcome of an experiment. The associated discrete probability distribution assigns a probability to the random variable taking each of these integer values.

Definition 4.1: Random variable

An experiment has an associated sample space Ω. A **random variable** is a rule that assigns a unique real number to every element in the sample space. If we denote an element in the sample space by ω, a random variable X is defined by

$$X(\omega) \;=\; x,$$

where x is a unique real number associated with ω.

Thus, X is a function defined on the sample space, Ω, and Ω is said to be the domain [1] of X. The image of X is the set of possible x values.

If the random variable X is discrete then its image is a discrete set of real numbers.

Example 4.1: Coin toss [A discrete random variable]

A coin is flipped once. The sample space is $\Omega = \{H, T\}$. A random variable X is defined by

$$
\begin{aligned}
X(T) &= 0 \\
X(H) &= 1.
\end{aligned}
$$

The domain of X is $\{H, T\}$ and its image is $\{0, 1\}$.

---

[1] A function, $X()$, between a set Ω and a set $\mathcal{X}$ is a rule that for each $\omega \in \Omega$ assigns a single value $x \in \mathcal{X}$, written as $x = X(\omega)$. The set Ω is the **domain** of the function and the set $\mathcal{X}$ is the **co-domain**. The subset of elements of $\mathcal{X}$ that are assigned to an $\omega \in \Omega$ is known as the **image** of the function (the image can equal the co-domain).

Example 4.2: Loss of separation [A discrete random variable]

Air Traffic Control at a large airport records the number of loss of separation incidents (aircraft passing too close) each month. A random variable is defined as that number. Formally it assigns "n loss of separation incidents" the number n.

4.1.1 Definition of a discrete probability distribution

A **discrete probability distribution** defines the probabilities that X takes each of the possible values in its image. The image of X is referred to as the **support** of the probability distribution[2].

Definition 4.2: Probability mass function (pmf)

The **probability mass function** is:

$$P(X = x), \qquad \text{for } x \text{ in image of } X.$$

The pmf $P(X = x)$ is usually abbreviated to $P(x)$, or $P_X(x)$ if there is any doubt about the random variable it refers to. The distinction between the upper case X and lower case x is that the former represents the random variable and the latter represents a particular value of that random variable. The pmf of any discrete random variable must satisfy

$$\sum P(x) = 1,$$

where the sum is over the support of $P(\cdot)$

Definition 4.3: Cumulative distribution function (cdf)

The **cumulative distribution function**, $F(x)$, is customarily defined[3] over a continuous domain by

$$F(x) = P(X \le x) \quad \text{for} \quad -\infty < x < \infty.$$

Although the cdf is formally defined over a continuous domain, in the case of a discrete random variable the cdf has steps at integer values. In particular, suppose the support of the pmf is the set of integers between L and M. Then we have

$$F(x) = P(X \le x) = \begin{cases} 0 & \text{for } x < L, \\ \sum_{i=0}^{\lfloor x \rfloor} P(i) & \text{for } L \le x \le M, \\ 1 & \text{for } M < x, \end{cases}$$

---

[2]The support of the pmf is its domain but "support" is generally used in a statistics context.
[3]The same definition then applies for a continuous random variable.

where $\lfloor x \rfloor$ is the floor function, defined as the largest integer not greater than its argument. Since $F(x)$ is a cumulative probability it can only take values between 0 and 1.

Any discrete distribution can be characterized by its pmf or by its cdf.

Example 4.3: Decagon spinner [A discrete uniform distribution]

Let the random variable X be the number obtained when a decagonal spinner, marked with the digits $\{0, 1, \ldots, 9\}$, is spun. If the spinner is fair, the pmf of X is:

$$P(x) \;=\; P(X = x) \;=\; \frac{1}{10}, \qquad \text{for } x = 0, \ldots, 9.$$

This is an example of a **discrete uniform distribution**[4]. The set of values $\{0, 1, 2, 3, 4, 5, 6, 7, 8, 9\}$ that X can take is the support, or domain, of the pmf. Note that

$$\sum_{x=0}^{9} P(x) \;=\; \frac{1}{10} + \ldots + \frac{1}{10} \;=\; 1.$$

4.1.2 Expected value

Expectation is averaging in the population, and in particular the expected value of a random variable X is its mean value. In Section 3.7.1 we defined the mean for discrete data $\{x_k\}$ with frequencies $\{f_k\}$ as $\bar{x} = \sum_{k=1}^{K} x_k f_k / n$, and this can be rearranged as

$$\bar{x} \;=\; \sum_{k=1}^{K} x_k \frac{f_k}{n}.$$

The mean of a discrete random variable X is defined by replacing the relative frequencies with probabilities.

Definition 4.4: Expected value of a discrete random variable

The mean of X, also known as the expected value of X, is defined by:

$$\mu \;=\; E[X] \;=\; \sum_{x} x P(x),$$

where μ denotes the population mean and $E[\cdot]$ is the expectation operator.

Expected value is a population average and we can average functions of the random variable X.

Definition 4.5: Expected value of a function of a discrete random variable

We define the **expected value** of any function of X, $\phi(X)$, by

$$E[\phi(X)] \;=\; \sum_{x} \phi(x) P(x).$$

---

[4]In general a discrete uniform distribution assigns the same probability to each element in its support.

In particular, this leads to the definition of the variance of X.

Definition 4.6: Variance of a discrete random variable

The variance of X is the expected value of the function $\phi(X) = (X - \mu)^2$, and is defined by

$$\sigma^2 \;=\; E\big[(X - \mu)^2\big] \;=\; \sum_x (x - \mu)^2 P(x),$$

where σ is the population standard deviation and σ^2 is the population variance. A useful identity is that the variance is the expected value of X^2 less the mean squared. This follows from the definition:

$$
\begin{aligned}
\sigma^2 &= E\big[(X - \mu)^2\big] \;=\; E\big[X^2 - 2\mu X + \mu^2\big] \\
&= E\big[X^2\big] - 2\mu E[X] + \mu^2 \\
&= E\big[X^2\big] - 2\mu\mu + \mu^2 \\
&= E\big[X^2\big] - \mu^2.
\end{aligned}
$$

Example 4.3: (Continued) Discrete uniform distribution

If X has a discrete uniform distribution with support $\{0, 1, \ldots, 9\}$, the mean of X is

$$\mu \;=\; E[X] \;=\; \sum_{x=0}^{9} x\left(\frac{1}{10}\right) \;=\; 0 \times \frac{1}{10} + \ldots + 9 \times \frac{1}{10} \;=\; 4.5$$

and the variance of X is

$$
\begin{aligned}
\sigma^2 &= E\big[(X - \mu)^2\big] \;=\; \sum_{x=0}^{9}(x - 4.5)^2\left(\frac{1}{10}\right) \\
&= (0 - 4.5)^2 \times \frac{1}{10} + (1 - 4.5)^2 \times \frac{1}{10} + \ldots + (9 - 4.5)^2 \times \frac{1}{10} \;=\; 8.25
\end{aligned}
$$

Hence, the standard deviation of X is $\sqrt{8.25} = 2.87$.

4.2 Bernoulli trial

4.2.1 Introduction

A Bernoulli trial is an experiment with only two possible outcomes, which are conventionally referred to as "success" and "failure", and is the elementary unit from which many probability distributions can be formulated[5]. An example of a Bernoulli trial is casting an aluminum cylinder head and determining whether or not it is of acceptable quality.

---

[5]It is named after Jacob Bernoulli (1654-1705) whose posthumous work Ars Conjectandi (1713) was of great significance in the theory of probability.

4.2.2   Defining the Bernoulli distribution

To provide a mathematical description of a Bernoulli trial, we define a random variable X as 1 if the trial results in a success and 0 if it results in a failure. We denote the probability of a success as p, and this probability is referred to as a **parameter**[6] of the distribution. The parameter enables us to give a general description of the distribution, and in applications the parameter may be substituted by a specific number. The Bernoulli trial can be defined by its pmf as

$$P(x) = \begin{cases} 1 - p & \text{for } x = 0, \\ p & \text{for } x = 1. \end{cases}$$

Example 4.4: Alloy casting [a Bernoulli trial]

Aluminum alloy castings are used for many high performance automotive engines. A limitation of the casting process is the tendency for gas bubbles to form within a casting as it cools, because the solubility of gas is lower in the solid than the liquid, a phenomenon known as porosity. Casting processes have been developed to reduce porosity as much as possible, and castings will typically be subject to inspection by X-ray tomography. A casting is acceptable provided: upper limits on the number of pores per unit volume and sizes of pores are not exceeded; and the locations of pores are away from critical surfaces.

A manufacturer of cast aluminum cylinder heads knows from experience of the process that the probability that a casting is of acceptable quality is 0.80. Unacceptable castings are designated as defective and are recycled by returning to the crucible which contains the molten alloy. Take a single casting from production and let X be 0 if the casting is acceptable and 1 if it is defective. Then X is a Bernoulli trial with a probability of success of 0.2. It may seem perverse to associate success with a casting being defective, but it is often more convenient to set p as the smaller probability.

4.2.3   Mean and variance of the Bernoulli distribution

The mean and variance of a Bernoulli distribution are:

$$\mu = p \text{ and } \sigma^2 = p(1 - p).$$

The proof follows directly from the definitions.

$$\mu = \text{E}[X] = 0 \times (1 - p) + 1 \times p = p \text{ and}$$

$$\sigma^2 = \text{E}[X^2] - \mu^2 = 0^2 \times (1 - p) + 1^2 \times p - p^2 = p - p^2 = p(1 - p).$$

---

[6]We define probability distributions in general terms using letters to represent characteristics that will take specific values in a particular application. The letters are the parameters of the distribution.

4.3   Binomial distribution

4.3.1   Introduction

Consider taking a random sample of three cast aluminum cylinder heads from a process which produces defective cylinder heads with a constant probability p. Let the random variable X be the number of defective cylinder heads in the sample of three. A sample space is shown in the tree diagram of Figure 4.1. The eight possible sequences of G, for

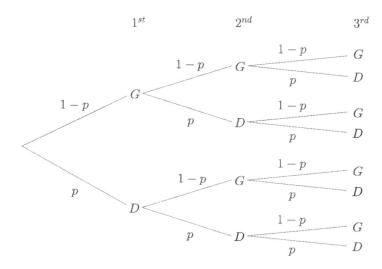

FIGURE 4.1: Tree diagram for a sample of three cylinder heads.

good, and D, for defective, are mutually exclusive and exhaustive. The probability that X takes the values $0, 1, 2, 3$ can be summarized by the formula

$$\mathrm{P}(x) \quad = \quad \binom{3}{x} p^x (1-p)^{3-x}, \quad \text{for } x = 0, 1, 2, 3.$$

Since there are just eight sequences, the formula can be verified quickly from the tree diagram. For example, if $x = 1$ there are $\binom{3}{1} = 3$ ways of choosing one of the three positions for $D : GGD, GDG, DGG$. Each sequence has probability $p(1-p)^2$. The formula for $\mathrm{P}(x)$ follows because $\binom{3}{x}$ is the number of ways of labelling x of the cylinder heads in the sequence of 3 as defective.

4.3.2   Defining the Binomial distribution

In general, imagine a sequence of n Bernoulli trials with a constant probability of success p. Let the random variable X be the number of successes in the n trials. X is a discrete random variable and its image is the set of integers from 0 up to n. A sample space is the 2^n possible sequences of Ss (successes) and Fs (failures), each sequence being of length n. The pmf is given by

$$\mathrm{P}(x) \quad = \quad \binom{n}{x} p^x (1-p)^{n-x}, \quad \text{for } x = 0, 1, \dots, n.$$

The justification for this formula is that $\binom{n}{x}$ is the number of ways of choosing x positions for the Ss in a sequence of length n with the remaining positions taken by Fs. An equivalent justification for the $\binom{n}{x}$ factor is that it is the number of ways of arranging x Ss and $(n-x)$ Fs. The probability of any specific arrangement of x Ss and $(n-x)$ Fs is $p^x(1-p)^{n-x}$, and the formula for the pdf follows.

The cdf is

$$F(x) \;=\; \mathrm{P}(X \le x) \;=\; \begin{cases} 0 & \text{if } x < 0 \\ \sum_{i=0}^{\lfloor x \rfloor} \mathrm{P}(i) & 0 \le x \le n \\ 1 & n < x. \end{cases}$$

The assumption of a constant probability of success p implies that the trials are independent. The binomial distribution is specified by the two **parameters** n and p, and we will write[7]

$$X \;\sim\; \mathrm{binom}(n,p),$$

The set of possible numbers of successes $\{0,1,\ldots,n\}$ is an alternative sample space. These outcomes are mutually exclusive by their definition, and they are exhaustive because precisely one must occur. A proof that

$$\sum_{x=0}^{n} \mathrm{P}(x) \;=\; 1$$

follows from the binomial expansion, which explains the choice of name for the distribution. First notice that $(p+(1-p))^n \;=\; 1^n \;=\; 1$. Now use the binomial expansion

$$(p+(1-p))^n \;=\; p^n + \binom{n}{1}p^{n-1}(1-p) + \binom{n}{2}p^{n-2}(1-p)^2 + \cdots + (1-p)^n$$

$$=\; \sum_{x=0}^{n}\binom{n}{x}p^x(1-p)^{n-x} \;=\; \sum_{x=0}^{n}\mathrm{P}(x).$$

In summary, a binomial distribution is appropriate when:

1. There is a fixed number of trials (n).

2. Each trial has just two possible outcomes $(S$ or $F)$.

3. The probability of a success p is the same for all trials.

Condition (3) implies that trials are independent. The converse is not necessarily true because trials could be independent with p changing over time. The computing syntax is

function	R	MATLAB
pmf	dbinom(x,n,p)	binopdf(x,n,p)
cdf	pbinom(x,n.p)	binocdf(x,n,p)

[7]The tilde ($\sim$) is read as "is distributed as".

Example 4.5: A binomial distribution

The random variable X has a binomial distribution with $n = 5$ and probability of success $p = 0.3$. The pmf, calculated from the formula, is

$$P(0) \;=\; (1-0.3)^5 = 0.16807$$

$$P(1) \;=\; \binom{5}{1}(1-0.3)^4(0.3) = 0.36015$$

$$P(2) \;=\; \binom{5}{2}(1-0.3)^3(0.3)^2 = 0.30870$$

$$P(3) \;=\; \binom{5}{3}(1-0.3)^2(0.3)^3 = 0.13230$$

$$P(4) \;=\; \binom{5}{4}(1-0.3)(0.3)^4 = 0.02835$$

$$P(5) \;=\; (0.3)^5 = 0.00243$$

These calculations can be checked in R by

```
> x = c(0:5)
> dbinom(x,5,0.3)
```

The cdf is

$$
\begin{aligned}
F(x) &= 0.00000 & x < 0\\
F(x) &= 0.16807 & 0 \le x < 1\\
F(x) &= 0.52822 & 1 \le x < 2\\
F(x) &= 0.83692 & 2 \le x < 3\\
F(x) &= 0.96922 & 3 \le x < 4\\
F(x) &= 0.99757 & 4 \le x < 5\\
F(x) &= 1.00000 & 5 \le x.
\end{aligned}
$$

These calculations can be checked in R by

```
> pbinom(x,5,0.3)
```

The pmf and cdf are plotted in Figure 4.2, where the filled circles include the integer values and the empty circles exclude the integers.

This can be achieved in R using the code[8]

---

[8]Graphs containing disjoint line segments with different symbols at the end are a bit intricate. Here we use the package *dplyr: A Grammar of Data Manipulation*, and, in particular, *mutate()* adds columns. The combination % > % (package magrittr, loaded with *dplyr*)is the equivalent of Unix's pipe operator and can be used to string together operators. For example imagine that you have a data frame of customers called *df*. The data frame has many columns of data including: location coded as domestic/OS; size; years (as a customer); revenue (annual in dollars). You could use
df % > % *filter(location == OS)*
% > % *select(size, years, revenue)*
% > % *mutate(rev_k = revenue/1000)*
This would take the data frame, select all the rows that are OS, keep the columns size, years, revenue and finally add a column with revenue in 1000 dollars, height from cm to m.

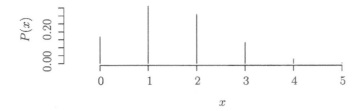

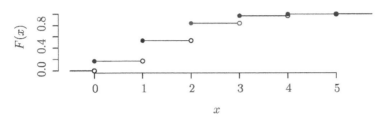

FIGURE 4.2: pmf (upper) and cdf (lower) of binomial distribution with $n = 5$ and $p = 0.3$.

```
> library(dplyr)
> n = 5
> p = 0.3
> df = data.frame(x = 0:5)
> df = df %>% mutate(y = dbinom(x,size = n,prob = p),
    F = pbinom(x,size = n,prob = p))
> with(df,plot(x,y,type='h',ylab = "\pr{x}",bty='n'))
> plot(df$x,df$F,ylim=c(0,1),xlim=c(-0.5,5.5),bty='n',
    ylab="F(x)",pch=16,type = 'n',xlab='x')
> for(i in 0:5){
    lines(c(i,i+1),rep(pbinom(i,size = n,prob = p),2))
    }
> lines(c(-0.5,0),c(0,0))
> points(df$x+1,df$F,bg='white',pch=21)
> points(df$x,df$F,pch=16)
> points(0,0,bg='white',pch=21)
```

Example 4.6: Process improvement [a binomial cdf]

Will Wrench is responsible for the process for casting aluminum cylinder heads at SeaDragon. The process has been producing 20% defective castings, almost all of which have been classed as defective due to porosity. Although the defective castings are recycled, and there is no evidence that recycling compromises the quality of the alloy, recycling reduces the efficiency of the process and wastes energy. The company metallurgist carries out research with the aim of reducing the number of defectives, and recently proposed a new degassing procedure. Wrench implemented the new procedure a week ago and since then only two of 28 cylinder heads cast have been defective. We can identify 28 trials with two possible outcomes, either good or defective. If we assume that the process continues to produce 20% defective castings and that defective castings occur independently, then the constant probability of a defective casting is 0.2. The expected number of defects is $28 \times 0.2 = 5.6$ and so only 2 defects in 28 castings is a promising result. But how surprising is it if the probability of a defect is 0.2? To answer this question we need to consider what is surprising. Is it that exactly 2 defects is unlikely, or is that doing as well as or better than 2 defectives (0 or 1 or 2 defectives) is unlikely? The probability of 2 or less defectives is the relevant probability [9] for quantifying this promising result (Exercise 4.14). Let X be the number of defective castings. The probability of 2 or less defectives is:

$$P(X \leq 2) \text{ where } X \sim \text{binom}(28, 0.2).$$

Using R we have

```
> pbinom(2,28,0.2)
[1] 0.06117165
```

Although the probability of 0.061 is quite small, it is still large enough to leave some doubt about the success of the new degassing procedure. Also, if there is a tendency for the defective castings to occur in clusters, the assumption of independent trials will not be realistic and the binomial probability will be an underestimate. Wrench decides to continue the new degas procedure for a few more weeks and then conduct a reassessment.

Example 4.7: Flooding in York [complementary binomial probability]

York in the north-east of England is a walled city that is renowned for its historic sites and buildings. York is built at the confluence of the River Ouse and River Foss and is prone to flooding. Given the architectural heritage, considerable effort has been made to provide non-intrusive flood defenses, which include wash-lands and the Foss Barrier. A typical design criterion for such defenses is that they will protect against a flood with a 100 year **average recurrence interval** (ARI) [10]. A 100 year ARI is equivalent to a probability of a flood of this magnitude, or greater, occurring in any one year of $1/100$.

It is often reasonable to model occurrences of floods in a sequence of years as independent provided *water years* are defined so that the main flood risk is mid-year. In the UK the water year runs from the beginning of September until the end of August. The assumption that the probability of a flood exceeding the current 100 year return period in any one year will remain constant is rather more questionable, given climate change predictions. Nevertheless we will calculate the probability that a flood with a return period of 100 years will occur at least once in the next 50 years. The answer is the complement of no such flood in 50 years and is given by

```
> 1-(1-1/100)^50
[1] 0.349939
```

The general result is given in Exercise 4.12.

---

[9]You are in a statistics class of 100 students, You score 94 in a quiz for which the class average is 50. Are you pleased because you scored exactly 94 or because only four people, say, scored 94 or more?

[10]The ARI is also known, more colloquially, as the return period. The use of "return period" is sometimes criticized because it might be mistakenly taken to imply that floods occur at regular intervals, which they do not.

4.3.3 A model for conductivity

[Forster, 2003] discusses a simple model for electrical conductivity that is based on a binomial distribution with three trials. At the atomic level individual circular atoms are conductive with a probability p_0. Atoms cluster randomly into threes, each cluster of three atoms is enclosed by a circle of minimum radius, tangent to all three atoms, and we refer to this circular entity as a first stage unit. First stage units randomly cluster into threes to form a second stage unit as shown in Figure 4.3. The process continues to build up a fractal

FIGURE 4.3: Second stage unit for model of conductive. Atoms (9 shown) randomly cluster into threes (first stage units, 3 shown) and three first stage units cluster to give a second stage unit.

model for a macroscopic piece of semi-conductor. A first stage unit is conductive if at least two of its three constituent atoms are conductive, and a nth stage unit is conductive if two or three $(n-1)^{th}$ stage units are conductive (Figure 4.4).

FIGURE 4.4: An nth stage unit is conductive if 2 or 3 of the $(n-1)^{th}$ stage units are conductive (filled circles).

The probability p_1 that a first stage unit is conductive is given by

$$p_1 = 3p_0^2(1-p_0) + p_0^3.$$

A second stage unit is conductive, with probability p_2, if at least two of its three constituent first stage units are conductive. Then

$$p_2 = 3p_1^2(1-p_1) + p_1^3.$$

The process continues in the same fashion, so the probability that an nth stage unit is conductive is

$$p_n = 3p_{n-1}^2(1-p_{n-1}) + p_{n-1}^3.$$

Now suppose that p_n tends to a limit p as $n \to \infty$. Then p satisfies

$$p = 3p^2(1-p) + p^3$$

and either $p = 0$ or $1 = 3p(1-p) + p^2$ which has solutions $p = \frac{1}{2}$ or $p = 1$. So, there are three possibilities:

- $p_0 = \frac{1}{2}$ and $p_i = \frac{1}{2}$ for $i = 1, \ldots, n$,

- $p_0 < \frac{1}{2}$ and $p_n \to 0$ or
- $p_0 > \frac{1}{2}$ and $p_n \to 1$.

The solution $p_0 = \frac{1}{2}$ is an unstable fixed point and both 0 and 1 are stable fixed points. You can confirm this with the following R-code:

```
> RGT=function(n,p){
+ for (i in 1:n) {
+ p= 3*p^2*(1-p)+p^3
+ }
+ return(p)
+ }
> RGT(10,.4)
[1] 1.584182e-35
> RGT(10,.49)
[1] 0.07541918
> RGT(10,.5)
[1] 0.5
```

With this model, the macroscopic behavior is quite different from the atomic behavior. The piece of semiconductor will not conduct if atoms conduct with any probability in the interval $[0, 0.5)$. Other applications of the model include phenomena such as the percolation of oil or water through rock.

4.3.4   Mean and variance of the binomial distribution

The mean of the binomial distribution is

$$\mathrm{E}[X] \;=\; \mu \;=\; \sum_{x=0}^{n} x \binom{n}{x} p^x (1-p)^{n-x}.$$

The first term in the sum is 0 so we can write

$$\mu \;=\; \sum_{x=1}^{n} x \binom{n}{x} p^x (1-p)^{n-x} \;=\; np \sum_{x=1}^{n} \frac{(n-1)!}{(n-x)!(x-1)!} P^{x-1}(1-p)^{n-x}.$$

Now substitute $y = x - 1$, and $m = n - 1$ in the summation to get

$$\mu \;=\; np \sum_{y=0}^{m} \frac{m!}{(m-y)!y!} p^y (1-p)^{m-y} \;=\; np \times 1 \;=\; np.$$

An alternative derivation is to explicitly write X as the sum of n Bernoulli random variables. That is

$$X = \sum_{i=1}^{n} W_i,$$

where W_i are Bernoulli variables. Then

$$\mathrm{E}[X] \;=\; \mathrm{E}\left[\sum_{i=1}^{n} W_i\right] \;=\; \sum_{i=1}^{n} \mathrm{E}[W_i] \;=\; \sum_{i=1}^{n} p = np.$$

The variance can be shown to be (Exercise 4.6)

$$\sigma^2 \;=\; np(1-p)$$

and the standard deviation is $\sigma = \sqrt{np(1-p)}$.

4.3.5 Random deviates from binomial distribution

In Chapter 2 we discussed PRNG for generating random sequences of digits. If we take these in consecutive runs of 4 and divide by 10 000 we have random numbers that are equally likely to take any value between 0 and 1 in steps of 0.0001. We can generate a random Bernoulli trial with probability p as: 0 if the uniform random number is less than $1 - p$ and 1 otherwise. A random deviate from a binomial distribution with parameters n and p is given by the sum of n Bernoulli trials, each with probability of success p. Kroese et al, [2011] give several more sophisticated algorithms, but we will generally rely on the inbuilt software functions.

The R code for simulating binomial deviates is:

$$\mathtt{rbinom(n, size, prob)},$$

where n is the number of deviates required, size is the number of trials, and prob is the probability of success. For example, to obtain 10 deviates from a binomial distribution with 20 trials with a probability of success of $1/2$:

```
> set.seed(2014)
> rbinom(10,20,0.5)
 [1]  9  8 11  9 10  7 13 11  7  8
```

4.3.6 Fitting a binomial distribution

In an application the number of trials n will be defined, but we may need to estimate the probability of a success from records of the process. The probability p is estimated by equating the sample mean to the population mean, an example of strategy known as **method of moments** (MoM).

Definition 4.7: Method of moments (MoM)

The population moments (typically mean and variance), expressed as functions of the parameters of the distribution, are equated to the sample moments. These equations are solved to give method of moments estimates of the parameters.

A common convention for denoting estimates of population parameters[11] is the parameter with a hat over it.

In the case of a binomial distribution, the population mean is np. Suppose we have N samples of n trials and the number of successes in the ith sample is x_i. The sample mean is $\bar{x} = \sum_{i=1}^{N} x_i/N$. Equate $n\hat{p}$ to $\bar{x}$ to obtain

$$\hat{p} = \bar{x}/n.$$

---

[11]In cases when it is not convenient to use equivalences between Greek and Roman alphabets

Example 4.8: Filter [estimating p from samples]

A company manufactures filters for the removal of oil and other contaminants from compressed air systems. There are several hand operations involved in the manufacture of filter elements and typical batch sizes are around one thousand elements. Random samples of size 40 are taken from each batch before shipping and tested against a stringent internal specification[12] which includes dimensional checks and appearance. The number of non-conforming filter elements in the past twenty samples are: $0, 3, 4, 3, 2, 0, 3, 1, 5, 1, 2, 2, 3, 0, 1, 4, 1, 0, 2, 1$. The non-conformances are slight and the non-conforming elements are within the advertised product specification. The estimate of p is calculated as follows.

```
> n = 40
> x = c(0, 3, 4, 3, 2, 0, 3, 1, 5, 1, 2, 2, 3, 0, 1,  4, 1, 0, 2, 1)
> N = length(x)
> N
[1] 20
> mean(x)
[1] 1.9
> phat=mean(x)/n
> phat
[1] 0.0475
```

and is $\widehat{p} = 0.0475$. We now compare the sample variance with the theoretical value for the binomial distribution.

```
> var(x)
[1] 2.2
> phat*(1-phat)*n
[1] 1.80975
```

The variance of a binomial random variable with $n = 40$ and $p = 0.0475$ is 1.81. The sample variance is rather higher but as we only have twenty samples this may just be sampling variability (Exercises 4.7). Alternatively, there may be a tendency for non-conforming elements to occur in clusters.

In many applications we have only one sample ($N = 1$) of size n with x successes and the objective is to estimate p. Then

$$\widehat{p} \;=\; \frac{x}{n}.$$

This estimator is covered in detail in Chapter 7.

4.4   Hypergeometric distribution

Suppose we take a random sample from a batch that contains a proportion p of defectives. The probability that the first item selected is a defective is p. However, we usually sample

---

[12]A stringent internal specification allows the effects of process modification to be assessed.

without replacement and probabilities for subsequent defectives will differ. The hypergeometric distribution allows for this.

4.4.1 Defining the hypergeometric distribution

A batch consists of N items of which B are classified as defective and the remaining $N - B$ are classified as good. We take a simple random sample of size n from the batch, which is considered as a finite population. The sample is taken without replacement, and a random variable X is defined as the number of defective items in the sample. The probability the first item is defective is B/N. The probability the second item is defective depends on whether or not the first was defective. If the first item was defective the probability the second is defective is $(B - 1)/(N - 1)$, and if the first item is good the probability the second is defective is $B/(N - 1)$. If both N and B are large then the difference in these two probabilities will be negligible. Moreover, if n is small in comparison with N and B the probabilities of the 3rd up to the nth being defective will not change much with the number of preceding defectives. In such cases the distribution of X is very well approximated by a binomial distribution with constant probability of success $p = B/N$.

Conversely, if B is small the exact distribution of X, the **hypergeometric distribution**, is more ppropriate. It is found by applying the equally likely definition of probability. Imagine that the items in the population are numbered from 1 to N. Then there are $\binom{N}{n}$ equally likely samples of size n. The number of samples of size n that contain exactly x defectives is the product of the number of ways of choosing x from the B defectives items with the number of ways of choosing $n - x$ good items from the $N - B$ good items: $\binom{B}{x} \times \binom{N-B}{n-x}$. It follows that the pmf of X is

$$\mathrm{P}(x) = \frac{\binom{B}{x}\binom{N-B}{n-x}}{\binom{N}{n}}, \qquad x = \max(0, n - N + B), \ldots, \min(n, B).$$

If the sample is small ($n \leq B$ and $n \leq (N - B)$), the support consists of the integers from 0 to n, (The hypergeometric distribution has three parameters n, N and B.) the mean and variance are

$$\mu = \frac{B}{N}, \qquad \sigma^2 = n\frac{B}{N}\left(1 - \frac{B}{N}\right)\left(\frac{N - n}{N - 1}\right).$$

The binomial distribution with $p = B/N$ gives the precise distribution of the number of defectives if the simple random sample is taken with replacement. The hypergeometric distribution has a smaller variance than the binomial distribution, because sampling without replacement is more informative.

Example 4.9: A hypergeometric distribution

A batch of size $N = 20$ contains $B = 6$ defective items. A random sample of $n = 5$ without replacement is selected. The number of defectives X has a hypergeometric distribution. The pmf is

$$P(0) = \frac{\binom{5}{0}\binom{14}{5}}{\binom{20}{5}} = 0.12913$$

$$P(1) = \frac{\binom{5}{1}\binom{14}{4}}{\binom{20}{5}} = 0.38738$$

$$P(2) = \frac{\binom{5}{2}\binom{14}{3}}{\binom{20}{5}} = 0.35217$$

$$P(3) = \frac{\binom{5}{3}\binom{14}{2}}{\binom{20}{5}} = 0.11739$$

$$P(4) = \frac{\binom{5}{4}\binom{14}{1}}{\binom{20}{5}} = 0.01354$$

$$P(5) = \frac{\binom{5}{5}\binom{14}{0}}{\binom{20}{5}} = 0.00039$$

The cdf is

$$
\begin{aligned}
F(x) &= 0.00000 \quad x < 0 \\
F(x) &= 0.12913 \quad 0 \le x < 1 \\
F(x) &= 0.51651 \quad 1 \le x < 2 \\
F(x) &= 0.86868 \quad 2 \le x < 3 \\
F(x) &= 0.98607 \quad 3 \le x < 4 \\
F(x) &= 0.99961 \quad 4 \le x < 5 \\
F(x) &= 1.00000 \quad 5 \le x.
\end{aligned}
$$

The pmf and cdf are plotted in Figure 4.5. You are asked to compare the pmf with that of the binom(5,0.3) in Exercise 4.18.

4.4.2 Random deviates from the hypergeometric distribution

The following R code simulates the number of defectives in 30 samples of size 7 from a population of 20 items of which 4 are defective. The draws within each sample of 7 are without replacement.

```
> N=20;B=4;n=7
> rhyper(30,B,N,n)
 [1] 3 2 0 1 0 3 1 0 2 4 2 3 1 0 1 0 1 2 1 1 2 1 0 1 0 1 0 0 3 0 1 1
```

4.4.3 Fitting the hypergeometric distribution

In an application N and n will be defined, which leaves the unknown number of defectives, B, to be estimated. If records of defectives in samples of size n from N are available B could be estimated as:

$$\widehat{B} = N\frac{\overline{x}}{n},$$

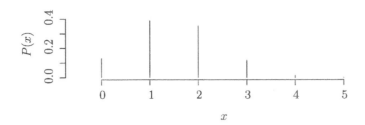

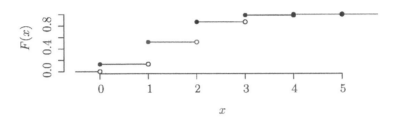

FIGURE 4.5: pmf (upper) and cdf (lower) of hypergeometric distribution with $N = 20, B = 6$, and $n = 5$.

where $\bar{x}$ is the average number of defects in these samples. Alternatively, $\bar{x}/n$ can be replaced by some other estimate (or assumed value) of the proportion of defectives. In the case of a single sample of size n with x defectives, the estimate of the number of defectives in the batch, B, is

$$\widehat{B} = N\frac{x}{n}$$

4.5 Negative binomial distribution

These distributions arise from a sequence of Bernoulli trials, but they differ from the binomial distribution because the number of trials is not fixed.

4.5.1 The geometric distribution

A random variable X is defined as the number of Bernoulli trials until the first success. The probability that $X = x$ is the probability of $x - 1$ failures followed by the success, so if the probability of a success on each trials is p, the pmf of X is

$$P(x) = (1-p)^{x-1}p, \quad x = 1, 2, \ldots$$

The cdf at integer values is

$$F(x) = 1 - (1-p)^x, \quad x = 1, 2, \ldots$$

The mean and variance of X are

$$\mu = \frac{1}{p}, \qquad \sigma^2 = \frac{1-p}{p^2}.$$

You are asked to prove these results in Exercise 4.22, but the mean follows from the following argument. If an event occurs on ν occasions in a long sequence of N Bernoulli trials, its probability is ν/N and the mean number of trials until an event occurs is N/ν.

Example 4.10: Peak flows [a geometric distribution]

The annual maximum river flow at a gage exceeds a flow F on average once every 10 years, and water years are defined so that annual maxima are approximately independent. The number of years until the next exceedance of F has a geometric distribution with $p = 0.1$. The mean number of years until an exceedance of F is 10 years and the standard deviation is 9.49 years. The probability of at least one exceedance within the next 7 years is

$$1 - (1 - 0.1)^7 \quad = \quad 0.522$$

The geometric distribution gives the number of years from now until an exceedance of F. Because exceedances of F are independent, "now" can be taken as a year with an exceedance of F and the times between exceedances of F have a geometric distribution. The geometric distribution is said to be **memoryless** because the probability distribution of the number of trials until the next success is independent of the past.

4.5.2   Defining the negative binomial distribution

The random variable X is the number of failures before the rth success in a sequence of Bernoulli trials with probability of success p. The probability that $X = x$ is the probability of $r - 1$ successes in $x + r - 1$ trials followed by a success at the xth trial. So the pmf is

$$P(x) \quad = \quad \binom{x + r - 1}{r - 1} p^r (1 - p)^x, \quad x = 0, \ldots$$

The mean and variance of X are

$$\mu \quad = \quad \frac{r(1 - p)}{p} \quad \text{and} \quad \sigma^2 \quad = \quad \frac{r(1 - p)}{p^2}.$$

The negative binomial distribution is also called the **Pascal distribution**. Although the given explanation for the pmf referred to r as integer, the distribution is well defined for any non-negative r. The expression $\binom{x+r-1}{r-1}$ is then interpreted in terms of the gamma function as $\Gamma(x + r)/(\Gamma(r)x!)$. The negative binomial random variable is sometimes defined as the number of trials until the rth success with support $\{r \leq x\}$, so that in the special case when $r = 1$, it corresponds to the geometric distribution (Exercise 4.21). You always need to check which form is being used.

Example 4.11: A negative binomial distribution

The random variable X has a negative binomial distribution with $r = 2$ and $p = 0.3$.

The pmf is

$$P(0) = \binom{1}{1}0.3^2(1-0.3)^0 = 0.09000000$$

$$P(1) = \binom{2}{1}0.3^2(1-0.3)^1 = 0.12600000$$

$$P(2) = \binom{3}{1}0.3^2(1-0.3)^2 = 0.13230000$$

$$P(3) = \binom{4}{1}0.3^2(1-0.3)^3 = 0.12348000$$

$$P(4) = \binom{5}{1}0.3^2(1-0.3)^4 = 0.10804500$$

$$P(5) = \binom{6}{1}0.3^2(1-0.3)^5 = 0.09075780$$

$$\vdots$$

The R code to check these values is `x = 0:5; pnbinom(x,2,0.3)` The cdf is

$$
\begin{aligned}
F(x) &= 0.0000000 & x < 0 \\
F(x) &= 0.0900000 & 0 \le x < 1 \\
F(x) &= 0.2160000 & 1 \le x < 2 \\
F(x) &= 0.3483000 & 2 \le x < 3 \\
F(x) &= 0.4717800 & 3 \le x < 4 \\
F(x) &= 0.5798250 & 4 \le x < 5 \\
F(x) &= 0.6705828 & 5 \le x < 6
\end{aligned}
$$

$$\vdots$$

The pmf and cdf are plotted in Figure 4.6.

4.5.3 Applications of negative binomial distribution

Example 4.12: Planetary exploration [negative binomial distribution]

A planetary lander deploys three exploration robots sequentially. The probability that a robot survives a (planetary) day in the harsh terrain is 0.8, and the probability of survival during one day is independent of survival on preceding days.

1. Find the expected number of days until the third robot fails and the standard deviation of this quantity.

2. Find the expected number of days of planetary exploration and the standard deviation of this quantity.

3. Calculate the probability that the number of complete days of exploration exceeds 20.

4. Comment on the assumption of independence.

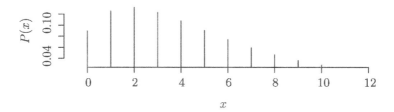

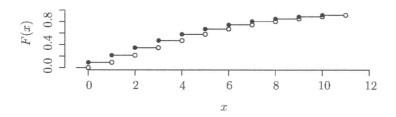

FIGURE 4.6: pmf (upper) and cdf (lower) of negative binomial distribution with $r = 2$ and $p = 0.3$.

We can model the situation as negative binomial with a probability that a robot fails (does not survive a day) as $1 - 0.8 = 0.2$. We require the distribution of the number of trials until the third failure.

1. The expected number of days until the third robot fails and its standard deviation are given by

$$3 \times \frac{1 - 0.2}{0.2} = 12 \quad \text{and} \quad \sqrt{3 \times \frac{1 - 0.2}{0.2^2}} = 7.75 \quad \text{respectively.}$$

2. If we assume a mean of half a day of exploration on days that a robot fails, the expected number of days of exploration is

$$12 + 3 \times 0.5 = 13.5$$

The standard deviation of the number of days of exploration is the same as the standard deviation of the number of days before the third robot fails, 7.75, if we ignore the uncertainty about the half days (see Exercise 6.5 for a more precise answer).

3. If we neglect any exploration on days that a robot fails, the probability of 20 complete days is

```
> 1-pnbinom(20,3,0.2)
[1] 0.1331855
```

If we allow one complete day of exploration to account for parts of the three days on which robots fail the probability is somewhat greater

```
> 1-pnbinom(19,3,0.2)
[1] 0.1544915
```

4. The assumption of independence would be unrealistic if, for example, meteor storms last for several days.

4.5.4 Fitting a negative binomial distribution

Example 4.13: Car pooling [negative binomial distribution]

The negative binomial distribution can be used as an empirical model of a discrete random variable with a variance that exceeds the mean and no fixed upper bound.
An engineer has been asked to set up a computer simulation for a cost benefit analysis of a car pooling system for a government organization. The number of passengers in cars has been observed on two mornings with the following results.

Number of passengers	0	1	2	3	4	5	total
Number of cars	160	96	28	13	2	1	300

The mean and variance of the number of passengers are 0.680 and 0.902 respectively. Method of moments estimates of the parameters in a negative binomial distribution are given by solving the equations

$$\bar{x} = \hat{r}\frac{(1 - \hat{p})}{\hat{p}}$$

and

$$s^2 = \hat{r}\frac{(1 - \hat{p})}{\hat{p}^2}.$$

These can be rearranged to give

$$\hat{p} = \frac{\bar{x}}{s^2}, \qquad \hat{r} = \bar{x}\frac{\hat{p}}{1 - \hat{p}}$$

and remember thatthe distribution is well defined for non-integer r. In this case

$$\hat{p} = \frac{0.680}{0.814} = 0.835, \qquad \hat{r} = 0.680\frac{0.835}{1 - 0.835} = 3.45.$$

The probability of more than 6 passengers in a car is

```
> 1- pnbinom(6,3.45,.835)
[1] 0.0001448711
```

4.5.5 Random numbers from a negative binomial distribution

Following the usual syntax, the number of passengers in the next 25 cars can be simulated:

```
> set.seed(101)
> rnbinom(25,3.45,.835)
 [1] 1 0 2 2 0 2 1 0 0 0 2 0 2 0 0 1 1 4 0 0 2 0 0 3 0
```

4.6   Poisson process

4.6.1   Defining a Poisson process in time

At the beginning of the twentieth century Ernest Rutherford and Johannes Geiger performed experiments with radioactive materials. Alpha particles (α-particles), which consist of two protons and two neutrons, are a type of ionizing radiation ejected by the nuclei of some unstable atoms, and in one experiment they recorded the number of α-particles emitted by a polonium source in each of 2608 periods of 7.5 seconds.

A polonium source contains millions of atoms and occasionally an atom will eject an α-particle and transmute to lead. The emissions of α-particles are events in time. If atoms transmute independently of each other the events occur as a Poisson process (Figure 4.7).

In a Poisson process:

- Events occur randomly and independently. By independence we mean that the numbers of events in disjoint time intervals are independent. This is a strong assumption, and one consequence is that an event you have just seen does not affect your chance of seeing another.

- Events themselves occupy a negligible amount of time.

- Events cannot occur simultaneously.

- Events occur at a constant average rate per unit of time.

The Poisson process can be thought of as the limit of a sequence of Bernoulli trials. Time is described as a sequence of small time increments. In each increment there is a small probability of an event occurring and this is defined as a success. No event occurring during a time increment is defined as a failure. The time increments are small enough for the probability of two or more events occurring to be negligible.

4.6.2   Superimposing Poisson processes

The Poisson process is fundamental for probabilistic models. Since events occur independently, if Poisson processes are superimposed the resulting process is also Poisson.

4.6.3   Spatial Poisson process

A Poisson process can also be used to model events occurring independentlyin volumes or over areas. The number of enteroviruses found in 10 liter samples of sea water might reasonably be modeled by a Poisson process [Mocé-Llivina et al., 2005]. The occurrences of flaws in woven cloth, such as stray threads in Harris Tweed, might be modeled with a Poisson distribution. The assumptions for a spatial Poisson process are the same as for a Poisson process in time with time replaced by volume or area:

- Events occur randomly and independently. That is, the number of events in disjoint volumes, or areas, are independent.

- Events occupy a negligible amount of volume or area.

- Events occur singly. That is, two, or more, events cannot occur at the same point.

- Events occur at a constant rate per unit of volume or area.

4.6.4 Modifications to Poisson processes

A point process is a random process in which events occur at isolated points in some continuum such as time, along a line transect, over an area or in a volume. A Poisson process is an example of a point process. Poisson processes can be modified, for example by allowing the rate to vary over time or allowing simultaneous events, and combined in other ways such as the Neyman- Scott cluster point process model.

4.6.5 Poisson distribution

Events occur in a Poisson process at an average rate of λ per unit of continuum. Let X be the number of events in a fixed length, or volume or area, of continuum t. The pmf of X is:

$$P(x) \quad = \quad \frac{e^{-\lambda t}(\lambda t)^x}{x!} \qquad x = 0, 1, \dots$$

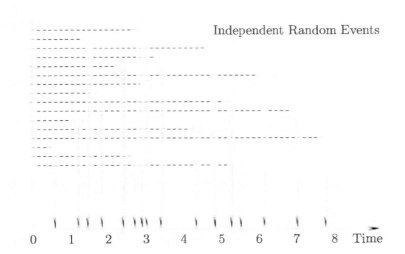

FIGURE 4.7: Atoms of polonium eject an α-particle (represented as a circle) independently of millions of other atoms in the polonium source leading to a Poisson distribution of the number of events in a fixed period of time.

The following derivation of the pmf of X shows the importance of the assumption that events occur independently. Divide t into n disjoint intervals of equal length [13] δt such that

$$n\delta t \quad = \quad t \Leftrightarrow \delta t \quad = \quad \frac{t}{n}.$$

Now suppose that n is very large so that δt is very small. Then the probability of more that one event in δt is negligible[14]and:

$$P(1 \text{ event in } \delta t) \quad \approx \quad \lambda \delta t$$
$$P(0 \text{ events in } \delta t) \quad \approx \quad 1 - \lambda \delta t.$$

---

[13]The derivation is described for t representing time, but "length" can be replaced by area or volume
[14]Formally the P(more than one event in δt) is of order $(\delta t)^2$ which vanishes relative to P(1 event in δt) as $\delta t \to 0$.

TABLE 4.1: Number of α particles omitted in 7.5 second intervals.

number of particles	0	1	2	3	4	5	6	7	8	9	10	11	12
number of 7.5s periods	57	203	383	525	532	408	273	139	45	27	10	4	2

This defines a binomial distribution for X with n trials and constant probability of success $\lambda \delta t$. The constant probability of success implies independence:

$$P(x) = \binom{n}{x}(\lambda \delta t)^x (1 - \lambda \delta t)^{n-x}.$$

The mean of the binomial distribution is $n(\lambda \delta t) = \lambda t$ which corresponds to λ being the average rate per unit time (often abbreviated to the "rate"). It remains to substitute t/n for δt in the binomial pmf to obtain

$$P(x) = \binom{n}{x}\left(\frac{\lambda t}{n}\right)^x \left(1 - \frac{\lambda t}{n}\right)^{n-x}$$

$$= \frac{n \times (n-1) \times \ldots \times (n-x+1)}{x!} \frac{(\delta t)^x}{n^x}\left(1 - \frac{\lambda t}{n}\right)^n \left(1 - \frac{\lambda t}{n}\right)^{-x}.$$

Now let $n \to \infty$ to obtain

$$P(x) = \frac{(\lambda t)^x}{x!} \lim_{n \to \infty}\left(1 - \frac{\lambda t}{n}\right)^n.$$

The remaining limit is $e^{-\lambda t}$ (Exercise 4.33). Although this derivative assumes λ is constant, the assumptions can be relaxed to λt being constant.

The probabilities can be shown to sum to 1 by recognizing the Taylor series expansion of $e^{\lambda t}$.

$$\sum_{x=0}^{\infty} P(x) = e^{-\lambda t}\sum_{x=0}^{\infty}\frac{(\lambda t)^x}{x!} = e^{-\lambda t}e^{\lambda t} = 1.$$

Examples of the pmf of the Poisson distribution for four values of λt are given in Figure 4.8. The mean μ and the variance σ^2 are both λt (Exercise 4.31).

4.6.6 Fitting a Poisson distribution

Rutherford and Geiger (1910) considered the numbers of α-particles emitted in 2608 time periods of 7.5 seconds. Their data iare given in Table 4.1.

The mean is

$$\bar{x} = \frac{0 \times 57 + \cdots + 12 \times 2}{2608} = 0 \times \frac{57}{2608} + \cdots 12 \times \frac{2}{2608} = 3.87$$

The method of moments estimate $\widehat{\lambda}$ of λ is obtained from

$$\bar{x} = \widehat{\lambda}t$$

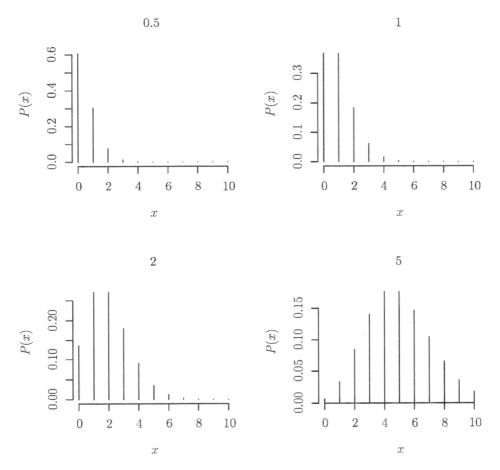

FIGURE 4.8: pmf of Poisson random variable with $\lambda t = 0.5, 1, 2$ and 5.

Here $\widehat{\lambda} = 3.87/7.5 = 0.530$ α-particles per second.

One test of how well the data are modeled by a Poisson distribution is to calculate the variance. If the variance is close to the mean this is at least consistent with a Poisson distribution. The variance of the number of particles in 7.5 periods is

$$s^2 = \frac{(0 - 3.87)^2 \times 57 + \ldots + (12 - 3.87)^2 \times 2}{2608 - 1} = 3.68.$$

This is remarkably close to the mean. We consider a more stringent test in Chapter 7.

4.6.7 Times between events

You may wonder why Rutherford and Geiger counted the number of particles emitted in 7.5 second intervals rather than record the times between emissions. We imagine that the reason is that particles are emitted too frequently for such timing to have been carried out with good precision using the instrumentation that was available at the time. In general, we can focus on the times between events, rather than the numbers of events in some fixed length of time. However, the times between events are measured on a continuous scale, and the distribution of times between events in a Poisson process, known as the exponential distribution is discussed in the next chapter.

4.7 Summary

4.7.1 Notation

$X(\omega)$ or X	random variable, where $\omega \in \Omega$
$P_X(x)$ or $P(x)$	probability mass function, pmf
$F_X(x)$ or $F(x)$	cumulative probability distribution, cdf
$E[\phi(X)]$	expected value of $\phi(X)$
μ_X or μ	expected value or mean of X
σ_X^2 or σ^2	variance of X

4.7.2 Summary of main results

A discrete random variable X assigns a non-negative integer number to each element in the sample space.

A random variable X has a probability distribution. In the case of a discrete random variable X the probability mass function (pmf) assigns a probability to each number in the image of X.

Taking the expected value of a random variable, or a function of a random variable, is equivalent to averaging in the population.

Some common pmf are described below with corresponding formulas for the support, $P(x)$, μ, and σ^2 given in Table 4.2.

- A Bernoulli trial is an experiment with two possible outcomes: success (S) and failure (F) and a probability p of success. A Bernoulli random variable X takes S to 1 and F to 0.

- A binomial random variable X is the number of successes in n Bernoulli trials.

- A hypergeometric random variable X is the number of defectives in a random sample of size n taken without replacement from a finite population of size N of which B are defective.

- A geometric random variable X is the number of Bernoulli trials until the first success.

- A negative binomial random variable X is the number of failures before the rth success in a sequence of Bernoulli trials.

- A Poisson random variable X is the number of events in time t if events occur randomly, independently and singly at an average rate λ per unit time.

TABLE 4.2: Discrete probability distributions.

Distribution	Support	pmf, $P(x)$
Binomial	$x = 0, \ldots, n.$	$\binom{n}{x} p^x (1-p)^{n-x}$
Hypergeometric	$x = \max(0, n\text{-}N\text{+}B),$ $\ldots, \min(n, B)$	$\frac{\binom{B}{x}\binom{N-B}{n-x}}{\binom{N}{n}}$
Geometric	$x = 0, 1 \ldots$	$(1-p)^{x-1} p$
Negative binomial	$x = 0, 1, \ldots$	$\binom{x+r-1}{x}(1-p)^x p^r$
Poisson	$x = 0, 1, \ldots$	$\frac{(\lambda t)^x e^{-\lambda t}}{x!}$

Distribution	Expected value, μ	Variance, σ^2
Binomial	np	$np(1-p)$
Hypergeometric	$n\frac{B}{N}\frac{(N-B)}{N}\frac{N-n}{N-1}$	
Geometric	$\frac{1}{p}$	$\frac{1-p}{p^2}$
Negative binomial	$\frac{pr}{1-p}$	$\frac{pr}{(1-p)^2}$
Poisson	λt	λt

4.7.3 MATLAB and R commands

In the following x is a value in the support of the corresponding distribution. The variables n, p, B and N are the parameters for the corresponding distribution and mu is the expected value (μ) for the corresponding distribution. For more information on any built in function, type help(function) in R or help function in MATLAB.

R command	MATLAB command
dbinom(x,n,p)	binopdf(x,n,p)
dgeom(x,p)	geopdf(x,p)
dhyper(x,B,N,n)	hygepdf(x,B+N,N,n)
dnbinom(x,r,p)	nbinpdf(x,r,p)
dpois(x,mu)	poisspdf(x,mu)

The above table gives the probabilities for the pmf. By way of example just using the binomial distribution, the cdf, inverse cdf, and random numbers are obtained by:

R command	MATLAB command
pbinom(x,n,p)	binocdf(x,n,p)
qbinom(prob,n,p)	binoinv(prob,n,p)
rbinom(numdev,n,p)	binornd(n,p,numdevrow,numdevcol)

4.8   Exercises

Section 4.1 Discrete random variables

Exercise 4.1: Sample space

A civil engineer will assess ground conditions for an airport construction on an ordinal descriptive scale: bad; poor; fair; good; ideal. Define a random variable X on this sample space.

Exercise 4.2: Indicator variables

An event A either occurs or it does not occur. The random variable X is defined as 1 if A occurs and as 0, if A does not occur. Similarly, a random variable Y is defined as 1 if an event B occurs and 0 if B does not occur. The random variables X and Y are referred to as **indicator variables**.

(a) Show that an indicator variable for the event $A \cap B$ is XY.

(b) Show that an indicator variable for $A \cup B$ is $\max(X, Y)$.

(c) Suggest an alternative indicator variable for $A \cap B$.

(d) Suggest an alternative indicator variable for $A \cup B$.

Exercise 4.3: Discrete moments 1

A discrete uniform random variable X is defined with support $0, 1, 2, 3, 4$. Find

(a) the mean, μ, of X.

(b) the variance σ^2, and standard deviation, σ, of X.

(c) the kurtosis, κ, of X, where

$$\kappa \;=\; \frac{\mathrm{E}\big[(X - \mu)^4\big]}{\sigma^4}.$$

Exercise 4.4: Discrete moments 2

A discrete uniform random variable X is defined with support $1, 2, \ldots, n$, so

$$\mathrm{P}(x) \;=\; \frac{1}{n}, \qquad x = 1, 2, \ldots, n,$$

(a) Find the mean of X.

(b) Find the variance, and standard deviation, of X.

Hint: Use the result of Definition 4.6, and use

$$\sum_{i=1}^{n}\left((i+1)^3 - i^3\right) \;=\; (n+1)^3 - 1$$

to deduce the sum of squared integers.

Section 4.2 Bernoulli trial

Exercise 4.5: Skewness of a Bernoulli random variable

A random variable X takes the values 0 or 1 with probabilities $1-p$ and p respectively. Find the skewness, γ, of X, where

$$\gamma \;=\; \frac{E\left[(X - \mu)^3\right]}{\sigma^3}$$

Section 4.3 Binomial distribution

Exercise 4.6: Variance of the binomial distribution

Suppose $X \sim binom(n, p)$.

(a) Obtain an expression for $E[X(X-1)]$, in terms of n and p.

(b) Show that $E\left[(X - E[X])^2\right] = E\left[X^2\right] - (E[X])^2$.

(c) Assume that $E[X] = np$ and deduce that $\sigma^2 = np(1-p)$.

Exercise 4.7: Collections of binomials

Generate $NS = 1\,000$ sets of $N = 20$ random samples of size $n = 40$ from a binomial distribution with $p = 0.0475$.

(a) For each set of 20 calculate the mean and variance of the number of successes in the samples of 40.

(b) Verify that the mean of the $1\,000$ means and the variance are close to np and $np(1-p)$ respectively.

(c) How many of the $1\,000$ variances exceed 2.2, the value obtained in Example 4.7?

Exercise 4.8: Corrupted images

A ground station receives images from a space telescope. The probability of a corrupted image is 3%. Assume corrupted images occur independently.

(a) What is the probability that exactly 3 out of 25 images are corrupted?

(b) What is the probability that 3 or more out of 25 images are corrupted?

Exercise 4.9: Robotic fish

A shoal of 20 robotic fish is released to monitor pollution levels in the ocean.
If the probability that a robotic fish is still returning data after one year is 0.2, and fish fail independently, what is the probability that 5 or more of the 20 robotic fish will still be returning signals in one year's time?

Exercise 4.10: Life of fluorescent light

The probability that a fluorescent light has a life of over 500 hours is 0.9. Find the probabilities that among eleven such lights:

(a) exactly 8 last for more than 500 hours,

(b) at least 8 last for more than 500 hours,

(c) at least 2 do not last for more than 500 hours.

Exercise 4.11: Defective items

A process produces some defective items, which do not necessarily occur independently. The probability that an item taken from the production line at random is defective is p. If this item is defective, the probability that the next item off the line is defective is θ. If this item is good, the probability that the next item off the line is defective is ϕ. Define the random variable X as the number of defectives when two items are taken from the line in sequence.

(a) What is the image of X?

(b) If $E[X] = 2p$, find an expression for ϕ in terms of θ and p.

(c) Calculate the variance of X in terms of θ and p.

(d) Calculate the coefficient of variation (CV) of X in terms of θ and p.

(e) Verify that you obtain the results for a binomial distribution if you set $\theta = p$.

(f) Calculate the numerical value of the CV if $p = 0.1$ and $\theta = 0.2$, and compare this with the value if $p = 0.1$ and $\theta = 0.1$.

(g) Calculate the numerical value of the CV if $p = 0.1$ and $\theta = 0.05$.

Exercise 4.12: Flood levels

The annual maximum flood level, at a given location by a river, exceeds a level c with a probability p. Assume that such annual exceedances occur independently.

(a) What is the ARI, T, in terms of p?

(b) Show that the probability that c will be exceeded in at least one of n years is:

$$1 - \left(1 - \frac{1}{T}\right)^n.$$

Exercise 4.13: Flood plain

A factory is built in a flood plain. The probability of flooding in any one year is 0.02. Assume that years with flooding occur independently.

(a) What is the expected number of years until the factory is next flooded?

(b) What is the probability that the number of years until the factory is next flooded exceeds the expected number?

Exercise 4.14:

(a) Suppose $X \sim \text{binom}(28, 0.2)$

(i) Find P(5)

      (ii) Find P(6)

      (iii) Find the greatest x such that $P(X \leq x) \leq 0.10$

      (iv) Find the greatest x such that $P(X \leq x) \leq 0.05$

(b) Suppose $X \sim \text{binom}(112, 0.2)$

      (i) What is $E[X]$

      (ii) Find P(22)

      (iii) Find P(23)

      (iv) Find the greatest x such that $P(X \leq x) \leq 0.10$

      (v) Find the greatest x such that $P(X \leq x) \leq 0.05$

(c) Suppose $X \sim \text{binom}(1000, 0.2)$

      (i) What is $E[X]$

      (ii) Find P(200)

      (iii) Find the greatest x such that $P(X \leq x) \leq 0.10$

      (iv) Find the greatest x such that $P(X \leq x) \leq 0.05$

Exercise 4.15: Nitrates content in water

An inspector wants to check that at least 95% of water samples from the public supply contain less than the maximum specified level for nitrates.

(a) The inspector takes a random sample of 15 jars from the water supply each week. If the nitrate content of any jar exceeds the specified level a complaint is filed. The legislation permits a maximum of 5% of jars above the specified level. Find the probabilities a complaint is filed if:

      (i) only 2% of the jars that could conceptually be filled from the supply exceed the level,

      (ii) 5% of all such jars exceed the level,

      (iii) 10% of all such jars exceed the level.

(b) Repeat the exercise if the inspector takes random samples of 30 and files a complaint if more than 2 jars exceed the limit.

Exercise 4.16: U.S. Coast Guard flares

[McHale, 1977] reported on the safety of distress flares and associated smokes to the U.S. Coast Guard. He found that 2 out of 18 handheld red flares from manufacturer A failed, but no failures in samples of 14 from manufacturer B and 10 from manufacturer C.

(a) For what probability p is the probability of obtaining 2 or more failures in a random sample of 18 equal to:

      (i) 0.10

      (ii) 0.05

      (iii) 0.01

(b) Suppose that there is no difference between manufacturers and the probability that a hand held flare fails is 2/42. Also suppose that the flares tested are a random sample from the population of all such flares. Under these suppositions, find the following probabilities.

      (i) 0 failures in a sample of 10

(ii) 0 failures in a sample of 14

(iii) 2 or more failures in a random sample of 18.

Exercise 4.17: Crushed rock

A container of finely crushed rock contains 28 small gold nuggets. A sample of one sixth part of the rock in the container is taken for assay. Suppose that the number of nuggets in the sample has a binomial distribution.

(a) Identify the number of trials and the probability of success in this context.

(b) What do we assume about the distribution of gold nuggets to justify use of the binomial distribution?

(c) What is the expected number of gold nuggets in the sample?

Section 4.4 Hypergeometric distribution

Exercise 4.18: Compare binomial and hypergeometric

Compare the pmf of *binom*(5, 0.3) with that of a hypergeometric distribution with $n = 5, N = 20, B = 6$.
Compare the probabilities of obtaining more than 2 defective items.

Exercise 4.19: Shipping containers

A consignment of 15 shipping containers includes 4 that contain motor bikes that have not been declared on the shipping documents. If customs officers check the contents of 3 randomly selected containers,

(a) what is the probability that they check at least one of the containers with the bikes?

(b) How many randomly selected containers should be checked if the probability of detecting at least one container with the bikes is to be 0.9?

Exercise 4.20: Corrupted Images

A ground station receives images from a space telescope. The probability of a corrupted image is p. Assume corrupted images occur independently. Let the random variable X be the number, in time order, of the first corrupted image. Then the probability mass function of X is:

$$P(x) = (1 - p)^{x-1} p \quad \text{for} \quad x = 1, \ldots.$$

(a) Prove that $\sum\limits_{x=1}^{\infty} P(x) = 1$.

(b) Prove that $E[X] = 1/p$.

(c) If $p = 0.1$ what is the probability that the next 10 images, taken of a supernova, are all uncorrupted?

Section 4.5 Negative binomial distribution

Exercise 4.21: Negative binomial distribution variants

A negative binomial random variable, Y, can be defined as the number of trials until the rth success.

$$P(y) \;=\; \binom{y-1}{r-1} p^r (1-p)^{y-r}, \quad y = r, r+1, \ldots$$

(a) Explain why this is equivalent to the distribution of X in Section 4.5.2.

(b) Given that $E[X] = r(1-p)/p$ what is $E[Y]$? What is the variance of Y given that the variance of X is $E[X]/p$?

(c) Show that Y has a geometric distribution, as defined in Section 4.5.1, when $r = 1$.

Exercise 4.22: The geometric distribution

Define S_n as the sum of the n terms of a geometric series with initial value a and common ratio r. That is,

$$S_n \;=\; a + ar + ar^2 + \ldots ar^{n-1}$$

(a) Consider $S_n - rS_n$ and show that $S_n = \dfrac{a(1 - r^n)}{1 - r}$.

(b) Under what conditions will $\lim_{n \to \infty} S_n = S_\infty$, for finite S_∞?

(c) Hence for a random variable X with probability mass function

$$P(x) \;=\; (1-p)^{x-1} p, \quad x = 1, 2, \ldots$$

(i) Show that the mean of X is given by $\dfrac{1}{p}$.

(ii) Show that the variance of X is given by $\dfrac{1-p}{p^2}$.

Section 4.6 Poisson process

Exercise 4.23: Faulty metal plate

Sheets of metal have plating faults which occur randomly and independently at an average rate of 1 per m^2. What is the probability that a sheet 1.5 m by 2 m will have at most one fault?

Exercise 4.24: Solar powered vehicle

A prototype small solar powered four wheel drive (4WD) with an auxiliary rotary engine is being tested. Breakdowns appear to occur according to a Poisson process with a constant mean rate of 0.0025 per hour.

(a) Calculate the probability of 4 or more breakdowns in 1 000 hours.

(b) Calculate the probability that the time until the first breakdown exceeds 500 h.

(c) Calculate the probability that the time until the second breakdown is less than 1 000 h.

Exercise 4.25: Hand woven cloth

Hand woven lengths of cloth are classified as top quality and second quality. A weaver produces 20 lengths per day. The first length of the day is of top quality with probability $1 - p$, and of second quality with probability p. The remaining 19 lengths have a probability p of being of second quality if the preceding length is of top quality, and a probability kp (k such that $kp < 1$) of being of second quality if the preceding length is of second quality. Write a computer script to simulate the numbers of second quality lengths produced each day with p and k as parameters.

(a) Consider the case with $p = 0.04$, $k = 3$ and simulate 10^6 days.

   (i) What are the mean and variance of the number of second quality lengths per day?

   (ii) Fit a negative binomial distribution to the numbers of second quality lengths per day by the method of moments.

   (iii) Compare the empirical distribution of the numbers of second quality lengths per day obtained from the computer simulation with probabilities calculated from the negative binomial distribution.

   (iv) Why can the negative binomial model only be an approximation to the model which has been simulated?

(b) Construct a table giving the mean and coefficient of variation of the number of second quality lengths per day as a function of p and k for: p from 0.01 to 0.09 in steps of 0.02, and $k = 0.1, 0.5, 2, 10$.

Exercise 4.26: Poisson

Assume that serious road traffic accidents occur in a rural area as a Poisson process with rate 28 accidents per 12 months.

(a) What is the expected number of accidents in a two month period?

(b) What is the probability of exactly 2 accidents in a two month period?

(c) What is the probability of 2 or less accidents in a two year period?

(d) What are the assumptions of the Poisson process in this context?

Exercise 4.27: Neutrino detector

A neutrino detector detects an average of 3.7 neutrinos per day (24 hours).

(a) Assume the detection of neutrinos is a Poisson process.

   (i) What is the probability that no neutrino is detected in a 24 hour period?

   (ii) What is the probability of detecting more than one neutrino in a 1 hour period?

(b) Model the detection of neutrinos by a binomial distribution.

   (i) Taking a one hour period as a trial, what is the probability of a success, defined as detecting a neutrino, if the mean of the binomial distribution is to match the mean of the Poisson process?

   (ii) What probability of no success in a day is obtained with this binomial distribution?

   (iii) Repeat sub-part(i) taking a 3 hour period as a trial.

(iv) Repeat sub-part(i) taking a 1 minute period as a trial.

Exercise 4.28: Supernovae

There have been 6 supernovae observed in our galaxy, the Milky Way, between 1001 and 2000. Assume, on the basis of observations on external galaxies, that supernovae events in our galaxy occur as a Poisson process with rate $1/80$ yr. What is the expected number of supernovae in the Milky Way over 1000 years? Calculate the probability of 6 or less supernovae in 1000 years.

Exercise 4.29: Road accidents

There were 43 road accidents involving pedestrians in the central business district (CBD) of a city in the ten year period 1992-2001. New safety measures were introduced at the beginning of the year 2002, and there was only one such accident in the year 2002. Assume that occurrences of accidents can be modeled reasonably as a Poisson process.

(a) Calculate the probability of 0 or 1 accident in the year 2002 if the underlying rate is unchanged at 43 accidents per 10 years.

(b) Do you think there is substantial evidence that the new safety measure has been successful?

Exercise 4.30: Alternative derivation of the Poisson distribution

An alternative derivation of the Poisson distribution follows from the following argument.
Assume a Poisson process with rate λ and let $P(x, t)$ denote the probability of exactly x events in time t.

(a) Explain why $\quad P(0, \delta t) = 1 - \lambda t$.

(b) Explain why $\quad P(0, t + \delta t) = P(0, t)(1 - \lambda t)$.

(c) Deduce that $\quad \dfrac{dP(0, t)}{dt} + -\lambda P(0, t)$.

(d) Solve the differential equation using the initial condition $P(0, 0) = 1$.

(e) Explain why $\quad P(1, t + \delta t) = P(1, t) P(0, \delta t) + P(0, t) P(1, \delta t)$ and hence deduce the pmf.

Exercise 4.31: Poisson mean

Follow the method used to obtain the mean and variance of a binomial random variable to show that the mean and variance of a Poisson random variable with rate λ over time intervals t are both λt.

Miscellaneous problems

Exercise 4.32: Zero inflated Poisson distribution

The zero-inflated Poisson (ZIP) model combines two zero generating processes. The first process generates a 0 with probability p in one time unit. The second process is a Poisson distribution with rate λ per time unit. If the random variable X has a ZIP distribution the pmf is

$$P(0) = p + (1 - p)e^{-\lambda} \quad \text{and}$$

$$P(x) = (1 - p)\frac{\lambda^x e^{-\lambda}}{x!}, \quad x = 1, 2, \ldots$$

(a) Show that the mean is $(1 - p)\lambda$ and that the variance is $\lambda(1 - p)(1 + \lambda p)$.

(b) Show that the method of moments estimators of the parameters are

$$\widehat{\lambda} = \frac{s^2 + \bar{x}^2 - \bar{x}}{\bar{x}} \quad \text{and}$$

$$\widehat{p} = \frac{s^2 - \bar{x}}{s^2 + \bar{x}^2 - \bar{x}},$$

where $\bar{x}$ is the sample mean and s^2 is the sample variance.

(c) Fit a ZIP to the number of passengers in cars recorded in the survey of Example 4.13 (car pooling).

(d) Calculate the probability of more than 6 passengers for the ZIP model. Compare this with the probability given by the negative binomial model and comment.

Exercise 4.33: Interest

(a) I invest $\$1,000$ at a fixed rate of interest for 1 year. What is the value at the end of 1 year if

(i) Interest is 6% per annum paid out at the end of the year.

(ii) Interest is 6/4% per quarter compounded over 4 quarters.

(iii) Interest is 46/12% per month compounded over 12 months (assume months are all 365/12 days).

(iv) Interest is 6/365% per day compounded daily.

(b) Use the generalized binomial expansion (Taylor series about 0) for $\left(1 + \frac{\alpha}{n}\right)^n$ to show that

$$\lim_{n \to \infty} \left(1 + \frac{\alpha}{n}\right)^n = 1 + \alpha + \frac{\alpha^2}{2!} + \ldots = e^\alpha.$$

(c) In compound interest calculations α is known as the **force of interest** and corresponds to interest being compounded continuously.

(i) What is the value of my investment if a force of interest of 0.06 is applied over 1 year?

(ii) What force of interest corresponds to 6% per annum paid at the end of the year?

Exercise 4.34: Multinomial distribution

Suppose that independent trials have r possible mutually exclusive and exhaustive outcomes. For each trial the probabilities of the r outcomes are $p_1, p_2, \ldots, p_r$. Let X_i be the total number of outcomes of type i in n trials.

(a) Explain why

$$P(x_1, x_2, \ldots, x_r) = \frac{n!}{x_1! x_2! \ldots x_r!} p_1^{x_1} p_2^{x_2} \ldots p_r^{x_r}$$

(b) Substitute $r = 2$ and comment.

(c) Explain why $E[X_i] = np_i$ and $var(X_i) = np_i(1 - p_i)$.

(d) If $r = 3$ find $E[X_i|X_1 = m]$ where $0 \leq m \leq n$, and $i = 2, 3$.

(e) In R the syntax for calculating probabilities, when for example $r = 3$, is
dmultinom(c(x1,x2,x3),prob=c(p1,p2,p3)).

In an aluminum casting process for cylinder heads, each cylinder head is classified as good, rework or recycle.

The probability that a cylinder head is good is 0.8, and the probability that it can be reworked by removing flash is 0.07.

Suppose the classification of consecutive cylinder heads is an independent process.

(i) What is the probability of 18 good, 1 rework, and 1 recycle in a sample of 20?

(ii) What is the probability of at least 18 good cylinder heads in a sample of 20?

(f) In R, the syntax for generating N random numbers from a multinomial distribution with n trials, when for example $r = 3$, is *rmultinom(N,n,c(p1,p2,p3))*.

Simulate the classification of cylinder heads for 10 samples, each of size 20, using the seed 1729.

5

Continuous probability distributions

A continuous random variable can be defined in terms of a probability density function (pdf) which is the limit of a histogram as the sample size tends to infinity. One reason for fitting a pdf to data is extrapolation into the tails to estimate probabilities associated with extreme events or to make computer simulations realistic. It is necessary to choose appropriate models for pdfs and a range of distributions that are used in engineering are described: uniform; normal and lognormal; exponential; gamma; and Gumbel. Procedures for fitting these distributions to data are given and a method for the graphical assessment of the goodness of fit is described. Algorithms for the generation of pseudo-random deviates from continuous distributions are explained and implemented with software functions.

5.1   Continuous random variables

Most of the concepts underlying discrete distributions are applicable to continuous distributions, with probabilities being replaced by areas under curves.

5.1.1   Definition of a continuous random variable

An experiment has an associated sample space Ω. Recall Definition 4.1 of a random variable, which is a rule that assigns a unique real number to every element in the sample space. If we denote an element in the sample space by ω a random variable X is defined by

$$X(\omega) \;=\; x,$$

where x is a unique real number associated with ω.

If the random variable X is continuous then the set of all possible x is the continuum of real numbers between some lower limit L and upper limit U. The limits can be finite or set at $-\infty$ and $+\infty$ respectively. Example 5.1 is typical inasmuch as the random variable takes a measurement, which has a physical unit, to the corresponding real number.

Example 5.1: Ball bearings [random variable]

Consider a process for the manufacture of nominally identical deep groove ball bearings with a specified inside diameter of 40 mm. If the process were to continue in its present state indefinitely, there would be an infinite population of such bearings. A deep groove ball bearing is selected from this process and its inner diameter is measured precisely, to the nearest micron, and found to be x mm. It is convenient to overlook the fact that the measurement will be rounded to three decimal places and to consider the sample space as the continuum of all possible lengths between 0 and some arbitrarily large upper value which we can set to $+\infty$. In practice, the lengths will be around 40 mm.

A random variable X is defined by

$$X(x \text{ mm}) = x,$$

where $x \in [0, \infty)$. From here onwards we dispense with the precise definitions and refer to what is, in this case, an inner diameter as a random variable.

5.1.2 Definition of a continuous probability distribution

A **continuous probability distribution** defines the probability that X lies within any interval[1] $[a, b]$ in its image. The image of X is referred to as the **support** of the probability distribution. We will define the **probability density function (pdf)**, often abbreviated to **density** as a limit of a histogram.

Suppose we take four samples of sizes $20, 100, 1\,000$, and $10\,000$ from an idealized process producing deep groove ball bearings. The inside diameters in mm of the bearings are measured to the nearest micron. Histograms showing the distributions of the inside diameters are displayed in Figure 5.1. As the sample size increases we can take more bins and also have more data in each bin. The histograms become smoother as the sample size increases and we imagine the histogram tending to a smooth curve as the sample size increases towards the entire infinite population (Figure 5.2 lower figure). This smooth curve is the pdf, and is defined as a function $f(x)$ of x. Areas under the histogram represent relative frequencies and areas under the pdf represent probabilities.

The two requirements for a function $f(x)$ to be a pdf are:

$$0 \leq f(x) \quad \text{and} \quad \int_{-\infty}^{\infty} f(x)dx = 1.$$

The first rules out negative probabilities, and the second then restricts probabilities to be between 0 and 1. In Chapter 2 we considered the cumulative frequency polygon as an alternative display to the histogram and saw that either one can be deduced from the other. Similarly, the population can be described by either the pdf, $f(x)$, or the **cumulative distribution function (cdf)**[2], $F(x)$, given in Definition 4.3 by:

$$F(x) = \mathrm{P}(X \leq x) \quad \text{for} \quad -\infty < x < \infty.$$

In geometric terms, $F(x)$ is the area under the histogram between $-\infty$ and a vertical line passing through x. The cdf can be obtained by integrating the pdf:

$$F(x) = \int_{-\infty}^{x} f(\theta)d\theta$$

and the pdf can be found by differentiating the cdf, which leads to the following definition of the density function in terms of the cdf.

Definition 5.1: Probability density function

The probability density function of a continuous random variable is defined by

$$f(x) = \frac{dF(x)}{dx}.$$

---

[1]The only probability that can be assigned to the probability that X equals a single number x is 0, because there is an uncountable infinity of real numbers in the interval $(x, x + \delta x)$ for arbitrarily small δx.

[2]For a continuous random variable $\mathrm{P}(X \leq x) = \mathrm{P}(X < x)$ but the distinction is crucial for a discrete random variable.

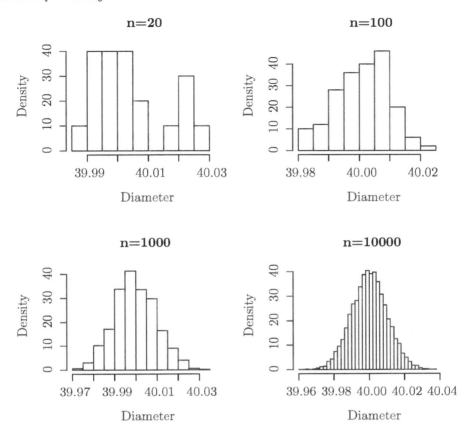

FIGURE 5.1: Histograms of inner diameters of deep groove ball bearings, of nominal diameter 40 mm, for samples of sizes: $20, 100, 1\,000, 10\,000$ (fictitious data).

5.1.3   Moments of a continuous probability distribution

Taking expected value is taking a mean value in the population. The population is modeled by a probability distribution. If you refer back to Section 3.7.2 you will see that we calculated the mean of continuous data that had been grouped into K bins by setting all the data in a bin to the mid point of the bin x_k, where $k = 1, \ldots, K$. If the number of data in bin k, the frequency, is f_k then

$$\bar{x} \;=\; \frac{\sum_{k=1}^{K} x_k f_k}{n},$$

where $n = \sum f_k$ is the number of data. This can be rewritten as

$$\bar{x} \;=\; \sum_{k=1}^{K} x_k \left(\frac{f_k}{n} \right).$$

Rewriting the mean in this fashion shows that the sample mean is equal to the sum of the mid-points of the bins multiplied by the area of the rectangle above the bin (Figure 5.2 upper frame). Now focus on the lower frame of Figure 5.2 which represents the population. The area of a typical rectangle is approximately equal to the product of the height of the pdf at the centre of the bin, $f(x_k)$, with the width of the bin δx.

This area is approximately

$$P\left(x_k - \frac{\delta x}{2} < X < x_k + \frac{\delta x}{2}\right) \;=\; \int_{x_k - \frac{\delta x}{2}}^{x_k + \frac{\delta x}{2}} f(x)dx \;\approx\; f(x_k)\delta x$$

and the approximations become more accurate as the bin width, δx, decreases.

So, the population mean μ is given by

$$\mu \;=\; \mathrm{E}[X] \;=\; \lim_{n,K \to \infty} \sum_{k=1}^{K} x_k \left(f(x_k)\delta x\right).$$

This limit is an integral, which is defined as a limit of a sum, and so we have the following definition.

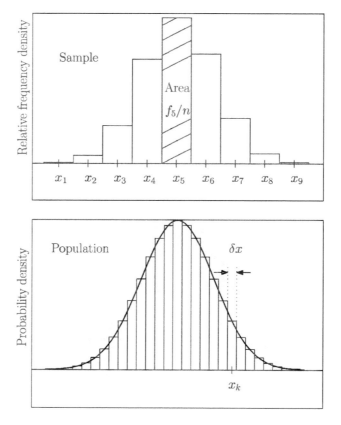

FIGURE 5.2: Relative frequencies of binned data in the histogram (areas of rectangles) tend towards probabilities of being in bins (areas under the pdf).

Definition 5.2: Expected value of a continuous random variable

The mean of a continuous random variable X with pdf $f(x)$, which is also known as the expected value of X, is defined as

$$\mu \;=\; \mathrm{E}[X] \;=\; \int_{-\infty}^{\infty} xf(x)dx.$$

A similar argument leads to the following definition.

Definition 5.3: Expected value of a function of a continuous random variable

The expected value of any function $\phi(X)$ of a continuous random variable X with pdf $f(x)$ is defined by

$$\mathrm{E}[\phi(X)] \;\; = \;\; \int_{-\infty}^{\infty} \phi(x)f(x)dx.$$

In particular, the expected value of a constant, a, is itself.

$$\mathrm{E}[a] \;\; = \;\; \int_{-\infty}^{\infty} af(x)dx \;\; = \;\; a\int_{-\infty}^{\infty} f(x)dx \;\; = \;\; a \times 1 \;\; = \;\; a,$$

since $f(x)$ is a pdf, and more generally:

$$\mathrm{E}[a\phi(X)] \;\; = \;\; \int_{-\infty}^{\infty} a\phi(x)f(x)dx \;\; = \;\; a\int_{-\infty}^{\infty} \phi(x)f(x)dx \;\; = \;\; a\mathrm{E}[\phi(X)].$$

Also, since the integral of a sum is the sum of the integrals of the summands, if $\psi(X)$ is some other function of X.

$$\mathrm{E}[\phi(X) + \psi(X)] \;\; = \;\; \int_{-\infty}^{\infty} (\phi(x) + \psi(x))f(x)dx$$

$$= \;\; \int_{-\infty}^{\infty} \phi(x)f(x)dx + \int_{-\infty}^{\infty} \psi(x)f(x)dx \;\; = \;\; \mathrm{E}[\phi(X)] + \mathrm{E}[\psi(X)].$$

Definition 5.4: Variance of a continuous random variable

The variance, σ^2, of a continuous random variable X is defined by

$$\sigma^2 \;\; = \;\; \mathrm{E}\big[(X - \mu)^2\big] \;\; = \;\; \int_{-\infty}^{\infty} (x - \mu)^2 f(x)dx.$$

The variance $\mathrm{E}\big[(X - \mu)^2\big]$ is the second central moment, where in general the rth central moment[3] (central refers to the subtraction of μ), is defined as follows.

Definition 5.5: Central moments of a continuous random variable

The rth central moment of a continuous random variable X is defined by

$$\mathrm{E}[(X - \mu)^r] \;\; = \;\; \int_{-\infty}^{\infty} (x - \mu)^r f(x)dx,$$

provided the integral is proper[4].

---

[3]Moments, as opposed to central-moments are discussed in Exercise 5.39.
[4]The Cauchy distribution has no moments (Exercise 5.40).

In a sample, the sum of deviations from the sample mean is 0, and the equivalent result in the population is that the first central moment is 0. This follows from

$$E[(X - \mu)] \;=\; E[X] - E[\mu] \;=\; \mu - \mu \;=\; 0.$$

There is a useful physical interpretation of this result. Imagine the pdf shown in Figure 5.3 being made of cardboard with uniform density of 1 per unit area. This cardboard cutout will balance on a fulcrum at the mean μ. The explanation is that that the clockwise turning moment, about μ, of the area of width δx above x is the product of its mass, which is approximately $f(x)\delta x$, and its distance from μ which is $x - \mu$. If points are to the left of μ the clockwise turning moment is negative corresponding to an anti-clockwise turning moment. In the limit as $\delta x \to 0$ the sum of these turning moments is

$$\int_{-\infty}^{\infty} (x - \mu)f(x)dx \;=\; 0$$

and the pdf balances at μ, which is the x-coordinate of its centre of gravity.

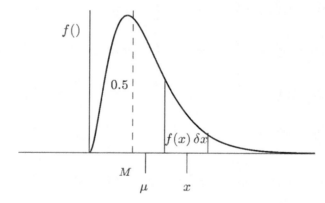

FIGURE 5.3: The sum of clockwise turning moments about μ equals 0 as δx tends to 0.

Definition 5.6: Skewness

The skewness of the distribution of X is defined as

$$\gamma \;=\; E\left[(X - \mu)^3\right]/\sigma^3.$$

If a distribution has a tail to the right the skewness will be positive, because cubing accentuates the difference from the mean and retains the sign (Exercise 5.1 for example).

Definition 5.7: Kurtosis

The kurtosis of the distribution of X is defined as

$$\kappa \;=\; E\left[(X - \mu)^4\right]/\sigma^4.$$

From its definition, kurtosis must be positive, and the kurtosis of a bell shaped curve (the normal distribution discussed in Section 5.4) is 3. A distribution is described as "heavy tailed" if the support is unbounded and the kurtosis is greater than 3. An example is the Laplace distribution defined in Exercise 5.41.

5.1.4 Median and mode of a continuous probability distribution

A useful notation is that x_α is the value of x such that

$$F(x_\alpha) \;=\; 1 - \alpha$$

and x_α eferred to as the **upper** α**- quantile**.

Definition 5.8: Median of a continuous probability distribution

The median, $x_{0.5}$, of a continuous probability distribution is the value of x such that

$$F(x_{0.5}) \;=\; 0.5$$

It follows from the definition of the median, that any pdf has one half of its area to the left of the median and one half of its area to the right. If a pdf is positively skewed, and so has a longer tail to the right, then the sum of clockwise moments about the median is greater than the sum of anti-clockwise moments. It follows that the centre of gravity of the pdf, the mean, must lie to the right of the median . An example is given in Exercise 5.1.

The mode of a continuous distribution is the value of x at which the pdf has a maximum. A distribution does not necessarily have a mode or can have more than one mode.

5.1.5 Parameters of probability distributions

In the following sections we describe various probability distributions. The pdf and cdf are functions of the variable x, but they include other letters, referred to as **parameters**, that define the mean and variance, and in some cases the skewness and kurtosis. If we fit a distribution to data we estimate the parameters of the distribution from the data. In many cases a convenient method of estimating the parameters is to equate the mean and variance of the distribution to the sample mean and variance. This is another example known as the method of moments or MoM. See also Section 4.3.6.

5.2 Uniform distribution

The pdf of a uniformly distributed random variable[5] is a rectangle of area 1.

---

[5] References to "random numbers" without further specification typically imply that they are uniformly distributed over $[0, 1]$.

5.2.1   Definition of a uniform distribution

The pdf of a random variable X which has a uniform distribution with support $[a, b]$ is:

$$f(x) = \begin{cases} \dfrac{1}{b-a} & \text{for } a \leq x \leq b, \\[2ex] 0 & \text{elsewhere} \end{cases}$$

and the cdf is

$$F(x) = \begin{cases} 0 & x < 0 \\[2ex] \dfrac{x-a}{b-a} & a \leq x \leq b, \\[2ex] 1 & b < x. \end{cases}$$

The pdf and cdf[6] are shown for the important special case of $[a, b]$ equal to $[0, 1]$ in Figure 5.4.

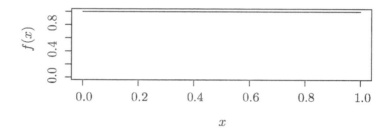

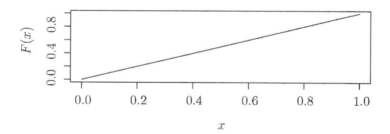

FIGURE 5.4: pdf and cdf of uniform distribution with support $[0, 1]$.

```
> x=c(0:100)/100
> y1=dunif(x,0,1)
> y2=punif(x,0,1)
```

---

[6]R syntax: the pdf, cdf, inverse cdf, and random deviates drawn from the distribution are obtained by preceding the name of the distribution with "d" for density, "p" for probability, "q" for quantile, and "r" for random respectively. The default for the quantile is the lower quantile, the upper quantile is obtained by including `lower=FALSE`.

```
> par(mfrow=c(2,1))
> plot(x,y1,type="l",ylab="f(x)",ylim=c(0,1))
> plot(x,y2,type="l",ylab="F(x)")
```

The mean and variance of a uniform distribution are (Exercise 5.6)

$$\mu \;=\; \frac{a+b}{2}, \qquad \sigma^2 \;=\; \frac{(b-a)^2}{12}.$$

Notation: $X \sim U[a, b]$

5.2.2 Applications of the uniform distribution

A uniform distribution is applicable when a variable is equally likely to be at any point between finite limits.

Example 5.2: Trams [uniform distribution]

Trams run from the north west of Adelaide to the beach at Glenelg along dedicated tram lines, and leave every 5 minutes at peak times. A computer model of commuter travel times incorporates a wait for a tram, that is uniformly distributed between 0 and 5 minutes.

Example 5.3: Flywheel [uniform distribution]

A flywheel has a radial line marked on on its circular face. It is given an arbitrary torque, and when it comes to rest the angle the line makes with an axis perpendicular to the axis of rotation is uniformly distributed with support $[0, 2\pi]$ radians.

5.2.3 Random deviates from a uniform distribution

The generation of pseudo-random deviates from $U[0, 1]$ is the basis for obtaining pseudo random deviates from any other distribution. A linear congruential PRNG with modulus m produces remainders between 0 and $m-1$ in an order that appears as if it is random. If m is large (10^8 say),then division by m gives a sequence of pseudo-random deviates from a uniform distribution. However, it is more convenient to use a software function. The following R commands generate 12 deviates from $U[0, 1]$ and prints them. The sequence is reproducible because a seed was set.

```
> set.seed(1)
> u=runif(12) ; print(u)
[1] 0.26550866 0.37212390 0.57285336 0.90820779 0.20168193 0.89838968
[7] 0.94467527 0.66079779 0.62911404 0.06178627 0.20597457 0.17655675
```

5.2.4 Distribution of $F(X)$ is uniform

If X is a random variable, so too is $F(X)$ where $F(\cdot)$ is the cdf of its probability distribution. If its probability distribution is continuous then

$$F(X) \;\sim\; U[0, 1].$$

The proof of this result follows. First notice that $F(x)$ is a probability and therefore must lie between 0 and 1. Since $F(\cdot)$ is a strictly increasing function,

$$F(x) \;=\; \mathrm{P}(X < x) \;=\; \mathrm{P}(F(X) < F(x)).$$

Now define $U = F(X)$, and write $u = F(x)$ to obtain

$$\mathrm{P}(U < u) \;=\; u, \qquad 0 \le u \le 1,$$

which is the cdf of $U[0,1]$. A corollary is that if $U \sim U[0,1]$ then $F^{-1}(U)$ has the distribution with cdf $F(\cdot)$.

This provides a method of generating random deviates from the distribution with cdf $F(\cdot)$. Generate a random deviate u from $U[0,1]$ and set $x = F^{-1}(u)$ as a random deviate from $F(\cdot)$. This result is quite general but it is only convenient when $F^{-1}(\cdot)$ can be written as a formula.

The general principle for generating random deviates is shown in Figure 5.5.

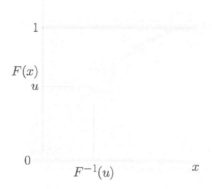

FIGURE 5.5: If u is a random deviate from $U[0,1]$, $F^{-1}(u)$ is a random deviate from the probability distribution with cdf $F(x)$.

5.2.5   Fitting a uniform distribution

It is unusual to estimate the limits for a uniform distribution from engineering data because the limits are generally known from the context. Method of moment estimation is not generally suitable (Exercise 5.5).

5.3   Exponential distribution

Times between events in a Poisson process have an exponential distribution.

5.3.1   Definition of an exponential distribution

Suppose events occur in a Poisson process at a mean rate of λ per unit time. Let the random variable T be the time until the first event. The time is measured from "now", and "now" can be any time because events in a Poisson process are assumed random and independent, and cannot occur simultaneously. In particular "now" can be the time at which the latest event occurred so T is also the time between events. The cdf of T follows directly from the Poisson distribution. If we consider some length of time t, the probability that T exceeds t is the probability of no event in time t. This probability is the complement of the cdf and is known as the **survivor function** or the **reliability function**.

$$P(t < T) \;=\; e^{-\lambda t}, \qquad 0 \le t.$$

Then the cdf is

$$F(t) \;=\; P(T \le t) \;=\; 1 - e^{-\lambda t}, \qquad 0 \le t.$$

We can check that $F(0) = 0$ and that $F(\infty) \to 1$. The pdf is obtained by differentiation with respect to t

$$f(t) \;=\; \lambda e^{-\lambda t}, \qquad 0 \le t.$$

The pdf and cdf for $\lambda = 2$ are shown in Figure 5.6.

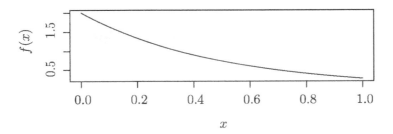

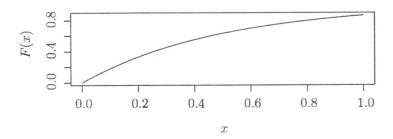

FIGURE 5.6: pdf and cdf of exponential random variable with $\lambda = 2$.

```
> x=c(0:100)/100
> y1=dexp(x,2)
> y2=pexp(x,2)
> par(mfrow=c(2,1))
> plot(x,y1,type="l",ylab="f(x)")
> plot(x,y2,type="l",ylab="F(x)")
```

The mean and variance of the exponential distribution are (Exercise 5.7)

$$\mu \;=\; \frac{1}{\lambda} \qquad \text{and} \qquad \sigma^2 \;=\; \frac{1}{\lambda^2}.$$

It follows that the CV (σ/μ) is 1. The median of the exponential solution $x_{0.5}$ is given by

$$F(x_{0.5}) \;=\; 1 - e^{-\lambda x_{0.5}} \;=\; 0.5 \Rightarrow x_{0.5} \;=\; \frac{\ln(2)}{\lambda}.$$

The exponential distribution is not symmetric about the mean, and it has a long tail to the right. Turning moments about the mean sum to 0, but if the deviation from μ is cubed the positive contributions will exceed the negative contributions. For the exponential distribution

$$\mathrm{E}\big[(X - \mu)^3\big] \;=\; \int (x - 1/\lambda)^3 \lambda e^{-\lambda t} dt \;=\; \frac{2}{\lambda^3}.$$

Hence the skewness is 2. The kurtosis is 9, much higher[7] than that of a bell-shaped distribution.

Notation: $T \sim Exp(\lambda)$.

5.3.2 Markov property

5.3.2.1 Poisson process

The exponential distribution has the Markov property. The **Markov property** is that the future depends on the present but not the past (given the present). So, the probability that the time until the next event, T, exceeds some time t measured from "now" is independent of how long "now" is from time 0. The exponential distribution is the only continuous distribution that has this property. Let τ be "now" and suppose the first event has not occurred so $\tau < T$. Then

$$P(\tau + t < T | \tau < T) \;=\; \frac{P((\tau < T) \cap (\tau + t < T))}{P(\tau < T)}$$

$$=\; \frac{P(\tau + t < T)}{P(\tau < T)} \;=\; \frac{e^{-\lambda(\tau+t)}}{e^{-\lambda\tau}}$$

$$=\; e^{-\lambda t} \;=\; P(t < T).$$

The Markov property is a defining characteristic of a Poisson process.

5.3.2.2 Lifetime distribution

An exponential distribution can be used to model lifetimes of components. However, given the Markov property the exponential distribution is only suitable if the failure mode is unrelated to the age of the component. Equivalently, the probability that the component fails in the next time interval, δt say, is independent of how long it has been working for. The proof of this claim follows from the definition of the exponential distribution. Suppose a component has been working for time τ, is working, and that the time until failure, T,

---

[7]The expression (kurtosis $- 3$) is defined as excess kurtosis and equals 6 for the exponential distribution.

has an $Exp(\lambda)$ distribution.

$$P(T < \tau + \delta\tau | \tau < T) = \frac{P((T < \tau + \delta\tau) \cap (\tau < T))}{P(\tau < T)}$$

$$= \int_{\tau}^{\tau + \delta t} \lambda e^{-\lambda\theta} \delta\theta / e^{-\lambda\tau}$$

$$\approx \lambda e^{-\lambda\tau} \delta t / e^{-\lambda\tau} = \lambda\delta t$$

which does not depend on τ. The approximation becomes exact as δt tends to 0.

5.3.3 Applications of the exponential distribution

The times between phone calls arriving at an exchange during business hours are often modeled as exponential, and the distribution is commonly used to model the time between arrivals in other queueing situations. The simplest queueing models assume the service time, length of phone call for example, is also exponential.

Example 5.4: Mobile phone calls [exponential distribution]

An engineer receives an average of 11.3 calls to her mobile phone each working day (8 hours). Assume calls follow a Poisson process with a constant mean and that the duration of calls is negligible. What are the mean and median times between calls? What is the probability of no call during a 2 hour period?

- The rate λ is 11.3/8 calls per hour. The mean time between calls is $8/11.3 = 0.71$ hours. The median time between calls is $\ln(2) \times 0.71 = 0.49$ hours.

- The probability of no call during a 2 hour period is $\exp^{-2 \times 11.3/8} = 0.059$.

Example 5.5: Computer networks [exponential distribution in queueing model]

- The average request rate from a university's browsers to a particular origin server on the Public Internet is 1.5 requests/sec (data from the University of New South Wales). If requests are well modeled as a Poisson process, the times between requests, T, have an exponential distribution with rate $\lambda = 1.5$ requests/sec. The mean time between requests is $1/\lambda = 0.667$ sec, and the median time between requests is $\ln(2)/\lambda = 0.462$ sec. The probability that the time between requests is less than 0.1 sec is

$$P(T < 0.1) = 1 - e^{-1.5 \times 0.1} = 0.139.$$

- Suppose that the mean object size, L, is 900 000 bits and that the access link can handle traffic at a rate $R = 1.5$ Mega bits/sec (Mbps), and that the mean amount of time it takes from when the router on the Internet side of the access link forwards a HTTP request until it receives the response (mean Internet delay) is 2 sec. The mean total response time is the sum of the mean Internet delay and the mean access delay. The mean access delay is defined as $A/(1 - AB)$ where A is the mean time required to send an object over the access link and B is the mean arrival rate of objects to the access link[8] The product AB is referred to as the mean traffic intensity. The time to transmit an object of size L over a link of rate R is L/R, so the average time is $A = 900000/150000 = 0.6$ sec. The mean traffic intensity on the link is $AB = (0.6$ msec/request$)(1.5$ requests/sec$) = 0.9$. Thus, the average access delay is $(0.6$ sec$)/(1 - 0.9) = 6$ sec. The total average response time is therefore 6 sec + 2 sec = 8 sec.

- Now suppose a cache is installed in the institutional LAN, and that the hit rate is 0.4. The traffic intensity on the access link is reduced by 40%, since 40% of the requests are satisfied within the university network. Thus, the arrival rate of the objects to the link also changes since only 60% of the objects need to be fetched from the origin servers, the rest being obtained from the cache. So, $B = 1.5 \times 0.6 = 0.9$ requests/sec and the average access delay is $(0.6$ sec$)/[1 - (0.6)(0.9)] = 1.3$ sec. The response time is approximately zero if the request is satisfied by the cache (which happens with probability 0.4) and the average response time is 1.3 sec + 2 sec = 3.3 sec for cache misses. The average response time for all requests is $0 \times 0.4 + 3.3 \times 0.6 = 1.96$ sec.

The exponential distribution is sometimes used to model the lifetimes of electronic or electrical components, especially under harsh test conditions designed to initiate early failure (accelerated life testing). According to the exponential model, the time until failure is independent of how long the component has been on test, and the failure mode is spontaneous rather than due to wear.

Example 5.6: Computer suite [exponential distribution]

The lifetimes of PCs in a computer room in a sorority house are assumed to have an exponential distribution with a mean of 5 years. At the beginning of a semester there are 20 machines in good working order. Six months later 2 have failed. What is the probability of 2 or more failures in six months if lifetimes are independent? The probability that a PC fails in 6 months is

```
> p=pexp(.3,rate=1/5)
> print(p)
[1] 0.05823547
```

If failures are independent we have a binomial distribution with 20 trials and a probability of success, defined as a PC failing, of 0.0582. The probability of 2 or more successes is

```
> 1-pbinom(1,20,p)
[1] 0.3263096
```

[8]We assume that the HTTP request messages are negligibly small and thus create no traffic on the network or the access link.

The probability is not remarkably small and is quite consistent with a mean lifetime of 5 years.

Example 5.7: Alpha particle emissions [exponential distribution]

Alpha particle emissions from independent sources are well modeled as a Poisson process as shown in Figure 5.7. (10 emissions shown in upper frame). The number of emissions per 500 time units, for example, has a Poisson distribution (foot of upper frame). The times between emissions have an exponential distribution (lower frame).

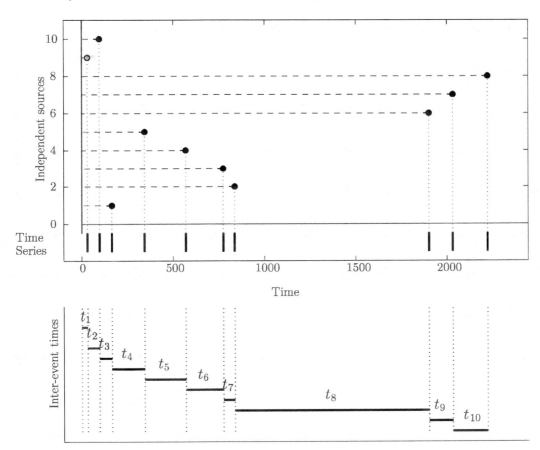

FIGURE 5.7: Alpha particle emissions from independent sources.

5.3.4   Random deviates from an exponential distribution

The general result of Section 5.2.4 is straightforward to implement in the case of an exponential distribution

$$F(t) \;=\; 1 - e^{-\lambda t} \;=\; u \quad\Longrightarrow\quad t \;=\; \frac{\ln(1-u)}{\lambda} \;=\; F^{-1}(u),$$

where u is a random deviate from $[0, 1]$.

This is the principle that R uses to provide pseudo-random deviates from $Exp(\lambda)$. The following R code gives 12 pseudo random deviates from $Exp(2)$.

```
> set.seed(1)
> x=rexp(12,2) ; print(x)
 [1] 0.37759092 0.59082139 0.07285336 0.06989763 0.21803431 1.44748427
 [7] 0.61478103 0.26984142 0.47828375 0.07352300 0.69536756 0.38101493
```

5.3.5   Fitting an exponential distribution

The exponential distribution has a single parameter which can be taken as either its mean $(1/\lambda)$ or its rate λ. The parameter can be estimated from the sample mean. However, the exponential distribution has a very distinctive shape and special properties, and we should ascertain whether or not it is a suitable model for an application before fitting it.

Example 5.8:   Insulators [exponential distribution]

The following data in Table 5.1 ([Kalkanis and Rosso, 1989]) show minutes to failure of mylar-polyurethane laminated DC HV insulating structure tested at five different voltage stress levels.

TABLE 5.1: Minutes to failure of mylar-polyurethane laminated insulation at five voltage levels.

361.4 kV/mm	219.0 kV/mm	157.1 kV/mm	122.4 kV/mm	100.3 kV/mm
0.10	15	49	188	606
0.33	16	99	297	1012
0.50	36	155	405	2520
0.50	50	180	744	2610
0.90	55	291	1218	3988
1.00	95	447	1340	4100
1.55	122	510	1715	5025
1.65	129	600	3382	6842
2.10	625	1 656		
4.00	700	1 721		

How can we assess whether an exponential distribution is a plausible model from a sample of size 10? A quantile-quantile plot (qq-plot) is the best we can do. Sort the data into ascending order, the **order statistics** (Definition 3.5), and plot them against the expected values of the corresponding order statistics for an exponential distribution. If the exponential distribution is plausible the plotted points should be scattered about a straight line. Clear curvature in the plot indicates that the exponential distribution is not a convincing model.

To understand the concept of expected values of the order statistics, imagine taking one million samples of size 10 from the exponential distribution. Sort each sample into ascending order and average the one million smallest values, average the one million 2nd smallest, and so on up to averaging the one million 10th smallest (that is largest) values. The following R code simulates this process for the exponential distribution with mean 1.

```
> mat = replicate(n = 1e6,expr = {sort(rexp(n = 10))})
> apply(mat,1,mean)
 [1] 0.1001165 0.2113083 0.3360231 0.4789103 0.6454660 0.8452678
 [7] 1.0957658 1.4285457 1.9285258 2.9284277
```

However, the following approximation is more convenient, and can be used for any continuous distribution. Denote the ith order statistic, as a random variable, by $X_{i:n}$. Then

$$F\left(\mathrm{E}[X_{i:n}]\right) \approx \frac{i}{n+1} \Rightarrow \mathrm{E}[X_{i:n}] = F^{-1}\left(\frac{i}{n+i}\right).$$

This is intuitively reasonable, the probability of being less than the expected value of the ith smallest order statistics out of n being approximated as $\frac{i}{n+1}$. The use of $n+1$ as the denominator gives a probability of $\frac{1}{n+1}$ of being less than the expected value of the smallest order statistics, and the same probability, $1 - \frac{n}{n+1} = \frac{1}{n}$ of being greater than the expected value of the largest order statistics.

Generally, the parameter λ of the exponential cdf will not be known, but this is a scaling parameter and does not affect the relative scatter of the points. If $X \sim Exp(\lambda)$, then division by the mean gives $\lambda X \sim Exp(1)$.

Define $Y = \lambda X$, so $Y \sim Exp(1)$. Then $\mathrm{E}[\lambda X_{i:n}] = \mathrm{E}[Y_{i:n}]$ and equivalently $\mathrm{E}[X_{i:n}] = \mathrm{E}\left[\frac{1}{\lambda}Y_{i:n}\right]$. If the data are from an exponential distribution, plotting $x_{i:n}$ against $\mathrm{E}[Y_{i:n}]$ will give points scattered about a straight line with gradient $\frac{1}{\lambda}$. Moreover, the parameter λ can be estimated by the reciprocal of a line drawn through the plotted points. However, the method of moments estimator $\widehat{\lambda}$, given by:

$$\bar{x} = \frac{1}{\lambda} \Rightarrow \widehat{\lambda} = \frac{1}{\bar{x}}$$

is more reliable than the graphical estimator (Exercise 8.21).

In the case of an exponential distribution with $\lambda = 1$:

$$F(x) = 1 - e^{-x} = p$$

$$x = -\ln(1-p)$$

$$\Rightarrow F^{-1}(p) = -\ln(1-p).$$

So, the approximation for the expected value of the ith order statistics is

$$E[Y_{i:n}] \approx -\ln\left(1 - \frac{i}{n+1}\right)$$

$$= -\ln\left(\frac{n+1-i}{n+1}\right)$$

A plot of the approximate quantiles of $Exp(1)$ against the more precise values obtained by the simulation is shown in the upper left frame of Figure 5.8. The approximation is adequate given the high variability of the greatest value in samples of size 10 from an exponential distribution. The qq-plots for the five samples of insulator are shown in the remaining frames.

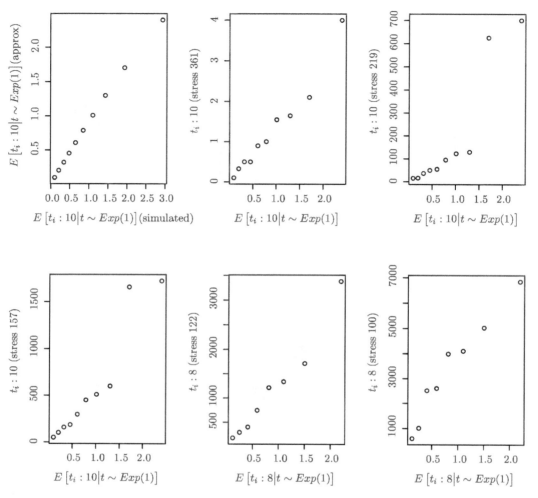

FIGURE 5.8: Plot of the approximate quantiles of $Exp(1)$ against the more precise values obtained by the simulation (upper left). The qq-plots of failure times for the five samples of insulator (Table 5.1) are shown in the remaining frames.

```
> mylar.dat=read.table("Mylar_insulation.txt",header=TRUE)
> print(head(mylar.dat));print(tail(mylar.dat))
   life stress
1 0.10       5
2 0.33       5
3 0.50       5
4 0.50       5
5 0.90       5
6 1.00       5
    life stress
41 2520       1
42 2610       1
43 3988       1
44 4100       1
```

```
45 5025      1
46 6842      1
> attach(mylar.dat)
> t_361=life[stress==5];t_219=life[stress==4];t_157=life[stress==3]
> t_122=life[stress==2];t_100=life[stress==1]
> n=10; i=1:n; p=i/(n+1)
> q10=-log(1-p)
> print(round(q10,2))
 [1] 0.10 0.20 0.32 0.45 0.61 0.79 1.01 1.30 1.70 2.40
> sim_q10=c(.10,.21,.34,.48,.65,.85,1.10,1.43,1.93,2.93)
> q8=-log(1-c(1:8)/9)
> par(mfrow=c(2,3))
> plot(sim_q10,q10,xlab="E[t_i:10|t~Exp(1)](simulated)",
  ylab="E[t_i:10|t~Exp(1)](approx)")
> xl10=c("E[t_i:10 | t~Exp(1)]")
> plot(q10,t_361,xlab=xl10,ylab="t_i:10   (stress 361)")
> plot(q10,t_219,xlab=xl10,ylab="t_i:10   (stress 219)")
> plot(q10,t_157,xlab=xl10,ylab="t_i:10   (stress 157)")
> xl8=c("E[t_i:8 | t~Exp(1)]")
> plot(q8,t_122,xlab=xl8,ylab="t_i:8   (stress 122)")
> plot(q8,t_100,xlab=xl8,ylab="t_i:8   (stress 100)")
```

The plots for voltage stresses of 361 and 122 are quite consistent with samples of size 10 from an exponential distribution. The plots for voltage stresses 219 and 157 show outlying points, but these are not unlikely to arise in such small samples from exponential distributions. In contrast, the plot for the voltage stress of 100 shows possible curvature which would indicate a CV less than 1. The exponential model for lifetimes of components has a unique feature that components do not wear and the probability that a component fails in the next minute is independent of how long it has been on test. If component lifetimes are well modeled by an exponential distribution then we should look for a failure mode that does not depend on wear. The CV of an exponential distribution is 1. Generally, components do tend to wear out and this tends to reduce the CV because lifetimes will tend to be relatively close to the expected lifetime. However, this is more apparent for mechanical components and solid state devices seem remarkably durable. In this case it seems that the failure mode at the lowest voltage, 100, is related to wear. One possible explanation for outlying values at 219 and 157 is that the test pieces of insulation are drawn from two populations, one of which is on average more durable (Exercise 5.11). The sample CV gives another indication of the suitability of an exponential distribution (Table 5.2).

TABLE 5.2: Statistics of lifetimes of mylar-polyurethane laminated insulation tested at different voltage levels.

Voltage	Mean	sd	$\widehat{CV}$
100	3 337.88	2 006.67	0.60
122	1 161.12	1 049.74	0.90
157	570.80	616.38	1.08
219	184.30	255.78	1.39
361	1.26	1.16	0.91

The sample CV for 100 is well below 1, again suggesting a failure mode that involves wear. The CV for 219 is substantially above 1 although the sample size is small and we can investigate the distribution of CVs in samples of size 10 from an exponential distribution by simulation (Exercise 5.10).

5.4   Normal (Gaussian) distribution

We consider a variable with a mean value μ. It is convenient to describe variation about μ as errors. Suppose the errors are the sum of a large number of small components of error. Assume that these small components of error have the same small magnitude[9] and are equally likely to be negative or positive. Then the errors have a normal distribution with mean 0 and the variable itself has a normal distribution with mean μ.

The normal distribution has a distinctive bell-shape, and many sets of data have histograms that are approximately bell shaped. For example, volumes of paint dispensed into cans with nominal contents of 1 liter. The normal distribution is central to statistical theory because of its genesis as the sum of random variables.

5.4.1   Definition of a normal distribution

Let the random variable X represent the measurement of some quantity which has a mean value in the population of μ. If the deviation of an individual X from μ is due to the sum of a large number of independent micro-deviations with mean 0, then X will be very well approximated by a normal distribution[10]. The normal distribution is often referred to as the Gaussian distribution[11], especially in an engineering context. The pdf of the normal distribution is commonly described as a bell curve. The pdf of the normal distribution is

$$f(x) \;\; = \;\; \frac{1}{\sqrt{2\pi}\sigma} e^{-\frac{1}{2}\left(\frac{x-\mu}{\sigma}\right)^2}, \qquad -\infty < x < \infty.$$

The parameters μ and σ are the mean and standard deviation of the distribution. Although the support of the distribution is the entire real line the area under the pdf beyond 6σ from μ is negligible (around 10^{-9}). It follows from the symmetry of the pdf that the median and mode are also μ, and that the skewness is 0. The kurtosis [12] of the normal distribution is 3. The cdf is defined by

$$F(x) \;\; = \;\; \int_{-\infty}^{x} \frac{1}{\sqrt{2\pi}\sigma} e^{-\frac{1}{2}\left(\frac{\xi-\mu}{\sigma}\right)^2} d\xi,$$

but it does not exist as a formula in terms of elementary functions and has to be calculated numerically. If X has a normal distribution with mean μ and standard deviation σ we

---

[9]The small components of error could have different magnitudes as long as no small component of error dominates.

[10]Abraham De Moivre (1667-1754) obtained the normal distribution as the limit of the distribution of a binomial random variable X, as $n \to \infty$, with X scaled as $\frac{X-np}{\sqrt{(np(1-p))}}$. Thomas Simpson drew attention to the physical significance of the result in 1757.

[11]After Carl Friedrich Gauss (1777-1855) who used the distribution in his *Theoria motus Corporum Celestium* 1809.

[12] **Excess kurtosis** is defined relative to a normal distribution as (kurtosis $- 3$).

write[13]

$$X \sim N(\mu, \sigma^2).$$

However, since we can scale any normal random variable X with mean μ and standard deviation σ to a normal variable Z with mean 0 and standard deviation 1 we can focus on the latter.

5.4.2 The standard normal distribution $Z \sim N(0,1)$

The **standard normal distribution** is a normal distribution with mean 0 and standard deviation 1. The pdf of a standard normal random variable, customarily denoted by Z is:

$$\phi(z) = \frac{1}{\sqrt{2\pi}} e^{-\frac{1}{2}z^2}, \qquad -\infty < z < \infty$$

and the cdf is

$$\Phi(z) = \int_{-\infty}^{z} \frac{1}{\sqrt{2\pi}} e^{-\frac{1}{2}\xi^2} d\xi.$$

The pdf, $\phi(z)$, is shown above its cdf, $\Phi(z)$, in Figure 5.9. The cdf, $\Phi(z)$, occurs in so many applications[14] that it is a standard mathematical function[15]. $\Phi(z)$ is described as an S-shaped curve and takes a real number to a probability. Although $\Phi(z)$ is defined for all real numbers, values less than -6 get taken to almost 0 and values above 6 get taken to almost 1. It follows from the symmetry about 0 of the standard normal distribution that $\Phi(0) = 0.5$, and that $\Phi(z)$ has a rotational symmetry of a half turn about $(0, 0.5)$:

$$\Phi(z) = 1 - \Phi(-z).$$

(Exercise-refbeginnormal). Also, since the area under a standard normal pdf between -1 and 1 is around 0.67, $\Phi(-1) \approx (1 - 0.67)/2 \approx 0.17$ and $\Phi(1) \approx 0.83$. Furthermore, since 0.95 of the area under a standard normal pdf lies between -2 and 2, $\Phi(-2) \approx 0.025$ and $\Phi(2) \approx 0.975$. More precisely, $\Phi(1.96) = 0.975$. The inverse function, $\Phi^{-1}()$, takes a probability to a real number. In terms of the quantiles of a standard normal distribution

$$\Phi(z_p) = 1 - p \Leftrightarrow z_p = \Phi^{-1}(1 - p).$$

A much used value is $z_{.025} = \Phi^{=1}(0.975) = 1.96$, and another common value that arises in quality control applications is $z_{.001} = \Phi^{-1}(0.999) = 3.09$.

If $X \sim N(\mu, \sigma^2)$ then the random variable X with its mean subtracted,

$$(X - \mu) \sim N(0, \sigma^2)$$

and if this random variable is scaled[16] by dividing by its standard deviation then

$$\frac{X - \mu}{\sigma} \sim N(0, 1).$$

---

[13]There are other conventions but specifying the variance rather than the standard deviation is slightly more convenient in later chapters.

[14]Including the heat equation in physics.

[15]$\Phi(\cdot)$ is available on scientific calculators as well as in Excel and other high level programming languages. In R, $\Phi(z)$ is pnorm(z).

[16]In general, a random variable is **standardized** by subtracting its mean and dividing by its standard deviation.

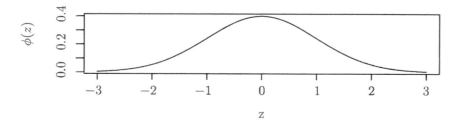

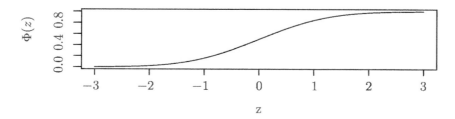

FIGURE 5.9: The pdf (upper) and cdf (lower) of the standard normal distribution.

It is conventional to use Z for the standard normal variable so we can write

$$\frac{X - \mu}{\sigma} \;=\; Z$$

and we say that X has been **standardized**.

The pdf and cdf of any normal distribution look the same if the scale is in a unit of standard deviations from the mean (0 becomes μ, 1 becomes $\mu + 1\sigma$, and so on). Three normal pdfs with different means or standard deviations are plotted on the same scale in Figure 5.10. The bell shapes are centered on their means, and are more spread out as the standard deviation increases. The peak of the pdf becomes lower as the standard deviation increases so that the area under the pdf remains at 1. The normal pdf has points of inflection when the argument is $\mu \pm \sigma$ (Exercise 5.18).

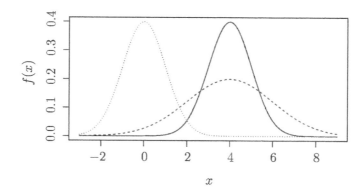

FIGURE 5.10: The pdfs of $N(0,1)$ (dotted line), $N(4,1)$ (solid line), and $N(4,2)$ (dashed line).

Typical questions involving the normal distribution are of the form: if $X \sim N(\mu, \sigma^2)$ find

$P(X < a)$, for some constant a, or find the value of b such that $P(X < b)$ equals some given probability. These can be answered using pnorm() or qnorm() with the mean and standard deviation specified. Nevertheless, you need to be familiar with the concept of standardizing a normal variable because:

- standardizing emphasizes that we are considering standard deviations from the mean;

- for some other questions it is necessary to work with the standardized variable;

- standardized variables are used in the construction of confidence intervals; (Chapter 7 onwards);

- you may only have access to, values for $\Phi(z)$.

Example 5.9: [Standard normal distribution]

Throughout this example $Z \sim N(0, 1)$.

1. Find $P(Z < -1.8)$. Using the value for $\Phi(1.8)$,

$$P(Z < -1.8) \quad = \quad \Phi(-1.8) \quad = \quad 1 - \Phi(1.8) \quad = \quad 1 - 0.9641 \quad = \quad 0.0359$$

Alternatively using R

```
> pnorm(-1.8)
[1] 0.03593032
```

2. We use R to calculate the areas of a standard normal distribution between -1 and 1, -2 and 2, and -3 and 3.

```
> z=c(1,2,3)
> area=round(1-2*(pnorm(-z)),3)
> print(cbind(z,area))
  z area
[1,] 1 0.683
[2,] 2 0.954
[3,] 3 0.997
```

The areas are approximately two thirds, 95% and all but 3 parts in a thousand respectively.

3. If $X \sim N(100, 20^2)$ find $P(X < 64)$. We standardize X and carry out the same subtraction of the mean and division by the standard deviation (which is positive) on the right hand side of the inequality to maintain the inequality.

$$
\begin{aligned}
P(X < 64) &= P\left(\frac{X - 100}{20} < \frac{64 - 100}{20}\right) \\
&= P(Z < -1.8) \quad = \quad \Phi(-1.8) \quad = \quad 0.0359.
\end{aligned}
$$

Alternatively, using R

```
> pnorm(64,mean=100,sd=20)
[1] 0.03593032
```

4. We use the notation z_α for $P(z_\alpha < Z) = \alpha$. If $\alpha = 0.025$, then $\Phi(z_{.025}) = 0.975$.

```
> qnorm(0.975)
[1] 1.959964
```

The upper $0.20, 0.10, 0.05, 0.025, 0.01, 0.001$ quantiles of the standard normal distribution are commonly used.

```
> p=c(0.20, 0.10, 0.05,0.025,0.01,0.001)
> z_p=round(qnorm(1-p),2)
> print(cbind(p,z_p))
    p    z_p
[1,] 0.200 0.84
[2,] 0.100 1.28
[3,] 0.050 1.64
[4,] 0.025 1.96
[5,] 0.010 2.33
[6,] 0.001 3.09
```

Notice that 95% of the distribution is between -1.96 and 1.96, which is approximately plus or minus two. Also, from the symmetry, about 0, of the standard normal distribution the lower quantiles are the negative of the corresponding upper quantiles. This can be shown formally as follows. Let ε be a number between 0 and 1. By definition both $P(Z < z_{1-\varepsilon}) = \varepsilon$ and $P(z_\varepsilon < Z) = \varepsilon$. Multiplying both sides of the second inequality by -1 and reversing the sign of the inequality gives $P(-Z < -z_\varepsilon) = \varepsilon$. But $-Z$ has the same probability distribution as Z so $P(Z < -z_\varepsilon) = \varepsilon$. Therefore $z_{1-\varepsilon} = -z_\varepsilon$.

5. If $X \sim N(100, 20^2)$ find b such that $P(X < b) = 0.98$.

$$P(X < b) \quad = \quad P\left(\frac{X - 100}{20} < \frac{b - 100}{20}\right) \quad = \quad P\left(Z < \frac{b - 100}{20}\right) \quad = \quad 0.98.$$

Since $\Phi^{-1}(0.98) = 2.054$, we require

$$\frac{b - 100}{20} \quad = \quad 2.054 \Rightarrow b \quad = \quad 100 + 2.054 \times 20 \quad = \quad 141.$$

Direct use of the **qnorm** function in R gives

```
> qnorm(.98,mean=100,sd=20)
[1] 141.075
```

6. Find σ such that if $X \sim N(100, \sigma^2)$ then $P(X < 80) = 0.01$. In this case we do need to standardize because the standard deviation has to be found. We require σ such that

$$P(X < 80) \quad = \quad P\left(\frac{X - 100}{\sigma} < \frac{80 - 100}{\sigma}\right) \quad = \quad P\left(Z < \frac{-20}{\sigma}\right) \quad = \quad 0.01$$

This reduces to

$$-\frac{20}{\sigma} \quad = \quad -z_{0.01}$$

Using R for the arithmetic

```
> sigma=-20/(-qnorm(.99))
> print(sigma)
[1] 8.597166
```

and $\sigma = 8.6$.

5.4.3   Applications of the normal distribution

In manufacturing industry many variables that are specified to lie within some specified interval, such as length or volume or capacitance, are assumed to be normally distributed. The assumption of a normal distribution is usually reasonable if deviations from the nominal value (typically the mid-point of the specification) are the sum of many small errors. Nevertheless, extrapolation beyond three standard deviations is placing a lot of trust in the model.

Example 5.10: Plates for ship building [meeting specification]

Plates in a shipyard have a specified length of 6 m and the specification is that cut plates should be within 2 mm of this length. Under the current process the lengths X are normally distributed with mean 6001 mm and standard deviation 1.2 mm. We calculate the proportion within specification. We can either scale to standard normal:

$$
\begin{aligned}
P(5998 < X < 6002) &= P\left(\frac{5998 - 6001}{1.2} < \frac{X - 6001}{1.2} < \frac{6002 - 6001}{1.2}\right) \\
&= P(-2.5 < Z < 0.833) \\
&= \Phi(0.833) - \Phi(-2.5) \\
&= 0.7977 - 0.0062 \ = \ 0.7915,
\end{aligned}
$$

using **pnorm(z)** in R for $\Phi(z)$, or rather more easily

```
> L=5998
> U=6002
>   pnorm(U,mu,sigma)-pnorm(L,mu,sigma)
[1] 0.791462
```

The proportion of plates outside the specification is $1 - 0.7915 = 0.21$, which is far too high. It can be reduced to a minimum by adjusting the mean to the middle of the specification (Figure 5.11).

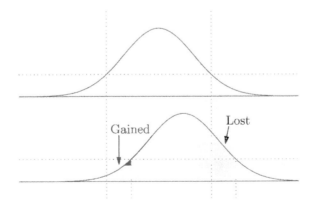

FIGURE 5.11: Proportion of items within specification maximized when mean set at centre of specification.

The proportion outside the specification is now

```
> 2*(1-pnorm(6002,6000,1.2))
[1] 0.0955807
```

where we have taken twice the proportion above the upper specification limit. This is an improvement but still too high. A common industrial requirement is that at most 1 in 1 000 items is outside specification. This is equivalent to an area of 0.0005 above the upper specification limit. The only way to achieve this tail area is to lower the standard deviation. We now determine a value of σ that will satisfy the criterion. In general, suppose $X \sim N(\mu, \sigma^2)$.

$$\Phi(z) \;=\; 0.9995 \;\leftrightarrow\; z \;=\; \Phi^{-1}(0.9995) \;=\; 3.29$$

using qnorm(0.9995) in R. It follows that

$$P(Z > 3.29) \;=\; 0.0005 \;\leftrightarrow\; P\left(\frac{X-\mu}{\sigma} > 3.29\right) \;\leftrightarrow\; P(X > \mu + 3.29\sigma) \;=\; 0.0005.$$

In this case we require

$$3.29\sigma \;=\; 6002 - 6000 \;=\; 2 \;\leftrightarrow\; \sigma \;=\; 0.608.$$

The standard deviation needs to be reduced to 0.61

Example 5.11: GPS receiver [accuracy depends on standard deviation]

According to GPS.GOV (updated 6 December 2016) data from the FAA show its high-quality GPS SPS receivers attaining better than 2.168 meter horizontal accuracy 95% of the time. What are the implied mean and standard deviation of the errors if they are normally distributed? If errors are normally distributed find the probabilities that a measurement is within: 1 meter, 1 foot (0.3018 m) and 1 inch (0.0254 m) of the actual value.

The implied mean is 0, and since 95% of a normal distribution is within two standard deviations of the mean, the implied standard deviation is around 1 meter. The probability that a measurement, X, is within a of the mean can be found in several ways, such as as $1 - 2P(X < -a)$ or $P(X < a) = P(X < a)$. We obtain precise probabilities using R.

```
> sigma=2.168/qnorm(.975)
> print(round(sigma,3))
[1] 1.106
> a=c(1,0.3018,0.0254)
> p=pnorm(a,mean=0,sd=sigma)-pnorm(-a,mean=0,sd=sigma)
> print(round(p,3))
[1] 0.634 0.215 0.018
```

So, to 2 decimal places, the implied standard deviation is 1.11 and the probabilities of being within 1 meter, 1 foot, and 1 inch of the actual value are 0.63, 0.22, and 0.02 respectively. The probability of being within 1 mm of the actual value is quite small (notice that the "mean=" and "sd=" can be omitted in the *pnorm()* function).

```
> p=1-2*pnorm(-0.001,0,sigma);p
[1] 0.0007213214
```

Example 5.12: Capacitors [proportion outside claimed tolerance]

A process has been set up to manufacture polypropylene capacitors with a 25 micro-Farad capacitance. The process mean is 25.08 and the standard deviation is 0.74. The capacitors are to be marketed with a tolerance of $\pm 10\%$. Assume that capacitances are normally distributed and calculate the proportion of production, in parts per million (ppm), that will lie outside the tolerance range. What would this reduce to if the process mean is adjusted to 25.00?

```
> mu=25.08;sigma=0.78
> L=25-0.10*25;U=25+0.10*25
> p=pnorm(L,mu,sigma)+(1-pnorm(U,mu,sigma))
> print(p*10^6)
[1] 1429.6
> print(pnorm(L,25,sigma)+(1-pnorm(U,25,sigma))*10^6)
[1] 675.0123
```

The process would produce 1430 ppm outside the tolerance, but this could be reduced to 675 ppm by resetting the mean to 25.00. If the process mean is reset, what does the standard deviation need to be reduced to for only 100 ppm to be outside the tolerance interval?

```
> p=100*10^(-6)/2;z=qnorm(p)
> newsigma=(25-L)/(-z);print(newsigma)
[1] 0.6425758
```

The standard deviation would need to be reduced to 0.64.

Example 5.13: Safes [probability of exceeding a critical value]

Safes are tested in a specially designed room that is heated to 1200 degrees Fahrenheit, which is the highest temperature in the average house fire. Once the room is heated, the safe is placed inside the room and sensors inside the safe record the internal temperature of the safe for the 1.5 hour duration of the test. Tests on a new design of safe had a mean maximum internal temperature of 346 degrees, and the standard deviation of maximum internal temperature was 2.7 degrees. What is the probability that the maximum internal temperature for a randomly selected safe exceeds 350 degrees Fahrenheit (the temperature at which most standard paper will char and combust), during this test if maximum internal temperatures are normally distributed?

```
> 1-pnorm(350,346,2.7)
[1] 0.06923916
```

The probability is about 0.07.

Example 5.14: Electric car [probability of being less than a critical value]

A design of electric car and battery set has a range of 485 km for driving outside city limits with an initial full charge. The standard deviation associated with this figure is around 10%. If you own such a car, what is the probability that you will run out of charge on a 400 km trip if the range is normally distributed and you start with a full charge?

```
> mu=485;sigma=0.1*mu
> pnorm(400,mu,sigma)
[1] 0.03983729
```

The probability is around 0.04, small but not negligible.

Example 5.15: Six Sigma [very high conformance with specification]

The name of the Six Sigma business improvement strategy, used by Motorola for example, is a reference to a process with a standard deviation that is one twelfth of the width of a specification with both a lower L and upper U limit. Then $U - L = 12\sigma$, and if the process mean is 1σ from the middle of the specification and the variable is precisely normally distributed about 0.3 parts per million (ppm) will lie outside specification ($\Phi(-5) = 2.97 \times 10^{-7}$). This may seem excessive precision for a manufacturing process, but might be expected of an ATM machine and would be unacceptable for safety critical processes. Moreover, if a manufacturing process consists of many stages, a few ppm out of specification at each stage will accumulate.

Example 5.16: Smoothing [approximating general pdfs using normal pdfs]

Kernel smoothing is a technique for producing a smooth curve, rather than a histogram, as an estimate, $\widehat{f}$, of a probability density $f(x)$ from a random sample, $\{x_i\}$ for $i = 1, \ldots, n$. The curve in Figure 3.23 fitted to the waiting times for Old Faithful was obtained by: superimposing a Gaussian distribution with a relatively small variance on each datum; summing the ordinates at a sufficiently fine scale to plot as a smooth curve, and then scaling the ordinates to give an area of 1. The equation is

$$\widehat{f}(x) \;=\; \frac{1}{nh} \sum_{i=1}^{n} K\left(\frac{x - x_i}{h}\right),$$

where $K()$ is the **kernel** and h is the **bandwidth**. The Gaussian, or normal, kernel is the standard normal pdf. $K() = \phi()$. You are asked to explain why the area under $\widehat{f}(x)$ is 1 in Exercise 5.35. The curve becomes smoother as h is increased, but detail is lost. A default value for h is $1.06sn^{-0.2}$, where s is the standard deviation of the data set, but you can try other values for the bandwidth.

5.4.4 Random numbers from a normal distribution

There are many algorithms for generating pseudo-random numbers from a normal distribution, and two are given in Exercise 5.45 and Exercise 5.46. The R syntax for random standard normal deviates is

```
> n=180
> set.seed(1)
> x=rnorm(n)
> head(x)
[1] -0.6264538  0.1836433 -0.8356286  1.5952808  0.3295078 -0.8204684
> tail(x)
[1] -0.33400084 -0.03472603  0.78763961  2.07524501  1.02739244  1.20790840
```

Random deviates from a normal distribution with mean μ and standard deviation σ can be obtained by scaling standard normal deviates or by specifying the parameters in the function call.

```
> mu=61
> sigma=4
> x1=61+sigma*x
> head(x1)
[1] 58.49418 61.73457 57.65749 67.38112 62.31803 57.71813
> set.seed(1)
> x2=rnorm(n,mu,sigma)
> head(x2)
[1] 58.49418 61.73457 57.65749 67.38112 62.31803 57.71813
```

5.4.5 Fitting a normal distribution

It is straightforward to fit a normal distribution. The mean μ and standard deviation σ are estimated by the sample mean $\bar{x}$ and sample standard deviation s respectively. But, it does not necessarily follow that the normal distribution is a good model for the data. Probability plots allow us to assess whether a chosen distribution is a suitable model for the data.

5.5 Probability plots

In Section 5.3.5 we drew quantile-quantile plots for an exponential distribution. The same principle can be used for any probability distribution. Suppose we have a random sample $\{x_i\}$, for $i = 1, \ldots, n$, from some distribution. We wish to assess whether it is reasonable to suppose the sample is from some probability distribution with cdf $F(\cdot)$. A probability plot provides a graphical assessment.

Any methods of assessment are limited by the sample size. If the sample size is small, 20 or less, many forms of distribution may seem plausible, If the sample size is very large, thousands, small discrepancies from models for distributions may be apparent, yet these discrepancies are not necessarily of practical importance. It all depends on the application.

Normal Q-Q Plot

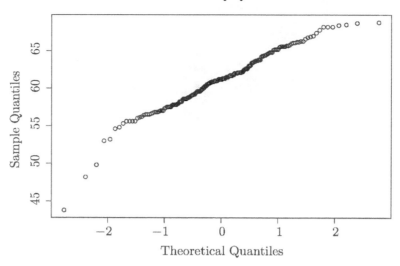

FIGURE 5.12: Normal qq-plot of the cubes dataset.

5.5.1 Quantile-quantile plots

The general procedure is to plot the sample order statistics, $x_{i:n}$, against the expected value of the order statistics from a specified reference probability distribution. The reference distribution is generally **reduced** so for a two-parameter distribution, $Y = a(X - b)$, and Y has a distribution with some specified parameter values, typically 0 for location and 1 for scale. The **quantile-quantile plot** (**qq-plot**) is

$$x_{i:n} \quad \text{against} \quad \mathrm{E}[Y_{i:n}].$$

- A convenient approximation to the expected values of order statistics in the reduced distribution is:

$$F\big(\mathrm{E}[Y_{i:n}]\big) \quad \approx \quad \frac{i}{n+1}.$$

- The standard practice is to plot the variable, here sample order statistics $x_{i:n}$, against the set values $\mathrm{E}[Y_{i:n}]$ as in Figure 5.12.

- The slope, and intercept in the case of a two-parameter distribution, provide rough estimates of the parameters of the distribution the data are drawn from but it is generally preferable to use method of moments estimates.

- There are various strategies for dealing with three-parameter distributions.

5.5.2 Probability plot

A probability plot is essentially identical to a qq-plot, but the display is different. Imagine a cdf drawn on a sheet of rubber and stretch the rubber so that the cdf is a straight line. The probability scale will now be non-linear, except for a cdf of a uniform distribution. The order statistics $x_{i:n}$ are plotted against probability $\frac{i}{n+1}$.

- Because the probability plot is a plot of a cdf it is usual to set the probability scale vertically and to use the horizontal scale for the order statistics. Therefore, the probability plot will be a reflection of the corresponding qq-plot.

- The probability of $\frac{i}{n+1}$ is known as a **plotting position**. The use of this plotting position corresponds to the approximation to the expected value of the order statistic given for the qq-plot. In particular i assigns a probability of $\frac{1}{n+1}$ to being less than the smallest value in the sample and a probability of $1 - \frac{n}{n+1} = \frac{1}{n+1}$ to being greater than the largest value.

- You can obtain a probability scale on a qq-plot by relabelling the $\mathrm{E}[Y_{i:n}]$ with $\frac{i}{n+1}$.

- A probability plot is a more intuitive presentation than a qq-plot but it needs a lot more programming in R. MATLAB offers nice probability plots, using the command `probplot(x)`, and probability graph paper is freely available off the internet.

Example 5.17: Concrete cubes [probability plots]

A normal quantile-quantile plot of the compressive strengths of 180 concrete cubes (Section 3.3.8) is shown in Figure 5.12 (obtained in R with `qqnorm()`).

A normal distribution provides a reasonable fit although there are a few outlying values at the lower end. These outliers correspond to cubes with low strengths. However, the variance of order statistics increases as i moves towards either 1 or n, so outliers can be expected. Simulation could be used to estimate the probability of as extreme or more extreme outliers in random samples of 200 from a normal distribution. A probability plot for the strengths of the cubes, drawn in MATLAB, is shown in Figure 5.13. In this case the strengths are along the horizontal axis. The probabilities associated with the outlying low compressive strengths are higher than expected with a normal distribution.

5.6 Lognormal distribution

One approach to deal with a variable that is not well approximated by a normal distribution is to find same transformation of the variable that is near normal. A common transformation of a non-negative variable is to take the logarithm, in some cases after adding a constant[17].

5.6.1 Definition of a lognormal distribution

A non-negative random variable X has a lognormal distribution if $\ln(X)$ has a normal distribution. A normal distribution arises as the sum of a large number of micro-variations, so the lognormal distribution arises as the product of a large number of positive micro-variations. This gives it physical credibility as a probability distribution for some applications involving non-negative variables. If we write

$$Y = \ln(X), \quad \text{then} \quad X = e^{Y}.$$

---

[17]A constant has to be added if there are 0s in the data set. In other cases it can lead to a better empirical fit.

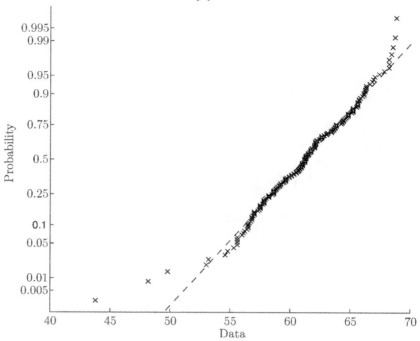

FIGURE 5.13: Probability plot for the cubes dataset.

If $Y \sim N(a, b^2)$ we can obtain the pdf of X by the following argument which can be used quite generally to find the probability distribution of strictly increasing, or strictly decreasing, functions of random variables.

$$F_Y(y) \;=\; \mathrm{P}(Y \le y) \;=\; \mathrm{P}\!\left(e^Y \le e^Y\right) \;=\; \mathrm{P}(X \le x) \;=\; F_X(x)$$

and it follows from the chain rule of calculus that

$$f_X(x) \;=\; \frac{dF_X(x)}{dx} \;=\; \frac{dF_Y(y)}{dy}\frac{dy}{dx}.$$

Now

$$\frac{dF_Y(y)}{dy} \;=\; \frac{d}{dy} \int_{-\infty}^{y} \frac{1}{\sqrt{2\pi}b} \exp\!\left(-\frac{1}{2}\left(\frac{u-a}{b}\right)^2\right)$$

$$=\; \frac{1}{\sqrt{2\pi}b} \exp\!\left(-\frac{1}{2}\left(\frac{y-a}{b}\right)^2\right)$$

and substituting $y = \ln(x)$

$$\frac{dy}{dx} \;=\; \frac{d}{dx}\ln(x) = \frac{1}{x},$$

we obtain the pdf of X as

$$f(x) \;=\; \frac{1}{x\sqrt{2\pi}b}\exp\!\left(-\frac{1}{2}\left(\frac{\ln(x)-a}{b}\right)^2\right).$$

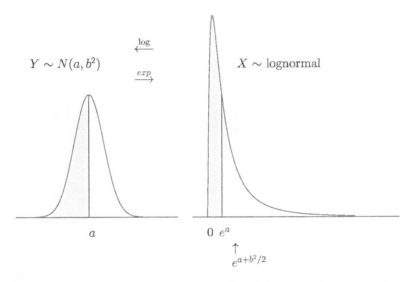

FIGURE 5.14: Normal and lognormal pdf showing the relationship.

The mean of X is most easily obtained as

$$E[X] \;=\; E[e^Y] \;=\; \int e^y \frac{1}{\sqrt{2\pi}b} \exp\left(-\frac{1}{2}\left(\frac{y-a}{b}\right)^2\right) dy \;=\; \exp\left(a + \frac{b^2}{2}\right).$$

The median of the lognormal distribution is $\exp(a)$ because the median of a normal distribution is also its mean and

$$P(Y < a) \;=\; P(\exp(Y) < \exp(a)) \;=\; P(X < \exp(a)) \;=\; 0.5.$$

It follows that the distribution is positively skewed, as shown in Figure 5.14. The variance of a lognormal distribution is

$$\exp\left(2a + b^2\right)\left(\exp\left(b^2\right) - 1\right),$$

so the coefficient of variation is

$$CV \;=\; \sqrt{\left(\exp\left(b^2\right) - 1\right)}$$

and the skewness is

$$\gamma \;=\; \left(\exp\left(b^2\right) + 2\right) \times CV.$$

Both the variance and the skewness, which are positive, are determined by the parameter b. A plot of the pdf and cdf for $a = 0$ and $b = 1$ is shown in Figure 5.15.

```
> y=seq(-3,3,.01)
> x=exp(y)
> f=dlnorm(x)
> F=plnorm(x)
> par(mfrow=c(2,1))
> plot(x,f,type="l",ylab="f(x)")
> plot(x,F,type="l",ylab="F(x)")
```

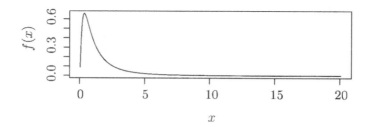

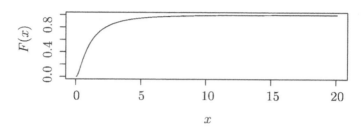

FIGURE 5.15: pdf and cdf of lognormal distribution with $a = 0$ and $b = 1$.

5.6.2   Applications of the lognormal distribution

A lognormal distribution is commonly used to model positive variables that are highly skewed, such as concentrations of rare elements in rock and outturn costs of engineering schemes.

Example 5.18:   Gold grades [three parameter lognormal distribution]

The lognormal distribution is commonly used in the mining industry. The gold grades (Example 3.11) are well modeled as lognormal. In general, a third parameter $0 < L$ is often introduced so that $Y = ln(X + L)$ is $N(a, b^2)$. The effect of taking logarithms is reduced as L increases, and if L is large relative to the range of X, the transformation is close to linear (Exercise 5.26). The introduction of the third parameter detracts from the physical interpretation of the variable being a product of positive micro-variations. Common logarithms, $\log_{10}$, are typically used in place of natural logarithms which introduces a factor of 0.4342945 (Exercise 5.27).

Example 5.19:   Out-turn costs [predicting cost]

The out-turn costs of engineering schemes are reasonably modeled by lognormal distributions. Models for costs are fitted to the logarithms of out-turn costs for past schemes and are then used to predict expected logarithms of costs for new projects. It is important to remember the mean logarithm of cost corresponds to a median cost and that the mean cost is higher. The difference between mean and median will lead to a substantial underestimate of costs if many new projects are being costed. If Y is logarithm of cost then the expected cost is

$$\mathrm{E}[X] \;=\; \exp\left(\mathrm{E}[Y]\right) \times \exp\left(\frac{\mathrm{var}(Y)}{2}\right),$$

provided the costs are well modeled by a lognormal distribution.

5.6.3 Random numbers from lognormal distribution

Random numbers can be obtained by either generating random numbers from a normal distribution and taking exponential or by using the **rlnorm()** function. For example, 5 random numbers from a lognormal distribution with $a = 1$ and $b = 3$ can be obtained by

```
> set.seed(1)
> x=rlnorm(5,1,3) ; print(x)
[1]    0.4150479    4.7158538    0.2215990 325.6562595    7.3047390
```

5.6.4 Fitting a lognormal distribution

The simplest method is to fit a normal distribution to the logarithms of the data. Figure 5.16 shows the histogram and normal quantile plots of the Canadian gold grades (Example 5.18).

```
> gold.dat=read.table("gold_grade.txt",header=TRUE)
> attach(gold.dat)
> x=gold
> y=log(gold)
> par(mfrow=c(2,2))
> hist(x,main="",xlab="gold grade",freq=FALSE)
> hist(x,main="",xlab="ln(gold grade)",freq=FALSE)
> qqnorm(x)
> qqnorm(y)
```

Notice the difference between the mean and the median of the distribution of gold grades. The sample mean of the gold grades is a more reliable estimate of the population mean than that based on the mean and variance of the logarithms of gold grades because the latter depends on the assumption of a lognormal distribution.

```
> print(mean(x))
[1] 0.7921547
> print(exp(mean(y)+var(y)/2))
[1] 0.7921111
> print(median(x))
[1] 0.65
> print(exp(mean(y)))
[1] 0.6336289
```

One method for setting a value for a third parameter L is described in Exercise 5.29.

5.7 Gamma distribution

A sum of independent exponential random variables has a gamma distribution. The distribution approaches a normal distribution as the number of variables in the sum increases. As a consequence the distribution is quite versatile for fitting positively skewed data.

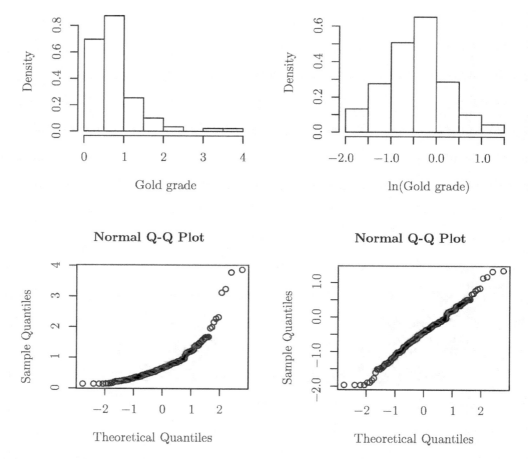

FIGURE 5.16: Histogram and normal qq-plots for gold grade (left) and log gold grade (right).

5.7.1   Definition of a gamma distribution

Consider a Poisson process with rate λ. Let the random variable T be the time until the kth event occurs. The cdf of T is

$$F(t) \;=\; 1 - \sum_{i=0}^{k-1} \frac{(\lambda t)^i \mathrm{e}^{-\lambda t}}{i!}, \qquad 0 \le t$$

and the pdf follows by differentiation

$$f(t) \;=\; \frac{\lambda^k t^{k-1} \mathrm{e}^{-\lambda t}}{\Gamma(k)}.$$

The random variable T is the sum of k independent exponential random variables with the same rate λ and its probability distribution is referred to as the **Erlang distribution**[18]. The mean and variance of T are $\frac{k}{\lambda}$ and $\frac{k}{\lambda^2}$ respectively [19] The skewness is $\frac{2}{\sqrt{k}}$ and the kurtosis

---

[18]Named after the Danish telecommunications engineer (1975-1932).

[19]These results are a most easily obtained by using the fact that the mean of a sum of random variables is the sum of the means and, if the variables are independent, the variance of the sum is the sum of the variances.

is $3 + \frac{6}{k}$. The parameter k is referred to as the shape parameter, and the reciprocal of the rate is referred to as the scale parameter. The distribution is exponential when $k = 1$ and tends to a normal distribution as $k \to \infty$ [20]. The distribution generalizes to non-integer positive values of k when it is known as the **gamma distribution**, and the support is given as $0 < t$ because the vertical axis is an asymptote to the pdf when $k < 1$. The pdf and cdf of gamma distributions with shape 1.5 and scale 2, and shape 3 and scale 1 are shown on the left and right panels respectively of Figure 5.17. The R syntax allows either the scale

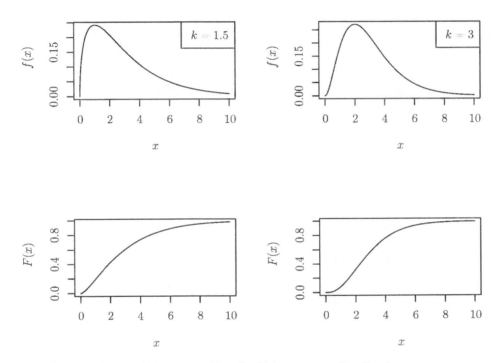

FIGURE 5.17: pdf and cdf for gamma distribution.

or rate (the default) to be specified. Notation: $T \sim Gamma(k, \lambda)$.

```
> x=seq(0,10,.01)
> pdf1=dgamma(x,shape=1.5,scale=2)
> cdf1=pgamma(x,shape=1.5,rate=1/2)
> pdf2=dgamma(x,shape=3,scale=1)
> cdf2=pgamma(x,shape=3,scale=1)
> par(mfcol=c(2,2))
> plot(x,pdf1,type="l",ylab="f(x)")
> legend(7,.20,"k=1.5")
> plot(x,cdf1,type="l",ylab="F(x)")
> plot(x,pdf2,type="l",ylab="f(x)")
> legend(7,.20,"k=3")
> plot(x,cdf2,type="l",ylab="F(x)")
```

---

[20] A consequence of the Central Limit Theorem (Section 6.4.3).

5.7.2   Applications of the gamma distribution

The gamma distribution is often chosen for modeling applications in which the skewness of the distribution can changes, for example, the modeling of internet traffic at different times of the day and different days of the week.

Example 5.20: Internet [gamma distribution]

The Erlang distribution is commonly used in models for queueing systems, including internet traffic.

Example 5.21: Rainfall [gamma distribution]

Stochastic rainfall models are used as the input to rainfall-runoff models that are used to design and optimize water storage reservoirs, storm sewer networks, flood defense systems and other civil engineering infrastructure. The advantages of rainfall models rather than historic records are: they can be used at locations without rain gage records; they can be used to simulate rainfall over far longer periods than historic records and so give more insight into extreme rainfall events; and they can be adapted to match various climate change scenarios. The gamma distribution is a popular choice for the distribution of rainfall on wet days, or shorter periods.

5.7.3   Random deviates from gamma distribution

Random deviates from Erlang distributions can be can be obtained by adding random deviates from exponential distributions. A method for obtaining random deviates from a gamma distribution with $k < 1$, and hence any non-integer k, is considered in Exercise 5.30, but in practice we can rely on the R function. For example, 5 random numbers from a gamma distribution with $k = 0.5$ and $\lambda = 0.1$ can be obtained by

```
> set.seed(1)
> x=rgamma(5,0.5,0.1) ; print(x)
[1]   0.9881357  5.5477635  2.0680270  3.4721073 39.5464914
```

5.7.4   Fitting a gamma distribution

Suppose we have a sample $\{x_i\}$ for $i = 1, \ldots, n$. The method of moments estimates of the parameters are obtained from the equations

$$\bar{x} = \frac{\widehat{k}}{\widehat{\lambda}}, \qquad s^2 = \frac{\widehat{k}}{\widehat{\lambda}^2}$$

and are

$$\widehat{\lambda} = \frac{\bar{x}}{s^2}, \qquad \widehat{k} = \widehat{\lambda}\bar{x}.$$

Example 5.22: Times between earthquakes [gamma distribution]

The data in `EARTHQUAKEIOP.txt` are the times in days between significant (Primary Magnitude $\geq$ 3) earthquakes, inter-occurrence periods (IOP), in the U.S. since 1944 (NOAA National Geophysical Data Center). The following R code estimates the parameters of a gamma distribution and superimposes the fitted pdf on the histogram (Figure 5.18).

```
> IOP=scan("data/EARTHQUAKEIOP.txt")
Read 120 items
> head(IOP)
[1] 574 374 400 334 860 335
> mean(IOP)
[1] 212.875
> sd(IOP)
[1] 267.9725
> lambda=mean(IOP)/var(IOP)
> k=lambda*mean(IOP)
> k
[1] 0.6310575
> lambda
[1] 0.002964451
> x=seq(1,1800,1)
> y=dgamma(x,shape=k,rate=lambda)
> hist(IOP,freq=FALSE,main="")
> lines(x,y)
```

The mean and standard deviation of the IOP are 213 and 268 respectively. Since the standard deviation exceeds the mean we know that the shape parameter will be less than 1. The shape and rate parameters are estimated as 0.63 and 0.0030 respectively. The fitted pdf is quite close to the histogram suggesting a reasonable fit (Figure 5.18). An explanation for the small shape parameter is that the earthquakes sometimes tend to cluster in time giving more variability than would be expected if earthquakes occurred randomly and independently as a temporal Poisson process.

5.8 Gumbel distribution

The Gumbel distribution has a theoretical justification as a model for extreme values such as annual maximum floods.

5.8.1 Definition of a Gumbel distribution

The **Gumbel distribution**[21] distribution is also known as the **extreme value Type I** distribution. It is obtained as the distribution of the maximum in a simple random sample

---

[21]Emile Gumbel (1891-1966)

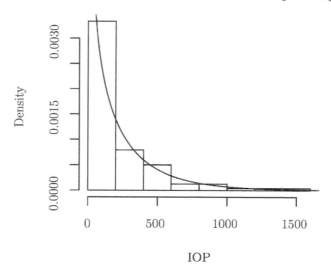

FIGURE 5.18: Histogram of IOP for earthquakes with fitted density.

of size n from a distribution of the exponential type as $n \to \infty$. A distribution of the exponential type has a pdf with an unbounded upper tail that decays at least as rapidly as an exponential distribution, and so includes the normal distribution and gamma distributions with $1 \leq k$. The derivation of the Gumbel distribution does not require a specification of the form of the distribution of exponential type. The cdf of a random variable X which has a Gumbel distribution is

$$F(x) \;=\; e^{-e^{-\frac{x-\xi}{\theta}}}, \qquad -\infty < x < \infty$$

and the pdf follows from differentiation as

$$f(x) \;=\; \frac{1}{\theta} e^{-\frac{x-\xi}{\theta}} e^{-e^{-\frac{x-\xi}{\theta}}}, \qquad -\infty < x < \infty.$$

The parameter ξ is the mode of the pdf and θ is a scale factor. The mean and standard deviation are given by:

$$\mu \;=\; \xi + (-\Gamma'(1))\theta \;\approx\; \xi + 0.577216\,\theta \qquad \text{and} \qquad \sigma = \frac{\pi}{\sqrt{6}}\,\theta.$$

The Gumbel distribution with mode 0 and scale factor 1 is known as the **reduced distribution** and the cdf is

$$F(x) \;=\; e^{-e^{-x}}, \qquad -\infty < x < \infty.$$

The following R code plots the pdf and cdf of the reduced distribution

```
> x=seq(-3,7,.01)
> pdf=exp(-x)*exp(-exp(-x))
> cdf=exp(-exp(-x))
> par(mfrow=c(2,1))
> plot(x,pdf,type="l",ylab="f(x)")
> plot(x,cdf,type="l",ylab="F(x)")
```

which are shown in Figure 5.19.

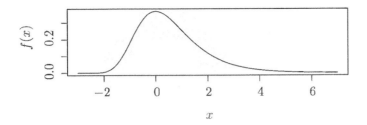

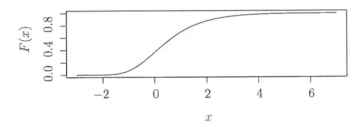

FIGURE 5.19: The pdf and cdf of the reduced Gumbel distribution.

5.8.2 Applications of the Gumbel distribution

Although the derivation of the Gumbel distribution uses an asymptotic argument[22] it often provides a good empirical fit to, for instance, the annual maximum daily flow of rivers. In such applications $n = 365$ and as daily flows are flows are typically highly dependent the equivalent random sample size is considerably smaller. However, daily flows are likely to be well modeled by some distribution of the exponential type, which is consistent with a Gumbel distribution of the annual maxima. Other applications include modeling of extreme wind speeds, maximum ice loading of electricity cables and annual maximum tidal levels.

The Gumbel distribution can also be used to model variables when there is no is explicit sample size from which a maximum is extracted. An example is the peak flows of the annual flood of the Mekong River at Vientiane [Adamson et al., 1999].

5.8.3 Random deviates from a Gumbel distribution

The inverse cdf of the Gumbel distribution can be found as an algebraic formula. Set

$$F(x) = e^{-e^{-\frac{x-\xi}{\theta}}} = u.$$

Then

$$x = \xi - \ln\left(-\ln(u)\right)\theta = F^{-1}(u).$$

So, random deviates from a Gumbel distribution can be generated by applying $F^{-1}()$ to random deviates from a uniform distribution. The following R code draws a random sample of $1,000$ from a Gumbel distribution with parameters 80 and 6 and plots histograms of the uniform deviates and corresponding Gumbel deviates (Figure 5.20).

---

[22]The size of the sample from which the maximum is extracted tends to infinity.

```
> set.seed(1)
> u=runif(1000)
> xi=80
> theta=6
> x=xi-log(-log(u))*theta
> par(mfrow=c(2,1))
> hist(u,freq=FALSE,main="")
> hist(x,freq=FALSE,main="")
```

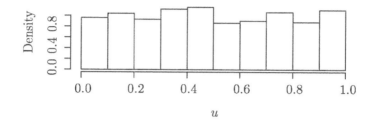

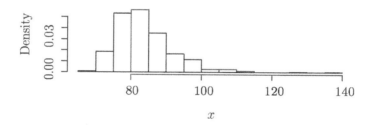

FIGURE 5.20: Random deviates from a uniform [0,1] distribution (upper) and corresponding random deviates from the Gumbel distribution (lower).

5.8.4   Fitting a Gumbel distribution

The methods of moments estimators of the parameters are the solutions of

$$\bar{x} \;=\; \widehat{\xi} + 0.5772\,\widehat{\theta} \quad \text{and} \quad s \;=\; \frac{\pi}{\sqrt{6}}\,\widehat{\theta}$$

given by

$$\widehat{\theta} \;=\; 0.7797s \quad \text{and} \quad \widehat{\xi} \;=\; \bar{x} - 0.5772\,\widehat{\theta}.$$

Example 5.23:   Animas River [annual peak streamflows]

We fit a Gumbel distribution to the annual peak streamflows of the Animas River at Durango from 1924 until 2012. The mean and standard deviation of the annual maximum flows are 5371 and 2061 respectively, and hence $\widehat{\theta}$ and $\widehat{\xi}$ are 2051 and 4187 respectively. The calculations in R are:

```
> animas1924.dat=read.table("animas1924.txt",header=T)
```

```
> attach(animas1924.dat)
> print(c("mean",round(mean(peak))))
> print(c("sd",round(sd(peak))))
> theta=sqrt(6)*sd(peak)/pi
> print(c("theta",round(theta)))
> xi=mean(peak)-0.5772*theta
> print(c("xi",round(xi)))
[1] "mean"  "5371"
[1] "sd"    "2631"
[1] "theta" "2051"
[1] "xi"    "4187"
```

The following R code provides: a time series plot of the peaks; histogram with the fitted pdf superimposed; box plot; and a plot of the order statistics against approximate expected values of order statistics from a reduced Gumbel distribution. These are shown in Figure 5.21.

```
> xp=seq(0,20000,1)
> yp=exp(-(xp-xi)/theta) * exp(-exp(-(xp-xi)/theta)) /theta
> par(mfrow=c(2,2))
> plot(as.ts(peak),main="",ylab="peak",xlab="time from 1924")
> hist(peak,freq=FALSE,main="")
> lines(xp,yp)
> boxplot(peak,ylab="peak")
> peak_sort=sort(peak)
> n=length(peak_sort)
> p=c(1:n)/(n+1)
> w=-log(-log(p))
>plot(w,peak_sort,main="",xlab="-ln(-ln(p))",ylab="peak (order
 statistic)")
```

The Gumbel distribution seems a reasonable fit to the data although the largest annual maximum over the 89 year period, 20 000, appears rather high. The variance of the largest order statistic, $X_{n:n}$, is high and sampling variability is a possible explanation. If the data are from a Gumbel distribution with the fitted parameter values, the probability of exceeding 20 000 is:

$$1 - \exp\left(-\exp\left(-\frac{20000 - 4187}{2051}\right)\right) = 0.000448.$$

The estimated average recurrence interval (ARI) of a peak of 20 000 is therefore $1/0.000448 = 2230.8$ years. If annual maxima are assumed to be independently distributed, the probability of an exceedance of 20 000 in an 89 year record is $1 - (1 - 0.000448)^{89} = 0.0391$. These calculations suggest that although the flow of 20 000 was extreme, it is not incompatible with a Gumbel distribution of annual maximum flows. A lognormal model of annual maximum flows is considered in Exercise 5.34. Another issue is that peak flows are typically derived from water level measured at a gage which can lead to considerable inaccuracies for very high flows.

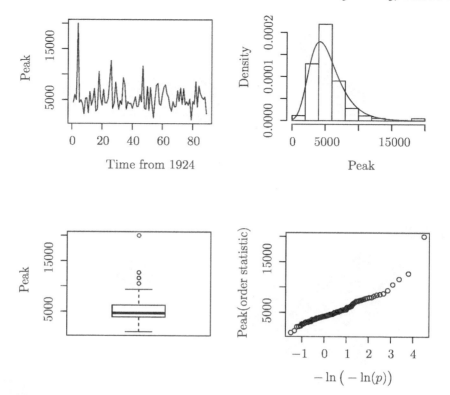

FIGURE 5.21: Annual peak streamflows in the Animas River 1924-2012: time series (top left); histogram and fitted Gumbel distribution (top right); box-plot (lower left); and qq-plot (lower right).

5.9   Summary

5.9.1   Notation

$X(\omega)$ or X	random variable, where $\omega \in \Omega$
$f(x)$	probability density function, pdf
$F(x)$	cumulative probability distribution, cdf
$\mathrm{E}[\phi(X)]$	expected value of $\phi(X)$
μ_X or μ	expected value or mean of X
σ_X^2 or σ^2	variance of X

5.9.2   Summary of main results

A random variable X can be said to have, or follow, a continuous distribution when X can take any value in the continuum of real numbers between some lower limit L and upper limit U, which can be infinite. Some of these distributions are described below with corresponding formulas for the support, $f(x)$, μ and σ^2 given in Table 5.3. For a given X with pdf $f(x)$, then $\mathrm{P}(a \leq X \leq b) = \int_a^b f(x)dx$. This may be used to calculate any probability required.

As for any mathematical model, a probability distribution is an approximation to reality. It follows that there is no correct distribution. However, some models are better than the others. Probability plots can be used to assess the goodness of fit, and to choose between probability distributions, for particular applications. Also, there may be conceptual reasons for choosing a particular model, such as a Gumbel distribution for extreme values.

TABLE 5.3: Continuous probability distributions.

Distribution	Support	PDF, $f(x)$	Expected value, μ	Variance, σ^2
Uniform	$[a,b]$	$\dfrac{1}{b-a}$	$\dfrac{a+b}{2}$	$\dfrac{(b-a)^2}{12}$
Exponential	$[0,\infty)$	$\lambda e^{-\lambda t}$	$\dfrac{1}{\lambda}$	$\dfrac{1}{\lambda^2}$
Normal	$(-\infty,\infty)$	$\dfrac{1}{\sqrt{2\pi}\sigma}e^{-\frac{1}{2}\left(\frac{x-\mu}{\sigma}\right)^2}$	μ	σ^2
Lognormal	$(0,\infty)$	$\dfrac{1}{x\sqrt{2\pi}b}e^{-\frac{1}{2}\left(\frac{\ln(x)-a}{b}\right)^2}$	$e^{a+\frac{b^2}{2}}$	$e^{2a+b^2}\left(e^{b^2}-1\right)$
Gamma	$[0,\infty)$	$\dfrac{\lambda^k t^{k-1}e^{-\lambda t}}{\Gamma(k)}$	$\dfrac{k}{\lambda}$	$\dfrac{k}{\lambda^2}$
Gumbel	$(-\infty,\infty)$	$\dfrac{1}{\theta}e^{-\frac{x-\xi}{\theta}}e^{-e^{-\frac{x-\xi}{\theta}}}$	$\xi-\Gamma'(1)\theta$	$\dfrac{\pi^2}{6}\theta^2$

5.9.3 MATLAB and R commands

In the following, **data** is a vector containing data points, **x** is a value (scalar or vector) in the support of the corresponding distribution. The variables **mu** and **sigma** are the expected value (μ) and standard deviation, respectively, for the corresponding distribution and **n** is some integer. The variables **a** and **b** are the parameters for the corresponding distribution and **lambda** is the rate for the exponential distribution. For more information on any built in function, type **help(function)** in R or **help function** in MATLAB.

R command	MATLAB command
qqplot(data)	qqplot(data)
dunif(x,a,b)	unifpdf(x,a,b)
punif(x,a,b)	unifcdf(x,a,b)
runif(n,a,b)	random('unif',a,b,[n,1])
	or a + (b-a)*rand(n,1)
dexp(x,lambda)	exppdf(x,1/lambda)
pexp(x,lambda)	exppdf(x,1/lambda)
rexp(n,lambda)	random('exp',1/lambda,[n,1])
	or exprnd(1/lambda, [n,1])
dnorm(x,mu,sigma)	normpdf(x,mu,sigma)
pnorm(x,mu,sigma)	normcdf(x,mu,sigma)
rnorm(n,mu,sigma)	random('norm',mu,sigma,[n,1])
	or mu + sigma*randn(n,1)
dlnorm(x,mu,sigma)	lognpdf(x,mu,sigma)
plnorm(x,mu,sigma)	logncdf(x,mu,sigma)
rlnorm(n,mu,sigma)	random('logn',mu,sigma,[n,1])
dgamma(x,a,b)	gampdf(x,a,b)
pgamma(x,a,b)	gamcdf(x,a,b)
rgamma(n,a,b)	random('gam',a,b,[n,1])

5.10   Exercises

Section 5.1 Continuous probability distributions

Exercise 5.1: A garden pump

The lifetime, T, of pond pumps made to a particular design has the pdf:

$$f(t) = \begin{cases} a - bt & 0 \le t \le 2 \\ 0 & \text{elsewhere,} \end{cases}$$

with $f(2) = 0$, where t is measured in 10 year units.

(a) Sketch the function $f(x)$ and find the values of a and b for it to be a pdf.
(b) Calculate the mean.
(c) Calculate $E[T^2]$ and hence find the variance and standard deviation.
(d) Write down a formula for the cdf, find the median, and verify that the median is less than the mean.
(e) What area lies within one standard deviation of the mean?
(f) Calculate the skewness.

Exercise 5.2: Moment generating function

(a) Use a Taylor expansion about 0 to show that

$$e^x \;=\; 1 + x + \frac{x^2}{2!} + \cdots + \frac{x^n}{n!} + \cdots$$

Denote the infinite series on the right by P(x). Show that

$$\frac{dP(x)}{dx} \;=\; P(x)$$

and that P(0) = 1.

(b) The **moment generating function (mgf)** of a random variable X is defined as

$$M(\theta) \;=\; E\big[e^{\theta x}\big],$$

where the argument θ is treated as a constant in the expectation. Show that

$$\left(\frac{d^n M(\theta)}{d\theta^n}\right)_{\theta=0} \;=\; E[X^n].$$

(c) Use the mgf to find the mean, variance, skewness and kurtosis of a uniform random variable with support $[0, 1]$. Write down the corresponding results for a uniform distribution with support $[a, b]$.

(d) Use the mgf to find the mean, variance, skewness and kurtosis of the exponential distribution.

(e) Use the mgf to find the mean and variance of the binomial distribution.

Exercise 5.3: Probability density function 1

A random variable X has probability density function given by $f(x) = k\big(1 - |x - 1|\big)$, for $0 \le x \le 2$ and 0 elsewhere.

(a) What is the value of k?

(b) Sketch the pdf.

(c) Find the mean, variance and standard deviation of X.

Exercise 5.4: Probability density function 2

A probability density function (pdf) has the form:

$$f(x) \;=\; a - a(1-x)^2, \qquad 0 \le x \le 2$$

and 0 elsewhere.

(a) For what value of x does the pdf take its maximum value and what is this value in terms of a?

(b) Show that the pdf can be expressed in the form:

$$f(x) \;=\; a(2x - x^2), \qquad 0 \le x \le 2.$$

(c) What is the value of a that makes $f(x)$ a pdf?

(d) Sketch the pdf.

(e) What is the mean of the distribution?

(f) If the random variable X has this distribution, what is $E[X^2]$?

(g) What is the variance, and the standard deviation, of the distribution?

(h) If the random variable X has this distribution, what is the probability that X lies within two standard deviations of its mean?

Section 5.2 Uniform distribution

Exercise 5.5: Fitting Uniform distribution

The following data are deviations from a target diameter (microns) for a computer controlled lathe: 57, -13, 21, 31, -50, -33, 12, -46, -38, and 12. Given the digital control an engineer thinks that the deviations may be well approximated by a uniform distribution. Estimate the a and b parameters of a uniform distribution for these data in two ways. Comment on the differences in the estimates.

(a) Solve

$$\bar{x} = \frac{\widehat{a} + \widehat{b}}{2}$$

$$s^2 = \frac{(\widehat{b} - \widehat{a})^2}{12}.$$

(b) Calculate a and b as

$$(x_{1:n} + x_{n:n})/2 \quad \pm \quad \frac{n+1}{n-1}(x_{n:n} - x_{1:n})/2.$$

Exercise 5.6: Uniform distribution

(a) Prove that a uniform distribution with support $[0, 1]$ has a mean of $1/2$ and a variance of $1/12$.

(b) Deduce the mean and variance for a uniform distribution with support $[a, b]$.

(c) Explain why the skewness of a uniform distribution is 0.

(d) Show that the kurtosis of a uniform distribution is $9/5$.

Section 5.3 Exponential distribution

Exercise 5.7: A random variable

Suppose T has an exponential distribution with rate λ. That is $f(x) = \lambda \exp^{\lambda t}$ for $0 \le t \le \infty$.

(a) Find $E[T]$ in terms of λ.

(b) Find $E[T^2]$ in terms of λ [hint: use integration by parts].

(c) Deduce an expression for the variance of T in terms of λ.

(d) What is the CV of T?

Exercise 5.8: Valve audio amplifier

The lifetime of a valve audio amplifier, T, has an exponential distribution with mean 10 years.

(a) Calculate the probability a valve audio amplifier lasts longer than 10 years.

(b) What is the probability that an amplifier will fail somewhere between 10 and 20 years hence?

(c) Now suppose the amplifier is 10 years old. What is the probability it fails within the next 10 years?

Exercise 5.9: LED screens

LED screens for a particular make of TV have lifetimes (T) with an exponential distribution. The mean lifetime of screens is 7.5 years.

(a) What is the probability that the screen lasts at least 7.5 years?

(b) What is the probability that the screen lasts another 7.5 years once it has reached 7.5 years and is still working.

(c) I have just bought one of these TVs and a screen. What is the probability I have a TV with a working screen in 15 years time if the spare screen has had the same lifetime distribution as the original screen since the purchase? Note that it is quite possible that the spare screen will have failed if I need to use it!

(d) I have just bought one of these TVs in a special promotion with a guarantee of one replacement screen. What is the probability I have a TV with a working screen in 15 years time if the replacement screen has a lifetime with an exponential distribution with mean 7.5?

(e) You are now told that the manufacturer put screens aside, in order to cover the guarantees at the time of the promotion, when the TVs were sold. Explain whether or not this will change your answer to (d).

Exercise 5.10: Random samples 1

For $k = 10^4$ generate random samples of size 10 from an exponential distribution with mean 1, and calculate the $\widehat{CV}$ for each sample.

(a) Draw the histogram of the $\widehat{CV}$.

(b) Calculate the mean and standard deviation of the $\widehat{CV}$.

(c) What proportion of $\widehat{CV}$ is less than or equal to 0.60 and what proportion of $\widehat{CV}$ is equal to or exceeds 1.39?

Exercise 5.11: Random samples 2

Generate $K = 10^4$ random samples from a mixture of two exponential distributions and investigate the distribution of the $\widehat{CV}$.

(a) $n = 10, \lambda_1 = 1.0, \lambda_2 = 2.0$ and a probability $p = 0.9$ of a draw from distribution 1.

(b) As (a) with $n = 30$.

Exercise 5.12: Arrival time

Packets arrive at a router at an average rate of 130 per millisecond (ms). Assume that arrivals are well modeled as a Poisson process.

(a) State the assumptions of a Poisson process.

(b) What is the probability of more than one packet arrival in the next 0.01 ms?

(c) What is the probability that the time until the next packet arrives exceeds 0.02 ms?

Exercise 5.13: Spatial Poisson process

A space telescope has a near planar dish of radius a. Meteors pass through the spatial plane containing the dish, randomly and independently, at a rate of λ per unit area per year.

(a) What is the probability, in terms of a and λ, that the dish is not hit by a meteor in one year?

Now let R be the distance from the centre of the dish to the nearest point at which a meteor crosses the spatial plane in the next m years.

(b) What is the probability R exceeds r?

(c) Use your result in (b) to write down the cdf of R, $F(r)$.

(d) Use your result in (c) to write down the pdf of R, $f(r)$ (your answers should include the parameters m and λ).

(e) Now take m and λ equal to 1 and plot $f(r)$ for r from 0 to 2.

Section 5.4 Normal (Gaussian) distribution

Exercise 5.14: Normal pdf

Sketch two standard normal pdfs. On one pdf shade the area $\Phi(z_p)$ where z_p is a positive value around 1.5. On the other shade the area $\Phi(-z_p)$, What is the sum of the two shaded areas?

Exercise 5.15: Packaging

A machine fills bags with cement. The declared mass on the bags is 50 kg.

(a) Assume the actual masses have a normal distribution.

  (i) If the standard deviation of the amount dispensed is 0.80 kg and the mean is set to 52 kg, what proportion of bags have a mass over 53 kg?

  (ii) If the standard deviation of the amount dispensed is 0.80 kg, to what value should the mean be set for 1% of the bags to have mass below 50 kg?

(b) (i) The manufacturer states that a proportion 0.01 of bags is underweight. A random sample of 20 bags will be taken. What is the probability of finding two or more underweight bags in the sample?

  (ii) If you took such a sample and found that it contained two underweight bags, what would you conclude?

Exercise 5.16: Normal probabilities

If $X \sim N\left(12, (2.5)^2\right)$

(a) Find $P(0 < X < 13)$.

(b) Find a such that $P(a < X) = 0.05$

(c) Find b such that $P(X < b) = 0.02$

Exercise 5.17: Manufacture of capacitors

A ceramic multilayer capacitor is rated at 100 microFarad with a tolerance of $\pm 20\%$. The manufacturer's internal specification is that the capacitance of capacitors should be within the interval [90, 110] microFarad. Assume that the capacitances have a normal distribution.

(a) Suppose the process mean is 104.0 and the standard deviation is 4.1. What proportion of capacitors will lie outside the internal specification? How many parts per million would be outside the tolerance?

(b) The manufacturer adjusts the process so that the mean is 100.0. What does the standard deviation need to be reduced to so that 0.997 of production is within the internal specification?

(c) Suppose the standard deviation is reduced to the value you calculate in (b), but the mean remains at 104.0. What proportion of capacitors would now lie outside the internal specification?

Exercise 5.18: Inflection point

If Z has pdf $\phi(z) = \frac{1}{\sqrt{2\pi}} e^{-z^2/2}$

(a) Show that $\dfrac{d^2\phi}{dz^2} = 0$ when $z = \pm 1$.

(b) Deduce that $\phi(z)$ has a point of inflection at -1 and at $+1$.

(c) Deduce that the pdf of a normal distribution has points of inflection when $x = \mu - \sigma$ and when $x = \mu + \sigma$.

Exercise 5.19: Brake cylinder pistons

The length of a piston in a brake cylinder is specified to be between 99 mm and 101 mm. The production process produces pistons with lengths which are normally distributed with mean 100.3 mm and standard deviation 0.8 mm. Each piston costs $3 to manufacture. Pistons longer than 101 mm can be shortened to be within spec at an additional cost of $1. Pistons shorter than 99 mm are scrapped and have no scrap value.

(a) What is the average cost of a within spec piston?

(b) Now suppose any pistons outside the spec are scrapped.

    (i) If the mean of the process can be altered without affecting the standard deviation, what value would minimize the scrap?

    (ii) What would the average cost of a within spec piston now be?

Exercise 5.20: Normal distribution

   (a) What is the upper quartile of the standard normal distribution?

   (b) What is the inter-quartile range of the standard normal distribution?

   (c) If $Z \sim N(0,1)$, what is the probability that Z is more than one and a half inter-quartile ranges above the upper quartile?

Exercise 5.21: Box plot

A box plot is constructed to show individually any points more than 1.5 inter-quartile ranges from a quartile.

   (a) What is the probability that a normal random variable is more than 1.5 inter-quartile ranges from a quartile?

   (b) What is the probability that the largest in a random sample of 30 is more than 1.5 inter-quartile ranges above the upper quartile?

Exercise 5.22: Wire rope

The specification for the breaking load of a wire rope is that it exceeds 8 tonnes. Assume that the breaking loads have a normal distribution with a mean of 8.45.

   (a) Suppose the standard deviation is 0.20. Find the probability that a rope has a breaking load below 8.00 in terms of the function $\Phi(z)$?

   (b) To what value must the standard deviation be reduced if the proportion of ropes with breaking load below 8.00 is to be 0.1%?

Section 5.5 Probability plots

Exercise 5.23: Normal plots

Plot either a normal probability plot or a normal quantile quantile plot for

   (a) the inflows to the Hardap Dam in Table 3.9 and the logarithms of inflow.

   (b) Comment on your plots.

Exercise 5.24: Gumbel plots

The cdf of a reduced Gumbel distribution is

$$F(y) \quad = \quad exp(-exp(-y)).$$

   (a) Explain how to draw a quantile-quantile plot for a Gumbel distribution.

   (b) Compare quantile-quantile plots for the Gumbel distribution with normal quantile plots for the logarithms of the data for

      (i) Canadian gold grades (see website)

      (ii) Animas River peak flows (see website)

     (iii) Annual maximum peak discharges on the River Uruguay (see website)

     (iv) Flood volumes on the River Uruguay (see website)

Section 5.6 Lognormal distribution

Exercise 5.25: Ice load

The logarithm of the depth (mm) of the annual maximum ice load on electricity cables has a normal distribution with mean 2.4 and standard deviation 0.85.

(a) What is the median annual maximum depth of ice load?

(b) What is the mean annual maximum depth of ice load?

(c) What depth of ice load has a probability of 0.01 of being exceeded in a year?

(d) What is the probability that the depth of ice load will exceed 50 mm in a year?

(e) What is the average recurrence interval, in years, for a depth of ice load of 50 mm, if annual maximum ice loads are independent?

Exercise 5.26: Taylor expansion

Use the Taylor expansion for $\ln(1 + x)$ to show that $\ln(X + L) \approx \ln(L) + X/L$.

Exercise 5.27: Logs

Show that

$$\log_{10}(x) = \frac{\ln(x)}{\ln(10)}.$$

Exercise 5.28: Lognormal 1

In the usual notation $Z \sim N(0, 1)$ and Z has *pdf* $\phi(z) = \frac{1}{\sqrt{2\pi}} e^{\frac{-z^2}{2}}$.

(a) Find $E[e^z]$

(b) Define $Y = e^z$.

    (i) Use the result in (a) to write down the mean of Y.

    (ii) What is the median value of Y?.

    (iii) Find the *pdf* of Y, $\psi(y)$ say, where the formula for $\psi()$ is expressed in terms of $\phi()$ and y.

    (iv) Plot $\psi(y)$ for y from 0 to 8.

    [Y has a lognormal distribution]

Exercise 5.29: Lognormal 2

Obtain a formula for L in terms of the median $x_{0.50}$ and quartiles, lower $x_{0.75}$ and upper $x_{0.25}$ of the original data by applying a symmetry criterion

$$ln(x_{0.50} + L) - ln(x_{0.75} + L) = ln(x_{0.25} + L) - ln(x_{0.50} + L).$$

Use this method to fit a 3 parameter lognormal distribution to the gold grades. Do you think it is a substantial improvement?

Section 5.7 Gamma distribution

Exercise 5.30: Animas peak flows

Consider the annual peak flows in the Animas River.

(a) Fit a gamma distribution to the peak flows by the method of moments. Draw a quantile-quantile plot.

(b) Fit a gamma distribution to (peak flows − least flow) by the method of moments. Draw a quantile-quantile plot.

(c) Fit a gamma distribution to ln (peak flows) by the method of moments. Draw a quantile-quantile plot.

(d) Fit a gamma distribution to ln (peak flows − least flow) by the method of moments. Draw a quantile-quantile plot.

(e) Compare estimates of the flow with an ARI of 100 years using the fitted distributions in (a) to (d).

Section 5.8 Gumbel distribution

Exercise 5.31: Random numbers from a Gumbel distribution

I want to draw a random number from the Gumbel distribution with cdf given by

$$F(x) \;=\; e^{-e^{-(x-40)/5}}.$$

Using the uniform $(0, 1)$ random number 0.57, calculate a corresponding random number from $F(\cdot)$.

Exercise 5.32: Gumbel versus lognormal

Fit a lognormal distribution to the annual maximum flows of the Animas River at Durango. Assume that annual maxima are independently distributed with the fitted lognormal distribution.

(a) Calculate the probability of exceeding 20 000 and the ARI corresponding to 20 000.

(b) What is the probability of exceeding 20 000 at least once in an 89 year record?

(c) Compare the results with those for the fitted Gumbel distribution given in Section 5.8.4. Comment on the physical justifications for the two probability distributions in the context of modeling annual maxima.

(d) Calculate the flow with an ARI of 1 000 years using the Gumbel and lognormal distributions.

Exercise 5.33: Fitting to data

The annual maximum flows in the Animas River at Durango for the years

$$1898, 1900, 1909, 1911, 1913 - 1923 \quad \text{are given by}$$

$$4\,680, \quad 3\,830, \quad 10\,000, \quad 25\,000, \quad 3\,700, \quad 8\,330, \quad 4\,430, \quad 6\,140,$$
$$8\,460, \quad 4\,400, \quad 5\,600, \quad 9\,260, \quad 9\,300, \quad 7\,800, \quad 4\,680.$$

Assume that these years are a random sample of additional years and augment the 1924 − 2012 data given on the website with these values. Fit a Gumbel and a lognormal distribution to the augmented data.

(a) Calculate the probability of exceeding 25 000 and the ARI corresponding to 25 000, for the two models.

(b) Assume that 25 000 was the highest recorded flow between 1898 and 2012. What are the probabilities of exceeding 25 000 at least once in a 115 year record according to the two models?

(c) Compare the results with those for the fitted Gumbel distribution given in Section 5.8.4. Comment on the physical justifications for the two probability distributions in the context of modeling annual maxima.

(d) Calculate the flow with an ARI of 1 000 years using the Gumbel and lognormal distributions.

Exercise 5.34: Animas river

(a) Fit a normal distribution to the logarithms of the annual peak streamflows in the Animas River. What is the estimated ARI of a peak of 20,000?

(b) Compare probability plots of a normal distribution for the logarithms of annual peaks and the Gumbel distribution for annual peaks.

Miscellaneous problems

Exercise 5.35: Kernel Smoother

Explain why the area under the kernel smoother $\widehat{f}(x)$ equals 1. Investigate the effect of changing the bandwidth for the waiting times for Old Faithful.

Exercise 5.36: Rayleigh distribution (see Exercise 6.31)

The distribution is named after Rayleigh (John William Strutt) who derived it in 1919 to solve a problem in acoustics. The cdf is

$$F(x) = 1 - exp(-x^2/(2\theta^2)), \qquad 0 \leq x.$$

The mean and variance are $\theta\sqrt{\pi/2}$ and $(2 - \pi/2)\theta^2$ respectively. The Rayleigh distribution is sometimes used to model wave heights, defined as the distance from a trough to the following peak. The significant wave height (H_s) is defined as four times the root mean squared height. Show that if wave heights have a Rayleigh distribution then about one third exceed H_s.

Exercise 5.37: Half normal distribution

A half normal distribution is obtained by folding a normal distribution with mean 0 about a vertical line through 0. It is a plausible model for run-out of circular components such as brake discs, CDs, and bicycle wheels. The pdf is

$$f(x) = \frac{2}{\sigma\sqrt{\pi}}e^{-x^2/(2\sigma^2)} \text{ for } x \geq 0.$$

Show that the mean and variance of the distribution are $\sigma\sqrt{2/\pi}$ and $(1 - 2/\pi)\sigma^2$ respectively.

Exercise 5.38: ROC curve

A manufacturer of medical instruments has developed an expert system for analyzing tomographic scans and giving a warning if a subject has condition C. A variable X has a standard normal distribution in the population of people without C and a normal distribution with a mean of 3 and a standard deviation of 1 in the population of subjects with condition C. Consider a rule in which you warn that a subject has condition C if X exceeds c. Consider values of c from 0 to 3 in steps of 0.1.

(a) For each value of c calculate the probability that X exceeds c if a subject does not have C.

(b) For each value of c calculate the probability that X exceeds c if a subject does have C.

(c) For each value of c plot the probability that X exceeds c if a subject does have C against the probability that X exceeds c if a subject does not have C.

(d) What does the plot look like if the distribution of X amongst subjects with C is identical to the distribution of X amongst subjects who do not have C?

Exercise 5.39: Moments

The rth moment about the origin of the distribution of a random variable X is

$$\mathrm{E}[X^r].$$

(a) What is the usual name for the 1st moment?

(b) Show that

$$\sigma^2 = \mathrm{E}[X^2] - (\mathrm{E}[X])^2.$$

(c) Obtain an expression for $\mathrm{E}[(X-\mu)^3]$ in terms of moments about the origin.

(d) Obtain an expression for $\mathrm{E}[(X-\mu)^4]$ in terms of moments about the origin.

Exercise 5.40: Cauchy distribution

A shielded radioactive source in a smoke detector emits α-particles into a half plane. Take the location of the sources as the origin and suppose that the angle between the particle path and the positive x-axis is uniformly distributed over $-\pi/2, pi/2]$. A screen is set up through $x = a$ and parallel to the y-axis. Let the intercept the particle path makes with the screen be Y. Show that Y has pdf

$$f(y) = \frac{a}{\pi(a^2 + y^2)}, \qquad -\infty < y < \infty.$$

The distribution of Y is known as the **Cauchy distribution**. The integral defining the mean is improper and does not converge, although its Cauchy principal value is 0. The median of the distribution is 0. Show that $\mathrm{E}[Y^2] = \infty$. The distribution is described as having infinite variance.

Exercise 5.41: Laplace distribution

The Laplace distribution can be described as back-to-back exponential distributions. A random variable X with mean μ has the Laplace distribution with pdf

$$f(x) = \frac{\lambda}{2}e^{-\lambda|x-\mu|}, \qquad \infty \leq x \leq \infty.$$

(a) Sketch the pdf. Does the distribution have a mode?

(b) Calculate the variance and standard deviation.

(c) Calculate the probability that X is more than $1, 2$, and 3 standard deviations away from the mean.

(d) Calculate the ratio of the inter-quartile range to the standard deviation and compare this with the value for a normal distribution.

(e) Explain why the difference of independent identically distributed exponential random variables has a Laplace distribution with mean 0.

(f) Explain how you can generate pseudo-random variates from a Laplace distribution given algorithms for generating pseudo-random variates from exponential and binomial distributions. Implement your method, construct a histogram, and check your answer to (e) when $\mu = 0$ and $\lambda = 2$.

(g) Use the result of (e), and a function for generating pseudo-random exponential variates, to generate a million pseudo-random variates from a Laplace distribution .with $mu = 0$ and $\lambda = 2$. Construct a histogram, and check your answer to (e) when $\mu = 0$ and $\lambda = 2$.

Exercise 5.42: Mekong river

Fit a Gumbel distribution to the peak flows during the annual flood of the Mekong river at Vientiane (see the website). Draw a histogram and superimpose the fitted pdf.

Exercise 5.43: Markov's inequality

If X is a random variable with support restricted to non-negative numbers and $E[X] = \mu$, Markov's inequality is:

$$P(x < X) \leq \frac{\mu}{x}.$$

Verify this inequality for $X \sim Exp(1)$ and $x = 3$. Prove the inequality in the general case.

Exercise 5.44: Weibull distribution

A chain has N links. The probability a single link can support a load x is

$$e^{(x/\theta)^\alpha}.$$

Assume that the strength of the chain is the strength of the weakest link and deduce that the load that can be supported X has a two parameter cdf of the form

$$F(x) = 1 - e^{-(x/\beta)^\alpha}, \qquad 0 \leq x.$$

Obtain an expression for the median of the distribution in terms of α and β. Obtain an expression for the quartiles of the distribution in terms of α and β, and suggest a method for estimating the parameters of the distribution from a set of data. Fit the distribution to the annual maximum flows in the Animas River from $1924 - 2012$. Assuming annual maximum flows have the fitted distribution, calculate the ARI of $20\,000$ and the flow with an ARI of $1\,000$ years.

Exercise 5.45: Box-Muller algorithm

Generate two different sequences $U_{1,i} \sim U(0,1]$ and $U_{2,i} \sim U[0,1]$, for $i = 1, 2, \ldots, 1\ 000$, and

(a) for each i set
  - $X_i = \sqrt{-2\ln\left(U_{1,i}\right)}\cos\left(2\pi U_{2,i}\right)$ and
  - $Y_i = \sqrt{-2\ln\left(U_{1,i}\right)}\sin\left(2\pi U_{2,i}\right)$.

(b) Verify graphically that the two sequences of random variables $\{X_i\}$ and $\{Y_i\}$ are normally distributed by creating histograms and qq-plots.

Note that two different streams of numbers must be used to generate the pair of U_is, because if U_1 and U_2 are generated by a single LCG in sequence, it has been shown by Bratley, Fox and Schrage (1987) that X and Y will be highly correlated.

Exercise 5.46: Polar Box-Muller algorithm

Generate two different sequences $U_{1,i} \sim U(0,1]$ and $U_{2,i} \sim U[0,1]$, for $i = 1, 2, \ldots, 1\ 000$, and

(a) for each i set
  - $V_{1,i} = 2U_{1,i} - 1$ and $V_{2,i} = 2U_{2,i} - 1$
  - Then if $V_{1,i}^2 + V_2^2 \leq 1$ (`acceptance/rejection`)
    - Let $W_i = \sqrt{\dfrac{-2\ln\left(V_{1,i}^2 + V_{2,i}^2\right)}{V_{1,i}^2 + V_{2,i}^2}}$

    - Return
      * $X_i = V_{1,i}W_i$
      * and $Y_i = V_{2,i}W_i$,
  - Otherwise, go back to Step 1 and try again.

(b) Verify graphically that the two sequences of random variables $\{X_i\}$ and $\{Y_i\}$ are normally distributed by creating histograms and qqplots.

As $V_{1,i}$ and $V_{2,i}$ are uniformly distributed across the square, we would expect to perform on average $\frac{4}{\pi} \approx 1.2732$ iterations to get two $N(0,1)$ random variates X_i and Y_i.

6

Correlation and functions of random variables

Covariance is a measure of linear association and the correlation coefficient is its non-dimensional form. We explain sample covariance and population covariance in the context of bivariate probability distributions. We derive formulae for the mean and variance of a linear combination of random variables in terms of their means, variances and pair-wise covariances. A special case of this result is that the mean of a simple random sample of size n from a population with mean μ and variance σ^2 has a mean of μ and a variance σ^2/n. We state the Central Limit Theorem and discuss the consequence that the sample mean has an approximate normal distribution. See relevant example in Appendix E:

Appendix E.3 *Robot rabbit.*

6.1 Introduction

Part of the threaded headset assembly on a bicycle is a crown race which is pressed onto a seat on the front fork and provides the inner race for the lower ball bearing. The seat diameter is designed to be slightly greater than the inside diameter of the crown race so that the crown race is held in position by tension. The interference is defined as:

$$(\text{diameter seat}) - (\text{inner diameter of crown race}).$$

The industry recommendation for the interference is 0.1 mm.

A production engineer in a company that manufactures bicycles was told that crown races made by supplier A were frequently fracturing when fitted to front forks for mountain bikes. Following discussion with other employees, the production engineer found that crown races made by supplier B also fractured in significant numbers during the pressing process whereas those manufactured by supplier C appeared satisfactory. The cost of a fracture is more than the cost of the crown race, because there is the labour cost of prizing the fractured race off the seat and repeating the pressing operation with another race. Moreover, if failures occur after bicycles leave the factory the costs are much higher. A highly stressed crown race, a consequence of too large an interference, could fracture in use and cause an accident. In contrast, too small an interference can lead to a loose fit which will soon be noticeable through play at the handlebars. The direct costs of customers' complaints and warranty repairs through dealers are increased by the hidden costs associated with losing a good reputation.

The engineer wanted the problem to be identified and rectified quickly. Forks were sourced from a single supplier and the specification for the diameter of seat was:

$$27.05 \pm 0.05.$$

The specification for the inside diameters of the crown races was

$$26.95 \pm 0.05.$$

The company expects numbers of out of specification items to be less than 3 per thousand so, if diameters are normally distributed, the standard deviations should be less than $0.05/3 = 0.017$. The engineer took random samples of 35 forks, and 35 crown races from stock supplied by A, B and C, and measured the diameters of the seats and the inside diameters of the crown races. A summary of these measurements is given in Table 6.1.

TABLE 6.1: Means and standard deviations of diameters of seats on front forks and of inner diameters of crown races.

Item	Sample size	Mean (mm)	Standard deviation (mm)
Fork seat	35	27.052	0.030
Crown race A	35	26.923	0.017
Crown race B	35	26.990	0.027
Crown race C	35	26.950	0.019

The sample means are close to the middle of the specification for the fork seats and for crown race C. However, the sample mean for crown races from A is more than one sample standard deviation less than the middle of the specification, and this suggests that they are being manufactured with a mean diameter that is too small. The crown races from manufacturer B appear to be manufactured with a mean diameter that is too large, and they are also too variable. The variability of the seat diameter on the forks is also too high, with an estimated standard deviation of nearly twice the maximum that would enable the specification to be met.

The production engineer would like to know how to combine the summary information in Table 6.1 to find the means and standard deviations of interferences. Let $\{x_i\}$ be the diameters of a random sample of n seats and $\{y_i\}$ be the diameters of a random sample of n crown races, where $1 \leq i \leq n$. The interferences are

$$d_i = x_i - y_i.$$

The mean interference $\bar{d}$ is the difference in the mean diameters $\bar{x} - \bar{y}$:

$$\bar{d} = \sum d_i/n = \sum (x_i - y_i)/n = \sum x_i/n - \sum y_i/n = \bar{x} - \bar{y}.$$

The variance of interferences, s_d^2, and hence the standard deviation of interferences, follows from the following argument.

$$
\begin{aligned}
s_d^2 &= \sum (d_i - \bar{d})^2/(n-1) \\
&= \sum ((x_i - y_i) - (\bar{x} - \bar{y}))^2/(n-1) \\
&= \sum ((x_i - \bar{x}) - (y_i - \bar{y}))^2/(n-1) \\
&= \sum (x_i - \bar{x})^2/(n-1) + \sum (y_i - \bar{y})^2/(n-1) - 2\sum (x_i - \bar{x})(y_i - \bar{y})/(n-1) \\
&= s_x^2 + s_y^2 - 2\sum (x_i - \bar{x})(y_i - \bar{y})/(n-1).
\end{aligned}
$$

The variance of the interferences is the sum of the variance of the seat diameters and the variance of the crown race diameters less twice a quantity known as the covariance of the two diameters. In the next sections we show that if the seats and crown races are paired at random then the expected value of this covariance term is 0, and so:

$$s_d^2 \approx s_x^2 + s_y^2.$$

The result for the standard deviation is

$$s_d \approx \sqrt{s_x^2 + s_y^2}.$$

This formula can be illustrated by applying Pythagoras' Theorem to the right angled triangle shown in Figure 6.1.

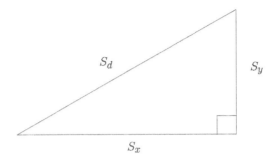

FIGURE 6.1: Relationship between standard deviation of the difference (S_d) and the standard deviations of seat diameters (S_x) and race diameters (S_y) for components assembled at random.

Using these formulae, the mean interference with crown race A is estimated as

$$27.052 - 26.923 = 0.129$$

and the estimated standard deviation of interference is

$$\sqrt{0.030^2 + 0.017^2} = 0.034$$

A summary of the statistics for interferences with all three crown races is given in Table 6.2.

TABLE 6.2: Means and standard deviations of interferences between seats on front forks and crown races.

Crown race	Mean (mm)	Standard deviation (mm)
A	0.129	0.034
B	0.062	0.040
C	0.102	0.036

The sample mean interference with crown race A, 0.129, is higher than the specified 0.1, and this may contribute to fractures, but the mean interference with crown race B is too low and these crown races also fracture. The main contribution to the variance of the interferences with crown races from A and C is the variance of the seats on the forks. The engineer continued the investigation with metallurgical tests of a few crown races, and finding that those from supplier C were less brittle, decided to use supplier C as the sole supplier of crown races. The engineer took a further sample of 20 forks, and the estimated standard deviation remained around 0.03 (Exercise 6.1), so the engineer requested that the supplier of forks reduce the standard deviation of the seat diameters so as to meet the specification.

6.2   Sample covariance and correlation coefficient

We consider covariance as a measure of linear association, and correlation as its non-dimensional form.

6.2.1   Defining sample covariance

Interlocking concrete paving blocks (pavers) are commonly used for exterior flooring in car parks, patios, driveways and walkways. Pavers are manufactured by pouring concrete, with some coloring agent added, into molds. A typical concrete mix is 17% cement, 28% aggregate and 55% sand. The cement content is important for strength and for frost resistance, but the strength also depends on the amount of water in the concrete mix and the amount and size of aggregate. Nevertheless, we expect to see a relationship between the compressive strengths and cement contents of pavers. The data in the pavers table on the website are the cement contents (percentage dry weight) and compressive strengths (MPa)[1], of a random sample of 24 pavers from approximately 1 million pavers used to construct a container park for a port authority. A scatter plot of these data is shown in Figure 6.2.

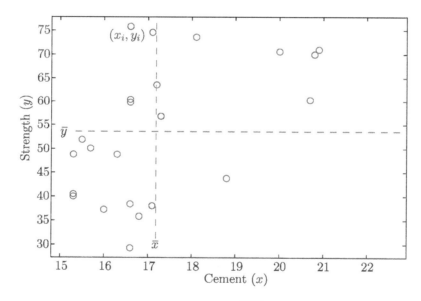

FIGURE 6.2: Calculating $\widehat{\text{cov}}(x, y)$ for 24 pavers.

Over this range of cement content there appears to be a tendency for the compressive strength to increase as the cement content increases, but there is considerable scatter about any straight line. We introduce **covariance** as a measure of the strength of linear association.

Definition 6.1:   Sample covariance

The **sample covariance** of a sample of n data pairs, (x_i, y_i), for $i = 1, \ldots, n$ is given

---

[1]One MPa is equivalent to 145 psi.

by

$$\widehat{\operatorname{cov}(x,y)} \;=\; \frac{\sum (x_i - \bar{x})(y_i - \bar{y})}{n-1}.$$

In the case of the pavers $n = 24$ and we take cement content and strength as x and y respectively. The reason for the definition of covariance becomes clear if we divide the set of points in Figure 6.2 into four quadrants by drawing the lines $x = \bar{x}$ and $y = \bar{y}$. Now look at a typical point, (x_i, y_i), in the upper left quadrant. Since $x_i - \bar{x} < 0$ and $y_i - \bar{y} > 0$ the product

$$(x_i - \bar{x})(y_i - \bar{y}) \;<\; 0,$$

and this point makes a negative contribution to $\widehat{\operatorname{cov}(x,y)}$. Similarly, all the other points in this quadrant, and in the lower right quadrant, make negative contributions to $\widehat{\operatorname{cov}(x,y)}$. In contrast, all the points in the lower left and upper right quadrants will make positive contributions to $\widehat{\operatorname{cov}(x,y)}$. In the case of the pavers, most of the points are in the lower left and upper right quadrants and we can deduce that $\widehat{\operatorname{cov}(x,y)}$ will be positive. The calculations, shown in the following R code, give $\bar{x} = 17.2$, $\bar{y} = 54$, and $\widehat{\operatorname{cov}(x,y)} = 13.4$ in a unit of $MPa \times \%$ cement. The `madx` column is mean-adjusted x, that is x less the mean of x. The column `mady` is mean-adjusted y and `madp` is their product.

```
> pavers.dat=read.table("pavers.txt",header=TRUE)
> attach(pavers.dat)
> x=cement
> mx=mean(cement) ; print(mx)
[1] 17.1875
> y=strength
> my=mean(y) ; print(my)
[1] 53.66667
> n=length(x) ; print(n)
[1] 24
> madx=x-mx
> mady=y-my
> madp=madx*mady
> pavers=data.frame(x,y,madx,mady,madp)
> head(pavers)
     x    y    madx       mady        madp
1 16.6 38.4 -0.5875 -15.266667    8.969167
2 16.6 75.8 -0.5875  22.133333 -13.003333
3 15.3 40.0 -1.8875 -13.666667   25.795833
4 17.1 38.0 -0.0875 -15.666667    1.370833
5 20.7 60.3  3.5125   6.633333   23.299583
6 20.8 70.0  3.6125  16.333333   59.004167
> scov=sum(madp)/(n-1) ; print(scov)
[1] 13.35783
```

The R function `cov(·)` calculates the covariance[2] directly.

---

[2]The MATLAB command `cov(cement,strength)` calculates a 2×2 matrix of variances of each of cement contents and compressive strength on the diagonals and the covariance between cement contents and compressive strength on the off diagonals.

```
> cov(cement,strength)
[1] 13.35783
```

The absolute magnitude of the sample covariance depends on the choice of units and we scale it to a non-dimensional quantity known as the **correlation coefficient** to give a summary measure of how closely the points are scattered about a straight line.

Definition 6.2: Sample correlation coefficient

The **sample correlation coefficient** (r) is a dimensionless quantity obtained from the covariance by making it non-dimensional. It is calculated by dividing the sample covariance by the product of the sample standard deviations of x and y, denoted by s_x and s_y.

$$r = \frac{\widehat{\text{cov}(x,y)}}{s_x s_y}.$$

It is shown in Section 6.4.1.1 that

$$-1 \leq r \leq 1.$$

The correlation coefficient will take its extreme value of -1 if the points lie on a straight line with a negative slope, and 1 if they lie on a line with a positive slope If points are scattered equally over all four quadrants, so that there is no apparent linear relationship, the correlation coefficient will be close to 0. Table 6.3 is a guide to interpretation of the correlation coefficient .

TABLE 6.3: Guide to interpretation of correlation coefficients.

Value around	Description
0	No linear relationship
$-0.3/+0.3$	Weak negative/positive linear relationship
$-0.5/+0.5$	Moderate negative/positive linear relationship
$-0.7/+0.7$	Strong negative/positive linear relationship
$-1/+1$	Points lie on a straight line with negative/positive slope

The correlation coefficient between the compressive strengths and cement contents is 0.53.

```
> sx=sd(cement) ; print(sx)
[1] 1.792815
> sy=sd(strength) ; print(sy)
[1] 14.17514
> correl=scov/(sx*sy) ; print(correl)
[1] 0.5256215
```

The R function cor(·) calculates the correlation coefficient [3] directly.

---

[3]The MATLAB command corr(cement,strength) calculates the correlation coefficient between cement contents and compressive strength (see also corrcoef(·,·) by typing help corrcoef).

```
> cor(cement,strength)
[1] 0.5256215
```

A non zero correlation coefficient does not imply that a change in one variable is a physical cause of a change in the other variable. There may be physical reasons to explain causation, but it does not follow from the statistical analysis alone.

Example 6.1: Cement pavers [interpreting the correlation coefficient]

In the case of the pavers the cement is known to contribute to their strength and there is a convincing physical explanation for this. The scatter plot shows that whilst strength does tend to increase with cement content over the range 15% to 21%, there is considerable variation about the theoretical relationship. This variation arises from variability in other factors, known or unknown, that affect the strength. The measurement error is probably negligible compared with these other sources of variability.

Example 6.2: Moonlight Gold Prospect [correlation and log transformation]

Arsenic occurs as a mineral compound, combined with sulphur and one of iron or nickel or cobalt, and can be associated with gold. The data in MoonlightAsAu.txt are measurements (ppm) of arsenic (As) and gold (Au) in 115 spot analyses of pyrite from the Moonlight epithermal gold prospect in Queensland, Australia [Winderbaum et al., 2012]. The data are plotted in Figure 6.3 (left panel), and the correlation coefficient is 0.65. However, the three outlying points in the upper right of the figure have a disproportionate effect on the calculation of the correlation coefficient . In the mining industry it is common to use logarithms of element concentrations, and a plot of the logarithm of Au against the logarithm of As is shown in Figure 6.3 (right panel).

The correlation coefficient is reduced to 0.53.

```
> Moonlight.dat=read.table("MoonlightAsAu.txt",header=T)
> head(Moonlight.dat)
        As      Au
1 16117.45 219.23
2 11030.52  85.98
3 34879.37 359.09
4  9822.49  13.72
5 26180.63 211.97
6  3460.68   3.74
> attach(Moonlight.dat)
> par(mfrow=c(1,2))
> plot(As,Au)
> cor(As,Au)
[1] 0.6510413
> plot(log10(As),log10(Au))
> cor(log10(As),log10(Au))
[1] 0.5333749
```

The association between As and Au is relevant to geologists and mining engineers but neither element causes the other.

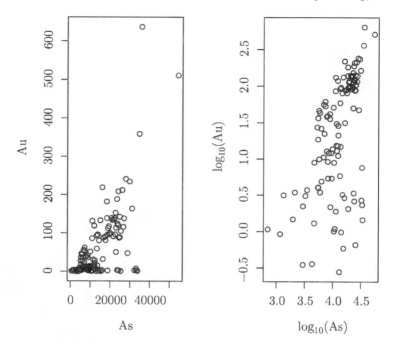

FIGURE 6.3: Scatter plots of ppm measurements of Au against As (left) and of $\log_{10}(\text{Au})$ against $\log_{10}(\text{As})$ (right).

Although there is no direct causal link between As and Au, it is reasonable to suppose that a common cause, certain physical properties of lava, accounts for the association. However, we can show high correlations between variables, particularly if they are measured over time, when there is no plausible common cause. We demonstrate this in Example 6.3.

Example 6.3: Sydney to Hobart yacht race [spurious correlation]

Consider the Sydney to Hobart Yacht Race from 1945 until 2013 and the Annual Mean Land-Ocean Temperature Index in 0.01 degrees Celsius relative to the base period 1951-1980 (NASA) for the same period. Time series plots are shown in Figure 6.4(a) and (b).

If we take the sampling unit as a year we have two variables, winning time and global temperature, and the scatter plot is shown in Figure 6.4(c). The correlation coefficient between winning time and global temperature is -0.73, but this arises because there is a tendency for the winning times to decrease over time, due to improved yacht design, and there has been an increasing trend in global temperature over the same period.

```
> SHYR.dat=read.table("Sydney2HobartYR.txt",header=T)
> head(SHYR.dat)
  year D  H  M  S
1 1945 6 14 22  0
2 1946 5  2 53 33
3 1948 5  3  3 54
4 1949 5 10 33 10
5 1950 5  5 28 35
```

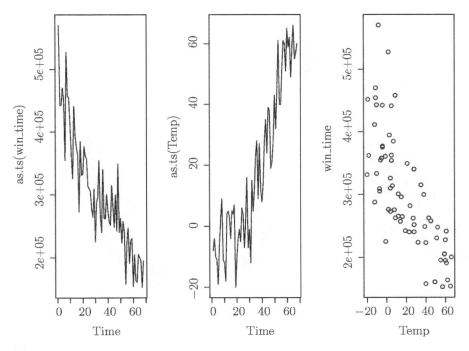

FIGURE 6.4: Time series plot of winning times in the Sydney-Hobart yacht race (left panel), global temperatures (middle panel), together with a scatter plot of time, temperature pairs (right panel).

```
6 1951 4  2 29  1
> attach(SHYR.dat)
> win_time=D*24*60*60+H*60*60+M*60+S
> par(mfrow=c(1,3))
> plot(as.ts(win_time))
> TEMP.dat=read.table("GlobalTemp.txt",header=T)
> head(TEMP.dat)
  Year Glob NHem SHem X24N90N X24S24N X90S24S X64N90N X44N64N
1 1880  -20  -33   -8     -37     -21      -3    -103     -46
2 1881  -12  -22   -2     -30      -6      -4     -69     -44
3 1882  -15  -24   -7     -25     -19      -2    -134     -25
4 1883  -18  -30   -7     -38     -16      -1     -38     -72
5 1884  -26  -41  -11     -56     -15     -12    -133     -67
6 1885  -24  -35  -14     -53     -12     -12    -121     -57
  X24N44N EQU24N X24SEQU X44S24S X64S44S X90S64S Year.1
1     -22    -26     -16      -6       2      60   1880
2     -14    -11       0      -7      -1      33   1881
3      -5    -23     -14      -6       4      49   1882
4     -16    -16     -16      -6       7      46   1883
5     -34    -19     -10     -18      -3      31   1884
6     -37     -7     -17     -22       2      48   1885
> Temp=TEMP.dat$Glob[67:134]
> plot(as.ts(Temp))
> plot(Temp,win_time)
```

```
> cor(Temp,win_time)
[1] -0.756081
```

The term **spurious correlation** is sometimes used to describe such irrelevant and potentially misleading correlation coefficients[4].

As well as being careful about interpreting substantially valued correlation coefficients, we should not take a zero correlation coefficient as evidence that variables are independent. The correlation coefficient is a measure of linear association and data can show non-linear patterns that have near zero correlation coefficients. The following examples demonstrate some of the limitations of correlation coefficients, and emphasize the need to plot the data.

Example 6.4: Pairs of Independent deviates [sampling variability]

We first look at some plots for data pairs when the two variables are just independent random numbers. In Figure 6.5 the left hand column shows pairs of independent normal deviates and the right hand column shows pairs of independent exponential deviates. The rows correspond to sample sizes of 10, 100 and 1 000 respectively.

```
> set.seed(8)
> par(mfcol=c(3,2))
> x11=rnorm(10) ; y11=rnorm(10)
> plot(x11,y11) ; print(c("Normal n=10",round(cor(x11,y11),2)))
 [1] "Normal n=10" "-0.37"
> x21=rnorm(100) ; y21=rnorm(100)
> plot(x21,y21) ; print(c("Normal n=100",round(cor(x21,y21),2)))
 [1] "Normal n=100" "0.04"
> x31=rnorm(1000) ; y31=rnorm(1000)
> plot(x31,y31) ; print(c("Normal n=1000",round(cor(x31,y31),2)))
 [1] "Normal n=1000" "-0.04"
> x12=rexp(10) ; y12=rexp(10)
> plot(x12,y12) ; print(c("Exponential n=10",round(cor(x12,y12),2)))
 [1] "Exponential n=10" "0.77"
> x22=rexp(100) ; y22=rexp(100)
> plot(x22,y22) ; print(c("Exponential n=100",round(cor(x22,y22),2)))
 [1] "Exponential n=100" "0.25"
> x32=rexp(1000) ; y32=rexp(1000)
> plot(x32,y32) ; print(c("Exponential n=1000",round(cor(x32,y32),2)))
 [1] "Exponential n=1000" "-0.01"
```

With samples of size 10 the correlation coefficient is highly influenced by any outlying points. The population correlation coefficient is 0 yet the sample correlation coefficients are -0.37 and 0.77 from the normal and exponential distributions respectively.

---

[4]A claim that global warming might lead to increased wind speeds which in turn tend lead to reduced passage times for yachts between Sydney and Hobart is not convincing. Changes in average wind speeds have been slight and the tendency has been for a decrease because of warming at the poles, despite evidence of an increase in extreme events. Even if average wind speeds had increased over the period, this would be a contributing factor rather than a sole cause.

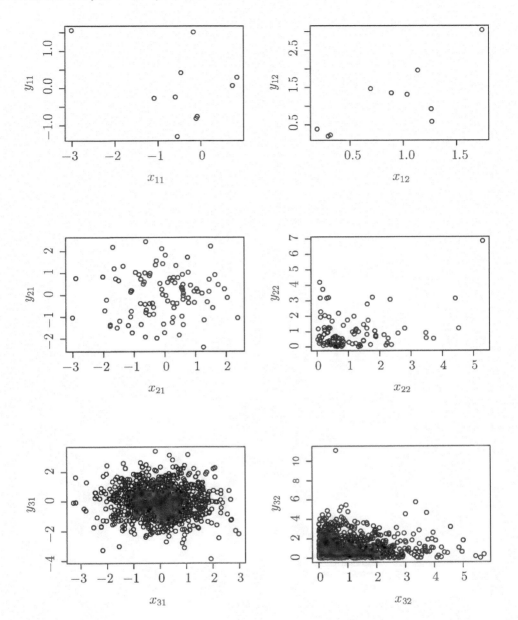

FIGURE 6.5: independent pairs of random deviates from normal (left column) and exponential (right column) distributions for sample sizes 10, 100, 1 000 (rows).

Example 6.5: Compression strength [curvature apparent in scatter plot]

The compressive strength of high-performance concrete is related to the plastic viscosity of the mix.

Laskar (2011) describes a test procedure program and results, viscosity (Pa) and compressive strength (MPa), from tests of 33 concrete cubes are shown in Figure 6.6.

```
> HPconcrete.dat=read.table("HPconcrete.txt",header=T)
> attach(HPconcrete.dat)
> with(HPconcrete.dat,plot(Viscosity,Strength))
> with(HPconcrete.dat,cor(Viscosity,Strength))
[1] 0.3492888
```

The correlation coefficient of 0.35 is weakly positive, but it can be seen from the scatterplot that the compressive strength tends to increase with plastic velocity from around 20 up to around 65 but then tends to decrease as the plastic viscosity increases towards 100.

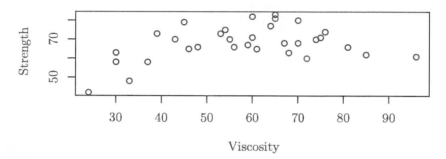

FIGURE 6.6: High performance concrete compressive strength against viscosity.

Example 6.6: Near zero correlation coefficient [non-linear relationship]

The data plotted in Figure 6.7 have a near zero correlation coefficient (0.06), but there is a clear pattern that would be missed without plotting the data.

```
> A=read.table("A.txt",header=T)
> plot(A$x,A$y)
```

Can you guess what these data represent?

To summarize, the correlation coefficient is a measure of linear association between two variables measured for each item. Do not rely on the correlation coefficient alone, always plot the data. If there is a substantial correlation coefficient use physical arguments to support an explanation such as: directly causal; some common cause; or just matching trends over time.

6.3   Bivariate distributions, population covariance and correlation coefficient

We often need to consider more than one variable for each member of a population, and any relationships between the variables may be of particular importance. An example is the data for sea states in the North Sea in Table 3.15.The cells are classed by significant wave

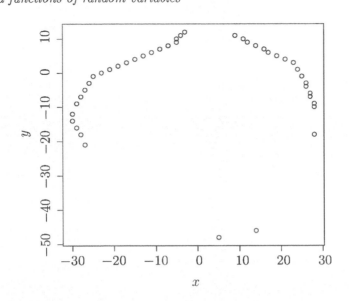

FIGURE 6.7: Plan view of engineered structure.

height from 0 m to 11 m in intervals of 0.5 m (rows), and by mean zero crossing period from 0 s to 13 s in intervals of 1 s (columns). If we add over the rows in each column we obtain the distribution of crossing period. This distribution is referred to as a **marginal distribution** because it was obtained by summing over other variables. We can also find the marginal distribution of the heights by adding over the columns in each row. However, these two marginal distributions would not tell us anything about the relationship between amplitude and frequency.

An offshore structure is relatively sensitive to specific frequencies, and if these tend to coincide with higher-amplitude waves then the safe operating life will be reduced. The approximate correlation coefficient can be calculated from grouped data, by taking midpoints of grouping intervals and multiplying by the frequencies (Section 6.3.3).

6.3.1 Population covariance and correlation coefficient

The sample covariance, which is defined for bivariate data, is an average value of the products of the deviations from their means. Taking expected value is averaging over the population, and the population definitions correspond to the sample definitions with expectation replacing division by $(n-1)$.

Definition 6.3: Population covariance

The covariance of a bivariate random variable $(X.Y)$ is

$$\mathrm{cov}(X,Y) \quad = \quad \mathrm{E}[(X - \mu_X)(Y - \mu_Y)],$$

where μ_X and μ_Y are the means of X and Y respectively.

Definition 6.4: Population correlation coefficient

The correlation coefficient of $(X.Y)$ is

$$\rho \;=\; \frac{\mathrm{cov}(X,Y)}{\sigma_X \sigma_Y},$$

where σ_X and σ_Y are the standard deviations of X and Y respectively.

The correlation coefficient ρ for any bivariate random variable (X,Y) satisfies

$$-1 \;\leq\; \rho \;\leq\; 1.$$

This is a consequence of its definition as we demonstrate in Section 6.4.1.1 If X and Y are independent then ρ is 0, but a correlation coefficient of 0 does not necessarily imply that X and Y are independent.

The expected value of a discrete random variable is obtained by replacing relative frequencies with probabilities. A continuous variable in a population is modeled with a probability density function (pdf), and replacement of relative frequencies in a histogram by areas under the pdf leads to the definition of expected value as an integral. The same principles apply to bivariate distributions.

6.3.2   Bivariate distributions - Discrete case

Prabhu (1996) classified a sample of 298 manufacturing companies in the north-east of England by size and world class status. The size was categorized by number of employees as: $1-49$; $50-149$; $150-249$; and 250 or more. We code these size categories by a random variable X which takes values $1, 2, 3, 4$ if a company is in the categories $1-49\ldots$, up to 250 or more. We represent world class status by Y which can take values $1, 2$ or 3 where:

- 1 corresponds to Prabhu's "poor practices and poor performance";

- 2 corresponds to both Prabhu's "promising - good practices but not yet good performance" and his "vulnerable - poor practices but good performance";

- 3 corresponds to "good practices and good performance".

The numbers of companies in the ordered categories are given in Table 6.4. [5]
    The row sums give us the distribution of these companies by world class in the right hand margin. When the distribution of one variable is obtained from a bivariate distribution in this way it is referred to as a **marginal distribution**. The column totals give us the marginal distribution of companies by size. To demonstrate the concept of a bivariate probability mass distribution we will divide the numbers of companies in categories by their total (298) to obtain relative frequencies, and define a probability mass function with probabilities equal to these relative frequencies. A bivariate pmf has the form

$$P_{XY}(x,y) \;=\; \mathrm{P}(X = x \cap Y = y).$$

Numerical values for the manufacturing companies are given in Table 6.5 The marginal

---

[5]Prabhu (1996) distinguishes promising (17 companies: 3, 6, 6, 2 by increasing size) from vulnerable (74 companies), but in the context of this section we want to define ordered categories and consider both promising and vulnerable companies as above 1 and below 3 on the world class scale.

TABLE 6.4: The numbers of companies in the ordered categories size (horizontal) and world class (vertical).

	1	2	3	4	
3	35	23	31	23	112
2	44	21	20	6	91
1	26	47	15	7	95
	105	91	66	36	298

TABLE 6.5: A bivariate distribution of companies in the ordered categories: size (horizontal); and world class (vertical).

	1	2	3	4	
3	0.12	0.08	0.10	0.08	0.38
2	0.14	0.07	0.07	0.02	0.30
1	0.09	0.16	0.05	0.02	0.32
	0.35	0.31	0.22	0.12	1.00

distribution of X is given by:

$$P_X(x) = \sum_{y=1}^{3} P_{XY}(x, y)$$

and is given along the foot of the table. The mean of X is

$$
\begin{aligned}
\mu_X &= \mathrm{E}[X] = \sum_{x=1}^{4} x P_X(x) \\
&= 1 \times 0.35 + 2 \times 0.31 + 3 \times 0.22 + 4 \times 0.12 \\
&= 2.11.
\end{aligned}
$$

The variance of X is

$$
\begin{aligned}
\sigma_X^2 &= \mathrm{E}\left[(X - \mu_X)^2\right] \\
&= \sum_{x=1}^{4} (x - \mu_X)^2 P_X(x) \\
&= (1 - 2.11)^2 \times 0.35 + \ldots + (4 - 2.11)^2 \times 0.12 \\
&= 1.0379.
\end{aligned}
$$

Hence the standard deviation

$$\sigma_X = \sqrt{1.0379} = 1.019.$$

Similar calculations give $\mu_Y = 2.06$ and $\sigma_Y = 0.835$. The covariance is

$$
\begin{aligned}
\text{cov}(X, Y) &= \text{E}[(X - \mu_X)(Y - \mu_Y)] \\
&= \sum_{x=1}^{4}\sum_{y=1}^{3}(x - \mu_X)(y - \mu_Y)\text{P}(x, y) \\
&= (1 - 2.11)(1 - 2.06) \times 0.09 + (1 - 2.11)(2 - 2.06) \times 0.14 + \ldots \\
&\qquad \ldots + (4 - 2.11)(3 - 2.06) \times 0.08 \\
&= 0.01404.
\end{aligned}
$$

The correlation coefficient is

$$
\rho = \frac{0.01404}{1.019 \times 0.835} = 0.017
$$

and although it is almost 0, X and Y are far from independent. The probabilities of a company being in the highest world class category conditioned on its size are:

$$
\text{P}(Y = 3 | x = 1) = (0.12/0.35) = 0.34, \qquad \text{P}(Y = 3 | x = 2) = (0.08/0.31) = 0.20,
$$

$$
\text{P}(Y = 3 | x = 3) = (0.10/0.22) = 0.45, \qquad \text{P}(Y = 3 | x = 4) = (0.08/0.12) = 0.67.
$$

According to this model companies in size category 2 ($50 - 149$ employees) are less successful than smaller or larger companies.

Although $\rho = 0$ does not imply X and Y are independent, if X and Y are independent then $\rho = 0$. To prove this fact, we first note that if X and Y are independent then

$$
\text{P}_{XY}(x, y) = \text{P}(X = x \cap Y = y) = \text{P}(X = x) \times \text{P}(Y = y) = \text{P}_X(x)\text{P}_Y(y).
$$

It follows that

$$
\begin{aligned}
\text{cov}(X, Y) &= \sum_x\sum_y(x - \mu_X)(y - \mu_Y)\text{P}_{XY}(x, y) \\
&= \sum_x\sum_y(x - \mu_X)(y - \mu_Y)\text{P}_X(x)\text{P}_Y(y) \\
&= \left(\sum_x(x - \mu_X)\text{P}_X(x)\right)\left(\sum_y(y - \mu_Y)\text{P}_Y(y)\right) \\
&= 0 \times 0 = 0.
\end{aligned}
$$

In general, the expected value of any function $\phi(\cdot, \cdot)$ of X and Y is

$$
\text{E}[\phi(X, Y)] = \sum_x\sum_y\phi(x, y)\text{P}_{XY}(x, y).
$$

The same principles hold for continuous distributions with integration replacing summation.

6.3.3 Bivariate distributions - Continuous case

6.3.3.1 Marginal distributions

Silver and lead often occur together in mineral deposits. The data in Table 6.6 are a summary of the log-silver content $\left(log_{10}(\text{Ag})\right)$, where Ag is silver content in ppm weight and log-lead

TABLE 6.6: Moonlight Gold Prospect.

		log-lead					
		$-1 \to 0$	$0 \to 1$	$1 \to 2$	$2 \to 3$	$3 \to 4$	
log silver	$3 \to 4$	0	0	0	0	12	12
	$2 \to 3$	0	11	24	15	7	57
	$1 \to 2$	0	8	49	20	0	77
	$0 \to 1$	2	1	6	3	0	12
	$-1 \to 0$	0	0	1	0	0	1
		2	20	80	38	19	159

content $(log_{10}(\text{Pb}))$, where Pb is lead content in ppm weight from 159 spot analyses of drill cores from the Moonlight Gold Prospect in Queensland [Winderbaum et al., 2012].

The row sums give us the distribution of log-silver in the right hand margin. The column totals give us the distribution of log-lead along the lower margin. We can construct histograms of log-lead and log-silver from the marginal distributions and calculate approximate means and standard deviations from the grouped data using the formulae.

$$\bar{x} = \sum_{k=1}^{K} x_k \frac{f_x}{n}, \qquad s = \sqrt{\sum_{k=1}^{K} (x_k - \bar{x})^2 \frac{f_x}{n-1}},$$

where K is the number of bins, x_k are mid points of the bins, f_k are the frequencies, and $n = \sum f_k$ is the number of data. The approximate sample mean of the log-lead values is

$$\bar{x} = -0.5 \times \frac{2}{159} + 0.5 \times \frac{20}{159} + 1.5 \times \frac{80}{159} + 2.5 \times \frac{38}{159} + 3.5 \times \frac{19}{159} = 1.86$$

The approximate variance of the log-lead values is

$$s^2 = (-0.5 - \bar{x})^2 \times \frac{2}{159 - 1} + \ldots + (3.5 - \bar{x})^2 \times \frac{19}{159 - 1} = 0.7309$$

and the standard deviation is 0.85. Similar calculations give the mean, variance, and standard deviation of log-silver as 1.95, 0.5953, and 0.77 respectively.

6.3.3.2 Bivariate histogram

We can construct a bivariate histogram in three dimensions (3D) by constructing blocks over the rectangular bins (which happen to be square for the log-lead, log-silver data) such that the volume of the block equals the proportion of data in the bin (see Figure 6.8). In general, we can suppose that the data are pairs (x, y), and that the range of x values is divided into K bins and that the range of y values is divided into L bins. There are then $K \times L$ rectangular bins.

Denote :

the number of data in the (k, ℓ) rectangular bin by $f_{k,\ell}$ and

the mid-point of the rectangular bin by (x_k, y_ℓ),

where $1 \le k \le K$ and $1 \le \ell \le L$ and $n = \sum_{k=1}^{K} \sum_{\ell=1}^{L} f_{k,\ell}$. Then, the heights of the blocks are relative frequency densities:

$$\frac{f_{k,\ell}/n}{\text{bin area}}$$

and the total volume of the histogram equals **1**. The volume of the histogram above some area gives the proportion of data in that area.

6.3.3.3 Covariate and correlation

The approximate sample covariance is:

$$\widehat{\text{cov}(x,y)} \;=\; \sum_{k=1}^{K}\sum_{\ell=1}^{L}(x_k-\bar{x})(y_k-\bar{y})\frac{f_{k,\ell}}{n-1}$$

and the approximate sample correlation coefficient is:

$$r \;=\; \frac{\widehat{\text{cov}(x,y)}}{s_x s_y}.$$

The height of the tallest block for the 3D histogram of log-lead and log-silver is:

$$\frac{49/159}{1\times 1} \;=\; 0.308$$

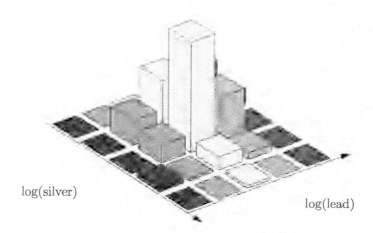

FIGURE 6.8: Bivariate histogram for log(lead) and log(silver). The vertical axis is relative frequency density.

The approximate covariance is

$$(-0.5-1.86)(0.5-1.95)\frac{2}{159}+\ldots+(3.5-1.86)(3.5-1.95)\frac{12}{159} \;=\; 0.257$$

and the approximate correlation coefficient is given by

$$r \;=\; \frac{0.257}{0.85\times 0.77} \;=\; 0.39.$$

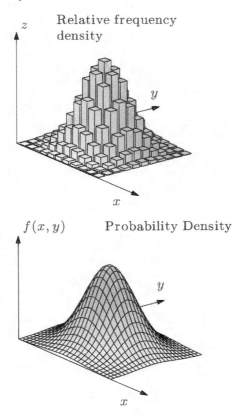

FIGURE 6.9: 3D histogram for sample and bivariate pdf for population.

6.3.3.4 Bivariate probability distributions

If you imagine the sample size increasing towards infinity, the bivariate histogram shown in Figure 6.9 (upper frame) will tend towards a smooth surface defined by a bivariate probability density function (pdf) $f_{XY}(x, y)$ shown in Figure 6.9 (lower frame). A bivariate pdf satisfies two conditions:

$$0 \leq f_{XY}(x, y), \qquad -\infty < x, y < \infty,$$

and

$$\int_{-\infty}^{\infty} \int_{-\infty}^{\infty} f_{X,Y}(x, y) dx dy = 1.$$

The summation of the volumes of the blocks of the histogram tends towards a bivariate integral. We define the marginal distributions by

$$f_X(x) = \int_{-\infty}^{\infty} f_{XY}(x, y) dy \quad \text{and} \quad f_Y(y) = \int_{-\infty}^{\infty} f_{XY}(x, y) dx.$$

The expected value of a general function of X and Y, $\phi(X, Y)$ is given by

$$\mathrm{E}[\phi(X, Y)] = \int_{-\infty}^{\infty} \int_{-\infty}^{\infty} \phi(x, y) f_{XY}(x, y) dx dy.$$

In particular, the covariance is

$$\mathrm{E}[(X - \mu_X)(Y - \mu_y)] \; = \; \int_{-\infty}^{\infty} \int_{-\infty}^{\infty} (x - \mu_X)(y - \mu_Y) f_{XY}(x, y) dx dy.$$

Also, taking $\mathrm{E}[X]$ in the bivariate distribution gives the mean of X in the marginal distribution.

$$\mu_X \; = \; \mathrm{E}[X] \; = \; \int_{-\infty}^{\infty} \int_{-\infty}^{\infty} x f_{XY}(x, y) dx dy$$

$$= \; \int_{-\infty}^{\infty} x \left(\int_{-\infty}^{\infty} f_{X,Y}(x, y) dy \right) dx \; = \; \int_{-\infty}^{\infty} x f_X(x) dx.$$

The bivariate cumulative distribution function (cdf) is defined by:

$$F_{XY}(x, y) \; = \; \mathrm{P}(X \le x, Y \le y) \; = \; \int_{-\infty}^{x} \int_{-\infty}^{y} f_{XY}(\xi, \eta) d\xi d\eta$$

and the bivariate pdf can be obtained from the cdf by partial differentiation:

$$f_{XY}(x, y) \; = \; \frac{\partial^2 F_{XY}(x, y)}{\partial x \partial y}.$$

The definition of independence of random variables X and Y follows from the definition of independent events that was given in Chapter 2. The random variables X and Y are **independent** if, and only if,

$$f_{XY}(x, y) \; = \; f_X(x) f_Y(y).$$

The equivalence with the definition of independent events can be demonstrated with the following argument. Consider a particular point (x_p, y_p). The height of the pdf $f(x_p, y_p)$ does not itself represent a probability, but if it is multiplied by a rectangular element of area $\delta x \delta y$ centered on (x_p, y_p), it represents the probability of (X, Y) being within this element. Then, if and only if X and Y are independent

$$\mathrm{P}(X \text{ within } \delta x/2 \text{ of } x_p \cap Y \text{ within } \delta y/2 \text{ of } y_p)$$
$$= \; \mathrm{P}(X \text{ within } \delta x/2 \text{ of } x_p) \times \mathrm{P}(Y \text{ within } \delta y/2 \text{ of } y_p).$$

These probabilities are now written in terms of the pdfs (see Figure 6.10) :

$$f_{XY}(x_p, y_p) \delta x \delta y \; \approx \; f_X(x_p) \delta x \times f_Y(y_p) \delta y.$$

The result becomes exact as δx and δy tend towards 0, and since δx and δy appear on both sides of the equation and cancel we are left with $f_{XY}(x, y) = f_X(x) f_Y(y)$. It follows from the definitions that if X and Y are independent then the covariance will be 0. However, a covariance of 0 does not necessarily imply that the variables are independent.
From hereon we will usually drop the subscripts on the pdfs and cdfs and rely on the context, and arguments of the functions, to distinguish distributions.

Example 6.7: Start-up phone company [bivariate distribution]

A start-up company sells a new design of mobile phone and also offers a repair service for all phones. Phones are delivered early on Monday morning, and the policy is to

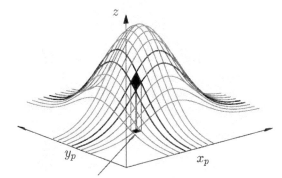

Rectangle area $\delta x \delta y$

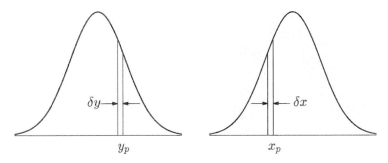

FIGURE 6.10: Upper frame: probability of being within an element entered on (x_p, y_p) as the volume $f(x_p, y_p)\delta x \delta y$. Lower frame: probability of being within $\delta y/2$ of y_p as an area and similarly for x.

start each week with a fixed number of phones in stock. Let X represent the number of mobile phones sold in a week as a proportion of the fixed number of phones at the start of the week. Let Y be the number of repairs as a proportion of the maximum number of repairs that the company can handle in one week. The weekly sales (X, Y) are modeled by the bivariate pdf:

$$f(x, y) = \frac{12}{7}\left(x^2 + xy\right), \qquad 0 \le x, y \le 1.$$

1. The cdf is

$$F(x, y) = \int_0^x \int_0^y \frac{12}{7}\left(\xi^2 + \xi\eta\right) d\xi d\eta$$

$$= \frac{12}{7}\left(\frac{1}{3}x^3 y + \frac{1}{4}x^2 y^2\right) = \frac{1}{7}\left(4x^3 y + 3x^2 y^2\right).$$

We can check that partial differentiation of the cdf $F(x, y)$ with respect to x and y does return the pdf $f(x, y)$, and that $F(1, 1) = 1$.

2. The marginal pdf of X is given by:

$$f(x) = \int_0^1 \frac{12}{7} (x^2 + xy) \, dy = \left[\frac{12}{7}x^2 y + \frac{12}{14}xy^2 \right]_{y=0}^{y=1}$$

$$= \frac{12}{7} \left(x^2 + \frac{1}{2}x \right) = \frac{6}{7} (2x^2 + x)$$

and the marginal cdf is $F(x) = \frac{1}{7} (4x^3 + 3x^2)$. Notice that $F(x) = F(x, 1)$.

3. The marginal pdf of Y is given by

$$f(y) = \int_0^1 \frac{12}{7} (x^2 + xy) \, dx = \left[\frac{12}{21}x^3 + \frac{12}{14}x^2 y \right]_{x=0}^{x=1} = \frac{4}{7} + \frac{6}{7}y.$$

and the marginal cdf is $F(y) = \frac{1}{7} (4y + 3y^2)$. Check that $F(y) = F(1, y)$.

4. We can calculate probabilities such as $P(((X < 0.4) \cap (Y < 0.6))$ by writing an R function for the cdf.

```
Fxy=function(x,y){(4*x^3*y+3*x^2*y^2)/7}
F(0.4,0.6)
Fxy(0.4,0.6)
[1] 0.04662857
Fxy(0.4,1)*Fxy(1,0.6)
[1] 0.05227102
```

The probability is 0.047, and since $P(X < 0.4) \times P(Y < 0.6) = 0.052 \neq 0.047$ we have demonstrated that X and Y are not independent.
The fact that $P((X < 0.4) \cap (Y < 0.6)) < P(X < 0.4) \times P(Y < 0.6)$, suggests that there is a negative correlation coefficient .

5. A general expression for the probability that X exceeds a and Y exceeds b is

$$P((a < X) \cap (b < Y)) = 1 - P((X < a) \cup (Y < b))$$

$$= 1 - \left(P(X < a) + P(Y < b) - P(X < a) \cap (Y < b)) \right).$$

In terms of the cdfs, $P((a < X) \cap (b < Y)) = 1 - F(a) - F(b) + F(a, b)$.

6. The $P((0.5 < X) \cap (0.5 < Y))$ is given by

```
1-(Fxy(0.5,1)+Fxy(1,0.5)-Fxy(0.5,0.5))
[1] 0.4910714
```

This is slightly less than $P(0.5 < X) \times P(0.5 < Y)$

```
(1-Fxy(0.5,1))*(1-Fxy(1,0.5))
[1] 0.4987245
```

which again suggests a negative correlation coefficient.

We can define a **conditional probability distribution** in the same way as we define conditional probability[6]. The conditional pdf of Y given that $X = x$ is defined by:

$$f_{Y|x}(y|x) = \frac{f_{XY}(x,y)}{f_X(x)}.$$

If we consider a particular value of x which we denote by x_p then we can write

$$f(y|x_p) = \frac{1}{f(x_p)} f(x_p, y),$$

which emphasizes that the conditional pdf is proportional to the bivariate pdf with x fixed at x_p. Geometrically, the conditional pdf corresponds to a scaled cross-section through the bivariate pdf cut by a plane through the point x_p and normal to the x-axis. The factor $1/f(x_p)$ scales the area of the cross-section to equal 1 and is known as the **normalizing factor** (see Figure 6.11)

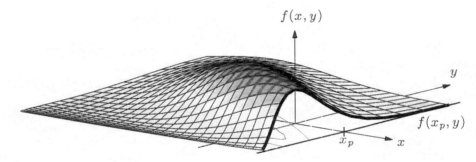

FIGURE 6.11: The conditional distribution of Y given that $x = x_p$ is the cross section of the bivariate pdf cut by a plane through x_p, parallel to the y–z plane, scaled to have an area of 1.

Example 6.7: (Continued) Start-up phone company

Given the bivariate distribution of phone sales and repairs, find the conditional distributions of Y given x and X given y.

$$f(y|x) = \frac{(\frac{12}{7})(x^2 + xy)}{\frac{6}{7}(2x^2 + x)} = \frac{2x + 2y}{2x + 1} \quad \text{for } 0 \le y \le 1.$$

The conditional cdf of Y given x is

$$F(y|x) = \frac{2xy + y^2}{2x + 1}.$$

The pdf and cdf of X given y are

$$f(x|y) = \frac{6x^2 + 6xy}{2 + 3y} \qquad \text{and} \qquad F(x|y) = \frac{2x^3 + 3x^2 y}{2 + 3y}.$$

---

[6]The justification again relies on products of pdfs with ordinates of pdfs with elements of area or length being probabilities (Figure 6.10).

6.3.4 Copulas

Any continuous random variable X can be transformed to a uniform random variable on $[0,1]$, as we now prove. The cumulative distribution function of a continuous distribution $F(\cdot)$ is defined by :

$$P(X < x) \;=\; F(x).$$

Since $F(\cdot)$ is an increasing function,

$$P(X < x) \;=\; P(F(X) < F(x)) \;=\; F(x).$$

Now define $U = F(X)$ and write $u = F(x)$ to obtain

$$P(F(X) < F(x)) \;=\; P(U < u) \;=\; u.$$

Since $0 \le F(x) = u \le 1$ it follows that U is $U[0,1]$. We used the inverse of this result when generating pseudo-random numbers.

A bivariate copula is a bivariate distribution with marginal distributions that are uniform on $[0,1]$. Since any continuous random variable can be transformed to $U[0,1]$, it follows that any bivariate distribution can be transformed to a copula. Conversely, we can take a bivariate copula and transform the uniform marginal distributions to any other distribution and we have constructed a bivariate distribution. The margins of the bivariate distribution can be of quite different types. These ideas extend to more than two variables.

Example 6.8: FMG copula [bivariate uniform distribution]

The cdf of the Farlie-Gumbel-Morgenstern (FMG) copula for the bivariate uniform random variable (U, V) is

$$C(u,v) \;=\; uv(1 - \alpha(1-u)(1-v)), \qquad 0 \le u, v \le 1,$$

where α is a dependency parameter constrained to $[-1, 1]$ although it is not the correlation coefficient . If α is set to 0, $C(u,v) = uv$ and U and V are independent. The marginal cdf of V is given by $C(1,v) = v$, for any value of α and is uniform as required. The pdf of the copula is given by

$$\frac{\partial^2 C(u,v)}{\partial u \partial v} \;=\; 1 + \alpha(1-2u)(1-2v).$$

The FMG copula is somewhat limited because it can only model a moderate degree of association. See Figure 6.12.

6.4 Linear combination of random variables (propagation of error)

We need to consider linear combinations of random variables in many applications. In particular we need expressions for the mean and variance of a linear combination.

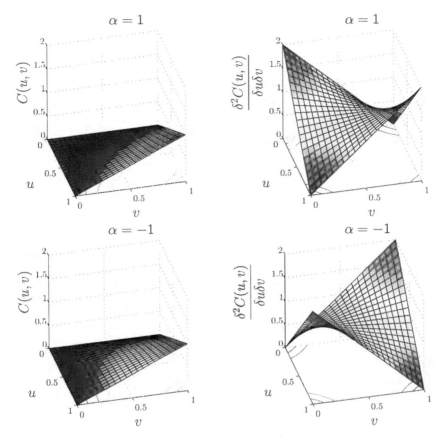

FIGURE 6.12: The cdf and pdf of the Farlie-Gumbel-Morgenstern (FMG) copula for the bivariate uniform random variable (U, V), with $\alpha = 1$ and $\alpha = -1$.

6.4.1 Mean and variance of a linear combination of random variables

If X and Y are random variables a linear combination is defined by,

$$W \quad = \quad aX + bY \ ,$$

where a and b are constants. A key result relates the mean and standard deviation of W, denoted by μ_W and σ_W, to those of X and Y, denoted by μ_X, μ_Y, σ_X and σ_Y, and the correlation coefficient between X and Y. The statement of the result is;

$$\mu_W \quad = \quad a\mu_X + b\mu_Y,$$

$$\sigma_W^2 \quad = \quad a^2\sigma_X^2 + b^2\sigma_Y^2 + 2ab\rho\sigma_X\sigma_Y,$$

$$\implies \sigma_W \quad = \quad \sqrt{(a^2\sigma_X^2 + b^2\sigma_Y^2 + 2ab\rho\sigma_X\sigma_Y)},$$

These formulae will be used in many places throughout this book.

The proof follows from the definitions of expected value. First the mean

$$\mu_W \quad = \quad \mathrm{E}[W] \quad = \quad \mathrm{E}[aX + bY] \quad = \quad a\mathrm{E}[X] + b\mathrm{E}[Y] \quad = \quad a\,\mu_X + b\,\mu_Y.$$

Then the variance

$$
\begin{aligned}
\sigma_W^2 &= \mathrm{E}\big[(W - \mu_W)^2\big] &= \mathrm{E}\Big[\big((aX + bY) - (a\mu_X + b\mu_Y)\big)^2\Big] \\[2mm]
&= \mathrm{E}\Big[\big((aX - a\mu_X) + (bY - b\mu_Y)\big)^2\Big] \\[2mm]
&= \mathrm{E}\big[(aX - a\mu_X)^2 + (bY - b\mu_Y)^2 + 2(aX - a\mu_X)(bY - b\mu_Y)\big] \\[2mm]
&= \mathrm{E}\big[a^2(X - \mu_X)^2 + b^2(Y - \mu_Y)^2 + 2ab(X - \mu_X)(Y - \mu_Y)\big] \\[2mm]
&= a^2\mathrm{E}\big[(X - \mu_X)^2\big] + b^2\mathrm{E}\big[(Y - \mu_Y)^2\big] + 2ab\mathrm{E}[(X - \mu_X)(Y - \mu_Y)] \\[2mm]
&= a^2\sigma_X^2 + b^2\sigma_Y^2 + 2ab\,\mathrm{cov}(X, Y),
\end{aligned}
$$

which can be expressed in terms of the correlation coefficient by substituting $\mathrm{cov}(X, Y) = \rho\sigma_X\sigma_Y$. The result extends to any number of variables by mathematical induction (Exercise 6.21). An important special case is: the variance of a sum of independent random variables is the sum of the variances.

Example 6.9: Marine survey vessel [mean and variance of linear combination]

A marine survey vessel has two instruments that provide measurements X and Y of the depth of the sea bed. The instruments have been carefully calibrated and give **unbiased** estimates of the depth. That is, if the depth is θ then

$$
\mathrm{E}[X] = \mathrm{E}[Y] = \theta.
$$

However the measurements are subject to error, and the first instrument is more precise with

$$
\sigma_Y = \sqrt{2}\sigma_X.
$$

A surveyor intends averaging the measurements but thinks that some weighted average will be better, inasmuch as it has a smaller standard deviation, than $(X + Y)/2$. The mean of

$$
0.5X + 0.5Y \quad \text{is} \quad 0.5\theta + 0.5\theta = \theta
$$

and if we assume that the errors are independent, the variance is

$$
0.25\sigma_X^2 + 0.25\sigma_Y^2 = (0.25 + 0.25(\sqrt{2})^2)\sigma_X^2 = 0.75\sigma_X^2,
$$

giving the standard deviation as $0.87\sigma_X$.
This is an improvement on X alone, but it seems sensible to give more weight to the more accurate measurement. If we try,

$$
0.8X + 0.2Y,
$$

the mean is

$$
0.8\theta + 0.2\theta = \theta
$$

and the variance, assuming that the errors are independent, is

$$
0.64\sigma_X^2 + 0.04\sigma_Y^2 = (0.64 + 0.04(\sqrt{2})^2)\sigma_X^2 = 0.72\sigma_X^2,
$$

giving the standard deviation as $0.85\sigma_X$.

This estimator is only a slight improvement on using equal weights, so we may have given too much weight to X. We can find the weights that minimize the variance by considering:

$$aX + (1-a)Y.$$

The sum of the weights must be 1 for the expected value to equal θ.

$$\mathrm{E}[aX + (1-a)Y] \;=\; a\mathrm{E}[X] + (1-a)\mathrm{E}[Y] \;=\; \big(a + (1-a)\big)\theta \;=\; \theta.$$

If we assume errors are independent

$$\mathrm{var}(aX + (1-a)Y) \;=\; a^2\sigma_X^2 + (1-a)^2 \times 2\sigma_X^2.$$

To find a stationary point of this expression, which will be a minimum, differentiate with respect to a and set the derivative to 0 to obtain $a = 2/3$. The variance is then

$$\left(\frac{4}{9} + 2\frac{1}{9}\right)\sigma_X^2 \;=\; \frac{2}{3}\sigma^2.$$

The standard deviation is now $0.82\sigma_X$, and this is the minimum. Now suppose that tests of the instruments at known depths have shown that the errors have a correlation coefficient of 0.4, because turbidity affects both instruments. How will this affect the mean, variance, and standard deviation of $\frac{2}{3}X + \frac{1}{3}Y$?

The mean is unaffected by the correlation coefficient

$$\mathrm{E}\left[\frac{2}{3}X + \frac{1}{3}Y\right] \;=\; \frac{2}{3}\mathrm{E}[X] + \frac{1}{3}\mathrm{E}[Y] \;=\; \frac{2}{3}\theta + \frac{1}{3}\theta \;=\; \theta.$$

The correlation coefficient is positive, so the variance will be greater than if the errors are independent.

$$\mathrm{var}\left(\frac{2}{3}X + \frac{1}{3}Y\right) \;=\; \frac{4}{9}\sigma_X^2 + \frac{1}{9} \times 2\sigma_Y^2 + 2 \times \frac{2}{3} \times \frac{1}{3}\mathrm{cov}(X,Y).$$

The covariance term is

$$\mathrm{cov}(X,Y) \;=\; \rho\sigma_X\sigma_Y \;=\; 0.4\sigma_X\sqrt{2}\sigma_X$$

and the variance is $0.92\sigma_X^2$. The standard deviation is $0.96\sigma_X$.

You are asked to find the minimum variance unbiased linear combination of X and Y, when they are correlated, in Exercise 6.14.

6.4.1.1 Bounds for correlation coefficient

We can now prove that $-1 \le \rho \le 1$. Since the variance of any quantity is non-negative,

$$0 \le \mathrm{var}(aX + bY) \;=\; a^2\sigma_X^2 + b^2\sigma_Y^2 + 2ab\,\rho\sigma_X\sigma_Y,$$

for any a, b. In particular, we can set $a = \sigma_Y$ and $b = \sigma_X$. Then

$$0 \;\le\; \sigma_Y^2\sigma_X^2 + \sigma_X^2\sigma_Y^2 + 2\sigma_Y\sigma_X\rho\sigma_X\sigma_Y$$

and it follows that

$$-2\sigma_X^2\sigma_Y^2 \;\leq\; 2\sigma_X^2\sigma_Y^2\rho.$$

Divide through by $\sigma_X^2\sigma_Y^2$, a positive quantity, to get $-1 \leq \rho$. Setting $a = \sigma_Y$ and $b = -\sigma_X$ leads to $\rho \leq 1$, and hence

$$-1 \;\leq\; \rho \;\leq\; 1.$$

A similar argument with population quantities replaced by the corresponding sample quantities leads to

$$-1 \;\leq\; r \;\leq\; 1.$$

6.4.2 Linear combination of normal random variables

The results for the mean and variance of a linear combination of random variables apply for any distributions for which means and variances are defined, but the form of the distribution is not specified. If the random variables have a normal distributions then any linear combination is also normally distributed, and this holds whether or not the variables are correlated. Probability distributions with this property are known as **stable distributions**, and the only stable distribution with a finite variance is the normal distribution[7].

Example 6.10: Timber framed houses [linear combination of variables]

A company manufactures timber framed houses. The side walls are composed of lower and upper sections. Lower sections have heights $X \sim N(3.00, (0.02)^2)$, and upper sections have heights $Y \sim N(2.50, (0.016)^2))$, where the measurements are in meters. If sections are despatched at random, what is the distribution of the difference in the heights of side walls when a house is assembled on a flat concrete slab? What is the probability that this difference exceeds 0.05?

Let X_1, X_2 be the heights of the two lower sections and Y_1, Y_2 be the heights of the upper sections. The difference

$$D \;=\; (X_1 + Y_1) - (X_2 + Y_2) \;=\; X_1 + Y_1 - X_2 - Y_2.$$

The mean of D

$$\mu_D \;=\; \mathrm{E}[D] \;=\; 3.00 + 2.50 - 3.00 - 2.50 \;=\; 0.$$

It is reasonable to suppose that heights of sections are independent and the variance of D is then

$$\sigma_D^2 \;=\; (0.02)^2 + (0.016)^2 + (0.02)^2 + (0.016)^2 \;=\; 0.001312 \;=\; (0.0362)^2$$

and hence $D \sim N(0, (0.362)^2)$. The probability the difference exceeds 0.05 is given by

```
> 1-2*(pnorm(0.05,0,0.0362)-0.5)
[1] 0.1672127
```

which rounds to 0.17

---

[7]A sum of independent gamma random variables with the same scale parameter will be gamma but this does not hold for any linear combination and is limited to independent variables.

Example 6.11: Shipyard designs [linear combination of variables]

[Kattan, 1993] aimed to reduce rework in shipyards by simplifying designs. The panel shown in Figure 6.13 is made up of 5 plates ($6m \times 1.5m$), 8 stiffeners and 3 webs. The specification for the width of the panel is within $[-7mm, +3mm]$ of 7.5m. Welders recorded process data and summarized the findings as shown in Table 6.7. The histograms of discrepancies from each operation looked close to normal distributions.

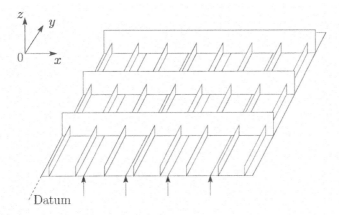

FIGURE 6.13: Built up stiffened panel of five plates. The eight stiffeners run along plates in the y-direction. The three webs run across plates in the x-direction.

TABLE 6.7: Stiffened plate data.

Operation	mean discrepancy(mm)	standard deviation of discrepancy(mm)
Plate cutting	+0.5	0.9
Plate alignment	+1.0	0.8
Butt welding shrinkage	-4.0	1.1
Weld shrinkage per stiffener	-1.0	1.0
Weld shrinkage per web	-1.5	1.2

There are 5 plate cuttings and 4 alignments for each panel, before a welder makes 4 butt welds, welds 8 stiffeners, and welds 3 webs. The mean error in the overall width of a panel will be

$$5 \times 0.5 + 4 \times 1 + 4 \times (-4) + 8 \times (-1) + 3 \times (-1.5) = -22.$$

This can be compensated for if the designer increases the width of the plates by 20/5 to 1.504 mm so that the mean error is in the middle of the specification at -2. The variance of the overall width, assuming individual discrepancies are independent, is

$$5 \times 0.9^2 + 4 \times 0.8^2 + 4 \times 1.1^2 + 8 \times 1.0^2 + 3 \times 1.2^2 = 23.8$$

The standard deviation is 4.9 mm. The probability of reworking a panel is

```
> 1-2*(pnorm(5,0,4.9)-0.5)
[1] 0.3075349
```

and 0.31 is far too high if the shipyard is to stay in business. The panel was re-designed with 3 plates of width 2.5 and 6 stiffeners, with the plates being slightly thicker to compensate for the reduced number of stiffeners. The variance is now

$$3 \times 0.9^2 + 2 \times 0.8^2 + 2 \times 1.1^2 + 6 \times 1.0^2 + 3 \times 1.2^2 \quad = \quad 17.0$$

The standard deviation is reduced to 4.12 mm, and the probability of reworking a panel is reduced to

```
>  1-2*(pnorm(5,0,4.12)-0.5)
[1] 0.2249035
```

A probability of 0.22 is an improvement, but is still too high. The next step is to investigate whether the welding variability can be reduced as this makes a greater contribution to the overall variability than the plate cutting or alignment.

6.4.3   Central Limit Theorem and distribution of the sample mean

Let $\{X_i\}$ for $1 \leq i \leq n$ be n independent random variables from a distribution with mean μ and finite variance σ^2. Define the sample total T

$$T \quad = \quad T_1 + \cdots + T_n.$$

The mean of T is

$$\mu_T \quad = \quad \mu + \cdots + \mu \quad = \quad n\,\mu.$$

A random sample justifies an assumption of independence and the variance of T is then

$$\sigma_T^2 \quad = \quad \sigma^2 + \cdots + \sigma^2 \quad = \quad n\sigma^2$$

and hence

$$\sigma_T \quad = \quad \sqrt{n}\,\sigma.$$

The sample mean $\overline{X}$ is defined by

$$\overline{X} \quad = \quad \frac{T}{n}.$$

The mean of the distribution of $\overline{X}$ is

$$\mu_{\overline{X}} \quad = \quad \frac{\mu_T}{n} \quad = \quad \frac{n\,\mu}{n} \quad = \quad \mu.$$

The variance of the distribution of $\overline{X}$ is

$$\sigma_{\overline{X}}^2 \quad = \quad \frac{\sigma_T^2}{n^2} \quad = \quad \frac{n\,\sigma^2}{n^2} \quad = \quad \frac{\sigma^2}{n}$$

and hence its standard deviation is

$$\sigma_{\overline{X}} \quad = \quad \frac{\sigma_T}{n} \quad = \quad \frac{\sqrt{n}\,\sigma}{n} \quad = \quad \frac{\sigma}{\sqrt{n}}.$$

We now have the mean and standard deviation of $\overline{X}$. The standard deviation of $\overline{X}$ is also commonly known as the **standard error** of $\overline{X}$.

The **Central Limit Theorem** (CLT) is a major theorem in statistics that gives us an approximation to the distribution of $\overline{X}$. In statistics an **asymptotic** result, is a result which holds as the sample size n tends to ∞. Asymptotic results are useful if they provide good approximations for small values of n, and the CLT is one such theorem. The distribution of

$$\frac{\overline{X} - \mu}{\sigma/\sqrt{n}} \rightarrow \quad \sim N(0,1),$$

as $n \rightarrow \infty$. If the X_i are normally distributed the result is exact for any value of n. In practice it will provide an excellent approximation for any population with finite variance [8] when the sample size n exceeds around 30, and may be adequate for n greater than two or three if the population is not dramatically different from normal. The practical consequence of the CLT is the approximation that

$$\overline{X} \quad \sim \quad N\left(\mu, \frac{\sigma^2}{n}\right).$$

The idea that $\overline{X}$ has a probability distribution may take a bit of getting used to. Imagine taking a large number of simple random samples of size n from the population and calculate the mean of each of these samples. The large number of sample means defines the distribution. We can simulate this process in R, and at the same time investigate the adequacy of the CLT for small sample sizes. The following code takes 10^6 random samples of sizes $4, 10, 20, 30$ from an exponential distribution with mean 1 and draws the histograms (see Figure 6.14).

```
> set.seed(1)
> N=1000000
> xbar4=rep(0,N);xbar10=rep(0,N);xbar20=rep(0,N);xbar30=rep(0,N)
> for (i in 1:N){
 x4=rexp(4)  ; bar4[i]=mean(x4)
 x10=rexp(10) ; bar10[i]=mean(x10)
 x20=rexp(20) ; bar20[i]=mean(x20)
 x30=rexp(30) ; bar30[i]=mean(x30)
}
> par(mfrow=c(2,2))
> xplot=c(1:1000)*4/1000
> y=exp(-xplot)
> hist(xbar4,breaks=30,freq=FALSE,main="",xlab="n=4");lines(xplot,y)
> hist(xbar10,breaks=30,freq=FALSE,main="",xlab="n=10");lines(xplot,y)
> hist(xbar20,breaks=30,freq=FALSE,main="",xlab="n=20");lines(xplot,y)
> hist(xbar30,breaks=30,freq=FALSE,main="",xlab="n=30");lines(xplot,y)
```

The histograms become closer to normal distributions as n increases.

---

[8]The Cauchy distribution (see Exercise 5.40) is an example of a distribution with infinite variance.

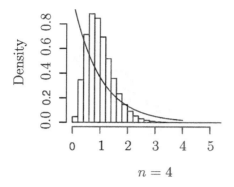

$$n = 4$$

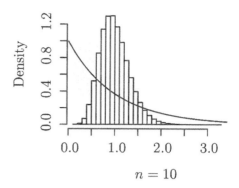

$$n = 10$$

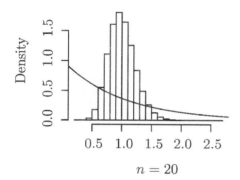

$$n = 20$$

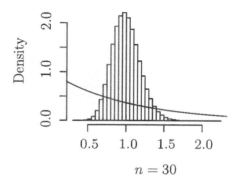

$$n = 30$$

FIGURE 6.14: Histograms of 1 000 000 samples of size n from exponential.

Example 6.12: The EPA [CLT]

The U.S. Environmental Protection Agency sets a Secondary Standard[9] for iron in drinking water as less than 0.3 milligrams per liter (mg/l). A water company will be fined, if the mean of hydrant tests at four randomly selected locations exceeds 0.5 mg/l. We calculate the probability that the company will be fined if the population mean is 0.3 mg/l, in terms of the coefficient of variation (CV). If the CV is c then the standard deviation of hydrant tests is, by definition of CV, $0.3 \times c$.

Let X_i be the iron concentration in water from a hydrant identified by i. Assume the four X_i are randomly and independently (idd) with a mean of 0.3. Define

$$\overline{X} = (X_1 + X_2 + X_3 + X_4)/4.$$

Then $\overline{X}$ has a mean of 0.3 and a standard deviation if $0.3c/\sqrt{4}$. We now assume that the distribution of X_i is near enough to a normal distribution for X_i to have a distribution that is close to a normal distribution, by the CLT. Then

$$P\big(\overline{X} > 0.5\big) \;\; = \;\; P\left(\frac{\overline{X} - 0.3}{0.3c/\sqrt{4}} > \frac{0.5 - 0.3}{0.3c/\sqrt{4}}\right) \;\; = \;\; P(Z > 1.333/c) \;\; = \;\; 1 - \Phi(1.333/c).$$

---

[9]Secondary drinking water contaminants do not pose health risks at levels usually found in water sources.

We calculate the probabilities for a values of c, from 0.4 up to 1 in steps of 0.1 using R

```
> c=seq(.4,1,.1)
> pnorm(1.333/c,lower.tail=FALSE)
[1] 0.0004303474 0.0038379846 0.0131529204 0.0284364919 0.0478318155
[6] 0.0692884838 0.0912659019
```

If the company just meets the standard with a mean of 0.3, and the standard deviation of hydrant tests is 0.15 (CV is 0.5), the probability that the company will be fined is about 0.004. If the standard deviation is as high as the mean value of 0.3 (CV is 1), the probability of a fine increases to 0.09.

Since $T = n\overline{X}$, an approximation to the distribution of T is that $T \sim N\left(n\mu, n\sigma^2\right)$.

Example 6.13: Oil exploration [CLT]

An oil exploration company owns a small helicopter to fly personnel out to remote sites. There are 7 passenger seats but the payload must be less than 800 kg. The distribution of the masses of passengers has mean 81 kg and a standard deviation of 12 kg. The distribution of the masses of equipment carried by by one passenger has a mean of 20kg and a standard deviation of 16 kg, and is positively skewed. What is the probability that the total mass of 7 randomly selected passengers and their equipment will exceed 800 kg?

The mean mass of a passenger plus equipment is $81 + 20 = 101$. The standard deviation of a passenger plus equipment, assuming independence of the mass of a passenger and mass of equipment, is $\sqrt{12^2 + 16^2} = 20$. The distribution of masses of passengers plus equipment has a noticeable positive skewness, but the total of 7 randomly selected passengers plus equipment will be approximately normal by the CLT. The total mass T has a mean of $7 \times 101 = 707$, and a standard deviation of $\sqrt{7} \times 20 = 52.9$. The required probability is

```
> 1-pnorm(800,701,52,9)
[1] 0.02846511
```

which is approximately 0.03.

6.5 Non-linear functions of random variables (propagation of error)

Many engineering quantities are non-linear functions of random variables. We can use a Taylor series expansion to obtain approximate results for the mean and variance of the non-linear function. We will demonstrate the method for an arbitrary differentiable function $\phi(,)$ of two random variables X and Y. We make a Taylor series expansion about the mean values of X and Y, and the usual approximation is to only consider the linear and quadratic terms. Then

$$\phi(X, Y) \approx \phi(\mu_X, \mu_Y) + \frac{\partial \phi}{\partial x}(X - \mu_X) + \frac{\partial \phi}{\partial y}(Y - \mu_Y)$$

$$+ \frac{1}{2!}\left(\frac{\partial^2 \phi}{\partial x^2}(X - \mu_X)^2 + 2\frac{\partial^2 \phi}{\partial x \partial y}(X - \mu_X)(Y - \mu_Y) + \frac{\partial^2 \phi}{\partial y^2}(Y - \mu_Y)^2\right).$$

If we take expectations we obtain

$$\mathrm{E}[\phi(X, Y)] \approx \phi(\mu_X, \mu_Y) + \frac{1}{2}\left(\frac{\partial^2 \phi}{\partial x^2}\sigma_X^2 + 2\frac{\partial^2 \phi}{\partial x \partial y}\mathrm{cov}(X, Y) + \frac{\partial^2 \phi}{\partial y^2}\sigma_Y^2\right).$$

The approximation for the variance is usually based on only the linear terms.

$$\mathrm{var}(\phi(X, Y)) \approx \left(\frac{\partial \phi}{\partial x}\right)^2 \sigma_X^2 + 2\left(\frac{\partial \phi}{\partial x}\right)\left(\frac{\partial \phi}{\partial y}\right)\mathrm{cov}(X, Y) + \left(\frac{\partial \phi}{\partial y}\right)^2 \sigma_Y^2.$$

The partial derivatives are evaluated at $X = \mu_X$ and $Y = \mu_Y$. The expressions simplify if X and Y are independent because $\mathrm{cov}(X, Y)$ will then be 0.

Example 6.14: River flow [mean and variance of non-linear function]

The flow in rivers is commonly measured from a weir. For a sharp crested rectangular weir the flow Q is calculated from the height of the water surface above the weir H, the weir discharge coefficient K, the length of the weir across the river L according to

$$Q \approx a\,KH^{3/2},$$

where

$$a = \frac{2}{3}\sqrt{2\mathrm{g}}L$$

and g is the acceleration due to gravity. A hydrologist considers that L is a known constant, but that K is uncertain and that H is subject to measurement error. More specifically, K and H are modeled as random variables with: means equal to the true values; CV of K is 10%; CV of H is 5%; and the errors are independent. Using the Taylor series approximations we obtain in H and K

$$\mathrm{E}[Q] \approx a\,\mu_K\mu_H^{3/2} + \frac{3}{8}a\,\mu_K\mu_H^{-1/2}\sigma_H^2$$

and

$$\mathrm{var}(Q) \approx \left(a\,\mu_H^{3/2}\right)^2\sigma_K^2 + \left(\frac{3}{2}a\,\mu_K\mu_H^{1/2}\right)^2\sigma_H^2.$$

Substituting the numerical values of the CVs of K and H leads to an approximate CV of Q as 12.5%. The difference between $\mathrm{E}[Q]$ and $a\,\mu_K\mu_H^{3/2}$ is approximately 0.1% which is negligible.

6.6 Summary

6.6.1 Notation

$\widehat{\text{cov}}(X,Y)$ covariance of sample data (x_i, y_i)

$\text{cov}(X,Y)$ covariance of random variables X and Y

r correlation of sample data (x_i, y_i)

ρ correlation of random variables X and Y

6.6.2 Summary of main results

TABLE 6.8: Summary table.

Sample	Population
$\bar{x}, \bar{y}$	μ_X, μ_Y
$s_x,\ s_y$	σ_X, σ_Y
$\widehat{\text{cov}}(X,Y) = \dfrac{\sum (x_i - \bar{x})(y_i - \bar{y})}{n-1}$	$\text{cov}(X,Y) = \text{E}[(X - \mu_X)(Y - \mu_Y)]$
$r = \dfrac{\widehat{\text{cov}}(X,Y)}{s_x s_y}$	$\rho = \dfrac{\text{cov}(X,Y)}{\sigma_X \sigma_Y}$

If X and Y have means μ_X, μ_Y and variances σ_X^2, σ_Y^2 and we define a random variable

$$W = aX + bY,$$

where a, b are constants, then

$$\mu_W = a\mu_X + b\mu_Y$$
$$\sigma_W^2 = a^2\sigma_X^2 + b^2\sigma_Y^2 + 2ab\,\text{cov}(X,Y).$$

If $f(x,y)$ is a bivariate pdf, then the marginal pdfs are:

$$f(x) = \int f(x,y)\,dy$$
$$f(y) = \int f(x,y)\,dx$$

The conditional pdf of Y given x is:

$$f(y|x) = \frac{f(x,y)}{f(x)}.$$

A consequence of the Central Limit Theorem is the approximation

$$\overline{X} \sim N\left(\mu, \frac{\sigma^2}{n}\right).$$

6.6.3 MATLAB and R commands

In the following, x and y are vectors containing data points.

R command	MATLAB command
cov(x,y)	cov(x,y)
cor(x,y)	corr(x,y)

6.7 Exercises

Section 6.1 Introduction

Exercise 6.1: Crown race

The sample mean and standard deviation of the diameters of the crown race seat on a random sample of 35 forks for a design of mountain bike were 27.052 and 0.030 (mm) respectively. A second random sample of 20 forks was taken, and the mean and standard deviation were 27.047 and 0.037 respectively.

(a) Estimate the population mean as a weighted mean of the two sample means.

(b) Estimate the population variance as a weighted mean of the two sample variances, with weights equal to degrees of freedom.

(c) Estimate the population standard deviation by the square root of the variance calculated in (b).

(d) Assume that diameters are normally distributed. Calculate the proportion of forks that are outside the specified diameter

$$27.050 \quad \pm \quad 0.05$$

if the process continues with a process mean and standard deviation equal to your combined estimates.

Section 6.2 Sample covariance and correlation coefficient

Exercise 6.2: Insulators

The following data are density $(x, \text{g cm}^{-3})$ and thermal conductivity $(y, \text{W m}^{-1})$ for 6 samples of an insulating material. (Materials Research and Innovation 1999, pp 2 through 8.)

x	0.1750	0.2200	0.2250	0.2260	0.2500	0.2765
y	0.0480	0.0525	0.0540	0.0535	0.0570	0.0610

(a) Draw a scatter plot and label the axes.

(b) Calculate the correlation coefficient .

(c) Comment on your results.

Exercise 6.3: Correlation between means

Let X and Y be random variables with: means of 0, standard deviations of σ_X and σ_Y respectively, and correlation ρ.

(a) (i) Define c as the covariance of X and Y and give a formula for c in terms of σ_X, σ_Y, and ρ.

(ii) Show that the covariance and correlation of $X + a$ and $Y + b$ are c and ρ respectively, for any constants a and b.

(iii) Explain why it suffices to assume random variables have a mean of 0 when proving general results about covariances and correlations.

(b) Define

$$ \overline{X} \;=\; \sum_{i=1}^{n} X_i, $$

where $\{X_i\}$ are an SRS of size n from the distribution of X (that is X_i are independent and identically distributed (iid) with the distribution of X).

(i) Define $\overline{Y}$ similarly relative to the distribution of Y.

(ii) What is the covariance between $\overline{X}$ and $\overline{Y}$?

(iii) What is the correlation between $\overline{X}$ and $\overline{Y}$?

Exercise 6.4: Clearances

Each of the following data pairs is the distance (feet) of a bicycle from the center line on a road and the clearance distance (feet) between the bicycle and a passing car.

Center	3.93	4.60	3.90	5.94	4.15	4.42	3.93	5.33	6.34	4.45
Clear	1.92	2.16	1.68	3.29	2.13	2.38	1.89	3.05	3.35	2.53

(a) Plot clearance against distance from center.

(b) Calculate the correlation and comment.

Exercise 6.5: Chemical reactor

The following data shows how the percentage explosive content varied with time (minutes) in a chemical reactor.

Time	0.0	1.5	3.0	4.5	6.0	7.5	9.0	10.5	12.0	13.5	15.0	16.5	18.0
Conc	21.0	19.0	15.0	12.5	10.5	9.0	7.8	7.0	6.2	5.7	5.4	5.0	4.7

(a) Plot percentage explosive content against time, calculate the correlation coefficient and comment.

(b) Plot the logarithm of percentage explosive content against time, calculate the correlation coefficient and comment.

Exercise 6.6: Water samples

The following data pairs are volatiles (ppb) and organic contaminant (ppb) in water samples.

Volatiles	12	7	15	2	5	19	7	13	7	19	12	10	14	13	14
Organic	21	9	28	3	18	32	7	20	15	31	20	17	21	17	22
Volatiles	4	16	7	7	10	17	17	16	6	13	2	14	0	0	0
Organic	4	19	8	24	13	20	18	22	7	18	7	18	0	0	0
Volatiles	0	0	1	1	2	2	3	3	10	7	7	20			
Organic	0	0	1	1	2	2	7	7	12	11	11	22			

(a) Plot volatiles against organic contaminant.

(b) Calculate the correlation coefficient and comment.

Exercise 6.7: Air samples

The following data pairs are carbon monoxide concentration (ppm) and benzo[a]pyren concentration (ppb) measured in 16 air samples from Herald Square in Manhattan.

CO	2.8	15.5	19.0	6.8	5.5	5.6	9.6	13.3
Benzo[a]pyren	0.5	0.1	0.8	0.9	1.0	1.1	3.9	4.0

CO	5.5	12.0	5.6	19.5	11.0	12.8	5.5	10.5
Benzo[a]pyren	1.3	5.7	1.5	6.0	7.3	8.1	2.2	9.5

(a) Plot benzo[a]pyren against CO.

(b) Calculate the correlation coefficient and comment.

Section 6.3 Bivariate distributions, population covariance and correlation coefficient

Exercise 6.8: Sea States

Let X and Y be zero crossing period and significant wave height respectively. The discrete scales correspond to zero crossing periods between 2 and 12 seconds and significant wave heights from 1 to 8 meters. The following probability mass function is based on data from the North Sea.

y	1	2	3	4
4	0.1	0.0	0.0	0.1
3	0.0	0.0	0.1	0.1
2	0.0	0.0	0.0	0.2
1	0.0	0.1	0.2	0.1

x

(a) Calculate the marginal distributions and demonstrate that X and Y are not independent.

(b) Calculate the conditional distribution of Y given $X = 3$.

Exercise 6.9: Bivariate pdf

A bivariate distribution has pdf

$$f(x,y) \quad = \quad x+y, \quad 0 \le x, y \le 1.$$

(a) Determine $P(X < 0.5 \text{ and } Y < 0.5)$.

(b) Determine $P(0.5 < X \text{ and } 0.5 < Y)$.

(c) Obtain an expression for the marginal pdf of X .

(d) Obtain an expression for the conditional pdf of X given y.

Exercise 6.10: Chemical manufacturer

A chemicals manufacturer stocks X kg of neodymium each day where X has a uniform distribution on $[0, 1]$. The daily demand, Y, conditional on $X = x$ has the pdf

$$f(y) \quad = \quad 2/x - 2y/x^2 \quad \text{for } 0 \le y \le x$$

(a) Find the conditional distribution of X, given that $Y = y$.

(b) If $y = 0.5$, what is the probability that X was less than 0.8?

Exercise 6.11: Bivariate cdf

Consider the bivariate cdf

$$F(x,y) \quad = \quad 1 - exp(-x) - exp(-y) + exp(-x - y - \theta xy) \quad 0 \le x, y$$

where $0 \le \theta \le 1$.

(a) Find an expression for the marginal distribution $f(x)$.

(b) Find an expression for the conditional density $f(y|x)$.

(c) What happens if $\theta = 0$?

Exercise 6.12: Marginal distributions

A bivariate pdf is defined by

$$f(x,y) \quad = \quad 8xy \quad \text{for } 0 \le x \le 1, 0 \le y \le x.$$

Find the marginal distributions of X and Y and hence show that they are not independent.

Exercise 6.13: Two component electronic system

An electronic system has two components in joint operation. Let X and Y denote the lifetimes of components of the first and second types, respectively. The joint pdf is given by

$$f(x,y) \quad = \quad \frac{1}{8} x \exp\left(-\frac{(x+y)}{2}\right) \quad \text{for } 0 < x, y.$$

(a) Find $P(1 < X$ and $1 < Y)$.

(b) Find $P(X + Y < t)$ for any $t \geq 0$.

(c) Hence write down the pdf of the random variable $T = X + Y$.

Section 6.4 Linear combination of random variables

Exercise 6.14: Marine survey vessel

A marine survey vessel has two instruments that provide unbiased measurements X and Y of the depth of the sea bed. The standard deviation of X is σ, the standard deviation of Y is $k\sigma$, and the correlation coefficient is ρ. Find a in terms of k, ρ such that $aX + (1 - a)Y$ has minimum variance.

Exercise 6.15: Rivet diameters

The diameters of rivets follows a normal distribution with mean 2.3 mm and standard deviation 0.05 mm. An independent process produces steel plates with holes whose diameters follow a normal distribution with mean 2.35 mm and standard deviation 0.1 mm. What is the probability that a randomly selected rivet willfit a hole selected at random?

Exercise 6.16: A chemical process

A chemical process produces a non-toxic pond coating material. The coating material is advertised with a drying time of 72 hours. The coating is produced by mixing solutions of chemicals A and B. The concentrations of the chemicals are normally distributed as:

$$A \sim N\left(40, (1.4)^2\right) \qquad B \sim N\left(44, (0.8)^2\right)$$

If the concentration of A exceeds the concentration of B the drying time of the product exceeds the advertised drying time. What is the probability that the concentration of A exceeds the concerntration of B?

Exercise 6.17: Battery lifetime

A battery for the capsule in a space flight has a lifetime, X, that is normally distributed with a mean of 100 hours and a standard deviation of 30 hours.

(a) What is the distribution of the total lifetime T of three batteries used consecutively,

$$T \;=\; X_1 + X_2 + X_3,$$

if the lifetimes are independently distributed?

(b) What is the probability that three batteries will suffice for a mission of 200 hours?

(c) Suppose the length of the mission is independent and normally distributed with a mean of 200 hours with a standard deviation of 50 hours. What is the probability that three batteries will now suffice for the mission?

Exercise 6.18: Gold plating

A gold plating solution is advertised with a specification of 3 g gold per liter. The gold is present as a salt, potassium gold cyanide. Let M represent the actual gold content in a liter bottle. Suppose M has a mean of 3.01 and a standard deviation of $3\,\mu$g. The assay method is subject to errors E which have a mean of 0 and a standard deviation of $2\,\mu$g. The errors are independent of the actual gold content. The measurement X is given by $X = M + E$.

(a) State the mean and standard deviation of X.

(b) Calculate the covariance and the correlation between X and M.

(c) Calculate the covariance and the correlation between X and E.

Exercise 6.19: Solar challenge

Let D be the distance traveled by a solar car, during a solar challenge event, in an 8 hour day. Assume D has a mean of 720 km and a standard deviation of 75 km. Let W be the total distance travelled in 5 such days.

(a) Assume daily distances are independent.

    (i) Write down the mean and standard deviation of W.

    (ii) Assume W is approximately normally distributed, and calculate the probability that W exceeds 4000 km.

(b) Assume daily distances are correlated with a correlation equal to 0.8^k, where k is the difference in day numbers. So, as examples, the correlation between the distance traveled on day 1 and day 2 is 0.8 and the correlation between the distance traveled on day 3 and day 5 is 0.64.

    (i) Write down the mean and standard deviation of W.

    (ii) Assume W is approximately normally distributed, and calculate the probability that W exceeds 4000 km.

Exercise 6.20: Lead content

Jars of water are taken from the public supply in an old quarter of the city and analysed for lead content. The variance of these measurements (M) is σ_M^2. The detection of minute amounts of lead is inevitably subject to errors (E) and the variance of measurements made on known standard solutions is σ_E^2.

(a) Express the variance σ_L^2 of actual (as opposed to measured) lead content (L) in terms of σ_M^2 and σ_E^2.

(b) A large sample of jars was taken and the mean and standard deviation of the measurements of lead content were 46 μgl^{-1} and 18 μgl^{-1}. The standard deviation of many measurements of lead content of known 40 μgl^{-1} solutions is 6 μgl^{-1}. Find the standard deviation of actual lead content and the correlation between errors and measurements.

Exercise 6.21: Linear combination of multiple variables

The random variables X_1, X_2, X_3 have means μ_i, standard deviations σ_i and covariances γ_{ij} for $1 \le i,j \le 3$.

(a) Find the mean, variance, and standard deviation of:

$$X_1 + X_2 + X_3$$

(b) Find the mean, variance, and standard deviation of:

$$\alpha_1 X_1 + \alpha_2 X_2 + \alpha_3 X_3.$$

(c) Write down the general result for the mean and variance of a linear combination:

$$\sum_{i=1}^{n} \alpha_i X_i.$$

Exercise 6.22: Lift capacity

A lift in a building is rated to carry 20 persons or a maximum load of 2,000 kg. If the mean mass of 20 persons exceeds 100 kg the lift will be overloaded and trip a warning buzzer. The people using the lift have masses distributed with a mean of 75 kg and a standard deviation of 24 kg. What is the probability that an individual has a mass above 100 kg, under an additional assumption that masses are normally distributed, and is this assumption likely to be realistic? What is the distribution of the mean mass of 20 randomly selected people, and what is the probability that the lift will be overloaded, whether or not the masses of people using the building are normally distributed?

Exercise 6.23: Reaction times

Let X represent driver reaction times ('thinking times') before applying brakes in an emergency. Assume X is distributed with mean 0.67 s and standard deviation 0.35 s. Let W represent the 'thinking distance' for cars braking from a speed of 50 kmh^{-1}.

(a) Remember that distance is the product of speed and time, and hence write down the mean and standard deviation of W.

(b) The 'braking distance' for cars braking from 50 kmh^{-1}, Y, is distributed with mean 14 m and standard deviation 8 m. Write down the mean and standard deviation of 'stopping distance', $W + Y$, for cars braking from 50 kmh^{-1} if W and Y are assumed independent.

(c) Assume 'stopping distance' is normally distributed and find the distance that will be exceeded by 1% of such cars.

(d) Repeat the calculation if 'stopping distance' is lognormal distributed.

Exercise 6.24: Geological survey

A large geological survey company transports employees to remote locations by a helicopter that can carry ten passengers. The company employs an equal number of males and females. The masses of males are distributed with mean 80 kg and standard deviation of 15 kg. The masses of females are distributed with mean 60 kg and standard deviation of 12 kg.

(a) Assume the distributions of masses are near enough to normal for the total masses of five males, and five females, to be close to a normal distribution by the CLT. Suppose five men and five women are randomly selected for a flight. What is the probability that the total mass will exceed 800 kg.

(b) Now suppose that ten employees are randomly selected for the flight. At each draw the employee is equally likely to be male or female. Let M, W and Y represent the masses of a randomly selected man, woman, and employee. Then:

$$\mathrm{E}[Y] = \frac{1}{2}\mathrm{E}[M] + \frac{1}{2}\mathrm{E}[W]$$

$$\mathrm{E}[Y^2] = \frac{1}{2}\mathrm{E}[M^2] + \frac{1}{2}\mathrm{E}[W^2].$$

(a) What is the mean and variance of Y?

(b) Assume that the total

$$T = \sum_{i=1}^{10} Y_i$$

is approximately normally distributed by the CLT. Calculate the probability that T exceeds 800.

Exercise 6.25: Measures of a sum of independent random variables

R_1 and R_2 are independent random variables with uniform distributions on $[0, 1]$.

(a) Define the variable W as their sum.

(b) What are the mean, variance, and standard deviation of W?

Exercise 6.26: Variance of product

The random variables X and Y are independent with means and variances $\mu_X, \mu_Y, \sigma_X^2, \sigma_Y^2$.

(a) What is $\mathrm{E}[XY]$?

(b) Use a Taylor series expansion to obtain an approximation to $\mathrm{var}(XY)$, and express this in terms of coefficients of variation (CV).

(c) Obtain the exact result for $\mathrm{var}(XY)$ and compare it with the approximation.

Exercise 6.27: Approximate variance of a ratio

The random variables X and Y can take only positive values and are independent. Their means are μ_X, μ_Y and their variances are σ_X^2, σ_Y^2. Define the ratio $Q = X/Y$.

(a) Use a Taylor series expansion as far as the linear terms to show that

$$\mathrm{E}[Q] \approx \frac{\mu_X}{\mu_Y}, \qquad \mathrm{var}(Q) \approx \frac{\sigma_X^2}{\mu_Y^2} + \sigma_Y^2 \left(\frac{-\mu_X}{\mu_Y^2} \right)^2$$

(b) Deduce that provided the coefficients of variation $CV()$ of X and Y are less than about 0.15

$$CV(Q)^2 \approx CV(X)^2 + CV(Y)^2.$$

(c) Deduce a similar result to (b) for $CV(P)$ where $P = XY$.

(d) Use a Taylor series expansion as far as the quadratic terms to find an improved approximation to $\mathrm{E}[Q]$.

Exercise 6.28: Resistors

Resistors are usually manufactured to a set of preferred values. Different resistance values can be obtained by combining resistors in series or in parallel. Suppose two resistors R_1 and R_2 are from populations with means of 1 kΩ and 2 kΩ respectively, and standrad deviations of 1%. Find the mean and standard deviation (in percentage terms) for:

(a) $R_S = R_1 + R_2$.

(b) $R_P = \left(R_1^{-1} + R_2^{-1}\right)^{-1}$.

Exercise 6.29: Approximation using Taylor series

Suppose that X and Y are independent positive random variables with means μ_X, μ_Y, and standard deviations σ_X and σ_Y respectively, and that $P(X > Y) \approx 0$.

(a) Use a Taylor series expansion, as far as the quadratic term, to approximate the mean and variance of $X/(1 - X/Y)$.

(b) Compare your approximate results with those obtained from a simulation if X and Y are exponential with means 1 and 10 respectively.

(c) Find $P(X > Y)$ if X and Y are independent exponential random variables with means μ_X and μ_Y respectively.

(d) Refer to Example 5.5 and explain why it is not appropriate to define the mean access delay as the mean value of the random variable of $X/(1-X/Y)$, where X is transmission time and Y is times between requests even if $P(X > Y)$ is negligible.

Exercise 6.30: Planetary exploration robots

A planetary exploration robot is powered by a battery which has a mean lifetime of 10.5 hours and a standard deviation of 2.2 hours in this application. The robot is equipped with one spare battery which is used as soon as the first is discharged. Assume battery lifetimes are independent.

(a) What are the mean and standard deviation of the total lifetime of the two batteries?

(b) If battery lifetimes are normally distributed, what is the probability that two batteries will suffice for a 15 hour mission?

(c) If battery lifetimes are exponentially distributed, what is the probability that two batteries will suffice for a 15 hour mission?

Section 6.5 Non-linear functions of random variables

Exercise 6.31: Rayleigh distribution (see Exercise 5.36)

Suppose X and Y and independent normal random variables with mean 0 and equal variances, Define

$$R \;=\; \sqrt{X^2 + Y^2}.$$

What are the approximate mean and variance of R?

Exercise 6.32: Helical gear

A helical gear wheel is held in a jig and measurements of distances from an origin to points on the profile are made in a rectangular cartesian coordinate system. The standard deviation of measurements along any axis is 8 μm. The distance between two points on the gear, with coordinates (x_1, y_1, z_1) and (x_2, y_2, z_2) is of the form

$$\left[(x_1 - x_2)^2 + (y_1 - y_2)^2 + (z_1 - z_2)^2 \right]^{\frac{1}{2}}.$$

What is the standard deviation of such distances?

7

Estimation and inference

We explain the difference between the scientific use of the terms accuracy and precision. We show how to augment estimates of population means, standard deviations and proportions with indications of their accuracy, known as confidence intervals. These ideas are extended to the comparison of two populations. The closely related procedures of hypothesis testing, that is assessing whether sample results are consistent with, or evidence against, some hypothesis about the population are described. The construction of prediction intervals and statistical tolerance intervals, for individual values rather than the population mean, is explained. See relevant example in Appendix E:

Appendix E.4 *Use your braking brains.*

7.1 Introduction

We aim to estimate a numerical characteristic of a population, generally referred to as a **parameter**, from a sample. We also aim to provide an indication of the reliability of this estimate. To do so, we need the notion of a sampling distribution. We might, for instance, aim to estimate the mean, μ, of a population. We would draw a simple random sample (SRS) of size n from the population and calculate the sample mean $\bar{x}$. We don't expect $\bar{x}$ to equal μ, but we can construct an interval centered on $\bar{x}$ that we are reasonably confident includes μ.

We will need the concept of a sampling distribution. Imagine that we draw very many SRSs of size n from the population, calculate the mean of each sample, and so obtain a probability distribution for the sample mean. A probability distribution that arises in this way is known as a **sampling distribution**. In general, we can use computer simulation to emulate the sampling distribution but probability theory provides theoretical results for the more common applications. For example, if SRSs are drawn from a normal distribution with a mean μ and standard deviation σ then the sampling distribution of $\overline{X}$ is normal with a mean μ and a standard deviation $\sigma/\sqrt{n}$. The term "sampling distribution" refers to the context in which the probability distribution is used, rather than to a class of probability distributions.

7.2 Statistics as estimators

We begin with some definitions.

7.2.1   Population parameters

Populations are modeled by probability distributions. A probability distribution is generally defined as a formula involving the variable x, say, and a few symbols which take specific values in an application. These symbols are known as **parameters** and in conjunction with the general formula, the parameters determine numerical characteristics of a population such as its mean, standard deviation, and quantiles, and in the case of binary data a proportion with a particular attribute. For example, a normal distribution has two parameters which are its mean μ and standard deviation σ. The exponential distribution for the time between events in a Poisson process is often defined by the rate parameter λ, events per unit time, in which case its mean is $1/\lambda$ time units. The uniform distribution is defined by two parameters a and b, which specify the range of the variable, and its mean is $(a + b)/2$.

Definition 7.1: Parameter

A parameter is a numerical characteristic of a population.

7.2.2   Sample statistics and sampling distributions

Definition 7.2: Statistic

A **statistic** is a function of a sample that does not include any unknown population parameters. It can be calculated for a particular sample, in which case it is a number, or it can be considered as a random variable in the context of imaginary repeated drawing of samples. When considered as a random variable, the statistic has a sampling distribution. Statistics are typically used as estimators of population parameters.

Definition 7.3: Estimator

An **estimator**, $\widehat{\theta}$, of an unknown population parameter, θ, is a statistic that can be used to provide a numerical value, referred to as an **estimate**, for that unknown parameter. The estimate is described as a **point estimate** when it is helpful to distinguish it from an interval estimate (Section 7.4).

Definition 7.4: Standard error

Suppose a population is modeled by a probability distribution that includes a parameter θ. A statistic $\widehat{\theta}$, considered as an estimator of θ, has a sampling distribution, and the mean and standard deviation of $\widehat{\theta}$ are defined as $E[\widehat{\theta}]$ and

$$\sqrt{E\left[\left(\widehat{\theta} - E\left[\widehat{\theta}\right]\right)^2\right]} \qquad \text{respectively.}$$

The standard deviation of $\widehat{\theta}$ is also commonly known as the **standard error** of $\widehat{\theta}$.

Example 7.1: Simple random sample [mean and variance]

Consider a simple random sample of size n from an infinite, or at least relatively large, population. The sample can be represented as $\{X_1, X_2, \ldots, X_n\}$ where the X_i have an identical probability distribution, with mean μ and standard deviation σ, and are independently distributed. The sample mean:

$$\overline{X} = \frac{\sum X_i}{n}$$

is a statistic as it is a function of the sample and does not include any unknown population parameters. The sample mean is used as an **estimator** of the population mean μ. The sampling distribution of the sample mean has mean, μ, and variance σ^2/n. Moreover, as a consequence of the Central Limit Theorem the sampling distribution is, approximately at least, $N(\mu, \sigma^2/n)$ provided either the population is near normal or the sample size n is reasonably large.

Once the sample is drawn we have data $\{x_1, x_2, \ldots, x_n\}$ and we calculate the sample mean $\bar{x}$ as an **estimate** of μ. Upper and lower case letters are used here to distinguish the random variables $\overline{X}$ and S from numerical values $\bar{x}$ and s respectively.

The sample variance depends only on the sample:

$$S^2 = \frac{\sum (X_i - \overline{X})^2}{n-1}.$$

It is a statistic and an estimator of the population variance σ^2, and the sample standard deviation S is an estimator of σ. The numerical values calculated from a specific sample are denoted by the sample variance, s^2, and sample standard deviation, s.

We use upper and lower case letters to distinguish the random variables $\overline{X}$ and S from numerical values $\bar{x}$ and s respectively. But, for other estimators of population parameters such as $\widehat{p}$ for a population proportion p, and $\widehat{\theta}$ for some population parameter θ, we rely on the context to distinguish estimators from estimates. We refer to sampling distributions of estimators, whereas estimates are numerical values. Using the same letter for a random variable and for the value taken by that variable in an application reduces the amount of notation.

Example 7.2: Sample of PCs [proportion]

A engineering school considers a suite of n nominally identical PCs as a random sample from the manufacturer's production process. Suppose that a proportion p of the manufacturer's production will fail within two years of student use. Let the random variable X be the number of PCs in the suite that fail within two years. Then the sample proportion:

$$\widehat{p} = \frac{X}{n}$$

is an estimator of p. The sampling distribution of X is $binom(n, p)$ and the sampling distribution of $\widehat{p}$ is usually approximated by a normal distribution with the same mean and variance as X/n, that is $N(p, p(1-p)/n)$.

At the end of the two years x PCs have been observed to fail. Then $\widehat{p} = x/n$ is an estimate of p.

7.2.3   Bias and MSE

Definition 7.5: Unbiased

An estimator $\widehat{\theta}$ of a population parameter θ is said to be **unbiased** if

$$\mathrm{E}\!\left[\widehat{\theta}\right] \;=\; \theta.$$

Example 7.3: Sample mean

The sample mean is an unbiased estimator of the population mean μ since

$$\mathrm{E}[\overline{X}] \;=\; \mu.$$

Example 7.4: Sample variance

The sample variance S^2 is an unbiased estimator of the population variance σ^2. We start by showing that

$$\sum_{i=1}^{n}(X_i - \overline{X})^2 \;=\; \sum\left((X_i - \mu) - (\overline{X} - \mu)\right)^2$$

$$=\; \sum(X_i - \mu)^2 - 2\sum(X_i - \mu)(\overline{X} - \mu) + \sum(\overline{X} - \mu)^2.$$

Now focus on the second term on the right hand side of the equation. The summation is over $i = 1, \ldots, n$ and as the factor $(\overline{X} - \mu)$ does not depend on i it is a common factor that can be moved outside the summation. Proceeding

$$\sum_{i=1}^{n}(X_i - \overline{X})^2 \;=\; \sum(X_i - \mu)^2 - 2(\overline{X} - \mu)\sum(X_i - \mu) + \sum(\overline{X} - \mu)^2$$

$$=\; \sum(X_i - \mu)^2 - 2(\overline{X} - \mu)n(\overline{X} - \mu) + n(\overline{X} - \mu)^2$$

$$=\; \sum(X_i - \mu)^2 - n(\overline{X} - \mu)^2.$$

If we now take expectation, using the assumption that the X_i are independent:

$$\mathrm{E}\!\left[\sum(X_i - \overline{X})^2\right] \;=\; \mathrm{E}\!\left[\sum(X_i - \mu)^2 - n(\overline{X} - \mu)^2\right]$$

$$=\; \sum \mathrm{E}\!\left[(X_i - \mu)^2\right] - n\mathrm{E}\!\left[(\overline{X} - \mu)^2\right]$$

$$=\; n\sigma^2 - n\left(\frac{\sigma^2}{n}\right) \;=\; (n-1)\sigma^2.$$

It follows that

$$\mathrm{E}[S^2] \;=\; \mathrm{E}\!\left[\frac{\sum(X_i - \overline{X})^2}{n-1}\right] \;=\; \frac{(n-1)\sigma^2}{n-1} \;=\; \sigma^2.$$

Amongst unbiased estimators we look for the one with minimum variance. You are asked to compare the sample mean and median as estimators of the mean of the normal and Laplace distribution, by simulation, in Exercise 7.47. But, we routinely use estimators that are not unbiased. A consequence of S^2 being unbiased for σ^2 is that S is not an unbiased estimator for σ. To understand the reason for this claim, consider any random variable X with mean $E[X] = \mu$ and

$$\text{var}(X) \quad = \quad E\big[(X - \mu)^2\big] \quad = \quad E\big[X^2 - 2X\mu + \mu^2\big] \quad = \quad E\big[X^2\big] - \mu^2.$$

It follows that

$$E\big[X^2\big] \quad = \quad \text{var}(X) + \mu^2.$$

In particular

$$E\big[S^2\big] \quad = \quad \text{var}(S) + (E[S])^2$$

and since $E\big[s^2 = \sigma^2\big]$

$$E[S] \quad = \quad \sqrt{\sigma^2 - \text{var}(S)} \quad = \quad \sigma\sqrt{1 - \text{var}(S)/\sigma^2} \quad < \quad \sigma.$$

Although $E[S]$ is less than σ, the ratio $\text{sd}(S)/\sigma$ becomes smaller as the sample size increases (Exercise 7.35), and its square $\text{var}(S)/\sigma^2$ is smaller still. Moreover the difference between $E[S]$ and σ is substantially less than the standard error of S (Exercise 7.34), and S is our usual estimator of σ.

Definition 7.6: Bias

The **bias** of an estimator $\widehat{\theta}$ of θ is

$$bias(\widehat{\theta}) \quad = \quad E\big[\widehat{\theta}\big] - \theta$$

If the bias is 0 then the estimator is unbiased.

Definition 7.7: Consistent

An estimator $\widehat{\theta}$, based on a sample of size n, of θ is **consistent** if

$$\text{var}\big(\widehat{\theta}\big) \quad \to \quad 0 \quad \text{as} \quad n \to \infty \quad \text{and}$$
$$E\big[\widehat{\theta}\big] \quad \to \quad \theta \quad \text{as} \quad n \to \infty.$$

In particular any unbiased estimator is consistent, provided its standard error decreases as the sample size increases.

Example 7.5: Sample standard deviation [consistent]

The sample standard deviation S is **consistent** for σ because its variance tends to 0, and so its expected value tends towards σ, as the sample size increases.

There is a distinction between the technical statistical meaning of "bias", and its everyday meaning of prejudiced or unrepresentative. Biased estimators can be useful when we know the extent of the bias. In contrast, the reason why a biased sample, believed to be unrepresentative is of little use for estimating population parameters is that the bias is unknown and could be large.

Example 7.6: Estimating a proportion [biased sample]

A graduate student in a large computer science school is developing a graphics package in his spare time. The student aims to produce a package that is easy for non-specialists to use. The student distributes a beta version to colleagues and asks whether they would have downloaded it, if the price was around one tenth of a basic PC. Colleagues' advice for improving the package will be valuable, but the proportion responding that they would download it is not a reliable estimate of the proportion in the target population. Apart from the fact that they haven't parted with any money, they might all say "no" because they routinely use more sophisticated products or all say "yes" because they wish to offer encouragement.

Biased estimators are useful provided their bias is small compared with their standard error[1]. In particular, an estimator with a small standard error and a known small bias will be preferable to an unbiased estimator with a relatively large standard error. We can compare estimators by calculating the mean squared error (MSE).

Definition 7.8: Mean square error

If $\widehat{\theta}$ is an estimator of θ, then the **mean square error**

$$MSE = \mathrm{E}\left[(\widehat{\theta} - \theta)^2\right].$$

We now show that the MSE is the sum of the variance and the square of the bias.

$$MSE = \mathrm{E}\left[(\widehat{\theta} - \theta)^2\right] = \mathrm{E}\left[\left(\left(\widehat{\theta} - \mathrm{E}\left[\widehat{\theta}\right]\right) + \left(\mathrm{E}\left[\widehat{\theta}\right] - \theta\right)\right)^2\right]$$

$$= \mathrm{E}\left[\left(\widehat{\theta} - \mathrm{E}\left[\widehat{\theta}\right]\right)^2\right] + 2\left(\mathrm{E}\left[\widehat{\theta}\right] - \theta\right)\mathrm{E}\left[\left(\widehat{\theta} - \mathrm{E}\left[\widehat{\theta}\right]\right)\right] + \left(\mathrm{E}\left[\widehat{\theta}\right] - \theta\right)^2,$$

but

$$\mathrm{E}\left[\widehat{\theta} - \mathrm{E}\left[\widehat{\theta}\right]\right] = 0$$

$$\text{so } MSE = \mathrm{var}\left(\widehat{\theta}\right) + \left(\mathrm{bias}(\widehat{\theta})\right)^2.$$

We usually choose the estimator with the minimum MSE.

---

[1]The bias often arises because the sampling distribution of the estimator is skewed, and the median of the sampling distribution may be closer to the population parameter than is the mean.

7.3 Accuracy and precision

Companies need to ensure that measuring instruments are accurate and regularly check them against internal standards. The internal standards will be checked against national standards which are themselves checked against the international standard. It is no use manufacturing to micron precision if your measurements are not consistent with the customer's measurements.

In scientific work, accuracy and precision are given specific and distinct meanings. A measurement system is **accurate** if, on average, it is close to the actual value. A measurement system is **precise** if replicate measurements are close together. Suppose three different designs of miniature flow meters were tested under laboratory conditions against a known steady flow of 10 liters per minute. Each meter was used on 10 occasions. Suppose the first meter gave 10 readings scattered between 8 and 12 with a mean around 10, the second meter gave 10 readings between 10.9 and 11.1, and the third meter gave 10 readings between 9.9 and 10.1. Then the first meter is accurate but not precise, the second meter is precise but not accurate, and the third meter is both accurate and precise. A graphical representation is shown in Figure 7.1.

An estimator is accurate if it is unbiased, or if the bias is small compared with the standard error. Precision is a measure of replicability and decreases as the variance of the estimator increases. Precision is commonly defined as the reciprocal of the variance.

FIGURE 7.1: You make eighyt shots at three targets using three different air rifles at a fairground. The rifle on the left may be accurate but is certainly not precise (two shots missed the target), the second rifle in the center is more precise but not accurate, and the third rifle on the right is both accurate and precise (after Moore, 1979).

We construct confidence intervals to indicate the accuracy and precision associated with an estimate.

7.4 Precision of estimate of population mean

In some applications it may be reasonable to suppose that the population standard deviation is known from experience of similar investigations. The focus is to estimate the population mean.

7.4.1 Confidence interval for population mean when σ known

Assume $\{X_i\}$, for $i = 1, \ldots, n$ is a simple random sample of n observations from a distribution with mean μ and variance σ^2 (see Figure 7.2).

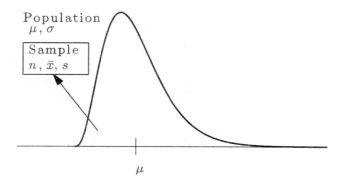

FIGURE 7.2: Population and sample.

The result for the variance of a linear combination of random variables, together with the Central Limit Theorem, justifies an approximation

$$\overline{X} \quad \sim \quad N\left(\mu, \frac{\sigma^2}{n}\right),$$

where n is the sample size, provided n is sufficiently large. If we standardize the variable the

$$\frac{\overline{X} - \mu}{\sigma/\sqrt{n}} \quad \sim \quad N(0,1).$$

It follows, from the definition of the quantiles of a standard normal distribution, that

$$P\left(-z_{\alpha/2} < \frac{\overline{X} - \mu}{\sigma/\sqrt{n}} < z_{\alpha/2}\right) \quad = \quad 1 - \alpha.$$

If we multiply throughout by the positive constant $\dfrac{\sigma}{\sqrt{n}}$ then we obtain

$$P\left(-z_{\alpha/2}\frac{\sigma}{\sqrt{n}} < \overline{X} - \mu < z_{\alpha/2}\frac{\sigma}{\sqrt{n}}\right) \quad = \quad 1 - \alpha.$$

Now subtract $\overline{X}$ throughout and multiply by -1, remembering to change the directions of the inequalities,

$$P\left(\overline{X} - z_{\alpha/2}\frac{\sigma}{\sqrt{n}} < \mu < \overline{X} + z_{\alpha/2}\frac{\sigma}{\sqrt{n}}\right) \quad = \quad 1 - \alpha.$$

That is, the probability that a random interval $\left[\overline{X} - z_{\alpha/2}\frac{\sigma}{\sqrt{n}}, \ \overline{X} + z_{\alpha/2}\frac{\sigma}{\sqrt{n}}\right]$ includes μ is $1 - \alpha$. Now consider a particular sample obtained in an experiment. Given the mean of this sample, $\overline{x}$, which is the value taken by the random variable $\overline{X}$, a $(1 - \alpha) \times 100\%$ confidence interval for μ is defined as

$$\left[\overline{x} - z_{\alpha/2}\frac{\sigma}{\sqrt{n}}, \ \overline{x} + z_{\alpha/2}\frac{\sigma}{\sqrt{n}}\right].$$

The formal interpretation of a $(1-\alpha) \times 100\%$ confidence interval is that in imagined repeated sampling, $(1 - \alpha) \times 100\%$ of such intervals include μ, and that we have one such interval. We are $(1 - \alpha) \times 100\%$ confident that μ is within the interval (see Exercise 7.43).

In this context the normal distribution is referred to as the **sampling distribution** because it describes the probability distribution of $\overline{X}$ in imagined repeated sampling. The normal distribution is an approximation, unless the distribution of $\{X_i\}$ is itself normal when it is exact, but the approximation improves rapidly as the sample size increases. The distribution of $\{X_i\}$, which represents the population, is known as the **parent distribution**. The normal approximation for the sampling distribution of $\overline{X}$ is generally good for any parent distribution with finite variance if the sample size n exceeds around 30. In practice, the approximation is commonly used for any n unless the parent distribution is thought to be substantially different from a normal distribution (see Exercise 7.44). The higher the confidence level, the wider is the confidence interval. For a given confidence level, the confidence interval becomes narrower as the sample size increases.

Example 7.7: Fireman's clothing [CI for mean, σ known]

A researcher is evaluating a new high performance fabric for use in the manufacture of firefighters' protective clothing. The arc rating is a measure of thermal protection provided against the heat generated by an electrical arc fault. A single layer of the fabric that is now used in the manufacture of the clothing (standard fabric) has an arc rating of 11 cal/cm^2. If there is evidence that the new fabric has a higher arc rating than the standard fabric, then it will be used in place of the standard fabric in prototype protective clothing, that will be evaluated further during firefighting exercises. Four test pieces of the fabric, taken from different rolls of fabric (bolts), are available. On the basis of experience with the standard material, the researcher assumes that the coefficient of variation of arc rating is about 0.10. The results of the test for the new fabric are: $11.3, 13.2, 10.5$ and 14.6. The sample mean, $\bar{x}$, is 12.4, but how reliable is this estimate of the mean arc rating of the notional population of all test pieces of this new fabric?

We construct a 95% confidence interval for the mean arc rating of the population of such test pieces test pieces of the fabric. The standard deviation σ is taken as the product of the coefficient of variation and the mean: $0.1 \times 12.4 = 1.24$.

```
> x=c(11.3,13.2,10.5,14.6)
> mean(x)
[1] 12.4
> n=length(x)
> z_.025=qnorm(.975)
> sigma=0.1*mean(x)
> U95=mean(x)+z_.025*sigma/sqrt(n)
> L95=mean(x)-z_.025*sigma/sqrt(n)
> print(round(c(L95,U95),2))
[1] 11.18 13.62
```

We are 95% confident that the population mean arc rating is between 11.18 and 13.62. We consider this sufficient evidence that the mean exceeds 11 to proceed with the manufacture of prototype clothing.

The construction of a confidence interval with known standard deviation provides a nice example of a function in R.

```
> zint=function(x,sigma,alpha){
```

```
+ n=length(x)
+ z=qnorm(1-alpha/2)
+ L=mean(x)-z*sigma/sqrt(n)
+ U=mean(x)+z*sigma/sqrt(n)
+ print(c(L,U))
+ }
> zint(x,1.24,0.10)
[1] 11.38019 13.41981
> zint(x,1.24,0.05)
[1] 11.18482 13.61518
> zint(x,1.24,0.01)
[1] 10.80299 13.99701
```

The 90% confidence interval (CI) for μ is $[11.38, 13.42]$, which is narrower than the 95% CI. The 99% CI is $[10.80, 14.00]$, which is wider than the 95% CI. Suppose the manager of the manufacturing company is only prepared to trial the new material in prototype garments if he is very confident it has an arc rating above 11. Although the 99% CI includes 11, it extends up to 14 and is sufficiently promising to justify further arc testing of the fabric. The increased sample size will lead to a narrower 99% CI which might have a lower limit that exceeds 11 (see Exercises 7.45 and 7.46).

7.4.2 Confidence interval for mean when σ unknown

In most experiments we will only have a rough idea of the standard deviation of the response, and the sample standard deviation s will be our best estimate of σ.

7.4.2.1 Construction of confidence interval and rationale for the t-distribution

In Section 6.2.1 we demonstrated that the sampling distribution of $\dfrac{\overline{X}-\mu}{\sigma/\sqrt{n}}$ is well approximated by $N(0,1)$, and that the result is exact if the parent distribution is normal. However, if we replace σ by s we are replacing a constant with a random variable, and if the parent distribution is normal the sampling distribution is known as **Student's t-distribution**[2] or simply the **t-distribution**. The t-distribution depends on a parameter ν known as the **degrees of freedom** that corresponds to the degrees of freedom in the definition of S^2, and in this context $\nu = n - 1$.

Assume $\{X_i\}$, for $i = 1, \ldots, n$, is a random sample of n observations from a distribution with mean μ and variance σ^2. Then

$$\frac{\overline{X}-\mu}{S/\sqrt{n}} \sim t_{n-1},$$

where t_ν is the t-distribution with ν degrees of freedom. Following the derivation in Section 7.4.1 a $(1-\alpha) \times 100\%$ confidence interval for μ is

$$\left[\overline{x} - t_{n-1,\alpha/2}\frac{s}{\sqrt{n}}, \ \ \overline{x} + t_{n-1,\alpha/2}\frac{s}{\sqrt{n}}\right].$$

The difference in the intervals is that σ has been replaced by the sample estimate s and the standard normal quantile has been replaced by a t-quantile.

---

[2]Student was the pseudonym of WS Gosset, who worked for the Guinness Brewery in Dublin. He published his work developing the t-distribution in **Biometrika** in 1907, under the name Student.

7.4.2.2 The *t*-distribution

The pdf of the *t*-distribution with ν degrees of freedom is

$$f(x) = \left(\frac{\Gamma((\nu+1)/2)}{\sqrt{\nu\pi}\Gamma(\nu/2)}\right)\left(1+\frac{x^2}{\nu}\right)^{-(\nu+1)/2}, \quad \text{for } -\infty < x < \infty.$$

The mean and variance are 0 and $\dfrac{\nu}{\nu-2}$ for $2 < \nu$ respectively. If $\nu \leq 2$, the variance is infinite[3]. The kurtosis[4] is $3 + \dfrac{1}{(\nu-4)}$ for $4 < \nu$, so the *t*-distribution has heavier tails than the standard normal distribution and the quantiles are larger in absolute value, but tend to those of the standard normal distribution as the degrees of freedom increase. This can be seen in the following short table of upper 0.025 quantiles, generated in R. For comparison $z_{0.025} = 1.96$.

```
> nu=c(1,2,4,5,10,15,20,30,50,100)
> t_.025=qt(.025,nu,lower.tail=FALSE);t_.025=round(t_.025,3)
> print(data.frame(nu,t_.025),row.names=FALSE)
   nu t_.025
    1 12.706
    2  4.303
    4  2.776
    5  2.571
   10  2.228
   15  2.131
   20  2.086
   30  2.042
   50  2.009
  100  1.984
```

The pdfs of t_5 and the standard normal distributions are plotted in Figure 7.3. You can see the heavier tails, and hence lower peak, of t_5 relative to the standard normal distribution. The R code for drawing Figure 7.3 is:

```
> x=seq(-4,4.01,.01)
> plot(x,dt(x,5),ylim=c(0,0.4),ylab="density",type="l")
> lines(x,dnorm(x),lty=2)
> legend(2.5,.3,c("t_5","N(0,1)"),lty=c(1,2))
```

Although the *t*-distribution was invented in the context of a sampling distribution, it can be rescaled to any mean and standard deviation and so used as a model for data that are heavier tailed than data from a normal distribution.

---

[3] A *t*-distribution with 1 degree of freedom is a Cauchy distribution.
[4] The kurtosis is infinite if $\nu \leq 4$.

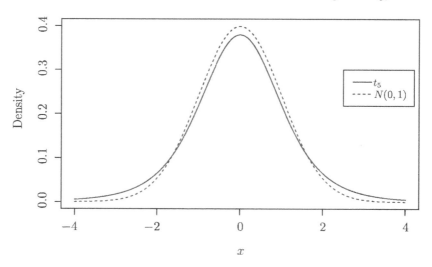

FIGURE 7.3: The standard normal distribution and the t-distribution with 5 degrees of freedom.

Example 7.8: Road stone [CI for mean, σ unknown]

The friction between a vehicle's tires and a bitumen road is due to the aggregate that is bound with the tar. A good crushed stone for use as aggregate, will maintain frictional forces despite the polishing action of tires. Samples of aggregate from a large road building project were sent to four independent laboratories for friction test readings (FTR) according to British Standard BS812:Part114:1989 [StandardsUK.com, 2014]. The FTR were:

$$62.15, 53.50, 55.00, 61.50$$

We calculate a 95% CI for the mean FTR μ of the notional population of all such aggregate samples using the formula, and we check the result with the R function `t.test()`.

```
> x=c(62.15, 53.50, 55.00, 61.50)
> n=length(x)
> xb=mean(x)
> s=sd(x)
> print(round(mean(x),2))
[1] 58.04
> print(round(sd(x),2))
[1] 4.42
> t_.025=qt(.025,(n-1),lower.tail=FALSE)
> print(t_.025)
[1] 3.182446
> L=xb-t_.025*s/sqrt(n)
> U=xb+t_.025*s/sqrt(n)
> print(round(c(L,U),2))
[1] 51.00 65.08
```

So, the 95% CI for μ is [51.00, 65.08]. We now compare the answer with `t.test()`.

```
> t.test(x)

        One Sample t-test

data:  x
t = 26.2372, df = 3, p-value = 0.0001215
alternative hypothesis: true mean is not equal to 0
95 percent confidence interval:
 50.99784 65.07716
sample estimates:
mean of x
  58.0375
```

The default confidence level in `t.test()` is 0.95, and it can be changed. If we round to two decimal places, then the 95% CI for μ is $[51.00, 65.08]$, as we obtained from explicit use of the formula. The 90% confidence interval will be narrower.

```
> t.test(x,conf.level=0.90)

        One Sample t-test

data:  x
t = 26.2372, df = 3, p-value = 0.0001215
alternative hypothesis: true mean is not equal to 0
90 percent confidence interval:
 52.83179 63.24321
sample estimates:
mean of x
  58.0375
```

The 90% CI for μ is $[52.83, 63.24]$.

7.4.3   Robustness

The construction of a confidence interval for μ when σ is unknown using the t-distribution assumes a SRS from a normal parent distribution. The assumption of a SRS is crucial, but how important is the assumption that the parent distribution is normal? In general, statistical techniques that are not sensitive to such distributional assumptions are described as **robust**. A nice feature of R is that we can easily check robustness by simulation. The following code calculates the coverage if the parent distribution is exponential. The mean is set at 1 but this is just a scaling factor and does not affect the results[5].

```
> n=20
> alpha=0.05
> t=qt(1-alpha/2,(n-1))
> K=10000
```

_____

[5]The `if()` function in R is very versatile. Here we use it in the simplest form, if the logical expression is TRUE then the action to the right is taken. If the action needs to extend over another line use { and } to enclose the action. Notice that the logical "and" is &. The `if()` function can be followed by `else()` functions.

```
> CI=0
> set.seed(17)
> for (k in 1:K) {
 x=rexp(n)
 xbar=mean(x)
 s=sd(x)
 L=xbar-t*s/sqrt(n)
 U=xbar+t*s/sqrt(n)
 if (L < 1 & 1 < U) CI=CI+1
}
> print(c("SRS from exponential, n=",n))
> print(c("nominal coverage",(1-alpha)*100,"%"))
> print(c("coverage %",round(CI/K*100,1)))
```

The results of running the code with $n = 20$ and $n = 100$ are shown below.

```
> print(c("SRS from exponential, n=",n))
[1] "SRS from exponential, n=" "20"
> print(c("nominal coverage",(1-alpha)*100,"%"))
[1] "nominal coverage" "95"                     "%"
> print(c("coverage %",round(CI/K*100,1)))
[1] "coverage %" "91.9"
> print(c("SRS from exponential, n=",n))
[1] "SRS from exponential, n=" "100"
> print(c("nominal coverage",(1-alpha)*100,"%"))
[1] "nominal coverage" "95"                     "%"
> print(c("coverage %",round(CI/K*100,1)))
[1] "coverage %" "94.2"
```

With a sample of 20 the coverage is about 92% rather than the claimed 95%, but the discrepancy decreases as the sample size increases. The coverage is 94.3% with sample size of 100. The exponential distribution is a substantial departure from normality and the CI based on a t-distribution is considered a reasonably robust procedure. For comparison, if the standard deviation of the exponential distribution was assumed known then the CI based on a normal distribution[6] had a coverage of 95.5%$(K = 10^6)$ and this is a very robust procedure.

7.4.4   Bootstrap methods

7.4.4.1   Bootstrap resampling

The **bootstrap** is a **resampling** method that avoids assumptions about the specific form of parent distribution. The bootstrap procedure estimates the parent distribution by the sample, and repeatedly samples with replacement from the sample. The statistic is calculated for each of these re-samples and an empirical sampling distribution is obtained.

Suppose θ is the parameter to be estimated and $\hat{\theta}$ is an estimate of θ calculated from a SRS of size n, $\{x_1, x_2, \ldots, x_n\}$ from the population. A bootstrap sample is a sample of size n drawn with replacement from the set $\{x_1, x_2, \ldots, x_n\}$ with equal probability $\frac{1}{n}$ of selection for each element at each draw, and is denoted $\{x_1^\star, x_2^\star, \ldots, x_n^\star\}$. The estimate of θ calculated from the bootstrap sample is $\hat{\theta}_b^\star$. If n is moderately large there are very many

---
[6]The code was modified by changing $t = qt()$ to $z = qnorm()$ and using z in place of t.

possible bootstrap samples and we draw a large number B of these, where $b = 1, \ldots, B$ (it doesn't matter if some are replicates) and so we have an empirical sampling distribution of $\widehat{\theta}_b^*$. The basis of the bootstrap method is that the sampling distribution of $\widehat{\theta}$ with respect to θ is approximated by the sampling distribution of $\widehat{\theta}_b^*$ with respect to $\widehat{\theta}$ So, the bias is approximated as

$$E[\widehat{\theta}] - \theta \approx \overline{\widehat{\theta^*_\cdot}} - \widehat{\theta},$$

where the dot subscript on $\overline{\widehat{\theta^*_\cdot}}$ indicates that the averaging was over b. The variance is approximated as the variance of the bootstrap estimates

$$\mathrm{var}(\widehat{\theta}) \approx \frac{\sum(\widehat{\theta}_b^* - \overline{\widehat{\theta^*_\cdot}})^2}{B - 1}.$$

The square root of the variance of the bootstrap estimates is an approximation to the standard error and can be given as a measure of precision of the estimator $\widehat{\theta}$. There are several procedures for constructing approximate confidence intervals for θ [Hesterberg, 2015], and two are described below.

7.4.4.2 Basic bootstrap confidence intervals

The starting point for the construction of the **basic bootstrap confidence interval** is the definition of the lower quantile of the sampling distribution of $\widehat{\theta}$,

$$P\left(\widehat{\theta}_{1-\alpha/2} < \widehat{\theta}\right) = 1 - \alpha/2.$$

Subtracting the unknown parameter from both sides of the inequality gives

$$P\left(\widehat{\theta}_{1-\alpha/2} - \theta < \widehat{\theta} - \theta\right) = 1 - \alpha/2.$$

Now use $\widehat{\theta}_{1-\alpha/2}^* - \widehat{\theta}$ as an approximation to the left hand side of the inequality and rearrange to obtain

$$P\left(\theta < 2\widehat{\theta} - \widehat{\theta}_{1-\alpha/2}^*\right) = 1 - \alpha/2.$$

A similar argument starting from the upper quantile leads to

$$P\left(2\widehat{\theta} - \widehat{\theta}_{\alpha/2}^* < \theta\right) = 1 - \alpha/2.$$

It follows that a $(1 - \alpha) \times 100\%$ confidence interval for θ is given by

$$\left[2\widehat{\theta} - \widehat{\theta}_{\alpha/2}^*, \; 2\widehat{\theta} - \widehat{\theta}_{1-\alpha/2}^*\right].$$

7.4.4.3 Percentile bootstrap confidence intervals

The **percentile bootstrap confidence interval** follows from the following argument.

$$P\left(\widehat{\theta}_{1-\alpha/2} < \theta < \widehat{\theta}_{\alpha/2}\right) = 1 - \alpha.$$

Write $\widehat{\theta} = \theta + W$ where W is a random variable that represents the estimation error. Then

$$P\left(\widehat{\theta}_{1-\alpha/2} - W < \theta < \widehat{\theta}_{\alpha/2} - W\right) = 1 - \alpha$$

and a $(1 - \alpha) \times 100\%$ confidence interval for θ is given by $\left[\widehat{\theta}_{1-\alpha/2} - w, \widehat{\theta}_{\alpha/2} - w\right]$. The percentile bootstrap replaces the unknown quantiles of $\widehat{\theta}$ with those of $\widehat{\theta}^*$ and assumes an observed value of 0 for the estimation error. So, the $(1 - \alpha) \times 100\%$ percentile bootstrap confidence interval is given by

$$\left[\widehat{\theta}^{\star}_{1-\alpha/2}, \widehat{\theta}^{\star}_{\alpha/2}\right].$$

Although the justification seems less convincing than that for the basic interval the percentile interval tends to give better results [Hesterberg, 2015].

Example 7.9: Camera flash [CI using bootstrap]

The following data are the number of camera flash pulses provided by notional 1.5 V dry cell batteries before the voltage dropped to 1.3 V ([Dunn, 2013])

$$37, 38, 37, 46, 34, 44, 47, 47, 44, 40.$$

We use R to construct confidence intervals for the mean number of pulses in the corresponding population, using both basic and percentile bootstrap intervals and an interval based on the t-distribution.

```
> set.seed(163)
> x=c(37, 38, 37, 46, 34, 44, 47, 47, 44, 40)
> n=length(x);alpha=0.10;B=10000;Bmean=rep(0,B)
> for (b in 1:B){
+ xx=sample(x,n,replace=TRUE)
+ Bmean[b]=mean(xx)
+ }
> BS=sort(Bmean)
> L=BS[(alpha/2)*B];U=BS[(1-alpha/2)*B]
> bias=mean(Bmean)-mean(x)
> Lbasic=2*mean(x)-U
> Ubasic=2*mean(x)-L
> print(c("basic bp 90,[",Lbasic,Ubasic,"]"),quote=FALSE)
[1] basic bp 90,[ 39.1           43.8              ]
> print(c("percen bp 90, [",L,U,"]"),quote=FALSE)
[1] percen bp 90, [ 39             43.7              ]
> t.test(x,conf.level=0.90)

        One Sample t-test

data:  x
t = 27.4714, df = 9, p-value = 5.443e-10
alternative hypothesis: true mean is not equal to 0
90 percent confidence interval:
 38.63746 44.16254
sample estimates:
mean of x
     41.4
```

The basic and percentile bootstrap intervals are very close because the sampling distribution of $\bar{x}^*$ is almost symmetric about $\bar{x}$. They are both narrower than the interval constructed using a t-distribution.

Example 7.10: Comparison of bootstrap CI with *t*-distribution CI

We now investigate whether the intervals constructed using the bootstrap procedure match the nominal confidence level. Our investigation is limited to confidence intervals for the population mean constructed from random samples from an exponential distribution, and for comparison a normal distribution.

To do this we need to run a simulation in which we take a large number, N, of random samples from the parent distribution. For each sample we construct bootstrap intervals, basic and percentile, and for comparison intervals based on the t-distribution and the normal distribution using the known standard deviation. The parent distributions are set to have a mean and variance of 1, but this does not affect the results. The number of nominal 90% confidence intervals for the mean that do not include the population mean, which has been set to 1, are counted. The code has two loops, the outer draws the samples and the inner re-samples each sample to construct bootstrap intervals. The code listing below uses $N = 10\,000$ samples of size $n = 10$ from an exponential distribution with a mean of 1, and $B = 1\,000$ bootstrap samples from each sample.

```
> set.seed(28)
> N=10000;Lb=rep(0,N);Ub=rep(0,N);Lp=rep(0,N);Up=rep(0,N)
> Lz=rep(0,N);Uz=rep(0,N);Lt=rep(0,N);Ut=rep(0,N)
> n=10;alpha=0.10
> for (j in 1:N) {
+ x=rexp(n)
+ B=1000;Bmean=rep(0,B)
+ for (b in 1:B)   {
+ xx=sample(x,n,replace=TRUE)
+ Bmean[b]=mean(xx)
+ }
+ BS=sort(Bmean)
+ L=BS[B*alpha/2];U=BS[B*(1-alpha/2)]
+ Lb[j]=2*mean(x)-U;Ub[j]=2*mean(x)-L
+ Lp[j]=L;Up[j]=U
+ Lz[j]=mean(x)-qnorm(1-alpha/2)*1/sqrt(n)
+ Uz[j]=mean(x)+qnorm(1-alpha/2)*1/sqrt(n)
+ Lt[j]=mean(x)-qt(1-alpha/2,(n-1))*sd(x)/sqrt(n)
+ Ut[j]=mean(x)+qt(1-alpha/2,(n-1))*sd(x)/sqrt(n)
+ }
> nfLb=length(which(Lb>1));nfUb=length(which(Ub<1))
> print(c("Basic","1<L",nfLb,"U<1",nfUb))
[1] "Basic" "1<L"    "251"    "U<1"    "1818"
> nfLp=length(which(Lp>1));nfUp=length(which(Up<1))
> print(c("Percentile","1<L",nfLp,"U<1",nfUp))
[1] "Percentile" "1<L"       "325"       "U<1"       "1581"
> nfLz=length(which(Lz>1));nfUz=length(which(Uz<1))
> print(c("z int","1<L",nfLz,"U<1",nfUz))
[1] "z int" "1<L"    "627"    "U<1"    "245"
> nfLt=length(which(Lt>1));nfUt=length(which(Ut<1))
> print(c("t int","1<L",nfLt,"U<1",nfUt))
```

```
[1] "t int" "1<L"    "138"    "U<1"    "1323"
```

The code was run again with random samples of sizes $n = 10, 30, 50$ from a normal distribution with mean and variance equal to 1 as well as from the exponential distribution with a mean of 1. The results are collated in Table 7.1.

TABLE 7.1: Coverage of nominal 90% bootstrap CI: number of intervals per 10 000 that do not include the population mean. The expected number if the procedure is accurate is 1 000.

	Exponential parent			Normal parent		
n	Basic	Percen	t	Basic	Percen	t
10	2 069	1 906	1 461	1 605	1 628	1 060
30	1 452	1 372	1 226	1 194	1 194	1 016
50	1 278	1 199	1 132	1 049	1 065	952

The results suggest that bootstrap intervals are too narrow in small samples. In this example the construction using the *t*-test appears more reliable even when sampling from an exponential distribution. However, an advantage of the bootstrap procedure is that it can be used for any statistic, for example an upper quantile of a Gumbel distribution (Example 7.11).

7.4.5 Parametric bootstrap

The basis for the bootstrap procedure is that the population distribution is estimated by the sample, which is resampled. In contrast, the parametric bootstrap avoids re-sampling the sample and instead assumes some probability distribution as a model of the population. The parameters of this parent distribution are estimated from the sample. The percentiles of the sampling distribution of $\widehat{\theta}$ are then obtained by Monte-Carlo simulation from the fitted probability distribution. Although the method does assume a specific parent distribution there are no restrictions on the choice of distribution and it circumvents assumptions of, for example, normality.

Our investigation of the parametric distribution is also limited to confidence intervals for the population mean constructed from random samples from an exponential distribution. The code replaces the bootstrap re-sampling of the sample with a random draw from an exponential distribution with a mean equal to the sample mean, and hence a rate equal to the reciprocal of the sample mean.

```
> #Parametric bootstrap
> set.seed(28)
> N=10000;Lb=rep(0,N);Ub=rep(0,N);Lp=rep(0,N);Up=rep(0,N)
> n=50;alpha=0.10
> for (j in 1:N) {
+ x=rexp(n)
+ B=1000;Bmean=rep(0,B)
+ for (b in 1:B)  {
+ xx=rexp(n,1/mean(x))
+ Bmean[b]=mean(xx)
+ }
```

```
+ BS=sort(Bmean)
+ L=BS[B*alpha/2];U=BS[B*(1-alpha/2)]
+ Lb[j]=2*mean(x)-U;Ub[j]=2*mean(x)-L
+ Lp[j]=L;Up[j]=U
+ }
> nfLb=length(which(Lb>1));nfUb=length(which(Ub<1))
> print(c("Basic","1<L",nfLb,"U<1",nfUb))
[1] "Basic" "1<L"    "185"    "U<1"    "925"
> nfLp=length(which(Lp>1));nfUp=length(which(Up<1))
> print(c("Percentile","1<L",nfLp,"U<1",nfUp))
[1] "Percentile" "1<L"         "305"         "U<1"         "783"
```

The parametric bootstrap performs rather well, around 11% of nominal 90% CIs did not contain the population mean, at least in the idealized case when the form of the assumed distribution is the same as the parent distribution. It is a useful procedure if it is reasonable to suppose some particular form of parent distribution.

Example 7.11: Maximum wind speeds [parametric bootstrap]

The data are annual maximum wind speeds (fastest mile wind speed 10 meters above ground level in mph) recorded at Jacksonville, FL 1948-1977. They are obtained from a National Institute of Standards and Technology website [NIST website, 2017].

The data are plotted in Figure 7.4. There is no apparent trend over this period and the histogram and quantile-quantile plot indicate that the data are compatible with an assumption of a SRS from a Gumbel distribution.

```
> Jacksonville.dat=read.table("Jacksonville.txt",header=TRUE)
> attach(Jacksonville.dat)
> print(speed)
 [1] 65 38 51 47 42 42 44 42 38 34 42 44 49 56 74 52 44 69 47 53 40 51
[23] 48 53 48 68 46 36 43 37
> summary(speed)
  Min. 1st Qu. Median  mean 3rd Qu.   Max.
 34.00   42.00  46.50  48.10   51.75  74.00
> par(mfrow=c(2,2))
> plot(as.ts(speed),ylab="speed");hist(speed,main="")
> n=length(speed);p=c(1:n)/(n+1)
> x=-log(-log(p));y=sort(speed)
> plot(x,y,xlab="-ln(-ln(i/(n+1)))",ylab="speed (order statistic)")
```

The annual maximum wind (fastest mile) with an average recurrence interval of 100 years is estimated by fitting a Gumbel distribution.

```
> #estimate parameters of Gumbel and upper 1% quantile
> theta=sqrt(6*var(speed)/pi^2);xi=mean(speed)-0.5772*theta
> print(c("xi",round(xi,2),"theta",round(theta,2),
+ "upper .01 quantile",round(xi-log(-log(.99))*theta,2)),quote=FALSE)
[1] xi                   43.6              theta            7.8
[5] upper .01 quantile 79.48
```

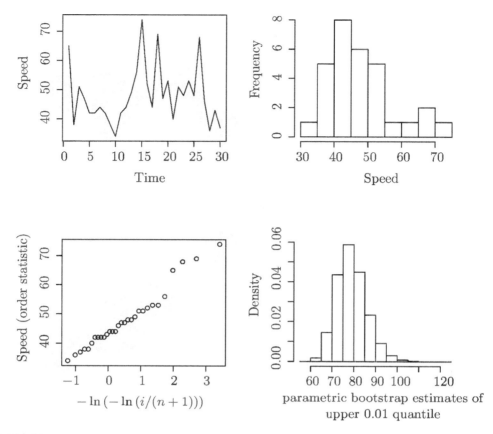

FIGURE 7.4: Jacksonville annual maximum wind speeds (mph): time series plot (upper left); histogram (upper right); Gumbel quantile-quantile plot (lower left); histogram of the parametric bootstrap estimates of the upper 0.01 quantile (upper right).

We now estimate the accuracy and precision of the estimator using the parametric bootstrap. It is neater to do this for the reduced variate and then make a linear transformation for the Jacksonville application.

```
> #simulate with reduced distribution
> B=100000;Bq=rep(0,B)
> for (b in 1:B) {
+ p=runif(n)
+ x=-log(-log(p))
+ thetab=sqrt(6*var(x)/pi^2);xib=mean(x)-0.5772*thetab
+ Bq[b]=xib-log(-log(.99))*thetab
+ }
> #now apply linear transform to Jacksonville application
> Bqlt=xi+theta*Bq
> bias=mean(Bqlt)-(xi-log(-log(.99))*theta)
> hist(Bqlt,main="",xlab="parametric bootstrap estimates of upper 0.01
  quantile",freq=FALSE)
> print(c("mean",round(mean(Bqlt),2),"bias",round(bias,2),
+ "standard error",round(sd(Bqlt),2)),quote=FALSE)
```

```
[1] mean 78.94
[3] bias -0.54
[5] standard error 6.94
> print(c("95% parametric bootstrap percentile CI"),quote=F)
[1] 95% parametric bootstrap percentile CI
> print(round(c(sort(Bqlt)[B*.025],sort(Bqlt)[B*0.975])))
[1] 67 94
```

A histogram of the bootstrap estimates is shown in the lower right frame of Figure 7.4. The estimated bias of the estimator is −0.54 which is slight. The standard error is 6.94 and the mean plus or minus two standard errors is close to the 95% CI.

Design codes, outside hurricane zones, are based on similar calculations made over a network meteorological stations. Moreover, the maximum annual 3 second gust speed is typically about 20% higher than the fastest mile.

7.5 Hypothesis testing

In some applications we focus on a specific value for the mean and ask whether the data are consistent with this value. One approach is to calculate a confidence interval for the mean and note whether it includes the specific value. Hypothesis testing is an alternative presentation which addresses the question directly. Hypothesis tests are widely used in statistics, and have more general application than confidence intervals.

Definition 7.9: Null hypothesis

A null hypothesis is made in the context of a general probability model which constitutes the assumptions behind the test. The null hypothesis defines a specific case of the general model.

Example 7.12: Water from well [null hypothesis]

The World Health Organization (WHO) recommends a maximum arsenic level in drinking water of 10 micrograms per liter (part per billion ppb). However, a more practical limit for countries with naturally high levels of arsenic, such as Bangladesh, is 40 ppb. In 1942 the U.S. limit was set at 50 ppb but it is now reduced to 10 ppb.

Assume that the arsenic contents of aliquots of well water taken for analysis have a normal distribution. A null hypothesis is that the mean of this distribution is 40 ppb.

Definition 7.10: Alternative hypothesis

The hypothesis to be accepted if the null hypothesis is rejected. It is the negation of the null hypothesis together with possible constraints as in the case of one-sided alternative hypotheses.

Example 7.13: Water from well [alternative hypothesis]

The most general alternative hypothesis is that the mean is not 40 ppb. A one-sided alternative hypothesis is that the mean exceeds 40 ppb.

You may notice that the null hypothesis specifies a precise single value, whereas the alternative hypothesis refers to a range of values. The null hypothesis is set up to provide a basis for argument, and in general we aim to provide evidence against it.

7.5.1   Hypothesis test for population mean when σ known

A hypothesis test sets up a **null hypothesis**, H_0, that the population mean is equal to a specific value. We denote the specific value by μ_0. If H_0 is true then

$$\frac{\overline{X} - \mu_0}{\sigma/\sqrt{n}}$$

is a statistic, the **test statistic**, because μ_0 is a specified number. We don't expect the null hypothesis to be precisely true, but we set it up as the basis for a decision. The reason for taking a specific value for H_0 is that it specifies the parameter μ of the sampling distribution. We write

$$H_0 : \mu = \mu_0$$

Having set up a null hypothesis, H_0, that we are going to test we also need to specify an alternative hypothesis, H_1, that we are testing it against. The most general alternative hypothesis is the **two-sided alternative hypothesis** that the mean is not equal to the specified value. That is

$$H_1 : \mu \neq \mu_0$$

We perform the test at some pre-selected **level of significance**, α, which is defined as

$$\alpha \;\; = \;\; \mathrm{P}(\text{reject } H_0 | H_0 \text{ true}).$$

The level of significance is typically kept low, and 0.05 is a common choice.

If H_0 is true, and we assume that the population standard deviation σ is known, the

$$\frac{\overline{X} - \mu_0}{\sigma/\sqrt{n}} \;\; \sim \;\; N(0,1).$$

From a practical point of view we want to reject H_0 if $\overline{X}$ is far from μ_0. We need to quantify "far from". By definition of the quantiles of a normal distribution

$$P\left(-z_{\alpha/2} < \frac{\overline{X} - \mu}{\sigma/\sqrt{n}} < z_{\alpha/2} \right) = 1 - \alpha$$

which is equivalent to

$$P\left(-z_{\alpha/2} \frac{\sigma}{\sqrt{n}} < \overline{X} - \mu < z_{\alpha/2} \frac{\sigma}{\sqrt{n}} \right) = 1 - \alpha$$

and we take "far from" as $z_{\alpha/2} \dfrac{\sigma}{\sqrt{n}}$.

So, we will reject H_0 if $\bar{x}$ is sufficiently far from μ_0, and for a given α, and a two-sided H_1, this corresponds to

$$\left| \frac{\bar{x} - \mu_0}{\sigma/\sqrt{n}} \right| > z_{\alpha/2} \quad \Longleftrightarrow \quad |\bar{x} - \mu_0| > z_{\alpha/2} \frac{\sigma}{\sqrt{n}}$$

If there is evidence against H_0 then we will conclude: there is evidence that $\mu < \mu_0$ if $\bar{x} < \mu_0$; or there is evidence that $\mu > \mu_0$ if $\bar{x} > \mu_0$. It is important to realize that no evidence against H_0 does not imply that $\mu = \mu_0$[7].

Example 7.14: Fluoridated water supply [hypothesis testing for μ, σ known]

In some cities the public water supply is fluoridated as a public dental health measure. This practice remains controversial and it is important to maintain the agreed target level. A city sets a target level of 0.8 ppm, and every week a public health inspector takes a random sample of 12 bottles from kitchen taps and sends them for fluoride analysis. The standard deviation of fluoride contents of bottles filled from kitchen taps has been estimated as the square root of the average of the variances of the weekly samples, taken over several years, and is 0.04. Moreover, the distribution of fluoride contents each week is reasonably modeled as a normal distribution.

Last week the analyses were:

$$0.78, 0.91, 0.91, 0.98, 0.83, 0.85, 0.87, 0.85, 0.95, 0.84, 0.90, 0.92.$$

The question is whether or not this sample is consistent with a population mean of 0.8. A confidence interval provides a succinct answer to the question (Section 7.5.3), but the same theory can be used to test a hypothesis about a population mean μ. In the case of the fluoridation:

$$H_0 : \mu = 0.8$$

and the alternative hypothesis is

$$H_1 : \mu \neq 0.8.$$

If testing is performed weekly, a value of $\alpha = 0.05$ would result in an expected $52 \times 0.05 = 2.6$ non-compliance orders per year if the water company maintains $\mu = 0.8$. This is too many and would result in non-compliance orders being seen as routine rather than a call for improvements, so in this application a lower value of $\alpha = 0.01$ is chosen.

If H_0 is true then the **test statistic** is

$$\frac{\bar{X} - 0.8}{0.04/\sqrt{12}} \sim N(0,1).$$

We will reject H_0 if $\bar{x}$ is sufficiently far from 0.8, and given $\alpha = 0.01$ this corresponds to the calculated value of the test statistic being less than $-z_{.005}$ or greater than $z_{.005}$ (Figure 7.5), with $\alpha = 0.01$.

---

[7] An analogy is the null hypothesis that a one year old used car, that is advertised as being "as new", is equivalent to a new car. The test is our inspection. The fact that we find nothing wrong does not prove that the car is equivalent to a new car.

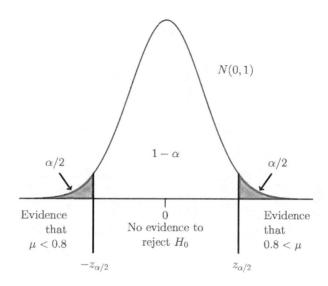

FIGURE 7.5: Distribution of $(\overline{X} - 0.8)/(0.04/\sqrt{12})$ if H_0 is true: test at α level with two sided alternative.

```
> x
[1] 0.78 0.91 0.91 0.98 0.83 0.85 0.87 0.85 0.95 0.84 0.90 0.92
> mean(x)
[1] 0.8825
> test_statistic=(mean(x)-0.8)/(0.04/sqrt(12))
> print(qnorm(.005,lower=FALSE))
[1] 2.575829
> print(test_statistic)
[1] 7.14471
```

The value of the test statistic is substantially higher than $z_{0.005}$, which is 2.58, so there is evidence to reject H_0, and the inspector would issue a non-compliance order stating that there is evidence that the level of fluoride is too high.

7.5.2   Hypothesis test for population mean when σ unknown

The theory is similar to the known case. Consider a null hypothesis

$$H_0 : \mu = \mu_0$$

and the two-sided alternative hypothesis

$$H_1 : \mu \neq \mu_0.$$

If H_0 is true then

$$\frac{\overline{X} - \mu_0}{S/\sqrt{n}} \quad \sim \quad t_{n-1}.$$

We reject at $\alpha \times 100\%$ level if

$$\left| \frac{\bar{x} - \mu_0}{s/\sqrt{n}} \right| > t_{n-1,\alpha/2}.$$

We illustrate this in Example 7.15.

Example 7.15: Inductors [hypothesis testing for μ, σ unknown]

An inductor is manufactured to a specified inductance of 470 microhenrys. A customer tests a sample of 20 inductors and finds the sample mean ($\bar{x}$) and standard deviation (s) are 465.8 and 8.7 respectively. If we assume the sample is an SRS from production is there evidence against a null hypothesis that the mean is 470 at the 5%, level of significance, against an alternative that it is not 470? We assume that inductances are approximately normally distributed.

$$H_0 : \mu = 470$$
$$H_1 : \mu \neq 470.$$

If H_0 is true then the test statistic is

$$\frac{\overline{X} - 470}{S/\sqrt{20}} \sim t_{19}.$$

The absolute value of the test statistic in the sample is

$$\left| \frac{465.8 - 470}{8.7/\sqrt{20}} \right| > 2.16.$$

This exceeds $t_{19,.025}$ which is 2.09 so there is evidence to reject H_0. Since $\bar{x}$ is 465.8 we have evidence, at the 5% level of significance, that the population mean is lower than the specified 470.

7.5.3   Relation between a hypothesis test and the confidence interval

Instead of the hypothesis test, we could construct a CI and issue a non-compliance order if this does not include 0.8. The $(1 - \alpha) \times 100\%$ CI is the set of all μ_0 such that we would not reject $H_0 : \mu = \mu_0$ at the α level against a two-sided H_1, as we now explain. The $(1 - \alpha) \times 100\%$ CI is

$$\bar{x} \pm z_{\alpha/2} \frac{\sigma}{\sqrt{n}}.$$

If μ_0 lies in this interval then

$$|\bar{x} - \mu_0| \leq z_{\alpha/2} \frac{\sigma}{\sqrt{n}} \qquad \text{so} \qquad \left| \frac{\bar{x} - \mu_0}{\frac{\sigma}{\sqrt{n}}} \right| \leq z_{\alpha/2}$$

and there is no evidence to reject H_0 at α level. Using our `zint()` function the 99% CI for μ is

```
> zint(x,0.04,0.01)
[1] 0.8527569 0.9122431
```

The lower limit of this interval is greater than 0.80 and so we have evidence that the mean is too high at the 0.01 level. The CI has the advantage that the effect of the sample size is clearly shown. If the sample size is very large we may have evidence to reject H_0 when the sample mean is very close to 0.8 and the discrepancy is quite acceptable, and this will be clear from the narrow CI which is close to, but doesn't include, 0.8. If the sample size is very small we may have no evidence against H_0 when the sample mean is far from 0.8, and the wide CI, which includes 0.8, will show this. An excessively large sample size is a waste of resources. Too small a sample is also a waste of resources if we have little chance of detecting a substantial non-compliance. A suitable sample size should be specified before the testing begins.

7.5.4 *p*-value

The *p*-**value** associated with a test of a null hypothesis, against a two sided alternative hypothesis, is defined as

$$P\left(|\text{test statistic}| > |\text{calculated value of test statistic in experiment}| \,\middle|\, H_0 \text{ is true}\right).$$

There is evidence to reject H_0 at the α level if

$$p\text{-value} \ < \ \alpha.$$

R provides *p*-values with the default that the null hypothesis about a single mean is that it is 0. In the road stone example (Example 7.43) the *p*-value is 0.0001215. If the sample had been drawn from a normal distribution with a mean of 0, then

$$P\left(\left|\frac{\overline{X}-0}{S/\sqrt{n}}\right| > 26.372\right) = 0.000125.$$

In this context a mean FTR of 0 is meaningless and we ignore the *p*-value. In general, if a *p*-value is reasonably small, typically less than 0.10, experimenters write that a result is statistically significant and give the *p*-value in brackets.

Example 7.15: (Continued) Inductors [*p*-value]

The calculated value of the test statistics is

$$\frac{465.8 - 470}{8.7/\sqrt{20}} = -2.16.$$

The R function **pt**(-2.16,19) gives $P(t_{19} < -2.16) = 0.022$ so the *p*-value with a two-sided alternative hypothesis is $2 \times 0.022 = 0.044$. As this is less than 0.05 there was evidence to reject H_0 at the 5% level.

7.5.5   One-sided confidence intervals and one-sided tests

In some applications we are only concerned that a variable is less than some critical value. An example is runout of discs (see Example 7.29) for which 0 is the ideal value and the specification is in terms of a maximum acceptable value.

The construction of the confidence interval is adapted slightly so that an upper $(1 - \alpha) \times 100\%$ CI for μ if σ is unknown is:

$$\left(-\infty, \ \bar{x} + t_{n-1,\alpha}\frac{s}{\sqrt{n}}\right].$$

If the variable is non-negative the formal $-\infty$ is replaced by 0. In other cases we may be concerned only that a variable exceeds some critical value. An example is the breaking loads of wire ropes for cranes. A lower $(1 - \alpha) \times 100$ CI for μ is given by

$$\left[\bar{x} + t_{n-1,\alpha}\frac{s}{\sqrt{n}} \ , \infty\right).$$

Example 7.16: Camera flash [one-sided CI and hypothesis testing]

The mean and standard deviation of the number of camera flash pulses before a voltage drop for ten batteries (Example 7.9) are 41.4 and 4.77 respectively. Given that $t_{9,.05} = 1.83$, the lower limit for a one-sided 95% CI for the mean lifetime in the population of all such batteries is

$$41.4 - 1.83 \times 4.77/\sqrt{10} \ = \ 38.64.$$

The 90% (two-sided) CI for the population mean is

```
> x=c(37, 38, 37, 46, 34, 44, 47, 47, 44, 40)
> t.test(x,conf.level=0.90)
        One Sample t-test
data:  x
t = 27.4714, df = 9, p-value = 5.443e-10
alternative hypothesis: true mean is not equal to 0
90 percent confidence interval:
 38.63746 44.16254
sample estimates:
mean of x
     41.4
```

So, we are 95% confident that the mean exceeds 38.64 and 90% confident that the mean lies between 38.64 and 44.16. Notice that the lower, or upper, limit of a $(1 - \alpha)$ two-sided CI is the limit of the $(1 - \alpha/2)$ one-sided CI.

7.6 Sample size

The precision of the estimate of the sample mean depends upon the standard deviation, and the sample size, n, which is set at the start of the investigation.

- The calculation of a suitable sample size depends on the population standard deviation, which is usually unknown. A value has to be assumed from: experience; a literature search; or a pilot study.

- The choice of sample size depends upon the required precision, which can be specified by the width of the confidence interval.

- A rationale for choosing the width of a confidence interval is to specify the smallest difference, δ, from some nominal value, μ_0, that is of practical importance. It would be embarrassing to present a confidence interval that includes both μ_0 and $\mu_0 + \delta$, and this cannot happen if the width of the confidence interval is less than δ. This requirement can be expressed as

$$2\, t_{n-1,\alpha/2}\, \frac{s}{\sqrt{n}} \;\; < \;\; \delta.$$

for a $(1 - \alpha/2)100\%$ confidence interval. Making n the subject of the formula

$$n \;=\; \left(\frac{2\, t_{n-1,\alpha/2}\, s}{\delta}\right)^2.$$

- Apply the formula with s replaced by the assumed value for σ and for a 95% CI, which is a customary choice, $t_{n-1,\alpha/2}$ replaced by 2.0.

- If the value for n is small repeat using a t-distribution with degrees of freedom set at $n - 1$ and iterate.

Example 7.17: Electric vehicle [sample size calculation]

A compact electric vehicle has an EPA certified range of 238 miles. A motoring organization will test the range under typical driving conditions and sets a route heading south out of Portland. A sample of n of these electric vehicles will be driven, each with a different driver, along the route and the range will be recorded. The coefficient of variation of ranges for similar tests has been around 15% so the motoring organization will assume a value of $0.15 \times 238 \approx 36$ for σ. A range of 200 miles under typical conditions would be substantially less than the certified 238 miles, so the motoring organization asks for a 95% confidence interval with a width less than 38, so that the confidence interval cannot include numbers less than 200 and numbers greater than 238. The formula gives

$$n \;=\; \left(\frac{2 \times 2.0 \times 36}{38}\right)^2 \;=\; 14.4.$$

The upper 0.025 quantile of the t-distribution is

```
> qt(.975,13)
[1] 2.160369
```

and repeating the calculation with 2.2 in place of 2.0 gives $n = 17.4$. The motoring organization would take a sample of 17 or 18 cars.

If the requirement for a 95% CI is relaxed to a 90% CI the sample size could be reduced to 12.

```
> (2*qt(.95,11)*36/38)^2
[1] 11.57857
```

The motoring organization decides to take a sample of 12. Sample size calculations are inevitably approximate, because the actual value of the sample standard deviation is not known in advance, but an approximate answer is far better than no answer[8].

7.7 Confidence interval for a population variance and standard deviation

The construction of a confidence interval for the variance, and hence the standard deviation, of a normal distribution depends on the **chi-square distribution**[9]. A chi-square distribution with ν degrees of freedom is defined as the distribution of the sum of ν squared standard normal variables, Z_i for $i = 1, \ldots, \nu$. That is:

$$W = \sum_{i=1}^{\nu} Z_i^2,$$

has a chi-square distribution with ν degrees of freedom, and we write $W \sim \chi_\nu^2$. The mean and variance of W are ν and 2ν respectively.

We now show that the sampling distribution of the variance of a random sample of size n from a normal distribution is proportional to a Chi-squared distribution with $n-1$ degrees of freedom. Assume $\{X_i\}$, for $i = 1, \ldots, n$ is a random sample of n observations from a normal distribution with mean μ and variance σ^2. Then it follows from the definition that

$$\sum_{i=1}^{n} \left(\frac{X_i - \mu}{\sigma} \right)^2 \sim \chi_n^2.$$

Replacing μ by $\overline{X}$ accounts for the loss of one degree of freedom (Exercise 7.14). So

$$\sum_{i=1}^{n} \left(\frac{X_i - \overline{X}}{\sigma} \right)^2 \sim \chi_{n-1}^2.$$

Now $\sum_{i=1}^{n} \left(X_i - \overline{X} \right)^2 / (n-1) = S^2$, so we have the sampling distribution:

$$\frac{(n-1)S^2}{\sigma^2} \sim \chi_{n-1}^2.$$

This result is sensitive to the assumption that the population is normal, even with a large sample size (Exercise 7.15). We can now construct a $(1-\alpha) \times 100\%$ CI for σ^2. Starting from

$$\mathrm{P}\left(\chi_{1-\alpha/2}^2 < \frac{(n-1)S^2}{\sigma^2} < \chi_{\alpha/2}^2 \right) = 1 - \alpha,$$

we rearrange the inequalities to have σ^2 in the middle.

$$\mathrm{P}\left(\frac{(n-1)S^2}{\chi_{\alpha/2}^2} < \sigma^2 < \frac{(n-1)S^2}{\chi_{1-\alpha/2}^2} \right) = 1 - \alpha.$$

---

[8] Sample size calculations are often made in the context of a hypothesis test with a specified probability of rejecting $H_0 : \mu = \mu_0$ when $H_1 : \mu = \mu_1$ for a specified μ_1 (Exercise 7.12).

[9] The chi-square distribution is a special case of the gamma distribution with parameters $(\nu/2, 1/2)$.

Given the results of an experiment we can calculate s^2 and the $(1 - \alpha) \times 100\%$ CI for σ^2 is

$$\left[\frac{(n-1)s^2}{\chi^2_{\alpha/2}}, \quad \frac{(n-1)s^2}{\chi^2_{1-\alpha/2}} \right].$$

The $(1 - \alpha) \times 100\%$ CI for σ is obtained by taking the square root of the CI for σ^2.

Example 7.18: Fluoridation of water supplies [CI for σ]

The standard deviation of fluoride content of liter bottles, filled from a random sample of kitchen taps, is expected to be 0.04 ppm. The standard deviation of the 12 analyses last week was 0.056. Is there evidence that the standard deviation is not 0.04?

We will answer this question by constructing a 90% CI for σ.

```
> x=c(0.78,0.91, 0.91, 0.98, 0.83, 0.85, 0.87, 0.85, 0.95, 0.84, 0.90, 0.92)
> n=length(x)
> print(mean(x))
[1] 0.8825
> s=sd(x) ; print(s)
[1] 0.05610461
> chisq_.05=qchisq(.05,(n-1),lower.tail=FALSE) ; print(chisq_.05)
[1] 19.67514
> chisq_.95=qchisq(.05,(n-1)) ; print(chisq_.95)
[1] 4.574813
> LSS90=(n-1)*s^2/chisq_.05
> USS90=(n-1)*s^2/chisq_.95
> LS90=sqrt(LSS90)
> US90=sqrt(USS90)
> print(c("90% CI for sigma^2",round(LSS90,5),round(USS90,5)))
[1] "90% CI for sigma^2" "0.00176"          "0.00757"
> print(c("90% CI for sigma",round(LS90,3),round(US90,3)))
[1] "90% CI for sigma" "0.042"          "0.087"
```

A 90% confidence interval for the standard deviation of fluoride is $(0.042, 0.087)$, and as the lower limit is above 0.040, this suggests the standard deviation may be too high. However, the inspector will only issue a non-compliance order if there is evidence against a null hypothesis that the standard deviation is 0.04 with a one-sided alternative that the standard deviation is greater than 0.04, at the 0.01 level. Should a non-compliance order be issued?

We can carry out the formal test by constructing a 99% one-sided lower confidence interval (sometimes known as a **lower confidence bound**) for μ. If the lower point of this interval exceeds 0.04 then there is evidence to reject the null hypothesis at the 0.01 level with a one-sided alternative.

```
> LSS99=(n-1)*s^2/qchisq(.01,(n-1),lower.tail=FALSE)
> LS99=sqrt(LSS99) ; print(LS99)
[1] 0.037422
```

The 99% lower confidence bound for the standard deviation is 0.038, and as this is less than 0.04, there is not sufficiently strong evidence to issue a non-compliance order. You are asked to perform a formal hypothesis test in Exercise 7.16.

7.8 Comparison of means

In this section we aim to compare the means of two populations. One strategy is to draw independent simple random samples, one from each population, and then compare the sample means. However, it may be possible to set up matched pairs, one item from each population, in which case we can analyze the differences of the items in each pair. If it is feasible matching will give a more precise comparison.

7.8.1 Independent samples

7.8.1.1 Population standard deviations differ

There are two populations, designated A and B with means μ_A and μ_B and standard deviations σ_A and σ_B. The objective is to compare the population means. An SRS of size n_A is drawn from population A and an independent SRS of size n_B is drawn from population B (see Figure 7.6).

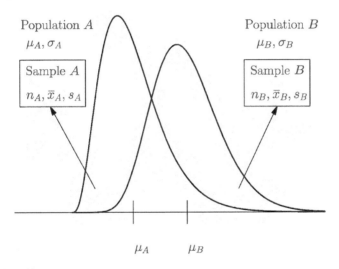

FIGURE 7.6: Independent samples from populations A and B.

The sample means $\bar{x}_A$ and $\bar{x}_B$, and sample standard deviations $\bar{s}_A$ and $\bar{s}_B$, are calculated. The difference in the population means is $\mu_A - \mu_B$. The estimate of the difference in population means is $\bar{x}_A - \bar{x}_B$.

We now quantify the precision of this estimate, and the following argument leads to a $(1 - \alpha) \times 100\%$ CI for $\mu_A - \mu_B$. We start from an approximation

$$\overline{X}_A \sim N\left(\mu_A, \frac{\sigma_A^2}{n_A}\right) \quad \text{and} \quad \overline{X}_B \sim N\left(\mu_B, \frac{\sigma_B^2}{n_B}\right),$$

where σ_A^2 and σ_B^2 are the variance of population A and B respectively, and approximate normality follows from the Central Limit Theorem. Since the samples are independent the variance of the difference in the means is the sum of the variances of the means and

$$\overline{X}_A - \overline{X}_B \sim N\left(\mu_A - \mu_B, \frac{\sigma_A^2}{n_A} + \frac{\sigma_B^2}{n_B}\right).$$

It follows that

$$\frac{(\overline{X}_A - \overline{X}_B) - (\mu_A - \mu_B)}{\sqrt{\frac{\sigma_A^2}{n_A} + \frac{\sigma_B^2}{n_B}}} \sim N(0,1).$$

If the population variances are known this result can be used to construct the CI, but the usual situation is that the population variances are replaced by sample estimates. The following is an approximate mathematical result[10], even when the populations are normal, but it is then an excellent approximation:

$$\frac{(\overline{X}_A - \overline{X}_B) - (\mu_A - \mu_B)}{\sqrt{\frac{S_A^2}{n_A} + \frac{S_B^2}{n_B}}} \sim t_\nu,$$

where the degrees of freedom (generally not an integer valued), are given by

$$\nu \approx \frac{\left(\frac{S_A^2}{n_A} + \frac{S_B^2}{n_B}\right)^2}{\left(\frac{(S_A^2/n_A)^2}{n_A-1} + \frac{(S_B^2/n_B)^2}{n_B-1}\right)}.$$

Since

$$\min(n_A - 1, n_B - 1) \le \nu \le n_A + n_B - 2$$

the lower bound can be used a conservative approximation. The $(1-\alpha) \times 100\%$ CI for $\mu_A - \mu_B$ is

$$\overline{x}_A - \overline{x}_B \pm t_{\nu,\alpha/2} \sqrt{\frac{s_A^2}{n_A} + \frac{s_B^2}{n_B}}.$$

The following example shows that a difference in population means needs to be viewed in the context of differences in standard deviations.

Example 7.19: Wire ropes [using the mean with the standard deviation]

Wire ropes with a mean breaking load of 11 tonne and a standard deviation of breaking load of 0.5 tonne would be preferable to ropes with a mean of 12 tonne and a standard deviation of breaking load of 1 tonne. If breaking loads are precisely normally distributed then 1 in a billion ropes will fail under a load equal to 6 standard deviations below the mean. This load is 8 tonne for the ropes with mean 11 and 6 tonne for the ropes with mean 12 tonne. The normal distribution model is unlikely to hold so precisely, but the same principle applies to any distributions.

Example 7.20: Steel cable [comparison of means]

A sample of 23 pieces of cable was taken from the George Washington Bridge in 1933, and a sample of 18 pieces of corroded cable was taken in 1962. The data are breaking loads (kN) (Stahl and Gagnon, 1995).

First we load the data into R:

---

[10]Due to B.L. Welch, 1947.

```
xA=c(3466, 3478, 3532, 3506, 3455, 3494, 3550, 3432, 3340, 3271, 3512,
  3328, 3489, 3485, 3564, 3460, 3547, 3544, 3552, 3558, 3538, 3551, 3549)
xB=c(2406, 3172, 2858, 2750, 2828, 3162, 2691, 2808, 3054, 3221, 3260,
  2995, 2651, 2897, 2799, 3201, 2966, 2661)
```

In this case, xA is the data on breaking load for the new cable, while xB is the data on the breaking load for the corroded cable.

Next we get the number of observations, sample means, and sample standard deviations for each sample:

```
> (nA=length(xA))
[1] 23
> (nB=length(xB))
[1] 18
> (xbarA=mean(xA))
[1] 3487
> (xbarB=mean(xB))
[1] 2910
> (sA=sd(xA))
[1] 79.33473
> (sB=sd(xB))
[1] 237.1557
```

Note the use of brackets around the R commands so that we can assign the values to variables for use later and also see the values.

Next we compare the distribution of each group using box plots:

```
boxplot(xA,xB,names=c("New 1933","Corroded 1962"))
```

The figure is given in Figure 7.7. As expected there is a larger median breaking load for the new cable, and a smaller IQR compared to the corroded cable. Finally we calculate the 95% confidence intervals for the difference in mean breaking load for the two groups using a lower bound of $18 - 1$ for the degrees of freedom:

```
> (diff=xbarA-xbarB)
[1] 577
> (sediff=sqrt(sA^2/nA + sB^2/nB))
[1] 58.29454
> (t=qt(.025,min((nA-1),(nB-1)),lower.tail=FALSE))
[1] 2.109816
> lwr=diff-t*sediff
> upr=diff+t*sediff
> c(lower=lwr,upper=upr)
   lower    upper
454.0093 699.9907
```

So we are 95% confident that the mean breaking load of the new steel is between 454 and 700 kN greater than the mean breaking load of the corroded steel cable. This is a substantial reduction in mean strength. Moreover, the variability of strength appears to have greatly increased (Figure 7.7).

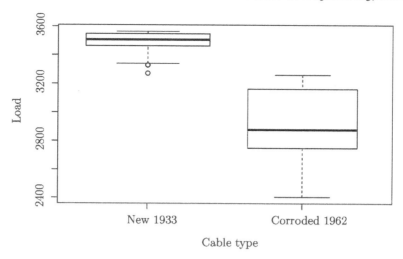

FIGURE 7.7: Box-plots of the breaking load of the new cable (1993) and corroded cable (1962) on the George Washington Bridge.

The R function `t.test()` is a more convenient way to calculate the confidence interval, We have explicitly specified 95% confidence and no assumption of equal population variances, although these are default values.

```
> t.test(xA,xB,conf.level=.95,var.equal=FALSE)

        Welch Two Sample t-test

data:  xA and xB
t = 9.898, df = 19.99, p-value = 3.777e-09
alternative hypothesis: true difference in means is not equal to 0
95 percent confidence interval:
 455.3957 698.6043
sample estimates:
mean of x mean of y
    3487      2910
```

The interval obtained with `t.test()` is slightly narrower because the conservative lower bound for the degrees of freedom, 17, was replaced by the more accurate 19.99.

7.8.1.2   Population standard deviations assumed equal

In some experiments we may be think it is reasonable to suppose that population standard deviations are equal. In this case we can use a pooled estimator S_p^2 of this common variance σ^2. This estimator is a weighted average of the sample variances, with weights equal to the degrees of freedom. Then

$$S_p^2 \;=\; \frac{(n_A - 1)S_A^2 + (n_B - 1)S_B^2}{n_A + n_B - 2}$$

is an unbiased estimator of σ^2 on $n_A + n_B - 2$ degrees of freedom. If the populations are normal then

$$\frac{(\overline{X}_A - \overline{X}_B) - (\mu_A - \mu_B)}{S_p\sqrt{\frac{1}{n_A} + \frac{1}{n_B}}} \sim t_{n_A+n_B-2}$$

and a $(1 - \alpha) \times 100\%$ CI for $\mu_A - \mu_B$ is

$$\overline{x}_A - \overline{x}_B \quad \pm \quad t_{n_A+n_B-2,\alpha/2}\, s_p \sqrt{\frac{1}{n_A} + \frac{1}{n_B}}.$$

Example 7.21: Circuit breakers [comparing means with equal population sds]

The data are from a manufacturer of high voltage switchgear. The open-time of circuit breakers is an important characteristic. A modified circuit breaker with a novel switch mechanism that will make it even more durable in remote settings with little, or no, maintenance, has been designed. Ten prototypes have been built and are compared with the last 38 from production. The designer does not expect the standard deviation of open-time to be affected, but the mean may be affected. A requirement of the new design is that the mean open-time should not be increased.

First as above, we read in the data.

```
> xA = c(22.41, 23.14, 22.22, 24.43, 23.28, 22.23, 23.42, 23.65,
  23.50, 22.70, 24.35, 23.33, 22.41, 20.97, 24.00, 22.94, 22.96,
  23.84, 23.72, 23.52, 23.81, 23.69, 23.05, 21.17, 23.54, 22.93,
  22.84, 21.64, 22.54, 23.36, 24.21, 22.88, 23.33, 22.93, 21.73,
  22.60, 22.62, 22.92)
> xB = c(23.68, 23.20, 21.88, 21.75, 23.11, 22.91, 21.14, 21.11,
  22.63, 23.21)
```

Again, we calculate the summary statistics:

```
> (nA=length(xA))
[1] 38
> (nB=length(xB))
[1] 10
> (xbarA=mean(xA))
[1] 23.02132
> (xbarB=mean(xB))
[1] 22.462
> (sA=sd(xA))
[1] 0.8058307
> (sB=sd(xB))
[1] 0.9224219
```

The box plots (Figure 7.8):

```
boxplot(xA,xB,names=c("production","prototype"))
```

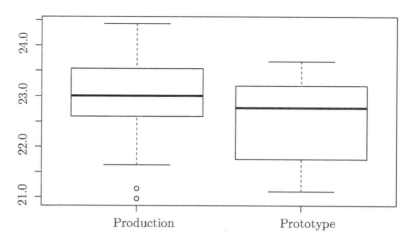

FIGURE 7.8: Box-plots of the open-times for production and prototype circuit breakers.

show that larger median time for the production compared to the prototype group, while the IQR is similar in both groups.

We can calculate the 95% confidence intervals for the differences using the formula:

```
> (diff=xbarA-xbarB)
[1] 0.5593158
> (dof=nA+nB-2)
[1] 46
> (sp=sqrt(((nA-1)*sA^2+(nB-1)*sB^2)/dof))
[1] 0.8299318
> (sediff=sp*sqrt(1/nA + 1/nB))
[1] 0.2949655
> (t=qt(.025,dof,lower.tail=FALSE))
[1] 2.012896
> lwr=diff-t*sediff
> upr=diff+t*sediff
> c(lower=lwr,upper=upr)
      lower        upper
-0.03441899   1.15305057
```

but using the `t.test()` function in R is far quicker.

```
> t.test(xA,xB,var.equal=TRUE)

  Two Sample t-test

data:  xA and xB
t = 1.8962, df = 46, p-value = 0.06422
alternative hypothesis: true difference in means is not equal to 0
95 percent confidence interval:
 -0.03441899  1.15305057
sample estimates:
mean of x mean of y
```

```
23.02132   22.46200
```

The conclusion is that the sample means for the production open times and prototype open times are 23.02 and 22.46 respectively. We are 95% confident that the difference in population means is between −0.03 and 1.15. Although the 95% CI does extend to −0.03, the lower limit of a 90% interval is greater than 0 and we can be reasonably confident that the mean open time has not increased.

7.8.2 Matched pairs

In Section 7.8.1 we considered a comparison of population means based on independent samples from each population. In some applications it may be possible to pair items from the two populations so that the items in a pair are as similar as possible, except for the feature under investigation. If this strategy is feasible, then a random sample of matched pairs will generally provide a more precise estimate of the difference in population means.

For example, suppose we want to compare masses recorded by a load cell weigh scale and spring loaded weigh scale. We would weigh the same objects on both scales are record the differences. This is the principle of the matched pairs procedure.

Example 7.22: Comparison of weigh-bridges [matched pairs]

A government inspector intends checking an old weigh-bridge at a quarry against a recently calibrated weigh-bridge at the public works department. The inspector will drive an unladen lorry to the quarry, where it will be weighed on the weigh-bridge, load it with building materials, drive to the exit where it is re-weighed, and note the recorded mass of building materials (exit mass less entry mass) paid for. The inspector will reverse, weighing the loaded and then unloaded lorry at the public works department. The discrepancy is the mass paid for at the quarry less the mass calculated at the public works department. The inspector will send colleagues with different lorries to buy building materials over the next four days. There will then be 5 measured discrepancies $\{d_i\}$, and the objective is to construct a 95% CI for the mean of all such discrepancies μ_D, which corresponds to a systematic over, or under, recoding of masses at the quarry. The discrepancies are: $15.8, 11.4, 10.4, -1.6, 6.7$. The analysis is that of a single sample from a population of discrepancies.

```
> x = c(15.8, 11.4, 10.4, -1.6,  6.7)
> t.test(x)

        One Sample t-test

data:  x
t = 2.9245, df = 4, p-value = 0.04305
alternative hypothesis: true mean is not equal to 0
95 percent confidence interval:
  0.4322098 16.6477902
sample estimates:
mean of x
     8.54
```

The 95% CI for μ_D is $[0, 17]$, and the inspector will tell the quarry to re-calibrate the weigh-bridge.

The experimental unit is a lorry load of building materials, and the pairing is calculating the mass of materials using the quarry weigh-bridge and calculating the mass on the public works weigh-bridge. The pairing removes the variability of masses of building materials loaded on the 5 occasions from the comparison. In general, if we can remove a component of variability when making comparisons we should do so.

Example 7.23: Gas cutting [matched pairs]

A company specializes in steel fabrication and uses oxy-propane gas cutting to cut steel plates. An engineer wants to investigate the use of oxy-natural gas as a more convenient alternative. An undesirable side-effect of any gas cutting is the hardening of the steel near the cut edge. The engineer will not consider natural gas instead of propane if the hardening side-effect is increased, and decides to perform an experiment to make a comparison. The engineer finds 8 plates of different grade steels and of different thicknesses. To remove the variability between plates from the comparison, the engineer decides to make two cuts on each plate, one with oxy-propane and the other with oxy-natural gas. It is possible that a first cut increases slightly the hardness of the plate, so the second cut might give systematically higher hardness readings near the cut edge. The engineer allows for this possibility by randomly selecting 4 of the 8 plates to be cut with oxy-propane first, the other 4 being cut with oxy-natural gas first. The variable to be analyzed is derived from Vickers hardness (VH10) measurements made in a fixed pattern alongside the cut edge.

To perform a matched pairs on this data, we first read in the data:

```
> oxy = read.csv("data/oxy.csv")
> oxy
  test.plate oxy.propane oxy.natural.gas
1          1         370             333
2          2         332             336
3          3         330             299
4          4         306             294
5          5         314             297
6          6         322             373
7          7         290             304
8          8         312             278
```

We then perform a matched-pairs using the R command t.test() with the extra argument paired = TRUE to let R know we are using matched-pairs.

```
> t.test(oxy$oxy.propane,oxy$oxy.natural.gas,paired = TRUE)

	Paired t-test

data:  oxy$oxy.propane and oxy$oxy.natural.gas
t = 0.7335, df = 7, p-value = 0.4871
alternative hypothesis: true difference in means is not equal to 0
95 percent confidence interval:
```

```
  -17.2339  32.7339
sample estimates:
mean of the differences
                7.75
```

The mean difference in hardness, oxy-propane less oxy-natural gas, is 7.75 which is encouraging. But, the 95% confidence interval for the difference is $[-17, 33]$ and we cannot be confident that the hardness of cut edges will not increase. Further testing is recommended.

7.9 Comparing variances

When comparing populations the most striking features are typically differences in location and spread. Location can be described by the mean and the spread can be described by standard deviations. We now consider the comparison of standard deviations.

Suppose we have independent simple random samples of sizes n_A and n_B from populations A and B. We know that

$$\frac{(n_A - 1)S_A^2}{\sigma_A^2} \sim \chi_{n_A-1}^2$$

and it follows that

$$\frac{S_A^2}{\sigma_A^2} \sim \frac{\chi_{n_A-1}^2}{n_A - 1}.$$

We have a similar result for S_B^2. The F-distribution[11] is defined as the ratio of two independent chi-square distributions divided by their degrees of freedom, to facilitate the comparison of population variances. It has two parameters ν_1 and ν_2, which are referred to as its degrees of freedom. The F-distribution has a mean of $\nu_2/(nu_2 - 2)$ and is asymmetric.

$$\frac{\frac{S_A^2}{\sigma_A^2}}{\frac{S_B^2}{\sigma_B^2}} = \frac{S_A^2 \sigma_B^2}{S_B^2 \sigma_A^2} \sim F_{n_A-1, n_B-1}.$$

A $(1 - \alpha) \times 100\%$ CI for σ_B^2/σ_A^2 follows from the following argument

$$\mathrm{P}\left(F_{n_A-1, n_B-1, 1-\alpha/2} < \frac{S_A^2 \sigma_B^2}{S_B^2 \sigma_A^2} < F_{n_A-1, n_B-1, \alpha/2} \right) = 1 - \alpha$$

and so

$$\mathrm{P}\left(\frac{S_B^2}{S_A^2} F_{n_A-1, n_B-1, 1-\alpha/2} < \frac{\sigma_B^2}{\sigma_A^2} < \frac{S_B^2}{S_A^2} F_{n_A-1, n_B-1, \alpha/2} \right) = 1 - \alpha.$$

Hence a $(1 - \alpha) \times 100\%$ CI for σ_B^2/σ_A^2 is given by

$$\left[\frac{s_B^2}{s_A^2} F_{n_A-1, n_B-1, 1-\alpha/2}, \quad \frac{s_B^2}{s_A^2} F_{n_A-1, n_B-1, \alpha/2} \right].$$

---

[11] Fisher-Snedecor distribution.

A $(1 - \alpha) \times 100\%$ CI for σ_B/σ_A is obtained by taking square roots of the end points of the interval for σ_B^2/σ_A^2.

Example 7.21: (Continued) Steel cable

The standard deviation of breaking load (kN) of 23 new pieces of cable from the Washington Bridge was 79, and the standard deviation of 18 corroded pieces was 237. A 95% CI for the ratio of the standard deviation of corroded pieces to the standard deviation of new pieces can be calculated in R.

```
> # Get ratio of corroded (grp A) to new (grp B)
> sB/sA
[1] 2.989305
> lwr = (sB/sA)^2*qf(.025,nA-1,nB-1)
> upr = (sB/sA)^2*qf(.025,nA-1,nB-1,lower.tail=FALSE)
> # 95% CI for variance
> c(lower=lwr,upper=upr)
    lower      upper
 3.650659 23.103698
> # 95% CI for std dev
> sqrt(c(lower=lwr,upper=upr))
    lower      upper
1.910670 4.806631
```

A 95% CI for $\sigma_{\text{corroded}}/\sigma_{\text{new}}$ is $[1.9, 4.8]$, and there is strong evidence that the standard deviation is greater for the corroded cable.

There is an R function `var.test()`, which is quicker than using the formula.

```
> var.test(xB,xA)

    F test to compare two variances

data:  xB and xA
F = 8.9359, num df = 17, denom df = 22, p-value = 5.229e-06
alternative hypothesis: true ratio of variances is not equal to 1
95 percent confidence interval:
  3.650659 23.103698
sample estimates:
ratio of variances
          8.935943
```

7.10 Inference about proportions

7.10.1 Single sample

The usual approach for making inferences about proportions is to approximate the sampling distribution of the sample proportion by a normal distribution. Let X be the number of successes in a simple random sample of size n, then to an excellent approximation $X \sim binom(n, p)$, and X has a mean np and variance $np(1-p)$. We approximate this distribution by a normal distribution with the same mean and variance.

$$X \quad \sim \quad N\left(np, np(1-p)\right).$$

This approximation is good provided the smaller of np and $n(1-p)$ exceeds about 5. The sample proportion (see Figure 7.9)

$$\widehat{p} \quad = \quad \frac{X}{n}$$

and the approximation is

$$\widehat{p} \quad \sim \quad N\left(p, \frac{p(1-p)}{n}\right).$$

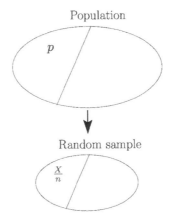

Population

p

Random sample

$\frac{X}{n}$

FIGURE 7.9: Single sample.

A $(1 - \alpha) \times 100\%$ CI for p follows from a similar argument to that used for the CI for a population mean[12]. That is:

$$P\left(-z_{\alpha/2} < \frac{\widehat{p} - p}{\sqrt{p(1-p)/n}} < z_{\alpha/2}\right) \quad = \quad 1 - \alpha.$$

We then rearrange, and replace the standard error, $\sqrt{p(1-p)/n}$, by the sample estimate, $\sqrt{\widehat{p}(1-\widehat{p})/n}$, to obtain

$$\widehat{p} \quad \pm \quad z_{\alpha/2} \sqrt{\frac{\widehat{p}(1-\widehat{p})}{n}}.$$

Although the standard error has been replaced with an estimate from the sample, it is usual to use $z_{\alpha/2}$ rather than $t_{\alpha/2}$ and refer to the interval as a large sample approximation. This further approximation is customarily accepted as reasonable provided the sample size, n, exceeds about 30.

---

[12]In fact, p is the population mean of X, where X takes the value 0 or 1.

Example 7.24: Pipes [CI for a proportion]

A water company is implementing a program of replacing lead communication pipes between properties and the water main by PVC pipes. The finance manager needs to estimate the cost of the program, and to provide an indication of the precision of this estimate. A random sample of 200 properties was taken from the billing register and 38 were found to have lead communication pipes to the water main. A 95% CI for the proportion in the corresponding population is

$$\frac{38}{200} \pm 1.96 \times \sqrt{\frac{(38/200)(1-38/200)}{200}},$$

which reduces to [0.14, 0.24]. You are asked to estimate costs in Exercise 7.53.

The R function `prop.test()` is quicker than making the calculations.

```
> prop.test(38,200)

        1-sample proportions test with continuity correction

data:  38 out of 200, null probability 0.5
X-squared = 75.645, df = 1, p-value < 2.2e-16
alternative hypothesis: true p is not equal to 0.5
95 percent confidence interval:
 0.1394851 0.2527281
sample estimates:
   p
0.19
```

R uses a slight modification of the approximation which accounts for the small difference in the confidence intervals (Exercise 7.53).

7.10.2   Comparing two proportions

We use subscripts 1 and 2 to distinguish the populations (Figure 7.10) and samples. As in Section 7.10.1 we use the approximation

$$\widehat{p_1} \sim N\left(p_1, \frac{p_1(1-p_1)}{n_1}\right).$$

for the distribution of $\widehat{p_1}$. We use a similar approximation for the distribution of $\widehat{p_2}$. Then

$$\widehat{p_1} + \widehat{p_2} \sim N\left(p_1 - p_2, \frac{p_1(1-p_1)}{n_1} + \frac{p_2(1-p_2)}{n_2}\right).$$

To construct a confident interval the unknown variance is replaced by its sample estimate

$$\widehat{p_1}(1-\widehat{p_1})/n_1 + \widehat{p_2}(1-\widehat{p_2})/n_2.$$

Then a $(1-\alpha/2) \times 100\%$ confidence interval for

$$\widehat{p_1} - \widehat{p_2} \pm z_{\alpha/2}\sqrt{\frac{\widehat{p_1}(1-\widehat{p_1})}{n_1} + \frac{\widehat{p_2}(1-\widehat{p_2})}{n_2}}.$$

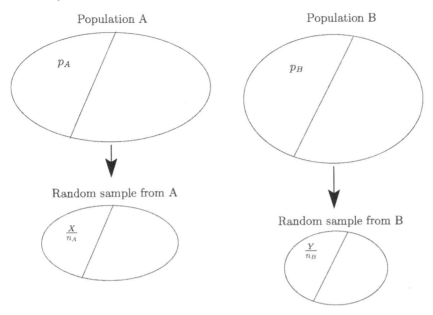

FIGURE 7.10: Comparing proportions.

The assumption is good provided: $\min(n_1 p_1, n_1(1 - p_1)) > 5$; $\min(n_2 p_2, n_2(1 - p_2)) > 5$; and $30 < n_1, n_2$.

Example 7.25: Plastic sheeting [comparison of proportions]

A company manufactures transparent plastic in 1m by 2m sheets. Sheets often contain flaws, in which case the flawed part of the sheet is cut away, and re-cycled into lower value products, while the remainder is sold as an off-cut. A production engineer finds that 38 out of the last 400 sheets contained flaws. The engineer implements changes to the manufacturing process and finds that 11 out of the next 200 sheets contain flaws. Is there evidence that the changes have been effective?

```
> prop.test(c(38,11),c(400,200))

	2-sample test for equality of proportions with continuity correction

data:  c(38, 11) out of c(400, 200)
X-squared = 2.3362, df = 1, p-value = 0.1264
alternative hypothesis: two.sided
95 percent confidence interval:
 -0.006457993  0.086457993
sample estimates:
prop 1 prop 2
 0.095  0.055
```

The decrease in the proportion of flawed sheets is promising but doesn't reach statistical significance at a 10% level ($p = 0.13$).

Now suppose that there are only 10 flawed sheets in the next 200 off production. There are now 21 flawed sheets out of 400 post change. There is now more convincing evidence of an improvement (p-value 0.03).

```
> prop.test(c(38,21),c(400,400))

    2-sample test for equality of proportions with continuity correction

data:  c(38, 21) out of c(400, 400)
X-squared = 4.6845, df = 1, p-value = 0.03044
alternative hypothesis: two.sided
95 percent confidence interval:
 0.003897438 0.081102562
sample estimates:
prop 1 prop 2
0.0950 0.0525
```

Example 7.26: Simpson's paradox [beware of combining subgroups]

Wardrop (1995) discusses the following statistics on the results of two free shots in basketball taken by Bird and Robey.

Larry Bird				Rick Robey		
1st shot	2nd shot			1st shot	2nd shot	
	Hit	Miss			Hit	Miss
Hit	251	34		Hit	54	37
Miss	48	5		Miss	49	31

For both players, there is a slightly higher proportion of misses on the second shot if the first was a hit. However, this observation is not statistically significant at a 20% level, even, for either player.

```
> prop.test(c(34,5),c(289,53))

    2-sample test for equality of proportions with continuity correction

data:  c(34, 5) out of c(289, 53)
X-squared = 0.0654, df = 1, p-value = 0.7982
alternative hypothesis: two.sided
95 percent confidence interval:
 -0.07487683  0.12149170
sample estimates:
    prop 1     prop 2
0.11764706 0.09433962

> prop.test(c(37,31),c(91,80))

    2-sample test for equality of proportions with continuity correction
```

```
data:  c(37, 31) out of c(91, 80)
X-squared = 0.0096, df = 1, p-value = 0.922
alternative hypothesis: two.sided
95 percent confidence interval:
 -0.1395592  0.1777460
sample estimates:
   prop 1    prop 2
0.4065934 0.3875000
```

But, look what happens if the results for the two players are combined.

```
> prop.test(c(34,5)+c(37,31),c(289,53)+c(91.80))

    2-sample test for equality of proportions with continuity correction

data:  c(34, 5) + c(37, 31) out of c(289, 53) + c(91.8)
X-squared = 2.1321, df = 1, p-value = 0.1442
alternative hypothesis: two.sided
95 percent confidence interval:
 -0.14747148  0.02313307
sample estimates:
   prop 1    prop 2
0.1864496 0.2486188
```

There is now a higher proportion of misses on the second shot if the first was a miss[13]. This reversal of trend occurs because in the combined set: most of the hits on the first shot were attributable to Bird whereas only half of the misses on the first shot were by Bird; and Bird is better at taking free shots. This is an an example of **Simpson's paradox**, which refers to situations in which a trend in different groups changes if the groups are combined[14]. The implications are: that we should be careful before combining sub-groups for analysis; and that we are aware of the possibility of unknown, or undeclared, variables [15] that might account for statistical summaries of data that have been obtained by combining smaller data sets that relate to different populations.

7.10.3   McNemar's test

McNemar's test is a paired comparison test for a difference in proportions, and the analysis is based on a normal approximation to the sampling distribution of a single sample proportion. Suppose we have n comparisons of two processes A and B, and the result of each comparison can be classified as: both successful; A successful but B not successful; A not successful but B successful; and both not successful. This classification is shown in Table 7.2.

We first focus on the $x+y$ comparisons with different outcomes for A and B, and define a random variable X as the number of successes for A in the $x+y$ comparisons. It is convenient to set $m = x + y$. Then using a normal distribution approximation $\frac{X}{m} \sim N(p, p(1-p)/m)$,

---

[13]This difference in proportions is statistically significant at a 20% level (p=0.14).

[14]Similar results can be found in game theory, Parrondo's games, and in physics, the Brownian Ratchet.

[15]Sometimes referred to as lurking variables.

TABLE 7.2: Summary of paired comparison of proportions.

	B success	B not success	
A success	u	x	u+x
A not success	y	w	w+y
	u+y	w+x	n=u+w+x+y

where p is the proportion of such comparisons for which A is successful. A $(1-\alpha) \times 100\%$ confidence interval for p is given by

$$\frac{x}{m} \pm z_{\alpha/2}\sqrt{\frac{(x/m)(1-x/m)}{m}}.$$

The formula for this interval does not include n. The difference in the proportions of successes for A and B is given by

$$\frac{u+x}{n} - \frac{u+y}{n} = \frac{x-y}{n}.$$

The difference $x - y$ is the observed value of the random variable $2X - m$ and we use a normal distribution approximation $X \sim N(mp, mp(1-p))$. Then a $(1-\alpha) \times 100\%$ confidence interval for the difference in the proportions of successes for A and B is given by

$$\frac{x-y}{n} \pm z_{\alpha/2}\frac{2\sqrt{m(x/m)(1-x/m)}}{n}.$$

The formula for the confidence interval for the difference in proportions includes both m and n.

Example 7.27: Cockpit layout [McNemar's test]

An aircraft manufacturer wishes to compare two prototype cockpit layouts, A and B, for a small seaplane. The manufacture asks 50 pilots to sit in the two prototype cockpits and assess them as either "clear" or "confusing". Twenty five pilots are randomly assigned to assess A first, and the other 25 pilots assess B first. Their assessments are given in Table7.3

TABLE 7.3: Pilots' assessment of cockpit layout for a seaplane.

	B clear	B confusing	
A clear	30	11	41
A confusing	4	5	9
	34	16	50

We begin with the 15 pilots who differed in their assessments. A 95% confidence interval for the proportion of such pilots who consider A clear and B confusing is

$$\frac{11}{15} \pm z_{.025}\sqrt{\frac{(11/15)(1-11/15)}{15}}$$

which gives $[0.51, 0.96]$. The lower point of this interval exceeds 0.5 so there is some evidence that A is the clearer layout ($p < 0.05$). A 95% confidence interval for the difference in the proportions of "clear" assessments for layout A and layout B is given by

$$\frac{11-4}{50} \pm z_{.025} \frac{2\sqrt{15(11/15)(1-11/15)}}{50}$$

which gives $[0.01, 0.27]$. The manufacturer decides to adopt layout A for production.

7.11 Prediction intervals and statistical tolerance intervals

7.11.1 Prediction interval

We may be more concerned about individual values in a population than the population mean. Suppose we have a sample from a population $\{X_i\}$ for $i = 1, \ldots, n$, with unknown mean μ and standard deviation σ, and want to provide an interval that has a $(1 - \alpha) \times 100\%$ chance of containing a single future observation X. This interval is referred to as a $(1 - \alpha) \times 100\%$ **prediction interval**. We first notice that

$$\mathrm{E}\big[X - \overline{X}\big] \;=\; \mu \;=\; 0$$

and that since X is independent of $\overline{X}$

$$var(X - \overline{X}) \;=\; \sigma^2 + \frac{\sigma^2}{n}.$$

If we now assume that the population has a normal distribution then

$$\frac{X - \overline{X}}{\sigma\sqrt{1 + \frac{1}{n}}} \;\sim\; N(0, 1)$$

and replacing the unknown σ by S gives

$$\frac{X - \overline{X}}{S\sqrt{1 + \frac{1}{n}}} \;\sim\; t_{n-1}.$$

There is therefore a probability of $(1 - \alpha)$ that

$$-t_{n-1, \alpha/2} \;<\; \frac{X - \overline{X}}{S\sqrt{1 + \frac{1}{n}}} \;<\; t_{n-1, \alpha/2}$$

and a $(1 - \alpha) \times 100\%$ prediction interval for X is

$$\left[\overline{x} - t_{n-1, \alpha/2}\, s\, \sqrt{1 + \frac{1}{n}}, \quad \overline{x} + t_{n-1, \alpha/2}\, s\, \sqrt{1 + \frac{1}{n}} \right].$$

As the sample size increases, this $(1 - \alpha) \times 100\%$ prediction interval for X from a normal distribution approaches

$$\left[\mu - z_{\alpha/2}\sigma, \;\; \mu - z_{\alpha/2}\sigma \right].$$

In contrast, the width of the CI for μ approaches 0 as $n \to \infty$.

Example 7.28: Electroplating [prediction interval]

A company sells gold cyanide solution for electroplating in 1 liter containers. The precise volumes (ml) of solution in a random sample of 25 containers are

```
> x
 [1] 1002.6 1003.0 1003.0 1003.5 1003.6 1001.0 1002.3 1002.9
 [9] 1002.7 1002.9 1003.1 1003.3 1003.6 1004.2 1003.3 1002.5
[17] 1003.6 1002.3 1004.6 1002.4 1003.4 1005.1 1002.0 1002.1 1004.3
> mean(x)
[1] 1003.092
> sd(x)
[1] 0.8953212
# A 95% prediction interval is given by
> n=length(x)
> L=mean(x)-qt(.025,24,lower.tail=FALSE)*sd(x)*sqrt(1+1/25)
> L
[1] 1001.208
> U=mean(x)+(.025,24,lower.tail=FALSE)*sd(x)*sqrt(1+1/25)
> U
[1] 1004.976
```

This 95% prediction interval, $[1001.21, 1004.98]$, is sensitive to the assumption that the gold volumes are normally distributed. The prediction interval also provides limits within which we expect $(1-\alpha) \times 100\%$ of the population to lie. However, this represents a best estimate rather than limits we are highly confident about. The next section deals with this limitation.

7.11.2   Statistical tolerance interval

We now calculate an interval that we are 90% confident includes 95% of volumes, based on the contents of the 25 containers.

An approximate construction for such intervals is outlined in Exercise 7.31, but the R package tolerance can be used for the calculation[16]. Like prediction intervals, tolerance intervals are highly sensitive to the assumed distribution and the package tolerance includes several options. We will assume that the population of gold volumes is normal.

```
> normtol.int(x,alpha=.10,P=.95,side=2)
  alpha    P    x.bar 2-sided.lower 2-sided.upper
1   0.1 0.95 1003.092      1000.871      1005.313
```

To summarize we have:

95% confidence interval for μ as $[1\,002.72, 1\,003.46]$
95% prediction interval for a single observation is $[1\,001.2, 1\,005.0]$
90% confidence that 95% of the population is within $[1\,000.9, 1\,005.3]$.

---

[16] In R click on Packages, Install package, then select tolerance. Now all you need do is load package tolerance.

The prediction interval and tolerance interval are sensitive to the assumption of normality, but the confidence interval for the mean of the distribution is not. The effect of changing the level of confidence in the limits for 95% of the population can be seen by changing alpha.

```
> normtol.int(x,alpha=.50,P=.95,side=2)
  alpha    P    x.bar 2-sided.lower 2-sided.upper
1   0.5 0.95 1003.092       1001.278       1004.906
> normtol.int(x,alpha=.01,P=.95,side=2)
  alpha    P    x.bar 2-sided.lower 2-sided.upper
1  0.01 0.95 1003.092        1000.42       1005.764
```

If alpha is set to 0.5, the tolerance interval is very close to the prediction interval. If alpha is reduced to 0.01, we are 99% confident that 95% of the population is within $[1\,000.4, 1\,005.8]$. A customer is mainly concerned that the volumes are not less than the nominal $1\,000$ ml, and a 1-sided tolerance interval is more appropriate.

```
> normtol.int(x,alpha=.01,P=.95,side=1)
  alpha    P    x.bar 1-sided.lower 1-sided.upper
1  0.01 0.95 1003.092       1000.734       1005.45
```

We are 99% confident that 95% of the volumes exceed $1\,003.1$ ml.

In some industrial situations the variable of interest is not well modeled by a normal distribution. The runout of automobile brake discs is an example. One consequence of runout is brake judder and a typical specification for runout of discs is less than 50 microns (0.050 mm). The variable is non-negative with an ideal value of 0 and unlikely to have a normal distribution. A half-normal distribution is more plausible.

Example 7.29: Runout [tolerance interval]

A manufacturer of brake discs has been asked to demonstrate to a potential customer that a specified runout of less than 50 microns can be met. In a sample of 50 discs the maximum runout is 46 microns. [Vardeman, 1992], and Exercise 7.32, suggests a distribution free one-sided tolerance interval for a fraction p of the population as

$$(-\infty, x_{n:n}],$$

with confidence level

$$(1 - p^n) \times 100\%.$$

In this case for $p = 0.95$ the tolerance interval is $(1 - 0.95^{50}) \times 100 = 93.2\%$

The manufacturer can claim 93.2% confidence that 95% of discs will have runout less than 0.46.

7.12 Goodness of fit tests

We consider statistical tests that can be used to test whether data are a random sample from some specified probability distribution. In these cases there is no population parameter for which to construct a confidence interval, but a goodness of fit test can be used to decide whether or not some model for the probability distribution is adequate. However, we need to be aware that if the sample is small we are unlikely to have evidence against a variety of distributions and if the sample is very large we may have evidence against distributions that are adequate for their purpose. A fit is more plausible if there is some physical justification for the choice of probability distribution.

7.12.1   Chi-square test

The chi-square distribution provides an approximation to a multinomial distribution in a similar way to the normal, and hence chi-square with one degree of freedom provides an approximation to the binomial distribution (Exercise 7.54). We demonstrate the test with the Rutherford and Geiger data.

Suppose we have a set of n observations of a discrete random variable X, that take values over a set of integers $\{i\}$, where $L \leq i \leq U$. Denote the observed frequencies by O_i where

$$\sum_{i=L}^{U} O_i = n.$$

The sum is over the number of categories $m = U - L + 1$. The null hypothesis to be tested, H_0, is that X has a pmf $P(x)$. If H_0 is true the probabilities that x takes the values i for $L < i < U$ are $P(i)$ and the expected frequencies are $E_i = P(i) \times n$. The expected frequency of values less than or equal to L is $E_L = F(L) \times n$, where $F(x)$ is the sdf of X. Similarly, the expected frequency of values greater than or equal to U is $E_U = (1 - F(U-1)) \times n$. The $\sum E_i = n$ because we have included values of x in the support of $P(x)$ for which there were no observations. Define

$$W = \sum_i \frac{(O-E)^2}{E_i}.$$

If H_0 is true, and the E_i exceed about 5 then, to a good approximation:

$$W \sim \chi^2_{m-p-1}.$$

The degrees of freedom for the chi-square distribution is m less the number of parameters of the distribution estimated from the data, p, and less 1 because $\sum E_i = \sum O_i = n$. Large values of W are evidence against H_0. If we test H_0 at the α level, then we reject it if the calculated value of W, w, exceeds $\chi^2_{m-p-1,\alpha}$. If $w < \chi^2_{m-p-1,\alpha}$ we have no evidence against H_0. Small values of W correspond to a good fit and are consistent with H_0. [17] If any of the E_i are substantially less than 5, adjoining categories should be combined. We have already done this by, for example, combining categories above U, in which there were no observations with U. It is also possible that the expected number above U exceeds 5, in which case this should be included as an additional category with an observed value of 0. In the case of the Rutherford and Geiger data the null hypothesis is

H_0: number of particles in 7.5 second periods have a Poisson distribution.

There is a single parameter to be estimated, λt where $t = 7.5$, from the sample mean.

---

[17]Extraordinarily small w might be taken to indicate that the fit is too good and that the data have been fabricated, because there is very little variability about the expected values.

```
> #Chi-square goodness of fit for Rutherford & Geiger data
> numpart=0:12
> observ=c(57,203,383,525,532,408,273,139,45,27,10,4,2)
> n=sum(observ)
> lamt=sum(numpart*observ)/n
> p4expect=dpois(numpart,lamt)
> expect=n*p4expect
> print(cbind(numpart,observ,expect))
      numpart observ      expect
 [1,]       0     57   54.376939
 [2,]       1    203  210.460438
 [3,]       2    383  407.282911
 [4,]       3    525  525.449093
 [5,]       4    532  508.424381
 [6,]       5    408  393.561020
 [7,]       6    273  253.873015
 [8,]       7    139  140.369972
 [9,]       8     45   67.910971
[10,]       9     27   29.204727
[11,]      10     10   11.303394
[12,]      11      4    3.977149
[13,]      12      2    1.282763
> E=expect[1:12]
> E[12]=E[12]+expect[13]
> O=observ[1:12]
> O[12]=O[12]+observ[13]
> chisqcomponent=(O-E)^2/E
> print(cbind(c(numpart[1:11],"11 or more"),chisqcomponent))
                      chisqcomponent
 [1,] "0"          "0.126532473118882"
 [2,] "1"          "0.264458917246348"
 [3,] "2"          "1.44778912787491"
 [4,] "3"          "0.000383832841227581"
 [5,] "4"          "1.09320051730064"
 [6,] "5"          "0.529737762433205"
 [7,] "6"          "1.44104151548125"
 [8,] "7"          "0.013370551924469"
 [9,] "8"          "7.72942260156628"
[10,] "9"          "0.166439469691283"
[11,] "10"         "0.150294273458119"
[12,] "11 or more" "0.104132964955602"
> chisqcal=sum((O-E)^2/E)
> Pv=1-pchisq(chisqcal,(12-1-1))
> print(c("calculated value of chisq",round(chisqcal,2),
"p-value",round(Pv,3)))
[1] "calculated value of chisq" "13.07"
[3] "p-value"                   "0.22"
```

We combined the 11 and 12 categories into 11 or more, to avoid an expected value as low as 1.28. The main contribution to w is the greater than expected frequency for 8 particles, but there is no evidence against H_0 at the 0.1 level (because the P=value of 0.22 exceeds

0.10). The sample size is quite large and the data suggest that the Poisson distribution is an excellent model for number of observations in 7.5 s intervals, and support the theory that the atoms disintegrate as a Poisson process.

7.12.2   Empirical distribution function tests

Although the chi-square goodness of fit test can be used if the data are grouped (Exercise 7.33), tests based on the **empirical cumulative distribution function** generally have higher power for a given sample size. The Kolmogorov-Smirnov test is an example of such an empirical cumulative distribution function test (EDF-test). Suppose we have a random sample of size n from some continuous probability distribution. The empirical (c)df is defined by

$$\widehat{F}(x) \quad = \quad \frac{number\ of\ data \le x}{n}.$$

This is a step function which increases by $\frac{1}{n}$ at each order statistic. The Kolmogorov-Smirnov test is a test of the null hypothesis

H_0: The data are a random sample from a probability distribution with cdf $F(x)$.

against an alternative.

H_1: The data are a random sample from some other probability distribution.

The test is based on the maximum discrepancy, measured as the vertical difference between the cdf, $F(x)$, and the ecdf. Since the ecdf is a step function, and $F(x)$ is a continuous increasing function, this maximum discrepancy must occur just before or just after a step. If the cdf lies above the ecdf the maximum discrepancy occurs before the step and will be obtained from

$$D^- \quad = \quad \max \left[F(x_{i:n}) - \frac{(i-1)}{n} \right].$$

If the cdf lies below the ecdf the the maximum discrepancy occurs before the step and will be obtained from

$$D^+ \quad = \quad \max \left[\frac{i}{n} - F(x_{i:n}) \right].$$

Then the maximum discrepancy D is obtained as

$$\max[D^-, D^+].$$

A relatively large value, d, of D is evidence against H_0, and the easiest way to determine whether d is relatively large is to use simulation and take a large number of random samples of n from the distribution with cdf $F(x)$. In most applications the parameters of this distribution are not specified and have to be estimated from the data, which tends to improve the fit. It is straightforward to allow for estimation of the parameters in the simulation.

Example 7.30: Gold grades [Goodness of fit with ecdf test]

The null hypothesis is that the gold grades (Example 3.11) are a random sample from a lognormal distribution, which is equivalent to a null hypothesis that the logarithms of the gold grades are from a normal distribution. The R code for plotting the ecdf and the cdf of the fitted normal distribution (Figure 7.11), and calculating d and its statistical significance is:[18]

```
> #gold grades
> gold_grades.dat=read.table("gold_grade.txt",header=T)
> x=gold_grades.dat$gold
> y=log(x)
> edf.y=ecdf(y)
> n=length(y)
> mu=mean(y)
> sigma=sd(y)
> #plot ecdf and normal cdf
> xp=seq(min(y),max(y),(max(y)-min(y))/1000)
> cdfy=pnorm(xp,mu,sigma)
> plot(edf.y,main="",xlab="ln(grade), y",ylab="F(y)")
> lines(xp,cdfy)
> #calculate KS statistic
> y.sort=sort(y)
> z=pnorm(y.sort,mu,sigma)
> astep=c(1:n)/n
> bstep=(c(1:n)-1)/n
> Dp=max(astep-z)
> Dn=max(z-bstep)
> D=max(Dp,Dn)
> print(c("KS statistic d=",round(D,4)))
[1] "KS statistic d="    "0.0465"
> #now to calculate significance
> K=1000000
> DD=rep(0,K)
> set.seed(1)
> for (k in 1:K){
+ xx=rnorm(n)
+ zz=pnorm(sort(xx),mean(xx),sd(xx))
+ DD[k]=max(max(astep-zz),max(zz-bstep))
+ }
> print(c("proportion exceeding d",1-max(which(sort(DD)<=D))/K))
[1] "proportion exceeding d" "0.445163"
```

---

[18]Notice the use of the which() function for obtaining the index in an array that corresponds to a particular value. This is one of the indexing features in R.

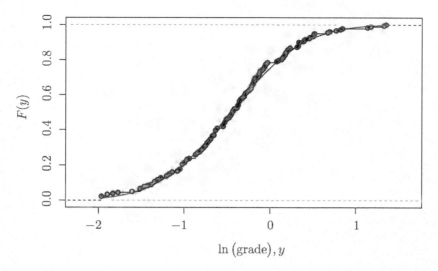

FIGURE 7.11: Empirical cdf (ecdf) for logarithms of gold grades and normal cdf.

The calculated value of D is 0.0465, and the proportion of d that exceeds 0.0465 in the one million draws from a normal distribution is 0.445. Formally, the p-value is 0.45 and there is no evidence against the hypothesis that the data are from a lognormal distribution at even a level of significance of 0.20.

Even with a reasonably large sample of 181 data, it is necessary to allow for the estimation of the parameters. If this is not done the proportion of d exceeding 0.0465 increases to about 0.80. Two other ecdf tests, one based on the Cramer-von Mises statistic and the other based on the Anderson-Darling statistic are covered in the exercises.

7.13 Summary

7.13.1 Notation

n	sample size
$\overline{X}, \bar{x}/\mu_X$	mean of sample/population
$S_X, s_x/\sigma_X$	standard deviation of sample/population
Z	standard normal random variable
z_α	upper α quantile standard normal
t_ν	t-distribution with ν dof
$t_{\nu,\alpha}$	upper α quantile t-distribution
χ_ν^2	chi-square distribution ν dof
$\chi_{\nu,\alpha}^2$	upper α quantile chi-square
F_{ν_1,ν_2}	F-distribution ν_1, ν_2 dof
$F_{\nu_1,\nu_2,\alpha}$	upper α quantile F

7.13.2 Summary of main results

Flow chart for confidence intervals

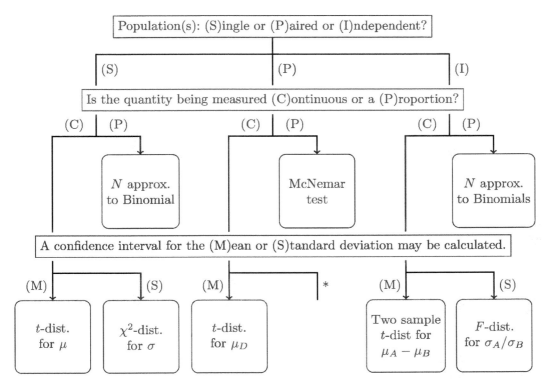

* In the case of paired comparisons we aim to match pairs so that the standard deviation of the differences is as small as possible. We are unlikely to require a confidence interval for the standard deviation of the differences as it does not provide information about the standard deviations of the populations.

Sample size

If the width of the confidence interval and degree of confidence are specified, the above formula can be used to determine the required sample size if a value for the population standard deviation is assumed. The assumption can be based on experience, or published data, or a pilot sample. In the case of comparisons with independent samples, equal sample sizes is most efficient. If the width of the confidence interval is to be reduced by a factor of $\frac{1}{k}$, the sample size has to be increased by a factor of about k^2.

Interval estimates

Suppose X has a normal distribution with a mean μ and a standard deviation σ. We are $(1-\alpha) \times 100\%$ confident that the $(1-\alpha) \times 100\%$ *confidence interval* for the population mean, μ, includes μ.

$$\bar{x} \ \pm \ t_{n-1,\alpha/2} \frac{s}{\sqrt{n}}.$$

Statistical Tests

Parameter	Random variable	Confidence interval	Null hypothesis	Test statistic
μ (σ known)	$\dfrac{\bar{X}-\mu}{\sigma/\sqrt{n}} \sim N(0,1)$	$\bar{x} \pm z_{\alpha/2}\dfrac{\sigma}{\sqrt{n}}$	$\mu = \mu_0$	$\dfrac{\bar{x}-\mu_0}{\sigma/\sqrt{n}}$
p	$\dfrac{\hat{p}-p}{\sqrt{p(1-p)}/n} \sim N(0,1)$	$\hat{p} \pm z_{\alpha/2}\sqrt{\dfrac{\hat{p}(1-\hat{p})}{n}}$	$p = p_0$	$\dfrac{\hat{p}-p_0}{\sqrt{p_0(1-p_0)}/n}$
μ	$\dfrac{\bar{X}-\mu}{s/\sqrt{n}} \sim t_{n-1}$	$\bar{x} \pm t_{n-1,\alpha/2}\dfrac{s}{\sqrt{n}}$	$\mu = \mu_0$	$\dfrac{\bar{x}-\mu_0}{s/\sqrt{n}}$
μ_D	$\dfrac{\bar{D}-\mu_D}{s_D/\sqrt{n}} \sim t_{n-1}$	$\bar{d} \pm t_{n-1,\alpha/2}\dfrac{s_D}{\sqrt{n}}$	$\mu_D = 0$	$\dfrac{\bar{d}}{s_D/\sqrt{n}}$
$\mu_A - \mu_B$	$\dfrac{(\bar{X}_A-\bar{X}_B)-(\mu_A-\mu_B)}{\sqrt{\frac{s_A^2}{n_A}+\frac{s_B^2}{n_B}}} \sim t_\nu$	$\bar{x}_A - \bar{x}_B \pm t_{\nu,\alpha/2}\sqrt{\dfrac{s_A^2}{n_A}+\dfrac{s_B^2}{n_B}}$	$\mu_A = \mu_B$	$\dfrac{\bar{x}_A-\bar{x}_B}{\sqrt{\frac{s_A^2}{n_A}+\frac{s_B^2}{n_B}}}$
$\mu_A - \mu_B$ ($\sigma_A = \sigma_B$)	$\dfrac{(\bar{X}_A-\bar{X}_B)-(\mu_A-\mu_B)}{s_p\sqrt{\frac{1}{n_A}+\frac{1}{n_B}}} \sim t_{n_A+n_B-2}$	$\bar{x}_A - \bar{x}_B \pm t_{n_A+n_B-2,\alpha/2}\, s_p\sqrt{\dfrac{1}{n_A}+\dfrac{1}{n_B}}$	$\mu_A = \mu_B$	$\dfrac{\bar{x}_A-\bar{x}_B}{s_p\sqrt{\frac{1}{n_A}+\frac{1}{n_B}}}$
σ^2	$\dfrac{(n-1)S^2}{\sigma^2} \sim \chi^2_{n-1}$	$\left[\dfrac{(n-1)s^2}{\chi^2_{n-1,\alpha/2}}, \dfrac{(n-1)s^2}{\chi^2_{n-1,1-\alpha/2}}\right]$	$\sigma = \sigma_0$	$\dfrac{(n-1)s^2}{\sigma_0^2}$
$\dfrac{\sigma_A^2}{\sigma_B^2}$	$\dfrac{S_A^2/\sigma_A^2}{S_B^2/\sigma_B^2} \sim F_{n_A-1,n_B-1}$	$\left[\dfrac{s_B^2}{s_A^2}F_{n_A-1,n_B-1,1-\alpha/2}, \dfrac{s_B^2}{s_A^2}F_{n_A-1,n_B-1,\alpha/2}\right]$	$\sigma_A = \sigma_B$	$\dfrac{s_A^2}{s_B^2}$

There is a probability of $(1 - \alpha)$ that a $(1 - \alpha) \times 100\%$ *prediction interval* includes a single value of X

$$\overline{x} \quad \pm \quad t_{n-1,\alpha/2}s\sqrt{1 + \frac{1}{n}}.$$

The *tolerance interval* is wider than the prediction interval because we are $(1 - \epsilon)100\%$ confident that it includes at least a proportion $(1 - \alpha)$ of the population where ϵ is usually chosen to be 0.1 or less.

7.13.3 MATLAB and R commands

In the following, `data1` and `data2` contain some number of data points (possibly different), `mu0` and `sigma` are some real values and `x`, `n` are integers and `prop`, `a` are values between 0 and 1. and For more information on any built in function, type `help(function)` in R or `help function` in MATLAB.

R command	MATLAB command
`t.test(data1,mu=mu0)`	`ttest(data1,mu0)`
`t.test(data1,data2)`	`ttest2(data1,data2)`
`var.test(data1,ratio=sigma)`	`vartest(data1,sigma^2)`
`var.test(data1,data2)`	`vartest2(data1,data2)`
`prop.test(x,n,p = prop)`	–
`ecdf(data1)`	`[F, X] = ecdf(data1)`
`normtol.int(data1, alpha = a, P = prop)`	–

7.14 Exercises

Section 7.2 Statistics as estimators

Exercise 7.1: Independence of $\overline{X}, S^2$ with SRS from normal distribution

If a SRS of size n is taken from a normal distribution then the sample mean and variance are independent random variables. Demonstrate this theoretical result by writing a script to make K draws of SRS of size n from a standard normal distribution. For each sample of size n calculate the mean and the variance. This will give K pairs, each pair consisting of a mean and a variance. Plot the variances against the means. Calculate the correlation between the means and the variances.

(a) What is the expected value of the correlation between the means and variances?

(b) Try different values of n with $K = 10000$. Are the correlations obtained from your simulations consistent with your answer to Part(a)?

(c) Would the results differ if you used standard deviations in place of variances? What is the expected value of the correlation between the means and standard deviations?

Exercise 7.2: $\overline{X}, S^2$ from exponential

Write a script to make K draws of SRS of size n from an exponential distribution with mean 1. For each sample of size n calculate the mean, the variance and the standard deviation. This will give you K triples, each triple consisting of a mean, variance and a standard deviation. Plot the variances against the means and plot the standard deviations against the means. Calculate the correlation between the means and the variances, and the correlation between the means and the standard deviations.

(a) Try different values of n with $K = 10000$. Comment on the correlations between the means and variances. Give an informal explanation for the results.

(b) Do the results differ if you use standard deviations in place of variances?

Exercise 7.3: $\overline{X}, S^2$ from t-distribution

Write a script to make K draws of SRS of size n from a t-distribution with ν degrees of freedom. For each sample of size n calculate the mean and the variance. This will give you K pairs, each pair consisting of a mean and a variance. Plot the variances against the means. Calculate the correlation between the means and the variances. Try different values of n for a t-distribution with 4.1 degrees of freedom, with $K = 10000$. Comment on the correlations between the means and variances. Give an informal explanation for the results.

Exercise 7.4: $\overline{X}, S^2$ from Gumbel

The reduced Gumbel distribution has a cdf:

$$F(y) \;=\; exp(-(exp(-y))), \qquad -\infty < y < \infty.$$

Write a script to make K draws of SRS of size n from a reduced Gumbel distribution. For each sample of size n calculate the mean, the variance, $\widehat{\theta} = s * \sqrt{6}/\pi$ where s is the sample standard deviation, and $\widehat{\xi} = \overline{y} - 0.577216\widehat{\theta}$ where $\overline{y}$ is the sample mean. This will give you K quadruplet, each quadruplet consisting of a mean, variance, $\widehat{\theta}$ and a $\widehat{\xi}$. Plot the variances against the means and plot the $\widehat{\theta}$ against the $\widehat{\xi}$. Calculate the correlation between the means and the variances, and the correlation between the $\widehat{\theta}$ and the $\widehat{\xi}$.

(a) Try different values of n with $K = 10\,000$, and comment on the results.

(b) Draw a histogram of the $\widehat{\theta}$ and calculate their mean value.

(c) Draw a histogram of the $\widehat{\xi}$ and calculate their mean value.

Section 7.3 Accuracy and precision

Exercise 7.5: SRS statistics

(a) Draw $K = 10\,000$ SRS off size $n = 5$ from $N(0,1)$. For each SRS calculate $s^2, \widehat{\sigma}_n^2, s, \widehat{\sigma}_n$ where $\widehat{\sigma}_n^2 = \sum_{i=1}^{n} (x_i - \overline{x})^2/n$.

    (i) Compare the bias, standard error, and root mean squared error of s^2 and $\widehat{\sigma}_n^2$ as estimates of $\sigma^2 = 1$.

    (ii) Compare the bias, standard error, and root mean squared error of s and $\widehat{\sigma}_n$ as estimates of $\sigma = 1$.

    (iii) Repeat (a) with SRS from an exponential distribution with mean 1.

(b) The following R script gives a pseudo random sample of 100 from a standard bivariate normal distribution with correlation 0.8. It requires the package MASS.

```
library(MASS)
SIG=matrix(c(1,.8,.8,1),nrow=2,byrow=TRUE)
X=mvrnorm(100,c(0,0),SIG)
cor(X)
```

Adapt it to investigate the accuracy and precision of r as an estimator of ρ when sampling from a bivariate normal distribution.

Section 7.4 Precision of estimate of population mean

Exercise 7.6: Bootstrap and bias correction

An SRS of size n is taken from an exponential distribution with mean 1. The objective is to compare three estimators of the mean: the sample median with a bootstrap correction for the bias (muhat1); the sample mean (muhat2); and the sample median multiplied by a factor equal to the distribution mean divided by the distribution median (muhat3). The following R script (or MATLAB equivalent) draws K SRS of size n. For each SRS of size n it calculates: a bootstrap estimate of the bias when using the sample median to estimate the population mean, the sample median and the three estimates of the population mean.

```
#sample size n, number bootstrap samples B, number of samples K
n=4
K=100
bias=rep(0,K);med=rep(0,K);muhat1=rep(0,K);
muhat2=rep(0,K);muhat3=rep(0,K);
for (k in 1:K){
x=rexp(n)
B=1000
m=rep(0,B)
for (b in 1:B){
xb=sample(x,n,replace=TRUE)
m[b]=median(xb)
}
med[k]=median(x)
bias[k]=mean(m)-mean(x)
muhat1[k]=med[k]-bias[k]
muhat2[k]=mean(x)
muhat3[k]=median(x)/log(2)
}
```

(a) For a sample size $n = 4$.

(i) What is the expected value and the standard error of the sample mean when $n - 4$?

(ii) What is the median of the exponential distribution with mean 1? What is the numerical value of the factor for muhat3?

(iii) Using the results from your simulation do you think the sample median is an unbiased estimator of the population median ?

(iv) Use the results of your simulation to estimate the mean, standard error and root mean squared error of the three estimators of the population mean. What are their relative advantages and disadvantages of the three estimators?

(v) Do you think the value of K is large enough to detect any bias in the estimators?

(b) Repeat (a) for $n = 5, 10, 20, 50, 100$.

Exercise 7.7: Bootstrap standard errors for estimation of quantiles

An SRS of size n is taken from a reduced Gumbel distribution ($\theta = 1, \xi = 0$). The objective is to compare two estimators of the upper 0.1 quantile. The first is equal to the upper 0.1 quantile of an assumed Gumbel distribution with its parameters replaced by estimates from the sample

$$\widehat{\xi} + (-ln(-ln(1 - 0.1)))\widehat{\theta},$$

where $\widehat{\theta} = s\sqrt{6}/\pi$, where s is the sample standard deviation, and $\widehat{\xi} = \bar{y} - 0.577216\widehat{\theta}$, where $\bar{y}$ is the sample mean. The second is a linear interpolation between order statistics

$$y_{0.9(n+1):(n+1)}.$$

The following R script (or MATLAB equivalent) draws K SRS of size n. For each SRS it computes the two estimates and makes bootstrap estimates of their standard errors.

```
n=20;p=0.9
K=100
ty1=rep(0,K);ty2=rep(0,K);sety1b=rep(0,K);sety2b=rep(0,K)
for (k in 1:K){
y=-log(-log(runif(n)))
thetahat=sd(y)*sqrt(6)/pi
xihat=mean(y)-0.577216*thetahat
ty1[k]=xihat-log(-log(p))*thetahat
m=(n+1)*p
intm=floor(m)
fracm=m-intm
ty2[k]=sort(y)[intm]+fracm*(sort(y)[intm+1]-sort(y)[intm])
#begin bootstrap
B=1000;ty1b=rep(0,B);ty2b=rep(0,B)
for (b in 1:B){
yb=sample(y,n,replace=TRUE)
thetahatb=sd(yb)*sqrt(6)/pi
xihatb=mean(yb)-0.577216*thetahatb
ty1b[b]=xihatb-log(-log(p))*thetahatb
ty2b[b]=sort(yb)[intm]+fracm*(sort(yb)[intm+1]-sort(yb)[intm])
}
```

```
#end bootstrap
sety1b[k]=sd(ty1b)
sety2b[k]=sd(ty2b)
}
print(c(mean(ty1),sd(ty1),mean(ty2),sd(ty2)))
print(c(mean(sety1b),mean(sety2b)))
```

Consider estimation of the upper 0.1 quantile from an SRS of 20.

(a) What is the value of the upper 0.1 quantile in the reduced Gumbel distribution?

(b) On the basis of your simulation results comment on the relative performance of the two estimators. What are their relative advantages?

(c) Draw box plots of the bootstrap estimates of the standard errors. Plot the bootstrap estimates of the standard error against the point estimates. Comment on the accuracy of the bootstrap estimates of standard errors.

(d) Consider applying bootstrap bias corrections to the estimators.

(e) Consider different sample sizes and different upper quantiles but the interpolated quantile estimator is only feasible if $(n = 1)p < n$.

Section 7.5 Hypothesis testing

Exercise 7.8: Resistor tolerances

A resistor is rated at 100 ohms, with a tolerance of ± 1. Assume the quoted tolerance is based on an assumed process standard deviation of 0.2.

(a) If the process mean is 100 and the process standard deviation is 0.20, what proportion (expressed as ppm) of production will be out of tolerance?

(b) If the process mean is 99.8 and the process standard deviation is 0.20, what proportion (expressed as ppm) of production will be out of tolerance?

(c) A sample of 20 was taken from the process. The sample mean and standard deviation were 100.25 and 0.37 respectively. Is there evidence against a hypothesis that the mean is 100.0, with a two sided alternative hypothesis, at the 5% level?

(d) Given the data in (c) is there evidence against a hypothesis that the standard deviation is 0.2, with a one-sided alternative that it is greater than 0.2, at the 10% level?

Exercise 7.9: Repair times

In a dispute between a local network provider (LNP) and a trunk route provider (TRP) the TRP claims that the times to respond to repairs are longer for TRP. Data from the past 7 repairs is: Test a null hypothesis of no difference in times H_0 using a **ran-**

$$
\begin{array}{llll}
\text{TRP:} & 3 & 12 & 15 & 27 \\
\text{LNP:} & 1 & 4 & 9 \\
\end{array}
$$

domization test. If H_0 is true every possible allocation of 3 of the 7 repair times to LNP is equally likely.

(a) In how many ways can we allocate 3 from 7 repair times to LNP? Call this N.

(b) We need to choose a test statistic. Any sensible choice will do, though the p-value may vary with the choice. Choose the t-statistics for comparison of two means based on independent samples. What is the value of this statistic for the observed data? Call this c.

(c) How many of the allocations of 3 from 7 repair times to LNP give an absolute value of the test statistic greater than or equal to $|c|$? Call this number n.

(d) Calculate the p-value as n/N.

(e) What are the advantages and drawbacks of randomization tests?

Exercise 7.10: Corrosion

Ten mild steel coupons were immersed at shallow depth in marine sediments, at different locations, for three years. For each coupon the corrosion (micrometers per year) was measured for the upper half and lower half of the coupon.

Coupon	1	2	3	4	5	6	7	8	9	10
Lower half	63	52	40	45	62	61	58	53	49	53
Upper half	37	21	42	28	56	49	36	42	51	47

(a) Plot the two corrosion rates (using different plotting characters) against coupon number.

(b) Calculate a 95% confidence interval for the difference in corrosion rates (lower half less upper half) in the corresponding population.

(c) Is there evidence against a hypothesis of no difference in corrosion rates between lower and upper half at the 0.05 level of significance?

Exercise 7.11: Lead content

Five containers of water were taken from each of two different stations A and B on a river. Determinations of the lead content of each sample were made and the results (ppm) are given below.

Station A	9.6	10.2	10.6	11.8	11.1	$\bar{x}_A = 10.66$	$s_A = 0.841$
Station B	9.7	11.8	11.9	12.3	11.7	$\bar{x}_B = 11.48$	$s_B = 1.021$

(a) Suppose the samples from station A and B are independent and the hypothesis that the corresponding population variances are equal against the hypothesis that they differ at the 10% level. State any assumptions you need to make. Construct a 95% confidence interval for the difference in the corresponding population means.

(b) You are now told that the five containers of water from station A were taken on the same day as the corresponding container from station B. If the data are reanalyzed, taking account of this, then do they provide any evidence that the lead contents of water from the two stations differ? State any assumptions you need to make.

Section 7.6 Sample size

Exercise 7.12:   Determining sample size

A type of capacitor has a nominal capacitance of 100 microF and has a manufacturer's tolerance interval of $\pm 10\%$. Assume that the manufacturing tolerance is set at $\pm 5\sigma$ so that $\sigma = 2$.

(a) We wish to test the null hypothesis that the mean capacitance (μ) is 100, against an alternative that is not at the 5% level. So $H_0 : \mu = 100$ and $H_1 \neq 100$. We also require that the probability of rejecting H_0 if $\mu = 101$ should be 0.8.

   (i) Write down the distribution of $\overline{X}$ if H_0 is true, with the standard deviation written as $2/\sqrt{n}$.
   (ii) We will reject H_0 and conclude that $\mu > 100$ if $\bar{x} > c$. What is the value of c as a function of n?
   (iii) Write down the distribution of $\overline{X}$ if $\mu = 101$. How much less than 101, as a function of n, must c be for the probability of rejecting H_0 to be 0.8?
   (iv) We now have a distance from 100 up to c and a distance from c up to 101 as fractions of n. What value of n is needed?

(b) What sample size is required if the probability of rejecting H_0 if $\mu = 101$?

(c) What size sample is required if the alternative hypothesis is $\mu < 100$ and we require the probability of rejecting H_0 if $\mu = 99$ to be 0.8?

Exercise 7.13:   Power of the test

We wish to test a null hypothesis that the mean of a normal distribution is μ_0, against an alternative hypothesis that it is not, at the α level of significance. Assume the population standard deviation is known to be σ. Suppose that $\mu = \mu_1$, with $\mu_1 > \mu_0$, and that we require the probability of rejecting H_0 to be $1 - \beta$.

(a) Explain why $\dfrac{\overline{X} - \mu_0}{\sigma/\sqrt{n}} \sim N\left(\dfrac{\mu_1 - \mu_0}{\sigma/\sqrt{n}}, 1\right)$. Deduce that $z_{\alpha/2} + z_\beta = \dfrac{\mu_1 - \mu_0}{\sigma/\sqrt{n}}$ and hence show that $n = \left(\dfrac{z_{\alpha/2} + z_\beta}{\mu_0 - \mu_1} \times \sigma\right)^2$.

(b) Explain why the same result holds if $\mu_1 < \mu_0$.

(c) For a given n, the probability of rejecting and H_0, $1 - \beta$, depends on population mean μ, $\mu - \mu_0$ which we will write as δ. A plot of $1 - \beta$ against δ is known as the power curve. Show that $1 - \beta = \phi\left(\dfrac{\sqrt{n}\delta}{\sigma} - z_{\alpha/2}\right)$.

(d) Plot the power curve for the case of $\alpha = 0.05, n = 25$ and $\sigma = 1$.

Section 7.7 Confidence interval for a population variance and standard deviation

Exercise 7.14:   Chi-squared distribution 1

Assume that X_i are iid $\sim N(\mu, \sigma_2)$

(a) Show $\displaystyle\sum_{i=1}^{n}\left(\dfrac{X_i - \mu}{\sigma}\right)^2 = \sum_{i=1}^{n}\left(\dfrac{X_i - \overline{X}}{\sigma}\right)^2 + n\left(\dfrac{\overline{X} - \mu}{\sigma}\right).$

(b) Explain why $n\left(\dfrac{\overline{X} - \mu}{\sigma}\right)^2 \sim \chi_1^2$.

(c) When taking independent draw from a normal distribution $\overline{X}$ and S are independent. Give an informal explanation for this result.

(d) Suppose W_1 and W_2 are independent random variables with chi-square distribution ν_1 and ν_2 degrees of freedom respectively.

   (i) Explain why $W_1 + W_2$ has a chi-square distribution with $\nu_1 + \nu_2$ degrees of freedom.

   (ii) Explain why $W_1 - W_2$ does not have a chi-square distribution with $\nu_1 - \nu_2$ degrees of freedom.

(e) Explain why $\displaystyle\sum_{i=1}^{n}\left(\dfrac{X_i - \overline{X}}{\sigma}\right)^2 \sim \chi_{n-1}^2$.

Exercise 7.15: Chi-squared distribution 2

(a) If $Z \sim N(0,1)$ explain why:

   (i) $Z^2 \sim \chi_1^2$.

   (ii) $\mathrm{E}[Z^2]=1$ and $\mathrm{var}(Z^2)=1$ given that a χ_ν^2 distribution has mean ν and variance 2ν.

   (iii) $\mathrm{E}[Z^4]=3$ given that the kurtosis of a normal distribution is 3.

(b) Let Y be a random variable with mean 0 and standard deviation 1. Show that the variance of Y^2 is $\tau - 1$ where τ is the kurtosis of the distribution of Y. Hint: $var(Y^2)=\mathrm{E}[(Y^2)^2]-(\mathrm{E}[Y^2])^2$.

Exercise 7.16: SRS from normal distribution

If we have an SRS from $N(\mu, \sigma^2)$ then

$$\frac{(n-1)s^2}{\sigma^2} \quad \sim \quad \chi_{n-1}^2.$$

Suppose we wish to test $H_0 : \sigma = \sigma_0$ against $H_1 : \sigma > \sigma_0$ at the α level.

(a) Explain why we would reject H_0 if

$$\frac{(n-1)s^2}{\sigma^2} \quad \sim \quad \chi_{n-1,\,\alpha}^2.$$

(b) Refer to Example 7.18. The values of n, σ_0 and s are 12, 0.04 and 0.056 respectively. What is the value of the test statistic? What is the critical value the test at 0.01 level? What is the conclusion?.

Exercise 7.17: Chi-squared distribution 3

Generate a random sample of 1000 random deviates chi-square distributions for $2, 5, 10$ and 100 degrees of freedom.

(a) Draw histograms of the four samples on one graph so that you can easily compare them.

(b) Draw normal quantile-quantile plots, or normal probability plots, for the four samples on one graph.

(c) Calculate the mean, variance, skewness and kurtosis of each sample.

Exercise 7.18: Studentized variate from exponential distribution

An SRS of size n is taken from an exponential distribution with mean 1. Define the random variable

$$\tau = \frac{\overline{X} - 1}{S/\sqrt{n}},$$

where $\overline{X}$ is the sample mean and S is the sample standard deviation. The objective is to compare the distribution of τ with a t-distribution with $n - 1$ degrees of freedom by simulation.

(a) Take $n = 5$ and simulate $K = 10\,000$ SRS from an exponential distribution with mean 1. Calculate τ from each SRS and consider the distribution of $10\,000$ values.

   (i) What is the skewness of this distribution of τ?
   (ii) What are the lower and upper 0.025 quantiles of this distribution of τ? What are the lower and upper 0.025 quantiles of a t-distribution with 4 degrees of freedom?
   (iii) What proportion of this distribution of τ lies below the lower 0.025, and above the upper 0.025, quantiles of a t-distribution with 4 degrees of freedom? Hence what proportion of τ values lie between the lower 0.025 quantile and upper 0.025 quantile of a t-distribution with 4 degrees of freedom?

(b) Repeat the simulation for $n = 10, 30, 50, 100, 1\,000$ and answer similar questions, making comparisons with a t-distribution with $(n - 1)$ degrees of freedom.

Section 7.8 Comparison of means

Exercise 7.19: Arsenic concentration in wells

A national guideline for drinking water standard is that the arsenic concentration is less than 50 micrograms per liter. The WHO guideline value is much lower and is set at 10 micrograms per liter. A sample of water from a well is analyzed for arsenic over seven days. The results in micrograms per liter (parts per billion) are: $39, 30, 43, 40, 47, 57, 50$.

(a) Calculate the mean and standard deviation.
(b) Estimate the standard error of the sample mean.
(c) Construct a 95% confidence interval for the population mean, and state the assumptions you are making.
(d) Given the interval in (c) are you confident that the population mean is below 50?
(e) Given the interval in (c) are you confident that the population mean exceeds 10?
(f) Construct a 90% confidence interval for the population mean.
(g) What size sample would you recommend if the width of a 95% confidence interval is to be around 6?

Exercise 7.20: Magnesium alloys

The objective of the following experiment was to detect any systematic difference in hardness of magnesium alloys. The results of Vickers hardness tests (MPa) on samples of two magnesium alloys A and B are summarized below. Assume the corresponding populations are near normal.

$$n_A = 9 \quad \bar{x}_A = 825.8 \quad s_A = 28.9$$
$$n_B = 7 \quad \bar{x}_B = 893.8 \quad s_A = 21.5$$

(a) Construct approximate 90% and a 95% confidence interval for the difference in the means of the corresponding populations.

(b) Assume the population variances are equal and construct approximate 90% and a 95% confidence interval for the difference in the means of the corresponding populations.

Section 7.8.2 Matched pairs

Exercise 7.21: Cement content

The objective of the following experiment was to detect any systematic difference in cement content (% by weight) of concrete paving blocks Eight concrete paving blocks were obtained from different sources Each paver was ground and one half was chosen at random and sent to Laboratory A. The other half was sent to Laboratory B.

Block	1	2	3	4	5	6	7	8
Lab A	20.3	18.8	17.9	21.5	20.5	19.3	19.8	18.2
Lab B	21.4	19.2	17.4	22.7	21.3	19.1	19.7	19.2

(a) Construct 90% and 95% confidence intervals for the difference in the corresponding population means.

(b) Is there evidence to reject a null hypothesis of a mean difference of 0, with a two-sided alternative hypothesis, at the 10% level?

(c) What size sample would you recommend if the width of a 90% confidence interval for the difference is to be around 0.5?

Section 7.9 Comparing variances

Exercise 7.22: Flow meters

A magnetic flow meter was compared with a mass flow meter for measuring flow through a laboratory flume with a constant head of water. Twelve measurements were made with each meter and the sample standard deviations were 0.53 and 0.37 liter/second. Construct a 90% and a 95% confidence interval for the ratio of the corresponding population standard deviations.

Exercise 7.23: Gold prospects

A mining engineer wishes to compare two gold field prospects, A and B, and can take 20 drill cores. The engineer assumes that the variance of drill cores in A is twice that in B.

(a) The engineer takes 10 cores from each prospect. The ratio of the sample variances turns out to be 2.16. Use the chi-square distribution to construct a 90% confidence interval for the ratio of the variances, and for the ratio of the standard deviations. State the assumptions you make.

(b) What allocation of the 20 cores between A and B will result in a minimum variance for the difference in sample means, if the variance of drill cores in A is twice that in B?

(c) Construct a 90% confidence interval for the ratio of the variances, and for the ratio of the standard deviations, using your allocation in (b), and given the sample ratio of variances is 2.16.

Section 7.10 Inference about proportions

Exercise 7.24: **Proportion 1**

The secondary maximum contaminant level for iron in drinking water is 0.3 parts per million (ppm). Fifty seven out of 500 bottles filled at an SRS of kitchen taps in a city exceeded the level. Construct an approximate 95% confidence interval for the proportion in the corresponding population.

Exercise 7.25: **Proportion 2**

The U.S. Environmental Protection Agency (EPA) action level for lead in drinking water is 15 parts per billion (ppb). A public health officer wants to estimate the proportion of houses in a particular quarter of the city with lead content of water drawn at the kitchen tap exceeding 15 ppb. The public health officer thinks this proportion may be around 0.20 and wants the width of a 90% confidence interval to be less than 0.05. What size SRS should be taken?

Exercise 7.26: **Proportion 3**

A fire chief claims that private fire hydrants are poorly maintained by comparison with municipal hydrants. Simple random samples of size 80 and 120 were taken from lists of private and municipal fire hydrants respectively. Fifteen of the private hydrants were found to be non-compliant and eight of the municipal hydrants were non-compliant. Assume the total numbers of fire hyrants in the city are large by comparison with the sample sizes.

(a) Construct an approximate 90% confidence interval for the differences in proportions in the corresponding populations.

(b) Is there evidence to support the fire chief's claim at the 0.10 level of significance?

Exercise 7.27: **Blurred printing**

A process prints colored patterns on steel sheets that will be made into biscuit tins. Three ink colors are used and the plate alignment is critical for a sharp image. A modification has been made to the process. Before the modification 18 out of 126 plates had a blurred image, whereas after the modification 8 out of 136 plates were blurred. Assume blurred plates occur randomly and independently.

(a) Construct a 90% and a 95% confidence interval for the difference in the corresponding population proportions.

(b) Is there evidence that the modification has been successful?

Exercise 7.28: Breath testing 1

After first introducing random breath tests for motorists a police found that 17 out of 250 drivers tested were above the maximum legal blood alcohol content.

(a) Estimate the proportion in the corresponding population and its standard error.

(b) Construct a 95% confidence interval for the proportion in the corresponding population.

Exercise 7.29: McNemar's test

Two self driving cars, using different control systems A and B, were presented with 100 challenging driving situations on a test track. The response of the cars was rated as either as good as an expert driver or below expert driver level. Treat the 100 driving situations as a random sample from the notional population of all possible challenging driving situations.

A|B	Expert	Below expert
Expert	30	36
Below expert	19	15

(a) Is there evidence of a difference between control systems at the 0.10 level?

(b) Construct a 90% confidence interval for the difference in proportions of expert responses for control system A and control system B.

Exercise 7.30: Breath testing 2

Northumbria Police Force breath-tested 407 motorists over the Christmas period in 1989 and found 189 positive. During the same period of 1990 they found 157 positive in 470 tests.

(a) Is there any evidence of a difference in the proportions in the corresponding populations? What assumptions are you making? Are they realistic?

(b) Discuss to what extent, if any, these data provide evidence of a decrease in drunken driving in Northumbria.

(c) Describe how you might, conceptually, design an experiment to investigate any changes in drinking habits over the Christmas period. Would there be any drawbacks to carrying out your proposed experiment in practice?

Section 7.11 Prediction intervals and statistical tolerance intervals

Exercise 7.31: Tolerance interval 1

A company manufacturers precision 0.1 micro-Farad capacitors with a tolerance of $\pm 1\%$. Assume the capacitances are normally distributed. A random sample of 50 capacitors is taken and the sample mean and standard deviation of the capacitances are 0.10008 and 0.00029 respectively.

(a) Calculate 90% confidence interval for the population mean.

(b) Calculate a 90% prediction interval for the capacitance of a randomly selected capacitor.

(c) Calculate an 80% statistical tolerance interval for 99% of the population.

Exercise 7.32: Tolerance interval 2

Refer to Example 7.29. Let p be the proportion of the population less than $x_{n:n}$. Let X_i be a random draw from the population. Then . Hence $P(X_1 \cap X_2 \cap \cdots \cap X_n) < x_{n:n} = p^n$. Derive Vardeman's tolerance interval.

Section 7.12 Goodness of fit tests

Exercise 7.33: Grouped data

Consider data for Gold grades $\{x_i\}$ for $i = 1, \ldots, 181$. Define $w_i = log_{10}(x_i)$. Then, $\widehat{w} = -0.1982$, $s_w = 0.2902$, and the grouped data follow in Table 7.4. The objective is

TABLE 7.4: Grouped data.

Class interval	Frequency
−0.9 to −0.8	6
−0.8 to −0.7	2
−0.7 to −0.6	8
−0.6 to −0.5	11
−0.5 to −0.4	15
−0.4 to −0.3	21
−0.3 to −0.2	24
−0.2 to −0.1	27
−0.1 to 0.0	28
0.0 to 0.1	15
0.1 to 0.2	10
0.2 to 0.3	6
0.3 to 0.4	4
0.4 to 0.6	4

to test the null hypothesis

$$H_0 \quad : \quad W \text{ has a normal distribution}$$
$$\text{against the alternative hypothesis}$$
$$H_1 \quad : \quad W \text{ does not have a normal distribution}$$

at the 0.10(10%) level. Assume the mean and standard deviation of the hypothesized normal distribution are equal to the corresponding sample statistics $\overline{w}$ and s_w.

(a) Calculate $P(W < -0.8)$.

(b) Calculate the probabilities that W lies in the next 12 class intervals.

(c) Calculate $P(0.4 < W)$.

(d) Hence calculate expected frequencies for all the bins (correct to 2 decimal places).

(e) Pool adjacent bins so that all expected values exceed 2.0 (the minimum in the SPSS statistical software).

(f) Calculate the value of the chi-square goodness of fit statistic.

(g) Is there evidence to reject the null hypothesis?

(h) Pool adjacent bins so that all expected values exceed 5.0 (text books typically recommend this to obtain a good approximation to the sampling distribution of the chi-square goodness of fit statistic).

(i) Is there now evidence to reject the null hypothesis?

(j) What is the disadvantage of insisting that all expected values exceed 5.0?

Miscellaneous problems

Exercise 7.34: Bias S

(a) Show that $1 > \sqrt{(1 - a^2)} > 1 - a$ for any $0 < a < 1$. Hence deduce that the bias of the estimator S of θ is small by comparison with its standard error.

(b) The variance of S when sampling from a normal distribution is approximately $\sigma^2/(2n)$. Plot the ratio of the bias of the estimator S of θ to its standard error against the sample size n for $n = 2, \ldots, 100$.

Exercise 7.35: sd(S)

(a) Suppose we have a SRS X_i for $i = 1, \ldots, n$ from $N(\mu, \sigma^2)$. Define $Y_i = X_i - \mu$. Assuming that the kurtosis of the normal distribution is 3, what are $E[Y_i^2]$ and $E[Y_i^4]$?

(b) Hence show that

$$\text{var}(Y_i^2) = 2\sigma^4.$$

(c) Why is this result sensitive to the assumption of normality?

(d) Define $W = \sum Y_i^2$, and show that $E[W] = n\sigma^2$ and $\text{var}(V) = 2n\sigma^4$. Use a Taylor series approximation to deduce that

$$E\left[\sqrt{W/n}\right] \approx \sigma \quad \text{var}\left(\sqrt{W/n}\right) \approx \sigma^2/(2n).$$

(e) Use (d) to justify the approximation

$$S \sim N(\sigma, \frac{\sigma^2}{2n})$$

and state the assumptions on which this is based.

Exercise 7.36: SRS sample

A SRS of size n is taken from a normal distribution. Assume the approximation $S \sim N(\sigma, \frac{\sigma^2}{2n})$.

(a) Explain why

$$s \pm z_{\alpha/2} \frac{s}{\sqrt{2n}}.$$

is an approximate $(1 - \alpha) \times 100\%$ confidence interval for σ.

(b) Obtain an improvement on the approximate interval in (a) from

$$\frac{S - \sigma}{\sigma/\sqrt{2n}} \quad \sim \quad N(0, 1).$$

(c) Suppose $n = 30$ and $s = 25.3$. Compare 90% confidence intervals obtained with the approximations (a), (b) with a precise interval obtained using the chi-square distribution. Repeat for $n = 10$.

Exercise 7.37: Road stone

The road stone FTR values in Example 7.8 were: $62.15, 53.50, 55.00, 61.50$.

(a) How many different bootstrap samples are there?
(b) What is the probability of a bootstrap sample being the same as the original sample?
(c) What is the probability that the four numbers in the bootstrap sample are the same?
(d) Obtain the sampling distribution of $\bar{x}^*$ and hence construct 80% basic and percentile confidence intervals for the population mean μ.
(e) Compare your results with those obtained using resampling from R.

Exercise 7.38: Gold grades

A mining engineer wishes to compare the mean gold grade (g/tonne) of two gold field prospects, A and B, using soil surface samples. The engineer expects mean values of around 15 and assumes, from past experience, that the coefficient of variation of gold in soil surface samples is around 2.5.

(a) The engineer requires that the width of a 90% confidence interval should be around 3. Recommend a number of soil surface samples to be taken from each prospect.
(b) A colleague states that the logarithms of grade, rather than the grades themselves, are often analyzed because this reduces the influence of outliers. He suggests that a confidence interval based on logarithms of grade would provide a more precise indication of the difference between the gold prospects. Comment on this suggestion.

Exercise 7.39: Difference in two population means

We aim to construct a confidence interval for the difference in two population means, based on two independent samples with a combined size of m items. The standard deviations of populations A and B are assumed to be known and are σ and $k\sigma$ respectively.

(a) What size sample n would you allocate to A, leaving a sample size of $m - n$ from B, in terms of k and m?
(b) A mining company has a budget for 150 drill cores from two gold field prospects, A and B. A geologist considers that the grade in B is likely to be more variable than in A and takes some soil surface samples from A and from B. The ratio of the sample standard deviations s_A/s_B is around 0.5.

Recommend a division of the 150 drill sites for cores between A and B.

(c) Grades from drill cores obtained from sites that are close together tend to be more similar than grades from drill cores from sites that are well separated site.

Suggest a strategy for locating the drill sites within A and B.

Exercise 7.40: Cylinder wall thickness

An engineer wishes to compare two methods for estimating the wall thickness of cast aluminum cylinder heads: ultrasound and sectioning. Results (mm) from 10 cylinder heads are given in in the table below. For each cylinder head, the ultrasound method was used before the destructive sectioning method.

Cylinder head	1	2	3	4	5	6	7	8	9	10
Ultrasound	19.3	21.8	21.2	23.1	22.8	22.3	21.5	21.9	19.8	20.4
Sectioning	20.7	21.6	20.9	23.6	22.5	23.5	22.4	22.7	21.1	19.9

(a) Construct 90% and 95% confidence intervals for the difference between the population means.

(b) On the basis of this interval, is there statistically significant evidence, at the 5% level, of a difference between population means obtained by the two methods?

(c) The engineer would like the width of a 90% confidence interval to be about 0.5. What size sample, in total, would you recommend?

Exercise 7.41: Wiper motors

A car manufacturer tested 10 of supplier A's windscreen wiper motors and 10 of supplier B's windscreen wiper motors in a test rig which had the wiper blades moving over dry glass. The times to failure of the motors are given in Windscreenmotor.txt

(a) Construct a 95% confidence interval for the difference between the population mean lifetimes.

(b) Assuming the price is similar, which supplier would you recommend the manufacturer use, and why?

Exercise 7.42: Copper smelter

An environmental engineer wishes to test a null hypothesis, H_0, that the mean arsenic level at a site in the vicinity of a copper smelter is 20 ppm against a one-sided alternative that it is higher. A mean level below 20 ppm is acceptable, but if the arsenic level is as high as 25 ppm then further cleanup is required. Assume the standard deviation of arsenic in soil samples at such sites is around 8 ppm.

(a) What sample size would you recommend if the width of a 90% confidence interval is to be less than 5?

(b) What sample size would you recommend if the width of a 95% confidence interval is to be less than 5?

(c) The test is to be performed at the 0.05 level and the probability of rejecting H_0 if the mean level is in fact 25 ppm is to be 0.8.

(d) Suppose a sample size of n is taken. Write down the distribution of the sample mean $\overline{X}$ in terms of the population mean μ and n.

(e) Write down the critical value c for the test in terms of the sample size n.

(f) Suppose the mean is 25. How much lower, in terms of n, than 25 does c need to be to satisfy the criterion that the probability of rejection of H_0 is 0.8?

(g) Sketch the distributions of $\overline{X}$ when μ equals 20 and when μ equals 25 and show the position of c.

(h) What value of n is required?

(i) What sample size would you recommend if the probability of rejecting H_0 when the mean level is 25 ppm is to be 0.05?

Exercise 7.43: Subjective probability

Refer to Section 7.4.1. Informally, it is common to interpret this as a subjective probability of $(1 - \alpha)$ that the interval contains the population mean. There is no random event because the calculated interval either contains the population mean or it does not contain the population mean. But, as we don't usually know which of these two possibilities is the fact, the subjective probability interpretation is reasonable for this construction of a confidence interval. However, such a subjective probability interpretation fails in the following application.

A component made on a digital controlled machine has a length that has a uniform distribution with support $[\theta - 0.5, \theta + 0.5]$. A simple random sample of size 2 will be taken from production with the aim of estimating θ. Let the random variable X be the length of a component.

(a) Show that

$$P(X_{1:2} \leq \theta \leq X_{2:2}) = 0.5$$

(b) Deduce a 50% confidence interval for θ.

(c) A sample is taken and the observations are 5.1 and 4.4. Explain why it is not reasonable to claim that the subjective probability that θ lies in the interval $[4.4, 5.1]$ is 0.5. Reconcile this result with the first footnote in this chapter.

Exercise 7.44: Quantile-quantile plots

Generate N SRS of size n from the following distributions and for each sample of n calculate the mean $\bar{x}$. Draw a histogram of the N $\bar{x}$ and a normal quantile-quantile plot or equivalent probability plot. Take $N = 10^5$ and $n = 5, 10, 20, 30, 50, 100$.

(a) $N(0, 1)$

(b) Exponential with $\lambda = 1$

(c) Reduced Gumbel distribution

(d) $U[0, 1]$

(e) t with $5, 4, 2$ and 1 degrees of freedom

(f) Poisson with mean 5 and 1

(g) Weibull with cdf $F(x) = 1 - e^{-\sqrt{x}}$ for $x \geq 0$

Exercise 7.45: Fireman's clothing 1

Refer to Example 7.7, fireman's clothing. Suppose that 8 test pieces had been taken and the sample mean was 12.1. Calculate 95% and 99% confidence intervals for the population mean, assuming that $\sigma = 1.21$.

Exercise 7.46: Fireman's clothing 2

Refer to Example 7.7, fireman's clothing. Assume a standard deviation of arc rating of 1.3. If the population mean is as high as 13, then it would be a valuable increase. The researcher would not want a 99% confidence interval to include both 11 and 13, because below 11 is not worthwhile and 13 is valuable and the researcher cannot distinguish between valuable and not worthwhile. What size SRS is required?

Exercise 7.47: Laplace distribution

Compare the sample mean and sample median as estimators of the mean of a normal distribution and the mean of a Laplace distribution for sample sizes of $n = 5, 10, 20, 50$ and 100. Sample from a standard normal distribution and a Laplace distribution with mean of 0 and standard deviation of 1. Draw $N = 10^5$ samples of each size from each distribution. Tabulate the results and comment on the findings.

Exercise 7.48: Exponential and normal distributions

Write an R program that takes random samples of size n from a standard normal distribution and an exponential distribution with mean 1, calculates a 95% CI for σ, and records whether or not the interval includes 1. Repeat this calculation $N = 10^5$ times and compare the proportions of intervals that include 1, for $n = 5, 10, 30, 100$.

Exercise 7.49: Chi squared distribution

If ν is large and $W \sim \chi^2_\nu$ then W is approximately $N(\nu, 2\nu)$. Use the normal distribution to approximate the upper and lower 0.025 percentiles for $\nu = 20$, 50 and 100. Compare the precise values.

Exercise 7.50: Acceleration in subway trains

In an experiment to study the effect of acceleration on passengers in subway trains, the acceleration at which loss of balance occurred was measured for 12 adults, randomly selected from a pool of volunteers on the subway staff. The mean and standard deviation of the 12 measurements were 1.62 and 0.36 ms^{-2} respectively. Calculate a 90% CI for the mean and standard deviation in the corresponding population. Do you think the sample is representative of all subway passengers?

Exercise 7.51: *F*-distribution 1

Plot the pdf of the F-distribution with 4 and 6 degrees of freedom using the R function *df(x,4,6)*.

Exercise 7.52: *F*-distribution 2

Show that

$$F_{n_B-1,n_A-1,1-\alpha/2} = \frac{1}{F_{n_A-1,n_B-1,\alpha/2}}.$$

Exercise 7.53: Lead pipes

An engineer considers that the cost of replacing a lead pipe with PVC is twice as likely to be within the range $[800, 1400]$ dollars as outside this range, Estimate the cost of the program if the water company has 3 million customers, and provide a 95% prediction interval for this cost.

Exercise 7.54: Bernoulli trials

Let X be the number of successes in a sequence of n Bernoulli trials with probability of success p. Then, denoting failure as 0 and success as 1, there are two categories 0 and 1 with $n - X$ and X being the corresponding frequencies. The expected numbers are $n(1 - p)$ and np respectively. Show that

$$\frac{((n - X) - n(1 - p))^2}{n(1 - p)} + \frac{(X - np)^2}{np} \sim [N(0, 1)]^2 = \chi_1^2.$$

Exercise 7.55: Platinum plating

An experiment was performed to compare two test laboratories A and B. Bottles of platinum plating solution, of nominal strength 50 grams/liter, were available from nine manufacturers. One half of each bottle was randomly assigned to A and the other half was sent to B. The results in grams/liter are: The mean of the assays from laboratories

Bottle	1	2	3	4	5	6	7	8	9
Lab A	61	60	56	63	56	63	59	55	61
Lab B	55	54	47	59	51	61	57	62	58

A and B are 59.33 and 56.00 respectively.

(a) The standard deviation of the 9 differences is 4.47. Construct a 95% confidence interval for the difference in the population means. You may refer to the following MATLAB output, but note that only one of the values is relevant.

```
>> tinv(.95,8)
ans = 1.8595
>> tinv(.95,9)
ans = 1.8331
>> tinv(.975,8)
ans = 2.3060
>> tinv(.975,9)
ans = 2.2622
```

(b) Consider the null hypothesis, H_0, that the mean difference in the corresponding population is 0. Is there evidence against H_0 at the 5% level of significance? Give a reason for your answer.

(c) What size sample would you recommend for a follow up experiment if the width of a 95% confidence interval should be around 4.0?

Exercise 7.56: Adhesive curing time

A materials engineer performed an experiment to investigate the difference in curing times for an adhesive, X, that is now used, and a new formulation, Y. The engineer had 14 nominally identical test joints to be glued, and randomly assigned 6 to Y, which would use up all the new formulation. The other 8 joints were assigned to X. The curing times (hours) and an analysis using MATLAB are given below.

```
>> x = [14    16    13    15    16    16    15    11]
>> y = [10    11    14    12    9    13]
>> mean(x) = 14.5; mean(y) = 11.5000
```

```
>> [h,p,ci,stats]=ttest2(x,y,'vartype','unequal','alpha',0.05)
h = 1
p = 0.0118
ci = 0.8147
stats = 5.1853
struct with fields:
tstat: 3.0364
df: 10.5761
sd: [1.7728 1.8708]
```

(a) Write down a 95% confidence interval for the difference in the means of the corresponding populations.

(b) Is there evidence of a difference between the population means at a 5% level of significance? Give a reason for your answer.

(c) Suppose that the two formulations provide equivalent bond strengths, have the same standard deviations in the corresponding populations, and cost the same. If a shorter curing time is preferred, which formulation would you recommend?

(d) Calculate the estimate of the standard error of the difference in the sample means using information given in the MATLAB output.

Exercise 7.57: Laterite

A chemical engineer compared an atomic absorption method (A) and a volumetric method (B) for determining the percentage of iron in laterite. Both methods were used on six specimens of laterite. Each specimen was halved and one half was randomly assigned to A and the other to B. The results (%) are: The means of the percentages

Specimen	Method A	Method B
1	30.99	30.05
2	31.47	31.75
3	30.00	28.50
4	35.25	35.12
5	30.62	30.55
6	27.21	27.26

from A and B are 30.923 and 30.538 respectively.

(a) Given that the objective is to discover any systematic difference in assays of iron content between the two methods, what is the rationale for the design of the experiment?

(b) The standard deviation of the 6 differences is 0.685. Construct a 95% confidence interval for the mean of the differences in the corresponding population. You may refer to the following MATLAB output, but note that only one of the values is relevant.

```
>> tinv(0.95,5)
ans = 2.0150
>> tinv(0.95,6)
ans = 1.9432
>> tinv(0.975,5)
ans = 2.5706
```

```
>> tinv(0.975,6)
ans = 2.4469
```

(c) Consider the null hypothesis,

H_0 : the mean difference in the corresponding population is 0.

Is there evidence against H_0 at the 5% level of significance? Give a reason for your answer.

Exercise 7.58: Joint geometries

A engineer is planning to perform an experiment to investigate the difference in cycles to failure of two joint geometries. The comparison will be made in terms of the logarithms of cycles to failure which are approximately normally distributed. From past experience the engineer thinks that the standard deviation (σ) of the logarithms of cycles to failure will be the same for both geometries, and around 0.18. Assume that the same number (n) of joints of each type will be tested.

(a) Write down an expression for the standard error of the difference of the means from the two samples in terms of σ and n.

(b) What is the minimum value of n if the width of the 95% confidence interval for the difference in population means should be less than 0.20? (You may assume $z_{0.025} = 1.96$.)

8

Linear regression and linear relationships

We consider a model for a response variable as the sum of a linear function of a predictor variable and independent random variation. This is known as the linear regression model. We discuss the linear regression model in the particular context of a bivariate normal distribution and relate it to the correlation coefficient. The linear model assumes the predictor variable is measured without error. We consider a measurement error model for cases when this assumption is unrealistic. Some functional relationships can be transformed to linear relationships, and one such example is presented. See relevant example in Appendix E:

Appendix E.5 *Predicting descent time from payload.*

8.1 Linear regression

8.1.1 Introduction

Mike Mandrel is the engineer responsible for the turbine division of SeaDragon. The division has been performing well and the turbines have a reputation for being durable and reliable. However, the larger customers are asking to see records of quality assurance procedures. To take an example, tungsten steel erosion shields are fitted to the low-pressure blading in the steam turbines, and the most important feature of an erosion shield is its resistance to wear. Customers would like to see records of wear measurements for a sample of the production of erosion shields. Direct measurement of resistance to wear in terms of abrasion loss is a lengthy, expensive and destructive process. However, abrasion loss is known to be largely dependent on the hardness of steel. For everyday quality control purposes it would be far more convenient to use hardness measurements, provided they give reasonably reliable estimates of the abrasion losses.

Mike decided to investigate the relationship between abrasion loss and hardness of erosion shields. During routine production, batches of 100 erosion shields, and 4 test coupons, are cast from single heats of the alloy. For the investigation, he randomly selects 1 erosion shield from each of the next 25 batches. The hardness of the alloy in each batch is taken as the average of Vickers hardness measurements [1] made on the 4 test coupons. The abrasion losses of the 25 erosion shields were measured after two weeks in a chamber designed to produce rapid wear. The data are listed in Table 8.1 and shown as a scatter plot in Figure 8.1.

The abrasion loss tends to decrease as the hardness increases, and the relationship appears to be approximately linear over the range of hardness values found in the experiment. However, the points do not lie precisely on a straight line. Even if the abrasion loss and hardness of the erosion shields could be measured exactly there would be deviations about

---

[1] Vickers hardness (VH) is measured by applying a pyramidal diamond indenter to the surface of the metal and calculating the ratio of the force to the area of indentation.

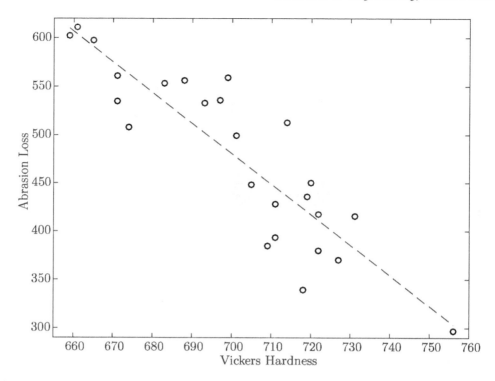

FIGURE 8.1: Abrasion loss plotted against Vickers hardness for 25 tungsten steel erosion shields (the dashed line is a fitted regression line).

TABLE 8.1: Vickers hardness $(9.8MPa)$ and abrasion loss (mg) for 25 erosion shields.

Vickers hardness	Abrasion loss	Vickers hardness	Abrasion loss
665	597	727	370
719	436	711	428
659	602	731	416
756	297	699	559
711	393	661	611
671	561	722	417
709	385	674	508
722	380	705	448
718	340	688	556
714	513	693	533
701	499	697	536
671	535	683	553
720	450		

the line, because many other factors, unmeasured or unknown, have some influence on the abrasion loss. We will refer to this as **inherent variability** in the population in order to distinguish it from **measurement error**. We propose a model which includes an intercept α, a slope β, and an unexplained variation as an **error** term:

$$\text{abrasion} \quad \text{loss} \quad = \quad \alpha + \beta \times \text{hardness} + \text{error}.$$

The errors allow for variations in abrasion loss of erosion shields of the same specified hardness, and accounts for the deviations of the points from the hypothetical line.

8.1.2 The model

We first define the model in general terms, and then discuss it in the context of Mike's experiment. Let Y be the variable we wish to predict, the **response variable**, and x be a **predictor variable**. An investigation, which can be a designed experiment or an observational study, has been performed to estimate a relationship between x and Y. The estimated relationship will be used to predict future values of Y when only values of x are known. The investigation provides n observations, each being a pair of x and y measurements:

$$(x_i, y_i) \qquad \text{for } i = 1, 2, \ldots, n.$$

Suppose that a plot of these data indicates that a linear relationship between x and y is a sensible approximation. Then we model the process that gives the observations by

$$Y_i = \alpha + \beta x_i + \epsilon_i \qquad \text{for } i = 1, 2, \ldots, n.$$

The intercept, α, and slope, β, are parameters of the model, and are referred to as the **coefficients**. The coefficients [2] are unknown constants, and β represents the change in the mean of Y due to a change of 1 unit in x. The x_i are known values from the investigation. The errors ϵ_i and hence Y_i are random variables, and y_i are the known values taken by the random variables Y_i in the investigation. The objectives of the investigation are:

to estimate the coefficients from the n data pairs;
to estimate the standard deviation of the errors;
to estimate the precision of the estimates of the coefficients;
to provide confidence intervals for the mean value of Y for given values of x;
to predict future values of Y for given values of x and to give limits within which future values of Y for given x are likely to lie.

We make five assumptions about the errors (ϵ_i).

A1: The mean value of the errors is zero.
A2: The errors are uncorrelated with the predictor variable.
A3: The errors are independent of each other.
A4: The errors all have the same variance, which we will denote by σ^2.
A5: The errors are normally distributed.

Model assumptions are never satisfied exactly, but it is important to check that they appear reasonable. To begin with, the underlying assumption of a linear relationship between x and y should be checked by plotting the data. Given that a linear model seems appropriate, the first assumption about the errors, **A1**, is crucial. But, we can not check that errors have a mean of 0 from the data because any other value of the mean would be indistinguishable from the intercept α. If the errors represent only inherent variation in the population, then the assumption of a mean of 0 is a consequence of their definition. However, the errors can include a component of measurement error in the response. If we wish to check that measurement error can reasonably be assumed to have a mean of 0, we will have to undertake a separate calibration exercise. In practice, this is usually unnecessary as

---

[2] A coefficient is a multiplier: α is the coefficient of 1 and β is the coefficient of x_i.

manufacturers ensure that their measuring instruments are correctly calibrated. They do so by implementing periodic checks against internal, and less frequently national, standards, and by training staff to use the instruments correctly. This will not eliminate measurement errors but it should make an assumption that measurement errors have a mean of 0 realistic.

The second assumption, **A2**, is also crucial. But, we can not detect a correlation between the predictor variable and the errors from the data, because any correlation would be incorporated into the slope. For example, if errors, defined as observed value less true value, tend to be positive for higher values of x and negative for lower values of x we will tend to overestimate the slope (Exercise 8.6). Correct calibration of measuring equipment should remove any such correlation between the predictor variable and measurement errors in the response. We assume that the predictor variables themselves are measured without error, or, in the case of a designed experiment, set precisely to target values (non-negligible measurement error in predictor variables is covered in Section 8.5).

For Mike's experiment the measurement errors for the abrasion loss are negligible compared with the inherent variation, because weighing is a precise operation. However, it wouldn't affect the analysis if measurement errors for abrasion loss were a substantial proportion of the error term in the model. The measurement of the VH of a batch is the mean of the measurements of VH on each of the four test coupons. This mean VH is defined as the predictor variable, and with this definition there is no measurement error in the model.

The remaining three assumptions are less critical because they can be checked from the data, to some extent, and if they are not realistic the analysis can be modified. Assumption **A3** is that the errors are independent of each other. In many cases there is no reason to doubt this, but it is possible for observations taken over time to have correlated errors. A positive correlation will result in underestimation of standard errors of the estimators of the coefficients.

If the assumption that the errors all have the same variance, **A4**, is infringed, the estimation of the standard errors of the coefficients may be inaccurate and prediction intervals should not have a constant width. For example, variation is sometimes nearer to being constant in percentage terms than in absolute terms so if Y tends to increase with x so will the standard deviation. Also, the higher variance variables will be given undue influence.

The fifth assumption, **A5**, is that the errors are normally distributed. If instead, errors have a heavy tailed distribution, such as the Laplace distribution, prediction intervals will be too narrow and outlying points will have undue influence.

To summarize, we are modeling the conditional distribution of Y given a particular value of x, x_p say, as

$$Y_p \sim N(\alpha + \beta x_p, \sigma).$$

The model for the process giving the observations is

$$Y_i = \alpha + \beta x_i + \epsilon_i, \quad \text{for } i = 1, 2, \ldots, n.$$

Notice that Y_i is short for $Y|x = x_i$, and since

$$\mathrm{E}[\epsilon_i] = 0, \qquad \mathrm{var}(\epsilon_i) = \sigma^2$$

it follows that

$$\mathrm{E}[Y_i] = \alpha + \beta x_i, \qquad \mathrm{var}(Y_i) = \sigma^2.$$

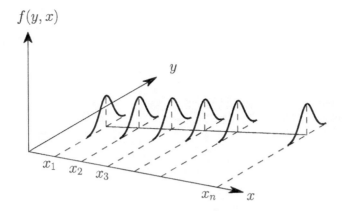

FIGURE 8.2: The linear regression model. The line in the xy plane is the mean values of Y given x (regression line). The bell shaped curves represent the normal distribution of Y given x, and all have standard deviation σ.

The unknown parameters which have to be estimated from the data are α, β and σ. The line on which the mean values of Y lie,

$$y = \alpha + \beta x,$$

is known as the **regression** line of y on x (Figure 8.2).

8.1.3 Fitting the model

8.1.3.1 Fitting the regression line

The model for the process giving the observations,

$$Y_i = \alpha + \beta x_i + \epsilon_i \qquad \text{for } i = 1, 2, \ldots, n$$

can be rearranged to give the errors as a linear combination of the response and predictor variable

$$\epsilon_i = Y_i - (\alpha + \beta x_i) \qquad \text{for } i = 1, 2, \ldots, n.$$

The sum of squared errors, Ψ, is given by:

$$\Psi = \sum \epsilon_i^2 = \sum (Y_i - (\alpha + \beta x_i))^2.$$

We have n data pairs, (x_i, y_i), and estimate the parameters α and β by the values, $\widehat{\alpha}$ and $\widehat{\beta}$, which minimize the sum of squared errors obtained in the investigation (ψ):

$$\psi = \sum (y_i - (\alpha + \beta x_i))^2.$$

The values $\widehat{\alpha}$ and $\widehat{\beta}$ are known as **least-squares estimates**[3].

---

[3]This is an application of the **principle of least squares**. The first explicit account was published by A.M. Legendre (1805) in his work on the estimation of orbits of comets. Legendre acknowledged Euler's contributions to the method, which had also been used by Karl Friedrich Gauss, working independently, from about 1795 [Plackett, 1972].

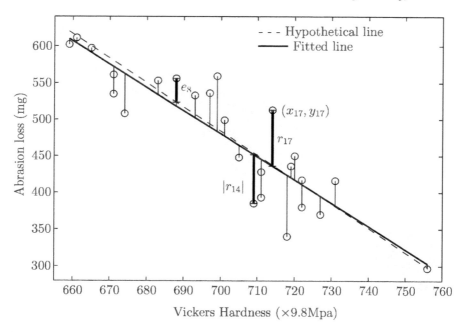

FIGURE 8.3: The estimated regression line (unbroken) is such that the sum of squared vertical distances from the points to the line (e.g. r_{17} and $\mid r_{14} \mid$) is a minimum. Also shown is a typical unmeasurable error (ϵ_8), its absolute magnitude is the vertical distance to the unknown hypothetical line (dashed line is a plausible scenario). The residual r_i is an estimate of the unknown ϵ_i.

In geometric terms we are finding the line

$$y = \widehat{\alpha} + \widehat{\beta}x,$$

such that the sum of squared vertical distances (that is distances measured parallel to the y-axis) from the points to the line is a minimum (Figure 8.3). Necessary conditions for ψ to have a minimum are that

$$\frac{\partial \psi}{\partial \alpha} = 0 \quad \text{and} \quad \frac{\partial \psi}{\partial \beta} = 0.$$

The partial differentiation is routine and leads to two linear simultaneous equations in two unknowns.

$$-2\sum (y_i - (\widehat{\alpha} + \widehat{\beta}x_i)) = 0$$
$$-2\sum x_i(y_i - (\widehat{\alpha} + \widehat{\beta}x_i)) = 0.$$

Rearrangement gives

$$n\widehat{\alpha} + \widehat{\beta}\sum x_i = \sum y_i$$
$$\widehat{\alpha}\sum x_i + \widehat{\beta}\sum x_i^2 = \sum x_i y_i .$$

Dividing the first equation by n and rearranging gives

$$\widehat{\alpha} = \bar{y} - \widehat{\beta}\bar{x} .$$

Substitution into the second gives

$$\widehat{\beta} = \frac{\sum x_i y_i - \bar{y} \sum x_i}{\sum x_i^2 - \bar{x} \sum x_i}.$$

Although this is a formula for the least squares estimate of the slope, the numerator and denominator are prone to rounding errors, which obscures the relatively simple form of the estimate. It is straightforward to express the formula in an equivalent form that is more convenient and instructive.

8.1.3.2 Identical forms for the least squares estimate of the slope

It is helpful to introduce abbreviations for mean-adjusted sums of squares and mean-adjusted sums of products.

$$S_{xx} = \sum (x_i - \bar{x})^2, \quad S_{xy} = \sum (x_i - \bar{x})(y_i - \bar{y}), \quad S_{yy} = \sum (y_i - \bar{y})^2.$$

The mean-adjusted sum of products

$$S_{xy} = \sum (x_i - \bar{x})(y_i - \bar{y}) = \sum x_i(y_i - \bar{y}) - \sum \bar{x}(y_i - \bar{y})$$

and since $\bar{x}$ does not depend on i, $\bar{x}$ is a common factor for the terms in $\sum \bar{x}(y_i - \bar{y})$ and

$$S_{xy} = \sum x_i(y_i - \bar{y}) - \bar{x} \sum (y_i - \bar{y}).$$

Now since $\sum (y_i - \bar{y}) = 0$,

$$S_{xy} = \sum x_i(y_i - \bar{y}) = \sum x_i y_i - \bar{y} \sum x_i.$$

So, the numerator in the formula for $\widehat{\beta}$ is equivalent to $\sum (x_i - \bar{x})(y_i - \bar{y})$. An identical argument with y replaced with x shows that the denominator in the formula for $\widehat{\beta}$ is equivalent to $\sum (x_i - \bar{x})^2$. Thus

$$\widehat{\beta} = \frac{\sum (x_i - \bar{x})(y_i - \bar{y})}{\sum (x_i - \bar{x})^2} = \frac{S_{xy}}{S_{xx}}.$$

An identical argument, interchanging x and y, gives

$$S_{xy} = \sum (x_i - \bar{x}) y_i$$

and similarly

$$\sum (x_i - \bar{x}) (Y_i - \overline{Y}) = \sum (x_i - \bar{x}) Y_i .$$

We use these forms in the theoretical development.

8.1.3.3 Relation to correlation

The estimate of the slope of the regression line is the correlation coefficient r scaled by the ratio of the standard deviation of y to the standard deviation of x. The scaling makes the estimate of the slope dimensionally consistent.

$$\widehat{\beta} = r \left(\frac{s_y}{s_x} \right).$$

Proof:

By definition

$$r \left(\frac{s_y}{s_x} \right) = \frac{S_{xy}/(n-1)}{\sqrt{(S_{xx}/(n-1))(S_{yy}/(n-1))}} \left(\sqrt{\frac{S_{yy}/(n-1)}{S_{xx}/(n-1)}} \right) = \frac{S_{xy}}{S_{xx}} = \widehat{\beta}.$$

8.1.3.4   Alternative form for the fitted regression line

The fitted regression line is,

$$y = \widehat{\alpha} + \widehat{\beta}x.$$

Since

$$\widehat{\alpha} = \bar{y} - \widehat{\beta}\bar{x},$$

the line also can be expressed as

$$y = \bar{y} + \widehat{\beta}(x - \bar{x}).$$

The latter form emphasizes that the line passes through the centroid of the data, $(\bar{x}, \bar{y})$, and is far more convenient for the theoretical development. Furthermore, we can use the fact that $\widehat{\beta} = rs_y/s_x$ to express the fitted regression in the non-dimensional form:

$$\frac{y - \bar{y}}{s_y} = r\frac{x - \bar{x}}{s_x}.$$

Example 8.1: Erosion shields [fitting a regression line]

For the erosion shields,

$$\bar{x} = 701.08, \qquad \bar{y} = 476.9, \qquad S_{xx} = 14\,672, \qquad S_{xy} = -46373.$$

The estimated regression line is

$$y = 476.9 - 3.161(x - 701.08).$$

The average abrasion loss is 476.9 and the estimated reduction in abrasion loss per unit increase in Vickers Hardness is 3.161, over a range of Vickers Hardness from 660 to 760. It can be written in an equivalent form

$$y = 2693.1 - 3.161x .$$

The intercept defines the fitted line but it has no physical interpretation because the model was fitted to erosion shields with Vickers Hardness in a range 660 to 760.

The R commands for the arithmetic are:

```
> shields.dat<-read.table("Erosion_shields.txt",header=T)
> attach(shields.dat)
#check the data read correctly
> head(shields.dat)
> x<-VH
> y<-loss
> Sxx<-sum((x-mean(x))^2)
> Sxy<-sum((x-mean(x))*(y-mean(y)))
> b<-Sxy/Sxx
> a<-mean(y)-b*mean(x)
> print(c(mean(x),mean(y),Sxx,Sxy))
[1]     701.08    476.92  14671.84 -46372.84
> print(c(a,b))
[1] 2692.80231   -3.16067
```

A much quicker approach is to use the `lm()` function in R which fits the model in a single command[4]. The "lm" is a mnemonic for "linear model". The syntax is the response, followed by $\sim$, followed by the predictor variable.

```
> shields.lm <- lm(loss~VH)
> summary(shields.lm)

Call:
lm(formula = loss ~ VH)\

Residuals:
    Min      1Q  Median      3Q     Max
-83.441 -24.995   6.043  30.542  76.916

Coefficients:
             Estimate Std. Error t value Pr(>|t|)
(Intercept) 2692.8023   242.8751  11.087 1.05e-10 ***
VH            -3.1607     0.3462  -9.129 4.15e-09 ***
---
Signif. codes:  0 *** 0.001 ** 0.01 * 0.05 . 0.1 1

Residual standard error: 41.94 on 23 degrees of freedom
Multiple R-squared: 0.7837,Adjusted R-squared: 0.7743
F-statistic: 83.34 on 1 and 23 DF,  p-value: 4.145e-09
```

For the moment, concentrate on the part of the R output that follows `Coefficients`: The first column describes the coefficients: α as (`intercept`), and β by the name of the variable, `VH`, for which it is the coefficient. The second column is the estimates of the coefficients. We explain the remainder of the R output later in the chapter. The fitted regression line can be added to a scatterplot with the `abline()` function, which takes the `lm()` function as its argument. Figure 8.1 was produced with

```
plot(VH,loss,xlab="Vickers hardness",ylab="Abrasion loss")
abline(lm(loss~VH),lty=2)
```

8.1.3.5 Residuals

The residuals (r_i) are defined as the differences between the observed y_i and the values predicted by the fitted regression line, $(\widehat{y}_i)$, which are known as **fitted values**. That is,

$$r_i = y_i - \widehat{y}_i,$$

where

$$\widehat{y}_i = \widehat{\alpha} + \widehat{\beta}x_i = \bar{y} + \widehat{\beta}(x_i - \bar{x}).$$

---

[4]In MATLAB, the 'polyfit' command gives us the slope and intercept of a fitted line

```
VH = importdata('Erosion_shields.txt');
p = polyfit(VH.data(:,1),VH.data(:,2),1)
p =
    -3.1607 2692.8
```

The residuals are our best estimates of the values taken by the errors. The residuals are not precisely equal to the values taken by the errors because the unknown α and β have been substituted by their least squares estimates $\widehat{\alpha}$ and $\widehat{\beta}$ respectively. In geometric terms, the absolute values of the residuals are the vertical distances from the points to the estimated regression line (Figure 8.3). If a point lies above the fitted regression line then the residual is positive, and if it lies below the line then the residual is negative.

Example 8.1: (Continued) Erosion shields [fitted value and residual]

Calculate the fitted values for the 5th and 10th datum, $(711, 393)$ and $(714, 513)$ respectively, and hence calculate the corresponding residuals. For the 5th datum, the fitted value is:

$$476.9 - 3.161 \times (711 - 701.8) \quad = \quad 445.57.$$

The residual is

$$393 - 445.57 \quad = \quad -52.57.$$

Similar calculations for the 10th datum give a fitted value of 436.08 and a residual of 76.92.

You can see the fitted values and residuals by[5]

```
> print(cbind(VH,loss,shields.lm$fit,shields.lm$res))
    VH loss
1  665  597 590.9570    6.043038
2  719  436 420.2808   15.719201
3  659  602 609.9210   -7.920981
4  756  297 303.3360   -6.336021
5  711  393 445.5662  -52.566157
...
```

8.1.3.6   Identities satisfied by the residuals

For any set of data: the sum of the residuals, and hence their mean, equals 0; and the correlation coefficient between the residuals and the predictor variable is 0. That is:

$$\sum r_i \;\; = \;\; 0 \quad \text{and} \quad \sum (x_i - \bar{x}) r_i \;\; = \;\; 0.$$

The proof of these results follow. For the first,

$$
\begin{aligned}
\sum r_i \;\; &= \;\; \sum (y_i - (\bar{y} + \widehat{\beta}(x_i - \bar{x})) \\
&= \;\; \sum (y_i - \bar{y}) - \widehat{\beta} \sum (x_i - \bar{x}) \\
&= \;\; 0 - \widehat{\beta} \times 0 \;\; = \;\; 0.
\end{aligned}
$$

---

[5]In MATLAB we use the output from 'polyval' to calculate a vector of residual values

```
yfit = polyval(p,VH.data(:,1));
yresid = VH.data(:,2) - yfit;
```

For the second,

$$
\begin{aligned}
\sum (x_i - \bar{x}) r_i &= \sum (x_i - \bar{x})(y_i - (\bar{y} + \widehat{\beta}(x_i - \bar{x}))) \\
&= \sum (x_i - \bar{x})(y_i - \bar{y}) - \widehat{\beta} \sum (x_i - \bar{x})^2 \\
&= S_{xy} - \widehat{\beta} S_{xx} \\
&= S_{xy} - \frac{S_{xy}}{S_{xx}} S_{xx} \;=\; 0.
\end{aligned}
$$

These results prove that it is impossible to detect that errors have a non-zero mean, or to detect a correlation between errors and the predictor variable, from the data alone. In fact the formulae for the least squares estimates can be derived by requiring the two conditions,

$$
\sum (y_i - (\alpha + \beta x_i)) \;=\; 0 \quad \text{and} \quad \sum (x_i - \bar{x})(y_i - (\alpha + \beta x_i)) \;=\; 0
$$

to hold (see Exercise 8.1).

8.1.3.7   Estimating the standard deviation of the errors

We estimate the variance of the errors, σ^2, by the variance of the residuals, s^2, calculated with the denominator $(n-2)$. Since $\bar{r} = 0$

$$
s^2 \;=\; \sum r_i^2 / (n-2).
$$

We divide by $n-2$ rather than n, and say we have lost two degrees of freedom by using estimates of two parameters, $\widehat{\alpha}$ and $\widehat{\beta}$. The term "loss of two degrees of freedom" is used because given any $n-2$ residuals, the remaining two residuals are determined by the constraints

$$
\sum r_i \;=\; 0 \quad \text{and} \quad \sum (x_i - \bar{x}) r_i \;=\; 0.
$$

Further motivation for division by $n-2$ is: if we have only two points: the fitted regression line passes through them; the sum of squared residuals is 0; and $s^2 = 0/0$ which is an undefined quantity. This is appropriate because we have no information about the variability of the errors.

The formal reason for defining s^2 with a denominator $n-2$ is that it gives an unbiased estimator of σ^2. That is, if s^2 is imagined to be averaged over repeated experiments it would equal σ^2. This is proven on the website, but a brief justification is that, on average, division by $n-2$ compensates for the fact that $\sum r_i^2$ must be slightly less than the sum of squared errors (Figure 8.3).

If the regression is a useful model, the standard deviation of the residuals, s, should be less than standard deviation s_y of the $\{y_i\}$. The relationship between s^2 and s_y^2 is that

$$
s^2 \;=\; s_y^2 (1 - r^2) \frac{n-1}{n-2},
$$

where r is the correlation coefficient (Exercise 8.7).

Example 8.1:   (Continued) Erosion shields [estimation of sd of errors]

For the tungsten shield data s^2 is 1759, so s is approximately 42 on $25 - 2 = 23$ degrees of freedom. The explicit calculation in R the follows

```
> SSR<-sum(shields.lm$res^2)
> s2<-SSR/(length(VH)-2)
> print(c(SSR,s2))
[1]  40450.610  1758.722
```

However, it is already available in the output from the R command `lm()` which refers to s as

```
Residual standard error: 41.94 on 23 degrees of freedom
```

We can find s as a single number, that we can use in subsequent calculations, in R by:

```
> summary(shields.lm)$sigma
[1]  41.93712
```

8.1.3.8   Checking assumptions A3, A4 and A5

We should now check that assumptions **A3**, **A4**, and **A5** are plausible. The reasonableness of assumptions of constant variance (**A4**) and normality (**A5**) can be assessed from the scatterplot and fitted line, Figure 8.1, and in this case they do seem reasonable. However, a plot of the residuals against the fitted values together with a normal qq plot enables a more rigorous assessment, and will be necessary in Chapter 9 when we have more than one predictor variable. If we have the time order, or some other relevant order for the observations, we can check the assumption that the errors are independent (**A3**) by plotting the acf of the residuals. These plots can be produced quickly in R since the fitted values and residuals are available in the `lm()` object. The acf is based on the order of observations given in Table 8.1, which is the order in which the VH measurements were made. We have added a histogram of the residuals to make up four panels:

```
> par(mfrow=c(2,2))
> plot(shields.lm$fit,shields.lm$res)
> hist(shields.lm$res)
> qqnorm(shields.lm$res)
> acf(shields.lm$res)
```

There is no evidence against any of assumptions **A3**, **A4** or **A5**. We again emphasize that **A1** and **A2** cannot be checked from the data.

The plot of the residuals against the fitted values (the pattern is the same if we plot residuals against x because the fitted values are a linear function of x), in the top left panel of Figure 8.4, gives us an opportunity to reassess the assumption of a linear relation between x and the mean value of Y. It may be easier to discern evidence of curvature in this plot than it is in the original plot of y against x. If there is evidence of curvature we can, for example, consider fitting a quadratic term in x (Chapter 9) or using logarithms of the data (Section 8.7 and Exercise 8.20).

8.1.4   Properties of the estimators

The regression is a model for the conditional distribution of the random variable Y given x. The x_i are fixed values, known from the investigation. To avoid excessive notation we rely on the context to distinguish the estimator from the estimate.

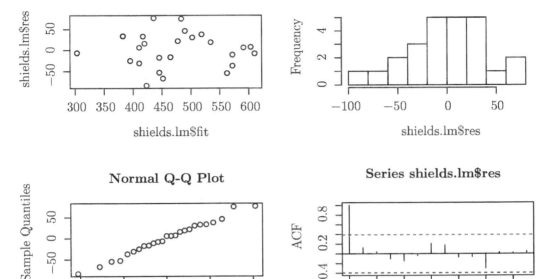

FIGURE 8.4: Residual plots for the Vickers data.

8.1.4.1 Estimator of the slope

Following the argument of Section 8.1.3.2 the estimator $\widehat{\beta}$ can be expressed in the form

$$\widehat{\beta} = \frac{\sum(x_i - \bar{x})Y_i}{S_{xx}},$$

which shows clearly that it is a weighted sum of the Y_i, the weights being $(x_i - \bar{x})/S_{xx}$. Taking expectations, and assuming **A2** (errors uncorrelated with x_i) holds, we obtain

$$\mathrm{E}\left[\widehat{\beta}\right] = \frac{\sum(x_i - \bar{x})\mathrm{E}[Y_i]}{S_{xx}}.$$

Now, assuming **A1** (errors have mean 0) holds

$$\mathrm{E}\left[\widehat{\beta}\right] = \frac{\sum(x_i - \bar{x})(\alpha + \beta x_i)}{S_{xx}} = \frac{0 + \beta S_{xx}}{S_{xx}} = \beta.$$

The variance of $\widehat{\beta}$ follows from the result for the variance of a linear function of random variables. If we assume **A3** and **A4** (errors are independent with the same variance) hold, then

$$\mathrm{var}\left(\widehat{\beta}\right) = \frac{\sum(x_i - \bar{x})^2\mathrm{var}(Y_i)}{S_{xx}^2} = \sigma^2 \frac{S_{xx}}{S_{xx}^2} = \frac{\sigma^2}{S_{xx}}.$$

Finally if **A5** (errors have a normal distribution) holds, then

$$\widehat{\beta} \sim N\left(\beta, \frac{\sigma^2}{\sum(x_i - \bar{x})^2}\right).$$

This seems reasonable enough: on average $\widehat{\beta}$ would equal the hypothetical β, and its variance decreases as the sample size increases because there will be more terms in $\sum (x_i - \bar{x})^2$ (Figure 8.5).

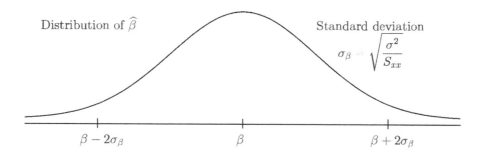

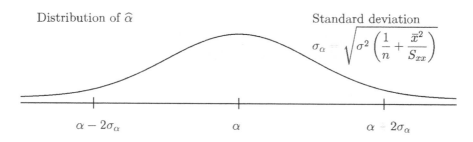

FIGURE 8.5: Normal distributions of estimators: slope upper and intercept lower.

By the usual argument, a $(1 - \alpha) \times 100\%$ (here α represents a tail probability rather than an intercept) confidence interval for β is given by

$$\widehat{\beta} \;\pm\; t_{n-2,\alpha/2}\sqrt{\frac{s^2}{\sum (x_i - \bar{x})^2}}\;.$$

If the sample size is large the distribution of $\widehat{\beta}$ is close to a normal distribution even if the distribution of the errors is not, as a consequence of the Central Limit Theorem. In Exercise 8.8 we ask you to compare this theoretical distribution of $\widehat{\beta}$ with approximations obtained using bootstrap methods, and to investigate the effect of different error distributions on the normality of the distribution of $\widehat{\beta}$ in small samples.

Example 8.1: (Continued) Erosion shields [CI for slope]

A 95% confidence interval for the slope of the regression of abrasion loss on hardness is:

$$-3.161 \;\pm\; 2.069\sqrt{(1759/14672)},$$

which gives $[-3.88, -2.44]$.

8.1.4.2 Estimator of the intercept

If the values of the predictor variable are far removed from 0, the intercept is of no practical interest. Be that as it may, the distributional result is,

$$\widehat{\alpha} \;\sim\; N\left(\alpha, \sigma^2\left(1/n + \bar{x}^2 / \sum (x_i - \bar{x})^2\right)\right),$$

which is a particular case of the general result that we prove in the next section. It is convenient to use the R `confint()` command

```
> confint(shields.lm)
                2.5 %       97.5 %
(Intercept) 2190.376840 3195.227774
VH            -3.876887   -2.444452
```

We should ignore the confidence interval for the intercept α. A Vickers hardness of 0 is physically meaningless and the approximate linear relationship has only been validated over a hardness range $[660, 760]$. If we want the estimate of the slope or its standard error for further calculations we can find them in R by:

```
> summary(shields.lm)$coef[2,]
     Estimate     Std. Error      t value      Pr(>|t|)
-3.160670e+00   3.462233e-01  -9.128991e+00  4.145057e-09
> summary(shields.lm)$coef[2,1]
[1] -3.160670e+00
> summary(shields.lm)$coef[2,2]
[1] 3.462233e-01
```

We can now explain the remaining columns under `Coefficients:` in the summary given by the command `summary(shields.lm)`. The t-value is the ratio of the estimate to its estimated standard deviation. $Pr(>|t|)$ is the probability that the absolute value of t would exceed the value obtained in the investigation if the coefficient in the corresponding population is 0. That is, $Pr(>|t|)$ is the p-value associated with a test of the hypothesis that the coefficient is 0 against a two sided alternative. If $Pr(>|t|)$ exceeds 0.05 the 95% confidence interval for β will include 0. In most engineering applications the confidence interval is more relevant than the result of a test in which the coefficient is 0.

8.1.5 Predictions

Predictions rely on the assumption that the assumed linear relationship is realistic. Unless we have information from other sources we can only check that a linear regression is a reasonable approximation within the range of x observed in the investigation. We should generally restrict predictions to values of x within this range, and if we must extrapolate it should not be far beyond the range (Exercise 8.9). The following results assume a linear relationship is valid.

8.1.5.1 Confidence interval for mean value of Y given x

Given any value of x, x_p say, the mean value of Y is $\alpha + \beta x_p$. The estimator of this conditional mean is $\widehat{\alpha} + \widehat{\beta} x_p$, which, for our present purposes, is more conveniently written in its equivalent form

$$\overline{Y} + \widehat{\beta}(x_p - \bar{x}).$$

We first show that this estimator is unbiased.

$$\mathrm{E}\left[\overline{Y} + \widehat{\beta}(x_p - \bar{x})\right] \;=\; \mathrm{E}\left[\overline{Y}\right] + (x_p - \bar{x})\mathrm{E}\left[\widehat{\beta}\right] \;=\; \alpha + \beta\bar{x} + (x_p - \bar{x})\beta \;=\; \alpha + \beta x_p.$$

We now notice that a consequence of the regression line passing through the point $(\bar{x}, \overline{Y})$ is that $\widehat{\beta}$ is independent of $\overline{Y}$ (for example, we could add any constant to all the Y_i and $\overline{Y}$ would increase by this constant whereas $\widehat{\beta}$ would be unchanged). So, the variance of the estimator of the conditional mean is

$$\mathrm{var}(\overline{Y}) + (x_p - \bar{x})^2 \mathrm{var}\left(\widehat{\beta}\right),$$

since $\overline{Y}$ and $\widehat{\beta}$ are independent. Now

$$\overline{Y} \;=\; \sum Y_i / n.$$

The variance of the Y_i, remember they are conditional on x_i, is σ^2. Furthermore, the errors are assumed to be independent (**A3**), so

$$\mathrm{var}(\overline{Y}) \;=\; n\sigma^2 / n^2 = \sigma^2 / n.$$

To summarize,

$$\widehat{\alpha} + \widehat{\beta} x_p \;\sim\; N\left(\alpha + \beta x_p, \sigma^2 \left(\frac{1}{n} + \frac{(x_p - \bar{x})^2}{\sum(x_i - \bar{x})^2}\right)\right)$$

and a $(1 - \alpha) \times 100\%$ confidence interval for the mean value of Y given that x equals x_p is

$$\bar{y} + \widehat{\beta}(x_p - \bar{x}) \;\pm\; t_{n-2,\alpha/2}\, s \sqrt{\frac{1}{n} + \frac{(x_p - \bar{x})^2}{\sum(x_i - \bar{x})^2}}.$$

A confidence interval for the intercept is obtained by putting x_p equal to 0, but it should be disregarded if 0 is well beyond the range of x values observed in the investigation.

Example 8.1: (Continued) Erosion shields [CI for mean Y given x]

A 95% confidence interval for the mean value of Y when x equals 670 is

$$476.92 - 3.1607 \times (670 - 701.08) \quad \pm \quad 2.048 \times 41.937 \times \sqrt{\frac{1}{30} + \frac{(670 - 701.08)^2}{14671.8}},$$

which gives an estimated mean of 575 with a 95% confidence interval of $[547, 603]$.

The arithmetic can be performed with the R function `predict()`.

```
> predict(shields.lm,newdata=data.frame(VH=670),
    interval=c("confidence"),level=0.95)
        fit      lwr      upr
1 575.1536 546.9303 603.377
```

The construction of the confidence intervals depends on assumptions **A1-A4**. If errors are positively correlated, or the variance of the errors is not constant, the calculated confidence intervals will be too narrow. The assumption of normality is less critical (Exercise 8.8).

8.1.5.2   Limits of prediction

The point prediction, $\widehat{Y}_p$, for Y given that x equals x_p, Y_p, is the same as the estimator of the conditional mean

$$\widehat{Y}_p \;=\; \overline{Y} + \widehat{\beta}(x_p - \overline{x}).$$

We will now write μ_p for $E[Y_p] = \alpha + \beta x_p$. Since $E\left[\widehat{Y}_p\right] = \mu_p$, the prediction error $Y_p - \widehat{Y}_p$ has mean 0. To construct limits of prediction we need the variance of the prediction error, which is:

$$
\begin{aligned}
E\left[\left(Y_p - \widehat{Y}_p\right)^2\right] &= E\left[\left((Y_p - \mu_p) + (\mu_p - \widehat{Y}_p)\right)^2\right] \\[2mm]
&= E\left[(Y_p - \mu_p)^2\right] + E\left[\left(\widehat{Y}_p - \mu_p\right)^2\right] - 2E\left[(Y_p - \mu_p)(\widehat{Y}_p - \mu_p)\right] \\[2mm]
&= \sigma^2 + \sigma^2\left(\frac{1}{n} + \frac{(x_p - \overline{x})^2}{\sum(x_i - \overline{x})^2}\right).
\end{aligned}
$$

The expectation of the product term is 0 because Y_p is independent of the estimator of its mean ($\widehat{Y}_p$). It follows that a $(1 - \alpha) \times 100\%$ prediction interval for Y given that x equals x_p is

$$\overline{y} + \widehat{\beta}(x_p - \overline{x}) \;\pm\; t_{n-2,\alpha/2}\, s\, \sqrt{1 + \frac{1}{n} + \frac{(x_p - \overline{x})^2}{\sum(x_i - \overline{x})^2}}\;.$$

This is equivalent to the confidence interval for the mean value with the addition of 1 under the square-root sign to allow for the variance of a single value of Y about its mean. If the parameters were known exactly a 95% prediction interval for Y would be:

$$\alpha + \beta x_p \;\pm\; 1.96\sigma$$

and if the sample size is reasonably large $\pm 2s$ is a useful approximation for the limits of prediction. The construction of limits of prediction does rely on assumption **A5**, that the errors are normally distributed.

Example 8.1:   (Continued) Erosion shields [PI for a single Y given x]

A 95% prediction interval for the mean value of Y when x equals 670 is

$$476.92 - 3.1607 \times (670 - 701.08) \;\pm\; 2.048 \times 41.937 \times \sqrt{1 + \frac{1}{30} + \frac{(670 - 701.08)^2}{14671.8}},$$

which gives a point prediction of 575 with a 95% prediction interval of $[484, 666]$, or more succinctly 575 ± 91.

The R function **predict()** gives

```
> predict(shields.lm,newdata=data.frame(VH=670),
    interval=c("prediction"),level=0.95)
        fit      lwr        upr
1 575.1536 483.9246 666.3826
```

8.1.5.3 Plotting confidence intervals and prediction limits

The width of a confidence interval for the mean value of Y, given x equals x_p, becomes noticeably wider as x_p moves further from $\bar{x}$. This is because of the $(x_p - \bar{x})^2$ term which reflects the increasing influence of the uncertainty in the estimate of the slope. There is a similar, but less marked, effect for limits of prediction. If the limits of prediction and confidence intervals are calculated for a few values of x they can be superimposed on the plot of the data and joined with smooth curves. This has been done for 95% intervals with the erosion shield data in Figure 8.6

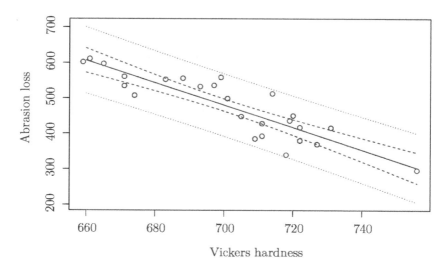

FIGURE 8.6: Abrasion loss against VH for 25 erosion shields with 95% CI for the mean (dashed line); and PI for individual shields (dotted line).

```
#abbreviate variable names
> x<-VH
> y<-loss
> s<-summary(shields.lm)$sigma
> n<-length(x)
#set length of y axis
> Uaxy<-max(y)+2.2*s
> Laxy<-min(y)-2.2*s
> Sxx<-sum((x-mean(x))^2)
> tc<-qt(0.975,n-2)
> xplot<-min(x)+(max(x)-min(x))*c(1:100)/100
> newdat<-data.frame(VH=xplot)
> yplot<-predict(shields.lm,newdata=newdat)
> Lp<-yplot-tc*s*sqrt(1+1/n+(xplot-mean(x))^2/Sxx)
> Lc<-yplot-tc*s*sqrt(  1/n+(xplot-mean(x))^2/Sxx)
> Uc<-yplot+tc*s*sqrt(  1/n+(xplot-mean(x))^2/Sxx)
> Up<-yplot+tc*s*sqrt(1+1/n+(xplot-mean(x))^2/Sxx)
> plot(x,y,ylim=c(Laxy,Uaxy),xlab="Vicker's Hardness",ylab="Abrasion Loss")
> lines(xplot,Lp,lty=3)
> lines(xplot,Lc,lty=2)
> lines(xplot,yplot)
```

```
> lines(xplot,Uc,lty=2)
> lines(xplot,Up,lty=3)
```

The upper prediction limits shown on the graph give Mike Mandrel, the engineer at SeaDragon, a value of abrasion loss that 2.5% of shields are expected to exceed for a given hardness. If a customer wanted assurance that most of the shields had abrasion losses below 700, Mike could recycle any batches with VH below 660. However, this seems rather wasteful and a better strategy would be to investigate reasons for low VH and to try to rectify the situation.

8.1.6 Summarizing the algebra

The analysis of variance (ANOVA) table is a convenient summary of the main results. The original mean-adjusted sum of squares of the y can be split into the sum of squared residuals, **residual sum of squares**, and a sum of squares that is attributed to the regression, **regression sum of squares**. The explanation follows.

$$\sum (y_i - \bar{y})^2 = \sum ((y_i - \hat{y}_i) + (\hat{y} - \bar{y}))^2$$

$$= \sum (y_i - \hat{y}_i)^2 + \sum (\hat{y}_i - \bar{y})^2 + 2 \sum (y_i - \hat{y}_i)(\hat{y}_i - \bar{y}).$$

The cross product term is equal to zero because it can be written

$$2 \sum r_i \left(\hat{\beta}(x_i - \bar{x}) \right) = 2\hat{\beta} \sum r_i (x_i - \bar{x}) = 0$$

and it follows that

$$\sum (y_i - \bar{y})^2 = \sum (\hat{y}_i - \bar{y})^2 + \sum r_i^2 .$$

That is, the mean-adjusted sum of squares of response = the regression sum of squares + the residual sum of squares.

TABLE 8.2: ANOVA.

Source of variation	Corrected sum of squares	Degrees of freedom	mean square	$E[$ mean square$]$
Regression	$\sum (\hat{y}_i - \bar{y})^2$	1	$\hat{\beta}^2 \sum (x_i - \bar{x})^2$	$\sigma^2 + \beta^2 \sum (x_i - \bar{x})^2$
Residual	$\sum r_i^2$	$n - 2$	s^2	σ^2
Total	$\sum (y_i - \bar{y})^2$	$n - 1$		

In Table 8.2, the mean square is the 'sum of squares' divided by the 'degrees of freedom'. One degree of freedom is allocated to the regression sum of squares because there was one predictor variable. The E[mean square] column includes the unknown parameters of the model and tells us, for example, that on average s^2 would equal σ^2. This final column is therefore algebraic rather than numerical and is not usually given by computer packages. The proof of the expected value of the regression mean square is straightforward, and follows.

$$E\left[\sum (\hat{y}_i - \bar{y})^2 \right] = E\left[\left(\hat{\beta} \right)^2 \sum (x_i - \bar{x})^2 \right].$$

Now from Exercise 8.10 and using the fact that $E\left[\widehat{\beta}\right] = \beta$, we write

$$E\left[\sum (\widehat{y}_i - \bar{y})^2\right] = \left(var\left(\widehat{\beta}\right) + \beta^2\right) S_{xx}$$
$$= \left(\frac{\sigma^2}{S_{xx}} + \beta^2\right) S_{xx} = \sigma^2 + \beta^2 S_{xx}.$$

If $\beta = 0$ then the regression mean square is an independent unbiased estimator of σ^2 and, given **A1-A5**, the ratio of the regression mean square to the residual mean square is distributed as $F_{1,n-2}$.

The arithmetic is handled by R.

```
> anova(shields.lm)
Analysis of Variance Table

Response: loss
          Df Sum Sq Mean Sq F value    Pr(>F)
VH         1 146569  146569  83.338 4.145e-09 ***
Residuals 23  40451    1759
---
Signif. codes:  0 *** 0.001** 0.01 * 0.05   .  0.1   1
```

The $Pr(> F)$ is the probability of such a large F value if $\beta = 0$, as given at the end of `summary(shields.lm)`. Also, as the F value is the square of the t value, $Pr(> F)$ equals $Pr(> |t|)$. In this case there is a clear and anticipated tendency for abrasion loss to reduce as VH increases and the hypothesis test is superfluous.

8.1.7   Coefficient of determination R^2

R^2 is the proportion of the variability in the data that is explained by the model. It is defined by

$$R^2 = \frac{\text{regression sum of squares}}{S_{yy}} = \frac{S_{yy} - \text{residual sum of squares}}{S_{yy}} = 1 - \frac{\sum r_i^2}{S_{yy}}.$$

The adjusted R^2 takes account of the loss of a degree of freedom that is a consequence of including the predictor variable and is defined by

$$R^2_{adj} = 1 - \frac{\sum r_i^2/(n-2)}{S_{yy}/(n-1)} = 1 - \frac{s^2}{s_y^2},$$

where s_y is the standard deviation of the $\{y_i\}$. R gives R^2 and R^2_{adj} at the end of `summary(shields.lm)` as 0.784 and 0.774 respectively.

8.2   Regression for a bivariate normal distribution

We defined conditional distributions from bivariate distributions in Chapter 6. The geometric interpretation is that the bivariate pdf $f(x, y)$ is a surface defined by $z = f(x, y)$, and the conditional distribution of Y given x is a scaling of the section of the pdf obtained by cutting it with a plane, parallel to the $y - z$ plane, that passes through x. In the case of the bivariate normal distribution the conditional distribution of Y given x, and the conditional distribution of X given y, are linear regressions satisfying all the assumptions **A1** through to **A5**.

8.2.1 The bivariate normal distribution

The bivariate normal distribution has the rather formidable looking pdf

$$f(x,y) \quad = \quad \frac{1}{2\pi\sigma_X\sigma_Y\sqrt{(1-\rho^2)}}\, e^{\Theta}, \quad \text{where}$$

$$\Theta \quad = \quad -\frac{1}{2(1-\rho^2)}\left[\left(\frac{x-\mu_X}{\sigma_X}\right)^2 - 2\rho\left(\frac{x-\mu_X}{\sigma_X}\right)\left(\frac{y-\mu_Y}{\sigma_Y}\right) + \left(\frac{y-\mu_Y}{\sigma_Y}\right)^2\right].$$

The parameters μ_X, μ_Y, σ_X and σ_Y are the means and standard deviations of the marginal distributions of X and Y. The parameter ρ is the correlation coefficient between X and Y, and therefore constrained to be between -1 and 1. The easiest way to find the conditional distribution of y on x is to use the standardized distribution which has means of 0 and standard deviations of 1. The general result then follows from a simple scaling argument. Suppose that (X, Y) has a bivariate normal distribution. Then if W and Z are defined by

$$W \quad = \quad (X - \mu_X)/\sigma_X \quad \text{and} \quad Z \quad = \quad (Y - \mu_Y)/\sigma_Y.$$

(W, Z) has a standardized bivariate normal distribution and the pdf is

$$f(w,z) \quad = \quad \frac{1}{2\pi\sqrt{(1-\rho^2)}}\exp\left\{\frac{-1}{2(1-\rho^2)}(w^2 - 2\rho wz + z^2)\right\}.$$

The marginal distributions are standard normal. To verify this, write $(w^2 - 2\rho wz + z^2)$ as $[(z-\rho w)^2 + (1-\rho^2)w^2]$ then substitute θ for $(z - \rho w)$ and integrate with respect to θ to obtain

$$f(w) \quad = \quad \int f(w,z)dz \quad = \quad \frac{1}{\sqrt{2\pi}}\exp(-\frac{1}{2}w^2).$$

Some straightforward algebra leads to the conditional distribution,

$$f(z|w) = \frac{f(w,z)}{f(w)} \quad = \quad \frac{1}{\sqrt{2\pi}\sqrt{(1-\rho^2)}}\exp\left\{\frac{-1}{2(1-\rho^2)}(z-\rho w)^2\right\}.$$

This is a normal distribution with mean, $\mathrm{E}[Z|w]$, equal to ρw and a variance of $(1-\rho^2)$. The regression line of z on w is

$$z \quad = \quad \rho w.$$

This result can be rescaled so it is explicitly in terms of Y and x. First, use the relationship between Z and Y to write

$$\mathrm{E}\left[\frac{Y-\mu_Y}{\sigma_Y}\middle| W = w_p\right] \quad = \quad \rho w_p.$$

If W is w_p then X equals $\mu_X + \sigma_X w_p$, and we define x_p by $\mu_X + \sigma_X w_p$. Then

$$\mathrm{E}\left[\frac{Y-\mu_Y}{\sigma_Y}\middle| X = x_p\right] \quad = \quad \rho\frac{x_p - \mu_X}{\sigma_X}$$

and finally

$$E[Y \mid X = x_p] \quad = \quad \mu_Y + \rho\frac{\sigma_Y}{\sigma_X}(x_p - \mu_X).$$

Furthermore, if the conditional distribution of Z given w has a variance of $(1 - \rho^2)$, the conditional distribution of Y given x has a variance $(1 - \rho^2)\sigma_Y^2$. Notice this variance does not depend on the value of x. The regression line of Y on x is

$$y \quad = \quad \mu_Y + \rho\frac{\sigma_Y}{\sigma_X}(x - \mu_X).$$

An identical argument leads to the regression of w on z:

$$w \quad = \quad \rho z$$

and the rescaled regression of X on y which is:

$$x \quad = \quad \mu_X + \rho\frac{\sigma_X}{\sigma_Y}(y - \mu_Y).$$

The two regression lines are not the same. Refer to Figure 8.7 which shows contours for, and a section of, the standardized bivariate normal distribution. Since $|\rho| < 1$, the regression line of z on w is less steep than the major axis of the elliptical contours, whereas the regression line of w on z is steeper.

The regression of z on w crosses contours at the points where their tangents are parallel to the z-axis. This is because $f(z|w_p)$ is a scaled section of the joint pdf $f(w, z)$ cut through w_p parallel to the plane containing the z-axis and the perpendicular axis. It is also a normal distribution so its mean coincides with its highest point. The highest point above the line through w_p parallel to the z-axis is where the line is tangent to a contour. The regression line consists of all these highest points. Although the diagram was drawn with a positive ρ, similar arguments hold for negative values. The fact that the regression lines predict values closer to the mean than the major axis of the ellipses is known as **regression towards the mean**.

8.3   Regression towards the mean

Imagine you are a male with one sister. Your sister can run 100 m in a time that is 3 standard deviations below the average. Will you be equally fast? Possibly, and you may be faster, but it is more likely that you will be slower. This is because factors other than genetic inheritance, which is itself highly variable, affect sprinting ability. These factors include the amount of training, diet, and enthusiasm. Now imagine the population of all brother and sister pairs in the U.S. Next, focus on the conditional distribution of all pairs for which the sister runs about 3 standard deviations faster than the average. The mean of this conditional distribution will be nearer to the mean for all brothers than 3 standard deviations (of the distribution of times for all brothers) because the other factors are unlikely to be as extreme as 3 standard deviations from their means. This phenomenon is a natural geometric property of a bivariate probability distribution, and of general multivariate distributions, and is known as **regression towards the mean**.

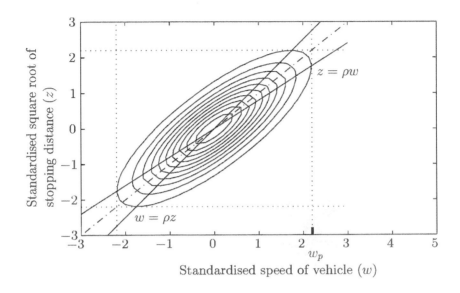

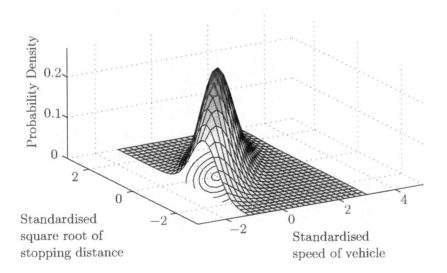

FIGURE 8.7: Contours of bivariate normal pdf (tangents to the contours shown as dotted lines) and regression lines, followed by a perspective plot.

Example 8.2: Mining engineering [regression to the mean]

Mining engineers and geologists are aware that gold (Au) and arsenic (As) tend to occur together. Suppose an assay of a sample of ore has an As content that is 3 standard deviations of As content above the mean As content of all such samples. We should not expect the Au content of the sample to be as far from the mean Au content of all such samples (in terms of standard deviations of Au content). Although the Au content of the sample could be as far or further from the mean Au content of all such samples, it is more likely to be closer to that mean.

Example 8.3: VIVDS [bivariate normal distribution]

A video image vehicle detection system (VIVDS) is set up to record speed and stopping distance for cars that brake when approaching a light controlled intersection, in response to the change from green light to orange, rather than continue over the intersection. The onset of braking is indicated by the car brake light. We consider the data as a random sample from the population of cars that brake on the green to orange light change. Published tables of stopping distances for a given speed have stopping distances increasing with the square of the speed, and we will suppose that the bivariate distribution of speed and square root of stopping distance is normal (Figure 8.7). For a bivariate normal distribution, if x, speed in this example, is σ_X above μ_X the mean value of Y, square root of stopping distance in this example, conditional on x is $|\rho|\,\sigma_Y$ rather than σ_Y.

Regression towards the mean is a feature of bivariate distributions in general. For example, suppose a car driven at two standard deviations above the population mean speed brakes as the lights change from green to orange. The stopping distance (as well as its square root) will not, on average, be as extreme as two standard deviations of stopping distance above the mean stopping distance. This is because other factors such as driver reaction time, standard of maintenance and potential efficiency of brakes are unlikely to be as far from their means as two standard deviations.

8.4   Relationship between correlation and regression

The errors in the regression model can include both inherent variation in the population and measurement errors. So, the measurements of the response Y can include measurement error. In contrast, it is assumed that the values of the predictor variable x are known with negligible error.

If we have a designed experiment in which we choose the values for x, a regression of Y on x will be appropriate, provided the assumptions are satisfied, but we should not consider a regression of x on Y. However, if we have an observational study, or a survey, in which we have a random sample of experimental units on which we record two variables X and Y, with negligible measurement error, we can consider regressing either variable on the other or a correlation analysis.

The measured abrasion loss of shields for turbine blades would vary even if they were all the same hardness. This variation is predominantly due to other differences in the test pieces, that is inherent variation in the population being sampled, as measurement of weight loss is quite precise. The VH measurement on the test coupons is also reasonably precise. So, we could consider a regression of abrasion loss on VH, a regression of VH on abrasion loss, or construct a confidence interval for the correlation coefficient between VH and abrasion loss. However, the relevant analysis for the application is that of abrasion loss on VH.

8.4.1 Values of x are assumed to be measured without error and can be preselected

The model for a linear regression of Y on x is

$$Y_i = \alpha + \beta x_i + \epsilon_i \,,$$

with ϵ_i having a zero mean and being independent of the x_i and each other. The errors, ϵ_i, may represent inherent variation or measurement error of the response, or both. In a designed experiment, the values of x are chosen in advance and there are considerable advantages to be gained by doing so. The x-values should be chosen to cover the range of values over which predictions may be required, and an even spacing of x-values over this range will allow us to assess whether a straight-line relationship is plausible (although this choice will not give the most efficient estimator of β (see Exercise 8.10)). Although the values of x may be chosen by the investigator, the experimental material should be randomly allocated to these values. Also, if the tests are made consecutively their order should be randomized, if it is practical to do so, to help justify the assumptions that the errors are independent of each other (**A3**).

8.4.2 The data pairs are assumed to be a random sample from a bivariate normal distribution

Assume that (X, Y) have a bivariate normal distribution and are measured without error. If a random sample from this distribution is available, either the regression of Y on x or the regression of X on y can be estimated. The choice depends on which variable the investigator wishes to predict.

An alternative to either of the regressions is a correlation analysis. This is appropriate if a measure of the association between the variable is required rather than an equation for making predictions. In a correlation analysis both X and Y are treated as random variables, whereas the regression of Y on x estimates the conditional distribution of Y for given x, specific values of which happen to have arisen at random. The following approximate distribution of the sample correlation coefficient (r), due to [Fisher, 1921], can be used to construct confidence intervals for ρ.

First construct a confidence interval for $\operatorname{arctanh}(\rho)$ using the normal approximation:

$$\operatorname{arctanh}(r) \sim N\left(\operatorname{arctanh}(\rho), 1/(n-3)\right).$$

If we are 95% confident that $\operatorname{arctanh}(\rho)$ is between L and U, we are equally confident that ρ is between $\tanh(L)$ and $\tanh(U)$.

The arctanh function, which is the inverse hyperbolic tangent function, can be expressed in terms of logarithms as

$$\operatorname{arctanh}(r) = \frac{1}{2}\ln\left(\frac{1+r}{1-r}\right).$$

The arctanh function, `atanh()` in both R and MATLAB, stretches the $[-1, 1]$ scale to $(-\infty, +\infty)$ as shown in Figure 8.8.

Example 8.4: Spot analyses [correlation influenced by outliers]

The data in Moonlight.txt are 159 multi-trace element spot analyses, using laser ablation inductively coupled plasma mass spectroscopy, of pyrite from the Moonlight epithermal gold prospect in Queensland [Winderbaum et al., 2012]. The concentrations (ppm) of 27 trace elements were measured with each spot analysis. Here we concentrate on As and Au.

The concentrations of trace elements in spot analyses vary considerably and tend to have highly positively skewed distributions. A consequence is that correlation coefficients calculated from samples of spot analyses can be dominated by outlying points, and it is a usual practice to use the common logarithms of concentration, with some constant added, in statistical analyses. After trying a few values for the additive constant, the histograms of $\log_{10}(As + 10\,000)$ and $\log_{10}(Au + 1)$ were as near to a normal distribution shape as can be achieved by this transform. Even if a perfect transformation could be found, variables with normal marginal distributions need not have a bivariate normal distribution. However, the transformed pairs are certainly closer to a random sample from a bivariate distribution than the untransformed pairs (Figure 8.8). The correlation coefficient between As and Au is 0.669 and it is reduced to 0.600 for

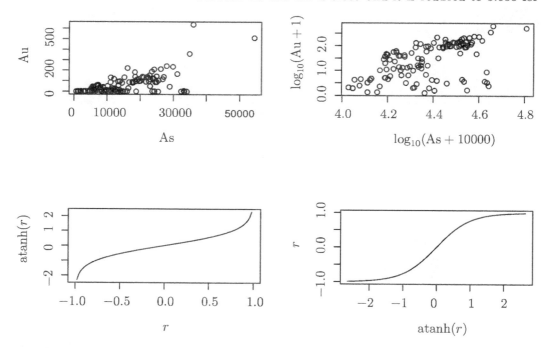

FIGURE 8.8: Upper left: Scatter plot of arsenic and gold pairs. Upper right: Scatter plot of transformed pairs. Lower left: arctanh(r) against r. Lower right: r against arctanh(r).

the transformed pairs. Using Fisher's construction, a 95% confidence interval for ρ in the corresponding population of transformed pairs is given by:

```
> r<-cor(log10(As+10000),log10(Au+1))
```

```
> n<-length(Au)
> L<-atanh(r)-qnorm(0.975)/sqrt(n-3)
> U<-atanh(r)+qnorm(0.975)/sqrt(n-3)
> print(tanh(L))
[1] 0.4898012
> print(tanh(U))
[1] 0.6908835
```

which is [0.49, 0.69] .

The two regression lines, drawn with the R code below, are shown in Figure 8.9. The regression towards the mean is apparent.

```
> elem.dat <- read.table("Moonlight.txt",header=T)
> head(elem.dat[,1:8])
     Au     Ag       As      Sb      Ti    V    Cr      Mn
1 12.04  65.25  8410.19 113.91 447.80 5.39 2.68 176.07
2 15.40  97.95 11465.80  78.31 130.09 0.86 0.00   0.94
3  9.61  31.30  9664.02  37.75  10.76 0.07 0.00   0.56
4 12.11  19.63  9315.57 149.18  18.39 0.25 0.00   9.87
5  1.07 119.84   693.04 381.16  43.71 0.45 2.28  21.79
6  5.40   8.65  9514.22 100.47 371.75 1.73 0.00  14.20
> attach(elem.dat)
> x<-log10(As+10000)
> y<-log10(Au+1)
> yonx.lm<-lm(y~x)
> xony.lm<-lm(x~y)
> plot(x,y,xlab="log10(As+10000)",ylab="log10(Au+1)")
> abline(lm(y~x),lty=2)
> abline(a=-xony.lm$coef[1]/xony.lm$coef[2],b=1/xony.lm$coef[2],lty=3)
> abline("v"=mean(x))
> abline("h"=mean(y))
```

8.5 Fitting a linear relationship when both variables are measured with error

In some cases we wish to estimate a linear relationship between two variables, when our measurements of both variables are subject to measurement error.

A researcher compared land and aerial survey methods. Let (X_i, Y_i) represent measurements of elevation at n locations, to be made by land and aerial survey, respectively. It is assumed that measurements are subject to independent errors, ϵ_i and ζ_i, respectively, with zero mean and constant variances σ_ϵ^2 and σ_ζ^2. Let u_i and v_i represent the unobservable error-free measurements. It is supposed that

$$v_i = \alpha + \beta u_i$$

for some α and β, and one of the main aims of the project is to see whether there is any

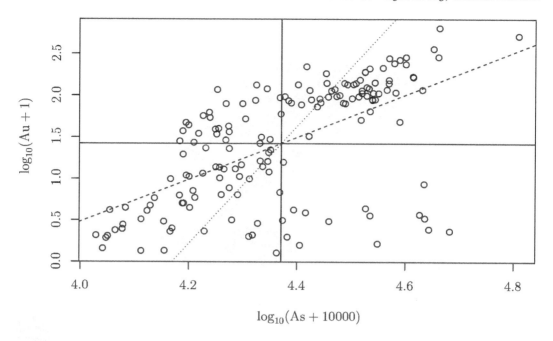

$$\log_{10}(\text{As} + 10000)$$

FIGURE 8.9: $(\log_{10}(As + 10\,000), \log_{10}(Au + 1))$ pairs. Regression of $\log_{10}(Au + 1)$ on $\log_{10}(As + 10\,000)$ (dashed); regression of $\log_{10}(As + 10\,000)$ on $\log_{10}(Au + 1)$ (dotted); lines through mean values (solid).

evidence that these parameters differ from 0 and 1, respectively.
Substituting

$$u_i = X_i - \epsilon_i \quad \text{and} \quad v_i = Y_i - \zeta_i$$

into the model relating v_i and u_i gives

$$Y_i = \alpha + \beta X_i + (\zeta_i - \beta \epsilon_i).$$

Despite first appearances, this is not the standard regression model because the assumption that the errors are independent of the predictor variable is not satisfied:

$$\begin{aligned}\text{cov}(X_i, (\zeta_i - \beta \epsilon_i)) &= \text{E}[(X_i - u_i)(\zeta_i - \beta \epsilon_i)] \\ &= \text{E}[\epsilon_i(\zeta_i - \beta \epsilon_i)] = -\beta \sigma_\epsilon^2.\end{aligned}$$

Maximum likelihood solution:
The usual approach to this problem is to assume the errors are normally distributed and that the ratio of σ_ζ^2 to σ_ϵ^2, denoted by λ in the following, is known. In practice, some reasonable value has to be postulated for λ, preferably based on information from replicate measurements. The (maximum likelihood) estimates of α and β are:

$$\widehat{\beta} = \left((S_{yy} - \lambda S_{xx}) + \left[(S_{yy} - \lambda S_{xx})^2 + 4\lambda S_{xy}^2\right]^{1/2}\right)\Big/(2S_{xy})$$

$$\widehat{\alpha} = \bar{y} - \widehat{\beta}\bar{x},$$

where S_{xy} is a shorthand for $\sum(x - \bar{x})(y - \bar{y})$, etc. These estimates follow from the likelihood function L (that is, the joint pdf treated as a function of the unknown parameters), which is proportional to

$$\sigma_\epsilon^{-1}\sigma_\zeta^{-1}\exp\left\{-\frac{1}{2}\sigma_\epsilon^{-2}\sum_{i=1}^{n}(x_i - u_i)^2 - \frac{1}{2}\sigma_\zeta^{-2}\sum_{i=1}^{n}(y_i - (\alpha + \beta u_i))^2\right\},$$

if $\alpha, \beta, \sigma_\epsilon$ and u_i are treated as the unknown parameters.
Estimates of σ_ϵ^2 and σ_ζ^2 are given by

$$\alpha_\zeta^2 = \lambda\widehat{\alpha}_\epsilon^2 = (S_{yy} - \widehat{\beta}S_{xy})/(n-2).$$

If $\lambda = 1$ the values of $\widehat{\alpha}$ and $\widehat{\beta}$ are the slope and intercept of the line such that the sum of squared perpendicular distances from the plotted points to the line is a minimum.
Upper and lower points of the $(1 - \alpha) \times 100\%$ confidence interval for β are given by

$$\lambda^{1/2}\tan(\arctan)(\widehat{\beta}\lambda^{-1/2}) \pm \frac{1}{2}\arcsin(2t_{\alpha/2}\theta)),$$

where

$$\theta^2 = \frac{\lambda(S_{xx}S_{yy} - S_{xy}^2)}{(n-2)((S_{yy} - \lambda S_{xx})^2 + 4\lambda S_{xy}^2)}$$

and $t_{a/2}$ is the upper $(a/2) \times 100\%$ point of the t-distribution with $n-2$ degrees of freedom. An approximation to the standard deviation of $\widehat{\beta}$ is given by one quarter of the width of such a 95% interval. Since $\widehat{\beta}$ is independent of $\bar{x}$

$$\text{var}(\widehat{\alpha}) \simeq \widehat{\sigma}_\zeta^2/n + \bar{x}^2\text{var}\left(\widehat{\beta}\right) + \widehat{\beta}^2\widehat{\sigma}_\epsilon^2/n.$$

The use of these formulae is demonstrated in the following example.

Example 8.5: Elevations above sea level [measurement error model]

The data pairs in Table 8.3, from the Department of Geomatics at the University of Newcastle upon Tyne, are the measured heights (in meters) above sea level of 25 points from a land survey (x) and an aerial survey (y). The points were equally spaced over a hilly area of 10 km by 10 km. Replicate measurements of the height of a single point suggest that errors in the aerial survey measurements have a standard deviation three times that of errors in the land survey. Calculations give

$$\bar{x} = 780.6, \quad \bar{y} = 793.0$$
$$S_{xx} = 177970, \quad S_{xy} = 179559 \quad \text{and} \quad S_{yy} = 181280,$$

where the ratio λ is assumed to be 9.
The formulae of this section lead to the following estimates

$$\widehat{\alpha} = 5.379, \quad \widehat{\beta} = 1.00899$$
$$\widehat{\sigma}_E = 0.716 \quad \text{and} \quad \widehat{\sigma}_H = 2.147$$

and the 95% confidence intervals for α and β are

$$[-7.6, \ 18.3] \quad \text{and} \quad [0.993, \ 1.025]$$

respectively. There is no evidence of any systematic difference between the results of the two surveys because the confidence intervals for α and β include 0 and 1 respectively.

TABLE 8.3: Heights in meters of 25 points determined by land and aerial surveys.

Land survey estimate	Aerial survey estimate	Land survey estimate	Aerial survey estimate
720.2	732.9	829.4	845.1
789.5	804.9	808.7	820.3
749.7	760.5	781.7	796.1
701.5	712.3	868.7	885.2
689.2	702.0	904.2	920.1
800.5	812.8	780.7	790.2
891.2	902.7	649.6	660.0
812.8	820.0	732.1	741.2
780.6	793.6	770.4	781.2
710.5	720.2	733.3	745.6
810.4	825.6	694.9	707.3
995.0	1 008.6	620.0	633.4
890.5	902.4		

8.6 Calibration lines

In the U.S. all states impose penalties for driving with a blood alcohol content (BAC) greater than 0.08 (percentage ethanol in the blood), and some have lower limits. Breathalyzer tests are generally used to screen drivers, and the amount of alcohol measured on the breath (BrAC) is generally accepted to be proportional to BAC in the ratio of 1 : 2100. In most jurisdictions court proceedings are based on BAC measurements following a driver's high BrAc reading. A police chief in Colorado, which imposes a limit of 0.05, decided to calibrate BrAC measurements from the police department's new issue of breathalyzer instruments against BAC.

A random sample of 21 men was taken from a pool of male volunteers in the department. The men were randomly assigned to one of 21 drinks that contained different measures of alcohol. After one hour each volunteer provided a blood sample and took the breathalyzer test.

The blood sample analysis (x) provided a measurement of BAC (milligrams of ethanol per 100 ml of blood) and the breathalyzer provided a measurement (y) of BrAC (micrograms of ethanol per liter). The measured BrAC for men with the same BAC varies considerably. We consider a measurement error model in which y is considered as a measurement of BAC with substantial error and x is considered a measurement of BAC with negligible error. Since the measurement error in x is negligible the measurement error model reduces to a linear regression of Y on x.

$$Y_i = \alpha + \beta x_i + \epsilon_i,$$

However, the objective is not to predict y from x. The purpose of fitting the model is to estimate α and β in the linear relationship

$$y = \alpha + \beta x$$

and hence provide a calibration

$$x = \frac{y - \widehat{\alpha}}{\widehat{\beta}}.$$

The calibration coefficients are not exactly unbiased because although $\widehat{\beta}$ is unbiased for β, $1/\widehat{\beta}$ is not unbiased for $1/\beta$. This is of little practical significance if the coefficient of variation of $\widehat{\beta}$ is small (see Exercise 8.15).

Example 8.6: Breathalyzer test [calibration line]

Twenty one male volunteers were randomly allocated one of 21 drinks containing from 0 to 20 units of 5.5 ml of 80 proof liquor. After one hour, they took the breathalyzer test and provided a blood sample. The results are given in Table 8.4 and the data are plotted in Figure 8.10

TABLE 8.4: Blood alcohol levels measured from blood samples and breathalyzer readings for 21 adult male volunteers.

Blood alcohol (x, mg alcohol per 100 ml blood)	Breathalyzer reading (y)	Blood alcohol (x, mg alcohol per 100 ml blood)	Breathalyzer reading (y)
0	64	52	248
1	114	54	321
6	98	58	325
13	27	63	352
21	153	70	389
23	160	72	377
30	76	77	403
37	268	78	320
36	230	89	466
38	216	93	483
45	154		

An R script to plot the data, fit the regression, and plot the calibration line is given below.

```
BA.dat=read.table("BloodAlcohol.txt",header=T)
BAC=BA.dat$BAC;BrAC=BA.dat$BrAC
M1=lm(BrAC~BAC)
par(mfrow=c(1,2))
plot(BAC,BrAC)
abline(M1)
calBA=(BrAC-M1$coef[1])/M1$coef[2]
plot(BrAC,calBA,ylab="Calibrated BAC",type="l")
```

The results of the regression are:

```
> summary(M1)

Call:
lm(formula = BrAC ~ BAC)

Residuals:
     Min      1Q   Median      3Q     Max
-104.375  -0.108   17.623  24.226  63.157
```

```
Coefficients:
            Estimate Std. Error t value Pr(>|t|)
(Intercept)  46.3768    20.3861   2.275   0.0347 *
BAC           4.4666     0.3819  11.695    4e-10 ***
---
Signif. codes:  0 *** 0.001 ** 0.01 * 0.05 . 0.1  1

Residual standard error: 48.78 on 19 degrees of freedom
Multiple R-squared: 0.878,      Adjusted R-squared: 0.8716
F-statistic: 136.8 on 1 and 19 DF,  p-value: 3.998e-10
```

The coefficients for the calibration line are

```
> 1/M1$coef[2]
0.223883
> -M1$coef[1]/M1$coef[2]
  -10.38298
```

The fitted calibration line is

$$\text{BAC} = -10.38 + 0.2239y.$$

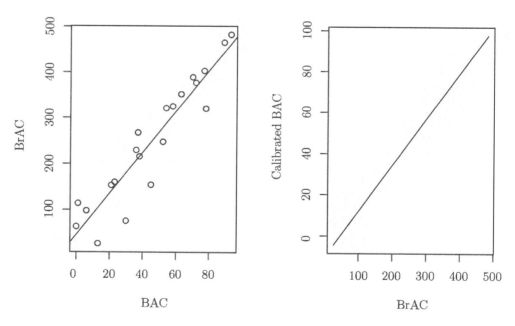

FIGURE 8.10: Breathalyzer reading plotted against blood alcohol determined by a laboratory test for 21 adult male volunteers.

8.7 Intrinsically linear models

Some non-linear relationships between two variables can be transformed to a linear relationship by taking a transform of one or other of the variables. For example, tests on steel specimens subject to fluctuating loading have indicated that the number of cycles to failure (N) is approximately inversely proportional to the stress range (S, equal to maximum stress less minimum stress) raised to some power between 3 and 4. That is,

$$N = kS^{-m}.$$

Appropriate values of k and m depend on the type and thickness of steel, and are obtained experimentally. The proposed nonlinear relationship can be transformed to a straight line by taking logarithms. Then

$$\ln(N) = \ln(k) - m\ln(S)$$

and if x and y are defined as $\ln(S)$ and $\ln(N)$, respectively, standard regression techniques can be used. If it is assumed that

$$\ln(N_i) = \ln(k) - m\ln(S_i) + \epsilon_i,$$

where the ϵ, satisfy the usual assumptions. Then

$$N_i = kS_i^{-m}\varphi_i ,$$

where φ_i has a lognormal distribution with a median value of 1.

TABLE 8.5: Amplitude of fluctuating stress and cycles to failure for 15 steel specimens.

Amplitude of stress (Nm^{-2})	Cycles to failure (x 1000)
500	20
450	19
400	19
350	40
300	48
250	112
200	183
150	496
100	1 883
90	2 750
80	2 181
70	3 111
60	9 158
50	15 520
40	47 188

Example 8.7: Fluctuating stress [intrinsically linear model]

Fifteen steel specimens with the same geometry were prepared and allocated to fluctuating loads of different amplitudes. The number of cycles to failure were recorded. The data are given in Table 8.5 and a plot of $\ln(S)$ against $\ln(N)$ is shown in Figure 8.11.

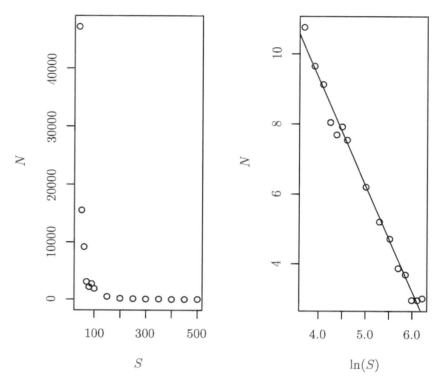

FIGURE 8.11: Natural logarithm of cycles to failure against natural logarithm of amplitude of fluctuating stress [predictions] for 15 steel specimens.

A linear relationship between the log variables is plausible and the fitted regression line is

$$\ln(N) = 28.4474 - 3.0582 \ln(S).$$

The estimates of k and m, for this steel in this geometry, are 2.26 x 10^{12} and 3.06, respectively.

The standard errors of the estimators of $\ln(k)$ and m are estimated as 0.4641 and 0.0914 respectively. A 95% confidence interval for $\ln(k)$ is given by

$$28.4474 \quad \pm \quad t_{13,0.025} 0.4641,$$

which gives [27.44, 29.45]. A 95% confidence interval for k is given by exponential of this interval, $[8.30 \times 10^{11}, \ 6.17 \times 10^{12}]$.

In this case the objective of the investigation is to estimate the parameters k and m. However, there is a snag if we use a regression of a transformed variable on some predictor variable to predict the mean of that variable. The snag is that the inverse transform of the mean of the transformed variable is not the mean of the (untransformed) variable. A way around this is to multiply the inverse transform of the mean of the transformed variable by an adjustment factor. If the errors in the transformed regression are normal then the adjustment factor, which is the ratio of the mean to the median in a lognormal distribution, is $exp(s^2/2)$. However, this adjustment is sensitive to the assumption of normality. An alternative adjustment factor is the ratio of the mean of the variable to the mean of the inverse transforms of the fitted values, in the regression of the transformed variable. The second adjustment factor is plausible and does not rely on a normality assumption, but it should not be used without checking that it is reasonable in the specific context (see Exercise 8.17). No adjustment is needed to compensate for transforming the predictor variable.

No adjustment is made to limits of prediction. This is because if g is some one-to-one transform of a random variable X and g^{-1} is the inverse transform:

$$P(L < g(X) < U) \;=\; P\big(g^{-1}(L) < g^{-1}(g(X)) < g^{-1}(U)\big) \;=\; P\big(g^{-1}(L) < X < g^{-1}(U)\big).$$

Example 8.7: (Continued) Fluctuating stress

A specimen of the same steel is subjected to a stress amplitude of 120.

Estimate:

(i) The expected value (mean value) of the natural logarithm of cycles to failure.

(ii) Exponential of the expected value of the natural logarithm of cycles to failure.

(iii) The expected value of the number of cycles to failure.

(iv) A 95% confidence interval for the mean number of cycles to failure.

(v) A 95% prediction interval for the mean number of cycles to failure.

An R script and the results follow.

```
AmpSt.dat<-read.table("AmplitudeStress.txt",header=T)
print(head(AmpSt.dat))
N=1000*AmpSt.dat$C;S=AmpSt.dat$S
y=log(N);x=log(S)
M1=lm(y~x)
print(summary(M1))
par(mfrow=c(1,2))
plot(S,N)
plot(x,y,xlab="ln(S)",ylab="ln(N)")
abline(M1)
newdata=data.frame(x=log(120))
PI=predict(M1,newdata,interval=c("prediction"),level=0.95)
print(PI)
#95% PI for N
print(round(exp(c(PI[2],PI[3]))))
CI=predict(M1,newdata,interval=c("confidenc"),level=0.95)
print(CI)
```

Statistics in Engineering, Second Edition

```
AF1=exp(summary(M1)$sigma^2/2)
AF2=mean(N)/mean(exp(M1$fit))
print(AF1)
print(CI[1])
print(exp(CI[1]))
print(AF1*exp(CI[1]))
#95% CI for N
print(round(AF1*exp(c(CI[2],CI[3]))))
> source("C:\\AndrewBook\\DATA\\AmplitudeStress.r")
     S   C
1 500  20
2 450  19
3 400  19
4 350  40
5 300  48
6 250 112


Call:
lm(formula = y ~ x)

Residuals:
    Min       1Q    Median       3Q      Max
-0.50401 -0.17574  0.06484  0.09631  0.50374

Coefficients:
            Estimate Std. Error t value Pr(>|t|)
(Intercept)  28.4474     0.4641   61.30  < 2e-16 ***
x            -3.0583     0.0914  -33.46 5.34e-14 ***
---
Signif. codes:  0 *** 0.001 ** 0.01 * 0.05 . 0.1     1

Residual standard error: 0.2941 on 13 degrees of freedom
Multiple R-squared: 0.9885,     Adjusted R-squared: 0.9876
F-statistic:  1120 on 1 and 13 DF,  p-value: 5.343e-14

        fit      lwr      upr
1 13.80608 13.14838 14.46377
[1]   513181 1912210
        fit      lwr      upr
1 13.80608 13.63628 13.97588
[1] 1.044201
[1] 13.80608
[1] 990610.4
[1] 1034396
[1]   872859 1225829
```

(i) The expected value of the natural logarithm of cycles to failure is 13.806.

(ii) Exponential of the expected value of the natural logarithm of cycles to failure is 990 610.

(iii) The expected value of the number of cycles to failure, using the adjustment factor based on the lognormal distribution, is 1 034 396.

(iv) A 95% confidence interval for the mean number of cycles to failure, using the first adjustment factor, in $1\,000$ s is $[873, 1\,226]$.

(v) A 95% prediction interval for the mean number of cycles to failure in $1\,000$ s is $[513, 1\,912]$.

8.8 Summary

8.8.1 Notation

n sample size
x_i/y_i observations of predictor/response variables
S_{xx} mean-adjusted sums of squares of predictor
S_{xy} mean-adjusted sums of products
S_{yy} mean-adjusted sums of squares of response
$\bar{x}/\bar{y}$ mean of observations of predictor/response variables
$\alpha/\widehat{\alpha}$ intercept for regression in the population/sample
$\beta/\widehat{\beta}$ coefficient of predictor variable in the population/sample
$\widehat{y}_i$ fitted values of response variable
ϵ_i/r_i errors/residuals
ρ/r correlation coefficient in population/sample
σ/s standard deviation of errors/ residuals

8.8.2 Summary of main results

The standard regression model is

$$Y_i \;=\; \alpha + \beta x_i + \epsilon_i \,,$$

where the $\epsilon_i \sim N(0, \sigma^2)$ and are independent of the x_i and each other.

Estimating the parameters:

$$S_{xy} \;=\; \sum (x_i - \bar{x})(y_i - \bar{y}), \quad \text{with} \quad S_{xx}, S_{yy} \text{ defined similarly and}$$

$$\widehat{\beta} \;=\; \frac{S_{xy}}{S_{xx}}, \quad \widehat{\alpha} \;=\; \bar{y} - \widehat{\beta}\bar{x} \quad s^2 \;=\; \frac{S_{yy} - S_{xy}^2/S_{xx}}{n-2}.$$

The estimated regression line is

$$y \;=\; \widehat{\alpha} + \widehat{\beta}x = \bar{y} + \widehat{\beta}(x - \bar{x}) \;=\; \bar{y} + r\frac{s_y}{s_x}(x - \bar{x}),$$

where r is the correlation coefficient . Expressing in terms of standardized variables

$$\frac{y - \bar{y}}{s_y} = r\frac{x - \bar{x}}{s_x}.$$

Residuals:

The residuals r_i, are the differences between the y_i and their fitted values $\widehat{y}_i$. That is,

$$\widehat{y}_i \;=\; \widehat{\alpha} + \widehat{\beta}x_i \quad \text{and} \quad r_i \;=\; y_i - \widehat{y}_i \,.$$

They are estimates of the values taken by the errors. The computational formula for s^2 is algebraically equivalent to $\sum r_i^2/(n-2)$, and

$$s^2 \;=\; s_y^2\left(1-r^2\right)\frac{n-1}{n-2}\;.$$

Confidence interval for slope:
$A(1-\alpha) \times 100\%$ confidence interval for β is given by

$$\widehat{\beta} \;\pm\; t_{n-2,\alpha/2}\; s\; \sqrt{s^2/S_{xx}}\;.$$

Confidence interval for mean of Y given x:
If $x = x_p$, a $(1-\varepsilon) \times 100\%$ confidence interval for the mean value of Y is

$$\widehat{\alpha}+\widehat{\beta}x_p \;\pm\; t_{n-2,\varepsilon/2}\; s\; \sqrt{1/n+(x_p-\bar{x})^2/S_{xx}}\;.$$

Limits of prediction for a single value of Y given x:
If $x = x_p$, $(1-\varepsilon) \times 100\%$ limits of prediction for a single value of Y are

$$\widehat{\alpha}+\widehat{\beta}x_p \;\pm\; t_{n-2,\varepsilon/2}\; s\; \sqrt{1+1/n+(x_p-\bar{x})^2/S_{xx}}\;.$$

Bivariate normal distribution:
A regression analysis assumes that the errors in measuring the x are negligible. Apart from that restriction, the x can be selected in advance or arise as a result of sampling some bivariate distribution. The bivariate normal distribution is relatively easy to work with, and, even if it is not directly appropriate, it is often plausible for some transformation of variables such as the natural logarithm. If we have a random sample from a bivariate normal distribution we can regress Y on x, regress X on y, or calculate the correlation coefficient r as an estimate of the population correlation coefficient ρ. A $(1-\alpha) \times 100\%$ confidence interval for ρ is given by

$$\tanh\left[\operatorname{arctanh}(r) \pm z_{a/2}/\sqrt{n-3}\right]\;.$$

Measurement error models:
Applicable when $v = \alpha + \beta u$ and the objective is to estimate α and β from measurements of u and v that include errors.

Intrinsically linear models:
Some non-linear relationships can be transformed to linear relationships.

8.8.3   MATLAB and R commands

In the following x is a column vector which contains observations of the predictor variable(s) and y is a column vector which contains the corresponding observations of the response variable. For more information on any built in function, type **help(function)** in R or **help function** in MATLAB.

R command	MATLAB command
lm(y~x)	lm = fitlm(x,y)
plot(x,y)	scatter(x,y)
coef(lm(y~x))	lm.Coefficients(:,1)
fitted(lm(y~x))	lm.Fitted
residuals(lm(y~x))	lm.Residuals

Note that a second MATLAB command **regstats** can be used instead of **fitlm**. It does not include all the features that **fitlm** has but can be easier to use for beginners as it includes a user interface (a dialog box).

8.9 Exercises

Section 8.1 Linear regression

Exercise 8.1: Regression on a single variable

Consider n data pairs (x_i, y_i).
Find a, b in terms of x_i, y_i and n such that

$$y_i = a + bx_i + r_i, \quad \sum r_i = 0, \quad \text{and} \sum x_i r_i = 0.$$

Comment on your results in the context of the regression model of Y on x.

Exercise 8.2: Baseflow and peakflow

The following pairs are baseflow (x) and peakflow (y) of the River Browney for 8 storms during 1983 in chronological order. The baseflow (x) is the minimum flow in the river before the storms. The units are $m^3 s^{-1}$ (cumecs).

x	2.03	2.35	4.14	1.27	2.52	0.74	0.46	0.31
y	23.90	31.55	17.27	30.97	38.82	3.21	1.42	1.58

(a) Plot the data in R or MATLAB.

(b) Write down the model for a regression of y on x and state all the usual assumptions.

(c) Calculate the linear regression of y on x.

(d) Estimate the values of y when x equals 0.30 and 4.20. Hence draw the regression line on your graph.

(e) Construct a 95% confidence interval for the slope of the regression in the corresponding population.

(f) Construct a 95% confidence interval for the standard deviation of the error in the model.

(g) Which of the usual assumptions can be checked by investigating the residuals, and what checks would you carry out?

Exercise 8.3: Sebacic acid concentration and viscosity

The following data are molar ratio of sebacic acid (x) and intrinsic viscosity of copolyester (y) for 8 minutes of a chemical process. [example from [Hsiue et al., 1995]]

x	1.0	0.9	0.8	0.7	0.6	0.5	0.4	0.3
y	0.45	0.20	0.34	0.58	0.70	0.57	0.55	0.44

(a) Plot the data, x horizontally and y vertically.
(b) Is random of Y variation about a straight line a plausible model for the data?
(c) Fit a regression of y on x and show the line on your graph.
(d) Calculate the ratio of s to s_y where s is the estimated standard deviation of the errors and s_y is the standard deviation of viscosity.
(e) Is their evidence to reject a null hypothesis that the slope is 0 against a two sided alternative that it is not 0 at a 0.10 level of significance.

Exercise 8.4: Pitch and encounter frequency

[Calisal et al., 1997] investigated the relationship between pitch (y, non-dimensional) and encounter frequency (x, non-dimensional) for model boats in a wave tank. Results of 13 tests with a wave height to model length ratio of 1/35 and a Froude number of 0.200 follow.

x	2.22	2.40	2.40	2.46	2.51	2.61	2.68	2.77	2.88	2.97	3.08	3.23	3.38
y	1.11	1.13	1.07	1.15	1.09	1.03	1.06	0.96	1.01	0.94	0.94	0.66	0.62

(a) Plot the data, x horizontally and y vertically.
(b) Does a linear relationship seem reasonable?
(c) Fit a regression of y on x.
(d) Use your regression in (c) to predict pitch for an encounter frequency of 2.6, and give approximate 95% limits for your prediction.
(e) Fit two regression lines, one for x from 2.22 up to 2.61 and the other for x from 2.68 up to 3.38.
(f) Is there evidence that either the slopes or the intercepts of the two lines differ?
(g) Make predictions of pitch for an encounter frequency of 2.6, and give approximate 95% limits for your prediction, using both the regressions in (e).

Exercise 8.5: Concrete permeability and strength

strength	31	45	34	25	22	12	53	48
permeability	330	310	349	356	361	390	301	312

strength	24	35	13	30	33	32	18
permeability	357	319	373	338	328	316	377

The following data are crushing strength (y, MPa) and permeability (x, mdarcy) of 15 concrete test cubes.

(a) Plot the data, x horizontally and y vertically.

(b) Fit the regression of y on x.

(c) Show the fitted regression line on your plot.

(d) Add curves showing 95% confidence intervals for $E[Y]$ and 95% prediction limits for Y over the range of porosity considered in the experiment.

Exercise 8.6: Slope of the regression equation

Suppose

$$y_i = 4 + 8x_i + \epsilon_i,$$

where x ranges from 0 up to 10 in steps of 0.5, so i runs from 1 up to 21.

(a) If $\epsilon_i \sim iid(0, 1)$ what are the expected values of $\widehat{\alpha}$ and $\widehat{\beta}$.

(b) Now suppose that ϵ_i are correlated with x_i so that

$$E[(x - 5)\epsilon_i] = 0.2$$

and that

$$E[\epsilon | x = 5] = 0.$$

The variance of ϵ_i is 1.

(i) What is the variance of x (use a denominator n as the mean is known to be 5)?

(ii) What is the correlation between x and ϵ?

(iii) What is the expected value of $\widehat{\alpha}$ and $\widehat{\beta}$?

Exercise 8.7: Correlation coefficient

Express s^2 in terms of S_{yy}, S_{xy}, S_{xx} and $n - 2$. Hence verify that

$$s^2 = s_y^2(1 - r^2)\frac{n - 1}{n - 2},$$

Exercise 8.8: Bootstrap

Consider the 13 VH abrasion loss pairs in Table 8.1. Investigate the bootstrap sampling distribution using two procedures

(i) Resample the n pairs with replacement.

(ii) Resample the residuals with replacement, to obtain r_i^b, and then take the bootstrap sample as $(x_i, \hat{y}_i + r_i^b)$ for $i = 1, \ldots, n$.

(a) Fit a regression. Find the sampling distribution of the slope by simulation assuming errors normal with standard deviation equal to that of the residuals.

(b) As for (a) but assume the errors have a Laplace distribution with standrad deviation equal to that of the residuals.

Exercise 8.9: Extrapolation

The film thickness $(x, 0.1 \text{ mm})$ and bond strength (y, N) for five test joints are:

x	16	15	10	12	11
y	44.7	43.0	51.0	48.1	48.6

(a) Plot the data with the x-axis ranging from 0 up to 20.

(b) Fit a regression of y on x, and show the line on your graph.

(c) Use your fitted line to predict the bond strength when $x = 0$. Is this prediction sensible?

(d) Sketch a possible relationship between x and y for $0 \le x \le 10$.

(e) Use your fitted line to predict the bond strength when $x = 20$.

(f) Construct a 95% prediction interval for bond strength when $x = 20$.

(g) What sources of uncertainty does the prediction interval allow for, and what source of uncertainty does it not allow for?

Exercise 8.10: Beta estimator

Assume

$$Y_i = \alpha + \beta x_i + \varepsilon_i,$$

where $\varepsilon_i \sim \text{iid}(0, \sigma^2)$. Compare the variance of the estimators of the slope, β, if

(a) $x = 1$ repeated five times and $x = 10$ repeated five times.

(b) x runs from 1 to 10 in steps of 1.

(c) $x = 1$ repeated four times and $x = 10$ repeated four times and $x = 5.5$ twice.

(d) What are the limitations of the design in (a)?

Section 8.2 Regression for a bivariate normal distribution

Exercise 8.11: Rio Parana flood peaks and volumes

Flood peaks $(x, 10^3 m^3/s)$ on the Rio Parana at Posadas, exceeding 35 during the period 1905-1993, and the associated volumes $(y, 10^9 m^3)$ [Adamson (1994)].

Flood peak (x)	37	38	41	36	38	48	54	50
Flood volume (y)	50	110	115	270	270	230	500	770

(a) Plot the data, x horizontally and y vertically. Does a linear relationship seem plausible?

(b) Assume the data are from a bivariate normal distribution and construct a 95% confidence interval for the correlation coefficient (ρ).

(c) Assume the logarithms of the data are from a bivariate normal distribution and construct a 95% confidence interval for the correlation coefficient (ρ).

Section 8.3 Regression towards the mean

Exercise 8.12: Two regressions 1

Consider the Rio Parana flood peaks and volumes (Exercise 8.11)

(a) Plot the data, x horizontally and y vertically.

(b) Fit a regression of y on x and show the line on your plot. Calculate a 95% confidence interval for the slope β.

(c) Fit a regression of x on y and show the line on your graph.

Section 8.4 Relationship between correlation and regression

Exercise 8.13: Two regressions 2

Suppose (X, Y) has a bivariate normal distribution. The means of X and Y are 1 and -1 respectively and the standard deviations are 2 and 5 respectively. The correlation between X and Y is -0.8.

(a) Write down the regression equations for Y on x and for X on y.

(b) Show the two lines on a sketch.

Section 8.5 Fitting a linear relationship when both variables are measured with error

Exercise 8.14: Measurement error (errors-in-variables) model

The income (x, euros) and ratio of actual to normative consumption for space heating (y, %) for ten percentiles of income of French households [Cayla et al., 2011] is:

x	7368	12588	16272	20263	25175	30088	34693	41140	62632
y	54	67	61	71	77	82	81	80	100

(a) Plot the data.

(b) Fit a regression of y on x and show as a broken line on your graph.

(c) Assume that both x and y are measured subject to error and that the standard deviation of the errors for x and the standard deviation of the errors for y are equal. Show the line on your graph.

Section 8.6 Calibration lines

Exercise 8.15: Calibration

A prototype drone expert system has been developed to measure the leaf area index (LAI) in vineyards. Ten vineyards agreed to take part in a test. The LAI as determined by experienced grape growers were

$$2.37, 1.62, 0.53, 0.87, 1.20, 2.13, 3.41, 0.94, 1.47, 2.81$$

The corresponding estimates from the drone were

$$2.56,\ 1.19,\ 1.44,\ 2.66,\ 2.20,\ 3.24,\ 3.13,\ 1.96,\ 1.52,\ 4.39$$

Compare the root mean squared errors obtained by predicting with a regression of LAI on drone, and by predicting using a formula based on the fitted regression of drone on LAI.

Exercise 8.16: Breathalyzer

Consider the calibration of the breathalyzer (Example 8.6)

(a) (i) What value of y corresponds to a BAC of 50?

    (ii) If you regress BAC on y, what is the value of y such that the point $(y, 50)$ lies on the line?

(b) The police chief is considering a policy of requiring male motorists with a breathalyzer reading above 250 to take a follow-up breath test on a truck mounted breath analyzer.

    (i) Estimate the probability that a man with a BAC of 80 will have a breathalyzer result below 250.

    (ii) Before administering a breathalyzer test, an experienced police officer considers that a male driver has the probability distribution of BAC shown below.

BAC	50	60	70	80
probability	0.1	0.1	0.3	0.5

The man's breathalyzer score is 250. What is the revised probability distribution?

Section 8.7 Intrinsically linear models

Exercise 8.17: Fluctuating stress

Consider the fifteen steel specimens subjected to fluctuating loads (Example 8.7). The R code includes the calculations of two adjustment factors. The first, based on the lognormal distribution (AF1) and the second (AF2), calculated as the ratio of the mean of the original data to the mean of the inverse transform of fitted variables.

(a) Why are AF1 and AF2 so different in this case?

(b) Do you agree that AF1 seems more appropriate?

Exercise 8.18: Michaelis-Menten formula

The Michaelis-Menten formula relates the rate of an enzymatic reaction v to the concentration of substrate $[S]$ by

$$v = \frac{m[S]}{k + [S]},$$

where m, k are constants. If data $([S]_i, v_i)$ are available:

(a) Explain how you might estimate m, k using linear regression.

(b) What are you assuming about the distribution of errors?

(c) What is the physical interpretation of m, k?

Exercise 8.19: Voltage and insulator breakdown

Times to dielectric breakdown of mylar-polyurethane insulation (minutes) at different voltages. [excerpt from [Kalkanis and Rosso, 1989]]

Voltage $(x, KV/mm)$	100.3	100.3	122.4	122.4	157.1	157.1	219	219	361.4	361.4
Lifetime $(z, minutes)$	1012	2612	297	744	99	180	16	50	0.33	0.50

(a) Plot

  (i) t against x.

  (ii) $\ln(t)$ against x.

  (iii) $\ln(t)$ against $\ln(x)$.

(b) In each case fit a straight line through the data. Which set of data is best described by a linear relationship?

(c) Use the model you choose in (b) to

  (i) Construct an approximate 95% confidence interval for the mean lifetime at a voltage of 361.4.

  (ii) Construct a 95% prediction interval for the lifetime of a single test specimen tested at a voltage of 100.3 and at a voltage of 361.4.

Exercise 8.20: Curvature

The following data [Hayter, 2012] are bulk stress (x, pounds per square inch) and resilient modulus (y, pounds per square inch 10^{-2}) for 16 samples of an aggregate made at different values of bulk stress.

x	8	11	12	16	21	25	30	35
y	99	99	117	129	139	138	146	156

x	38	41	47	51	55	61	68	75
y	159	168	176	172	187	177	190	195

(a) Plot y against x.

(b) Plot $\ln(y)$ against $\ln(x)$.

(c) Fit the model

$$y = kx^c$$

by fitting a linear regression to $(\ln(x), \ln(y))$.

(d) Superimpose the fitted model on your graph in (a).

(e) Calculate the differences between the observed y and the fitted values using the model in (c). For these differences calculate

Statistics in Engineering, Second Edition

(i) the mean,

(ii) standard deviation and

(iii) root mean squared error.

(f) Compare the root mean squared error $\big(e(iii)\big)$ with

(or we could simulate data around the model Hayter fitted.)

Miscellaneous

Exercise 8.21: Methods of Moments

Compare the method of moments estimator of the mean, and the rate parameter, of an exponential distribution, with an estimator based on the quantile-quantile plot. For a random sample of size n, the method of moments estimators of the mean and rate are $\overline{X}$ and $\dfrac{1}{\overline{X}}$ respectively. The graphical estimator of the mean (Section 5.3.5) is the gradient of a regression line fitted to the pairs $\left(-\ln\left(\dfrac{n+1-i}{n+1}\right), X_{i:n}\right)$. The graphical estimator of the rate is the reciprocal of the estimator of the mean.

(a) Take $K = 10^4$ random samples of size $n = 10$ from an exponential distribution with rate $\lambda = 1$. Compare the estimators in terms of their estimated bias and standard error.

(b) As for (a) with $n = 30$.

Exercise 8.22: Predicting costs

The following data are sizes (x, thousand properties) and out-turn costs of meeting European Community standards (y, monetary units) for 8 water supply zones.

size	1.0	2.3	4.5	5.1	6.7	6.8	7.2	9.3
cost	11	4	41	36	45	87	80	81

(a) Plot the data. Fit a regression of y on x and show the line on your graph. Predict the cost for a zone of 8 400 properties. Construct a 90% prediction for the cost.

(b) Take $w = \ln(y)$. Plot the data.

(c) Fit a regression of w on x and show the line on your graph.

(d) Predict the cost for a zone of 8 400 properties, as $\exp(\widehat{w}) + s^2/2$.

(e) Construct a 90% prediction for the cost.

(f) Fit the model

$$Y_i = bx_i + \epsilon_i,$$

where ϵ_i are iid$(0, \sigma^2)$, by least squares.

(g) State, with reasons, which of the models is the more plausible.

9

Multiple regression

Multiple regression models a response as a linear combination of several predictor variables. The predictor variables, and the response, can be non-linear functions of observations so the model is very versatile. We show how interactions, quadratic terms, categorical variables, sinusoidal functions, and past values of the response can be used as predictor variables. Apart from prediction, regression models can be used as empirical evidence to support, or suggest, explanations of relationships between predictor variables and the response.

There are many variants on the standard multiple regression model. We show how to fit models that are not linear in the unknown coefficients using the principle of least squares. We consider two models for a discrete response, logistic regression and Poisson regression, which are examples of the generalized linear model. The generalized least squares algorithm is included as an exercise. See relevant example in Appendix E.

Experiment E.6 *Company efficiency, resources and teamwork.*

9.1   Introduction

In Chapter 8 we modeled a response on a single predictor variable, we now extend the ideas to modeling a response on several predictor variables. We will begin with a case study[1] of a company, *Conch Communications*, that retails a wide range of mobile phones.

Example 9.1:   Conch [regression of response on two predictor variables]

Conch has 10 stores and the company CEO wants to expand by buying one of two stores that have been offered for sale: a smaller store in a busy high street and a larger store in a quieter street. The larger premises would allow a better display, but the number of walk-in customers is likely to be less in the quieter street. The operating costs would be approximately the same for the two stores, because rates are higher in the high street off-setting any savings in running a smaller store. The CEO is restricted to a single purchase and has to choose one of the stores on the basis of higher expected sales. Neither store has been used for selling mobile phones, the smaller had been a music store and the larger had been a stationery store, so the CEO cannot obtain relevant sales data. However, the CEO does have sales information for the 10 shops that *Conch* already owns, together with the mean numbers of pedestrians per hour passing down the street during shopping hours (pedestrians), and the floor areas (area) (Table 9.1). The aim is to fit a model for sales based on pedestrians and areas for these 10 shops, and to use this model to predict sales for the two options using their known pedestrians and areas.

---

[1]The *Conch* case is based on a consultancy by John Turcan at the University of Glasgow.

TABLE 9.1: Pedestrian traffic, area and sales for 10 *Conch* stores.

Pedestrians/hour	Floor area (square meters)	Sales (monetary units)
564	650	980
1072	700	1160
326	450	800
1172	500	1130
798	550	1040
584	650	1000
280	675	740
970	750	1250
802	625	1080
650	500	876

A simple model that allows for random variation about a linear relationships between sales and pedestrians and area has the form

$$Sales \;=\; \beta_0 + \beta_1 \times pedestrians + \beta_2 \times area + error.$$

In the next section we consider a general model that includes this model for sales on two predictor variables, and regression on a single predictor variable, as special cases.

9.2 Multivariate data

There are several variables defined for each member of a population, and a sample of size n is obtained from the population. The aim of a multiple regression analysis is to fit a model which can be used to predict one variable, which we call the **response** and denote by Y, from functions of the others, which we call the **predictor variables**. The response Y is a continuous variable, but the predictor variables can be continuous or discrete.

To fit the model, a random sample of size n is obtained from the population. The data are then of the form

$$(x_{1i}, \ldots, x_{mi}, y_i) \quad \text{for} \quad i = 1, \ldots, n.$$

Each datum relates to an item i in the sample. The datum has $m + 1$ components that are the value of the response y_i and the values of m variables which we think might affect the response, or be associated with the response. These variables are the **covariates** and are denoted by $x_{1i}, \ldots, x_{mi}$. It follows that each datum consists of $m + 1$ elements, for some positive integer m. We then fit a model in which y, or some function of y, is a linear combination of predictor variables which can be some or all of the covariates, and functions of the covariates, plus random variation. We can use the model to predict values of the response for items when we know only the values of the covariates.

Example 9.1: (Continued) Conch [data]

In the *Conch* case the notional population is all small shops suitable as retail outlets for mobile phones. The shops that the company now owns are a sample of size 10. The data have the form

$$(x_{1i}, x_{2i}, y_i) \quad \text{for} \quad i = 1, \ldots, 10,$$

where x_{1i} is pedestrian, x_{2i} is area, and y_i is sales. The data are shown in Table 9.1.

9.3 Multiple regression model

Here we define the multiple regression model, which is formally known as **the linear model** because it is linear in the unknown coefficients. The predictor variables can be nonlinear function of the covariates.

9.3.1 The linear model

The response is denoted by Y and k predictor variables[2] are denoted by $x_1, \ldots, x_k$. We model n observations by

$$Y_i = \beta_0 + \beta_1 x_1 + \ldots + \beta_k x_k + \varepsilon_i \quad i = 1, \ldots, n,$$

where β_j are unknown **coefficients** and the ε_i are errors. The assumptions about the errors are identical to those made for regression on a single predictor variable.

A1: The mean value of the errors is zero.

A2: The errors are uncorrelated with the predictor variables.

A3: The errors are independent of each other.

A4: The errors have the same variance σ^2.

A5: The errors are normally distributed.

The estimators of the coefficients are unbiased provided assumptions **A1** and **A2** hold, although these two assumptions cannot be checked from the data. This multiple regression model is known as the linear model because it is linear in the coefficients β_j, $(j = 0, \ldots, k)$. The model can be succinctly written in matrix terms and you will find that this makes the mathematical development remarkably neat.

---

[2]In general k will not be the same as m. For example x_{1i} might be x_i with $x_{2i} = x_i^2$.

9.3.2   Random vectors

Definition 9.1: Random vector

A **random vector** (an $n \times 1$ matrix) is a vector of random variables. For example,

$$\boldsymbol{Y} = \begin{pmatrix} Y_1 \\ Y_2 \\ \vdots \\ Y_n \end{pmatrix}.$$

Definition 9.2: mean vector

For the random vector $\boldsymbol{Y}$ we define the **mean vector, $\boldsymbol{\mu}$,** by

$$\boldsymbol{\mu} \;=\; \mathrm{E}[\boldsymbol{Y}] \;=\; \begin{pmatrix} \mathrm{E}[Y_1] \\ \mathrm{E}[Y_2] \\ \vdots \\ \mathrm{E}[Y_n] \end{pmatrix} \;=\; \begin{pmatrix} \mu_1 \\ \mu_2 \\ \vdots \\ \mu_n \end{pmatrix}.$$

Definition 9.3: Variance-covariance matrix

The **variance-covariance matrix**, which has size $n \times n$, is defined by

$$\mathrm{var}(\boldsymbol{Y}) \;=\; \mathrm{E}[(\boldsymbol{Y}-\boldsymbol{\mu})(\boldsymbol{Y}-\boldsymbol{\mu})'] \;=\; \Sigma \;=\; [\sigma_{ij}],$$

where

$$\sigma_{ij} \;=\; \begin{cases} \mathrm{cov}(Y_i, Y_j) & \text{for } i \neq j, \\ \mathrm{var}(Y_i) & \text{for } i = j. \end{cases}$$

This definition follows directly from the definitions of matrix multiplication, variance, and covariance (Exercise 9.2).

9.3.2.1   Linear transformations of a random vector

Suppose $\boldsymbol{Y}$ is a random vector of size $n \times 1$ with $\mathrm{E}[\boldsymbol{Y}] = \boldsymbol{\mu}$ and $\mathrm{var}(\boldsymbol{Y}) = \Sigma$ and let A and $\boldsymbol{b}$ be constant matrices of sizes $m \times n$ and $m \times 1$ respectively[3]. Then

$$\mathrm{E}[A\boldsymbol{Y} + \boldsymbol{b}] \;=\; A\boldsymbol{\mu} + \boldsymbol{b}$$

$$\mathrm{var}(A\boldsymbol{Y} + \boldsymbol{b}) \;=\; A\,\Sigma\,A'.$$

You are asked to prove this for the case of $\boldsymbol{Y}$ having two components in Exercise 9.3. the result includes the result for the variance of a linear combination of random variables, which we considered in Chapter 6.

---

[3] A common mathematical convention is to use lower case letters in bold type face for $n \times 1$ or $1 \times n$ matrices, considered as vectors, lower case symbols for 1×1 matrices, which are scalars, and upper case symbols for matrices of other sizes.

9.3.2.2 Multivariate normal distribution

We will use the notation,

$$\boldsymbol{Y} \ \sim \ N_n(\boldsymbol{\mu}, \Sigma)$$

to indicate that the n-dimensional random vector $\boldsymbol{Y}$ has the n-dimensional **multivariate normal distribution** with mean vector $\boldsymbol{\mu}$ and variance-covariance matrix Σ. The pdf of the multivariate normal distribution is:

$$f(\boldsymbol{Y}) \ = \ (2\pi)^{-n/2}|\Sigma|^{-1/2}e^{-\frac{1}{2}(\boldsymbol{Y}-\boldsymbol{\mu})'\Sigma^{-1}(\boldsymbol{Y}-\boldsymbol{\mu})},$$

provided that Σ has an inverse[4]. Since a linear combination of normal random variables is also normal, we have the following result.

If $\boldsymbol{Y} \sim N_n(\boldsymbol{\mu}, \Sigma)$ and A, of size $m \times n$, and $\boldsymbol{b}$, of size $m \times 1$, are fixed constant matrices then

$$A\boldsymbol{Y} + \boldsymbol{b} \ \sim \ N_m\left(A\boldsymbol{\mu} + \boldsymbol{b}, A\Sigma A'\right).$$

All marginal and conditional distributions arising from a multivariate normal distribution are also normal.

9.3.3 Matrix formulation of the linear model

Let

$$\boldsymbol{Y} \ = \ \begin{pmatrix} Y_1 \\ Y_2 \\ \vdots \\ Y_n \end{pmatrix}, \ X \ = \ \begin{pmatrix} 1 & x_{11} & x_{12} & \cdots & x_{1k} \\ 1 & x_{21} & x_{22} & \cdots & x_{2k} \\ \vdots & & & & \vdots \\ 1 & x_{n1} & x_{n2} & \cdots & x_{nk} \end{pmatrix}, \ \boldsymbol{\beta} \ = \ \begin{pmatrix} \beta_0 \\ \beta_1 \\ \vdots \\ \beta_k \end{pmatrix} \ \text{and} \ \boldsymbol{\varepsilon} \ = \ \begin{pmatrix} \varepsilon_1 \\ \varepsilon_2 \\ \vdots \\ \varepsilon_n \end{pmatrix}.$$

Then the multiple regression model is

$$\boldsymbol{Y} \ = \ X\boldsymbol{\beta} + \boldsymbol{\varepsilon},$$

with

$$E(\boldsymbol{\varepsilon}) \ = \ \boldsymbol{0} \quad \text{and} \quad \text{var}(\boldsymbol{\varepsilon}) \ = \ \sigma^2 I.$$

where I is the $n \times n$ identity matrix. The additional assumption of normality is expressed as

$$\boldsymbol{\varepsilon} \ \sim \ N_n(\boldsymbol{0}, \sigma^2 I).$$

The matrix $\boldsymbol{\beta}$ contains the coefficients to be estimated. The leading column of $1s$ in the matrix X gives the β_0, when pre-multiplying $\boldsymbol{\beta}$, and the matrix X has size $n \times (k+1)$.

9.3.4 Geometrical interpretation

The equation

$$y \ = \ \beta_0 + \beta_1 x_1 + \beta_2 x_2$$

---

[4] A variance-covariance matrix must be positive semi-definite and if it has an inverse it is positive definite.

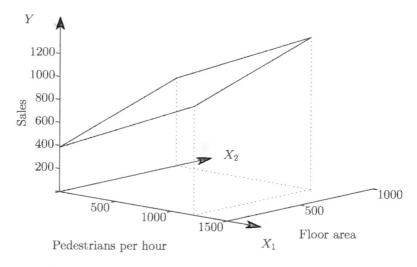

FIGURE 9.1: Equation $y = \beta_0 + \beta_1 x_1 + \beta_2 x_2$ defines a plane.

defines a plane in 3D, usually described with respect to a right handed coordinate system with the y corresponding to the vertical axis. Figure 9.1 shows a plane corresponding to the equation

$$Sales \;=\; 386 + 0.46 \times pedestrians + 0.47 \times area.$$

When we fit a model of the form

$$Y_i \;=\; \beta_0 + \beta_1 x_{1i} + \beta_2 x_{2i} + \varepsilon_i$$

using the principle of least squares the fitted regression equation represents the plane, in the volume containing the data points, such that the sum of squared vertical (parallel to the y-axis) distances from the points to the plane is minimized (Figure 9.2).

It is sometimes useful to refer to a hyperplane when we have more than two predictor variables but the pictorial representation is lost[5].

9.4   Fitting the model

Having obtained the data and defined the model, we now fit the model.

9.4.1   Principle of least squares

We use the principle of least squares, as we did for regression on a single predictor variable, to obtain estimators of the coefficients. The sum of squared errors is

$$\Psi \;=\; \sum_{i=1}^{n} \varepsilon_i^2 \;=\; \sum_{i=1}^{n} \left(Y_i - (\beta_0 + \beta_1 x_1 + \ldots + \beta_k) \right)^2.$$

---

[5]Unless the additional predictor variables are functions of the first two predictors, see, for example, quadratic surfaces in Section 9.7.4.

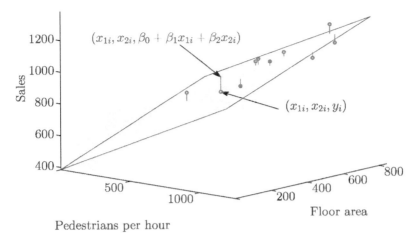

FIGURE 9.2: The fitted regression plane is such that the sum of squared vertical distances from the points to the plane is a minimum. The solid circles represent the data, and the other ends of the vertical lines touch the plane.

In matrix terms

$$\Psi \;=\; \varepsilon'\varepsilon \;=\; (Y - X\beta)'(Y - X\beta)'.$$

We consider Ψ as a scalar function of $\beta_0, \ldots, \beta_k$, that is a function of β, and we wish to minimize Ψ with respect to β. We need just three basic results from multivariate calculus[6].

9.4.2 Multivariate calculus - Three basic results

Rule 1: A necessary condition for a function $\psi(\beta)$ to have a minimum with respect to β is that the partial derivatives of ψ with respect to the β_j for $j = 0, \ldots, k$ all equal 0 (see Figure 9.3). With the following definition

$$\frac{\partial \psi}{\partial \beta} \;=\; \begin{pmatrix} \frac{\partial \psi}{\partial \beta_0} \\ \vdots \\ \frac{\partial \psi}{\partial \beta_k} \end{pmatrix},$$

a necessary requirement for a minimum, or in general any stationary point, is that

$$\frac{\partial \psi}{\partial \beta} \;=\; \mathbf{0}.$$

We now need two more results which follow directly from the matrix definitions and usual calculus results for differentiating linear and quadratic terms.

Rule 2: If c is a constant matrix of size $1 \times (k+1)$ then

$$\frac{\partial (c\beta)}{\partial \beta} \;=\; c'.$$

---

[6]The equations for the least squares estimates are identical to those obtained by imposing constraints on the residuals in the manner of Exercise 8.7. (see Exercise 9.17.)

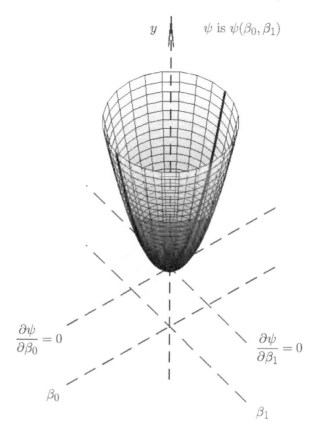

FIGURE 9.3: A function ψ of β_0 and β_1 has a minimum at coordinates $(0,0)$. The surface intersects the plane containing the y and β_0 axes in a curve (shown as bold) and $\dfrac{\partial \psi}{\partial \beta_0}$ is the gradient of a tangent to this curve. At a minimum this gradient must be 0. Similarly $\dfrac{\partial \psi}{\partial \beta_1}$ must be 0.

Convince yourself that this does follow directly from the definitions by completing Exercise 9.4(a).

Rule 3: If M is a constant matrix of size $(k+1) \times (k+1)$ then

$$\frac{\partial(\boldsymbol{\beta}'M\boldsymbol{\beta})}{\partial \boldsymbol{\beta}} = M\boldsymbol{\beta} + M'\boldsymbol{\beta}$$

Convince yourself that this also follows directly from the definitions by completing Exercise 9.4(b).

9.4.3   The least squares estimator of the coefficients

The sum of squared errors

$$\Psi \;=\; \boldsymbol{\varepsilon}'\boldsymbol{\varepsilon} \;=\; (\boldsymbol{Y} - X\boldsymbol{\beta})'(\boldsymbol{Y} - X\boldsymbol{\beta}) \;=\; \boldsymbol{Y}'\boldsymbol{Y} - 2\boldsymbol{Y}'X\boldsymbol{\beta} + \boldsymbol{\beta}'X'X\boldsymbol{\beta}$$

using the matrix transposition identity that $(AB)' = B'A'$. Now differentiate with respect to $\boldsymbol{\beta}$.

$$\frac{\partial \Psi}{\partial \boldsymbol{\beta}} = -2X'\boldsymbol{Y} + 2X'X\boldsymbol{\beta},$$

where we have used the fact that $X'X$ is symmetric, which follows from $(X'X)' = X'(X')' = X'X$. Set the derivative equal to $\mathbf{0}$ to obtain the least squares estimator

$$\widehat{\boldsymbol{\beta}} = (X'X)^{-1}X'\boldsymbol{Y},$$

provided $|X'X| \neq 0$. This last condition will be satisfied provided that no one predictor variable is a linear combination of some of the others (Exercise 9.5). The matrix $X'X$, which has size $(k+1) \times (k+1)$, has sums of squares of predictor variables down the leading diagonal and sums of cross products of two predictor variables elsewhere. The elements of $X'\boldsymbol{Y}$, which has size $(k+1) \times 1$, are of the form $\sum_{i=1}^{n} x_{ji}Y_i$ for $j = 0, \ldots, k$ (Exercise 9.6).

9.4.4 Estimating the coefficients

Once we have the data we obtain a specific estimate of the coefficients using the observed values of the response $\boldsymbol{y}$ in place of the random vector[7] $\boldsymbol{Y}$.

Example 9.1: (Continued) Conch

We could set up the matrices for the *Conch* data and perform the matrix calculations, see Exercise 9.7 for R code to do this, but it is much quicker to use the R linear model function lm(). First, however, we should draw some plots from the data (see Figure 9.4).

```
> #10 shops, Pedest, Area, Sales
> conch.dat=read.table("conch.txt",header=T)
> attach(conch.dat)
> head(conch.dat)
  Pedest Area Sales
1    564  650   980
2   1072  700  1160
3    326  450   800
4   1172  500  1130
5    798  550  1040
6    584  650  1000
> par(mfcol=c(2,2))
> plot(Pedest,Sales)
> plot(Area,Sales)
> plot(Pedest,Area)
> #If area below median plot with 'o' above median plot with 'X'
> Size=1
> cond=median(Area)
> for (i in 1:10){
+ if (Area[i] < cond) Size[i]=1 else Size[i]=4}
```

---

[7]Although we distinguish the random vector, $\boldsymbol{Y}$, from the data, $\boldsymbol{y}$, by upper and lower case respectively, we rely on the context to distinguish the estimator of the coefficients, a random variable, from the numerical estimates. Both are denoted by $\widehat{\boldsymbol{\beta}}$.

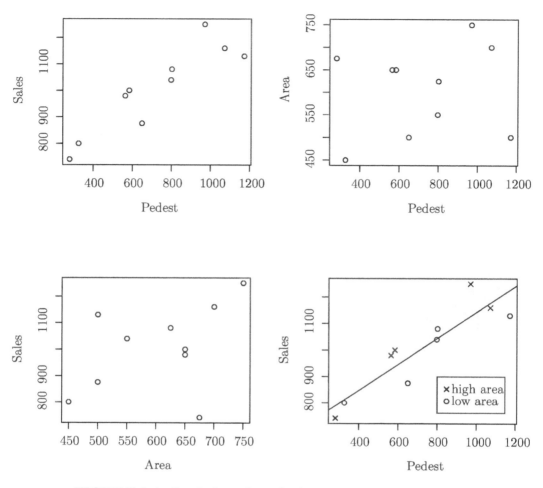

FIGURE 9.4: Conch data plots of sales versus pedestrian traffic.

```
> plot(Pedest,Sales,pch=Size)
> legend(800,900,c("high area","low area"),pch=c(4,1))
> abline(lm(Sales~Pedest))
```

The upper left frame of Figure 9.4 shows a clear tendency for sales to increase with pedestrian traffic. The upper right frame shows that there is little association between pedestrian traffic and area, and the lower left frame hints that there might be a slight increase of sales with area. The correlations between the three pairs of variables can be calculated with the R function cor().

```
> round(cor(cbind(Pedest,Area,Sales)),2)
        Pedest Area Sales
Pedest   1.00 0.17  0.90
Area     0.17 1.00  0.43
Sales    0.90 0.43  1.00
```

In this case there is little correlation between pedestrians and areas for the 10 shops considered. There are positive correlations between pedestrians and sales and between area and sales, as expected. But in general correlations can be misleading if they are considered in isolation. For example, suppose there had happened to be a substantial negative correlation between pedestrians and area for the 10 shops. Then there might be a negative correlation between area and sales because the larger area is associated with lower pedestrians, and pedestrians is more influential on sales. The multiple regression model allows for the effect of pedestrians when considering the influence of area.

The lower right frame of Figure 9.4 shows the effect of floor area once a regression of sales on traffic has been fitted, and is an example of a **pseudo-3D plot**. The crosses represent points that are above median area, and the circles represent points that are below median area. Three of the crosses are above the line which is consistent with the claim that sales increase with area for a given pedestrian traffic. However, one cross is well below the line and another is slightly below the line. With only 10 shops we are unsure about the effect of area on sales, for a given pedestrian traffic. A convenient package for a 3D plot[8] is `scatterplot3d`, which is implemented in the following R code to yield Figure 9.5.

```
library(scatterplot3d)
scatterplot3d(Pedest,Area,Sales,type="h")
```

We now fit a regression of sales on both pedestrian traffic and floor area (**m3**), and for comparison regressions of sales on pedestrian traffic only (**m1**) and floor area only (**m2**).

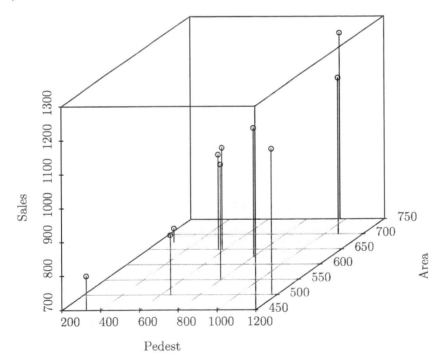

FIGURE 9.5: Conch: 3D scatter plot of sales on pedestrian traffic and area.

---

[8] Quick-R Scatterplots is a useful resource.

```
> m1=lm(Sales~Pedest)
> m2=lm(Sales~Area)
> m3=lm(Sales~Pedest+Area)
> summary(m1)

Call:
lm(formula = Sales ~ Pedest)

Residuals:
    Min      1Q  Median      3Q     Max
-96.273 -41.097  -7.271  47.583 122.740

Coefficients:
             Estimate Std. Error t value Pr(>|t|)
(Intercept) 651.79718   64.15972  10.159 7.54e-06 ***
Pedest        0.49017    0.08276   5.923 0.000353 ***
---
Residual standard error: 74.03 on 8 degrees of freedom
Multiple R-squared:  0.8143,    Adjusted R-squared:  0.7911
F-statistic: 35.08 on 1 and 8 DF,  p-value: 0.0003526

> summary(m2)

Call:
lm(formula = Sales ~ Area)

Residuals:
    Min      1Q  Median      3Q     Max
-314.73  -56.86   11.59   84.04  198.10

Coefficients:
             Estimate Std. Error t value Pr(>|t|)
(Intercept) 580.9379   319.2242   1.820    0.106
Area          0.7019     0.5214   1.346    0.215

Residual standard error: 155.1 on 8 degrees of freedom
Multiple R-squared:  0.1847,    Adjusted R-squared:  0.0828
F-statistic: 1.812 on 1 and 8 DF,  p-value: 0.2151

> summary(m3)

Call:
lm(formula = Sales ~ Pedest + Area)

Residuals:
   Min     1Q Median     3Q    Max
-93.45 -43.92  25.70  34.83  60.91

Coefficients:
             Estimate Std. Error t value Pr(>|t|)
(Intercept) 385.67909  125.64787   3.070 0.018080 *
```

```
Pedest        0.46424    0.06742   6.885 0.000234 ***
Area          0.47080    0.20272   2.322 0.053205 .
---
Residual standard error: 59.47 on 7 degrees of freedom
Multiple R-squared:  0.8951,    Adjusted R-squared:  0.8651
F-statistic: 29.87 on 2 and 7 DF,  p-value: 0.0003737
```

For the moment, there are just two details of these results to notice. The first, from summary (**m3**), is that the fitted regression model for predicting sales from pedestrian traffic and floor area is

$$Sales \; = \; 385.68 + 0.46424 \times pedestrians + 0.47080 \times area.$$

The physical interpretation of the coefficients of pedestrians and area is that sales increase by 0.46424 per week for an increase of one in the average pedestrians, and that sales increase by 0.47080 for an increase of one in area.

The intercept, which is the coefficient of 1, needs to be considered in the context of the ranges of values for Pedest and Area[9].

```
> summary(cbind(Pedest,Area,Sales))
      Pedest              Area            Sales
Min.    : 280.0   Min.    :450.0   Min.    : 740
1st Qu.: 569.0    1st Qu.:512.5    1st Qu.: 902
Median : 724.0    Median :637.5    Median :1020
Mean    : 721.8 mean     :605.0 mean     :1006
3rd Qu.: 928.0    3rd Qu.:668.8    3rd Qu.:1118
Max.    :1172.0   Max.    :750.0   Max.    :1250
```

The intercept makes Sales equal its mean value when Pedest and Area equal their mean values.

$$385.6791 + 0.4642 \times 721.8 + 0.4708 \times 605 \; = \; 1005.573.$$

The regression model provides a linear approximation to the relationship between sales and pedestrians and floor area over a range from 280 up to 1172 for pedestrians and a range from 450 up to 750 for floor area.

The second detail, referring to coefficients in summary (**m1**) and summary (**m2**), is that the coefficients in the regression on the two predictor variables are different from the coefficients in the regressions on single predictor variables. This is generally the case for observational studies, and is a consequence of the correlations between the predictor variables. In contrast, in a designed experiment we can choose the values of predictor variables and we typically choose them to be uncorrelated (Exercise 9.8).

---

[9]Sales will not be 386 if pedestrians and area are both 0, sales could only be 0.

9.4.5   Estimating the standard deviation of the errors

The standard deviation of the errors is also a parameter of the model. We estimate it as the square root of the variance of estimates of the errors. The errors are

$$\varepsilon_i = Y_i - (\beta_0 + \beta_1 x_1 + \ldots + \beta_k x_k)$$

and the values that the errors take in a particular case are[10]

$$\varepsilon_i = y_i - (\beta_0 + \beta_1 x_1 + \ldots + \beta_k x_k).$$

If we knew the values of β_j we would estimate the variance of the errors by

$$\widehat{\sigma}^2 = \frac{\sum \varepsilon_i^2}{n}.$$

But, as we don't know the values of β_j we replace β_j by our estimates $\widehat{\beta}_j$ and define **residuals**, as estimates of the errors.

Definition 9.4: Fitted value

The fitted values $\widehat{y}_i$ are the predicted values of the responses for the items in the sample.

$$\widehat{y}_i = \widehat{\beta_0} + \widehat{\beta_1} x_1 + \ldots + \widehat{\beta_k} x_k, \quad \text{for } i = 1, \ldots, n.$$

Definition 9.5: Residual

A residual is the signed difference between the observed value and the fitted value, defined as the observed value less the fitted value.

$$r_i = y_i - \widehat{y}_i,$$

The residuals are our best estimates of the values taken by the errors, but we know that

$$\sum r_i^2 < \sum \varepsilon_i^2,$$

because the estimates of the coefficients were obtained by minimizing the sum of squared differences between the observations and the fitted values. The minimum distances are the absolute values of the residuals. We allow for the sum of residuals being less than the sum of errors by using the estimate

$$s^2 = \frac{\sum r_i^2}{n - k - 1}$$

for the variance of the errors σ^2, and s as an estimate of the standard deviation of the errors. If we consider the residuals r_i as random variables[11], by replacing y_i with Y_i, then it can be shown that S^2 is unbiased for σ^2. That is, $\mathrm{E}[S^2] = \sigma^2$ where

$$S^2 = \frac{\sum r_i^2}{n - k - 1} = \frac{\sum (Y_i - \widehat{Y}_i)^2}{n - k - 1}.$$

[10]We rely on the context to distinguish between the random variable ε_i and the value it takes in a particular case.

[11]Again we rely on the context to distinguish between the random variable r_i and the value it takes in a particular case.

The denominator $n - k - 1$ is known as the **degrees of freedom** and equals the number of data, n, less the number of coefficients in the model that are estimated from the data. If the number of data is equal to the number of coefficients, then all the residuals are 0 and there are no degrees of freedom for estimating the error.

Example 9.1: (Continued) Conch

For the *Conch* case the marginal standard deviation of the sales, s_y, is

```
> sd(Sales)
[1] 161.9542
```

The estimated standard deviation of the errors in the regression of sales on pedestrians and area (**m3**) can be read from the R summary of the regression, where it is referred to as "residual standard error". The seven degrees of freedom come from 10 data less 2 coefficients of predictor variables and less 1 estimated intercept (coefficient of 1).

```
Residual standard error: 59.47 on 7 degrees of freedom
```

It can also be obtained from

```
> summary(m3)$sigma
[1] 59.4745
```

which is useful for further calculations. The estimated standard deviation of the errors in the regression on pedestrian traffic only (**m1**) is rather higher

```
> summary(m1)$sigma
[1] 74.02597
```

and this provides some justification for using the model **m3**, with the two predictor variables traffic and area, in preference to **m1**.

9.4.6 Standard errors of the estimators of the coefficients

The estimator $\widehat{\beta}$ is unbiased for β.

$$\mathrm{E}\left[\widehat{\beta}\right] = \mathrm{E}\left[(X'X)^{-1}X'Y\right] = (X'X)^{-1}X'\mathrm{E}[Y] = (X'X)^{-1}X'X\beta = \beta.$$

The variance-covariance matrix is

$$
\begin{aligned}
\mathrm{var}\left(\widehat{\beta}\right) &= \mathrm{E}\left[(\widehat{\beta} - \beta)'(\widehat{\beta} - \beta)\right] \\
&= \mathrm{E}\left[\left((X'X)^{-1}X'Y - (X'X)^{-1}X'\mathrm{E}[Y]\right)\left((X'X)^{-1}X'Y - (X'X)^{-1}X'\mathrm{E}[Y]\right)'\right] \\
&= \mathrm{E}\left[\left((X'X)^{-1}X'(Y - E\mathrm{E}[Y])\right)\left((X'X)^{-1}X'(Y - \mathrm{E}[Y])'\right)\right] \\
&= \mathrm{E}\left[(X'X)^{-1}X'(Y - \mathrm{E}[Y])(Y - \mathrm{E}[Y])'X(X'X)^{-1}\right], \quad \text{since } X'X \text{ is symmetric} \\
&= (X'X)^{-1}X'\mathrm{E}[(Y - \mathrm{E}[Y])(Y - E\mathrm{E}[Y])']X(X'X)^{-1}.
\end{aligned}
$$

Now

$$E[(Y - E[Y])(Y - E[Y])'] \;=\; I\,\sigma^2,$$

so the expression reduces to

$$(X'X)^{-1}\sigma^2.$$

The estimator of the variance-covariance matrix is $(X'X)^{-1}S^2$ and the corresponding estimates are calculated by replacing the estimator S^2 with its estimate s^2. The variances of estimators of coefficients are on the leading diagonal and the co-variances between two estimators of coefficients are given by the off-diagonal terms. The estimated variance-covariance matrix of the estimators can be calculated from the formulae using R (Exercise 9.10) but the estimated standard deviations of the estimators (standard errors) are available in the `lm()` summary.

Example 9.1: (Continued) Conch

```
Coefficients:
              Estimate Std. Error t value Pr(>|t|)
(Intercept) 385.67909  125.64787   3.070 0.018080 *
Pedest        0.46424    0.06742   6.885 0.000234 ***
Area          0.47080    0.20272   2.322 0.053205 .
```

The t-value is the ratio of the estimate to its standard error and `Pr(>|t|)` is the p-value for a test of a null hypothesis that the coefficient is 0 with a two-sided alternative hypothesis that it is not 0. We can calculate a 90% confidence interval for the coefficient of Area, without retyping any numbers, as follows.

```
> summary(m3)$coef
                 Estimate    Std. Error   t value     Pr(>|t|)
(Intercept) 385.6790856 125.64787099 3.069523 0.0180796177
Pedest        0.4642399   0.06742308 6.885475 0.0002343706
Area          0.4707976   0.20271969 2.322407 0.0532054974
> L=m3$coef[3]-qt(.95,m3$df.res)*summary(m3)$coef[3,2]
> U=m3$coef[3]+qt(.95,m3$df.res)*summary(m3)$coef[3,2]
> print(round(c(L,U),2))
Area Area
0.09 0.85
```

The 90% confidence interval for the coefficient of Area is $[0.09, 0.85]$. It is quite wide, but this is partly a consequence of the small sample size. We can deduce that the 95% confidence interval will just include 0 from the information that $P(> |t|) = 0.053$, which is the right hand column of the R output coefficients.

9.5   Assessing the fit

The residuals are our estimates of the errors and so are the basis for our assessment of the fit of the model to the data. There are two aspects to the assessment. The first is whether the model is consistent with the data inasmuch as the residuals seem to be realizations of random variation. The second is to quantify the predictive power of the model by the proportion of the original variability in the response that is explained by the model.

9.5.1 The residuals

If a multiple regression model is fitted to any set of data[12] then the residuals have mean 0 and the correlations between predictor variables and the residuals are 0. These results follow succinctly from the matrix formulation. First convince yourself that if the residuals are in the column matrix r then $X'r$ is a column matrix containing the sum of the residuals and the covariances between the residuals and the predictor variables. Once you have done so, the results are obtained in a few steps.

$$X'r \;=\; X'(y - \widehat{y}) \;=\; X'(y - X\widehat{\beta}) \;=\; X'(y - X(X'X)^{-1}X'y) \;=\; 0.$$

It follows that we cannot test the assumptions that errors have mean 0 (**A1**) and that errors are uncorrelated with the predictor variables (**A2**) from the data. We can however:

- plot residuals against the predictor variables to see if there is evidence of curvature,

- plot residuals against fitted values to see if there is, for example, a tendency for residuals to become more variable as fitted values get larger and

- draw a normal qqplot to see if the assumption of normality is reasonable and to identify any outlying observations.

If the observations can be placed in time order, or other physically meaningful order, then we can calculate auto-correlations to check whether the assumption that the errors are independent of each other is reasonable.

Example 9.1: (Continued) Conch

For the *Conch* analysis.

```
> r=m3$res
> par(mfrow=c(2,2))
> plot(Pedest,r)
> plot(Area,r)
> plot(m3$fit,r)
> qqnorm(r)
```

From the upper plots in Figure 9.6, there is no suggestion that inclusion of quadratic terms would improve the model. There is no indication that the assumptions of equal variance and normality are unreasonable from the lower plots.

---

[12]Conversely, the estimated coefficients in the regression can be obtained by imposing these constraints (Exercise 9.17).

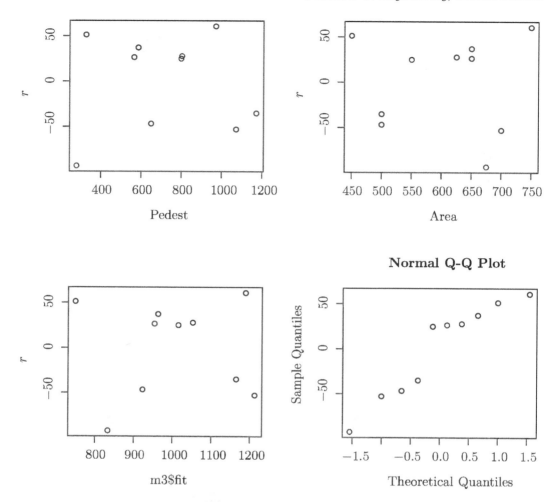

FIGURE 9.6: Conch residual plots.

9.5.2   R-squared

The statistic **R-squared**, also written as R^2 and known as the **coefficient of determination** is a measure of the proportion of the variance in the response that can be attributed to the regression model. It is defined by

$$R^2 \;=\; 1 - \frac{\sum r_i^2}{\sum (y_i - \bar{y})^2}$$

and you can see that $0 \le R^2 \le 1$. It would equal 0 if the estimates of the coefficients are all 0, corresponding to predicting the response by the mean of its past values. In contrast, it would equal 1 if the model is a perfect fit. This may sound impressive but you can obtain a perfect fit to n data by introducing $n-1$ predictor variables, and R^2 close to 1 may be a consequence of an excessive number of predictor variables. The adjusted R^2, given by

$$R^2_{adj} \;=\; 1 - \frac{\sum r_i^2 / (n-k-1)}{\sum (y_i - \bar{y})^2 / (n-1)} \;=\; 1 - \frac{s^2}{s_y^2},$$

penalizes addition of predictor variables by dividing the sums of squares in the definition by their degrees of freedom. This is equivalent to using s^2, relative to s_y^2, as a measure of fit. If the addition of variables causes R_{adj}^2 to increase, then s^2, and so s, must decrease.

Example 9.1: (Continued) Conch

If you refer back to the summaries for **m1** and **m3** you will see that R_{adj}^2 increases from 0.7911 to 0.8651 when Area is added as a second predictor variable and s decreases from 74.03 to 59.47. For comparison s_y is 161.95. The R^2 increased from 0.8143 to 0.8951 between **m1** and **m3**. The R_{adj}^2 for **m3**, about 0.87, indicates that our model will provide useful predictions for *Conch* management.

An alternative interpretation of R^2 is that it is the square of the correlation between $\mathbf{Y}$ and $\widehat{\mathbf{Y}}$. You are asked to prove this result in Exercise 9.15.

9.5.3 F-statistic

We have now explained everything in the `lm()` summary except the final F-statistic. This relates to a null hypothesis that all the coefficients in the model, except for the intercept, are 0. Informally, the null hypothesis is that the model is no improvement on predicting future values of the response by its mean value. We need a small p-value for a credible model. Formally

$$H_0 \;:\; \beta_1 = \ldots = \beta_k = 0$$

$$H_1 \;:\; \text{not all } \beta_j = 0, \quad \text{for} \quad 1 \leq j \leq k.$$

For any regression the total sum of squared deviations of the response variable from its mean (TSS) can be split up into two components: the sum of squared residuals (RSS) and the sum of squares attributed to the regression $(REGSS)$[13]. This is an example of an analysis of variance (ANOVA).

$$TSS \;=\; RSS + REGSS$$

and more explicitly in terms of the data

$$\sum (y_i - \bar{y})^2 \;=\; \sum (y_i - \widehat{y_i})^2 + \sum (\widehat{y_i} - \bar{y})^2.$$

The proof of this result follows from the fact that the residuals are uncorrelated with the predictor variables (Exercise 9.16). There are $n-1$ degrees of freedom for the TSS, and these are split as $n-k-1$ for RSS and k for $REGSS$. If H_0 is true then $RSS/(n-k-1)$ and $REGSS/k$ are independent estimators of the variance of the errors. Their ratio has an F-distribution:

$$\frac{REGSS/k}{RSS/(n-k-1)} \;\sim\; F_{k,n-k-1}$$

There is evidence to reject H_0 at the $\alpha\%$ level if the calculated value of this ratio exceeds $F_{k,n-k-1,\alpha}$.

---

[13] R^2 is often defined as $\frac{REGSS}{TSS}$.

Example 9.1: (Continued) Conch

Calculations for the *Conch* case are

```
> n=length(Sales)
> TSS=var(Sales)*(n-1)
> RSS=summary(m3)$sigma^2*m3$df
>   REGSS=TSS-RSS
> F=(REGSS/2)/(RSS/m3$df) ; print(F)
[1] 29.86839
> 1-pf(F,2,m3$df)
[1] 0.0003737359
```

We have strong evidence to reject a null hypothesis that sales depend on neither pedestrian traffic nor floor area. We would not usually make these calculations because the R summary of lm() gives us the result.

```
F-statistic: 29.87 on 2 and 7 DF,  p-value: 0.0003737
```

9.5.4   Cross validation

If we have a large data set, we can split it into **training data** and **test data**. The model is fitted to the training data, and the model is then used to make predictions for the test data. If the model is reasonable, the mean of the prediction errors should be close to 0 and their standard deviation should be close to s (Exercise 9.17).

A variation on this strategy is to leave out one datum at a time, fit the model to the remaining $n - 1$ data, and predict the value of the response for the removed datum. The prediction errors with this leave-one-out-at-a-time technique are known as the PRESS[14] residuals. The PRESS residuals can be obtained from the original analysis and it is not necessary to explicitly perform n regression, each with $(n - 1)$ data (Exercise 9.21).

9.6   Predictions

Suppose we want to predict the response when the predictor variables have values

$$x_p \;=\; (1, x_{1p}, \ldots, x_{kp})'.$$

The predicted value of the response $\widehat{y}_p$ is

$$\widehat{y}_p \;=\; x_p'\widehat{\beta}$$

The variance of this quantity is given by

$$\mathrm{var}\left(x_p'\widehat{\beta}\right) \;=\; x_p'\mathrm{var}\left(\widehat{\beta}\right)x_p \;=\; x_p'(X'X)^{-1}x_p\,\sigma^2$$

---

[14]PRESS = prediction error sum of squares.

and the estimated variance is obtained by replacing σ^2 with s^2. Therefore a $(1 - \alpha) \times 100\%$ confidence interval for the mean value of Y when x equals x_p is

$$\widehat{y}_p \quad \pm \quad t_{n-k-1,\alpha/2} \; s \; \sqrt{x'_p(X'X)^{-1}x_p}$$

and a $(1 - \alpha) \times 100\%$ prediction interval for a single value of Y when x equals x_p is

$$\widehat{y}_p \quad \pm \quad t_{n-k-1,\alpha/2} \; s \; \sqrt{1 + x'_p(X'X)^{-1}x_p}$$

If n is large relative to k, then a reasonable approximation to the predicted interval is

$$\widehat{y}_p \quad \pm \quad z_{\alpha/2}s.$$

Example 9.1: (Continued) Conch

Having set up a model for predicting sales from pedestrians and area we can advise *Conch* management about the two options for purchase. The pedestrians and area are known for the two stores, which we'll call A and B, and are given in Table 9.2.

TABLE 9.2: Options for purchase.

Option	Pedestrians/hour	Floor area (square meters)
A	475	1 000
B	880	550

```
> m3=lm(Sales~Pedest+Area)
> newdat=data.frame(Pedest=c(475,880),Area=c(1000,550))
> predm3=predict(m3,newdata=newdat,interval="prediction",level=0.95)
> print(predm3)
        fit      lwr      upr
1 1076.991 828.7505 1325.231
2 1053.149 900.4812 1205.817
```

The predicted sales for Shop A, 1 077, are slightly higher than for Shop B, 1 053, but notice how much wider the prediction interval is for Shop A. This is a consequence of the area for Shop A, 1 000, being so far, 4 standard deviations of area, from the mean Area of the 10 shops,

```
> summary(Area)
   Min. 1st Qu.  Median mean 3rd Qu.    Max.
  450.0   512.5   637.5  605.0   668.8   750.0
> sd(Area)
[1] 99.16317
> (1000-mean(Area))/sd(Area)
[1] 3.983334
```

and the lack of precision in the estimate of the coefficient of Area. The prediction, and prediction interval, rely on the assumed linear relationship between Sales and Area. The model was fitted to floor areas from 450 up to 750. Our prediction for Shop A is based on extrapolation to an area of 1 000, and we have no evidence to support the linear approximation over this range. In contrast the prediction for Shop B is based on interpolation, as both the Pedest and Area for Shop B are within the range of the data used to fit the model.

```
> summary(Pedest)
   Min. 1st Qu.  Median mean 3rd Qu.    Max.
  280.0   569.0   724.0 721.8   928.0  1172.0
```

We recommended purchase of Shop B, because the prediction for shop A is dubious and not substantially higher than the prediction for shop B.

9.7   Building multiple regression models

We now have most of the theoretical basis we need for using multiple regression. The following applications show how versatile the multiple regression model can be.

9.7.1   Interactions

Definition 9.6: Interaction

Two predictor variables interact if the effect on the response of one predictor variable depends on the value of the other predictor variable.

It is important to allow for the possibility of an interaction between predictor variables in regression models.

Example 9.1: (Continued) Conch

In the *Conch* case, it is possible that the effect of increasing floor area depends on the pedestrian traffic. It may be that a larger floor area is particularly effective for increasing sales when the pedestrian traffic is high. The interaction is allowed for by including a cross product in the model,

$$Sales = \beta_0 + \beta_1 \times pedestrians + \beta_2 \times area + \beta_3 \times pedestrians \times area + error,$$

which can be written as

$$Sales = \beta_0 + \beta_1 \times pedestrians + (\beta_2 + \beta_3 \times pedestrians) \times area + error$$

to show that an interpretation of the cross-product, interaction term, is that the coefficient of area depends on pedestrians[15].

```
> PAint=Pedest*Area
> cor(cbind(Pedest,Area,PAint))
            Pedest       Area      PAint
Pedest 1.0000000 0.1655823 0.9229291
Area   0.1655823 1.0000000 0.4997097
```

[15]The model could also be written with the coefficient of pedestrians depending on area, but we have chosen to take pedestrians as the leading predictor variable.

```
PAint   0.9229291 0.4997097 1.0000000
> m4a=lm(Sales~Pedest+Area+PAint) ; print(summary(m4a))

Call:
lm(formula = Sales ~ Pedest + Area + PAint)

Residuals:
   Min     1Q Median     3Q    Max
-78.53 -37.21  18.94  33.62  47.51

Coefficients:
              Estimate Std. Error t value Pr(>|t|)
(Intercept)  7.271e+02  2.791e+02   2.605   0.0404 *
Pedest      -5.205e-03  3.530e-01  -0.015   0.9887
Area        -1.120e-01  4.717e-01  -0.237   0.8203
PAint        7.934e-04  5.867e-04   1.352   0.2251
---
Residual standard error: 56.24 on 6 degrees of freedom
Multiple R-squared:  0.9196,    Adjusted R-squared:  0.8794
F-statistic: 22.88 on 3 and 6 DF,  p-value: 0.001102

> newdata=data.frame(Pedest=c(475,880),Area=c(1000,550),
+ PAint=c(475*1000,880*550))
> predm4a=predict(m2,newdata=newdat,interval="prediction",level=0.95)
> print(predm4a)
        fit      lwr      upr
1 1076.991 828.7505 1325.231
2 1053.149 900.4812 1205.817
```

When the interaction is included (**m4**) the estimated standard deviation of the errors, 56.2 on 6 degrees of freedom, is slightly less than for **m3**, 59.5 on 7 degrees of freedom. It follows that R^2_{adj} is slightly higher, 0.880 compared with 0.865 Both Pedest and Area are highly correlated with PAint, which makes interpretation of the coefficients awkward - in particular the coefficients of Pedest and Area are now negative, though Sales does increase with increasing Pedest through the interaction term. When adding cross-product, and squared terms, it makes interpretation of the model easier if we mean adjust predictor variables (subtract the means) and, in some cases at least, standardize by dividing the mean-adjusted variables by their standard deviations. The scaling does not affect predictions. Another consequence of such scaling is that it makes the $X'X$ matrix better conditioned for inversion (Exercise 9.26).

```
> x1=(Pedest-mean(Pedest))/sd(Pedest)
> x2=(Area - mean(Area))/sd(Area)
> x3=x1*x2
> cor(cbind(x1,x2,x3))
           x1         x2          x3
x1  1.0000000 0.16558228 -0.11661009
x2  0.1655823 1.00000000  0.01882527
x3 -0.1166101 0.01882527  1.00000000
> m4b=lm(Sales~x1+x2+x3) ; print(summary(m4b))
```

```
Call:
lm(formula = Sales ~ x1 + x2 + x3)

Residuals:
   Min    1Q Median    3Q    Max
-78.53 -37.21  18.94  33.62  47.51

Coefficients:
            Estimate Std. Error t value Pr(>|t|)
(Intercept)  1002.10      17.97  55.761 2.23e-09 ***
x1            141.56      19.15   7.392 0.000315 ***
x2             45.68      19.02   2.401 0.053188 .
x3             23.46      17.35   1.352 0.225057
---
Residual standard error: 56.24 on 6 degrees of freedom
Multiple R-squared: 0.9196,    Adjusted R-squared: 0.8794
F-statistic: 22.88 on 3 and 6 DF,  p-value: 0.001102

> standPed=(c(475,880)-mean(Pedest))/sd(Pedest)
> standArea=(c(1000,550)-mean(Area))/sd(Area)
> newdata=data.frame(x1=standPed,x2=standArea,x3=standPed*standArea)
> predm4b=predict(m2,newdata=newdat,interval="prediction",level=0.95)
> print(predm4b)
       fit      lwr      upr
1 1076.991 828.7505 1325.231
2 1053.149 900.4812 1205.817
```

The predictor variables $x1$ and $x2$ are now approximately uncorrelated with their interaction $x3$ and the coefficients of $x1$ and $x2$ in model **m4b** are close to their values in model **m3b**, which excludes their interaction (shown below). The coefficient of the interaction is positive which corresponds to the suggestion that an increase in area is more beneficial as the pedestrian traffic increases. The p-value of 0.22 associated with the coefficient of $x3$, and also with the coefficient of PAint, tells us that a 90% confidence interval for the interaction includes positive and negative values. Given this lack of precision we may choose to omit the interaction despite its being positive, as we thought it should be, and the slight increase in R^2_{adj}. No model is correct, but good models provide useful approximations.

```
> m3b=lm(Sales~x1+x2)
> summary(m3b)

Call:
lm(formula = Sales ~ x1 + x2)

Residuals:
   Min    1Q Median    3Q    Max
-93.45 -43.92  25.70  34.83  60.91

Coefficients:
            Estimate Std. Error t value Pr(>|t|)
(Intercept)  1005.60      18.81  53.468 2.1e-10 ***
```

```
x1              138.41       20.10    6.885 0.000234 ***
x2               46.69       20.10    2.322 0.053205 .
---
Residual standard error: 59.47 on 7 degrees of freedom
Multiple R-squared: 0.8951,    Adjusted R-squared:  0.8651
F-statistic: 29.87 on 2 and 7 DF,  p-value: 0.0003737
```

Definition 9.7: Centering, mean adjustment and standardizing

Centering, mean adjustment (also known as mean correction), and standardizing are all linear transformations of data $\{x_i\}$, for $i = 1, \ldots, n$. These are defined in Table 9.3, where c is any constant. mean adjustment is centering with $c = \bar{x}$.

TABLE 9.3: Centering, mean adjustment and standardizing.

Linear transform	Data	Transformed data
centering	x_i	$x_i - c$
mean adjustment	x_i	$x_i - \bar{x}$
standardizing	x_i	$\dfrac{x_i - \bar{x}}{s_x}$

Linear transforms, such as centering and standardizing, of the predictor variables do not affect predictions. For example:

$$y = \widehat{\beta}_0 + \widehat{\beta}_1 x_1 + \widehat{\beta}_2 x_2$$

is equivalent to

$$y = \widehat{\gamma}_0 + \widehat{\gamma}_1 \frac{x_1 - \bar{x}_1}{s_{x_1}} + \widehat{\gamma}_2 \frac{x_2 - \bar{x}_2}{s_{x_2}},$$

where

$$\frac{\widehat{\gamma}_1}{s_{x1}} = \widehat{\beta}_1, \qquad \frac{\widehat{\gamma}_2}{s_{x2}} = \widehat{\beta}_2, \qquad \widehat{\gamma}_0 = \bar{y} \quad \text{and} \quad \widehat{\beta}_0 = \bar{y} - \widehat{\beta}_1 \bar{x}_1 - \widehat{\beta}_2 \bar{x}_2.$$

It follows that models **m3** and **m3b** are equivalent, as are models **m4a** and **m4b** (Exercise 9.27).

The advantages of standardizing predictor variables include the following.

- It usually facilitates the assessment of whether or not quadratic terms and interactions improve the model.

- If all the predictor variables are standardized, the relative influence of predictor variables on the response is given by the magnitudes of their estimated coefficient s.

- If all predictor variables in the model are standardized the intercept is estimated by the mean response.

- The matrix $X'X$ is better conditioned for the numerical calculations.

The disadvantage is that the predictor variables are no longer measured in their physical units. It is straightforward, if somewhat inconvenient, to calculate the coefficients in the physical model from the coefficients in the standardized model [16]. The calculation requires the means and standard deviations of the predictor variables.

9.7.2 Categorical variables

The predictor variables in a multiple regression model are not restricted to continuous variables. Categorical variables can be introduced through the use of **indicator variables**.

Example 9.2: Elysium Electronics

An engineer in an electronics company ran an experiment to compare the bond strength of four formulations of adhesive: A, B, C and D. She thinks the bond strength will also depend on the film thickness which is supposed to be 12 (0.1 mm) but typically varies between 10 and 16 (see Figure 9.7). Although the film thickness is not tightly controlled it can be measured[17]. She tested 5 joints for each formulation and measured

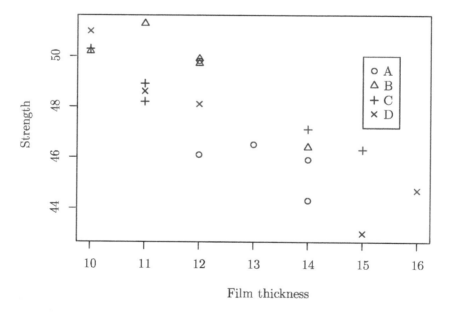

FIGURE 9.7: Film thickness versus strength for four different glue formulations.

the strength y (Newtons) and the film thickness. The results are given in Table 9.4. For the moment we will just consider the results for glue formulations A and B. We

---

[16]Similarly you can calculate coefficients in the standardized model from coefficients in the physical model, although you are unlikely to have reason to do so.

[17]In an experimental situation variables that cannot be set to precise values but can be monitored are called **concomitant variables**.

TABLE 9.4: Glue formulation, film thickness and strength for 20 test joints.

Glue	Film thickness (0.1 mm)	Bond strength (N)	Glue	Film thickness (0.1 mm)	Bond strength (N)
A	13	46.5	C	15	46.3
A	14	45.9	C	14	47.1
A	12	49.8	C	11	48.9
A	12	46.1	C	11	48.2
A	14	44.3	C	10	50.3
B	12	49.9	D	16	44.7
B	10	50.2	D	15	43.0
B	11	51.3	D	10	51.0
B	12	49.7	D	12	48.1
B	14	46.4	D	11	48.6

can set up a single indicator variable, *gluenp*[18] say, coded as

$$gluenp = \begin{cases} -1 & \text{for } A, \\ +1 & \text{for } B. \end{cases}$$

The R regression analysis is

```
> glue.dat=read.table("glue.txt",header=TRUE)
> head(glue.dat)
  glue film_thick strength
1    A         13     46.5
2    A         14     45.9
3    A         12     49.8
4    A         12     46.1
5    A         14     44.3
6    B         12     49.9
> attach(glue.dat)
> y=strength[1:10]
> thickness=film_thick[1:10]
> gluenp=c(rep(-1,5),rep(1,5))
> mnp=lm(y~thickness+gluenp)
> summary(mnp)

Call:
lm(formula = y ~ thickness + gluenp)

Residuals:
   Min     1Q  Median     3Q    Max
-1.5997 -0.9064  0.2080  0.6169  2.1003

Coefficients:
            Estimate Std. Error t value Pr(>|t|)
(Intercept)  62.6381     4.5520  13.761 2.52e-06 ***
```

---

[18]The *np* stands for negative or positive.

```
thickness     -1.1797      0.3656  -3.227   0.0145 *
gluenp         0.7822      0.4682   1.671   0.1387
---
Residual standard error: 1.308 on 7 degrees of freedom
Multiple R-squared:  0.7697,     Adjusted R-squared:  0.7039
F-statistic:  11.7 on 2 and 7 DF,  p-value: 0.005865
```

With this coding the difference between the mean strengths of joints with glue A and B is estimated as twice the coefficient of *glue*. That is B is stronger than A by $(2 \times 0.7822 = 1.56)$, although the precision of this estimate is low as the standard error is $(2 \times 0.4682 = 0.94)$. The intercept, 62.638, is an estimate of the average of the mean strengths with glue A and the mean strengths with glue B when the film thickness is at its mean value, less the product of the coefficient of film thickness and its mean[19]. An alternative indicator variable, *glue01*, can be coded as

$$glue01 \quad = \quad \begin{cases} 0 & \text{for } A, \\ 1 & \text{for } B. \end{cases}$$

and with this coding the coefficient of *glue01*, 1.56, is an estimate of the mean strength of joints with glue B relative to the mean with glue A. The intercept, 61.856, is now an estimate of the mean strength with glue A when the film thickness is at its mean value, less the product of the coefficient of film thickness and its mean.

```
> glue01=c(rep(0,5),rep(1,5))
> m01=lm(y~thickness+glue01)
> summary(m01)

Call:
lm(formula = y ~ thickness + glue01)

Residuals:
    Min      1Q  Median      3Q     Max
-1.5997 -0.9064  0.2080  0.6169  2.1003

Coefficients:
             Estimate Std. Error t value Pr(>|t|)
(Intercept)  61.8559     4.7884  12.918 3.87e-06 ***
thickness    -1.1797     0.3656  -3.227   0.0145 *
glue01        1.5644     0.9363   1.671   0.1387
---
Residual standard error: 1.308 on 7 degrees of freedom
Multiple R-squared:  0.7697,     Adjusted R-squared:  0.7039
F-statistic:  11.7 on 2 and 7 DF,  p-value: 0.005865
```

We can choose any coding and the estimate of the difference in strength with glues A and B will be the same.

In the general case of c categories we will require $c-1$ indicator variables. A convenient coding for our 4 glue formulations is given in Table 9.5, another option is given in Exercise 9.29.

---

[19]It might be more convenient to mean adjust the film thickness before performing the regression (Exercise 9.28), so that the intercept is the average of mean strength with glue A and mean strength with glue B when the film thickness is at its mean value.

TABLE 9.5: Coding of 3 indicator variables for glues B, C, D relative to A.

Glue	x_1	x_2	x_3
A	0	0	0
B	1	0	0
C	0	1	0
D	0	0	1

R sets up the coding in Table 9.5 with the `factor()` function, so all we need do is to add the name of the categorical variable as the argument.

We now analyze the results from the entire experiment. The following R code plots strength against film thickness using different plotting characters for the different glues.

```
psym=rep(1,length(glue))
psym[glue=="B"]=2
psym[glue=="C"]=3
psym[glue=="D"]=4
plot(film_thick,strength,pch=psym)
legend(15,50,pch=1:4,c("A","B","C","D"))
```

The most striking feature is the decrease in strength as the film thickness increases. The glue formulation effects are relatively small although it seems that the B points tend to lie above the A points. The regression analysis is

```
> m2=lm(strength ~ film_thick + factor(glue))
> summary(m2)

Call:
lm(formula = strength ~ film_thick + factor(glue))

Residuals:
     Min      1Q   Median      3Q      Max
-1.85815 -0.93808  0.09603  0.78135  2.27007

Coefficients:
               Estimate Std. Error t value Pr(>|t|)
(Intercept)     59.6491     2.0793  28.687 1.61e-14 ***
film_thick      -1.0099     0.1547  -6.529 9.54e-06 ***
factor(glue)B    1.7681     0.7704   2.295   0.0366 *
factor(glue)C    0.8321     0.7578   1.098   0.2895
factor(glue)D    0.3580     0.7483   0.478   0.6392
---
Residual standard error: 1.182 on 15 degrees of freedom
Multiple R-squared:  0.803,     Adjusted R-squared:  0.7505
F-statistic: 15.29 on 4 and 15 DF,  p-value: 3.59e-05
```

The first conclusion is that the joint strength decreases with film thickness over the range 10 to 16. It follows that the most effective way to increase the strength of the joint is to better control the process so that the film thickness is close to 10. We have fitted a linear relationship between film thickness and strength over the range of thickness values encountered in the experiment, 10 to 16, and we cannot infer that this linear relationship extends beyond this range. Moreover, we know that strength cannot continue to increase as film thickness decreases towards 0 because the bond strength with no glue, film thickness 0, will be 0. The second conclusion is that glue B gives a significantly[20] stronger joint than glue A. The 95% confidence interval for the increase is $[0.13, 3.41]$, which was calculated using R by

```
> m2$coef[3] - qt(.975,15)*summary(m2)$coef[3,2]
factor(glue)B
     0.1261029
> m2$coef[3] + qt(.975,15)*summary(m2)$coef[3,2]
factor(glue)B
     3.410062
```

You may think it is rather convenient that the comparison of glue formulations was made against glue A which happens to have the lowest sample mean strength. The `factor()` function default is to work with alphabetical order of the category names and estimate coefficients relative to the first. The default can be changed by specifying the order of categories. For example, if we wish to compare the other three glues against D, which might be the company's standard formulation:

```
> m2d=lm(strength~film_thick + factor(glue,levels=c("D","A","B","C")))
> summary(m2d)

Call:
lm(formula = strength ~ film_thick + factor(glue, levels = c("D",
    "A", "B", "C")))

Residuals:
     Min       1Q   Median       3Q      Max
-1.85815 -0.93808  0.09603  0.78135  2.27007

Coefficients:
                                         Estimate Std. Error t value
(Intercept)                               60.0071 2.0494 29.280
film_thick                                -1.0099 0.1547 -6.529
factor(glue, levels=c("D", "A", "B", "C"))A  -0.3580 0.7483 -0.478
factor(glue, levels=c("D", "A", "B", "C"))B   1.4101 0.7635  1.847
factor(glue, levels=c("D", "A", "B", "C"))C   0.4740 0.7534  0.629
                                          Pr(>|t|)
(Intercept)                               1.19e-14 ***
film_thick                                9.54e-06 ***
factor(glue, levels=c("D", "A", "B", "C"))A  0.6392
factor(glue, levels=c("D", "A", "B", "C"))B  0.0846 .
factor(glue, levels=c("D", "A", "B", "C"))C  0.5387
```

---

[20] Here we ignore the issue of multiple comparisons - see Section 9.7.3 and Chapter 12.

```
Residual standard error: 1.182 on 15 degrees of freedom
Multiple R-squared:  0.803,     Adjusted R-squared:  0.7505
F-statistic: 15.29 on 4 and 15 DF,  p-value: 3.59e-05
```

There is some weak evidence that B is stronger than D: the 90% confidence interval for the difference is all positive, but the 95% confidence interval includes 0. You might also reflect that the A versus B difference was the greatest of the $\binom{4}{2}$ possible comparisons and question whether there is strong evidence against a hypothesis that glue formulation makes no difference to the strength. We can answer this question using an F-test for an added set of variables. The test is a more stringent criterion for choosing a more complex model than is a reduction in the estimated standard deviation of the errors.

9.7.3 F-test for an added set of variables

There are n data and, in our usual notation, we fit a model **m1** with k predictor variables

$$Y_i \ = \ \beta_0 + \beta_1 x_1 + \ldots + \beta_k x_k + \varepsilon_{1i},$$

where $\varepsilon_{1i} \sim N\left(0, \sigma_1^2\right)$ and are independently distributed. We then fit a model **m2** with l predictor variables, which include the original k predictor variables and with an additional set of $\ell - k$ predictor variables.

$$Y_i \ = \ \beta_0 + \beta_1 x_1 + \ldots + \beta_k x_k + \beta_{k+1} + \ldots + \beta_\ell + \varepsilon_{2i},$$

where $\varepsilon_{2i} \sim N\left(0, \sigma_2^2\right)$ and are independently distributed. For the model **m2** to be a statistically significant improvement on model **m1**, at the α level, we need to test the hypothesis

$$H_0 : \beta_{k+1} \ = \ \ldots \ = \ \beta_\ell \ = \ 0$$

at the α level and reject it in favor of the alternative hypothesis

$$H_1 : \text{not all } \beta_{k+1}, \ldots, \beta_\ell \text{ equal } 0.$$

In the case that H_1 holds the $\beta_0, \ldots, \beta_k$ in the two models are not generally the same, but it follows from the definition of H_0 that they are identical if H_0 is true. The test follows from the fact that if the null hypothesis, H_0, is true then

$$\frac{(RSS_{m1} - RSS_{m2})/(\ell - k)}{RSS_{m2}/(n - \ell - 1)} \ \sim \ F_{\ell-k, n-\ell-1}.$$

A proof of this result is given on the website, but it is intuitively plausible. Whether or not H_0 is true, the denominator is an unbiased estimator of σ_2^2 with $n - \ell - 1$ degrees of freedom. If H_0 is true then $\sigma_1^2 = \sigma_2^2$, and the numerator is an independent estimator of this common variance with $\ell - k$ degrees of freedom.

Example 9.2: (Continued) Elysium Electronics

We test the hypothesis that there is no difference in the glue formulations. More formally if we define model **m1** as

$$Y_i \;=\; \beta_0 + \beta_4 x_4 + \varepsilon 1i,$$

for $i = 1, \ldots, 20$ where x_4 is film thickness, and model **m2** as

$$Y_i \;=\; \beta_0 + \beta_1 x_1 + \beta_2 x_2 + \beta_3 x_3 + \beta_4 x_4 + \varepsilon 2i,$$

where x_1, x_2, x_3 are the indicator variables for glues defined in the previous section, then the null hypothesis is

$$H_0 \;:\; \beta_1 \;=\; \beta_2 = \beta_3 = 0$$

and the alternative hypothesis is

$$H_1 \;:\; \text{not all } \beta_1, \beta_2, \beta_3 \text{ equal } 0.$$

The R code to test H_0 is

```
> m2=lm(strength~film_thick+factor(glue))
> print(c(summary(m1)$sigma,m1$df))
[1]   1.272665 18.000000
> print(c(summary(m2)$sigma,m2$df))
[1]   1.182152 15.000000
> RSSm1=summary(m1)$sigma^2*m1$df
> RSSm2=summary(m2)$sigma^2*m2$df
> F= F=((RSSm1-RSSm2)/(m1$df-m2$df))/summary(m2)$sigma^2
> print(1-pf(F,m1$df-m2$df,m2$df))
[1] 0.164329
```

The p-value is 0.16 and there is no evidence to reject H_0 at the 10% level.

The estimated standard deviation of the errors in **m2**, 1.18 on 15 degrees of freedom, is less than that for **m1**, 1.27 on 18 degrees of freedom, so model **m2** offers some improvement on **m1**. Also the confidence interval for the difference between glues A and B suggested that B gives stronger joints than A does. However, the p-value for the test of H_0 is 0.16 and the statistical significance of the improvement doesn't reach the customary 0.10, or more stringently 0.05, level. Overall, there is weak evidence that B is better than A and possibly D. We would suggest a follow up experiment to confirm these tentative findings.

R facilitates comparison of models with the *anova()* function. We first see how *anova()* works for one fitted regression model. For example

```
> anova(m1)

Analysis of Variance Table
Response: strength
            Df Sum Sq Mean Sq F value   Pr(>F)
film_thick   1 77.251  77.251  47.696 1.86e-06 ***
Residuals   18 29.154   1.620
---
```

gives, in the notation of Section 9.5.3, RSS as 77.251 and REGSS as 29.154. The TSS is $77.251 + 29.154 = 106.405$ as can be checked with

```
> 19*var(strength)
[1] 106.4055
```

Similarly

```
> anova(m2)

Analysis of Variance Table
Response: strength
             Df Sum Sq Mean Sq F value    Pr(>F)
film_thick    1 77.251  77.251  55.279 2.097e-06 ***
factor(glue)  3  8.192   2.731   1.954    0.1643
Residuals    15 20.962   1.397
---
```

The comparison of two models, by testing H_0, is given by

```
> anova(m1,m2)

Analysis of Variance Table
Model 1: strength ~ film_thick
Model 2: strength ~ film_thick + factor(glue)
  Res.Df    RSS Df Sum of Sq      F Pr(>F)
1     18 29.154
2     15 20.962  3    8.1919 1.954 0.1643
```

We see the p-value of 0.16 at the foot of the right hand column.

Example 9.3: Port throughput [indicator variables]

The data in Table 9.6 is the throughput in million twenty-foot equivalent units (TEU) during the year 1997 for 28 ocean ports in China [Frankel, 1998] together with: the total number of ocean berths (*tob*); the number of general cargo berths (*gcb*); and the region of China classified as north (N), east (E), and south (S). The number of specialist berths is the difference, $tob - gcb$. It seems reasonable to suppose that TEU will increase with the total number of ocean berths (see Figure 9.8), although the coefficient may be larger for ports that offer a higher proportion of specialist berths.

A model for this is:

$$TEU_i = \beta_0 + (\beta_1 - \delta \times gcb_i/tob_i) \times tob_i + \varepsilon_i$$

for $i = 1, \ldots, 28$. The model can be rewritten in the form

$$TEU_i = \beta_0 + \beta_1 \times tob_i + \beta_2 \times gcb_i + \varepsilon_i.$$

Notice that we include an intercept β_0 although throughput would inevitably be 0 if $tob = 0$. We do not wish to restrict the approximate linear relationship, over the range of *tob* for which we have data, to a proportional relationship.

TABLE 9.6: Ocean ports in China.

Region	tob	gcb	TEU
N	48	34	64.12
N	10	8	11.35
N	6	2	5.50
N	38	18	83.82
N	6	6	7.86
N	62	46	57.87
N	4	3	2.25
N	24	15	73.02

Region	tob	gcb	TEU
E	13	8	14.52
E	20	8	17.15
E	17	15	8.78
E	11	8	16.15
E	9	4	16.09
E	68	34	165.67
E	3	2	6.38
E	22	8	68.52
E	6	0	10.05
E	13	10	10.27

Region	tob	gcb	TEU
S	11	2	13.13
S	2	0	0.88
S	6	1	10.00
S	6	0	5.25
S	32	16	73.00
S	23	16	6.82
S	4	0	2.02
S	8	8	4.63
S	2	0	4.64
S	6	4	2.82

```
> Chinaports.dat=read.table("Chinaports.txt",header=TRUE)
> print(head(Chinaports.dat))
  region tob gcb    TEU
1      N  48  34 64.12
2      N  10   8 11.35
3      N   6   2  5.50
4      N  38  18 83.82
5      N   6   6  7.86
6      N  62  46 57.87
> attach(Chinaports.dat)
> n=length(TEU)
> rat=gcb/tob
> rat.bin=rep(1,n)
> rat.bin[rat > median(rat)] = 2
> plot(tob,TEU,pch=rat.bin)
> position=list(x=.9,y=150)
> legend(position,c("Below median gcb/tob","Above median gcb/tob"),
+ pch=c(1,2))
> #plot(tob,log(TEU),pch=rat.bin)
> #position=list(x=.9,y=150)
> #legend(position,c("Below median","Above median"),pch=c(1,2))
> m1=lm(TEU~tob+gcb) ; print(summary(m1))

Call:
lm(formula = TEU ~ tob + gcb)
```

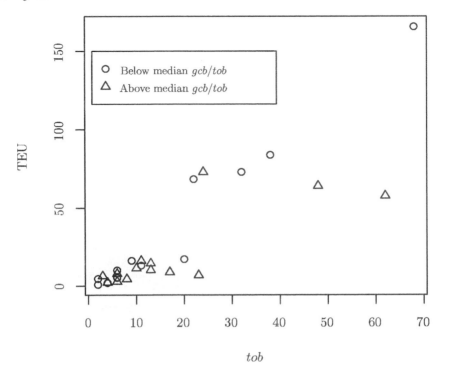

FIGURE 9.8: Twenty-foot equivalent units (TEU) vs total number of ocean ports (*tob*) for the China ports data.

```
Residuals:
    Min      1Q   Median      3Q      Max
-29.339  -5.389    0.572   7.393   36.496

Coefficients:
            Estimate Std. Error t value Pr(>|t|)
(Intercept)  -7.5277     3.6900  -2.040    0.052 .
tob           4.2392     0.4840   8.759 4.33e-09 ***
gcb          -3.8459     0.7418  -5.185 2.31e-05 ***
---
Residual standard error: 13.7 on 25 degrees of freedom
Multiple R-squared:  0.8782,    Adjusted R-squared:  0.8684
F-statistic: 90.09 on 2 and 25 DF,  p-value: 3.736e-12
```

The estimated increase in throughput for an additional specialist ocean berth is 4.24 TEU whereas the estimated increase for an additional general cargo is 4.24−3.85, which is only 0.39. The model suggests that the best strategy for increasing TEU is to build additional specialist berths. It may also be worth considering conversion of some general cargo berths to specialist berths. However, the model has been fitted to data that includes ports offering a high number of general cargo berths. If most of these were converted to specialist berths the remaining general cargo berths might see a dramatic increase in throughput.

The next step is to investigate whether there is evidence of a difference in the regions.

```
> m2=lm(TEU ~ tob + gcb + (factor(region)))
> summary(m2)

Call:
lm(formula = TEU ~ tob + gcb + (factor(region)))

Residuals:
    Min      1Q  Median      3Q     Max
-31.196  -6.810   0.257   7.210  31.782

Coefficients:
                  Estimate Std. Error t value Pr(>|t|)
(Intercept)        -5.4507     5.2586  -1.037    0.311
tob                 4.3670     0.4988   8.755 8.82e-09 ***
gcb                -4.1928     0.7900  -5.308 2.18e-05 ***
factor(region)N     4.7733     6.9149   0.690    0.497
factor(region)S    -6.1939     6.2163  -0.996    0.329
---
Residual standard error: 13.59 on 23 degrees of freedom
Multiple R-squared:  0.8896,    Adjusted R-squared:  0.8704
F-statistic: 46.34 on 4 and 23 DF,  p-value: 1.108e-10
```

There is a small reduction in the estimated standard deviation of the errors, from 13.7 to 13.6, and the fitted coefficients have the north with higher TEU, by 4.8, and the south with lower TEU, by 6.2, than the east. Before proceeding we should check that the residuals look reasonable. This step is particularly important in multiple regression because initial plots do not show the residuals in the way that a plot of y against a single predictor x does. We plot residuals against fitted values (see Figure 9.9a) to check whether an assumption that the errors have a constant variance ($A4$) is reasonable. The normal quantile-quantile plot (see Figure 9.9, right) indicates whether an assumption of normality ($A5$) is reasonable.

```
> par(mfrow=c(1,2))
> plot(m2$fit,m2$res)
> qqnorm(m2$res)
```

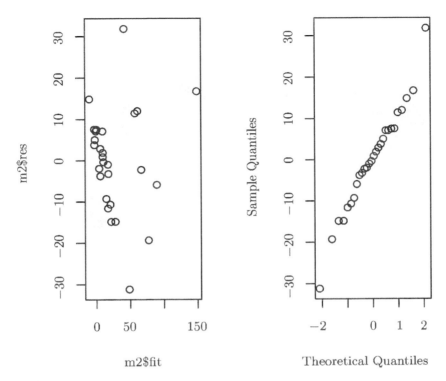

FIGURE 9.9: Residuals vs fitted values (left) and Q-Q plot (right) for the China ports data.

The original plot of the data (Figure 9.8) shows that the majority of ports are relatively small, in terms of total number of ocean berths, and that the larger ports will have a considerable influence on the estimated coefficients. It also suggests that the variance of the errors may increase with the size of the port. The fact that an increase in variance is less apparent in the plot of the residuals against the total number of ocean berths is due to the inclusion of the number of specialist berths in the model. The negative coefficient of general cargo berths is strong evidence that providing specialist berths increases throughput. The coefficients of the indicator variables for N and S relative to E correspond to an estimate that average TEU is 4.77 higher in N and -6.19 lower in S. Neither of these differences is statistically significant when compared with E. However, the difference between N and S is estimated as 10.97 and if the indicator variables are set up relative to N the coefficient for S will have a t-value around 1.6 (roughly $10.97/6.91$) and a p-value around 0.10. You are asked to find a more precise value in Exercise 9.30. There is some weak evidence that throughput is higher in the north than in the south for a port offering the same numbers of ocean berths.

9.7.4   Quadratic terms

Quadratic functions of the predictor variables can be added to regression models to allow for curvature in the response over the range of the predictor variables.

Example 9.4: Plastic sheet manufacture [squared terms]

Flexible plastic sheet can be manufactured by a bubble blowing process. The data in Table 9.7 is the tensile strength and extrusion rate for 35 sheets. Over this range of extrusion rates the tensile strength tends to increase with extrusion rate, and a process engineer is investigating whether this is well modeled as a linear relationship or better modeled by including a quadratic term. We read the data, plot strength against

TABLE 9.7: Flexible plastic sheet manufacture.

Extrusion rate	Strength	Extrusion rate	Strength
40	173	179	197
65	179	180	203
75	171	190	263
75	151	228	222
85	192	229	197
95	217	236	217
105	186	245	233
115	211	255	246
120	187	290	254
130	183	343	330
140	189	380	284
145	203	385	321
145	181	415	333
160	241	498	321
165	187	500	329
170	254	510	290
178	235	520	316
		750	337

extrusion rate, fit a regression of strength on extrusion rate, and add the fitted line to the plot in the top left of Figure 9.10.

```
> bubble.dat=read.table("bubble.txt",header=TRUE)
> print(head(bubble.dat))
  exrate strength
1     40      173
2     65      179
3     75      171
4     75      151
5     85      192
6     95      217
> attach(bubble.dat)
> par(mfrow=c(2,2))
> m1=lm(strength~exrate) ; print(summary(m1))
```

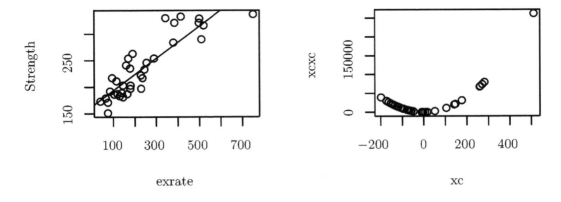

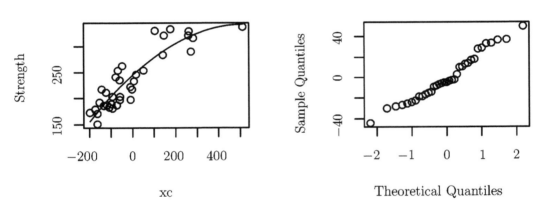

FIGURE 9.10: Plots of plastic sheet extrusion data.

```
Call:
lm(formula = strength ~ exrate)

Residuals:
    Min      1Q  Median      3Q     Max
-53.267 -16.984  -3.955  13.808  63.052

Coefficients:
            Estimate Std. Error t value Pr(>|t|)
(Intercept) 163.02032    7.92719   20.57  < 2e-16 ***
exrate        0.30300    0.02751   11.02 1.36e-12 ***
---
Residual standard error: 26.37 on 33 degrees of freedom
Multiple R-squared:  0.7862,    Adjusted R-squared:  0.7797
F-statistic: 121.4 on 1 and 33 DF,  p-value: 1.359e-12

> plot(exrate,strength)
```

```
> abline(m1)
```

It is generally good practice to center[21] predictor variables before squaring them because variables with a large mean and relatively small standard deviation (a low CV) are
highly correlated with their square. Generally, high correlations lead to ill-conditioned
matrices for inversion and make it more difficult to interpret the coefficients. In this
case it is not necessary to center exrate but there is nothing to lose by doing so, and
we subtract the mean.

```
> print(c("correlation between exrate and exrate^2",
+   round(cor(exrate,exrate^2),3)))
[1] "correlation between exrate and exrate^2"
[2] "0.96"
> xc=exrate-mean(exrate)
> xcxc=xc^2
> plot(xc,xcxc)
> print(c("correlation between xc and xcxc",round(cor(xc,xcxc),3)))
[1] "correlation between xc and xcxc" "0.711"
```

The centering considerably reduces the correlation between the linear and quadratic
terms describing extrusion rate, but it remains quite high because most of the deviations
from the mean are relatively small and negative (see the top right of Figure 9.10). The
regression including a quadratic term follows

```
> m2=lm(strength~xc+xcxc) ; print(summary(m2))

Call:
lm(formula = strength ~ xc + xcxc)

Residuals:
    Min     1Q  Median     3Q     Max
-44.247 -17.701  -4.954  15.656  49.839

Coefficients:
              Estimate Std. Error t value Pr(>|t|)
(Intercept)  2.448e+02  5.193e+00  47.136  < 2e-16 ***
xc           3.760e-01  3.530e-02  10.653 4.72e-12 ***
xcxc        -3.636e-04  1.249e-04  -2.911   0.0065 **
---
Residual standard error: 23.81 on 32 degrees of freedom
Multiple R-squared:  0.831,    Adjusted R-squared:  0.8204
F-statistic: 78.66 on 2 and 32 DF,  p-value: 4.437e-13
```

[21]Subtracting the mean from predictor variables is an example of centering. The term centering refers to
subtraction of any constant near the center of the data such as the mean, median or average of the greatest
and least values.

The estimate of the coefficient, -3.636×10^{-4}, of the quadratic term relative to its standard deviation, 1.249×10^{-4}, is nearly 3 in absolute value and the corresponding p-value is 0.0065. The coefficient of the quadratic term and its standard error do not change with the centering of extrusion rate, but the coefficient of the linear term does change. There is strong evidence that a quadratic curve is a better fit over the range of extrusion rates, from 50 to 750, but the curve is highly influenced by the single sheet manufactured at the high extrusion rate of 750. The fitted curve is shown in in the lower left of Figure 9.10. It would be advisable to carry out some more runs at high extrusion rates. The normal quantile-quantile plot of the residuals, lower right of Figure 9.10, does not show any particular outlying values.

```
> plot(xc,strength)
> xcp=1:1000;xcp=min(xc)+xcp*(max(xc)-min(xc))/1000
> newdat=data.frame(xc=xcp,xcxc=xcp^2)
> predm2=predict(m2,newdata=newdat)
> lines(xcp,predm2,lty=1)
> qqnorm(m2$res)
```

We have already seen that the predictor variables in a multiple regression analysis can be extended to include functions of the original predictor variables. A very important example is the product of two predictor variables that allows for their interaction. In this section we consider quadratic terms. As a general rule we retain linear terms if we have quadratic terms or interactions and we include an intercept term. In principle, we could include cubic and higher order terms but convincing practical applications are uncommon.

The next example is also taken from the chemical engineering industry (both data sets were from the erstwhile Imperial Chemical Industries in the UK).

Example 9.5: Calcium metal production [quadratic surface]

The data in Table 9.8 are obtained from 27 runs of a process that produces calcium (Ca) metal. The response variable is the percentage of calcium (Ca) in the mix, and the predictor variables are the percentage of calcium chloride, CaCl in the mix, and the temperature of the mix in degrees Celsius. We standardize the predictor variables by subtracting their means and dividing by their standard deviations. This not only centers the predictor variables but also makes them non-dimensional and measured on the same scale so that coefficients of linear terms are directly comparable.

```
> CaCl.dat=read.table("CaCl.txt",header=TRUE)
> print(head(CaCl.dat))
   Ca Temp CaCl
1 3.02  547 63.7
2 3.22  550 62.8
3 2.98  550 65.1
4 3.90  556 65.6
5 3.38  572 64.3
6 2.74  574 62.1
> attach(CaCl.dat)
> x1=(CaCl-mean(CaCl))/sd(CaCl)
> x2=(Temp-mean(Temp))/sd(Temp)
> m1=lm(Ca~x1+x2) ; print(summary(m1))
```

TABLE 9.8: Calcium metal production.

Ca	Temp	CaCl	Ca	Temp	CaCl
3.02	547	63.7	2.36	579	62.4
3.22	550	62.8	2.90	580	62.0
2.98	550	65.1	2.34	580	62.2
3.90	556	65.6	2.92	580	62.9
3.38	572	64.3	2.67	591	58.6
2.74	574	62.1	3.28	602	61.5
3.13	574	63.0	3.01	602	61.9
3.12	575	61.7	3.01	602	62.2
2.91	575	62.3	3.59	605	63.3
2.72	575	62.6	2.21	608	58.0
2.99	575	62.9	2.00	608	59.4
2.42	575	63.2	1.92	608	59.8
2.90	576	62.6	3.77	609	63.4
			4.18	610	64.2

```
Call:
lm(formula = Ca ~ x1 + x2)

Residuals:
     Min       1Q    Median       3Q       Max
-0.67278 -0.17656  0.01842  0.21991  0.66398

Coefficients:
            Estimate Std. Error t value Pr(>|t|)
(Intercept)  2.94778    0.07102  41.506  < 2e-16 ***
x1           0.48339    0.08606   5.617  8.8e-06 ***
x2           0.20981    0.08606   2.438   0.0225 *
---
Residual standard error: 0.369 on 24 degrees of freedom
Multiple R-squared:  0.5719,    Adjusted R-squared:  0.5362
F-statistic: 16.03 on 2 and 24 DF,  p-value: 3.788e-05
```

The linear model shows that the mean value of Ca is 2.95 and that x_1, the standardized concentration of CaCl on the mix, has a dominant effect on the response. The coefficient is positive indicating that an increase in CaCl will lead to an increase in Ca.

The next step is to compare this with a quadratic model that includes an interaction. This has the general form

$$Y_i = \beta_0 + \beta_1 x_1 + \beta_2 x_2 + \beta_3 x_1^2 + \beta_4 x_2^2 + \beta_5 x_1 x_2 + \varepsilon_i$$

where $i = 1, \ldots, n$, and $\varepsilon_i \sim N(0, \sigma^2)$. The deterministic part of the model represents a quadratic surface that is either: a paraboloid with vertex pointing up to give a maximum (upper right Figure 9.11); a paraboloid with vertex pointing down to give a minimum (lower left Figure 9.11); or a saddle point (lower right Figure 9.11); if β_3 and β_4 are 0, then y is a linear function of x_1 for fixed x_2 and similarly a linear function of x_2 for fixed x_1 and the surface is a warped plane (upper left Figure 9.11).

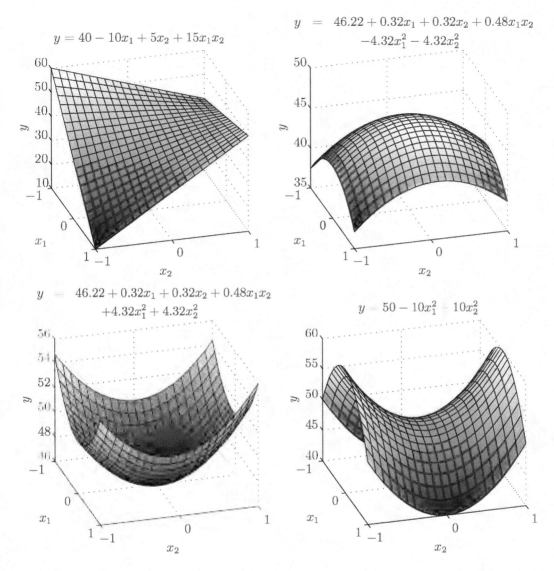

FIGURE 9.11: Quadratic model forms: warped plane (top left); maximum (top right); minimum (lower left); and saddle point (lower right).

```
> x1x1=x1*x1
> x2x2=x2*x2
> x1x2=x1*x2
> m2c=lm(Ca~x1+x2+x1x1+x2x2+x1x2) ; print(summary(m2c))

Call:
lm(formula = Ca ~ x1 + x2 + x1x1 + x2x2 + x1x2)

Residuals:
      Min       1Q    Median       3Q      Max
 -0.56103 -0.15307  0.04411  0.12568  0.62669
```

```
Coefficients:
             Estimate Std. Error t value Pr(>|t|)
(Intercept)   2.80736    0.08282  33.898  < 2e-16 ***
x1            0.41245    0.08615   4.788 9.91e-05 ***
x2            0.04822    0.08244   0.585  0.56485
x1x1          0.21806    0.06677   3.266  0.00369 **
x2x2          0.13920    0.07058   1.972  0.06189 .
x1x2          0.39081    0.10366   3.770  0.00112 **
---
Residual standard error: 0.2959 on 21 degrees of freedom
Multiple R-squared:  0.7592,    Adjusted R-squared:  0.7019
F-statistic: 13.24 on 5 and 21 DF,  p-value: 6.717e-06
```

The quadratic surface is a clear improvement on the plane given by **m1**. The estimated standard deviation of the errors has reduced from 0.37 to 0.30. Notice that we retain x_2 despite the standard error of its coefficient being nearly twice the value of the coefficient itself because its interaction with x_2, and to a lesser extent its square x_2^2 make a substantial contribution to the model. We now check the residuals from the model (see Figure 9.12).

```
par(mfrow=c(1,2))
plot(m2c$fit,m2c$res)
qqnorm(m2c$res)
```

Figure 9.12 (left) indicates that the assumption of constant variance (**A4**) seems plausible, as there are no clear outlying points. The assumption of normal errors (**A5**) (see Figure 9.12, right), is also plausible, which is mainly required for prediction intervals. When the response is some function of two predictor variables we can show the value of the response by contours in the plane. One function in R for drawing contours is **contour()**. The tricky bit is setting up a grid of (x, y) values at which to calculate z and this is facilitated by the **expand.grid()** function as shown below.

```
> x <- min(x1)+(max(x1)-min(x1))*c(1:100)/100
> y <- min(x2)+(max(x2)-min(x2))*c(1:100)/100
> X <- expand.grid(x)
> Y <- expand.grid(y)
> z <- matrix(m2c$coef[1]
+ + m2c$coef[2]*X
+ + m2c$coef[3]*Y
+ + m2c$coef[4]*X^2
+ + m2c$coef[5]*Y^2
+ + m2c$coef[6]*X*Y,100,100)
> contour(x,y,z,xlab="CaCl (standardized)",
+ ylab="Temperature (standardized)")
```

From the contour plot in Figure 9.13, we see that the process is operating around a saddle point. To maximize the percentage Ca, set a low temperature if the CaCl percentage is low and set a high temperature if the CaCl percentage is high[22].

---

[22]If the contours represented elevation, the pass would run from north-west to south-east, and this direction is determined by interaction.

Normal Q-Q Plot

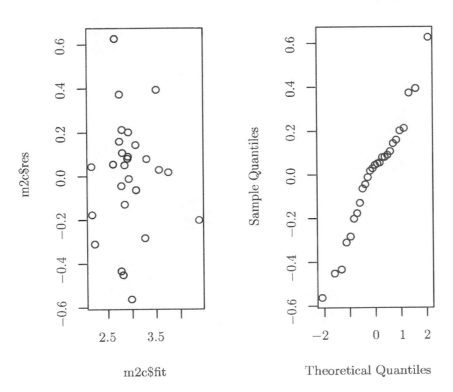

FIGURE 9.12: Residuals of fitted model for calcium metal regressed on linear, quadratic and intersection of calcium chloride and temperature: residuals against fitted values (left) ordered residuals against expected values of normal order statistics (right).

9.7.5 Guidelines for fitting regression models

We have aimed to demonstrate regression modeling through a variety of practical applications. In the case of *Conch communications*, the objective was to provide predictions to support management decisions. The main purpose of the analysis of the throughput for the China ports was to quantify the effect of providing specialist berths designed to handle specific material rather than general cargo berths, and to investigate whether there is a systematic difference between sea-board regions of China. The model of inputs to the Fontburn Reservoir, which is fitted in the next section, was used in a computer simulation to investigate the consequences of different operating policies. Although there is usually some subjectivity behind the choice of regression model there are some general principles to follow.

- There is no true model, but the lack of a true model does not imply that any model will do. We aim to ensure that the model we fit: is a reasonable approximation to the physical situation; is consistent with the assumptions we make about errors; makes full use of the available data; and provides answers to the questions posed.

- Statistical significance does not equate to practical significance. If we have a small data set the standard errors of the estimated coefficients of predictor variables that we know

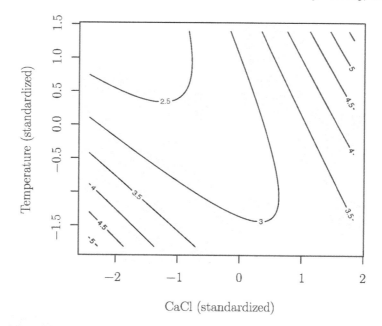

FIGURE 9.13: Contour plot of percentage Ca metal in a mix.

to have some effect on the response may be large. It follows that a 95%, for example, confidence interval may include 0 and the effect is not statistically significant at the 5% level. In contrast, if we have a very large data set predictor variables with a negligible practical effect on the response may be statistically significant. Ideally, the sample size is chosen to lie between these two scenarios but you may not have any say in this. For example, Conch owns only 10 shops. We discuss the choice of sample size in the context of defined experiments in Chapter 11.

- Include an intercept whether or not it is statistically significant[23] (Exercise 9.31).

- Include predictor variables that you know have an effect on the response from physical reasoning. For example an experiment to compare four different treatments of steel to reduce corrosion in sea water was carried out from oil rigs. The steel specimens were all from the same roll and were immersed in sea water for one year. However the precise number of days of immersion varied somewhat around 365 days, and is known for each specimen. The response is a measure of corrosion, and the regression for corrosion should include the number of days of immersion, whether or not it is statistically significant, as well as three indicator variables to represent the four treatments.

- Use indicator variables for categorical variables. If numbers in each category differ make the comparisons relative to the largest category. Include indicators for all the categories, whether or not they are statistically significant.

- If you intend comparing the relative effects of predictor variables, then consider standardizing the predictor variables. Their effects are then proportional to their estimated coefficients.

---

[23]An exception is if you want to model Y as directly proportional to x, for simplicity, and expect this relationship to hold for x from 0 up to at least its greatest value in the sample.

- Investigate interactions by including product terms in a model.

- Center predictor variables before adding their squares to the model.

- If you are squaring one predictor variable it may be appropriate to square them all and include all their interactions, fitting a quadratic surface rather than a plane, or hyper-plane. The statistical significance of the change can be assessed by an F-test.

- Plot residuals against predictor variables and the fitted values to check for evidence of curvature or of non-constant variance, and to identify particularly influential data. A normal qq-plot will identify outlying residuals.

- Predictor variables are generally correlated amongst themselves with the consequence that estimated coefficients will depend on which other variables are included in the model. It follows that the statistical significance of a predictor variable depends on which other predictor variables are in the model. There are automated systematic procedures for variable selection including the R function step() and the R package leaps, and these do allow for chosen predictor variables to be included in any model. However, the applications in this book are better suited to intervention by the modeler.

- Multicollinearity generally refers to high correlations between predictor variables. A consequence is that estimators of the coefficients are highly correlated and their standard errors are high. In the case of x and x^2 centering x can reduce, or remove, the correlation. In the case that two predictors are different measures of the same feature, either one can be chosen. For example, we might have 5-day, 7-day and 30-day strengths of concrete cubes made from the same batch, for 20 batches, with the intention of fitting a regression to predict the 30-day strength of concrete from batches using 5-day and 7-day measurements. It is likely that 5-day and 7-day strengths will be highly correlated, that either on its own is a useful predictor of 30-day strength, but including both does not give a reduced estimate of the standard deviation of the errors. Moreover including both results in high standard errors for both coefficients, one of the two coefficients could even be negative. We would then choose to use only 5-day strength as it is of more practical value, because the sooner potential sub-standard concrete is detected the better.

- In some applications we get an improved model if we take the response in the regression as some transformation of the measured response. In particular, a logarithmic transformation[24] is often effective, particularly if the standard deviation of the errors is increasing in proportion to the magnitude of the response. However, the model is for the transformed response and an improved R^2 does not necessarily imply that the sum of squared prediction errors in terms of the measured response will be reduced. Furthermore, the model is predicting the mean value of the transformed response and this will not transform back to the mean value of the measured response[25]. If the errors in the model for the transformed response have a symmetric distribution, then back transformation of a prediction for the mean value of the transformed response corresponds to the median of the distribution of measured responses. If you are predicting costs of engineering schemes (as in Chapter 14) then this would lead to systematic under prediction of out-turn costs.

[24]$\ln(y)$ or more generally $\ln(y+a)$ where a is a constant chosen to make all the $y+a$ positive or to reduce the effect of the transformation (Exercise 5.29).
[25]Unless the transform is a linear scaling as in centering or standardizing.

9.8   Time series

Multiple regression can be used for modeling time series. The covariates are time and past values of the variable.

9.8.1   Introduction

A time series is a sequence of observations of some variable over time. It is usual to choose a constant time step, and the variable is either sampled or aggregated. Examples of sampled time series are: the continuous electrical signal from an accelerometer during landing of an aircraft sampled with an analog-to-digital converter at 1 million samples per second (the sampling interval needs to be considerably shorter than the wavelength of the highest frequencies in the signal (Section 9.8.2)); and noon-day temperature measured by the National Weather Service at Zephyr Hills, Florida. Examples of aggregated series are: rainfall over 6 minute intervals; and inflows of water to a reservoir per month.

There are many possible models for time series, including multiple regression models. As with any mathematical modeling there is no correct model but a good model will give a close match to reality. There are at least three reasons for fitting time series models: to obtain some insight into the underlying process; to make short term forecasts; and for simulation studies. We illustrate general principles with two examples: monthly inflows of water to a reservoir; and water levels at the center of a wave tank when waves are generated by pseudo-random wave makers.

9.8.2   Aliasing and sampling intervals.

Once a continuous signal is sampled any frequencies, measured in cycles per second (Hz), higher than half the sampling frequency, measure in seconds, will be indistinguishable from their lower frequency aliases. For example, a sine function of frequency 1 Hz sampled at 0.2 second intervals will be indistinguishable from a sine function that makes an additional cycle within the sampling interval and has a frequency of 6 Hz (1 Hz + 5 Hz). Positive frequency is conventionally represented by anti-clockwise rotation of a radius in a unit circle centered on the origin, and negative frequency corresponds to clockwise rotation. Therefore, the sine function is also indistinguishable from 4 Hz (1 Hz - 5 Hz) Figure 9.14.

Definition 9.8: Nyquist frequency

The Nyquist frequency Q is related to the sampling interval Δ by

$$Q = \frac{1}{2\Delta}.$$

Definition 9.9: Alias frequencies

If a continuous signal is sampled at Δ second intervals. A sine function with frequency f cycles per second will be indistinguishable from sine functions with frequencies

$$f \pm k\frac{1}{\Delta},$$

for any integer k. These frequencies for $k \neq 0$ are known as the alias frequencies of f.

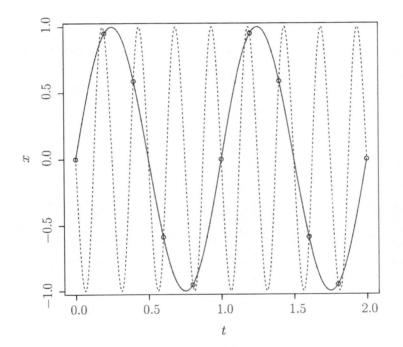

FIGURE 9.14: Two cycles of a sine function with frequency 1 Hz (solid line) and an alias frequency 4 Hz (broken line) with $\Delta = 0.2$. The Nyquist frequency is 2.5 Hz. Sampled values shown as circles.

```
> t=c(0:10)/5
> tc=c(0:2000)/1000
> x=sin(2*pi*t)
> xc=sin(2*pi*tc)
> xa=sin(-4*2*pi*tc)
> plot(t,x)
> lines(tc,xc)
> lines(tc,xa,lty="dashed")
```

Any signal higher than the Nyquist frequency, Q, will be indistinguishable from its lower frequency alias. In practice the cut-off is not so sharply defined and electronic measuring equipment uses analogue filters, known as anti-aliasing filters, to remove frequencies above around $0.5Q$.

9.8.3 Fitting a trend and seasonal variation with regression

Definition 9.10: Deterministic trend

A deterministic trend is a function of time t and the simplest example is a linear trend, $y = a + bt$, where a and b are constants.

A linear trend is often an adequate approximation over the time period for which there are data, and it can be extrapolated slightly to make short term forecasts.

Definition 9.11: Seasonal variation

Seasonal variation is a repeating deterministic pattern with a known period.

An example is expected electricity demand which varies in a consistent manner within the 24 hour day and throughout the year.

Definition 9.12: Stationarity in time series models

A time series model is stationary in the mean if its expected value is constant over time. That is, there is no trend or seasonal variation in the model.

If we identify and remove any trend and seasonal effects in a time series, the resultant time series can generally be considered a realization of a stationary time series model.

We now discuss seasonal variation in more detail in the context of monthly data, the period being a year of 12 time steps. A time series model for a sequence of random variables $\{Y_y\}$, where $t = 1, \ldots, n$ is month, with a linear deterministic trend and sinusoidal seasonal variation of frequency 1 cycle per year is

$$Y_t = \beta_0 + \beta_1 t + \beta_2 \cos\left(\frac{2\pi t}{12}\right) + \beta_3 \sin\left(\frac{2\pi t}{12}\right) + X_t,$$

where $\{X_t\}$ is a stationary time series. More complex seasonal patterns can be obtained by adding higher frequency sinusoidal curves or using an indicator variable for month of the year. Curvature in the trend can be modeled by including a quadratic term in t. The model can be fitted by least squares and the estimators of the coefficients are unbiased, but estimators of standard errors of the coefficients will be biased[26] unless $\{X_t\}$ is just a sequence of independent random variables.

Example 9.6: Reservoir inflows

The data in the file font.txt are monthly effective inflows $(m^3 s^{-1})$ to the Fontburn Reservoir, Northumberland, England for the period January 1909 until December 1980 (data courtesy erstwhile Northumbrian Water Authority). The objective is to fit a time series model to: investigate whether there is evidence of a change in the mean; to provide one month ahead forecasts of inflow to the control algorithm for releases from the reservoir; and to generate long term simulations to evaluate the consequences of different release policies. The marginal distribution of inflows is highly skewed by a few extreme values as seen in the left hand side of Figure 9.15, so it might be easier to work with the logarithm of inflow shown in the middle of Figure 9.15.

---

[26] If $\{X_t\}$ is positively autocorrelated at small lags, an equivalent number of independent variables would be lesser than the number of observations in the time series, so standard errors tend to be underestimated.

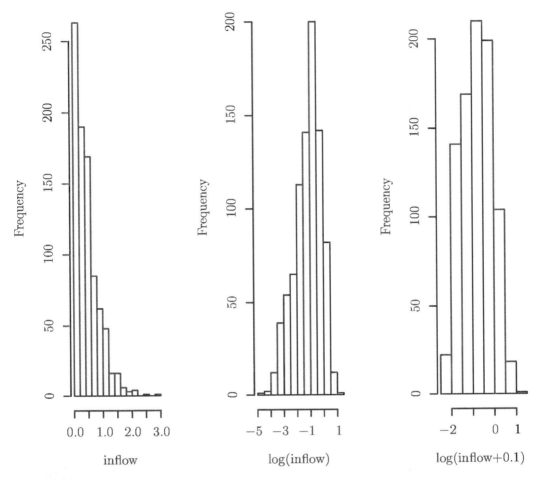

FIGURE 9.15: Histogram of the effective monthly inflows and transformed inflows into Fontburn Reservoir.

```
> source("C:\\Users\\Andrew\\Documents\\R scripts\\Font.r")
> Font.dat=read.table("font.txt",header=T)
> attach(Font.dat)
> par(mfrow=c(1,3))
> hist(inflow,main="")
> hist(log(inflow),main="")
> hist(log(inflow+0.1),main="")
```

An advantage of using logarithms is that simulations cannot generate negative flows, but a drawback is that simulations can generate the occasional extreme flow that is physically implausible. One compromise is to use $\log(\text{inflow} + a)$ for some choice of constant a. In principle, we should get similar results provided we model the errors in the model with sufficient accuracy. Here, we choose to take $a = 0.1$ as this gives a near symmetric distribution (right hand side of Figure 9.15), although simulations can then produce negative numbers in the range $(-0.1, 0)$. Any negative flows would then be set to 0. The time series of inflows and the transformed inflows are shown in Figure 9.16.

```
> Font.ts=ts(inflow, start=1909,freq=12)
> par(mfrow=c(2,1))
> plot(Font.ts)
> plot(log(Font.ts+0.1))
```

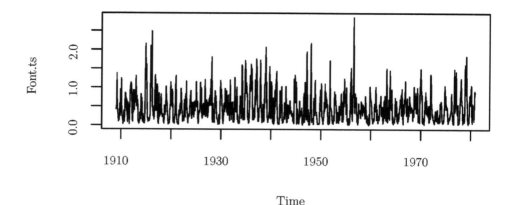

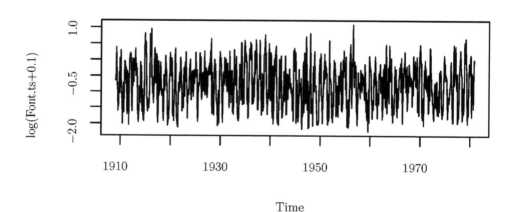

FIGURE 9.16: Monthly inflows and the transformed inflows to Fontburn Reservoir.

We now fit a regression (**m1**) that includes a linear trend and a sinusoidal seasonal term with a frequency of 1 cycle per year.

```
> y=log(inflow+0.1) ; print(sd(y))
[1] 0.6918156
> n=length(y)
> t=1:n
> C=cos(2*pi*t/12)
> S=sin(2*pi*t/12)
> m1=lm(y~t+C+S) ; print(summary(m1))

Call:
lm(formula = y ~ t + C + S)

Residuals:
```

```
     Min       1Q    Median      3Q       Max
 -1.44432 -0.42930  0.00713  0.36754   2.37059
```

```
Coefficients:
              Estimate Std. Error t value Pr(>|t|)
(Intercept) -7.196e-01  3.812e-02 -18.876   <2e-16 ***
t           -1.602e-04  7.636e-05  -2.098   0.0362 *
C            5.147e-01  2.693e-02  19.111   <2e-16 ***
S            2.531e-01  2.693e-02   9.397   <2e-16 ***
---
Residual standard error: 0.5598 on 860 degrees of freedom
Multiple R-squared:  0.3476,    Adjusted R-squared:  0.3453
F-statistic: 152.7 on 3 and 860 DF,  p-value: < 2.2e-16
```

The inflows are known to be seasonal with the summer being relatively dry. The R^2-value is 0.35 and this is mainly due to the seasonal component of the model. The evidence for a decreasing linear trend is equivocal because the p-value of 0.0362 is based on a dubious assumption of independent errors.

We compare **m1** with a regression model **m2** that includes a linear trend and an additive seasonal effect that is estimated separately for each month.

```
> month= t %% 12
> m2=lm(y~t+factor(month)) ; print(summary(m2))
```

```
Call:
lm(formula = y ~ t + factor(month))
```

```
Residuals:
    Min      1Q  Median      3Q     Max
 -1.5288 -0.4271 -0.0113  0.3806  2.2488
```

```
Coefficients:
                  Estimate Std. Error t value Pr(>|t|)
(Intercept)     -2.225e-01  7.388e-02  -3.011  0.00268 **
t               -1.605e-04  7.627e-05  -2.105  0.03559 *
factor(month)1   3.983e-02  9.319e-02   0.427  0.66922
factor(month)2  -6.267e-02  9.319e-02  -0.672  0.50149
factor(month)3  -1.594e-01  9.319e-02  -1.711  0.08751 .
factor(month)4  -5.460e-01  9.319e-02  -5.859 6.65e-09 ***
factor(month)5  -8.208e-01  9.319e-02  -8.808  < 2e-16 ***
factor(month)6  -1.057e+00  9.319e-02 -11.337  < 2e-16 ***
factor(month)7  -1.083e+00  9.319e-02 -11.620  < 2e-16 ***
factor(month)8  -8.517e-01  9.319e-02  -9.139  < 2e-16 ***
factor(month)9  -8.448e-01  9.319e-02  -9.065  < 2e-16 ***
factor(month)10 -4.983e-01  9.319e-02  -5.347 1.15e-07 ***
factor(month)11 -8.059e-02  9.319e-02  -0.865  0.38737
---
Residual standard error: 0.5591 on 851 degrees of freedom
Multiple R-squared:  0.3559,    Adjusted R-squared:  0.3468
F-statistic: 39.18 on 12 and 851 DF,  p-value: < 2.2e-16
```

```
> num=(.3559-.3476)/9;den=(1-.3559)/851;F=num/den
> print(c("F=",round(F,3),"p-value=",round(1-pf(F,9,851),3)))
[1] "F="        "1.218"     "p-value=" "0.28"
```

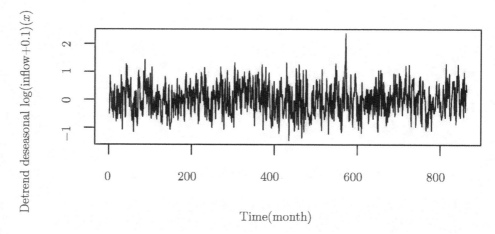

FIGURE 9.17: detrended and deseasonalized logarithms of inflows plus 0.1 to Fontburn Reservoir.

The standard deviation of the response (logarithm of inflows plus 0.1) is 0.6918. Model **m1** has an estimated standard deviation of the errors of 0.5598 and **m2** is a very slight improvement with an estimated standard deviation of the errors of 0.5591. The F-test for comparing **m2** with **m1** is not statistically significant (**m1** is a special case[27] of **m2**). The estimated trend is negative but the p-value is unreliable (it is rather too small and we return to this later). There is little to choose between the models. We take the residuals from **m1** as the detrended and deseasonalized time series of logarithms of inflows plus 0.1 and refer to this time series as $\{x_t\}$ (Figure 9.17).

9.8.4   Auto-covariance and auto-correlation

When modeling time series we imagine that the observed time series is a realization of an underlying model that could produce an infinite number of such time series. This infinite number of possible time series is known as the **ensemble** and expected values are taken relative to this ensemble. The mean at time t, μ_t, is defined by

$$\mathrm{E}[X_t] = \mu_t.$$

With only one time series $\{x_t\}$ we cannot estimate μ_t with any sensible precision so we typically assume that μ_t is some function of t, commonly a linear function. The notion of "lag", meaning time ago, is also needed to describe time series models.

Definition 9.13: Lag

Time ago is referred to as the lag, so that the variable at lag k is $\{X_{t-k}\}$.

---

[27]Using an indicator variable for month is equivalent to fitting $C_1, S_1, \ldots, C_6$, where $C_j = \cos(j \times 2\pi t/12)$ and $S_j = \sin(j \times 2\pi t/12)$ where $j = 1, 2, \cdots, 6$.

Definition 9.14: Second order stationarity

A time series model $\{X_t\}$ is second order stationary if the mean is constant for all t, $E[X_t] = \mu$, and if the variance and covariances at all lags are constants.

This requirement is discussed in Section 9.8.4.1.

9.8.4.1 Defining auto-covariance for a stationary times series model

The auto-covariance at a lag k is the covariance between X_t and X_{t-k}.

Definition 9.15: The auto-covariance function

The auto-covariance function (acvf) for a stationary times series model $\{X_t\}$ is defined as

$$E[(X_{t-k} - \mu)(X_t - \mu)] \quad = \quad \gamma(k) \quad \text{for all} \quad t,$$

where the expectation, $E[\cdot]$, is taken across the imagined infinite ensemble of all possible time series. In particular, $\gamma(0)$ is the variance of X_t.

Definition 9.16: The auto-correlation function

The auto-correlation function (acf) for a stationary times series model $\{X_t\}$ is defined as

$$\rho(k) \quad = \quad \frac{\gamma(k)}{\gamma(0)}.$$

Definition 9.17: Discrete white noise

Discrete white noise (DWN)[28] is defined as a sequence of independent random variables, with constant mean μ and constant variance σ^2.

For DWN

$$\rho(k) \quad = \quad \begin{cases} 1 & k = 0 \\ 0 & k \neq 0. \end{cases}$$

The best prediction for future values of DWN is the mean μ. When we fit time series models we aim for residuals that can plausibly be considered as a realization of DWN with mean 0.

---

[28]White noise has a flat spectrum (Exercise 9.35).

9.8.4.2 Defining sample auto-covariance and the correlogram

Assume we have a time series $\{x_t\}$, of length n, which is a realization of a second order stationary time series model $\{X_t\}$. The mean μ is estimated by

$$\bar{x} = \sum_{t=1}^{n} x_t/n$$

and the acvf is estimated by the sample acvf

$$c(k) = \sum_{t=1}^{n-k} (x_t - \bar{x})(x_{t+k} - \bar{x})/n.$$

The sample acf is

$$r(k) = \frac{c(k)}{c(0)}$$

and a plot of $r(k)$ against k, `acf()` in R, is known as the correlogram[29].

Example 9.6: (Continued) Reservoir inflows

We compare the correlogram of an independent sequence of normal random variables, also known as Gaussian white noise (GWN), with the deseasonalized and detrended inflow data $\{x_t\}$ for the Fontburn Reservoir in Figure 9.18.

```
> gwn=rnorm(n)
> x=m1$res
> set.seed(1)
> gwn=rnorm(n)
> par(mfrow=c(1,2))
> acf(gwn)
> acf(x)
```

It can be shown that, if a sequence is DWN then the approximate sampling distribution of $r(k)$ is

$$r(k) \sim N\left(\frac{-1}{n}, \frac{1}{n}\right)$$

and the dotted lines on the acf are drawn at

$$-1/n \pm \frac{2}{\sqrt{n}}$$

If a time series is a realization of DWN around 1 in 20 of the $r(k)$ are expected to be outside these lines. There happen to be 2 out of 30 $r(k)$ outside these lines for the realization of white noise in Figure 9.18.

In contrast, there is clear evidence that the inflows are positively autocorrelated, that is if this month is above the mean then next month is expected to be slightly above the mean. An explanation is that a portion of rainfall in the catchment runs off into the river and a portion is stored as groundwater which provides a slow recharge to the reservoir.

---

[29]The denominator n in the definition of c(k) ensures that $-1 \leq r(k) \leq 1$.

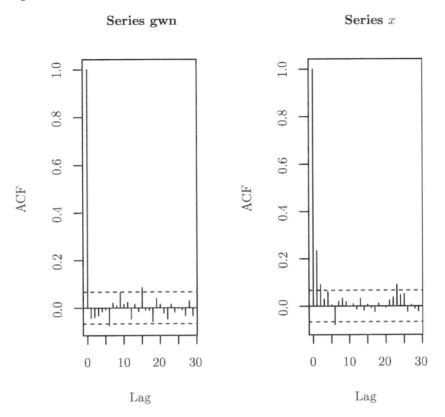

FIGURE 9.18: Comparison of correlograms for white noise (left panel) and x_t (right panel).

9.8.5 Auto-regressive models

We consider auto-regressive models for a stationary time series model $\{X_t\}$ in which the distribution of the variable now depends on its past values. The auto-regressive model of order p (AR(p)) is defined as

$$X_t - \mu \;\; = \;\; \alpha_1(X_{t-1} - \mu) + \ldots + \alpha_p(X_{t-p} - \mu) + \varepsilon_t,$$

where $\{\varepsilon_t\}$ is DWN with mean 0. An assumption of normality is often not realistic. The coefficients have to satisfy certain conditions for the model to be second order stationary (Exercise 9.32). An alternative expression for the model is

$$X_t \;\; = \;\; \alpha_0 + \alpha_1 X_{t-1} + \ldots + \alpha_p X_{t-p} + \varepsilon_t,$$

where $\alpha_0 = \mu(1 - \alpha_1 - \ldots - \alpha_k)$. An AR($p$) model can be fitted by a multiple regression model with x_t as the response and $x_{t-1}, \ldots, x_{t-p}$ as predictor variables. In practice many detrended and deseasonalized time series are modeled quite well by AR(1) or AR(2) models so we will focus on these.

9.8.5.1   AR(1) and AR(2) models

Definition 9.18: AR(1)

The AR(1) model has the form

$$(X_t - \mu) \quad = \quad \alpha(X_{t-1} - \mu) + \varepsilon_t,$$

where $-1 < \alpha < 1$ for the model to be stationary in the mean (Exercise 9.32), and $\{\varepsilon_t\}$ is DWN with mean 0.

Definition 9.19: AR(2)

The AR(2) model has the form

$$(X_t - \mu) \quad = \quad \alpha_1(X_{t-1} - \mu) + \alpha_2(X_{t-2} - \mu) + \varepsilon_t,$$

where $\alpha_1 + \alpha_2 < 1$, $\alpha_1 - \alpha_2 > -1$ and $\alpha_2 > -1$ for the model to be stationary in the mean (Exercise 9.32), and $\{\varepsilon_t\}$ is DWN with mean 0.

The following code generates times series of length 10 000 from two AR(1) models with coefficients $\alpha = 0.8, -0.8$ and an AR(2) model with coefficients $\alpha_1 = 1$ and $\alpha_2 = -0.5$. Realizations are shown in the upper row of Figure 9.19.

```
set.seed(11)
n=10000;a=0.8
x1p8=rep(0,n)
x1n8=rep(0,n)
x2=rep(0,n)
for (t in 2:n){
x1p8[t]=a * x1p8[t-1] + rnorm(1)
x1n8[t]= -a * x1n8[t-1] + rnorm(1)
}
a1=1;a2=-.5
x2=rep(0,n)
for (t in 3:n){
x2[t]=a1*x2[t-1] + a2*x2[t-2] + rnorm(1)
}
par(mfrow=c(2,3))
plot(as.ts(x1p8[5501:5550]))
plot(as.ts(x1n8[5501:5550]))
plot(as.ts(x2[5501:5550]))
acf(x1p8,main="")
acf(x1n8,main="")
acf(x2,main="")
```

An AR(1) with $0 < \alpha < 1$ has an acf that shows a geometric decay with lag, as seen in the lower left panel of Figure 9.19. The theoretical result (Exercise 9.33) is that

$$\rho(k) \quad = \quad \alpha^k \quad \text{for } k = 0, 1, \dots \text{ and } -1 < \alpha < 1.$$

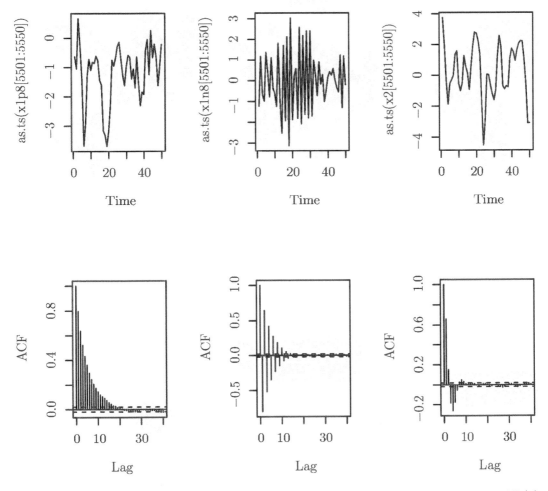

FIGURE 9.19: From left to right AR(1) with $\alpha = 0.8$, AR(1) with $\alpha = -0.8$ and AR(2) with $\alpha_1 = 1$, $\alpha_2 = -0.5$.

Features of realizations of AR(1) models with $\alpha < 0$, that become more noticeable as α tends towards 1, are that consecutive values tend to be relatively close and variation is on a slow timescale (upper left Figure 9.19). In contrast, if $-1 < \alpha < 0$ then the sign of the acf alternates (center bottom of Figure 9.19), the absolute value of the acf shows a geometric decay with lag (bottom left and center Figure 9.19)), and in realizations of AR(1) models with $\alpha < 0$, consecutive values tend to alternate either side of the mean (upper center Figure 9.19). The acf of the AR(2) model with the chosen parameters is a damped sinusoidal curve (Exercise 9.34) with a value of 1 at lag 0. This corresponds to a difference equation for a mass-spring-damper system forced by random noise when the damping is sub-critical.

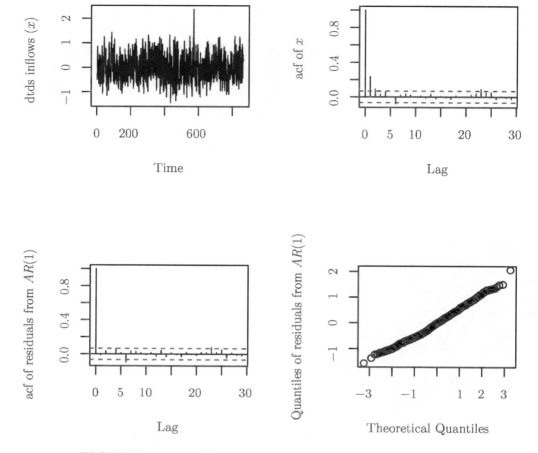

FIGURE 9.20: AR(1) model of the Fontburn reservoir inflows.

Example 9.6: (Continued) Reservoir inflows

The acf of the detrended deseasonalized logarithms of inflows plus 0.1 ($\{x_t\}$) is consistent with a realization from an AR(1) model with a small positive α. In the following code we fit an AR(1) model to $\{x_t\}$ and check that the acf of the residuals is consistent with a realization of DWN.

```
> m3=lm(x[2:n]~x[1:(n-1)]) ; print(summary(m3))

Call:
lm(formula = x[2:n] ~ x[1:(n - 1)])

Residuals:
     Min       1Q   Median       3Q      Max
-1.57033 -0.41767  0.00048  0.37822  2.06251

Coefficients:
            Estimate Std. Error t value Pr(>|t|)
(Intercept) 0.0006609  0.0185048   0.036    0.972
```

```
x[1:(n - 1)] 0.2343370   0.0331204    7.075 3.09e-12 ***
---
Residual standard error: 0.5436 on 861 degrees of freedom
Multiple R-squared:  0.05495,   Adjusted R-squared:  0.05385
F-statistic: 50.06 on 1 and 861 DF,  p-value: 3.09e-12
```

```
> par(mfrow=c(2,2))
> plot(as.ts(x),ylab="dtds inflows (x)")
> acf(x,ylab="acf of x",main="")
> acf(m3$res,ylab="acf of residuals from AR(1)",main="")
> qqnorm(m3$res,ylab="quantiles of residuals from AR(1)",main="")
```

The acf of the residuals after fitting the AR(1) model (Figure 9.20 lower left) has no systematic pattern and there is no evidence of auto-correlations at low lags. So the acf of the residuals is consistent with a realization of DWN. With the chosen transform of inflows, the residuals seem to be reasonably modeled by a normal distribution (Figure 9.20 lower right). Alternatively, a scaled t-distribution would be better if the residuals have a kurtosis greater than 3, as the t-distribution has heavier tails.

We now consider how the standard deviation of the original time series has been reduced by the models.

```
> print(c(round(sd(y),3),round(sd(x),3),round(sd(m3$res),3)))
[1] 0.692 0.559 0.543
```

The original variable has a standard deviation of 0.69, the detrended and deseasonalized variable has a standard deviation of 0.56, and the residuals from the AR(1) model have a standard deviation of 0.54. The dominant feature of the inflows time series is the seasonal variation.

The regression of y on t, C, S, and then the AR(1) for the residuals can be combined into a single step (Exercise 9.36). Although the interpretation of the model, in the context of time series analysis, is obscured the standard error for the coefficient of time is accurate so facilitating a test of significance of the trend. An explanation for a trend in the inflows is that it is a consequence of the increasing urbanization, of what was originally a rural catchment, over the 72 year period.

Example 9.7: Wave heights

The time series is the wave heights at the centre of a wave tank mm (known as a data wave) with random wave-makers, sampled at 0.1 second intervals over 39.7 seconds. The objective is to model the wave height for computer simulations of the performance of wave energy devices. We start by plotting the entire series and check for outlying or missing values (Figure 9.21).

We then plot a segment in more detail (Figure 9.22 upper left) and the acf of the entire series (Figure 9.22 upper right). The acf is similar to a damped sinusoid and suggests an AR(2) model. The residuals from the AR(2) show some slight auto-correlation (Figure 9.22 lower left) and the ar() function in R can provide a somewhat better fit (Exercise 9.37). The residuals are reasonably modeled as a realization from a normal distribution (Figure 9.22 lower right).

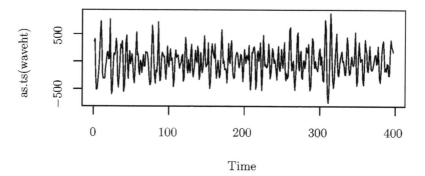

FIGURE 9.21: Time series plot of wave heights from a wave tank.

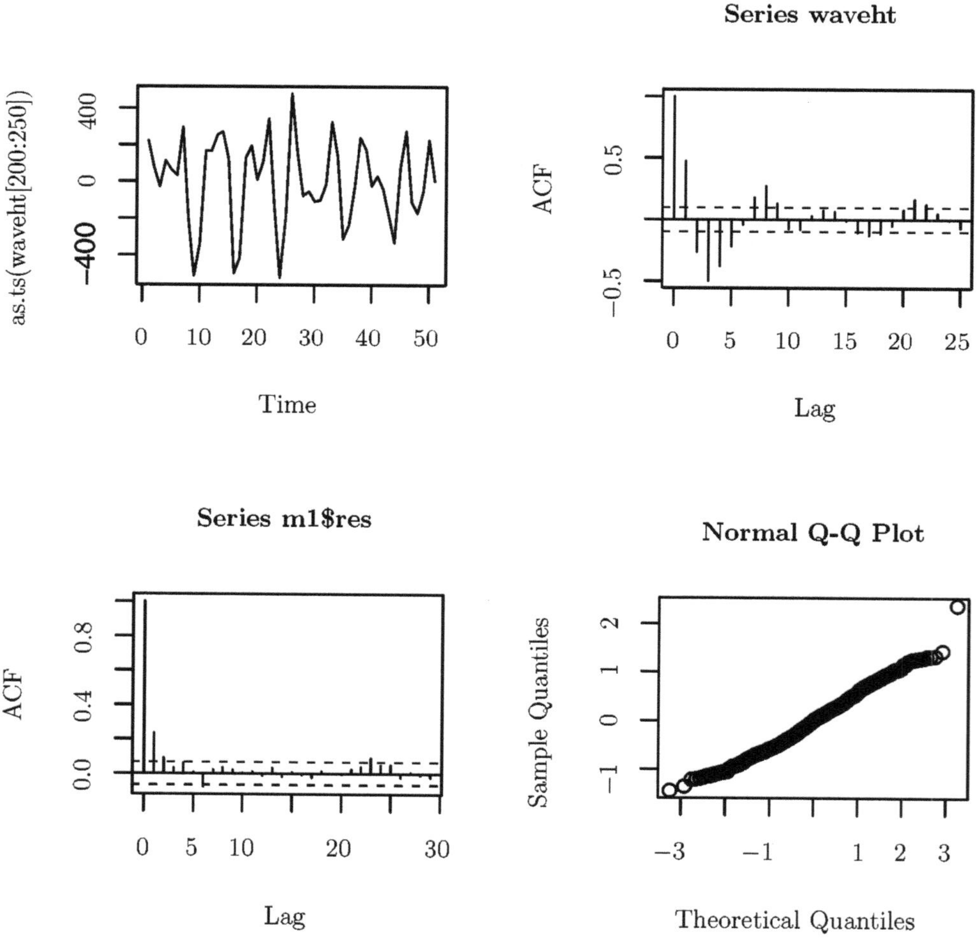

FIGURE 9.22: Residuals from an AR(2) model fitted to the wave height data.

```
wave.dat=read.table("wave.txt",header=T)
attach(wave.dat)
n=length(waveht)
```

```
plot(as.ts(waveht))
print(c("sd waveht",round(sd(waveht),1)))
m1=lm(waveht[3:n]~waveht[2:(n-1)]+waveht[1:(n-2)])
print(summary(m1))
par(mfrow=c(2,2))
plot(as.ts(waveht[200:250]))
acf(waveht)
acf(m1$res)
qqnorm(m1$res)
```

9.9 Non-linear least squares

The standard regression model is known as the linear model because it is linear in the unknown coefficients. This feature combined with the principle of least squares led to explicit formulae for the least squares estimators of the coefficients and their standard errors. But, we often need to fit models that cannot be written as linear in the unknown coefficients. We can apply the principle of least squares but the minimization has to be performed numerically. One approach is to use the Nelder-Mead algorithm and bootstrap methods to obtain estimates of standard errors. A more efficient procedure is to linearize the model around the last set of estimates of the coefficients using a Taylor series approximation. We skip the detail and rely on the **non-linear least squares** routine in R.

Example 9.8: Angle of draw

In open cast mining, and many construction projects, a pit is excavated and there will be subsidence near the edge of the pit. If the cross section of the pit is taken as a rectangle, the angle of draw is the angle between a line from the bottom of the pit to the point where subsidence begins to occur and the vertical. The angle of draw depends on the geology, but within a given soil type it tends to increase with the ratio of width to depth of the excavation. Following [Myers et al., 2010], we fit a model for angle of draw, y, as a function of the ratio of width to depth, x, of the form

$$y_i = a\left(1 - e^{-bx_i}\right) + \varepsilon_i, \qquad \text{for} \quad i = 1, \dots, n$$

to data from $n = 16$ mining excavations in West Virginia[30]. The ε_i are iid with mean 0 and variance σ_ε^2. In the model, the angle of draw approaches the upper threshold a as the ratio, x, increases.

The R code for fitting the model using the function `nls()` follows. The syntax is straightforward, and you need to provide the model formula and starting values. If the model is a convincing fit any plausible starting values should do, and we chose 30 and 1 for a and b respectively, based on inspection of the plot given in Figure 9.23 and the measurement data in Table 9.9.

```
> subsid.dat=read.table("subsid.txt",header=TRUE)
> attach(subsid.dat)
```

---

[30]Data provided by the Department of Mining Engineering at Virginia Polytechnic Institute.

```
> print(head(subsid.dat))
width depth angle
1    610    550   33.6
2    450    500   22.3
3    450    520   22.0
4    430    740   18.7
5    410    800   20.2
6    500    230   31.0
> ratiowd=width/depth
> plot(ratiowd,angle,xlab="width/depth",ylab="angle of draw")
> ratio=ratiowd
> m1=nls(angle~a*(1-exp(-b*ratio)),start=list(a= 30,b= 1))
> newdat=data.frame(ratio=c(0:2200)/1000)
> y=predict(m1,newdata=newdat)
> lines(newdat$ratio,y,lty=1)
> summary(m1)

Formula: angle ~ a * (1 - exp(-b * ratio))

Parameters:
Estimate Std. Error t value Pr(>|t|)
a  32.4644     2.6478  12.261 7.09e-09 ***
b   1.5111     0.2978   5.075 0.000169 ***
---
Residual standard error: 3.823 on 14 degrees of freedom

Number of iterations to convergence: 6
Achieved convergence tolerance: 7.789e-07
```

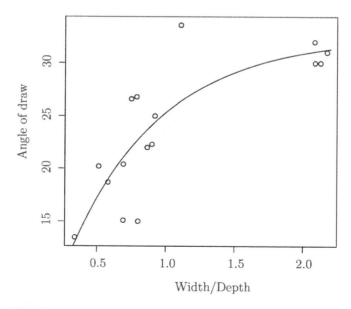

FIGURE 9.23: Angle of draw against width/depth ratio.

TABLE 9.9: Mining excavations: width (ft) and depth (ft) of excavation and angle of draw (degrees).

Width	Depth	Angle of draw
610	550	33.6
450	500	22.3
450	520	22.0
430	740	18.7
410	800	20.2
500	230	31.0
500	235	30.0
500	240	32.0
450	600	26.6
450	650	15.1
480	230	30.0
475	1400	13.5
485	615	26.8
474	515	25.0
485	700	20.4
600	750	15.0

You are asked to construct a 95% prediction interval for a in Exercise 9.39.

This is a simple example with just two parameters, and a model that seems to fit the data well with parameters that have a clear physical interpretation. The parameter a is the maximum angle of draw and b governs the rate at which this maximum is reached.

In general we need to consider the following points.

- Convergence may be dependent on finding initial values that are fairly close to the values that minimize the sum of squared errors (optimum values).

- If the expected value of a parameter, b say, is small, 0.01 say, it will be better to redefine $b = c/100$, and optimize with respect to c rather than b, because the default step size in optimization routines will typically be around 0.1

- There may be local minima. Trying different initial values may show that there are local minima.

- Parameters may be constrained to a particular interval, in particular to being positive. We can either use a constrained optimization procedure, or optimize on some function of the parameter that is not constrained. For example, if $0 < \theta < 1$ we can optimize with respect to $-\infty < \phi < \infty$ where

$$\theta = \frac{e^\phi}{1 + e^\phi}.$$

- The parameters may not all be identifiable from the data.

9.10 Generalized linear model

In the ordinary regression model the response Y_i is normally distributed with a mean equal to a linear combination of predictor variables,

$$\mathrm{E}[Y_i] \;=\; \beta_0 + \beta_1 x_{1i} + \ldots + \beta_k x_{ki},$$

and constant variance. The unknown coefficients, β_j, are estimated from the data. The **generalized linear regression** (glm) includes other distributions and allows for the mean to be some known function of the linear combination of predictor variables. In this section we consider logistic regression, which has a binomial response, and Poisson regression.

9.10.1 Logistic regression

Consider binomial experiments in which the probabilities of success p_i depend on the values of predictor variables, x_{ki}. The logistic regression model has the form

$$\ln\left(\frac{p_i}{1-p_i}\right) \;=\; \beta_0 + \beta_1 x_{1i} + \ldots + \beta_k x_{ki}.$$

The right hand side is the same as in the standard regression model for $\mathrm{E}\left[Y_i \big| x_{1i}, \ldots, x_{ki}\right]$, and the left hand side is known as the **logit** of p_i. The logit of p has domain $(0,1)$ and range $(-\infty, \infty)$ (Figure 9.24 left panel). Its inverse function, which is obtained by making p the subject of the formula

$$\ln\left(\frac{p}{1-p}\right) \;=\; \theta \iff p \;=\; \frac{e^\theta}{1+e^\theta},$$

is shown in Figure 9.24 (right panel). The data are the numbers of successes, x_i, in a sequence of binomial experiments with n_i trials and associated values of predictor variables $x_{1i}, \ldots, x_{ki}$. The model is fitted by maximizing the likelihood[31]. The likelihood is

$$\mathcal{L}(\beta_0, \ldots, \beta_k) \;=\; \prod_i^m \binom{n_i}{x_i} p_i^{x_i}(1-p_i)^{n_i-x_i},$$

where

$$p_i \;=\; \frac{\exp(\beta_0 + \beta_1 x_{1i} + \ldots + \beta_k x_{ki})}{1+\exp(\beta_0 + \beta_1 x_{1i} + \ldots + \beta_k x_{ki})}.$$

There are very efficient algorithms for finding the maximum likelihood estimates, based on linearization about the current estimates, that were originally implemented by hand calculation[32].

Example 9.9: Prototype gas tank

A motor vehicle manufacturer tested a prototype design of plastic gas tank by subjecting tanks to high impacts that exceed those anticipated in crashes, and recording whether or not they fail. The results are shown in Table 9.10 The analysis in R is im-

[31]A regression of logit(x_i/n_i) on the predictor variables is not very satisfactory because it ignores the change in variance of logit(x_i/n_i) with p and n, Exercise 9.47. Moreover, if x_i is 0 or n_i, some arbitrary small number has to be added or subtracted in order to calculate the logit.

[32]This was known as probit analysis, the probit being the inverse normal cdf of p with 5 added to keep it positive (Exercise 9.43).

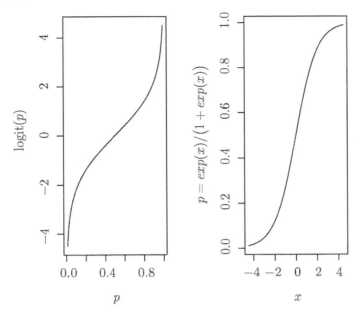

FIGURE 9.24: Logit p against p (left panel) and its inverse function (right panel).

TABLE 9.10: Gas tanks: number tested and number failing by impact.

Number tested	Number failed	Impact (coded units)
5	0	10
8	2	12
8	5	14
8	4	16
8	6	18
8	7	20
1	1	22

plemented with the generalized linear model function, glm(). The response, Y, is set up as an array with two columns, the number of failures and the number of successes. The right hand side of the model follows the same syntax as for lm(). The **family=binomial** argument specifies logistic regression.

```
> n=c(5,8,8,8,8,8,1)
> F=c(0,2,5,4,6,7,1)
> x=c(10,12,14,16,18,20,22) ; S=n-F ; Y=cbind(F,S)
> m1=glm(Y~x,family=binomial)
> summary(m1)

Call:
glm(formula = Y ~ x, family = binomial)

Deviance Residuals:
      1        2        3        4        5        6        7
-1.17121  0.03757  1.19197 -0.62839 -0.16384 -0.07023  0.34203
```

```
Coefficients:
             Estimate Std. Error z value Pr(>|z|)
(Intercept)  -5.8560     1.8908   -3.097  0.00195 **
x             0.3939     0.1224    3.217  0.00130 **
---

(Dispersion parameter for binomial family taken to be 1)

Null deviance: 17.7229  on 6  degrees of freedom
Residual deviance:  3.3376  on 5  degrees of freedom
AIC: 19.001

Number of Fisher Scoring iterations: 4
> m1$res
1              2            3            4            5            6
-1.14702718  0.03099881   0.86301309  -0.46111704  -0.14032052  -0.07802011
7
1.06023688
> xx=seq(10,22,.001)
> logitp=m1$coef[1]+m1$coef[2]*xx
> p=exp(logitp)/(1+exp(logitp))
> plot(xx,p,type="l",xlab="impact",ylab="probability of failure")
```

The residuals are defined as

$$r_i = \frac{x_i - n_i \widehat{p}_i}{\sqrt{n_i \widehat{p}_i (1 - \widehat{p}_i)}}$$

and in a logistic regression the **deviance** is approximately equal to the sum of squared residuals. If the data are from a sequence of binomial experiments the expected value of the deviance will equal the degrees of freedom. In this case the deviance is 3.3376 which is less than the number of degrees of freedom, 5, which is the number of data, 7, less the number of coefficients estimated from the data, 2, and the model provides a good fit. The probability of failure is plotted against impact in Figure 9.25. If the deviance is substantially greater than the degrees of freedom the data are said to be **over-dispersed** and an example is given in Exercise 9.48.

In some applications, all the binomial experiments are single trial ($n_i = 1$ for $i = 1, \ldots, m$). An example is given in Exercise 9.44.

9.10.2   Poisson regression

In a Poisson regression the responses, Y_i, are assumed to come from Poisson distributions with means μ_i that depend on values of predictor variables, x_k, according to the formula

$$\ln(\mu_i) \;=\; \beta_0 + \beta_1 x_{1i} + \ldots + \beta_k x_{ki}.$$

The right hand side is the same as in the standard regression model for $\mathrm{E}\big[Y | x_1, \ldots, x_k\big]$, and the left hand side is the natural logarithm of μ. The likelihood is

$$\mathcal{L}(\beta_0, \ldots, \beta_k) \;=\; \prod_i^m \frac{e^{-\mu_i} \mu_i^{y_i}}{y_i!},$$

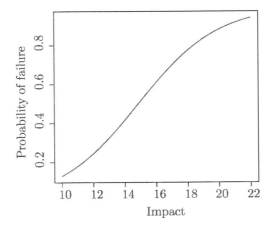

FIGURE 9.25: Gas tanks: probability of failure against impact.

where

$$\mu_i \;=\; \exp\left(\beta_0 + \beta_1 x_{1i} + \ldots + \beta_k x_{ki}\right)$$

and is maximized by the same algorithm as is used for logistic regression implemented with the R function `glm()`.

Example 9.10: Road traffic

[Aljanahi et al., 1999] investigated the effect of road traffic speed on the number of personal injury accidents on dual carriageways under free flow conditions. They monitored nine sites in the Tyne and Wear district of the UK and ten sites in Bahrain over a five year study period. The data from Tyne and Wear, length of monitored road section (km), vehicle flow through the site in a unit of 10^5 vehicles per dual carriageway per year, coefficient of upper spread of speed ($CUSS$), and the number of personal injury accidents over the five year period are given in Table 9.11 and plotted in Figure 9.26. The coefficient of upper spread of speed, defined as

$$CUSS \;=\; \frac{85\% \text{ quantile speed} - \text{median speed}}{\text{median speed}},$$

is a commonly used statistic of the speed distribution on roads. The $CUSS$ focuses on differences in speeds of the faster vehicles, which is believed to be more highly associated with accidents than the average speed.

According to the model accidents occur randomly and independently with the site means μ_i of the Poisson distributions being proportional to the length of carriageway (L), some power of flow (F), and some power of $CUSS$ (S). That is

$$\mu_i \;=\; k\, L_i F_i^a S_i^b,$$

where k, a, b are unknown constants to be estimated. The interaction between vehicles under free flow conditions is complex and there is empirical evidence that the accident rate is not necessarily directly proportional to the traffic flow, so a is estimated rather than assumed equal to 1. The logarithm of μ has the form of the Poisson regression

$$\ln(\mu_i) = \ln(k) + \ln(L_i) + a\,\ln(F_i) + b\,\ln(S_i)$$

TABLE 9.11: Accidents (1989-93) at nine dual carriageway sites in Tyne and Wear.

Length (km)	Flow (10^5)	CUSS	Accidents (5 years)
44.00	39.63	0.1588	142
18.00	50.00	0.2041	94
7.00	47.67	0.1869	37
0.64	46.91	0.2821	4
1.16	33.33	0.1441	3
0.47	52.51	0.1639	3
0.62	44.78	0.1593	2
1.05	54.21	0.1799	2
1.03	24.45	0.1214	1

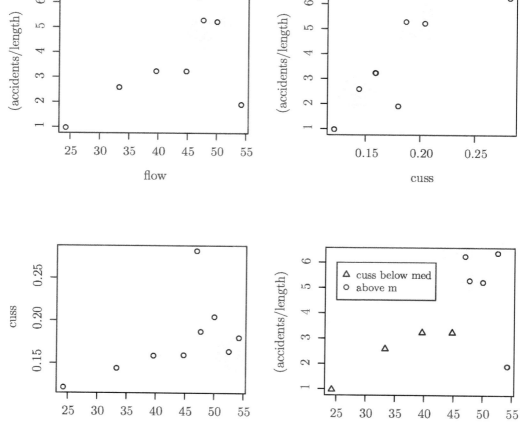

FIGURE 9.26: Accidents (1989-93) at nine dual carriageway sites in Tyne and Wear.

with the modification that $\ln(L_i)$ are known and referred to as an **offset**. The model can be fitted in R with the `glm()` function[33].

---

[33]When we introduced the Poisson distribution we assumed a constant rate of events per time λ. But, this assumption can be relaxed by defining a time dependent Poisson process $\lambda(t)$ for which the expected number of events in time t, $\mu = \lambda t$, is replaced by the more general $\mu = \int_0^t \lambda(\theta)d\theta$. This is a consequence of the general result that a superposition of Poisson processes is itself a Poisson process. In the context of Aljanahi et al's [1999] study, the rate could vary within days and seasonally.

```
> roadlinks.dat=read.table("roadlinks.txt",header=T)
> attach(roadlinks.dat)
> print(head(roadlinks.dat)
      L     F      S    accid
1 44.00 39.63 0.1588    142
2 18.00 50.00 0.2041     94
3  7.00 47.67 0.1869     37
4  0.64 46.91 0.2821      4
5  1.16 33.33 0.1441      3
6  0.47 52.51 0.1639      3
> lnF=log(F)
> lnS=log(S)
> lnL=log(L)
> m1=glm(accid~offset(lnL)+lnF+lnS,family=poisson)
> print(summary(m1))

Call:
glm(formula = accid ~ offset(lnL) + lnF + lnS, family = poisson)

Deviance Residuals:
    Min       1Q    Median       3Q      Max
-1.63227  -0.21693  -0.11538  0.08538  0.82767

Coefficients:
             Estimate Std. Error z value Pr(>|z|)
(Intercept)  -0.5498     5.1602   -0.107    0.915
lnF           0.9896     0.9823    1.007    0.314
lnS           1.0426     0.9160    1.138    0.255

(Dispersion parameter for poisson family taken to be 1)

Null deviance: 22.4345  on 8  degrees of freedom
Residual deviance:  4.0438  on 6  degrees of freedom
AIC: 45.152

Number of Fisher Scoring iterations: 4
```

Residuals can be defined as

$$r_i = \frac{y_i - \widehat{\mu}_i}{\sqrt{\widehat{\mu}_i}}$$

and the deviance is approximately the sum of these residuals squared[34]. If the data are generated by the model the expected value of the deviance is equal to the degrees of freedom. The model is a good fit although the standard error of the estimate of the coefficient of $CUSS$ is quite large. The coefficient of the logarithm of $CUSS$, 1.0426, is close to 1.0 so the accident rate is estimated to increase in proportion to the increase in $CUSS$. The coefficient of $CUSS$ is only slightly changed and the standard error is reduced considerably if the logarithm of flow is set as an offset, that is if a is assumed equal to 1, Exercise 9.40.

---

[34]The residuals from glm() are defined somewhat differently.

```
> n=length(accid)
> cussbin=rep(1,n)
> cussbin[S < median(S)]=cussbin[S < median(S)]+1
> AR=accid/L
> par(mfrow=c(2,2))
> plot(F,AR,xlab="flow",ylab="accidents/length")
> plot(S,AR,xlab="CUSS",ylab="accidents/length")
> plot(F,S,xlab="flow",ylab="CUSS")
> plot(F,AR,xlab="flow",ylab="accidents/length",pch=cussbin)
> legend(25,6,c("S low","S high"),pch=cussbin)
```

The Poisson regression can be approximated by an ordinary regression model with $\ln(y_i)$, or $\sqrt{y_i}$, as the response, Exercise 9.41, but the Poisson regression avoids the approximations, which may be dubious for small observed numbers of events, and also gives a model for μ rather than the expected value of some function of μ.

9.11   Summary

9.11.1   Notation

n	sample size
x_i/y_i	observations of predictor/response variables
$\bar{x}/\bar{y}$	mean of observations of predictor/response variables
$\beta_0/\widehat{\beta_0}$	constant term (coefficient of 1) for the model for the population/sample
$\beta_i/\widehat{\beta_i}$	coefficient for predictor i in the model for the population/sample
$\widehat{y}_i$	fitted values of response variable

9.11.2   Summary of main results

Standard regression model

$$Y_i \;=\; \beta_0 + \beta_1 x_1 + \ldots + \beta_k x_k + \varepsilon_i \qquad i = 1, \ldots, n,$$

or in matrix notation

$$\boldsymbol{Y} \;=\; X\boldsymbol{\beta} + \boldsymbol{\varepsilon},$$

where

$$\boldsymbol{Y} \;=\; \begin{pmatrix} Y_1 \\ Y_2 \\ \vdots \\ Y_n \end{pmatrix}, \; X \;=\; \begin{pmatrix} 1 & x_{11} & x_{12} & \ldots & x_{1k} \\ 1 & x_{21} & x_{22} & \ldots & x_{2k} \\ \vdots & & & & \vdots \\ 1 & x_{n1} & x_{n2} & \ldots & x_{nk} \end{pmatrix}, \; \boldsymbol{\beta} \;=\; \begin{pmatrix} \beta_0 \\ \beta_1 \\ \vdots \\ \beta_k \end{pmatrix} \; \text{and} \; \boldsymbol{\varepsilon} \;=\; \begin{pmatrix} \varepsilon_1 \\ \varepsilon_2 \\ \vdots \\ \varepsilon_n \end{pmatrix}.$$

and ε_i are independently distributed with mean 0 and variance σ^2.

Estimating the parameters

The estimates for the coefficients are found by minimizing the sum of the squared errors

$$\Psi = \sum_{i=1}^{n} \left(Y_i - (\beta_0 + \beta_1 x_1 + \ldots + \beta_k)\right)^2,$$

which gives

$$\widehat{\beta} = (X'X)^{-1}X'\boldsymbol{y}.$$

The variance of the errors is estimated by $\sum r_i^2/(n-k-1)$ where $r_i = y_i - \widehat{y}_i$.

Inference

We usually assume the errors have a normal distribution $\varepsilon \sim N(0, I\sigma^2)$.

Predictions

$$\widehat{y}_p = x'_p \widehat{\beta}$$

where

$$x_p = (1, x_{1p}, x_{,p}, \cdots, x_{kp}).$$

Logistic regression

$$logit(p_i) = ln\left(\frac{p_i}{1-p_i}\right) = \beta_0 + \beta_1 x_{1i} + \cdots + \beta_k x_{ki}.$$

Poisson regression

$$Y_i \sim P(\mu_i)$$

where

$$ln(\mu_i) = \beta_0 + x_{0i} + \beta_1 x_{1i} + \cdots + \beta_k x_{ki}$$

and x_{0i} allows for an offset variable.

9.11.3   MATLAB and R commands

The MATLAB and R commands for multiple regression are the same as linear regression. These commands are displayed again below, with additional predictor variables for multiple regression. In the following x1 and x2 are column vectors which contain observations of the predictor variables and y is a column vector which contains the corresponding observations of the response variable. For more information on any built in function, type help(function) in R or help function in MATLAB.

R command	MATLAB command
lm(y~x1 + x2)	lm = fitlm(x1,x2,y)
coef(lm(y~x1 + x2))	lm.Coefficients(:,1)
fitted(lm(y~x1 + x2))	lm.Fitted
r = residuals(lm(y~x1 + x2))	lm.Residuals
qqnorm(r)	qqplot(lm.Residuals)

9.12   Exercises

Section 9.3 Multiple regression model

Exercise 9.1:

Let $c = \begin{bmatrix} a \\ b \end{bmatrix}$ and $W = \begin{bmatrix} X \\ Y \end{bmatrix}$, where a, b are constants and X, Y are random variables each with mean 0, variances $\mathrm{var}(X)$ and $\mathrm{var}(Y)$, and covariance $\mathrm{cov}(X, Y)$.

(a) Explain carefully why $\mathrm{var}(c'W) = c'\mathrm{E}[WW']\,c$.

(b) Give the entries of the matrix $\mathrm{E}[WW']$ in terms of $\mathrm{var}(X)$, $\mathrm{var}(Y)$ and covariance $\mathrm{cov}(X, Y)$.

(c) Write the equation for $\mathrm{var}(c'W)$ in terms of $a, b, \mathrm{var}(X), \mathrm{var}(Y)$ and covariance $\mathrm{cov}(X, Y)$.

(d) Explain why there is no loss of generality in assuming that X and Y have means of 0.

Exercise 9.2:

If $\boldsymbol{Y} = (Y_1, Y_2)'$ and $\mathrm{E}[\boldsymbol{Y}] = \begin{pmatrix} \mu_0 \\ \mu_1 \end{pmatrix}$, then show that

$$\mathrm{E}[(\boldsymbol{Y} - \mu)(\boldsymbol{Y} - \mu)'] = \begin{pmatrix} \mathrm{var}(Y_1) & \mathrm{cov}(Y_1, Y_2) \\ \mathrm{cov}(Y_1, Y_2) & \mathrm{var}(Y_2) \end{pmatrix}.$$

Exercise 9.3:

Prove the result of Section 9.3.2.1 for the case of

$$\boldsymbol{Y} = \begin{pmatrix} Y_1 \\ Y_2 \end{pmatrix}, \quad \boldsymbol{b} = \begin{pmatrix} b_1 \\ b_2 \end{pmatrix} \text{ and}$$

(a) $A = \begin{pmatrix} a_1 & a_2 \end{pmatrix}$

(b) $A = \begin{pmatrix} a_{11} & a_{12} \\ a_{21} & a_{22} \end{pmatrix}$

Section 9.4 Fitting the model

Exercise 9.4:

Let

$$\boldsymbol{\beta} = \begin{pmatrix} \beta_0 \\ \beta_1 \end{pmatrix}, \quad M = \begin{pmatrix} m_{11} & m_{12} \\ m_{21} & m_{22} \end{pmatrix}, \quad \text{and } \mathbf{c} = \begin{pmatrix} c_1 & c_2 \end{pmatrix}.$$

(a) (i) Verify **Rule 2** by writing $c\beta$ in component form and partially differentiating with respect to β_0 and β_1.

(ii) Write down a corresponding results for $\dfrac{\partial \boldsymbol{\beta}' \mathbf{c}'}{\partial \boldsymbol{\beta}}$.

(b) Verify **Rule 3** by writing $B'MB$ in component form and partially differentiating with respect to β_0 and β_1.

Exercise 9.5:

Consider fitting a regression model

$$Y_i = \beta_0 + \beta_1 x_{1i} + \beta_2 x_{2i} + \varepsilon_i$$

to n data. Suppose that

$$x_2 = g + mx_i,$$

where g and m are constants.

(a) What is the correlation between x_1 and x_2?

(b) Substitute x_2 in terms of x_1 into the model, and explain why it is not possible to obtain unique estimates of β_1 and β_2.

(c) Show the form of the matrix X for the case of x_1 having mean 0, and $m = 0$.

(d) Explain why, with the X in (c), that $|X'X| = 0$.

(e) Explain why this result holds for x_1 with any mean and any value of m.

[Assume that a determinant is 0 if any one row is a linear combination of other rows.]

Exercise 9.6:

(a) Consider fitting a regression model

$$Y_i = \beta_0 + \beta_1 x_{1i} + \beta_2 x_{2i} + \varepsilon_i$$

to 5 data.

(i) Write down the matrices X, $X'X$ and $X'Y$ in terms of x_{1i}, x_{2i} and y_i.

(ii) How does $X'X$ simplify if x_1 and x_2 have means of 0?

(b) Show the form of the matrices $X'X$, and $X'Y$ in the general case of k predictor variables and n data.

Exercise 9.7:

Either read the data in from the website or type the Pedest, Area and Sales into x_1, x_2 and Y respectively. Then create a vector ones as the first column of X. Then check the matrix calculations against the lm() output. A short script to do this using R follows.

```
X=matrix(c(ones,x1,x2),nrow=4,ncol=3,byrow=FALSE)
print(X)
print(t(X)%*%X)
Bhat=solve(t(X)%*%X)%*%t(X)%*%Y
print(Bhat)
```

Exercise 9.8:

The yield of a chemical product in a reactor depends on temperature and pressure. An engineer runs the reaction on nine occasions with the nine possible combinations of temperature set low, middle, and high ($-1, 0, 1$ in scaled units) and pressure set low, medium and high ($-1, 0, 1$ in scaled units). Denote temperature and pressure, in scaled units, as x_1 and x_2 respectively. Set up an appropriate 9 by 3 matrix X for fitting the model

$$Y_i = \beta_0 + \beta_1 x_{1i} + \beta + 2x_{2i}.$$

(a) What is the correlation between x_1 and x_2?

(b) Show that $X'X$ is a diagonal matrix.

(c) Write down the inverse of $X'X$.

(d) Describe the estimators of the coefficients in terms of the yields $\{Y_i\}$.

Exercise 9.9:

The Bass formula for sales per unit time at time t is:

$$S(t) = m\frac{(p+q)^2 e^{-(p+q)t}}{p\left(1 + (q/p)e^{-(p+q)t}\right)^2},$$

where m is the total number of people who eventually buy the product and p and q are the coefficients of innovation and imitation respectively. Both coefficients lie in the range $[0, 1)$. Fit the model to sales of VCRs in the U.S. between 1980 and 1989 [data from Bass website]

$$840, 1470, 2110, 4000, 7590, 10950, 10530, 9470, 7790, 5890.$$

Exercise 9.10:

(a) Refer to the *Conch* data. Calculate $(X'X)^{-1}s^2$.

(b) Verify that the square-root of the leading diagonal in the matrix (a) corresponds to the standard errors given in Example 9.1 of Section 9.4.6.

(c) What is the covariance and correlation between $\widehat{\beta}_1$ and $\widehat{\beta}_2$?

Exercise 9.11:

The 57 data in urbtv.txt are from a study carried out by the Transportation Centre at Northwestern University, Chicago, in 1926. Each datum consists of five variables for a traffic analysis zone. They are in five columns:
column 1: trips per occupied dwelling unit;
column 2: average car ownership;
column 3: average household size;
column 4: socio-economic index; and
column 5: urbanization index.

(a) Fit the best regression you can find for the number of trips in terms of the variables in column 3 to 5, including their interactions if appropriate.

(b) Fit the best regression you can find for the average car ownership in terms of the variables in columns 3 to 5.

(c) Fit the regression of average car ownership in terms of all the other variables.

Exercise 9.12:

A manager works for a company that specializes in setting up computer systems for insurance companies. The manager has kept a record of the time taken (x_1) and the number of snags encountered (y) for the latest 40 contracts, together with the assessment of difficulty before the tasks started (x_2). The data are given in compsys.txt.

(a) Regress y on x_1 only.

(b) Regress y on x_2 only.

(c) Regress y on x_1 and x_2. Does anything surprise you? Can you explain your finding?

Exercise 9.13: Linear and quadratic functions

An engineer designed an experiment to investigate the effect of adding a compound to a varnish used in the electronics industry with the aim of reducing drying time. The amount of compound added, x, ranged from 1 up to 10 in unit steps (coded units). For each value of x, two samples of varnish were mixed with the compound and poured into Petri dishes to reach a 1 mm depth. The amounts of additive and drying times in minutes are shown in Table 9.12

TABLE 9.12: Varnish drying times

x	y	x	y
1	16.0	1	13.9
2	14.1	2	13.9
3	14.2	3	14.0
4	11.1	4	14.9
5	11.8	5	11.1
6	12.7	6	11.9
7	11.3	7	11.8
8	11.1	8	12.9
9	11.6	9	11.4
10	13.0	10	12.1

(a) Show the fitted line and the fitted quadratic curve on your plot.

(b) Fit a regression of y on x (model 1) and on $(x - 5.5)$ (model 2). What is the estimated standard deviation of the errors? Why is the intercept different in model 1 and model 2?

(c) Fit a regression of y on x, and x^2 (model 3). What is the estimated standard deviation of the errors?

(d) Fit a regression of y on $(x - 5.5)$, and $(x - 5.5)^2$ (model 4). What is the estimated standard deviation of the errors?

(e) Explain why the t-value associated with the coefficient of the linear term is smaller in model 3 than it is in model 4.

(f) Calculate the correlation between x and x^2, and the correlation between $(x - 5.5)$ and $(x - 5.5)^2$.

(g) Estimate the variance of observations at the same value of x by averaging the 10 estimates of variance based on the pair of observations at each x. How many degrees of freedom is this estimate based on?

(h) Compare your estimate in (g) with the estimated standard deviation of the errors in your quadratic model (model 2), and comment.

Exercise 9.14: Hole-drilling determination of stress (polynomial regression)

[Ostertagová, 2012] measured the strain (ε_a, ε_b, ε_c micro-m/m) in three directions as a function of depth (h) of a hole drilled in a metal beam. The data are given in the table below. Define $x = h - \bar{h}$.

Drilling stage (mm) h	Strain values in particular directions (μm/m)		
	ε_a	ε_b	ε_c
0.50	−16.00	−9.00	−6.00
1.00	−38.00	−22.00	−8.00
1.50	−50.00	−32.00	−5.00
2.00	−65.00	−48.00	−2.00
2.50	−72.00	−55.00	2.00
3.00	−80.00	−63.00	4.00
3.50	−85.00	−67.00	5.00
4.00	−89.00	−81.00	6.00
4.50	−93.00	−83.00	8.00
5.00	−100.00	−85.00	9.00

(a) Plot ε_a against x. Regress ε_a on x and ε_a on x^2. Superimpose the fitted line and the fitted quadratic curve on a scatter plot. Is the quadratic curve a convincing improvement on the line?

(b) As for (a) except with ε_b as the response.

(c) As for (a) except with ε_c as the response. Also fit, and plot the fitted curve, for a regression of ε_c on x, x^2, and x^3. Do you think the cubic curve is a convincing improvement on the line?

Section 9.5 Assessing the fit

Exercise 9.15:

Refer to Section 9.5.2. Consider the regression of y on two predictor variables x_1 and x_2. Show that the correlation between y and $\hat{y}$ is equivalent to R^2.

Exercise 9.16:

Refer to Section 9.5.3. Assume without loss of generality that the predictor variables $x_i, \cdots, x_k$ have mean values of 0 and so $\hat{\beta}_0 = \hat{y}$.

(a) Write $\sum(y_i - \bar{y})^2 = \sum((y_i - \hat{y}_i) + (\hat{y}_i - \bar{y}))^2$ and expand the right hand side.

(b) Use the result of Section 9.5.1 to show that the cross product term is 0.

(c) Hence demonstrate the ANOVA.

Exercise 9.17:

Consider fitting the model

$$y_i = \beta_0 + \beta_1 x_{1i} + \beta_2 x_{2i} + \varepsilon_i$$

to the *Conch* data set. Define

$$r_i = y_i - (\widehat{\beta}_0 + \widehat{\beta}_1 x_{1i} + \widehat{\beta}_2 x_{2i})$$

and show that imposing the constraints $\sum r_i = 0$, $\sum x_{1i} r_i = 0$, $\sum x_{2i} r_i$ leads to the same estimates as $\widehat{\beta} = (X'X)^{-1}X'Y$.

Exercise 9.18:

Fit a regression of sales on pedestrians and area for the first six shops in the *Conch* data file `head(conch.dat)`. Use this model to predict sales for the remaining four shops. Compare the estimated standard deviation of the errors in the fitted regression to the square root of the sum of the four prediction errors.

Exercise 9.19: Contingency tables

Madsen (1976) found the following proportions of high satisfaction among tenants in Copenhagen:
tower blocks with low level of contact with other residents 100/219;
tower blocks with high level of contact with other residents 100/181;
terraced houses with low level of contact with other residents 31/95;
and terraced houses with high level of contact with other residents 39/182.
Fit a logistic regression with tower block or terraced housing, and low or high contact as predictor variables. Comment on the fitted model, and improve on it.

Exercise 9.20:

When fitting the multiple regression model

$$\widehat{\beta} = (X'X)^{-1}X'Y,$$
$$\widehat{Y} = X\widehat{\beta} = X(X'X)^{-1}X'Y.$$

Define

$$H = X(X'X)^{-1}X'.$$

Then

$$\widehat{Y} = HY.$$

(a) Show that $H^2 = H$.

(b) The residuals are defined by $R = Y - \widehat{Y} = (I - H)Y$. Show that $R = (I - H)E$.

(c) Show that the variance-covariance matrix of the residuals is $(I - H)\sigma^2$.

(d) H is known as the projection matrix. Explain why this is an appropriate name for H.

Exercise 9.21: PRESS

In the notation of Exercise 9.20 the variance of the i-th residual is $(1 - h_{ii})\sigma^2$ where h_{ii} is the i-th element o the leading designed of H.

(a) Explain why the variance of the i-th residual is less than σ^2.

(b) Studentized residuals are defined by

$$\frac{r_{ii}}{\sqrt{(1 - h_{ii})s_{ii}^2}},$$

where r_i is the i-th residual. Calculate the studentized residuals for the *Conch* data and plot them against the corresponding residuals.

(c) The *PRESS* residuals (r_{ii}), corresponding to leaving out datum i, are given by

$$r_{(i)} = \frac{r_i}{1 - h_{ii}}.$$

Prove this formula for a regression on a single predictor variable (x_i) with mean 0, that is $\sum x_i = 0$.

(d) Calculate the *PRESS* residuals for the *Conch* data and plot them against the residuals.

Exercise 9.22: Cook's statistics

A datum in a multiple regression is said to be influential if the values of the predictor variables are far from the means for these variables. A measure of influence is $\dfrac{h_{ii}}{(1 - h_{ii})(1 + k)}$. The Cook's statistics are defined by

$$D_i = \frac{r_i^2}{(1 - h_{ii})s^2} \times \frac{h_{ii}}{(1 - h_{ii})(1 + k)}$$

which is a composite measure that identifies points of particular interest because they are either influential or outlying or both. In general, values of D_i that exceed about 2 are investigated.

(a) Calculate $\dfrac{h_{ii}}{(1 + h_{ii})(1 + k)}$ for the point corresponding to the extrusion rate of 750 in Example 9.4 on the plastic sheet manufacture

  (i) for a linear model and

  (ii) for a model that includes a quadratic term.

(b) Calculate the Cook's statistic for the point corresponding to the extrusion rate of 750 in Example 9.4

  (i) for a linear model and

  (ii) for a model that includes a quadratic term.

Section 9.6 Predictions

Exercise 9.23:

Refer to your linear (model 1) and quadratic model (model 2) for drying time of varnish (Exercise 9.13).

(a) Use model 1 to give approximate ($\hat{y} \pm s$) limits of prediction for the drying time when $x = 0$ and when $x = 5$.

(b) Use model 2 to give approximate ($\hat{y} \pm s$) limits of prediction for the drying time when $x = 0$ and when $x = 5$.

(c) Use model 2 to give precise limits of prediction for the drying time when $x = 0$ and when $x = 5$. What assumptions does your limit of prediction rely on?

(d) Use model 1 and model 2 to predict drying time if $x = 10$, and comment on these predictions.

Exercise 9.24:

Predict costs for outstanding schemes, given that there are $28, 5, 4, 2, 1, 1$ schemes with $1, 2, 3, 5, 6, 8$ properties respectively, using $m1, m2$ and $m3$.

Exercise 9.25:

Predict costs, and give 90% prediction limits for the costs, of schemes with 11 and 12 properties using $m1, m2$ and $m3$. Comment on their plausibility.

Section 9.7 Building multiple regression models

Exercise 9.26:

Consider the *Conch* data and a regression model sales on predictions x_1 and area x_2. Calculate $X'X$. Now define mean-adjusted pedestrians $u_{1i} = x_{1i} - \bar{x}_1$ and mean-adjusted areas $u_{2i} = x_{2i} - \bar{x}_2$. Calculate UU. Compare $X'X$ and $U'U$, then compare the determinants of the two matrices. Comment.

Exercise 9.27:

Refer to the *Conch* data and Section 9.7.

(a) Given model **m3b**, and a pedestrian mean of 721.8 and an area mean of 605.0. Deduce model **m3**.

(b) Given model **m4b** and the means of pedestrians and area, deduce model **m4a**.

Exercise 9.28:

Refer to the regression of joint strength on glue A or B and film thickness. Refit the model using mean-adjusted film thickness in place of film thickness. Verify that the intercept is the mean strength of all 10 joints.

Exercise 9.29:

Another option for coding the glue formation in Example 9.2 is the Helmert contrast matrix shown below.

Glue	x_1	x_2	x_3
A	−1	−1	−1
B	1	−1	−1
C	0	2	−1
D	0	0	3

(a) What are the means of X_1, X_2 and X_3?

(b) What are the correlations between X_1 and X_2, X_1 and X_3, and X_2 and X_3?

(c) Write down a Helmert contrast matrix for five categories A, B, C, D, E.

(d) Give an advantage and a drawback to using the Helmert contrast matrix.

Exercise 9.30: China ports with indicator variables

Refer to the China ports in Section 9.7.3. Refit the model with indicator variables for the regions set up relative to North. Give the precise t-value and p-value for the South region compared with the North.

Exercise 9.31:

R will fit a regression without an intercept by adding −1 to the model formula.

(a) Consider the *Conch* data in Table 9.1. Fit the model

$$m_0 = lm(Sales \sim Pedest + Area - 1).$$

Comment on the output.

(b) Consider the strength of joints in Example 9.2. Fit the model

$$m_0 = lm(Sales \sim Pedest + Area - 1).$$

Comment on the output.

(c) Run the following R script

```
y=rep(10,10)
x=1:10
plot(x,y)
m=lm(y~x-1)
summary(m)
```

You will see an impressive $R^2 = 0.7897$. How do you think this value of R^2 is calculated?

Section 9.8 Time series

Exercise 9.32:

Consider the time series model

$$X_t = \alpha X_{t-1} + \varepsilon_t$$

where $\varepsilon_t \sim$ indep $0, \sigma_2$ with $X_0 = 0$, and hence $X_1 = \varepsilon_1$.

(a) Show that $X_2 = \varepsilon_2 + \alpha\varepsilon_1$ and that $X_3 = \varepsilon_3 + \alpha\varepsilon_2 + \alpha^2\varepsilon_1$.

(b) Deduce that $X_t = \varepsilon_t + \alpha\varepsilon_{t-1} + \alpha^2\varepsilon_{t-2} + \alpha^3\varepsilon_{t-3}$.

(c) Explain why $E[X_t]=0$ for all t.

(d) Express $var(X_t)$ in terms of σ^2.

(e) Let S_n be the sum of the first n terms of a geometric series with initial term a and common ratio r. The $S_n = a + ar + \cdots + ar^{n-1}$. Show that $S_n = \dfrac{a - ar^2}{1 - r}$ and deduce that $S_\infty = \dfrac{a}{1-r}$ for $|r| < 1$.

(f) Use the result in (v) to show that $var(X_t) \simeq \dfrac{\sigma^2}{1 - \alpha^2}$ for large t provided $|\alpha| < 1$.

(g) Explain why $\{X_t\}$ is not second order stationary id $\alpha \geqslant 1$.

(h) Write down the model for the special case of $\alpha = 1$. This is known as a ransom walk.

Exercise 9.33:

We can assume that $E[X_t]=0$ without any loss of generality. Consider

$$X_t = \alpha X_{t-1} + \varepsilon_t$$

which is second order stationary ($|\alpha| < 1$).

(a) Multiple both sides by X_{t-k} to obtain

$$X_t X_{t-k} = \alpha X_t X_{t-k} + \varepsilon_t X_{t-k}$$

and explain why this implies $\gamma(k) = \alpha\gamma(k-1) + 0$.

(b) Assume $\gamma(k) = \theta^k$ and solve for θ.

(c) Deduce that $\rho(k) = \alpha^k$.

Exercise 9.34:

Assume without any loss of generality that $E[X_t]=0$. Consider

$$X_t = \alpha_1 X_{t-1} + \alpha_2 X_{t-2} + \varepsilon_t$$

and assume the model is stationary. Multiple both sides by X_{t-k}, take the expectation and solve the difference equation. Show that $\rho(k)$ has the form of a damped sinusoid if $\alpha_1 = 1$ and $\alpha_2 = -0.5$.

Exercise 9.35:

(a) The periodogram of a time series of length n, which is even, is calculated as follows.

   (i) Calculate $n/2$ frequencies from a lowest at 2π radians per record length which is $(2\pi/n)$ radians per sampling interval at multiples from 1 up to n/2. This gives: $2\pi/n, 2 \times 2/0i/n, 3 \times 2\pi/n, \ldots, (n/2) \times 2\pi/n$ radians per sampling interval. What is the highest frequency and why is it the highest frequency that can be fitted usefully?

   (ii) Calculate the cosine and sine functions at the sample points for these frequencies. What is the sine function of the highest frequency?

   (iii) Any pair from these cosine and sine functions are uncorrelated (they are orthogonal). Fit a regression with all the functions except the sine function at the highest frequency. Try this for the following example.

```
set.seed(7);y=rnorm(6,4,10)
n=length(y);t=c(1:n)
f1=2*pi/n;f2=2*f1;f3=3*f1
C1=cos(f1*t);S1=sin(f1*t);C2=cos(f2*t);S2=sin(f2*t);C3=cos(f3*t)
print(round(cor(cbind(C1,S1,C2,S2,C3)),2))
m=lm(y~C1+S1+C2+S2+C3)
```

(iv) The variance of the time series is equal to one half of the sum of the squared amplitudes of the sinusoidal functions plus another half of the amplitude of the cosine function at the highest frequency. This is known as Parseval's Theorem. Verify Parseval's theorem for the example.

```
P1=m$coef[2]^2+m$coef[3]^2
P2=m$coef[4]^2+m$coef[5]^2
P3=m$coef[6]^2
print(c(var(y)*(n-1)/n,0.5*(P1+P2)+P3))
plot(c(f1,f2,f3),c(.5*P1,.5*P2,P3))
```

(b) The R function `spectrum()` calculates the periodogram using highly efficient algorithms and plots a moving average of the ordinates. The number in the average is controlled with span.

(i) Try

```
n=1000000
y=rnorm(n)
spectrum(y,span=2*sqrt(length(y)))
```

Why is the the term "white noise" used for the spectrum?

(ii) Repeat (i) using pseudo-random numbers from an exponential distribution.

(iii) Generate a time series of length 100000 from an AR(1) time series model with $\alpha = 0.9$. Calculate the spectrum and comment.

(iv) Generate a time series of length 100000 from an AR(1) time series model with $\alpha = -0.9$. Calculate the spectrum and comment.

(v) Draw a spectrum for the wave tank.

Exercise 9.36:

Fit the model

$$Y_t = \beta_0 + \beta_1 Y_{t-1} + \beta_2 t + \beta_3 \cos\left(\frac{2\pi t}{12}\right) + \beta_4 \sin\left(\frac{2\pi t}{12}\right) + \varepsilon_t$$

to the logarithms of inflows to the Fontburn reservoir plus 0.1.

(a) Construct a 90% confidence interval and a 95% confidence interval for the coefficient of time (β_2).

(b) Explain why the estimated amplitude of the sinusoidal term $\sqrt{\widehat{B}_3^2 + \widehat{B}_4^2}$ is less than that in the regression

$$Y_t = \beta_0 + \beta_2 t + \beta_3 \cos\left(\frac{2\pi t}{12}\right) + \beta_4 \sin\left(\frac{2\pi t}{12}\right) + \varepsilon_t.$$

(c) Give an interpretation of the model that include the Y_{t-1} term as a physical system.

Exercise 9.37:

Fit an AR(1) model to the wave-height data using `ar()`.

(a) Compare the estimated standard deviation of the errors for the AR(2) and AR(p by Akaike information criterion (AIC)) model.

(b) Compare the ACF of the residuals for the two models.

(c) Compare the 20 most extreme values over a simulation of length 1 million for the two models.

Section 9.9 Non-linear least squares

Exercise 9.38:

Use a bootstrap procedure, resampling the 16 data, to estimate the standard error of the coefficients a and b in Example 9.8. Why is this rather unsatisfactory? Implement a bootstrap procedure in which you resample residuals from the fitted model.

Exercise 9.39: Confidence intervals

The standard errors given in the R `nls()` function are based linearization of the function about the parameter estimates using a Taylor series expansion. Ordinary least squares is then used with the linearized model. Then a $(1 - \alpha) \times 100\%$ confidence interval for a coefficient is given by the estimate plus or minus the product of its estimated standard error and $t_{\nu, \alpha/2}$, where ν is the degrees of freedom for the estimator of the error (number of data less the number of parameters estimated from the data). Calculate 95% confidence intervals for a and b in Example 3.1.

Section 9.10 Generalized linear model

Exercise 9.40:

Reanalyze the Tyne and Wear accident data with the logarithm of the product of length and flow as an offset.

Exercise 9.41: Machine breakdown

The data is the file machbrk.txt are: the number of machine breakdowns (y_t); and the number of person days lost to absenteeism (x_t) for a period of 72 weeks $t = 1, 2, \cdots, 72$. The data are from a large manufacturing company and the number of breakdowns was considered a serious problem. A manager had responded to the problem by implementing routine preventative maintenance and a training program for employees to take responsibility for this.

(a) Plot the number of breakdowns and the days lost to absenteeism against t on the same graph.

(b) Fit a model of the form $ln(y_t) = \beta_0 + \beta_1 I_t(t - 36) + \beta_2 x_t + \varepsilon_t$ where ε_t are independent with mean 0 and $I_t = 0$ for $t \leq 36$ and $I_t = 1$ for $36 < t$. Comment on the results.

(c) Fit a model of the form $\sqrt{y_t} = \beta_0 + \beta_1 I_t(t-36) + \beta_2 x_t + \varepsilon_t$ and comment on the results.

(d) Use the `glm()` function to fit a model of the form $Y_t = P(\mu_t)$ where $\mu_t = \beta_0 + \beta_1 I_t(t-36) + \beta_2 x_t$.

(e) Comment on the different analyses. Is there evidence that the number of breakdowns increases with days lost to absenteeism? Suggest how the manager might continue the quality improvement initiative.

Exercise 9.42: Generalized least squares

Consider the linear model

$$Y = XB + E, \qquad (9.1)$$

but now suppose that E has a variance-covariance matrix V rather than $I\sigma^2$.

(a) There is a matrix Q such that $Q^2 = V$. Is Q unique? If not how might a unique Q be defined?

(b) Show that the variance-covariance of $Q^{-1}E$ is I.

(c) Pre-multiply both sides of (9.1) by Q^{-1} and hence show that a generalised least squares estimator of B is

$$\widehat{B} = (X'V^{-1}X)X'V^{-1}Y.$$

Exercise 9.43: Probit

An option with R is to use a probit, defined as the inverse cdf of the standard normal distribution. The R code is

```
m2=glm(Y~x,family=binomial(link="probit"))
print(summary(m2))
```

(a) Plot the probit of p, $\Phi^{-1}(p)$. against the logit of p, $ln(p/(1-p))$, for values of p from 0.001 up to 0.999.

(b) Repeat the analysis of the fuel tanks using probit instead of logit. Are the conclusions any different?

Exercise 9.44: Shuttle

Data for 23 launches of the space shuttles are given in shuttle.txt. Use a logistic regression to model the probability of a field joint failing (1) against launch temperature and engine pressure [Dalal et al., 1989].

Exercise 9.45:

Small test panes of three types of toughened glass were subjected to the same impact from a pendulum. The results are:

Glass type	1	2	3
Undamaged	30	40	45
Broke	70	40	75

Fit a logistic regression using indicator variables for the glass type.

Miscellaneous problems

Exercise 9.46: Lifetimes

The following data are lifetimes (mins) of AA/lithium rechargeable dry-cell batteries under a constant load: $58, 55, 63, 64$.

(a) Calculate the mean and the median .

(b) Write down the 4 deviations from the mean. What is the sum of these deviations?

(c) Without using a calculator, calculate the variance, s^2 (NB this is defined with the denominator $n - 1$). Show all the steps in your calculation.

(d) Calculate the standard deviation, s.

Exercise 9.47: Bernoulli trials

If X is the number of successes in n Bernoulli trials with a probability of success p, then $E[X] = np$ and $\text{var}(X) = p(1 - p)/n$. Use a Taylor series expansion to obtain and approximation to the expected value and variance of $\text{logit}(X/n)$.

Exercise 9.48: Hand assembly

A company manufactures products which require a dexterous hand assembly operation. Employees work in small groups around tables to do this work. Some of the product is non-conforming and has to be sold as second quality ("second"). The company aims to keep the proportion of seconds below 0.02 and has provided additional training to the employees over a 26 week period. The data is the number of items in a batch at the end of each week and the number of seconds.

(a) Fit a logistic regression model for the proportion of seconds (p) as a function of time: $logit(p) = \beta_0 + \beta_1 t$ where $t = 1, 2, \cdots, 26$.

(b) Is there evidence of over-dispersion? If so, what might account for this?

(c) Is there evidence that the additional training has reduced the proportion of seconds? Test the null hypothesis of no effect against a hypothesis that training reduces the proportion of seconds at the 10% level. Adjust the z-score in the computer output by the square root of the ratio of the deviance to the number of battles if this exceeds 1.

10

Statistical quality control

To implement statistical quality control, we first need to understand the causes of variation in our process. We aim to identify and remove what is known as special cause variation and to keep common cause variation, to a minimum level. The next step is to establish that our process can meet the customer's specification, and that our product will continue to meet that specification during its design lifetime. Statistical quality control charts are used to monitor the process and so provide early warning of any new sources of special cause variation and any increase in the common cause variation.

Experiment E.8 Weibull analysis of cycles to failure

Experiment E.9 Control or tamper?

10.1 Continuous improvement

10.1.1 Defining quality

Being fit for purpose and within specification are requirements for a product or service, to be of high quality. In addition there are likely to be aesthetic features, that will help attract customers, and if the product or service exceeds customer expectations[1] it will help retain them. All this has to be achieved whilst keeping the price at the market level, so processes need to be efficient. Statistical quality control focuses on the requirement of being within specification.

Definition 10.1: Satisfactory quality

A product or service is of satisfactory quality if it is within specification.

---

[1] It is however crucial that we meet, rather than attempt to improve upon, the specification. A manufacturing company supplied a fabric that exceeded the specification for durability to a customer. The fabric was fit for purpose and was used, but the customer requested that future supplies should be less durable and demanded a pro-rata decrease in the price.

10.1.2 Taking measurements

It is hard to argue against the aim of continuous improvement, and there is plenty of advice for managers on how to achieve this goal[2]. Peter F. Drucker, whose ideas have influenced management teaching and practice since the 1950s is often quoted as saying "If you can't measure it you can't improve it", and "What gets measured gets improved". These aphorisms are sound advice for improving processes, but they can be counter productive when managing individuals [Caulcutt, 1999].

Example 10.1: PC sales

Benito has run a small business selling PC systems (PCs plus peripherals and home set up) for several years and has noticed a decline in sales over the period. He is inclined to believe that this is because his two sales people Angelo and Bill are a bit slack, despite the commission they earn on sales. Benito decides to set a target sales of 15 PC systems per week. If Angelo or Bill sell more than 19 systems in a week he will praise them and if they sell less than 11 he will blame them for the decline in sales.

In the computer simulation (see book website) we set weekly sales using independent draws from a normal distribution with mean of 15 and standard deviation of 2 that are rounded to the nearest integer. If a PC system is ordered in one week, Benito always arranges for the home set up in the next week.

1. Angelo ignores both blame and praise and declares his sales to Benito at the end of the week.

   (a) How long do customers wait for home set up?
   (b) What does Benito usually notice about Angelo's sales in the week following a week in which he has blamed Angelo for the decline in sales?
   (c) What does Benito usually notice about Angelo's sales in the week following a week in which he praises Angelo?

   The answers to these questions are 1 week, the week following blame will usually be closer to the mean and so an improvement (see Exercise 10.16 for the precise probability) and the week following praise will usually be closer to the mean and so a deterioration.

2. Bill dislikes the blame and is embarrassed by the praise so he attempts to evade both by being less truthful when he declares his sales to Benito. The plots in Figure 10.1 are obtained from a computer simulation and show Angelo's sales, Bill's reported sales and how many of Bill's customers wait 2 weeks for home setup. The + and * indicate when the actual weekly PC sales fall below 11 and above 19 respectively for Angelo and Bill.

   (a) What evasive strategy do you think Bill is using?
   (b) What is the likely effect on the business?
   (c) How would you advise Benito?

W. Edwards Deming, began his career as an electrical engineer before becoming a professor of statistics at New York University's graduate school of business administration (1946-1993). He is particularly renowned for his work with Japanese industry following the

---

[2]In his book *Images of Organization* Gareth Morgan describes organizations from different perspectives, which provide context for management theories.

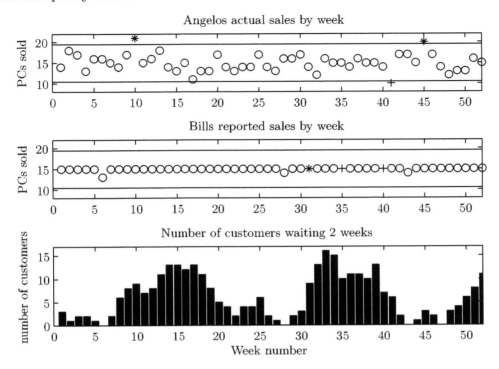

FIGURE 10.1: Angelo's actual and reported sales (top), Bill's reported sales (center) and the number of Bill's customers who wait two weeks for home setup (bottom).

second world war, and his book, *Out of the Crisis* [Deming, 2000], draws on his experiences working with industry in the U.S. and Japan. In this book he promotes 14 points, and Benito, in Example 10.1, might consider these in particular:

8: Drive out fear, so that everyone may work effectively for the company.

11: Eliminate work standards (quotas) on the factory floor, substitute leadership.

10.1.3 Avoiding rework

If quality is improved, all products will meet the specification and most rework will be eliminated. Quality can often be improved without investing in expensive new machinery by carefully studying the process, checking that the machines are set up correctly and removing unnecessary sources of variation.

An example was a production line in a factory that made fruit pastilles. The production line was frequently stopped because the pastilles became stuck in a chute. Also, when the pastilles were packed into tubes, some tubes failed to meet specification because the ends of the foils gaped. The cause of both problems was that some pastilles were slightly oversized as a consequence of one filling head dispensing too much syrup into the mold. Once this had been rectified:

- line stoppages were rare,

- all packed tubes were within specification, and

- there was a saving of thousands of dollars per year on syrup costs because less syrup was used.

The futility of manufacturing scrap was pointed out in an entertaining fashion by the Enterprise Initiative (UK in the 1980s) with a cartoon, and a short film *The Case of the Short Sighted Boss* featuring well known British actors [The Enterprise Initiative, 2015a, The Enterprise Initiative, 2015b]. The elimination of waste is the key principle of *lean manufacturing*. In the opening paragraph of *Out of the Crisis*, Deming emphasizes that productivity increases as quality improves, because there is less rework.

10.1.4 Strategies for quality improvement

Drucker's influence can be seen in the Six Sigma suite of techniques for process improvement. Deming's ideas, amongst many others, have been incorporated into Total Quality Management (TQM), that was taken up by many companies in the 1980s and continues in other guises. Deming's first point "Create constancy of purpose toward improvement of product and service, with the aim to become competitive and to stay in business and provide jobs" is just as pertinent now as it was in 1982.

Some of the general advice offered by well known management consultants may appear contradictory, but any general advice needs to be considered in the specific context. For example, it is neither efficient nor effective to produce a high quality product by 100% inspection, but automobile manufacturers may have every finished vehicle test driven around a track to check safety critical features such as brakes, steering and lights. Moreover, while much of the advice for managers may be common sense, we might recall Voltaire's remark that *"Le sens commun n'est pas si commun"*[3].

A process manager is not expected to improve quality on his or her own. The people working on the process can contribute and need to be given encouragement to do so. Formal structures for facilitating contributions from employees include *suggestion schemes*, *quality circles*, and *process improvement teams*. Also, a colleague from another department or division may be able to offer helpful advice, because they are somewhat removed from a problem and unconstrained by habit.

A common feature of any strategy for quality improvement is the *Shewhart cycle* [Shewhart, 1939], also known as the *PDSA cycle* and the *Deming cycle* despite Deming's attribution of the cycle to Walter A Shewhart. The P is for plan, the D is for do, the S is for study of the results, and the A is for act. It is presented as a circular process because improvement is on-going. Deming points out that statistical methodology provides valuable guidance for all steps in the cycle.

10.1.5 Quality management systems

The ISO9000 series of quality management system standards [ISO, 2015a] was designed so that companies could replace assessment of their processes, and procedures for ensuring satisfactory quality of products, by individual customers with a single third-party certification. Some early implementations were criticized for excessive documentation, which is not a requirement of the standard, and also on the ground that documented procedures do not necessarily assure high quality. A succinct rejoinder to the second criticism from a manager in a successful company that manufactures filters is that "It's pointless to set up procedures unless they assure high quality". He also emphasized the need to set up a

---

[3]Common sense is not so common.

system in which procedures can be changed easily and efficiently when improvements to processes are identified. The automotive industry tends to rely on its interpretation of ISO9000 known as QS9000. Apart from ISO9000, there is a more recent ISO standard, ISO18404 [ISO, 2015b] which integrates ISO9000 with Six Sigma, and the environmental standard ISO14000 [ISO, 2015c].

10.1.6   Implementing continuous improvement

We consider four aspects of managing our process:

1. **Stability:**
   To begin with,

   - is the process running as it should or are there frequent malfunctions?
   - are we sometimes manufacturing scrap?

   These issues are typically caused by what Deming refers to as **special cause variation**. Special cause variation has a specific cause, that can in principle be identified and corrected. For example:

   - the employee who usually sets up a process is absent due to illness and a colleague takes over the role but does so incorrectly, due to a lack of a clear procedure;
   - an employee makes mistakes because he or she has had inadequate training;
   - a bearing on a lathe is excessively worn due to a lack of preventative maintenance;
   - a batch of raw material is contaminated.

   In contrast **common cause variation** becomes an intrinsic part of the process. For example:

   - no machine is perfect and there will be slight variation between one item and the next even when the machine is well maintained, generally the higher the precision that is required the more expensive the machine will be;
   - raw material will vary, even if it is all within the specification, and this variability will be greater if there is a policy of buying from whichever supplier offers the lowest price at the time.

   Generally, employees can identify special cause variation and correct it, given support from management, whereas taking action to reduce common cause variation is a management decision.
   A process is stable if there is no special cause variation, and it is said to be in statistical control. The variation is a random process and in the context of time series the process is stationary in the mean and variance. It is often assumed that the random variation is independently distributed, in which case any corrective action, which Deming refers to as tampering, will be counter-productive and potentially de-stabilizing. Deming's funnel experiment is the basis for Experiment E.9, which shows that tampering is counter-productive (Exercise 10.2).

2. **Capability and Reliability:**
   Suppose our process is stable. Is all the product within specification? If not, we are likely to lose customers because we are not supplying product that meets their requirements. Attempting 100% inspection to ensure all the shipped product does meet the

specification is generally too inefficient to be viable. We need to consider whether our process is capable of meeting the specification, and capability indices are a customary measure. Whether or not reliability is part of the specification, our products need to be reliable, inasmuch as they continue to function correctly over their design life, if we expect customers to return or to recommend our products to others. We discuss lifetime testing.

3. **Quality control:**
   Now suppose we have dealt with all the special cause variation and our process is running as it should. There will inevitably be some variation but it has been reduced to a minimum. Deming refers to this background variation as common cause variation. Can we rely on this satisfactory state of affairs persisting? The answer is almost certainly "no", so the third aspect of managing our process is to monitor the performance and take remedial action when there is a substantial change[4]. However, we need to avoid tampering with the process if changes can reasonably be attributed to common cause variation.

4. **Research and development:**
   The fourth aspect of managing our process is research and development to improve the process so that we continue to impress our customers and attract more of them[5].

10.2 Process stability

A stable process has a constant mean, and common cause variation introduces variability about the mean.

10.2.1 Runs chart

This is a plot of the quality variable measured on items from production against time. Start by looking at recent records if they are available. Then

- Take n items from production, well separated by intervening items so that the selected items can plausibly be considered random draws.

- Measure the value of the quality variable for each item $\{x_t\}$ for $t = 1, \ldots, n$.

- Plot x_t against t.

- Calculate the moving lag 1 ranges

$$R_t = |x_t - x_{t-1}| \quad \text{for} \quad t = 2, \ldots, n.$$

- Estimate the process standard deviation by

$$\widetilde{\sigma} = 1.047 \times \text{median} (R_t).$$

---

[4]We may be able to learn something from the special cause variation as Alexander Fleming did in 1928 when he discovered penicillin.

[5]Improving the process does not necessarily imply changing the product, for example with traditional items when the challenge is to improve supporting processes so that the tradition can be upheld.

If $\{x_t\}$ is a random sample from a normal distribution with standard deviation σ then $\tilde{\sigma}$ is an unbiased estimator of σ (Exercise 10.3). The reason for using this estimator of the process standard deviation is that it is insensitive to changes (exercises 10.4, 10.5, 10.6) in the mean of the process.

- Draw lines at

$$\bar{x}, \quad \bar{x} \pm \tilde{\sigma}, \quad \bar{x} \pm 2\tilde{\sigma}, \quad \bar{x} \pm 3\tilde{\sigma}.$$

Label zones as:

C within $1\tilde{\sigma}$ of the mean

B between $1\tilde{\sigma}$ and $2\tilde{\sigma}$ from the mean

A between $2\tilde{\sigma}$ and $3\tilde{\sigma}$ from the mean.

The following R script defines a function, that we name `nelson()`, that draws a runs chart with the zones marked for a sequence of observations.

```
nelson=function(x)
{
n=length(x);t=c(1:n);mx=mean(x)
ml=t-t+mx\
R=abs(x[2:n]-x[1:(n-1)])
sg=median(R)*1.047
print(c('mean',mx))
print(c('s',sd(x)))
print(c('sigtilde',sg))
al3=ml-3*sg;al2=ml-2*sg;al1=ml-sg
au1=ml+sg;au2=ml+2*sg;au3=ml+3*sg
tpx=c(n,n,n,n,n);tpyn=c(-2.5,-1.5,0,1.5,2.5)
tpy=tpyn*sg+mx;names=c('A','B','C','B','A')
plot(t,x)
lines(t,al3,lty=5);lines(t,al2,lty=4);lines(t,al1,lty=3)
lines(t,ml);lines(t,au1,lty=3);lines(t,au2,lty=4)
lines(t,au3,lty=5);text(tpx,tpy,names)
}
```

In some cases it might be more useful to use the target value or the median in place of the mean as the centre line when defining the zones. Lloyd S. Nelson, who was Director of Statistical Methods for the Nashua Corporation, suggested the following 8 rules as a guide to deciding whether or not the process is stable. There is evidence that a process may not be stable if

R1: A point beyond Zone A. This rule aims to detect outlying values, and it can also indicate a change in the mean.

R2: Nine or more consecutive points the same side of the centre line. If the center line is set at the sample mean, the rule indicates step changes in the process mean. If the center line is set at the target value this rule would indicate that the process is off-target.

R3: Six consecutive points increasing or decreasing. This rule aims to detect a trend in the process mean.

R4: Fourteen consecutive points oscillating. This rule aims to detect repeated over-correction of the process.

R5: Two out of three in or beyond Zone A on the same side. This rule aims to detect a change in mean level.

R6: Four out of five in or beyond Zone B on the same side. This rule aims to detect a change in mean level.

R7: Fifteen consecutive points in Zone C. This rule indicates suggests that the process standard deviation changes from time to time and that this is a period of low standard deviation.

R8: Eight consecutive points avoid Zone C. This might indicate repeated over-correction.

All of the events described in the 8 rules are unlikely in a short record if the process has a constant mean and standard deviation, and zones are defined about the mean. For example, for a symmetric distribution, the probability of the next 9 consecutive points being on the same side of the mean is $(\frac{1}{2})^8 \approx 0.004$ if the distribution is symmetric. In a long record we'd expect to see some of these events a few times, for example 8 instances of 9 consecutive points being on the same side of the mean. We use the function, **nelson()**, with zones defined about the mean, for the following example

Example 10.2: Moisture content

Zirconium silicate $(ZrSiO_4)$, occurs naturally as zircon, and it is used in the manufacture of refractory materials, the production of ceramics, and as an opacifier in enamels and glazes. A company supplies zirconium silicate in 5 tonne batches, and part of the specification is that the water content should be below 1% by weight. Batches are tumbled in a drier before being packed in a moisture proof wrapper. The moisture contents of 17 batches randomly selected from the last 17 shifts are:

$0.18, 0.24, 0.11, 0.14, 0.38, 0.13, 0.26, 0.18, 0.16, 0.15, 0.27, 0.14, 0.21, 0.23, 0.18, 0.29, 0.13$

```
> x=c(.18,.24,.11,.14,.38,.13,.26,.18,
+ .16,.15,.27,.14,.21,.23,.18,.29,.13)
> print(x)
 [1] 0.18 0.24 0.11 0.14 0.38 0.13 0.26 0.18 0.16 0.15 0.27 0.14 0.21
[14] 0.23 0.18 0.29 0.13
> n=length(x);R=abs(x[2:n]-x[1:(n-1)])
> print(R)
 [1] 0.06 0.13 0.03 0.24 0.25 0.13 0.08 0.02 0.01 0.12 0.13 0.07 0.02
[14] 0.05 0.11 0.16
> nelson(x)
[1] "mean"                "0.198823529411765"
[1] "s"                   "0.0715788335457117"
[1] "sigtilde" "0.099465"
```

From Figure 10.2, the process seems stable. The distribution of moisture contents seems to be positively skewed, see the histogram and box plot in Figure 10.3, which is typical for non-negative variables that are within one or two standard deviations of 0. The

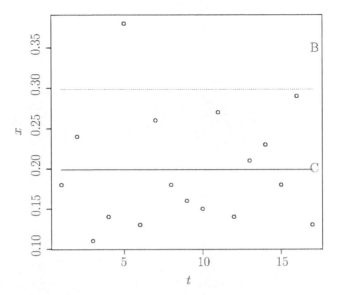

FIGURE 10.2: Moisture content (1% by weight) of zirconium silicate from 17 shifts.

$\tilde{\sigma}$ may be slightly biased [6] but the runs chart is not a precise statistical procedure. The acf of moisture has a statistically significant negative auto-correlation at lag 1 as shown in Figure 10.3, which suggests there may be a tendency to tamper with the process. Also, the zirconium silicate is being dried more than is required to meet the specification. Provided the moisture-proof wrapper is effective, a shorter tumbling time could be implemented and this would yield savings in costs through: reduced electricity cost operating the tumble drier and less wear on the drier.

10.2.2   Histograms and box plots

We may wish to check for stability between different suppliers, or between different shifts, and so on. Graphical displays provide a useful visual assessment of the situation. Also, for many manufacturing processes it is reasonable to suppose that common cause variation is errors made up of a sum of a large number of independent small components of error, which are equally likely to be positive or negative. In such cases the distribution of the errors will be well approximated by a normal distribution. An important exception is processes for which a variable, that cannot take negative values, has a target value of zero. In these cases the distributions are likely to be positively skewed, as we see in Figure 10.3.

Example 10.3: Tensile strengths

The data in Table 10.1 are tensile strengths (kg) of the 12 wires taken from each of 9 high voltage electricity transmission cables (Hald, 1952). Cables 1 − 4 were made from one lot of raw material, while cables 5 − 9 were made from a second lot. The box plots

---
[6]The bias could be estimated using a bootstrap procedure.

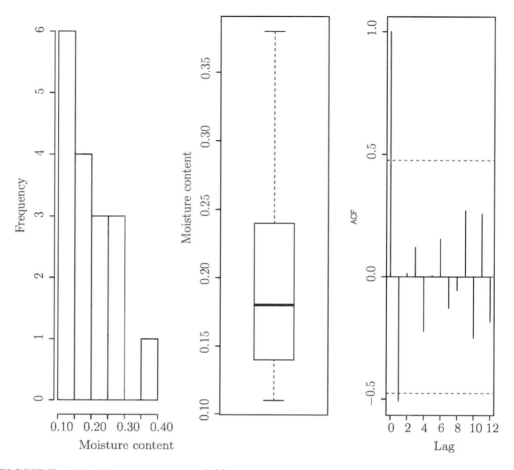

FIGURE 10.3: Moisture content (1% by weight) of zirconium silicate : histogram (left); box plot (center); auto-correlation function (right).

TABLE 10.1: Tensile strength of 12 wires taken from 9 cables.

Cable

1	2	3	4	5	6	7	8	9
345	329	340	328	347	341	339	339	342
327	327	330	344	341	340	340	340	346
335	332	325	342	345	335	342	347	347
338	348	328	350	340	336	341	345	348
330	337	338	335	350	339	336	350	355
334	328	332	332	346	340	342	348	351
335	328	335	328	345	342	347	341	333
340	330	340	340	342	345	345	342	347
337	345	336	335	340	341	341	337	350
342	334	339	337	339	338	340	346	347
333	328	335	337	330	346	336	340	348
335	330	329	340	338	347	342	345	341

show a clear difference in means of wires made from lot 1 and lot 2. The wires made from lot 2 have a higher mean strength[7]. This accounts for the rather flat appearance of the histogram shown in Figure 10.4. We will investigate the variations of strengths of wires within cables and the variation between cables.

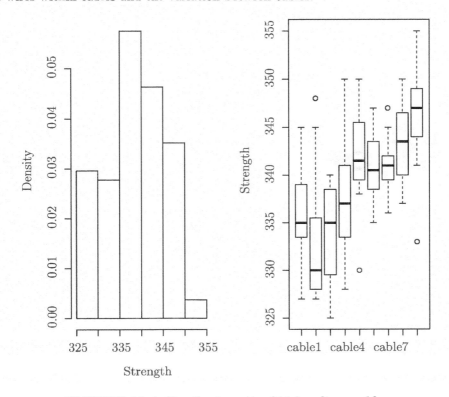

FIGURE 10.4: Tensile strength of high voltage cables.

10.2.3 Components of variance

If we are to reduce, or at least control variation, we need to know its sources. Suppose we have equal size samples of size J from I different batches. It is convenient to introduce a double subscript notation, so y_{ij} is the observation on item j from batch i, where $i = 1, \ldots, I$ and $j = 1, \ldots, J$. A model for the observations is

$$Y_{ij} \;=\; \mu + \alpha_i + \epsilon_{ij}, \qquad \text{where}$$

- μ is the mean of all batches in the hypothetical infinite population of all possible batches;

- α_i, referred to as between-batch errors, are the differences between the means of batches and μ, with $\alpha_i \sim 0, \sigma_\alpha^2$; and ϵ_{ij} are within batch errors with $\epsilon_{ij} \sim 0, \sigma_\epsilon^2$.

All errors are assumed to be independently distributed. Then a batch mean

$$\overline{Y}_{i.} \;=\; \mu + \alpha_i + \bar{\epsilon}_{i.} \,,$$

---

[7]This can be verified with a two sample t-test.

where the "." indicates that we have averaged over j. The overall mean

$$\overline{Y}_{..} \;=\; \mu + \overline{\alpha}. + \overline{\epsilon}_{..} \,,$$

where "." again indicates averaging over the corresponding subscript. From the first of these equations we have

$$\mathrm{var}(Y_{ij}|\text{within batch } i) \;=\; \sigma_\epsilon^2.$$

We can estimate σ_ϵ^2 by calculating the sample variance within each batch and then taking the mean of the I sample variances, which we denote by $\widehat{\sigma}_\epsilon^2$.

$$s_i^2 \;=\; \frac{\sum_{j=1}^{J}(y_{ij} - \overline{y}_{i.})^2}{J - 1} \qquad \widehat{\sigma}_\epsilon^2 \;=\; \frac{\sum_{i=1}^{I} s_i^2}{I}.$$

This estimate of σ_ϵ^2 is based on $I(J-1)$ degrees of freedom. From the second equation we have

$$\mathrm{var}(\overline{Y}_{i.}) \;=\; \sigma_\alpha^2 + \frac{\sigma_\epsilon^2}{J} \,.$$

We estimate σ_α^2 from this equation by replacing the left hand side with the sample variance of the batch means and σ_ϵ^2 with its estimate. That is,

$$s_{\overline{Y}_{i.}}^2 = \widehat{\sigma}_\alpha^2 + \frac{\widehat{\sigma}_\epsilon^2}{J} \,,$$

where

$$s_{\overline{Y}_{i.}}^2 \;=\; \frac{\sum_{i=1}^{I}(\overline{Y}_{i.} - \overline{Y}_{..})^2}{I - 1}$$

and the estimate $\widehat{\sigma}_\alpha^2$ is based on $I - 1$ degrees of freedom.

We can test a null hypothesis that the variance between batches is 0,

$$H_0 : \sigma_\alpha^2 \;=\; 0,$$

against an alternative hypothesis

$$H_1 : \sigma_\alpha^2 \;>\; 0,$$

with an F-test if we assume normal distributions for the random variation. If H_0 is true then

$$\frac{J s_{\overline{Y}_{i.}}^2}{\widehat{\sigma}_\epsilon^2} \;\sim\; F_{I-1, I(J-1)}.$$

It is possible that the estimate of the variance σ_α^2 turns out to be negative, in which case the estimate is replaced by 0. This eventuality is quite likely if σ_α^2 is small compared with σ_ϵ^2.

Example 10.3: (Continued) Tensile strengths

We focus on the 5 ropes from lot 2 and estimate the standard deviation of strengths of wires within ropes (σ_ϵ) and a standard deviation that represents an additional component of variability which arises between ropes in the same lot(σ_α).

```
> strength.dat=read.table("cableHV.txt",header=TRUE)
> attach(strength.dat)
> B=cbind(cable5,cable6,cable7,cable8,cable9)
> I=5
> J=12
> xbars=apply(B,2,mean)
> varin=mean(apply(B,2,var))
> varmean=var(xbars)
> varbet=var(xbars)-varin/12
> print(c("variance e",round(varin,2),"variance a",round(varbet,2)))
[1] "variance e" "19.79"       "variance a" "3.42"
> print(c("sd e",round(sqrt(varin),2),"sd a",round(sqrt(varbet),2)))
[1] "sd e" "4.45" "sd a" "1.85"
> F=varmean*J/varin
> print(c("F_calc",round(F,2),"p-value",round(1-pf(F,(I-1),I*(J-1)),3)))
[1] "F_calc" "3.07"       "p-value" "0.023"
```

The standard deviation of strengths of wires within a rope is estimated as 4.45 and the additional component of variability arising between ropes in the same lot has an estimated standard deviation of 1.85. If reasons for this additional component of variability could be identified the variation in the strengths of ropes from the same lot would be reduced[8]. It is also possible that the additional component has a standard deviation of 0, which is the ideal situation, in which case the chance of estimating a standard deviation as high as 1.85 is 0.023.

Example 10.4: Inter-laboratory trial

Laboratory tests on supposedly identical materials will rarely give identical results. Many factors contribute to this variability, including differences in: test specimens from the material; test equipment; calibration; technicians carrying out the test; and environmental conditions.

The term **precision** is a measure of how close together are tests on the same material, and is defined as the reciprocal of the variance of the test results. In contrast, **accuracy** is a measure of how close together are the average of all test results and the true value, and the lack of accuracy is measured by the bias. The true value is established by, or derived from, international conventions and accuracy depends on careful, and traceable, calibration against these standards. For example, the kilogram is defined as being equal to the mass of the *International Prototype of the Kilogram* held at the International Bureau of Weights and Measures (BIPM) near Paris, and the second is defined in terms of exactly 9 192 631 770 periods of a particular frequency of radiation from the cesium atom under specified conditions. In the U.S. the National Institute for Standards and Technology (NIST) is part of the U.S. Department of Commerce.

Inter-laboratory trials are undertaken by trade associations to review the **repeatability** and **reproducibility** of the test method. Repeatability quantifies the precision of tests on specimens from the same material that are as constant as possible: same

---

[8] Although the standard deviation between ropes (1.84) is smaller than the standard deviation of individual wires within the same rope (4.45), the latter is reduced by a factor of $1/\sqrt{12}$ when a rope is manufactured from 12 wires. So the estimated standard deviation of strength of ropes due to variation in the strength of wires within a rope is 1.28.

laboratory; same technician; same equipment; and separated by short time periods. Reproducibility relates to tests on specimens from the same material at different laboratories, with different equipment and technicians. Since reproducibility is based on a comparison of different laboratories it accounts for inaccurate calibrations[9]. Let Y_{ij} represent the jth test result from the ith laboratory, and model this by

$$Y_{ij} = \mu + \alpha_i + \epsilon_{ij},$$

where: μ is the mean of all such test results and ideally the true value; α_i are between laboratory errors with $\alpha_i \sim 0, \sigma_\alpha^2$; and ϵ_{ij} are within laboratory errors with $\epsilon_{ij} \sim 0, \sigma_\epsilon^2$. The repeatability *(rep)* is defined by

$$rep = 1.96\sqrt{\sigma_\epsilon^2 + \sigma_\epsilon^2} = 2.8\sigma_\epsilon,$$

which is half the width of an approximate 95% confidence interval for the difference between two test results under repeatability conditions. A similar rationale leads to the definition of reproducibility *(Rep)* as

$$Rep = 2.8\sqrt{\sigma_\alpha^2 + \sigma_\epsilon^2}.$$

Our example is an inter-laboratory trial for measurements of polished-stone value (PSV).

The friction between a vehicle's tires and a tarmacadam road surface is due to the aggregate that is bound with the tar. A good road-stone will maintain frictional forces despite the polishing action of tires. British Standard BS812:Part 114: 1989 is a method for measuring the friction between rubber and polished stone in a laboratory, and the result of the test is the PSV. The test is based on comparison with a control stone which has been stockpiled and is supplied to the 30 accredited laboratories in the UK[10] that have the test equipment for PSV measurements. Sixteen of these laboratories agreed to take part in an inter-laboratory trial that was run according to the British Standard on the precision of test methods (BS5497:Part1:1987). As part of this trial, the 16 laboratories were sent specimens of a particular road-stone. The test procedure involves polishing a sub-sample of the specimen. The entire procedure including the polishing was repeated on two separate occasions. The results are in Table 10.2, the plot in Figure 10.5 and analysis is performed using R as follows.

TABLE 10.2: Polished stone values obtained on two different occasions (Run 1, Run2) for the same road-stone from 16 laboratories.

Lab	Run1	Run2	Lab	Run1	Run2
1	62.15	60.00	9	58.00	58.50
2	53.50	54.20	10	54.15	56.00
3	55.00	55.15	11	54.65	55.00
4	61.50	61.50	12	54.80	53.50
5	62.30	62.85	13	64.15	61.15
6	56.50	54.65	14	57.20	60.70
7	59.00	57.30	15	64.15	61.00
8	56.00	55.00	16	59.15	61.50

---

[9] At least relative to each other.
[10] In the year 1994.

```
> PSV.dat=read.table("PSV.txt",header=TRUE)
> attach(PSV.dat)
> head(PSV.dat)
1   1 62.15 60.00
2   2 53.50 54.20
3   3 55.00 55.15
4   4 61.50 61.50
5   5 62.30 62.85
6   6 56.50 54.65
> plot(c(Lab,Lab),c(Run1,Run2),xlab="Laboratory",ylab="PSV")
> x=cbind(Run1,Run2)
> I=length(Run1)
> J=2
> xbars=apply(x,1,mean)
> varin=mean(apply(x,1,var))
> varmean=var(xbars)
> varbet=var(xbars)-varin/J
> print(c("variance within",round(varin,2),"variance between",
+ round(varbet,2)))
[1] "variance within"  "1.72"          "variance between" "9.99"
> print(c("sd within",round(sqrt(varin),2),"sd between",
+ round(sqrt(varbet),2)))
[1] "sd within"  "1.31"     "sd between" "3.16"
> print(c("repeatability",round(2.8*sqrt(varin),2)))
[1] "repeatability" "3.67"
> print(c("Reproducibility",round(2.8*sqrt(varin+varbet),2)))
[1] "Reproducibility" "9.58"
```

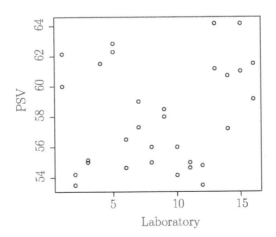

FIGURE 10.5: Polished stone value for a road-stone: two PSV results from 16 laboratories.

The estimates of σ_α and σ_ϵ are 3.16 and 1.31 respectively, and the repeatability and reproducibility are 3.7 and 9.6 respectively. These values were considered reasonable for this particular PSV test procedure. You are asked to analyze results from another part of this trial when laboratories were supplied with specimens of stone from the control stockpile in Exercise 10.14.

In the cases of the wire ropes, and the inter-laboratory trial, the number of observations from each batch, and laboratory, was the same (12 and 2 respectively). If the number of observations from each batch is the same the design of the investigation is said to be **balanced**. See Figure 10.6. If there are unequal numbers of observations from each batch, or laboratory or other group definition, then the same principle applies but the formulae are not so neat.

The model for components of variance can be extended to three or more components. For example, the variability of strengths of concrete cubes within batches, between batches of concrete made from the same delivery of cement, and between deliveries of cement used in making the batches[11]. We describe the model in this context and assume equal numbers of

Concrete delivery

Concrete batch

Concrete test cubes

FIGURE 10.6: Hierarchical structure for concrete test cubes.

batches sampled from each delivery and equal numbers of cubes sampled from each batch. We also assume that the mass of concrete used to manufacture the test cubes made from a batch is small compared with the mass of the batch and that the number of batches sampled from a delivery of cement is small by comparison with the total number of batches made from the delivery. There are I deliveries of cement, and from each delivery J batches of concrete are sampled. Then K test cubes are made from each batch. Let Y_{ijk} be the compressive strength of a concrete cube k made from batch j made using cement from delivery i, where $i = 1, \ldots, I$, $j = 1, \ldots, J$ and $k = 1, \ldots, K$. The model is

$$Y_{ijk} = \mu + \alpha_i + \beta_{ij} + \epsilon_{ijk},$$

where

$$\alpha_i \sim 0, \sigma_\alpha^2, \quad \beta_{ij} \sim 0, \sigma_\beta^2 \quad \text{and} \quad \epsilon_{ijk} \sim 0, \sigma_\epsilon^2.$$

All the errors are assumed to be independent, and it follows that $\text{var}(Y_{ijk})$, which we will refer to as σ^2, is $\sigma_\alpha^2 + \sigma_\beta^2 + \sigma_\epsilon^2$ and is an application of Pythagoras' Theorem in 3D (Figure 10.7). We now consider the variance within batches. From the model

$$\text{var}(Y_{ijk}|\text{within delivery } i \text{ and batch } j) = \sigma_\epsilon^2$$

and since

$$\overline{Y}_{ij\cdot} = \mu + \alpha_i + \beta_{ij} + \epsilon_{ij\cdot},$$

it follows that

$$\text{var}(\overline{Y}_{ij\cdot}|\text{within delivery } i) = \sigma_\beta^2 + \sigma_\epsilon^2/K$$

and since

$$\overline{Y}_{i\cdot\cdot} = \mu + \alpha_i + \beta_{i\cdot} + \epsilon_{i\cdot\cdot},$$

---

[11]Concrete is primarily a mixture of cement, sand, aggregate, and water.

it follows that

$$\mathrm{var}\left(\overline{Y}_{i\cdot\cdot}\right) \;=\; \sigma_\alpha^2 + \sigma_\beta^2/J + \sigma_\epsilon^2/(JK).$$

Estimators of the components of variance by replacing population variances with their sample estimators, beginning with the variance of cubes within batches.

$$\widehat{\sigma}_\epsilon^2 \;=\; \frac{\sum_{i=1}^{I}\sum_{j=1}^{J}\left(\sum_{k=1}^{K}(Y_{ijk}-\overline{Y}_{ij\cdot})^2/(K-1)\right)}{IJ}.$$

The estimator of the variance of batches within deliveries is

$$s^2_{\overline{Y}_{ij\cdot}|\text{within delivery }i} \;=\; \frac{\sum_{i=1}^{I}\left(\sum_{j=1}^{J}(Y_{ij\cdot}-\overline{Y}_{i\cdot\cdot})^2/(J-1)\right)}{I}$$

and so

$$\widehat{\sigma}_\beta^2 \;=\; s^2_{\overline{Y}_{ij\cdot}|\text{within delivery }i} \;-\;\widehat{\sigma}_\epsilon^2/K.$$

Finally

$$\widehat{\sigma}_\alpha^2 \;=\; \sum_{i=1}^{I}(\overline{Y}_{i\cdot\cdot}-\overline{Y}_{\ldots})^2/(I-1)-\widehat{\sigma}_\beta^2/J-\widehat{\sigma}_\epsilon^2/(JK).$$

All of these estimators of components of variance are unbiased for the corresponding population variances and the unbiased estimator[12] of the variance of strengths of concrete cubes, σ^2 say, is

$$\widehat{\sigma}^2 \;=\; \widehat{\sigma}_\alpha^2 + \widehat{\sigma}_\beta^2 + \widehat{\sigma}_\epsilon^2.$$

Compressive strengths (MPa) of concrete cubes tested during the construction of the main

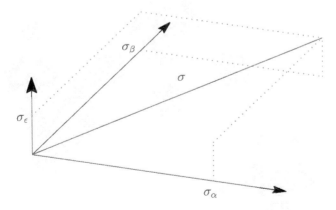

FIGURE 10.7: Pythagoras' Theorem in 3D ($\sigma^2 = \sigma_\alpha^2 + \sigma_\beta^2 + \sigma_\epsilon^2$). independent components of variance are additive.

runway at Heathrow Airport are in Table 10.3 (excerpt from Graham and Martin, 1946).

---

[12] $\sum(Y_{ijk}-\overline{Y}_{\ldots})^2/(IJK-1)$ is not unbiased because we do not have an SRS from the population of all cubes. The sampling is random but multistage: deliveries; batches within deliveries; cubes within batches. It follows that $\mathrm{var}(\overline{Y}_{\ldots}) \geq \sigma^2/(IJK)$.

We have reproduced data relating to six deliveries of cement. A large number of batches of concrete were made using the cement from each delivery and two batches were randomly selected from each delivery. Four test cubes were made from each batch of concrete and they were tested after 28 days. The following R code makes the calculations, and it is convenient to set up a 3-subscript array and use the `apply()` and `tapply()` functions, after first plotting the data.

TABLE 10.3: Compressive strengths (MPa) of concrete cubes by delivery and batch.

delivery	batch	cube	delivery	batch	cube	delivery	batch	cube
1	1	35.6	3	5	30.0	5	9	33.2
1	1	33.6	3	5	35.0	5	9	35.2
1	1	34.1	3	5	35.0	5	9	37.8
1	1	34.5	3	5	32.6	5	9	35.4
1	2	38.6	3	6	27.9	5	10	35.8
1	2	41.6	3	6	27.7	5	10	37.1
1	2	40.7	3	6	29.0	5	10	37.1
1	2	39.9	3	6	32.8	5	10	39.5
2	3	30.7	4	7	34.3	6	11	39.5
2	3	30.5	4	7	36.4	6	11	42.1
2	3	27.2	4	7	33.4	6	11	38.5
2	3	26.8	4	7	33.4	6	11	40.2
2	4	31.7	4	8	38.7	6	12	38.7
2	4	30.0	4	8	38.5	6	12	36.1
2	4	33.8	4	8	43.3	6	12	35.9
2	4	29.6	4	8	36.7	6	12	42.8

```
> DBC.dat=read.table("Heathrow_DBC.txt",header=TRUE)
> attach(DBC.dat)
> print(head(DBC.dat))
  delivery batch cube
1        1     1 35.6
2        1     1 33.6
3        1     1 34.1
4        1     1 34.5
5        1     2 38.6
6        1     2 41.6
> bid=rep(c(rep(1,4),rep(2,4)),6)
> boxplot(cube~bid/delivery)
> I=6;J=2;K=4
> y=array(0,dim=c(6,2,4))
> for (i in 1:I){for (j in 1:J){for (k in 1:K){
+ y[i,j,k]=cube[(i-1)*(J*K)+(j-1)*K + k]
+ }}}
> batchmean=tapply(cube,batch,mean)
> withinbatchvar=tapply(cube,batch,var)
> var_e=mean(withinbatchvar)
> y_ijdot=apply(y,1:2,mean)
> withindelivvar=apply(y_ijdot,1,var)
> var_b=mean(withindelivvar)-var_e/K
> delmean=apply(y,1,mean)
```

```
> var_a=var(delmean)-var_b/J-var_e/(J*K)
> print(c("variances a,b,e",round(var_a,2),round(var_b,2),round(var_e,2)))
[1] "variances a,b,e" "9.88"          "6.01"          "4.18"
> print(c("sd a,b,e",round(sqrt(var_a),2),round(sqrt(var_b),2),
+  round(sqrt(var_e),2)))
[1] "sd a,b,e" "3.14"      "2.45"      "2.04"
> sigsqu=var_a+var_b+var_e
> print(c("variance strengths cubes",round(sigsqu,1),"sd strengths cubes",
+  round(sqrt(sigsqu),1)))
[1] "variance strengths cubes" "20.1"
[3] "sd strengths cubes"        "4.5"
```

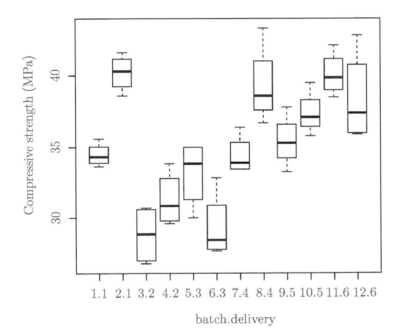

FIGURE 10.8: Box plots of compressive strengths (MPa) of 4 cubes within batches (2 batches from each of 6 deliveries).

TABLE 10.4: Components of variance of compressive strengths of concrete cubes.

Source	Variance	StdDev
Delivery	9.88	3.14
Batches within delivery	6.01	2.45
Cubes within batches	4.18	2.04
Cubes	20.07	4.48

The components of variance are set out in Table 10.4 The most effective way to reduce the variance of cubes is to reduce the largest component of variance. So the first action is to request the supplier to reduce the variability of the cement. The second action would be to aim for less variation between batches. The variation of the strength of cubes within batches is likely to be common cause variation and seems to be reasonably low.

10.3 Capability

Manufactured components are usually marketed with a statement of engineering tolerances, which are permissible limits of variation, on critical features. Some examples from the web are

- A bearing manufacturer (SKF) gives a tolerance for the inner ring diameter on a radial bearing of nominal diameter 120 mm as $0, -25$ μm.

- An electronics manufacturer gives a tolerance on a 10 μF capacitor as $\pm 5\%$.

- The IEC/EN60062 E3S tolerance for capacitors is $-20\%, +50\%$.

- An electronics manufacturer sells a wide range of resistors, some with tolerances as wide as $\pm 5\%$ through to a narrower tolerance of $\pm 0.1\%$.

- A manufacturer (ATE) advertises brake discs for cars with a maximum runout of 30 μm.

- Dry cell zinc chloride 1.5 volt batteries have a capacity of at least 0.34 amp-hour.

- A manufacturer of wire rope (SWR) states that the minimum breaking load for a wire rope of nominal diameter 20 mm is 25 708 kg.

In the case of electronic components, and high precision steel components, it is necessary to specify a temperature at which measurements will be made (typically 20 degrees Celsius). From a process engineer's point of view the tolerance interval is the specification that needs to met, and in the case of safety critical features the internal company specification may be more demanding than the advertised tolerance interval. In other cases the customer, which can include the next stage of the manufacturing process, will specify limits within which critical features should lie. It is customary to use the term "specification" when discussing the capability of processes.

10.3.1 Process capability index

A two sided specification (spec) for a feature of our product is that it be between within the interval $[L, U]$ where L and U are the lower and upper specification limits respectively. Assume that the process is stable, and that this feature has a probability distribution with mean μ and standard deviation σ. The **process capability index** is defined as

$$C_p = \frac{U - L}{6\sigma}.$$

If the feature has a normal distribution and the mean is at the center of the specification then a C_p of 1 corresponds to 3 in a 1 000, or 3 000 parts per million (ppm), outside specification, and a process is customarily described as capable if its C_p exceeds 1. But, a C_p of 1 may not be good enough.

- The 3 ppm outside specification is for a normal distribution, and as we are dealing with tail probabilities it is sensitive to the assumed distribution. The ppm outside specification will be higher for distributions with heavier tails (kurtosis above 3).

- Many industries expect higher values than 1 for C_p. The automotive industry generally expects suppliers to demonstrate that $C_p > 1.33$, and companies practicing Six Sigma aim for C_p around 2.

- Is our process mean at the center of the specification? If not the ppm outside specification will increase as in Figure 5.11.

- How precise is our estimate of C_p? If we have an active quality management system we will have extensive records of the process mean and standard deviation of established processes. If this is not the case we will have to take a sample from production.

Example 10.5: Carbon film resistors

An electronics company produces carbon film resistors with a tolerance of $\pm 2\%$. For the 100 ohm resistor, the process mean is 100.00 and the standard deviation is 0.35. The process capability relative to a specification of $[98, 102]$ is

$$C_p = \frac{102 - 98}{6 \times 0.35} = 1.90$$

If resistances are precisely normally distributed the proportion outside specification is $2(1 - \Phi(2/0.35))$, which is 11 parts per billion. If the mean was as far as one standard deviation from its target value of 100.00, 100.35 say, 12 ppm wold be outside the specification. Could these resistors be marketed as being with a $\pm 1\%$ tolerance? The C_p relative to a specification of $[99, 101]$ is 0.95, and even if the process mean is precisely on target there would be 4 275 ppm outside the specification. A claim of $\pm 1\%$ tolerance would not be justified.

10.3.2   Process performance index

It some situations the process mean may not be at the center of the specification, either because the process has not been set up precisely or because it is not practical to adjust the mean for a particular customer. The process performance index C_{pk} measures how the process is performing when the mean is not at the center of the specification, rather than how it would perform if the mean was at the center of the specification.

$$C_{pk} = \min\left(\frac{\mu - L}{3\sigma}, \frac{U - \mu}{3\sigma}\right).$$

Example 10.6: Tack weld

A consultant in a shipyard (Isao Ohno, 1990) found that the Sumi-Auto welding had a tack weld length distributed with a mean of 2.23 mm and a standard deviation of 0.63 mm. The specification for the length of the tack weld was between 1.5 and 3.0 mm. The process performance index

$$C_{pk} = \min\left(\frac{2.23 - 1.5}{3 \times 0.63}, \frac{3.0 - 2.23}{3 \times 0.63}\right) = \min(0.407, 0.386) = 0.39$$

is far too low. The shipyard implemented a quality improvement program and the standard deviation of tack weld length was reduced to 0.26 mm with the same welding equipment. The process capability is

$$C_p = \frac{3.0 - 1.5}{6 \times 0.26} = 0.962$$

The mean value 2.23 mm was already close, within 10% of the process standard deviation, to the middle of the specification, 2.25 mm. The improved process is on the borderline of being considered capable, and would be acceptable in the ship building industry.

In general, if we cannot demonstrate a sufficiently high C_p we have a choice. Either we reduce the standard deviation, which may not be feasible without investing in new equipment, or we decline the contract unless the customer can relax the specification. Customers will expect to see some evidence that a process is capable. The rationale for the ISO9000 quality management standard is that a company can demonstrate procedures for calculating performance indices once, to a third party certifier, rather than repeatedly to potential customers. If we have extensive past records from a stable process the sample size may be large enough to ignore sampling errors when estimating the mean and standard deviation, but if the process has been recently set up or recently modified we rely on a sample from recent production and sampling error may not be negligible. Suppose we have a sample of size n from a stable process and it is plausible to treat this as a random sample from a normal distribution. If the sample mean and standard deviation are $\bar{x}$ and s respectively an approximate $(1 - \alpha) \times 100\%$ confidence interval for C_{pk} is

$$\widehat{C}_{pk} \left(1 \pm z_{\alpha/2} \frac{1}{\sqrt{2n}} \right)$$

where

$$\widehat{C}_{pk} = \min \left(\frac{\bar{x} - L}{3s}, \frac{U - \bar{x}}{3s} \right)$$

Alternatively, we can use a procedure for calculating statistical tolerance intervals, or a bootstrap procedure, but the above construction is the most convenient. The process capability is greater than the process performance index unless $\bar{x}$ is equal to the middle of the specification when they are equal.

Example 10.7: Control actuators

The diameters (mm) of a random sample of 50 pistons for control actuators were measured. The specification is $[100.3, 101.5]$, and the sample mean and standard deviation were 100.93 and 0.14 respectively. An approximate 80% confidence interval for C_{pk}, assuming diameters are normally distributed, is:

$$1.357 \times \left(1 \pm 1.282 \frac{1}{\sqrt{2 \times 100}} \right),$$

which reduces to $[1.18, 1.53]$.

10.3.3   One-sided process capability indices

If the specification has only an upper limit the process capability is defined as

$$C_{pU} = \frac{U - \mu}{3\sigma}$$

and an approximate confidence interval is calculated in a similar way to that for for C_{pk} as

$$\widehat{C}_{pU}\left(1 \pm z_{\alpha/2}\frac{1}{\sqrt{2n}}\right),$$

where

$$\widehat{C}_{pU} = \frac{U - \bar{x}}{3s}$$

Example 10.8: Brake discs

A manufacturer of brake discs for trucks claims that the mean runout is 12 μm with a standard deviation of 6 μm, but a potential customer has asked for this to be validated. What size sample is required if a 95% confidence interval for C_{pU} is to be approximately $\pm 10\%$ of $\widehat{C}_{pU}$? The requirement is that

$$1 + \frac{z_{.025}}{\sqrt{2n}} = 1.1 \Rightarrow n = 192$$

The runout for a sample of 200 brake discs is measured and the mean and standard deviation are 14.8 and 8.3 respectively. The potential customer has specified an upper limit for runout of 50, and requires a demonstration that a 95% confidence interval for C_{pU} exceeds 1.3. Will the sample provide this assurance? The estimate of C_{pU} is $\frac{50-14.8}{3\times 8.3} = 1.41$, which is well above 1.3, although the 95% confidence interval is $[1.28, 1.55]$ and doesn't quite exceed 1.3. This confidence interval would be probably acceptable. An alternative is to calculate a one-sided confidence interval for C_{pu} (Exercise 10.12).

When there is only a lower limit for the specification the process capability is defined as

$$C_{pL} = \frac{L - \mu}{3\sigma}$$

and an approximate confidence interval is obtained in a similar way to that for C_{pk}.

Example 10.9: Batteries

The specification for a 1.5 volt alkaline cell is that it should power a test circuit that draws 700 milli-amps for at least 60 minutes. A sample of 40 cells from production was tested. The mean time a battery powered the circuit was 71 minutes and the standard deviation of time was 2.9 minutes. The estimated C_{pL} is 1.26. If we assume a random sample from a normal distribution an approximate 95% confidence interval for C_{pL} is

$$\frac{71 - 60}{3 \times 2.9}\left(1 + 1.96\frac{1}{\sqrt{(2 \times 40)}}\right), \quad \text{which reduces to } [0.99, 1.54].$$

10.4   Reliability

10.4.1   Introduction

Here we consider the reliability of components. The reliability of systems, repairable and non-repairable, is considered in the next Chapter 13.

10.4.1.1   Reliability of components

The reliability measure is lifetime, defined as the length of time that the component continues to perform its specific function under controlled test conditions. In normal operation the lifetimes of components, that are not designed to be renewed, should exceed the design life of the equipment they are installed in. If we were to test the lifetimes of such items in ordinary use, typical tests would run for years. So, tests are often conducted under extreme adverse conditions in order to obtain results more quickly and this strategy is known as **highly accelerated lifetime testing** (HALT).

Example 10.10: Wiper blades

Automobile windshield wiper blades are designed to move over a wet surface. If the wipers are used over a dry windshield the wiper motors will be overloaded and will typically burn out after a few hours. An automobile manufacturer implements HALT by mounting wiper motors in a test rig so that they continuously move the blades over a dry windshield for up to 24 hours. Most motors fail within this period and the lifetimes are recorded. The lifetimes of the motors that do not fail are said to be **censored** because we only know that they exceed 24 hours.

Definition 10.2: Censored observation

An observation of a variable is censored if the precise value taken by the variable is unknown, but a range within which the value lies is known. In particular, the range can be greater than, or less than, some known value.

Some renewable items such as batteries and machine tools have relatively short lifetimes when in continual use and lifetime testing can be performed under typical conditions of use, others, such as truck tires and automobile brake pads, have longer lifetimes. Lifetimes are usually measured in time or number of cycles to failure, or when testing tensile strength tests of materials the load that causes fracture. Lifetimes of renewable components such as batteries under a test load can often be modeled realistically by a normal distribution provided that the coefficient of variation is less than around 0.2. In contrast component lifetimes in HALT testing are not usually well modeled by normal distributions. For example, under extreme voltage stress mylar insulation appears to fail spontaneously rather than wear out and the lifetimes are better modeled by an exponential distribution.

10.4.1.2 Reliability function and the failure rate

Suppose the lifetime of a component T is modeled by the cumulative distribution function (cdf) $F(t)$. The reliability function[13] $R(t)$, is the complement of the cdf

$$R(t) = 1 - F(t).$$

The failure rate, also known as the hazard function [14], and usually written as $h(t)$ is proportional to the conditional probability that a component fails within the time interval $(t, t + \delta t]$ given that it has not failed by time t, where δt is a small time increment, which is the constant of proportionality. A more succinct definition is

$$h(t) = \frac{f(t)}{R(t)}, \quad 0 \le t,$$

where $f(t)$ is the pdf. We now show that the two definitions are equivalent.

$$P(\text{fail in } (t, t+\delta t] \mid \text{not failed at } t) = \frac{(\text{fail in } (t, t+\delta t]) \cap (\text{not failed at } t)}{\text{not failed at } t}$$

$$= \frac{\text{fail in } (t, t+\delta t]}{\text{not failed at } t} = \frac{\int_t^{t+\delta t} f(t)\, dt}{R(t)}$$

$$\approx \frac{f(t)\delta t}{R(t)} = h(t)\delta t.$$

The approximation becomes exact as δt tends to 0.

A unit of measurement used in electronics applications is "failures in time" (FIT), which is defined as $10^{-9} h(t)$ where $h(t)$ is expressed as failures h^{-1}.

Example 10.11: Exponential distribution

The reliability function of the exponential distribution is

$$R(t) = 1 - F(t) = 1 - \left(1 - e^{-\lambda t}\right) = e^{-\lambda t}.$$

The hazard function is constant,

$$h(t) = \frac{\lambda e^{-\lambda t}}{e^{-\lambda t}} = \lambda.$$

This is a consequence of the memoryless property of an exponential distribution. The parameter λ is the failure rate and its reciprocal is the mean lifetime.

Example 10.12: Normal distribution

The hazard function for a normal distribution, $N(\mu, \sigma^2)$, is

$$h(t) = \frac{\phi((t-\mu)/\sigma)}{1 - \Phi((t-\mu)/\sigma)}$$

and it is plotted for $\mu = 6, \sigma = 1$ in Figure 10.9. The hazard function starts off as negligible, but starts increasing around two standard deviations below the mean.

---

[13]The reliability function is also known as the survivor function, usually written as $S(t)$.
[14]Failure rate is typically used in engineering and hazard function is more common in mathematics.

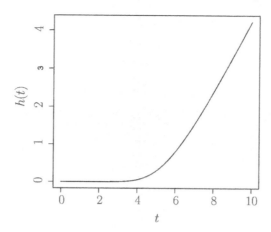

FIGURE 10.9: Hazard function for a normal distribution with mean 6 and standard deviation 1.

The hazard function is an alternative specification of a probability distribution to the cdf or pdf. In principle, if we know one of the three we can derive the other two, although numerical integration will often be needed.

Example 10.13: Constant hazard function

A hazard function is constant at λ. Then

$$\frac{f(t)}{R(t)} \;=\; \lambda,$$

which can be rewritten as the differential equation

$$\frac{dF(t)}{dt} \;=\; \lambda(1 - F(t)).$$

The boundary condition is $F(0) = 0$. The solution is the cdf of the exponential distribution

$$F(t) \;=\; 1 - e^{-\lambda t}, \quad 0 \le t.$$

The cumulative hazard function, $H(t)$, is

$$H(t) \;=\; \int_0^t h(u)du$$

and the average hazard rate (AHR) between times t_1 and t_2 is given by:

$$AHR(t_1, t_2) \;=\; \frac{H(t_2) - H(t_1)}{t_2 - t_1}.$$

10.4.2 Weibull analysis

10.4.2.1 Definition of the Weibull distribution

The Weibull distribution, named after the Swedish engineer Waloddi Weibull (1887-1979) who applied it to materials testing, is a versatile distribution which includes the exponential distribution as a special case, can be close to normal, and can have negative skewness. The cdf is

$$F(t) \;=\; 1 - e^{-(t/b)^a}, \quad 0 \leq t,$$

where a is the shape parameter and b is the scale parameter[15] and where both parameters are restricted to be positive numbers. A rationale for the form of the distribution is that it represents the reliability of a chain of v links each of which has reliability $e^{-(t/\theta)^a}$, if the chain fails when the weakest link fails $\left(b = \theta/v^{1/\alpha}\right)$. It is straightforward to verify that the hazard function is

$$h(t) \;=\; \frac{a}{b}\left(\frac{t}{b}\right)^{a-1}.$$

If $a = 1$ we get an exponential distribution with rate parameter $\lambda = 1/b$, and if $1 < a$ the failure rate increases with time. If $0 < a < 1$ then the failure rate decreases with time, which can be plausible in some applications. The pdfs and hazard functions of Weibull distributions with $a = 0.95, 1, 2, 4$ and $b = 1$ are plotted in Figure 10.10, with the following R code.

```
> t=seq(from=0,to=4,by=.01)
> a=.95;b=1
> f1=dweibull(t,shape=a,scale=1)
> h1=(a/b)*(t/b)^(a-1)
> a=1;b=1
> f2=dweibull(t,shape=a,scale=1)
> h2=(a/b)*(t/b)^(a-1)
> a=2;b=1
> f3=dweibull(t,shape=a,scale=1)
> h3=(a/b)*(t/b)^(a-1)
> a=4;b=1
> f4=dweibull(t,shape=a,scale=1)
> h4=(a/b)*(t/b)^(a-1)
> par(mfrow=c(1,2))
> plot(t,f4,type="l",lty=4,ylab="f(t)")
> lines(t,f1,lty=2)
> lines(t,f2,lty=1)
> lines(t,f3,lty=3)
> plot(t,h3,type="l",lty=3,ylab="h(t)")
> lines(t,h1,lty=2)
> lines(t,h2,lty=1)
> lines(t,h4,lty=4)
> legend(2,3.5,c(".95","1","2","4"),lty=c(2,1,3,4))
```

The mean and variance are

$$\mu \;=\; b\Gamma(1 + a^{-1}) \qquad \sigma^2 \;=\; b^2\left[\Gamma(1 + 2a^{-1}) - \Gamma^2(1 + a^{-1})\right].$$

---

[15]It is also commonly defined with a parameter λ, which is the reciprocal of b and is known as the rate parameter.

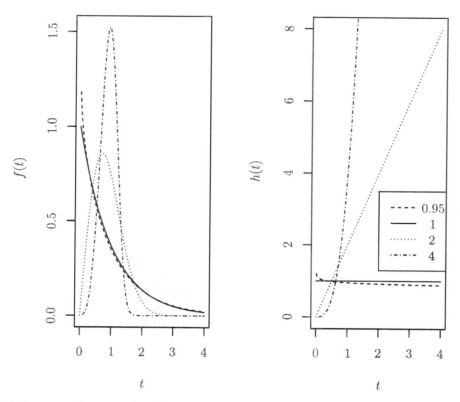

FIGURE 10.10: Densities (pdfs) (left frame) and hazard function (right frame) for Weibull distributions.

The skewness is positive for a less than about 5 and becomes negative for larger values of a. The R syntax for the pdf is

```
dweibull(t,a,b)
```

and, as for other distributions, the cdf, inverse cdf, and random deviates are obtained by changing the leading letter. A Weibull distribution with $a = 5$ is close to a normal distribution, as you can verify by taking a large random sample and plotting a normal quantile-quantile plot.

```
> qqnorm(rweibull(1000,5,1))
```

The method of moments estimators of the parameters are not particularly convenient to implement as the estimates have to be obtained using some numerical method. The following graphical method is an alternative that also provides an assessment of the suitability of the Weibull distribution.

10.4.2.2   Weibull quantile plot

We begin with the approximation that we have used for other quantile plots.

$$F(E[T_{i:n}]) \quad \approx \quad \frac{i}{n+1}$$

and in the case of a Weibull distribution this can be rearranged as

$$E[T_{i:n}] \quad = \quad F^{-1}\left(\frac{i}{n+1}\right) \quad = \quad b\left[-\ln(1-p_i)\right]^{1/a}$$

where $p_i = i/(n+1)$, and then linearized by taking logarithms of both sides.

$$\ln\left(E[T_{i:n}]\right) = b + \frac{1}{a}\ln\left(-\ln(1-p_i)\right)$$

If the failure times are a random sample from a Weibull distribution, a scatter plot of the logarithms of the order statistics $\ln(t_{i:n})$ against $\ln\left(-\ln(1-p_i)\right)$ should show random scatter about a straight line with intercept $\ln(b)$ and slope $1/a$. Then a line drawn through the points can be used to estimate the parameter a as the reciprocal of the slope and b as exponential of the intercept. It is convenient to draw the line using lm(), although this is not the optimum procedure[16], as more precise estimators can be obtained by using the maximum likelihood method introduced in the next section. The graphical estimates are used as initial values for the numerical optimization.

Example 10.14: HV insulators

Samples of mylar-polyurethane laminated DC HV insulating structure were tested at five different voltage stress levels (Kalkanis and Rosso, 1989), and the minutes to failure are shown in Table 10.5 The Weibull quantile plots for the three higher voltages and the lowest are shown in Figure 10.11.

TABLE 10.5: mylar-polyurethane laminated DC HV insulation structure: minutes to failure by voltage stress.

361.4 kV/mm	219.0 kV/mm	157.1 kV/mm	122.4 kV/mm	100.3 kV/mm
0.10	15	49	188	606
0.33	16	99	297	1012
0.50	36	155	405	2520
0.50	50	180	744	2610
0.90	55	291	1218	3988
1.00	95	447	1340	4100
1.55	122	510	1715	5025
1.65	129	600	3382	6842
2.10	625	1656		
4.00	700	1721		

The R code for the plot at the lowest voltage follows[17] (HHV,HV,MV,LV,LLV are from the highest to lowest voltage stress).

```
> m=read.table("mylar.txt",header=TRUE)
> attach(m)
```

---

[16]The variance of order statistics increases as i moves further from $(n+1)/2$, so ordinary least squares is overly influenced by the first few and last few order statistics.

[17]The mylar.txt file (headers HHV, HV, MV, LV, LLV for the highest to the lowest voltage stress) had NAs inserted at the end of the LV and LLV columns to make columns of equal lengths. The third and fourth lines in the R script remove the NAs to give columns of length 8 for LV and LLV. See for example, Tutorial on Reading and Importing Excel Files into R - DataCam, and Short reference card on the web.

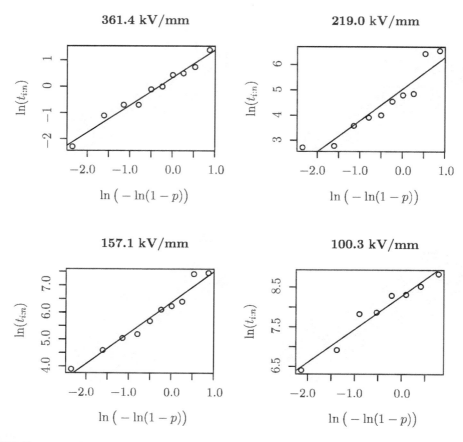

FIGURE 10.11: Weibull quantile plots for lifetimes of mylar-polyurethane laminated DC HV insulation structures at the three higher voltages and the lowest voltage.

```
> LV=as.numeric(na.omit(LV))
> LLV=as.numeric(na.omit(LLV))
> print(LLV)
[1]   606 1012 2520 2610 3988 4100 5025 6842
> a=rep(0,5)
> b=rep(0,5)
> x=HHV
> x=LLV
> n=length(x)
> lnmylar=log(x)
> i=c(1:n)
> p=i/(n+1)
> pp=log(-log(1-p))
> plot(pp,lnmylar,xlab="ln(-ln(1-p))",ylab="ln(t_i:n)",main="100.3 kV/mm")
> m1=lm(lnmylar~pp)
> abline(m1$coef[1],m1$coef[2])
> m1=lm(lnmylar~pp)
> a[5]=round(1/m1$coef[2],2)
> b[5]=round(exp(m1$coef[1]),2)
```

The graphical estimates of the a and b parameters from the highest to lowest voltage stress are

```
> a
[1] 0.96 0.80 0.88 1.01 1.20
> b
[1]    1.40  154.06  578.61 1285.98 3937.56
```

The shape parameters of the first four distributions seem consistent with an exponential distribution [18], but the shape parameter of the distribution at the lowest voltage stress may be higher[19]. We now compare the sample means and median s with the means and median s of the fitted distributions. A lower α quantile $t_{1-\alpha}$ of the Weibull distribution is obtained by rearranging

$$F(t_{1-\alpha}) \;=\; 1 - e^{-(t_{1-\alpha}/b)^a} \;=\; \alpha$$

to get

$$t_{1-\alpha} \;=\; b\,(-\ln(1-\alpha))^{1/a}\,.$$

The median is $t_{0.5}$ and is estimated by replacing a and b by their estimates from the Weibull quantile plot.

```
> meanlife=apply(m,2,mean,na.rm=TRUE)
> meanlife
      HHV        HV        MV        LV       LLV
    1.263   184.300   570.800  1161.125  3337.875
> meanW=round(b*gamma(1+1/a),2)
> meanW
[1]    1.43   174.55   616.53  1280.65  3703.89
> medianlife=apply(m,2,median,na.rm=TRUE)
> medianlife
     HHV       HV       MV       LV      LLV
    0.95    75.00   369.00   981.00  3299.00
> medianW=round(b*(log(2))^(1/a),2)
> medianW
[1]    0.96    97.44   381.51   894.61  2901.23
```

The sample means, meanlife, and means of the fitted distributions, meanW, are generally within 10%. The discrepancy is reasonable, given the small samples. The standard error expressed as a percentage of a sample mean from an exponential distribution sample of size n is $100/\sqrt{n}\%$. The agreement between sample median s, medianlife, and the median s of the fitted distributions, medianW, is with the exception of HV, slightly closer than that between the means.

---

[18] The 0.80 is noticeably is lower than 1, but a decreasing hazard function is not very plausible and the sampling error with a sample of 10 will be substantial.

[19] A bootstrap could be used to estimate its standard error.

10.4.2.3 Censored data

The Weibull quantile plot is directly applicable to censored data.

Example 10.15: Electrolytic capacitors

Tantalum electrolytic capacitors were subject to a HALT test and 18 out of 174 failed within the 12 500 hour duration of the test ([N.D. Singpurwalla, 1975]). The failure times for these 18 capacitors are in Table 10.6 The Weibull quantile plot is a scatter

TABLE 10.6: Times to failure of 18 capacitors that failed within 12 500 h in a HALT test of 174 capacitors.

25	50	165	500	620	720	820	910	980
1270	1600	2270	2370	4590	4880	7560	8750	12500

plot of the logarithms of the first 18 order statistics against $\ln(-\ln(1-p_i))$, for $i = 1, \ldots, 16$, where $p_i = 1/(n+1)$ with $n = 174$.

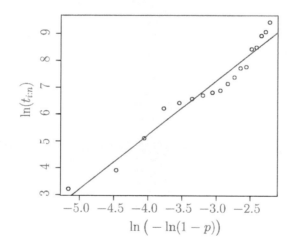

FIGURE 10.12: Weibull quantile plot for the electrolytic capacitors data.

The R code for Figure 10.12 follows.

```
> x=c(25,50,165,500,620,720,820,910,980,1270,
+ 1600,2270,2370,4590,4880,7560,8750,12500)
> n=174
> logt=log(x)
> i=c(1:length(x))
> p=i/(n+1)
> pp=log(-log(1-p))
> plot(pp,logt,xlab="ln(-ln(1-p))",ylab="ln(t_i:n)")
> m1=lm(logt~pp)
> abline(m1$coef[1],m1$coef[2])
> print(round(1/m1$coef[2],2))
 pp
```

```
0.5
> print(round(exp(m1$coef[1]),2))
(Intercept)
   591560.9
```

Of the 18 capacitors that failed, 9 failed before 1 000 hours whereas the remaining 9 failed between 1 000 and 12 500 hours. The estimated shape parameter of 0.5 is less than 1, which indicates a decreasing failure rate over the first 12 500 hours. However, it is unrealistic to extrapolate and predict that the failure rate will continue to decrease and that lifetimes have a Weibull distribution. The test indicates that some of the capacitors are relatively weak and fail early.

Censored observations can also arise during the study period as we show in the next example.

Example 10.16: Solar exploration vehicles

Eleven planet surface explorer vehicles of identical design were powered by a solar charged battery. The battery lifetime is defined as the number of planetary days over which the battery provides sufficient power to propel the vehicle. Seven batteries failed after: $9, 13, 18, 23, 31, 34, 48$ days. Three of the vehicles were destroyed by meteor showers on the 13th, 28th, and 45th day, so the battery lifetimes are censored at these values. A signal relay device failed on the 11th vehicle after 72 days, so its battery life is censored at 72. In the Weibull quantile plot, we need to allow for the reduction in the relevant sample size if an item is removed from the study before it fails. The first of 11 batteries to fail, fails after 9 days, but it is the third of 10 batteries that fails after 18 days because a meteor strike removed one battery after 13 days. It is the seventh of 8 that fails after 48 days. We don't know whether the second battery to fail failed before or after the meteor strike on day 13 so the sample size could be 11 or 10. In the following code to produce Figure 10.13 we assume the battery failed before the strike, but it makes very little difference if 10 is used in place of the 11.

```
> x=c(9,13,18,23,31,34,48)
> n=c(11,11,10,10,9,9,8)
> logt=log(x)
> i=c(1:length(x))
> p=i/(n+1)
> pp=log(-log(1-p))
> plot(pp,logt,xlab="ln(-ln(1-p))",ylab="ln(t_i:n)")
> m1=lm(logt~pp)
> abline(m1$coef[1],m1$coef[2])
> print(round(1/m1$coef[2],2))
  pp
1.68
> print(round(exp(m1$coef[1]),2))
(Intercept)
     36.91
```

A Weibull distribution is a good fit to the data, and the estimated parameters, a and b, are 1.68 and 36.91 respectively.

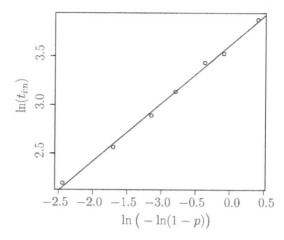

FIGURE 10.13: Weibull quantile plot for censored data.

10.4.3   Maximum likelihood

The **likelihood** of a set of parameter values given a sample of observations is defined as the probability of the observations given the parameter values. The **maximum likelihood estimators** (MLE) of the parameters are the values of the parameters, as a function of the observations, that maximize the probability. The principle is more easily appreciated for discrete distributions. In the HALT test of 174 tantalum electrolytic capacitors 9 failed within 1 000 hours. We will obtain the maximum likelihood estimate of the probability of failure within 1 000 hours under the same HALT conditions. The capacitors are considered to be a random sample from production so we have a binomial experiment with $n = 174$ trials and 9 observed failures.

In general suppose there are x failures in n trials with probability of failure p. Then the likelihood $\mathcal{L}(p)$ is

$$\mathcal{L}(p) \;=\; \mathrm{P}(\,x\,|\,p\,) \;=\; \binom{n}{x} p^x (1-p)^{n-x}$$

considered as a function of p. Although it is straightforward to differentiate $\mathcal{L}$ with respect to p and set the derivative equal to 0 as a necessary condition for a maximum, it is usually more convenient to work with the log-likelihood $\ell(p) = \ln(\mathcal{L}(p))$. If $\mathcal{L}(p)$ has a maximum at some value of p then $\ell(p)$ also has a maximum at that value of p, because $\mathtt{ln()}$ is a strictly increasing function of its argument.

$$\ell(p) \;=\; \ln\binom{n}{x} + x\ln(p) + (n-x)\ln(1-p).$$

A necessary condition for a maximum is that

$$\frac{d\ell(p)}{dp} \;=\; 0 + \frac{x}{p} + \frac{-(n-x)}{1-p}$$

equals 0. The value of p, $\widehat{p}$, for which $\frac{d\ell(p)}{dp} = 0$ is the maximum likelihood estimator. For the binomial distribution the MLE [20] of p follows from

$$\frac{x}{\widehat{p}} + \frac{-(n-x)}{1-\widehat{p}} = 0 \quad \Rightarrow \quad \widehat{p} = \frac{x}{n}.$$

In the case of the capacitors $\widehat{p} = 9/174 = 0.052$. The MLE and MoM estimators of the probability p of a failure in a binomial distribution based on observing x failures in n trials are identical.

If the distribution is continuous, the probability of observing the precise observations in the sample[21] can only be 0 which is not helpful, so instead we work with the probability of being within $\delta/2$ of the precise observed values, where δ is an arbitrary small quantity[22]. We illustrate this for a sample from an exponential distribution. The lifetime of a component has an exponential distribution with failure rate λ and pdf

$$f(t) = \lambda e^{-\lambda t}.$$

A random sample of n components fail at times t_i for $i = 1, \ldots, n$. The probability of being within $\delta t/2$ of each of the observed values is

$$\prod_{i=1}^{n} (f(t_i)\delta t) = \prod_{i=1}^{n} \left(\lambda e^{-\lambda t_i} \delta t\right),$$

where δt is arbitrarily small. Now the δt is irrelevant when finding the value of λ that maximizes the probability and the likelihood is defined with values of the pdf rather than probabilities:

$$\mathcal{L}(\lambda) = \prod_{i=1}^{n} f(t_i) = \prod_{i=1}^{n} \lambda e^{-\lambda t_i}.$$

It is more convenient to work with its logarithm:

$$\ell(\lambda) = \sum_{i=1}^{n} \ln(\lambda) - \sum_{i=1}^{n} \lambda t_i = n \ln(\lambda) - \sum_{i=1}^{n} \lambda t_i.$$

If we differentiate $\ell(\lambda)$ with respect to λ and set the derivative equal to 0 we get the MLE

$$\frac{n}{\widehat{\lambda}} - \sum_{i=1}^{n} t_i = 0 \quad \Rightarrow \quad \widehat{\lambda} = \frac{1}{\bar{t}}.$$

The MLE of the rate parameter of the exponential distribution is the same as the MoM estimate.

We now consider fitting an exponential distribution to censored data. The lifetime of a component in a HALT test has an exponential distribution with failure rate λ. A random sample of n components is tested in a HALT test of duration U, and m components fail at times t_i for $i = 1, \ldots, m$. The remaining $n - m$ components are still working after time U, so we only know that their lifetimes exceed U. The probability of being within $\delta t/2$ of each of the observed failure times and of the $n - m$ components lasting longer than U is

$$\left(\prod_{i=1}^{m} (f(t_i)\delta t)\right) (R(U))^{n-m},$$

---

[20] In this section we rely on the context to distinguish the MLE as an estimator when, for example, x is a random variable and the MLE as an estimate when x is the observed value.

[21] Here we imagine the observations are precise to an infinite number of decimal places.

[22] We have used this argument at several places throughout the book, one example being the definition of conditional probability in continuous multivariate distributions.

where $R(t) = 1 - F(t)$ is the reliability function. As before the δt is irrelevant when finding the value of λ that maximizes the probability and on substituting for $f(t)$ and $R(t)$ the likelihood becomes

$$\mathcal{L}(\lambda) \;\; = \;\; \left(\prod_{i=1}^{m} \lambda e^{-\lambda t_i} \right) e^{-(n-m)\lambda U}.$$

The log-likelihood is

$$\ell(\lambda) \;\; = \;\; m \ln(\lambda) - \sum_{i=1}^{m} \lambda t_i - (n-m)\lambda U.$$

If we differentiate $\ell(\lambda)$ with respect to λ and set the derivative equal to 0 we get the MLE

$$\frac{m}{\widehat{\lambda}} - \sum_{i=1}^{m} t_i - (n-m)U \;\; = \;\; 0 \;\; \Rightarrow \;\; \widehat{\lambda} \;\; = \;\; \frac{m}{\sum_{i=1}^{m} t_i + (n-m)U}.$$

This is a nice intuitive result, the estimate of the failure rate is the number of items that fail divided by the total time on test for all the items. The same result can also be obtained with a MoM approach. However, for a Weibull distribution with $a \neq 1$, the ML and MoM approaches give different results.

Example 10.17: Short wave radios

Ten short wave radios were tested under HALT conditions for 24 hours, and 6 failed after $5.5, 8.9, 9.5, 14.1, 19.8$, and 23.1 hours. If lifetimes are assumed to have an exponential distribution, then the estimate of λ is

$$\widehat{\lambda} \;\; = \;\; \frac{6}{5.5 + \cdots + 23.1 + 4 \times 24} \;\; = \;\; 0.0339$$

failures/hour.

The MLE argument becomes more useful when we have distributions which are not so amenable to MoM arguments, in particular the Weibull distribution. Let $F(t)$, $R(t)$ and $f(t)$ be the cdf, reliability function, and pdf of some lifetime probability distribution with two parameters a, b, such as the Weibull distribution. If we assume a random sample of n of which m fail before U the log-likelihood is

$$\ell(a, b) \;\; = \;\; \sum_{i=1}^{m} \ln(f(t_i)) + (n-m) \ln(R(U))$$

and it can be maximized with respect to a, b by a numerical search, such as the Nelder-Mead algorithm. The likelihood remains valid if no components outlast the test $m = n$. The argument generalizes to more than two parameters and more general censoring.

Example 10.18: Insulation

Consider the lifetimes of 8 pieces of mylar-polyurethane insulation tested at 100.3 kV/mm in Table 10.7.

TABLE 10.7: Lifetimes of 8 pieces of mylar-polyurethane insulation.

606	1 012	2 520	2 610
3 988	4 100	5 025	6 842

The mean time to failure is 3 337.9. The graphical estimates of the parameters a and b of a Weibull distribution were 1.20 and 3 937.6 The following R code computes the MLE of a and b using the Nelder-Mead algorithm with the graphical estimates as initial values[23]. It has been written to allow data censored at an upper point U, which can be set at any value greater than the observed data if all items failed.

```
> U=10000;n=8;m=8
> t=c(606,1012,2520,2610,3988,4100,5025,6842)
> a0=1.2
> b0=3938
> #fit Weibull by Nelder-Mead
> f=function(p){
+ a=exp(p[1]/100)
+ b=exp(p[2]/100)
+  -sum( log(a) + (a-1)*log(t) - a*log(b) - (t/b)^a ) + (n-m)*(U/b)^a
+ }
> par=c(100*log(a0),100*log(b0))
> avm=optim(par,f)
> a=exp(avm$par[1]/100)
> b=exp(avm$par[2]/100)
> print(c(a,b))
[1]     1.733746 3738.586835
```

The mean and standard deviation of the fitted Weibull distribution are 3 331.5 and 1 981.4 respectively, slightly different from the sample mean and standard deviation which are 3 337.9 and 2 077.1 respectively. In cases where MLE and MoM differ, the MLE are generally considered preferable because the MLE is asymptotically the optimum estimator under certain conditions. However, MLE are not always the best estimators in small samples. The MLE of the shape parameter a is 1.73, but the sample is small and we might ask whether an exponential distribution would be plausible for the lifetimes. The fact that the sample standard deviation is substantially less than the mean, when they are equal for an exponential distribution, suggests that it will not be. There are theoretical results that give asymptotic[24] expressions for standard errors of MLE, but we will use a bootstrap approach.

---

[23] In R optim() minimizes a function using the Nelder-Mead algorithm. Maximizing the log-likelihood is equivalent to minimizing log-likelihood. The reason for setting a and b at exp($\cdot$/100) of the optimization parameters is to ensure they are positive and to keep changes small. An alternative is to use the constrained optimization algorithm constrOptim.

[24] In statistics asymptotic results are results that are obtained by letting the sample size n tend to infinity. They often provide good approximations for small n, and the Central Limit Theorem is a renowned known example.

```
> t=c(606,1012,2520,2610,3988,4100,5025,6842)
> n=8
> BN=1000
> BSa=rep(0,BN)
> BSb=rep(0,BN)
> set.seed(1)
> for (i in 1:BN) {
+ yb=sample(t,n,replace=TRUE)
+ a0= 1.73
+ b0=  3739
+ f=function(p){
+ a=exp(p[1]/100)
+ b=exp(p[2]/100)
+  -sum( log(a) + (a-1)*log(yb) - a*log(b) - (yb/b)^a )
+ }
+ par=c(100*log(a0),100*log(b0))
+ avm=optim(par,f)
+ a=exp(avm$par[1]/100)
+ b=exp(avm$par[2]/100)
+ BSa[i]=a
+ BSb[i]=b
+ }
> a_B=mean(BSa)
> b_B=mean(BSb)
> sda_B=sd(BSa)
> sdb_B=sd(BSb)
> print(c("a_B=",round(a_B,2),"sd",round(sda_B,3)))
[1] "a_B="  "2.07"  "sd"     "0.831"
> print(c("b_B=",round(b_B,2),"sd",round(sdb_B,3)))
[1] "b_B="      "3733.95" "sd"           "777.924"
> print(c("correlation",round(cor(BSa,BSb),2)))
[1] "correlation" "0.5"
> BSa=sort(BSa)
> aL=BSa[50]
> aU=BSa[950]
> print(c("90% percentile int",round(aL,2),round(aU,2)))
[1] "90% percentile int" "1.24"                "3.5"
```

The mean and standard deviation of the bootstrap distribution of $\hat{a}_{bs}$ are 2.07 and 0.831 respectively. The bootstrap estimate of the bias of the estimator is the difference $2.07 - \hat{a} = 2.07 - 1.73 = 0.34$. The distribution of $\hat{a}_{bs}$ is highly positively skewed and a 90% bootstrap percentile interval for a is $[1.24, 3.5]$. The bootstrap percentile interval does not allow for bias, so the interval may tend to favor larger estimates, but the results of the bootstrap investigation nevertheless indicate that an exponential distribution would not be a convincing model.

We finish off our section on reliability by considering empirical reliability functions.

10.4.4   Kaplan-Meier estimator of reliability

The Kaplan-Meier product-limit estimator is a distribution free estimator of the reliability function $R(t)$. It is particularly useful if a large number of components have been tested and the form of the failure rate is expected to change with time as is likely for the tantalum capacitor of Example 10.15. Initially there are n components on test. The numbers that are still functioning at times

$$t_1 < t_2 < t_3 < \ldots < t_m$$

are recorded. These times can be either the times at which components fail or fixed inspection times. Denote the number of components failing during the interval $(t_{i-1}, t_i]$ by d_i and the number of items at risk during the interval by n_i. The number of items at risk during the interval is the number of items functioning, and therefore at risk, at the beginning of the interval less the number of items removed before they fail during the interval. Let T be the lifetime of a component. An estimate of

$$P(T > t_i | T > t_{i-1}) \;=\; p_i$$

is

$$\widehat{p}_i \;=\; 1 - \frac{d_i}{n_i}.$$

Also

$$
\begin{aligned}
P(T > t_i) \;&=\; P(T > t_i \,|\, T > t_{i-1})\, P(T > t_{i-1}) \\
&=\; P(T > t_i \,|\, T > t_{i-1})\, P(T > t_{i-1} \,|\, T > t_{i-2}) \cdots P(T > t{-}!)
\end{aligned}
$$

and so

$$\widehat{R}(t) \;=\; \prod_{i=1}^{max(m)\,t_m \le t} \left(1 - \frac{d_i}{n_i} \right).$$

Greenwood's $(1 - \alpha)100\%$ confidence bounds are given by

$$\widehat{R}\left(1 \pm z_{\alpha/2} \sqrt{ \sum_{i=1}^{max(m)\,t_m \le t} \frac{d_i}{n_i(n_i - d_i)} } \right).$$

Example 10.19:   Solar batteries

The solar batteries on the 11 planet explorer vehicles (refer previous example) provided power for: $9, 13, 13+, 18, 23, 28+, 31, 34, 45+, 48, 72+$ days, where the $+$ indicates that the vehicle was destroyed or failed for some unconnected reason. The calculation of $\widehat{R}(t)$ for the solar batteries is shown in Table 10.8

The plot in Figure 10.14 is generated using the survival package in R.

```
> #Kaplan-Meier plot
> library(survival)
> #days is number of days battery powers vehicle
> #status: 1 battery fails 2 operating at end of study 3 meteor shower
```

TABLE 10.8: Calculation of reliability function for solar batteries.

i	t_i	d_i	n_i	$\widehat{R}(t_i)$
1	9	1	11	0.91
2	13	1	10	0.82
3	18	1	8	0.72
4	23	1	7	0.61
5	31	1	5	0.49
6	34	1	4	0.37
7	48	1	2	0.18

```
> days=c(9,13,13,18,23,28,31,34,45,48,72)
> status=c(1,1,3,1,1,3,1,1,3,1,2)
> solarb=data.frame(days,status)
> #create survival object
> msurv<-with(solarb,Surv(days,status == 1))
> print(msurv)
 [1]  9  13  13+ 18  23  28+ 31  34  45+ 48  72+
> #Compute Kaplan-Meier estimator
> mfit<-survfit(Surv(days, status == 1)~1,data = solarb)
> print(summary(mfit))
Call: survfit(formula = Surv(days, status == 1) ~ 1, data = solarb)

 time n.risk n.event survival std.err lower 95% CI upper 95% CI
    9     11       1    0.909  0.0867       0.7541        1.000
   13     10       1    0.818  0.1163       0.6192        1.000
   18      8       1    0.716  0.1397       0.4884        1.000
   23      7       1    0.614  0.1526       0.3769        0.999
   31      5       1    0.491  0.1642       0.2549        0.946
   34      4       1    0.368  0.1627       0.1549        0.875
   48      2       1    0.184  0.1535       0.0359        0.944
> #plot the KM estimator
> plot(mfit,xlab="Days",ylab="P(survival)",lty=c(1,3,3))
```

10.5 Acceptance sampling

Inspecting all items (100% inspection), is generally inefficient:

- testing can be destructive, in which case 100% inspection is not an option;

- 100% inspection incurs high inspection costs;

- 100% inspection takes an excessive amount of time and will increase inventory;

- 100% inspection may not detect all defective items (e.g. Hill 1962);

- if the process capability is high 100% inspection is a waste of resources because it does not improve the quality.

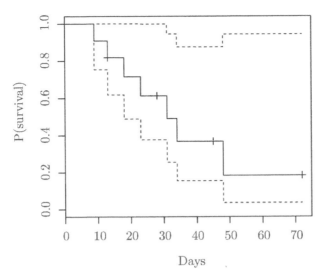

FIGURE 10.14: Kaplan-Meier reliability function for solar batteries.

Acceptance sampling is a more efficient alternative to 100% inspection. There are many different acceptance sampling plans and ANSI and ISO provide standards. Here we consider a single stage sampling plan based on attributes, typically whether an item is good or fails to meet the specification in some way (defective). We suppose that the proportion of defective items in batches from a particular supplier is p. The average outgoing quality (AOQ) is the proportion of defective items in batches leaving the inspection process. The AOQ generally depends on the proportion of defective items in batches arriving at the inspection process and the AOQ limit (AOQL) is the maximum theoretical value of the AOQ.

A random sample of n components is taken from each batch of items delivered. Provided the sample size is less than around 10% of the batch size N, a binomial approximation to the hypergeometric distribution is adequate. Suppose the proportion of defective items in the batch is p. The batch will be accepted if the number of defects in the sample is less than or equal to a critical value c. If the number of defects in the sample, X, exceeds c the entire batch will be inspected at the producer's expense and all defective items will be replaced with satisfactory items. If the proportion of defectives in batches is p then the AOQ is

$$P(X \leq c) \times p + P(c < X) \times 0,$$

because a proportion $P(X \leq c)$ of batches will be accepted, and these batches contain a proportion p of defectives, whereas all defectives in the rejected batches are replaced with satisfactory items.

Example 10.20: Bicycle wheels

A bicycle manufacturer buys in spoked wheels from a supplier. The specification for a wheel is that:

- the rim runout should be less than 0.4 mm,
- the radial true less than 0.4 mm,
- the dish less than 0.4 mm and
- the spoke tensions within 4% of each other.

The supplier uses a robotic machine to lace the spokes to the wheel and a small proportion of wheels do not meet the specification. However, the specification is more stringent than the standard for bike shop repairs in Barnett's manual (The 0.4 mm tolerances are replaced with 0.5 mm, Barnett Bicycle Institute 10e) and the manufacturer is willing to accept a notional maximum of 2% of wheels outside the specification (defective). The manufacturer and supplier agree on the following acceptance sampling plan. A random sample of 50 wheels will be taken from each batch of 1 000 wheels delivered. The batch will be accepted provided no more than one defective is found in the sample. The AOQ is

$$\left((1-p)^{50} + 50p(1-p)^{49}\right) p$$

It is plotted in Figure 10.15, and the AOQL is found to be 0.017, from the following R code.

```
> p=seq(0,.1,.0001)
> AOQ=((1-p)^50+50*p*(1-p)^49)*p
> AOQL=max(AOQ)
> tf=(AOQ == AOQL)
> indexpm=which(tf)
> print(c("incoming p",p[indexpm],"AOQL",round(AOQL,3)))
[1] "incoming p" "0.0318"     "AOQL"       "0.017"
> plot(p,AOQ,type="l")
> lines(p,AOQ-AOQ+AOQL,lty=2)
> text(0.002,0.017,"AOQL")
```

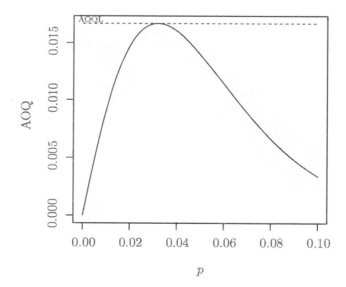

FIGURE 10.15: Proportion of defective wheels outgoing from the inspection process (AOQ) against proportion of defective wheels in incoming batches.

Although acceptance sampling is a substantial improvement on 100% inspection, it is open to the same criticisms. In particular, if the supplier's process capability is high, then acceptance sampling is itself a waste of resources because it does not improve the quality.

Example 10.21: Air filters

A company that manufactures filters for compressed air generators, and for use in the food and beverage industry, fabricates the filter membrane in-house, but buys in components such as housings and end caps. Some of these components may have cosmetic defects such as flash on a cast component or paint run on a filter housing.

Acceptance sampling could be used to control the AOQL, but the company aims to deal with regular suppliers who can be relied upon to deliver high quality components. Once a supplier has demonstrated an ability to consistently meet the specification, acceptance sampling is only occasionally implemented, or dispensed with. However, a single item is inspected from every delivery to ensure that the delivery corresponds to the order.

Another criticism of acceptance sampling is that it is based on the notion that a small proportion of defects is acceptable, and if the proportion is as small as 1 in 1 000 (1 000 ppm) the sample size for an effective acceptance sampling scheme is unreasonably large. [Deming, 2000] says that acceptance sampling techniques "guarantee that some customers will get defective products" and also bemoans the resources used to implement them. Nevertheless, acceptance sample may be useful when dealing with new suppliers if defects can be defined so that a small proportion is acceptable.

10.6 Statistical quality control charts

Statistical quality control (SQC) charts are used to monitor processes that are generally in statistical control, and to provide early warning of any special cause variation that affects the process. They plot statistics that represent the quality of the items produced, obtained from samples over time. The plot has a region which corresponds to the process appearing to be in statistical control, and a region or regions which indicate that some action needs to be taken. The action need not be as drastic as stopping a production line, and it may just be to monitor the process more closely, but there is little benefit to be obtained from SQC charts if points in the action region are ignored. Walter A Shewhart is credited with introducing the idea at the Western Electric Company's factory in Cicero, Illinois in 1924.

A common feature of SQC charts is that there are target values for the variables that are being monitored and that the standard deviations of the variables are known. The standard deviations are estimated from records of the process when it has been deemed to be in statistical control, and have already been used to demonstrate capability.

10.6.1 Shewhart mean and range chart for continuous variables

Samples of size n are taken from the process at unequally spaced times $t = 1, \ldots,$ and a continuous variable is measured. The data for each sample are denoted as $\{x_{t1}, \ldots, x_{tn}\}$. The sample means and ranges are plotted against t.

10.6.1.1 Mean chart

- The sample means, $\bar{x}_t = \sum_{i=1}^{n} x_{ti}/n$, are plotted against t.

534 Statistics in Engineering, Second Edition

- There is a target value, τ, for the variable X.

- The standard deviation of the variable, σ, is known.

- The sample size n is generally small, 4 or 5 is typical, because the emphasis is on monitoring the process over time rather than establishing a precise value of the mean at some specific time.

- The sample is assumed to be a random sample from the process at the time it is taken. This assumption is more plausible if the items are not consecutive items from production. For example, if they are traveling along a conveyor belt leave short gaps between removing the items.

- The distribution of $\overline{X}$ is approximately normal as a consequence of the Central Limit Theorem.

- The frequency of taking samples depends on the application. The cost of resources to take the samples, make the measurements, and record the measurements has to be offset against the loss incurred if the process produces sub-standard product or scrap. It might be as often as every half hour if a machine is manually operated or once a shift or even just weekly for reliable automated processes.

- Samples should not be taken at precise time intervals in case these correspond to periodicities in the process.

- Sample ranges are usually used in preference to standard deviations because they are quicker to calculate.

- The target value τ is shown as a line on the chart and lower and upper action lines (LAL) and (UAL) are drawn at

$$\tau \pm 3.09 \frac{\sigma}{\sqrt{n}}.$$

If the process mean is on target $\overline{X} \sim N(\tau, \sigma^2/n)$ and the probability a point plots beyond an action line is $2(1 - \Phi(3.09)) = 2/1000$, then the average run length (ARL) is the average number of points plotted between a point lying beyond an action line and is the reciprocal of the probability of a point lying beyond an action line. if the process is on target the ARL is $1000/2 = 500$. A value of 3 is sometimes used instead of 3.09, in which case the ARL is 370.4.

- If the process mean is $\tau + k\sigma$ the probability that a point plots above the UAL is

$$P\left(\tau + 3.09 \frac{\sigma}{\sqrt{n}} < \overline{X}\right) = P(3.09 - k\sqrt{n} < Z) = 1 - \Phi(3.09 - k\sqrt{n})$$

For $n = 5$ and $k = 1$

```
> 1-pnorm(3.09-sqrt(5))
[1] 0.1965713
```

and the ARL[25] is 5.09.

---
[25]The probability of a point lying below the LAL is negligible.

- Lower and upper warning lines, (LWL) and (UWL), are often added to the plot at

$$\tau \pm 1.96 \frac{\sigma}{\sqrt{n}}$$

and action is sometimes taken if two consecutive points lie below the LWL or above the UWL (see Exercise 10.10).

10.6.1.2 Range chart

- The sample ranges, $R_t = \max_i(x_{ti}) - \min(x_{ti})$, are plotted against t.

- Sample ranges are usually used in preference to standard deviations because they are easier to calculate.

- Action lines corresponding to the upper and lower 0.001 quantiles of the distribution of ranges for random samples of size n from normal distributions are shown. These quantiles are tabled, Table 10.9, but they can be quickly obtained by simulation. For example, with $n = 5$

```
> n=5
> m=matrix(rnorm(1000000),ncol=n)
> r=sort(apply(m,1,max)-apply(m,1,min))
> L=length(r)
> print(c(r[L*.001],r[L*.999]))
[1]  0.3792035 5.4903572
```

runs in a few seconds on a PC and is close enough to tabled values of 0.37 and 5.48 respectively. The action lines are drawn at 0.37σ and 5.48σ. The values are sensitive to the assumed distribution of the variable and simulation can allow for this.

- Warning lines can be shown corresponding to 0.025 quantiles.

TABLE 10.9: Quantiles of the distribution of the sample range in random samples of size n from a standard normal distribution.

n	Lower 0.001	Lower 0.025	Upper 0.025	Upper 0.001
2	0.00	0.04	1.17	4.65
3	0.06	0.30	3.68	5.06
4	0.20	0.59	4.20	5.31
5	0.37	0.85	4.36	5.48
6	0.53	1.07	4.49	5.62
7	0.69	1.25	4.60	5.73

Example 10.22: Burners

A small engineering company supplies burners for gas cookers in response to orders. The burners are made from perforated plates that are rolled into tubes and have flanges fixed at the ends in a purpose built machine. The company has just received an order for one thousand burners with a specification that: the length from the base to the first gas outlet hole is between 67.50 mm and 68.50 mm. Although the machine is old it has been well maintained and once set up the standard deviation of the length is 0.09.

The capability index for this burner is high, $C_p = 1/(6 \times .09) = 1.85$, but setting the machine up at the beginning of a run is a tricky operation and the run will typically be made with a mean that is not at the centre of the specification. The machine is set up for the order and the mean length of the first 25 burners produced is 68.10 mm. A mean of 68.10 corresponds to a process performance index of $C_{pk} = 0.40/(3 \times 0.09) = 1.48$ which is satisfactory. Samples of size $n = 5$ will be taken at approximate half hour intervals during the 7 hour run, and Shewhart mean and range charts are drawn with $\tau = 68.10$ and $\sigma = 0.09$. If a point lies outside an action line, the action will be to take another sample of size 20 immediately. It turned out that no action was required and R code for plotting the charts shown in Figures 10.16 and 10.17 from the 14 samples follows[26].

```
> burner.dat=read.table("burner.txt",header=T)
> attach(burner.dat)
> print(head(burner.dat))
  t    x1     x2     x3     x4     x5
1 1 68.076 67.950 68.147 68.109 68.243
2 2 68.111 68.037 68.087 68.314 68.274
3 3 68.336 67.946 68.116 68.198 67.892
4 4 68.144 68.003 68.134 68.029 68.288
5 5 68.111 68.223 68.122 68.048 68.175
6 6 68.243 68.146 68.070 68.043 67.994
> mb=matrix(as.matrix(burner.dat)[,2:6],ncol=5)
> xbar=apply(mb,1,mean)
> ran=apply(mb,1,max)-apply(mb,1,min)
> n=5;T=length(t);sigma=0.09;tau=rep(68.1,T)
> #mean chart
> LAL=tau-3.09*sigma/sqrt(n);UAL=tau+3.09*sigma/sqrt(n)
> LWL=tau-1.96*sigma/sqrt(n);UWL=tau+1.96*sigma/sqrt(n)
> yaxll=min(min(xbar),LAL[1])-.01;yaxul=max(max(xbar),UAL[1])+.01
> plot(t,xbar,ylim=c(yaxll,yaxul),ylab="mm",main="Means")
> lines(t,tau,lty=1);lines(t,LAL,lty=2);lines(t,UAL,lty=2)
> text(1.1,yaxll,"LAL");text(1.1,yaxul,"UAL")
> lines(t,LWL,lty=3);lines(t,UWL,lty=3)
> text(1.1,LWL[1],"LWL");text(1.1,UWL[1],"UWL")
> #range chart
> LAL=rep(0,T)+.37*sigma;UAL=rep(0,T)+5.48*sigma
> LWL=rep(0,T)+.85*sigma;UWL=rep(0,T)+4.20*sigma
> yaxll=0;yaxul=max(max(ran),UAL[1])+.01
> plot(t,ran,ylim=c(yaxll,yaxul),ylab="mm",main="Ranges")
> lines(t,LAL,lty=2);lines(t,UAL,lty=2)
> text(1.1,LAL[1],"LAL");text(1.1,yaxul,"UAL")
> lines(t,LWL,lty=3);lines(t,UWL,lty=3)
> text(1.1,LWL[1],"LWL");text(1.1,UWL[1],"UWL")
```

[26]SQC charts are updated after each sample. If the data are being entered into Excel it is more convenient to read the spreadsheet directly than to save it as a *.txt file each time. There are several R packages that enable this. For example if you install and load the package xlsx you can read data from sheet m of Data.xlsx with the command `read.xlsx("Data.xlsx",m)`. You do need JAVA-HOME for the package xlsx to run.

Means

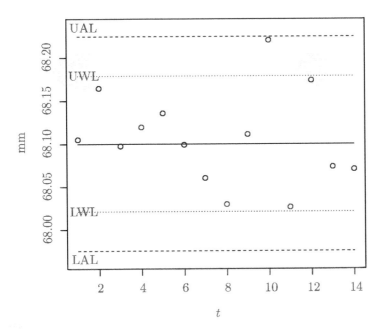

FIGURE 10.16: Shewhart mean chart for samples of 5 lengths of gas burners.

Ranges

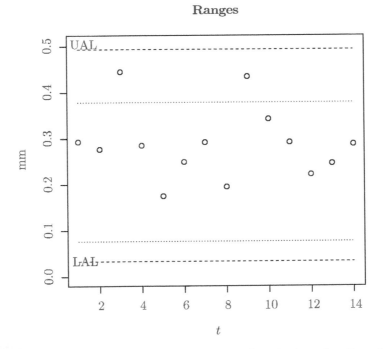

FIGURE 10.17: Shewhart range chart for samples of 5 lengths of gas burners.

Example 10.23: Concrete

The quality of concrete used in large civil engineering projects such as bridges, dams and tall buildings is closely monitored. For example, on a bridge project 6 samples of concrete were taken during each working day for a slump test, and for making cubes that are tested for compressive strength after 7 and 28 days. The times of taking the 6 samples were spaced throughout each day, with enough random variation in the spacing to avoid predictability. The slump test is an instant measure of the wetness (consistency) of the concrete and the specification is that the slump (drop of apex of a concrete cone tested according to the ASTM standard) should be be between 50 mm and 90 mm. The target slump is 70 mm, and from past experience the standard deviation is about 4.5 mm. The means and ranges of the first 5 days samples are shown in Table 10.10 You

TABLE 10.10: Mean and range of the first 5 days samples of 6 slump measurements on concrete.

Day	Mean	Range
1	72.3	13.4
2	68.4	7.3
3	70.8	15.7
4	74.1	21.4
5	75.9	24.7

are asked to set up Shewhart mean and range charts in Exercise 10.9. The supplier is given copies of these charts and will be told to adjust the mean consistency if a point plots outside the action lines on the mean chart, or to reduce its variability if a point plots outside the action lines on the range chart.

10.6.2   p-charts for proportions

The p-chart is a modification of the Shewhart mean chart for proportions. Suppose that random samples of size n are taken from production and denote the number of failures in each sample by X. Define $\widehat{p} = \frac{X}{n}$. If the proportion of failures produced by the process at the time of sampling is p, then provided $\min(np, n(1-p))$ exceeds about 5

$$\widehat{p} \sim N(p, p(1-p)/n),$$

is a good approximation. If a proportion p of failures is acceptable the chart is set up with action lines at

$$p \pm 3.09 \sqrt{\frac{p(1-p)}{n}}.$$

Warning lines can also be shown. For the chart to be useful p needs to be substantial, as it may be in tests that simulate extreme conditions.

Example 10.24: Windshields

A manufacturer of front windshields for automobiles subjects random samples of 30 to extreme impact tests each week. The test is more extreme than any in the customer's specification.

A windshield is mounted in a dummy automobile housing in the recommended manner and is considered a failure if any glass leaves the housing after the impact. The probability of a failure has been 0.40 over the past year and the purpose of the chart is to monitor any changes in the process, whether due to special cause variation or intentional modifications to the windshield manufacturing process or the recommendations for mounting it in the housing. The numbers of failures for 26 weeks after setting up the chart are given in Table 10.11.

TABLE 10.11: Number of windscreen failures.

Week	1	2	3	4	5	6	7	8	9	10	11	12	13
Failures	8	11	12	21	17	16	11	18	19	11	14	14	19

Week	14	15	16	17	18	19	20	21	22	23	24	25	26
Failures	11	15	8	11	12	13	5	6	7	14	8	11	6

The R code for plotting the p-chart shown in Figure 10.18 is

```
> x=c(8,11,12,21,17,16,11,18,19,11,14,14,
+ 19,11,15,8,11,12,13,5,6,7,14,8,11,6)
> n=30;T=length(x);t=1:T;p=rep(.4,T)
> phat=x/n
> LAL=p-3.09*sqrt(p*(1-p)/n);UAL=p+3.09*sqrt(p*(1-p)/n)
> LWL=p-1.96*sqrt(p*(1-p)/n);UWL=p+1.96*sqrt(p*(1-p)/n)
> yaxll=min(min(phat),LAL[1])-.01;yaxul=max(max(phat),UAL[1])+.01
> plot(t,phat,ylim=c(yaxll,yaxul),ylab="Proportion fail")
> lines(t,p,lty=1);lines(t,LAL,lty=2);lines(t,UAL,lty=2)
> text(1.1,yaxll,"LAL");text(1.1,yaxul,"UAL")
> lines(t,LWL,lty=3);lines(t,UWL,lty=3)
> text(1.1,LWL[1],"LWL");text(1.1,UWL[1],"UWL")
```

After 4 weeks there were 21 failures and the point plotted above the upper action limit. The process was checked but nothing amiss was found. After 8 and 9 weeks two points plotted above the upper warning line and the production engineer thought that modifying the plastic laminating layer might decrease the proportion of failures. Following discussions with design engineers the modification was implemented for week 14. Since then there does seem to be a downwards step change, and although no points have plotted below the lower action limit several are below the lower warning limit. The modification seems to have been successful in reducing failures.

The p-chart is sometimes set up for the number of failures, rather than the proportion of failures. An advantage of plotting the proportion of failures is that it can be used when the sample sizes differ. The action lines are sensitive to an assumption of independent draws from the population.

10.6.3   c-charts for counts

The Shewhart chart can be applied to count data over fixed period of time. If the events that are counted can reasonably be assumed to occur randomly and independently a normal approximation to the Poisson distribution can be used. Alternatively the standard deviation

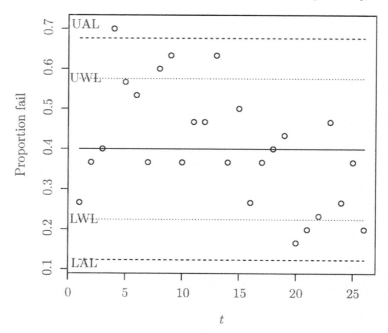

FIGURE 10.18: Proportion of windscreen failures per week.

of the counts can be estimated from past data. A c-chart is a plot of the number of events in each time period against time (measured in unit of the time period). If the mean count per period is μ and its standard deviation is σ action limits for the c-chart, based on a normal approximation to the distribution of counts, are set at

$$\mu \pm 3.09\sigma$$

If a Poisson distribution is assumed then $\sigma = \sqrt{\mu}$. Warning limits can also be set at $\pm 1.96\sigma$.

Example 10.25: Components

A large manufacturing company of small mechanical components had a mean number of 54.7 machine breakdowns per week over a one year period. The standard deviation was 4.6, which is considerably lower than the 7.4 which is consistent with a Poisson distribution with mean 54.7 and suggests that some breakdowns were regular rather than purely random occurrences. This high number of breakdowns each week was a serious problem and the production manager decided to implement a program of total productive maintenance (TPM). As part of this program the manager asked the machine operators to undertake routine preventative maintenance, and provided training and a daily time allowance so that they were able to do so. The number of breakdowns per week over the 30 weeks of the program are given in Table 10.12.

A c-plot generated from the following R script is shown in Figure 10.19.

```
x=c(46,60,52,52,52,51,48,42,40,41,42,27,29,39,36,35,
+   30,29,30,32,31,19,26,25,24,32,18,25,29,30)
> T=length(x);t=1:T;mu=rep(54.7,T);sigma=4.6
> LAL=mu-3.09*sigma;UAL=mu+3.09*sigma
```

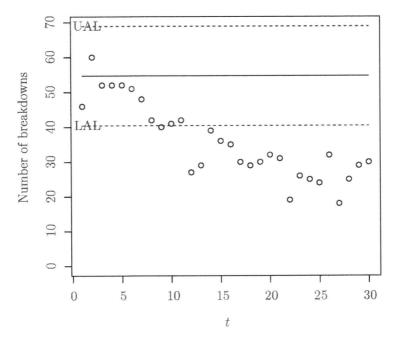

FIGURE 10.19: Number of machine breakdowns per week.

TABLE 10.12: Number of breakdowns per week.

Week	1	2	3	4	5	6	7	8	9	10
Breakdowns	46	60	52	52	52	51	48	42	40	41

Week	11	12	13	14	15	16	17	18	19	20
Breakdowns	42	27	29	39	36	35	30	29	30	32

Week	21	22	23	24	25	26	27	28	29	30
Breakdowns	31	19	26	25	24	32	18	25	29	30

```
> yaxll=0;yaxul=max(max(x),UAL[1])+.01
> plot(t,x,ylim=c(yaxll,yaxul),ylab="Number of breakdowns")
> lines(t,mu,lty=1);lines(t,LAL,lty=2);lines(t,UAL,lty=2)
> text(1.1,LAL[1],"LAL");text(1.1,yaxul,"UAL")
```

The number of breakdowns has reduced dramatically since the TPM program was introduced, and the level of absenteeism that had been identified as unreasonably high had also reduced. The next stage of the TPM process is to consolidate the gains and aim for further improvements. Further improvements will be harder to achieve because the more straightforward problems have already been tackled. A c-chart could be set up around 25 breakdowns per week to monitor progress.

10.6.4 Cumulative sum charts

The cumulative sum chart is particularly useful for monitoring processes when we have one-at-a-time data rather than samples on n items, although it is also used for sample means as an alternative to the Shewhart mean chart. We describe it in terms of one-at-a-time observations x_t from a process with a target value for the variable of τ. The cumulative sum (CUSUM) is defined as

$$S_t \;=\; \sum_{i=1}^{t}(x_t - \tau)$$

and the CUSUM chart is a plot of S_t against time. It is the slope of the chart, rather than its current value, that indicates the process mean has changed from the target value. A steep upwards slope indicates a positive change and a steep downwards slope indicates a negative change.

A V-mask can be used to decide whether action is justified. Assume that the standard deviation of the variable that is being plotted is known to be σ. The V-mask is centered on the latest observation, the width of the front of the mask is 10σ, the gradients of the arms are $\pm 0.5\sigma$ per sampling interval and action is indicated if any points in the CUSUM plot beyond the arms. The false alarm rate with this V-mask is about 1 in 440 if the variable has a normal distribution. An equivalent procedure to the V-mask which is easier to program, but lacks the visual impact, is

- $H = 5\sigma$

- $K = 0.5\sigma$

- $SH(t) = \max[0, (x_t - \tau) - K + SH(t-1)]$

- $SL(t) = \max[0, -(x_t - \tau) - K + SL(t-1)]$

- Take action if either of $SH(t)$ or $SL(t)$ exceeds H.

Example 10.26: High voltage switch gear

An engineering company manufactures high voltage switch gear and has recently modified a particular design of circuit breaker by replacing a moving part with an electronic device. The aim of the modification is to reduce maintenance requirements, without affecting the operating characteristics. Tests on prototypes indicated that the modification had little affect on operating characteristics. A production engineer uses CUSUM charts to monitor critical operating characteristics, which are measured on every circuit breaker before shipping. One of these is the open-time, and open-times (10^{-4} s) for the latest 44 circuit breakers are given in Table 10.13. The modification was implemented after the 20th circuit breaker.

TABLE 10.13: Open times (10^{-4} s) for 44 circuit breakers.

521	505	506	481	498	499	487	501	527	500	502
505	483	507	499	497	511	489	501	520	515	495
490	504	504	489	487	492	509	490	483	488	487
473	498	487	503	500	500	484	494	497	478	486

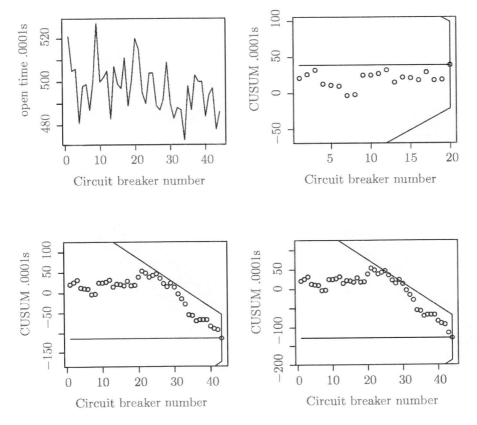

FIGURE 10.20: CUSUM charts for circuit breaker characteristics.

The following R script plots the data (Figure 10.20 upper left) and then draws a CUSUM, using the function **cumsum()**, and shows the V-mask set at the 44th observation (Figure 10.20 lower right).

```
> cb.dat=read.table("circuitbreaker.txt",header=T)
> attach(cb.dat)
> head(opent)
[1] 521 505 506 481 498 499
> par(mfrow=c(2,2))
> plot(as.ts(opent),xlab="circuit breaker number",
+ ylab="open time .0001s")
> n=44
> x=opent[1:n]
> tau=500
> sigma=12
> K=0.5*sigma
> H=5*sigma
> cx=cumsum((x-tau))
> yll=min(cx)-H
> ylu=max(cx)+H
> sampno=c(1:n)
```

```
> plot(sampno,cx,ylim=c(yll,ylu),xlab="circuit breaker number",
+ ylab="CUSUM .0001s")
> LV=cx[n]-H
> UV=cx[n]+H
> lines(c(n,n),c(LV,UV))
> LA=LV-(n-1)*K
> UA=UV+(n-1)*K
> lines(c(1,n),c(LA,LV))
> lines(c(1,n),c(UA,UV))
> lines(c(1,n),c(cx[n],cx[n]))
```

In Figure 10.20 the top right panel shows the CUSUM and V-mask after 20 observations, and there is no suggestion that the process mean is off target. The lower left panel shows the CUSUM and V-mask after 43 observations and there is a clear decreasing slope. However, no points lie outside the V-mask and there is no signal to take action. After 44 observations the CUSUM extends outside the upper arm of the V-mask and there is a signal to take action. The action is to calculate the means of the 20 open-times before the modification and the 24 open-times following the modification.

```
> mean(opent[1:20])
[1] 501.95
> mean(opent[21:44])
[1] 493.0417
```

A reduction in open time from the target vale of 500 down to around 493 does not affect the functionality of the design and the target value will be reset at 493.

The CUSUM chart can be used for means of n items, in which case $\sigma = \sigma_{\overline{X}} = \sigma_X/\sqrt{n}$, but the Shewhart range chart still needs to be used to monitor the process variability. The CUSUM chart will, on average, indicate a small shift in the mean more quickly than the Shewhart mean chart.

10.6.5 Multivariate control charts

For many manufacturing processes there will be more than one variable to be monitored. Relying on individual Shewhart mean charts for each variable is not ideal because

- There will be more charts to monitor and they may all get overlooked.

- If the process is in statistical control and we take action if a point on any chart lies beyond the action lines we will reduce the average run length (ARL).

- Points on several charts might be simultaneously between warning and action lines, indicating a change in the process, yet be ignored because none lies beyond an action line.

A better strategy is to plot Hotelling's statistic. Suppose there are m variables to monitor and that we take random samples of n items at unequally spaced times t. The vector of sample means at time t is

$$\bar{\boldsymbol{x}}_t = (\bar{x}_{1t}, \ldots, \bar{x}_{mt})'$$

The population or target mean, depending on the application, is μ, and the population variance-covariance matrix, Σ, is also assumed known.

$$H_t = n(\bar{x}_t - \mu)'\Sigma^{-1}(\bar{x}_t - \mu)$$

If $\bar{x}_t$ is the mean of a random sample of n from a multivariate distribution with mean $\mathbf{0}$ and variance-covariance Σ then $(\hat{x} - \mu)'(\Sigma^{-1/2}/\sqrt{n})$ has mean $\mathbf{0}$ and variance-covariance matrix $\boldsymbol{I}$. It follows that H_t is the sum of m independent chi-square variables and has a chi-square distribution with m degrees of freedom. If we set the action line at $\chi^2_{m,.005}$ the ARL will be 500. Although H_t neatly solves the ARL problem it loses the information on individual variables and whether they are above or below the mean. The first step of the action is to consider the mean charts for individual variables. It is important to monitor variability as well as the mean level. The sample **generalized variance** can be calculated at time t

$$\left|\widehat{\Sigma}_t\right|$$

A plot of the ratio of the sample generalized variance to the population generalized variance can be used to monitor variability. Theoretical results are available for the sampling distribution but they are sensitive to an assumption of multivariate normality and Monte-Carlo simulation provides an easily implemented, if less elegant, alternative (Exercise-SQCgeneralizedvar). The trace of the sample variance-covariance matrix (that is the sum of the variances, which lie along the leading diagonal) might also be used to monitor variability but it ignores information from covariances (Exercise-SQCtrace).

Example 10.27: Bank Bottom Engineering - robot arms

We demonstrate the use of Hotelling's statistic with an excerpt from the data on robot arms. For the purpose of this demonstration we consider the first 5 variables only, referred to here as $V_1, \ldots, V_5$, and assume that mu and Σ are known to be

$$\mu = (0.010, -0.030, 0.008, 0.000, 0.005)'$$

and

$$\Sigma = \begin{bmatrix} 1.00 & 0.11 & 0.00 & 0.05 & -0.06 \\ 0.11 & 1.00 & -0.41 & 0.33 & 0.15 \\ 0.00 & -0.41 & 1.00 & -0.22 & -0.20 \\ 0.05 & 0.33 & -0.22 & 1.00 & -0.20 \\ -0.06 & 0.15 & -0.20 & -0.20 & 1.00 \end{bmatrix}$$

from extensive past records. In the following R script the mu (μ), Sigma (Σ) and the corresponding correlation matrix, CorMat, and the means of 17 samples of size 6, xbar, had been entered in advance[27]. The Hotelling plot is shown in Figure 10.21 upper left and the Shewhart mean plots for the 5 variables are shown below and to the right. Notice that the Hotelling chart flags action after the 8th and 15th sample. No point on individual charts lies beyond action lines after the 8th sample. Points on two of the individual charts do lie beyond action lines after the 15th sample, but not so strikingly as on the Hotelling chart. One point lies below the lower action line on the chart for V_1 after the 7th sample but the Hotelling chart does not indicate that action is required.

---

[27]For the purpose of constructing data for this example this we took every other robot arm, to make an assumption of independence more plausible, to obtain 5 columns of length 103, and grouped these into 17 samples of size 6 with 1 left at the end.

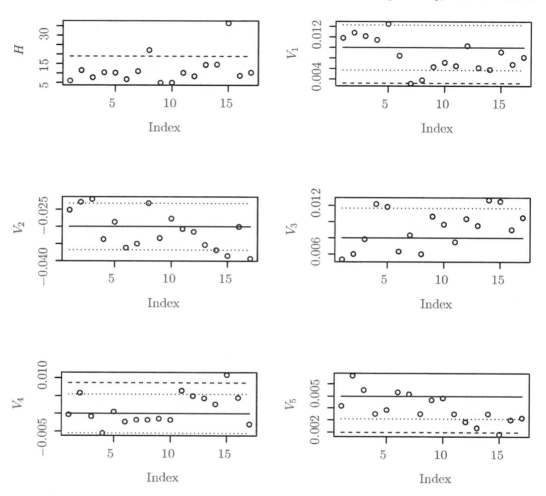

FIGURE 10.21: Hotelling statistic for 17 samples of 6 robot arms (top left) and Shewhart mean charts for $V_1, \ldots, V_6$. Action lines are shown as dotted lines. The mean (solid line) and warning lines (dotted lines) are also shown on the Shewhart mean charts.

```
> print(mu)
[1]  0.010 -0.030  0.008  0.000  0.005
> print(Sigma)
          V1         V2        V3        V4        V5
V1  2.88e-05   5.10e-06  0.00e+00   1.9e-06  -8.0e-07
V2  5.10e-06   7.17e-05 -1.59e-05   1.9e-05   3.1e-06
V3  0.00e+00  -1.59e-05  2.10e-05  -6.9e-06  -2.1e-06
V4  1.90e-06   1.90e-05 -6.90e-06   4.7e-05  -3.3e-06
V5 -8.00e-07   3.10e-06 -2.10e-06  -3.3e-06   5.7e-06
> print(CorMat)
      V1    V2    V3    V4    V5
V1  1.00  0.11  0.00  0.05 -0.06
V2  0.11  1.00 -0.41  0.33  0.15
V3  0.00 -0.41  1.00 -0.22 -0.20
```

```
V4   0.05   0.33  -0.22   1.00  -0.20
V5  -0.06   0.15  -0.20  -0.20   1.00
> print(xbar)
                [,1]           [,2]           [,3]            [,4]           [,5]
 [1,]  0.011833333  -0.02516667  0.005333333  -0.0003333333  0.004166667
 [2,]  0.012833333  -0.02283333  0.006000000   0.0058333333  0.006666667
 [3,]  0.012166667  -0.02200000  0.007833333  -0.0008333333  0.005500000
 [4,]  0.011500000  -0.03366667  0.012166667  -0.0055000000  0.003500000
 [5,]  0.014500000  -0.02866667  0.011833333   0.0005000000  0.003833333
 [6,]  0.008500000  -0.03616667  0.006333333  -0.0023333333  0.005333333
 [7,]  0.003166667  -0.03500000  0.008333333  -0.0018333333  0.005166667
 [8,]  0.003833333  -0.02316667  0.006000000  -0.0018333333  0.003500000
 [9,]  0.006333333  -0.03333333  0.010666667  -0.0015000000  0.004666667
[10,]  0.007166667  -0.02766667  0.009666667  -0.0018333333  0.004833333
[11,]  0.006500000  -0.03066667  0.007500000   0.0065000000  0.003500000
[12,]  0.010333333  -0.03150000  0.010333333   0.0050000000  0.002833333
[13,]  0.006166667  -0.03533333  0.009500000   0.0043333333  0.002333333
[14,]  0.005833333  -0.03683333  0.012666667   0.0026666667  0.003500000
[15,]  0.009166667  -0.03850000  0.012500000   0.0108333333  0.001833333
[16,]  0.006833333  -0.03000000  0.009000000   0.0045000000  0.003000000
[17,]  0.008166667  -0.03933333  0.010500000  -0.0030000000  0.003166667
> T=17
> H=rep(0,T)
> for (j in 1:T){
+ H[j]=n*t(xbar[j,1:5]-mu)%*%solve(Sigma)%*%(xbar[j,1:5]-mu)}
> varlab=c("V1","V2","V3","V4","V5")
> par(mfrow=c(3,2))
> plot(H)
> lines(t,rep(qchisq(.998,m),T),lty=2)
> for (i in 1:5) {
+ plot(xbar[,i],ylab=varlab[i])
+ lines(t,rep(mu[i]-3.09*sqrt(Sigma[i,i]/n),T),lty=2)
+ lines(t,rep(mu[i]-1.96*sqrt(Sigma[i,i]/n),T),lty=3)
+ lines(t,rep(mu[i],T))
+ lines(t,rep(mu[i]+1.96*sqrt(Sigma[i,i]/n),T),lty=3)
+ lines(t,rep(mu[i]+3.09*sqrt(Sigma[i,i]/n),T),lty=2)
+ }
#To monitor the variability
```

10.7   Summary

10.7.1   Notation

R_t moving lag
$\widetilde{\sigma}$ moving lag standard deviation
y_{ij} observation on item j from batch i
μ mean of all batches in the infinite population
I number of batches
J number of observations in each batch
α_i between batch errors
ϵ_{ij} within batch errors
σ_ϵ^2 variance of within batch errors
s_i^2 sample variance within batch
$\widehat{\sigma}_\epsilon^2$ mean of sample variance within batch
$R(t)$ reliability function

10.7.2   Summary of main results

Runs chart: Draw lines at

$$\bar{x}, \quad \bar{x} \pm \widetilde{\sigma}, \quad \bar{x} \pm 2\widetilde{\sigma}, \quad \bar{x} \pm 3\widetilde{\sigma}$$

and label the zones as:

C within $1\widetilde{\sigma}$ of the mean,

B between $1\widetilde{\sigma}$ and $2\widetilde{\sigma}$ from the mean, and

A between $2\widetilde{\sigma}$ and $3\widetilde{\sigma}$ from the mean.

Components of variance: A model for the observations is

$$Y_{ij} = \mu + \alpha_i + \epsilon_{ij},$$

where $\alpha_i \sim 0, \sigma_\alpha^2$ and $\epsilon_{ij} \sim 0, \sigma_\epsilon^2$. Within batch variance σ_ϵ^2 can be estimated by calculating the sample variance within each batch and then taking the mean of the I sample variances, which we denote by $\widehat{\sigma}_\epsilon^2$.

$$s_i^2 = \frac{\sum_{j=1}^{J}(y_{ij} - \bar{y}_{i.})^2}{J - 1} \qquad \widehat{\sigma}_\epsilon^2 = \frac{\sum_{i=1}^{I} s_i^2}{I}.$$

Capability: The process capability index is defined as

$$C_p = \frac{U - L}{6\sigma}.$$

where L and U are the lower and upper specification limits respectively. The process performance index is defined as

$$C_{pk} = \min\left(\frac{\mu - L}{3\sigma}, \frac{U - \mu}{3\sigma}\right).$$

If the sample mean and standard deviation are $\bar{x}$ and s respectively an approximate $(1 - \alpha) \times 100\%$ confidence interval for C_{pk} is

$$\widehat{C}_{pk} \left(1 \pm z_{\alpha/2} \frac{1}{\sqrt{2n}} \right),$$

where

$$\widehat{C}_{pk} = \min \left(\frac{\bar{x} - L}{3s}, \frac{U - \bar{x}}{3s} \right).$$

Reliability: The reliability function is the complement of the cdf

$$R(t) = 1 - F(t),$$

while the failure rate, also known as the hazard function is defined as

$$h(t) = \frac{f(t)}{R(t)}, \quad 0 \le t,$$

where $F(t)$ and $f(t)$ are the cdf and pdf respectively.

acceptance sampling : The average outgoing quality (AOQ) is the proportion of defective items in batches leaving the inspection process and is given by

$$\text{AOQ} = P(X \le c) \times p + P(c < X) \times 0,$$

wherein X is the number of defective items in the sample, c is the critical value for defective items below which the batch will be accepted, and p is the proportion of defective item in the batch.

Statistical quality control charts: Statistical quality control (SQC) charts are used to monitor processes that are generally in statistical control, and to provide early warning of any special cause variation that affects the process.

Shewhart mean chart The sample means are plotted against t. The target value τ is shown as a line on the chart and lower and upper action lines (LAL) and (UAL) are drawn at $\tau \pm 3.09 \frac{\sigma}{\sqrt{n}}$.

Shewhart range chart The sample ranges, $R_t = \max_i(x_{ti}) - \min_i(x_{ti})$, are plotted against t. Action lines corresponding to the upper and lower 0.001 quantiles of the distribution of ranges for random samples of size n from normal distributions are shown.

p-charts for proportions The p-chart is a modification of the Shewhart mean chart for proportions, where $\widehat{p}$ is used as the sample failure proportion and can be approximated by $\widehat{p} \sim N(p, p(1 - p)/n)$. Action lines can be included at $p \pm 3.09 \sqrt{\frac{p(1-p)}{n}}$.

c-charts for counts A c-chart is a plot of the number of events in each time period against time (measured in unit of the time period) and action lines are set at $\mu \pm 3.09\sigma$. If a Poisson distribution is assumed then $\sigma = \sqrt{\mu}$.

Cumulative sum charts The CUSUM chart is a plot of S_t against time where $S_t = \sum_{i=1}^{t}(x_t - \tau)$ and x_t are one-at-a-time observations with a target value of τ. Action lines (V-masks) are centered on the latest observation, the width of the front of the mask is 10σ, the gradients of the arms are $\pm 0.5\sigma$ per sampling interval and action is indicated if any points in the CUSUM plot beyond the arms.

Multivariate control charts Used when more than one variable needs to be monitored. Hotelling's statistic is plotted against time, given by $H_t = n(\bar{x}_t - \mu)'\Sigma^{-1}(\bar{x}_t - \mu)$, where n is the number of samples, $\bar{x}_t$ is the vector of samples means at time t, μ is the vector of means for all variables and Σ is the variance-covariance matrix.

10.7.3   MATLAB and R commands

In the following R and MATLAB commands, `data` contains the time series, `x` is is an integer, `m`, `tm` and `tsd` are real numbers. For more information on any built in function, type `help(function)` in R or `help function` in MATLAB.

R command
***
`cusum(data, decision.interval=x,center=tm,se.shift=m,std.dev=tsd)`

MATLAB command
***
`cusum(data,x,m,tm,tsd)`

10.8   Exercises

Section 10.1 Continuous improvement

Exercise 10.1:   Process performance

An amplifier is specified to have a gain between 190 and 210. The manufacturing process produces amplifiers with a mean of 202 and a standard deviation of 2.5.

(a) Calculate C_{pk}.

(b) What proportion, in ppm (parts per million), will be outside spec if gains have a normal distribution?

(c) Prove that the variance of a variable with a uniform distribution over $[0,1]$ is $1/12$.

(d) What proportion, in ppm, will be outside spec if gains have a uniform distribution?

Section 10.2 Process stability

Exercise 10.2:   Tamper

Let x_t be a process variable measured at time t, and suppose the unit of measurement is scaled so that the target value is 0. The process is stable with a mean of 0 and errors ϵ_t which have a mean of 0 and a standard deviation σ, so without any operator intervention

$$x_t = \epsilon_t$$

But, the process mean can be adjusted and a new operator, in an attempt to reduce the process variability, makes an adjustment equal to $-x_t$ after each measurement. For example, if the last measurement was 5 above target the operator reduces the mean level by 5. This introduces a deviation δ_t from the target value of 0. The deviation is unknown to the operator but it can be modeled by

$$\delta_t = \delta_{t-1} - x_t$$

Then the process with operator intervention is modeled by

$$x_t = \delta_{t-1} + \epsilon_t.$$

What is the effect of the operator intervention on the process variability?

Exercise 10.3: $\tilde{\sigma}$

Let X_1 and X_2 be independent $N(0, \sigma^2)$.

(a) What is the distribution of $D = X_1 - X_2$?

(b) What is the median of the distribution of $R = |D|$?

(c) Hence, justify the estimator $\tilde{\sigma}$ of σ for use in a runs chart given in Section 10.2.1.

Exercise 10.4: $\tilde{\sigma}$ **step change**

Write an R script to generate a sequence of 100 pseudo-random standard normal variates and calculate $\tilde{\sigma}$ for use in a runs chart as defined in Section 10.2.1 and also the sample variance (s^2) and standard deviation (s). Repeating the generation of 100 variates 10 000 times and compare the sampling distributions of $\tilde{\sigma}$ with that of S^2 and S. Now introduce some arbitrary step changes by for example:

```
> x=rnorm(100)
> sc=c(rep(0,25),rep(2,25),rep(-1,20),rep(3,10),rep(-2,20))
> y=x+sc
> plot(y)
```

and compare the sampling distributions of $\tilde{\sigma}$ and S.

Exercise 10.5: $\tilde{\sigma}$ **trend**

Suppose you have a sum of a sequence of n independent standard normal variables and a deterministic trend βt where $t = 1, \ldots, n$.

(a) What is the expected value of $\tilde{\sigma}$, as defined in Section 10.2.1, in terms of σ, β, n?

(b) What is the expected value of S^2 in terms of σ, β, n?

(c) Comment on the difference between $\tilde{\sigma}$ and S.

Exercise 10.6: $\tilde{\sigma}$ **fold** n

Write an R script to generate pseudo-random numbers from a folded normal distribution and so compare the sampling distributions of $\tilde{\sigma}$, as defined in Section 10.2.1, and S when sampling from a folded normal distribution.

Exercise 10.7: Inter-laboratory trial

A company produces a gold solution that is advertised as containing 20 g of potassium aurocyanide per liter. Customers sometimes send aliquots of the solution to test laboratories for assay. Following a customer complaint about the gold content, a manager decides to carry out an inter-laboratory trial.

The ideal for an international standard is a measurement procedure that will give the same result, for identical material, in any accredited laboratory with any competent technician. In practice the committees responsible for international standards realize that there will be slight differences between assays made by the same technician using the same apparatus on the same day, and somewhat larger, but still small, differences between assays made by different technicians in different labs. There is a standard for running inter-laboratory trials to investigate these discrepancies (ISO5725). The standard deviation of the assays by the same technician is taken as a measure of replicability, and the standard deviation of assays from different labs is a measure of reproducibility. The manager is particularly concerned about the reproducibility.

The laboratories work with 50 ml samples. The manager draws a random sample of 15 laboratories from a list of accredited laboratories in the country. She prepares 1.5 l of a thoroughly mixed gold solution and dispenses it into 50 ml bottles, 10 ml at a time. Two bottles are randomly assigned to each of the 15 labs. The results of the assays are given in the following table

Lab	Replicate1	Replicate2
1	21207	21276
2	21019	20996
3	21050	21045
4	21255	21255
5	21299	21281
6	21033	21093
7	21119	21174
8	21045	21077
9	21158	21142
10	21077	21017
11	21045	21033
12	20996	21038
13	21244	21341
14	21229	21012
15	21239	21341

(a) Estimate the within laboratory standard deviation.

(b) Estimate the between laboratory standard deviation.

(c) Hence estimate the overall standard deviation.

Section 10.3 Capability

Exercise 10.8: Zirconium silicate

A mineral processing company sells zirconium silicate in 7 tonne batches. The specification is that the moisture content is below 1%, and each batch is dried in a tumbler drier before being packaged in a moisture proof wrap.

Exercise 10.9: Slump

Set up a Shewhart mean and range chart for monitoring slump in Example 10.23. Plot the results for the five days and comment.

Exercise 10.10: Shewhart

A process produces brushes for electric motors. The brushes are made 4 at a time in a mold with 4 wells. The critical dimension is the length of the brush. The following data in Table 10.14 were recorded (0.1 mm above lower specification limit) from a random sample of 20 sets from the mold, but you don't know which well the four brushes are from. The upper specification limit on the scale is 150 (0.1 mm above the lower specification limit).

TABLE 10.14: Lengths (mm) of 20 sets of brushes made using a mold simultaneously producing 4 at a time.

sample	brush1	brush2	brush3	brush4	sample	brush1	brush2	brush3	brush4
1	69	33	59	36	11	104	8	25	44
2	88	7	61	62	12	68	25	18	32
3	58	7	54	25	13	77	28	33	58
4	71	38	59	15	14	65	33	62	43
5	61	25	31	27	15	62	0	52	54
6	70	1	65	31	16	45	31	29	60
7	78	22	42	46	17	64	5	60	77
8	76	37	26	65	18	121	35	70	48
9	52	56	54	46	19	71	35	58	39
10	78	23	80	58	20	27	29	33	39

(a) Plot Shewhart mean and range charts and comment on these.

(b) Assuming the process is in control estimate C_p and C_{pk}.

(c) You are now told that the four columns correspond to wells so brush 1 is from well 1 and similarly for brushes 2, 3 and 4. Analyze the data taking account of this additional information and advise the production engineer.

Exercise 10.11: Capacitors

A company manufactures 270 pF porcelain chip capacitors to tolerance code K. The capacitances are normally distributed with a mean of 273 and a standard deviation of 12.

(a) What proportion of the capacitors are within the specification of $[243, 297]$?

(b) Calculate the process performance index C_{pk} .

(c) Suppose the mean is adjusted to 270. To what must the standard deviation be reduced if 60 parts per million fall outside the specification?

(d) If the changes in the part above are made to the process, what is the process capability index C_p?

Exercise 10.12: Brake discs

Refer to Example 10.8

(a) Calculate a two-sided 90% confidence interval for C_{pu}.

(b) Calculate a one-sided 95% confidence interval for C_{pu} and give an explanation why this is appropriate.

(c) Calculate 95% confidence intervals for the process mean and standard deviation and comment in the context of the manufacturer's claim.

Section 10.4 Reliability

Exercise 10.13: Lifetimes

The lifetimes (T) of a certain manufacturer's starter motors have the following probability density function

$$f(t) \;=\; t/2, \quad \text{for } 0 \le t \le 2,$$

where t is measured in years (ignore leap years).

(a) Find the mean, median and standard deviation of the distribution.

(b) Find expressions for the cdf, survivor function, and the hazard.

(c) Today, your starter motor is one year old and still working. Use the hazard function to compute an approximation to the probability it is still working tomorrow and give the percentage error in this estimate.

Exercise 10.14: PSV

In the inter-laboratory trial the 16 laboratories were provided with specimens of the control roadstone from the stockpile. The laboratories were not told the source of the roadstone and the results are in Table 10.15. The PSV test requires that the stones be polished and polishing is carried out separately before the two tests so the entire test procedure is replicated.

(a) Plot the data.

(b) Calculate the replicability and the reproducibility.

(c) The PSV of the stockpiled roadstone is 52.5. Is there any evidence of bias in the PSV results from the laboratories?

TABLE 10.15: Roadstone specimen data.

Lab	Control-Run1	Control-Run2	Lab	Control-Run1	Control-Run2
1	55.50	54.70	9	52.00	52.00
2	50.35	48.65	10	51.00	52.50
3	52.15	52.15	11	49.00	49.15
4	53.50	53.20	12	48.00	49.15
5	59.00	58.20	13	54.15	55.00
6	54.35	52.35	14	53.00	53.80
7	55.50	51.15	15	56.35	53.80
8	49.50	53.00	16	49.15	51.15

Section 10.5 Acceptance sampling

Exercise 10.15:

Polycarbonate roofing sheets are of acceptance quality provided no more than 1% have flaws. Compare the following sampling schemes in terms of AOQL and the probability that a batch with 0.01 flawed sheets is rejected. Assume batches are large.

Scheme 1: Inspect a random sample of $n = 50$. If none is flawed then accept the batch, but if any are flawed then reject the batch.

Scheme 2: Inspect a random sample of $n = 50$. If none is flawed then accept the batch. If more than one is flawed reject the batch. If one is flawed then take a second sample of $n = 50$, and accept the batch if none is flawed in the second sample.

Section 10.6 Statistical quality control charts

Exercise 10.16: Benito

Two consecutive independent standard normal variables are denoted by Z_t and Z_{t+1}. Plot a graph of $P(Z_{t+1} > z_t)$ against z_t for $-3 < z_t < -1$. Comment on the graph in the context of Example 10.1. Can you think of other situations in which the same principle applies?

Exercise 10.17: Shewhart mean and range charts

A process produces capacitors. The target capacitance is 500 microfarads. From extensive past records, the standard deviation of capacitance can be assumed to be 12.0 microfarads (correct to one decimal place). The following data are random samples of 3 capacitors taken from 7 shifts.

503	508	519
505	492	506
526	492	502
505	521	511
517	483	515
511	518	510
513	536	518

(a) Draw a Shewhart mean and range chart.

(b) Comment on your charts.

Miscellaneous problems

Exercise 10.18: E[median(moving range)]

Suppose X_1 and X_2 are independent normal variables with mean μ and standard deviation σ. Define the difference

$$D \;=\; X_1 - X_2.$$

(a) Write down the mean, variance, and standard deviation of D.

(b) Is the distribution of D necessarily normal? You may quote a general result in your answer. A proof is not required.

(c) Now define the range (R) as the absolute value of D. That is:

$$R \;=\; |D|.$$

Sketch the distribution of D, and hence sketch the distribution of R.

(d) Use tables of areas under the standard normal distribution to find the median of R as a multiple of C.

(e) Hence express σ in terms of the median of R.

Exercise 10.19: Box plot

(a) Calculate the inter-quartile range (IQR) for a standard normal distribution.

(b) What is the probability a randomly drawn variate from a standard normal distribution will lie above the upper fence?

(c) Calculate the inter-quartile range (IQR) of an exponential distribution with mean 1.

(d) What is the probability a randomly drawn variate from this exponential distribution will lie above the upper fence?

Exercise 10.20: Autocorrelation

Ethernet data in the following table is from a LAN at Bellcore, Morristown, $\ln{(bytes/ms + 1)}$ in chronological order (first 24 data from a long time series). Call these data $\{x\}$.

4858	5020	562	726	466	516	832	470	600	4076	5986	670
726	3978	6190	450	762	742	446	580	644	446	644	696

(a) Plot x_{t+1} against x_t.

(b) Calculate: $\bar{x}, s$ and s^2.

(c) Calculate: $c(0), c(1)$ and $r(1)$.

Exercise 10.21: Process performance

BHC makes aluminium electrolytic capacitors. The ALS34 series includes a capacitor with a nominal capacitance of 680 000 microfarad with a specification from -10% to $+30\%$. Assume capacitance has a Weibull distribution, where the cdf of a Weibull distribution can be written

$$F(x) \;=\; 1 - exp(-((x - L)/b)^2) \quad \text{for } L \leq x.$$

Assume L is equal to 600 000, that a proportion of 0.1 of production is below 652 800 (-4% from nominal), and that a proportion 0.1 of production is above 761 600 ($+12\%$ from nominal).

(a) Determine the implied values of a and b.

(b) Plot the pdf of the distribution.

(c) How many parts per million (ppm) are below and above the lower and upper specification limits respectively?

(d) The mean and variance of a Weibull distribution are:

$$\mu = L + b\Gamma(1 + 1/a)$$
$$\sigma^2 = b^2 \left\{ \Gamma(1 + 2/a) - (\Gamma(1 + 1/a))^2 \right\}.$$

Calculate the mean, variance and standard deviation of the Weibull distribution describing capacitances.

(e) Assume the process runs with the mean and standard deviation you calculated in (d) and calculate the process performance index (C_{pk}).

11

Design of experiments with regression analysis

In a designed experiment we investigate the effects of predictor variables, that we can control and set to specific values, on some response. The design of the experiment is the choice of values for these predictor variables. The predictor variables are known as factors, and the chosen values for the factors are refered to as levels. This chapter deals with factorial experiments and their extensions including the central composite design. Multiple regression analysis is used for the analysis of these experiments. The strategy of experimenting by making small changes to a process during routine production, known as evolutionary operation (EVOP), is introduced.

Experiment E.7 Factorial experiment
Experiment E.10 Where is the summit?

11.1 Introduction

In an observational study, or survey, we take a sample from a population with the aim of investigating the distributions of variables and associations between variables. In contrast, in a designed experiment we typically think of a system with inputs and outputs. We can control some of the inputs and set them to specific values, either to investigate their effects on the outputs or, once we know these effects, to optimize the outputs. The design of an experiment is a set of values for the inputs, chosen to provide as much information about their effects on the outputs as is possible given practical constraints.

A statistical approach is required because the net effect of other inputs to the system, which cannot be measured and may not even be known, is modeled as independent random variation. Randomization helps make this modeling assumption reasonable.

The first steps in designing an experiment are to define the system, identify relevant responses (outputs), and list all the predictor variables (inputs) that we can think of. At this stage it is valuable to get contributions from all employees with relevant experience and to draw fish-bone diagrams. Then we plan and perform the experiment. We have already considered some simple experiments, which include:

- Comparison against some specified standard, based on a random sample from the population.

- Comparison of means of two populations: based on independent random samples; and based on a paired comparison.

- Comparison of variances of two populations: based on independent random samples.

- Comparison of proportions in two populations: based on independent random samples; and based on a paired procedure.

- Effect of a predictor variable on a response based on regression when we choose the values of the predictor variable.

- Comparison of means of several populations, based on independent samples from the populations analyzed by multiple regression using indicator variables.

- Estimation of components of variance.

In this chapter we extend these ideas. We define the following terms in the context of the design of experiments.

Definition 11.1: System or process

The system, or process, is the physical entity that we intend to study. It has input variables and output variables.

We can control the values of some of the input variables and these are known as factors or control variables. We can measure some of the other inputs but cannot control them, and these we refer to as concomitant variables. There will also be inputs to the system we know of but cannot measure and inputs to the system that we are unaware of, and their effects are attributed to random errors.

Example 11.1: Cement kiln [system inputs and response]

A cement kiln is an example of a system. The outputs are the quality of the cement produced and a particular response is the percentage of free lime (CaO). The response is affected by:

- factors, or control variables, include fuel and oxidant rates, rotation speed, feed rate of limestone meal into the kiln, speeds of fans, and setting of vents.
- concomitant variables are the water content and the chemical composition of the limestone meal, and ambient temperature and wind speed.
- variables that we know of but cannot measure are the precise temperature at all points in the kiln.

We return to this case in Example 11.10.

Definition 11.2: Response

A response is a variable that we need to control. The objective of the experiment is to investigate how other variables affect the response.

Definition 11.3: Factor

A factor is a variable which may have an effect on the response, and for which we can set specific values in an experiment. The values of the factor that we choose to consider in the experiment are known as **levels**.

Definition 11.4: Concomitant variable

A concomitant variable is a variable which may affect the response, but cannot be set at specific values in an experiment. It is however possible to monitor its value during an experiment.

Definition 11.5: Block

A block is a grouping of experimental material so that the material within blocks is relatively homogeneous.

Definition 11.6: Confounding

Predictor variables are confounded when it is not possible to distinguish their individual effects on the response.

Example 11.2: Alloy strength [overlooking confounding]

Anderson and McLean (1974) describe the case of a CEO who vowed to never use a designed experiment again. The reason was that the company had been misled into adopting a new process for making an alloy that had turned out to be inferior to the standard process. We now consider whether the CEO's decision is justified.

The tensile strength of test pieces made from ingots cast from one heat of the new process were compared with test pieces made from ingots cast from one heat of the standard process. The design was to take random sample of 10 ingots from each heat, make 5 test pieces from each ingot, and measure the tensile strength of the test pieces. The mean strength of the test pieces from the heat from the new process was statistically significantly higher than the mean strength of the test pieces from the heat from the standard process. The new process was adopted on the basis of this evidence.

The statistical analysis was correct, but the conclusion was seriously flawed. Variation between heats was confounded with the change in the process. There should have been at least two heats from each process for the comparison. Also, a check of records from past production would have indicated the extent of variability between heats of the standard process. The CEO's decision is not justified, as a well designed experiment would have estimated and allowed for the variation between heats.

There are some general principles (which we label from **P1** up to **P10**) for the design of an experiment which need to be considered at the start of the investigation. Anyone with knowledge or experience of the system, or of experimentation in general, may be able to offer useful advice at this stage.

P1: State the aims of the experiment.

P2: Define the response variable, or variables.

P3: Choose factors and levels.

P4: Keep everything else as constant as is possible subject to randomization, replication and blocking.

P5: Identify any concomitant variables and arrange to monitor them.

P6: Randomize experimental material to factor combinations. Randomize the order in which factor combinations are investigated, subject to any constraints.

P7: The design should include enough design points, replicated as necessary, to estimate effects with reasonable precision.

P8: Arrange for the experimental conditions and material to be representative of typical operating conditions. Blocking can be used to emulate the variation in typical operating conditions without disregarding the advice to keep everything other than the factors under investigation as constant as is possible.

P9: Consider possible confounding variables and ensure that they will not invalidate conclusions from the experiment.

P10: Allow for the possibility that the effect of one factor depends on the level of other factors (interactions between factors).

11.2   Factorial designs with factors at two levels

The experiment design consists of a list of combinations of levels for factors that will be set for the inputs to the process.

Definition 11.7: Run

Operating the process once with specific settings for factors is known as a run of the process.

11.2.1   Full factorial designs

If there are k factors, each at two levels, there are 2^k possible factor combinations. The reason for designing an experiment with a run at every factor combination is to investigate interactions. It is not satisfactory to design an experiment in which just one variable is changed at a time because it will give misleading results if there are interactions.

11.2.1.1   Setting up a 2^k design

There is a standard notation for these designs. Factors are denoted by capital Roman letters: $A, B, C, \ldots$. The two levels for each factor are coded as -1 and $+1$ and are typically referred to as low and high. The 2^k factor combinations are known as the design points.

Definition 11.8: Design points

The factor combinations that will be run in a designed experiment are known as design points.

The convention for a design point is that the lower case letter is used if the factor is at the high level, and 1 represents all factors at the low level (Table 11.1). The upper case letters serve both as the variable name and also as the variable[1] that takes values from $\{-1, +1\}$. The lower case letter combinations represent both design points and the values of the response at those design points. A run provides a value of the response at a particular design point.

Definition 11.9: Standard order

A list of the 2^k runs in a full factorial design with k factors in which A alternates between -1 and $+1$, B alternates between $-1-1$ and $+1+1$, C alternates between $-1-1-1-1$ and $+1+1+1+1$ and so on is known as the standard order.

It is convenient to enumerate runs in a standard order but the order of performing the runs within a replicate should be randomized if it is feasible to do so[2].

Example 11.3: Three factor full factorial design [runs in standard order]

Given three factors A, B and C, the full factorial design has the 2^3 points shown in standard order in Table 11.1. A single replicate of the 2^3 design has one run at each design point.

TABLE 11.1: Single replicate of 2^3 design in a standard order.

A	B	C	Design point
-1	-1	-1	1
$+1$	-1	-1	a
-1	$+1$	-1	b
$+1$	$+1$	-1	ab
-1	-1	$+1$	c
$+1$	-1	$+1$	ac
-1	$+1$	$+1$	bc
$+1$	$+1$	$+1$	abc

Standard order is a convenient way of setting up the design points in a 2^k factorial design. For example, design points for a 2^4 design are obtained by writing down the 2^3 design

---

[1]That is A represents the factor A and the variable that takes the value -1 if A is low and $+1$ if A is high, rather than introducing x_1, say, that takes the value -1 if A is low and $+1$ if A is high.

[2]An example where it might not be feasible is with kilns that have high thermal inertia and take many hours to reach a steady state if the temperature is changed.

for A, B, and C, and then adding a column for D as 8 −1s followed by 8 +1s.

Example 11.4: Emu Engineering [interaction]

Emu is a small engineering company which specializes in precision machining of ceramic circuit boards. In one process, it uses routers for cutting notches. An engineer is investigating the effects of drill speed and router diameter on the width of the notches.

P1: The aim is to control variation in the width of notches cut in the ceramic circuit boards.

P2: The main cause of variation in width is known to be vibration. The response is vibration (0.01g) measured with an accelerometer mounted on the ceramic circuit boards.

P3: Factor A is Bit Size and the two levels are 2 mm diameter and 4 mm diameter. Factor B is Rotation Speed and the two levels are 40 rpm and 90 rpm.

P4: A single replicate of the full factorial design has 2^2 design points. There will be 4 replicates. The first replicate will be performed by one operator at the start of the first shift of the week using new bits. The other three replicates will be by different operators on different days.

P5: No concomitant variables were identified.

P6: The most common job is to cut groves in a particular design of printed ceramic circuit board. A random sample of 16 of these boards was randomly allocated to the 16 runs.

P7: The experiment was replicated four times. If this did not provide sufficient precision it could be replicated more times. The order of the four runs within each day was randomized.

P8: The printed ceramic circuit boards used were a random sample from the process. The results of the experiment will be specific to the bit sizes, 2 mm and 4 mm, and rotation speeds, 40 rpm and 90 rpm, considered. The bit sizes are those used in routine production, but the rotation speed could be set anywhere between the two limits. It is unadvisable to assume a linear relationship between rotation speed and vibration, with a particular bit size, because there could be a resonant frequency in this range. This could be investigated with a follow up experiment.

P9: No confounding variables were known or anticipated.

P10: The analysis of factorial designs allows for interactions.

In this experiment A and B represent bit size and rotation speed respectively. The low level is coded −1 and the high level is coded +1. A full replicate includes all possible factor combinations, and in the case of two factors this equals 4. The overall design consists of four replicates of the full factorial design. The full factorial design is given in standard order. The order of performing the runs within each replicate was randomized as given in the columns titled "run order" in Table 11.2. The results of the experiment are given in the columns titled "vibe" of Table 11.2.

TABLE 11.2: Router vibration (0.01g) of printed ceramic circuit boards.

A	B	rep	run order	vibe	A	B	rep	run order	vibe
-1	-1	1	2	18.2	-1	-1	3	1	12.9
$+1$	-1	1	4	27.2	$+1$	-1	3	4	22.4
-1	$+1$	1	1	15.9	-1	$+1$	3	3	15.1
$+1$	$+1$	1	3	41.0	$+1$	$+1$	3	2	36.3
-1	-1	2	2	18.9	-1	-1	4	3	14.4
$+1$	-1	2	1	24.0	$+1$	-1	4	1	22.5
-1	$+1$	2	3	14.5	-1	$+1$	4	2	14.2
$+1$	$+1$	2	4	43.9	$+1$	$+1$	4	4	39.9

11.2.1.2 Analysis of 2^k design

We use a regression model to analyze the results from 2^k experiments. The predictor variables are categorized as main effects and interactions.

- Single factors A, B, C, D, E and so on. The **main effect** of A, for example, is defined as the mean of responses from all the runs when A is $+1$ less the mean of responses from all the runs when A is -1. The main effect of A is twice the estimated coefficient of A in the regression model. A itself is referred to as a main effect term.

- 2-factor products AB, AC, BC, and so on are known as the **2-factor interaction** terms. For example, the 2-factor interaction term AB allows the effect of A to depend on the value of B and vice-versa. The 2-factor interaction of AB is defined as the mean response when $AB = +1$ less the mean response when $AB = -1$, and is twice the coefficient of AB in the regression.

- 3-factor products ABC, ABD, ACD, and so on are known as the **3-factor interaction** terms. The 3-factor term ABC allows for the 2-factor interaction effects to depend on the level of third factor. The 3-factor interaction ABC, for example, is defined as the mean response when $ABC = +1$ less the mean response when $ABC = -1$, and is twice the coefficient of ABC in the regression.

- Higher order interactions are defined in a similar fashion. The k-factor interaction is the highest order interaction that can be included as a predictor variable in a 2^k design. The higher order interactions are typically seen to be, or assumed to be, negligible.

Definition 11.10: Design matrix

The matrix X in the regression model is known as the design matrix .

It follows from the patterns of the -1s and $+1$s in the definitions of the predictor variables that all the columns in the design matrix, other than the first which is a column of 1s, have mean 0 and that the sum of the products of row entries in any two columns of the design matrix is 0. So, the predictor variables all have mean 0, and any two are uncorrelated.

Definition 11.11: Orthogonal design

An experimental design is orthogonal if $X'X$ is a diagonal matrix. If $X'X$ is diagonal so too is $(X'X)^{-1}$ and estimators of coefficients are uncorrelated.

In particular, 2^k designs are orthogonal. The practical consequence of this is that estimates of main effects and interactions are unchanged when other main effects or interactions are added to, or removed from, the regression model. However, the associated standard errors of these estimates will in general change.

Definition 11.12: Saturated model

A saturated model is one in which the number of parameters to be estimated equals the number of data.

In the case of a linear model the saturated model will give an exact fit to the data and there will be no degrees of freedom for estimating the standard deviation of errors. Nevertheless, since all the predictor variables are standardized it is possible to check whether or not the higher order interactions are negligible compared with main effects and 2-factor interactions, by comparing the absolute magnitude of coefficients. If high order interactions are omitted from the model, their effects are confounded with the errors.

Example 11.5: Regression coefficients for 2^2 design [regression analysis]

We demonstrate the relationship between the estimates of the regression coefficients and differences in mean values for a 2^2 design.

$$Y = \begin{pmatrix} 1 \\ a \\ b \\ ab \end{pmatrix}, \quad X = \begin{pmatrix} +1 & -1 & -1 & +1 \\ +1 & +1 & -1 & -1 \\ +1 & -1 & +1 & -1 \\ +1 & +1 & +1 & +1 \end{pmatrix} \quad \text{and} \quad \beta = \begin{pmatrix} \beta_0 \\ \beta_1 \\ \beta_2 \\ \beta_3 \end{pmatrix}.$$

The saturated model is

$$Y = X\beta$$

and the estimates of the coefficients are

$$\widehat{\beta} = (X'X)^{-1}X'y = \begin{pmatrix} \frac{1}{4} & 0 & 0 & 0 \\ 0 & \frac{1}{4} & 0 & 0 \\ 0 & 0 & \frac{1}{4} & 0 \\ 0 & 0 & 0 & \frac{1}{4} \end{pmatrix} \begin{pmatrix} 1 + a + b + ab \\ -1 + a - b + ab \\ -1 - a + b + ab \\ 1 - a - b + ab \end{pmatrix}.$$

Notice that $\widehat{\beta}_0$ is the mean of the 4 observed responses. Next

$$\widehat{\beta}_1 = \frac{-1 + a - b + ab}{4} = \frac{1}{2}\left(\frac{a + ab}{2} - \frac{1 + b}{2}\right),$$

which is one half of the difference between the mean of the two observations when A

is at $\boxed{+1}$ and the mean of the two observations when A is at $\boxed{-1}$.

$$
Y = \begin{pmatrix} \boxed{1} \\ \boxed{a} \\ \boxed{b} \\ \boxed{ab} \end{pmatrix}, \qquad X = \begin{pmatrix} +1 & \boxed{-1} & -1 & +1 \\ +1 & \boxed{+1} & -1 & -1 \\ +1 & \boxed{-1} & +1 & -1 \\ +1 & \boxed{+1} & +1 & +1 \end{pmatrix}.
$$

That is, $\widehat{\beta}_1$ is one half the main effect of A. Similarly $\widehat{\beta}_2$ is one half the main effect of B and $\widehat{\beta}_3$ is one half the interaction effect of AB, where the interaction effect is defined as the difference between the mean of the two observations when AB is at $+1$ and the mean of the two observations when AB is at -1. That is

$$
\hat{\beta}_3 = \frac{1}{2}\left(\frac{1+ab}{2} - \frac{1+b}{2}\right).
$$

The interaction effect can also be defined as one-half of the difference between the estimate of the effect of A when B is high, and the estimate of the effect of A when B is low (and vice-versa).

$$
\frac{1}{2}\left((ab - b) - (a - 1)\right) = \frac{1}{2}\left((ab - b) - (b - 1)\right).
$$

If we fit a saturated model we have no degrees of freedom for error. We can estimate the standard deviation of the errors if we either replicate the 2^k design, or assume the higher order interactions are negligible. We use both strategies in the following examples.

Example 11.4: (Continued) Emu Engineering

The results from the experiment are in Table 11.2. We begin by reading the data and plotting vibration against bit size and rotation

```
router.dat=read.table("Router.txt",header=TRUE)
#A is bit size B is rotation speed
attach(router.dat)
print(head(router.dat))
plot(B,vibe,pch=(A+2),xlab="Speed",ylab="Vibration")
legend(0.5, 42, inset=.05, title="Bit size",c("-1","+1"),pch=c(1,3))
```

The practical conclusions follow from the plot (Figure 11.1). The vibration is around 15 at either speed with the small bit. With the larger bit, vibration is around 25 at the lower speed and 40 at the higher speed. The interaction effect between bit size and rotation speed is particularly noticeable. The regression analysis in R follows. Within the `lm()` function, the syntax $(A + B)^2$ is an instruction to fit all the main effects and 2-factor interactions of the factors in the sum in the bracket[3].

---

[3]Within `lm()`, $(A + B)^2$ does not include quadratic terms, even when A and B are at more than two levels and quadratic terms can be fitted. If a factor is at more than two levels then A^2 can be defined as a variable, $AA = A^2$ say, and included in the model formula, or the inhibit interpretation AsIs function can be used `I(A^2)`.

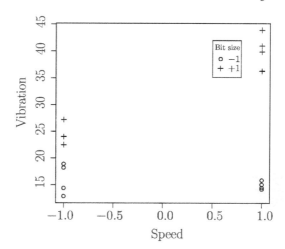

FIGURE 11.1: Router experiment: vibration against speed with bit size.

```
> m1=lm(vibe~(A+B)^2)
> summary(m1)
Call:
lm(formula = vibe ~ (A + B)^2)
Residuals:
   Min    1Q Median    3Q    Max
-3.975 -1.550 -0.200  1.256  3.625
Coefficients:
            Estimate Std. Error t value Pr(>|t|)
(Intercept)  23.8312     0.6112  38.991 5.22e-14 ***
A             8.3187     0.6112  13.611 1.17e-08 ***
B             3.7687     0.6112   6.166 4.83e-05 ***
A:B           4.3562     0.6112   7.127 1.20e-05 ***
---
Residual standard error: 2.445 on 12 degrees of freedom
Multiple R-squared:  0.9581,    Adjusted R-squared:  0.9476
F-statistic: 91.36 on 3 and 12 DF,  p-value: 1.569e-08
> m1$fit
     1      2      3      4      5      6      7      8      9     10
16.100 24.025 14.925 40.275 16.100 24.025 14.925 40.275 16.100 24.025
    11     12     13     14     15     16
14.925 40.275 16.100 24.025 14.925 40.275
```

A summary of the experiment is shown in Table 11.3. The estimated mean vibration is below 20 for the 2 mm bit at both 40 rpm and 90 rpm. The estimated mean vibration with the 4 mm bit is substantially higher at 90 rpm than at 40 rpm. The lower rotation speed is recommended with the larger bit size and the estimated mean vibration is the 24. The 2^2 design had to be replicated to investigate the interaction which turned out to be crucial. With 4 replicates there were 16 data, 4 parameters to be estimated and 12 degrees of freedom for error. The estimated standard deviation of the vibration of individual ceramic circuit boards, for a specific bit size and rotation speed, is 2.445.

TABLE 11.3: Router experiment: summary of 4 replicates of 2^2 design.

Bit size (mm)	Rotation speed rpm	Vibration (0.01g)
2	40	16.1
4	40	24.0
2	90	14.9
4	90	40.3

Our next case investigates 6 factors, each at 2 levels, in a 2^6 factorial design.

Example 11.6: SeaDragon aluminum wheels [case study for replicated 2^6]

SeaDragon manufactures aluminum wheels for the automotive sector (Figure 11.2). The process begins with melting the aluminum in a furnace at a temperature around 650° Celsius. A degassing procedure, in which a flux composed of chlorine and fluorine salts is added to the melt to reduce the hydrogen content, is applied. A robot then moves a measure of molten aluminum into the lower part of molds that are in two presses either side of the furnace. After the lower parts of the molds are filled the upper parts are lowered and pressure is applied. The molds are cooled on the exterior and open automatically when the aluminum solidifies. The wheels are conveyed to a water batch for quenching and then pass through an X-ray tomography station, where the porosity is assessed robotically, before any finishing processes such as painting. The quality of the wheels supplied is high, but this is achieved at the expense of recycling nearly 20% of the wheels produced. The main reason for recycling wheels is excessive porosity, and SeaDragon is keen to reduce the proportion of the wheels that are recycled. The production engineer decides to investigate possible causes of the porosity.

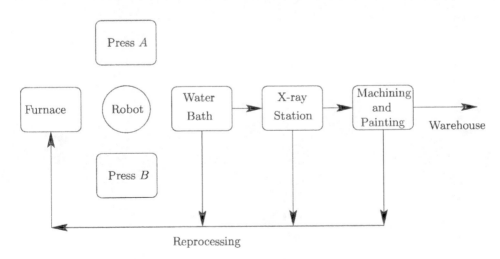

FIGURE 11.2: Flow chart for aluminum wheels.

P1: The objective of the experiment is to identify factors that influence porosity of wheels and so reduce the proportion of wheels recycled due to excessive porosity.

P2: The response variable is the porosity measurement provided by the X-ray tomography. The measurement is on a scale from 0 to 10. A wheel is recycled if the porosity exceeds 3.60. The process mean and standard deviation of porosity are 3.35 and 0.25 respectively.

P3: Before choosing factors and levels, the engineer asked other workers who were familiar with the process for their opinions about the likely causes of high porosity. Their contributions were added to a cause and effect diagram (Figure 11.3).

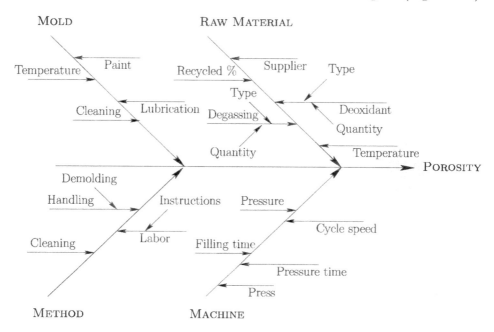

FIGURE 11.3: Cause and effect diagram.

She also reviewed the plant records on wheel quality. A list of factors was drawn up and is reproduced in Table 11.4. There are two presses and two types of lining for the mold. One lining is made in-house, and the other is from an external supplier. The production engineer and the company metallurgist agreed that the temperature and pressure could be safely and feasibly set at the low and high values shown in the table. The process could be run with or without the degas procedure and with or without recycled material.

TABLE 11.4: Porosity of aluminum wheels: 6 factors at 2 levels.

Label	Description	Low (-1)	High ($+1$)
A	Press	Left of furnace	Right of furnace
B	Lining of mold	In-house	External
C	Degas	No	Yes
D	Temperature	635°C	685°C
E	Pressure	800 kPa	1000 kPa
F	Recycled material	0%	20%

P4: Keep everything else as constant as is possible subject to randomization, replication and blocking.

P5: No concomitant variables were identified.

P6: It is not practical to continually adjust the furnace temperature between pressing wheels or to apply degas and different amounts of recycled material for single wheels so all the low temperature runs without degas and without recycled material were performed before the low temperature without degas with recycled material, and so on before the high temperature runs. Also, the two presses were in use at the same time. The pressure and mold lining levels were randomized.

P7: If the $2^6 = 64$ experiment is replicated once, then an effect will be estimated from a comparison of independent means of 32 wheels. The standard deviation of a wheel is assumed to be 0.25, so the standard deviation of an estimated effect is

$$\sqrt{\frac{0.25^2}{32} + \frac{0.25^2}{32}} = 0.0625$$

This was considered adequate as the engineer anticipated effects greater than 0.1. Also, a confirmatory experiment would be performed before making changes to the established process.

P8: Aluminum ingots were obtained from the usual smelter.

P9: The effects of temperature, degas, and recycle are confounded with time over the period of the experiment, but there is no reason to suppose that time will have any effect and a confirmatory experiment is planned,

P10: The analysis of the full factorial experiment allows for interactions between factors.

The results are shown in Table 11.5. First a saturated model is fitted[4] and a quantile-quantile plot of the coefficients, other than the intercept, is drawn (Figure 11.4 left frame). It is apparent from the plot that there are four substantial effects and these can be identified from the model m0 summary as C, D, E and the CE interaction. None of the higher order interactions has a noticeably high coefficient (this can be inferred from the plot and we only show an excerpt from the R output) so the next stage is to fit a model with just the main effects and 2-factor interactions.

```
> wheels.dat=read.table("WheelsFullFac.txt",header=TRUE)
> attach(wheels.dat)
> head(wheels.dat)
  Press Lining Degas Temp Pressure Recycle Porosity
1   -1     -1    -1   -1       -1      -1     5.13
2   -1     -1     1   -1       -1      -1     3.76
3   -1      1    -1   -1       -1      -1     5.28
4   -1      1     1   -1       -1      -1     3.97
5    1     -1    -1   -1       -1      -1     5.41
6    1     -1     1   -1       -1      -1     4.04
> A=Press; B=Lining; C=Degas; D=Temp; E=Pressure; F=Recycle;
+ y=Porosity
> m0=lm(y~A*B*C*D*E*F)
> summary(m0)

Call:
```

---

[4]Within `lm()` the syntax $A * B * C * D * E * F$ gives all possible products from single terms up to $ABCDEF$.

TABLE 11.5: Sea Dragon aluminum wheels experiment

Press	Lining	Degas	Temp	Pressure	Recycle	Porosity		Press	Lining	Degas	Temp	Pressure	Recycle	Porosity
−1	−1	−1	−1	−1	−1	5.13		−1	−1	−1	+1	−1	−1	4.65
−1	−1	+1	−1	−1	−1	3.76		−1	−1	+1	+1	−1	−1	2.93
−1	+1	−1	−1	−1	−1	5.28		−1	+1	−1	+1	−1	−1	4.91
−1	+1	+1	−1	−1	−1	3.97		−1	+1	+1	+1	−1	−1	3.10
+1	−1	−1	−1	−1	−1	5.41		+1	−1	−1	+1	−1	−1	4.32
+1	−1	+1	−1	−1	−1	4.04		+1	−1	+1	+1	−1	−1	2.96
+1	+1	−1	−1	−1	−1	5.45		+1	+1	−1	+1	−1	−1	4.56
+1	+1	+1	−1	−1	−1	3.80		+1	+1	+1	+1	−1	−1	3.10
−1	−1	−1	−1	−1	+1	5.45		−1	−1	−1	+1	−1	+1	4.19
−1	−1	+1	−1	−1	+1	3.88		−1	−1	+1	+1	−1	+1	2.50
−1	+1	+1	−1	−1	+1	3.58		−1	+1	−1	+1	−1	+1	4.09
−1	+1	−1	−1	−1	+1	5.35		−1	+1	+1	+1	−1	+1	3.01
+1	−1	−1	−1	−1	+1	5.64		+1	−1	−1	+1	−1	+1	4.39
+1	−1	+1	−1	−1	+1	3.68		+1	−1	+1	+1	−1	+1	3.08
+1	+1	−1	−1	−1	+1	5.56		+1	+1	−1	+1	−1	+1	4.66
+1	+1	+1	−1	−1	+1	3.63		+1	+1	+1	+1	−1	+1	2.94
−1	−1	−1	−1	+1	−1	3.61		−1	−1	−1	+1	+1	−1	3.00
−1	−1	+1	−1	+1	−1	3.43		−1	−1	+1	+1	+1	−1	2.66
−1	+1	−1	−1	+1	−1	3.97		−1	+1	−1	+1	+1	−1	2.95
−1	+1	+1	−1	+1	−1	3.91		−1	+1	+1	+1	+1	−1	3.02
+1	−1	−1	−1	+1	−1	3.63		+1	−1	−1	+1	+1	−1	3.05
+1	−1	+1	−1	+1	−1	3.69		+1	−1	+1	+1	+1	−1	2.83
+1	+1	−1	−1	+1	−1	3.90		+1	+1	−1	+1	+1	−1	2.95
+1	+1	+1	−1	+1	−1	3.64		+1	+1	+1	+1	+1	−1	2.83
−1	−1	−1	−1	+1	+1	3.72		−1	−1	−1	+1	+1	+1	2.88
−1	−1	+1	−1	+1	+1	3.63		−1	−1	+1	+1	+1	+1	3.01
−1	+1	−1	−1	+1	+1	3.86		−1	+1	−1	+1	+1	+1	3.22
−1	+1	+1	−1	+1	+1	3.72		−1	+1	+1	+1	+1	+1	3.33
+1	−1	−1	−1	+1	+1	3.45		+1	−1	−1	+1	+1	+1	3.37
+1	−1	+1	−1	+1	+1	4.08		+1	−1	+1	+1	+1	+1	2.79
+1	+1	−1	−1	+1	+1	3.65		+1	+1	−1	+1	+1	+1	3.27
+1	+1	+1	−1	+1	+1	3.71		+1	+1	+1	+1	+1	+1	3.28

```
lm(formula = y ~ A * B * C * D * E * F)

Residuals:
ALL 64 residuals are 0: no residual degrees of freedom!

Coefficients:
Estimate Std. Error t value Pr(>|t|)
(Intercept)  3.766250        NA      NA      NA
A            0.025625        NA      NA      NA
B            0.052500        NA      NA      NA
C           -0.406250        NA      NA      NA
D           -0.396562        NA      NA      NA
```

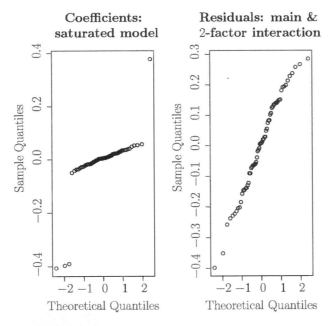

FIGURE 11.4: Wheel quality test results.

E	−0.390000	NA	NA	NA
F	0.002500	NA	NA	NA
A:B	−0.036250	NA	NA	NA
A:C	−0.005625	NA	NA	NA
B:C	−0.001875	NA	NA	NA
A:D	0.003438	NA	NA	NA
B:D	0.029062	NA	NA	NA
C:D	−0.002813	NA	NA	NA
A:E	−0.019375	NA	NA	NA
B:E	0.021875	NA	NA	NA
C:E	0.377500	NA	NA	NA
D:E	0.047812	NA	NA	NA
A:F	0.029375	NA	NA	NA
B:F	−0.017500	NA	NA	NA
C:F	0.003125	NA	NA	NA
D:F	0.003437	NA	NA	NA
E:F	0.056875	NA	NA	NA
A:B:C	−0.028125	NA	NA	NA
.				
.				
.				
A:B:C:D:E:F	0.040313	NA	NA	NA

```
Residual standard error: NaN on 0 degrees of freedom
Multiple R-squared:     1,     Adjusted R-squared:    NaN
F-statistic:   NaN on 63 and 0 DF,  p-value: NA

> par(mfrow=c(1,2))
```

```
> qqnorm(m0$coef[2:64],main="coefficients: sat model")
```

Since any 2^k design is orthogonal the coefficients of the main effects and 2-factor interactions will not change, but we now have their standard deviations (standard errors) and residuals from the model which are shown in the normal quantile-quantile plot (Figure 11.4 right frame).

```
> m1=lm(y~(A+B+C+D+E+F)^2)
> summary(m1)

Call:
lm(formula = y ~ (A + B + C + D + E + F)^2)

Residuals:
Min        1Q     Median      3Q       Max
-0.39813 -0.12391  0.00781  0.13500   0.28500

Coefficients:
Estimate Std. Error t value Pr(>|t|)
(Intercept)  3.766250   0.024387 154.440   <2e-16 ***
A            0.025625   0.024387   1.051   0.2994
B            0.052500   0.024387   2.153   0.0371 *
C           -0.406250   0.024387 -16.659   <2e-16 ***
D           -0.396562   0.024387 -16.262   <2e-16 ***
E           -0.390000   0.024387 -15.992   <2e-16 ***
F            0.002500   0.024387   0.103   0.9188
A:B         -0.036250   0.024387  -1.486   0.1446
A:C         -0.005625   0.024387  -0.231   0.8187
A:D          0.003438   0.024387   0.141   0.8886
A:E         -0.019375   0.024387  -0.794   0.4314
A:F          0.029375   0.024387   1.205   0.2351
B:C         -0.001875   0.024387  -0.077   0.9391
B:D          0.029062   0.024387   1.192   0.2401
B:E          0.021875   0.024387   0.897   0.3748
B:F         -0.017500   0.024387  -0.718   0.4770
C:D         -0.002812   0.024387  -0.115   0.9087
C:E          0.377500   0.024387  15.480   <2e-16 ***
C:F          0.003125   0.024387   0.128   0.8986
D:E          0.047812   0.024387   1.961   0.0566 .
D:F          0.003437   0.024387   0.141   0.8886
E:F          0.056875   0.024387   2.332   0.0246 *
---
 Residual standard error: 0.1951 on 42 degrees of freedom
Multiple R-squared:  0.9619,    Adjusted R-squared:  0.9428
F-statistic: 50.45 on 21 and 42 DF,  p-value: < 2.2e-16

> qqnorm(m1$res,main="residuals: main & 2-factor int")
```

The standard errors of the coefficients are 0.024, so the standard error of effects is $2 \times 0.024 = 0.048$. The standard error is smaller than the 0.0625 calculated before performing the experiment because the standard deviation of the residuals, 0.195, turned out to be lower than the assumed standard deviation of porosity 0.250. This may be a consequence of the more carefully controlled experimental conditions.

The conclusion is that the estimated response will be lowest when C (degas), D (temperature), E (pressure) are set at $+1$, but the $C : E$ interaction effect is almost as large as the two main effects and acts against them, the lining $B = -1$, and recycled F is set to the opposite sign of E to take account of the interaction. The effect of A (press) is not statistically significant at even a 0.20 level. The practical advice is to set the temperature to the high value of 685°C, the pressure to the high value of 1000 kPa, consider whether the expense of the degas procedure is worth the estimated reduction of 0.06, and use the in-house lining for the mold. Both presses are in continuous use so there is no decision to be made about factor A. The proportion of recycled material will naturally be reduced if the new settings are as successful as the analysis suggests. The production engineer decided to change to high temperature and pressure, and to monitor the process closely with and without degas and with the two types of lining for the mold.

```
> #future no degas
> newdat=data.frame(A=0,B=-1,C=-1,D=1,E=1,F=0)
> predict(m1,newdata=newdat,interval=c("confidence"))
fit      lwr      upr
1 2.95375 2.790526 3.116974
> #future degas
> newdat=data.frame(A=0,B=-1,C=1,D=1,E=1,F=0)
> predict(m1,newdata=newdat,interval=c("confidence"))
fit      lwr      upr
1 2.894375 2.731151 3.057599
```

If a mean porosity of 3.00 can be achieved, and the standard deviation is unchanged at 0.25, there would be around 1% of wheels recycled due to porosity.

```
> 1-pnorm(3.60,3.00,0.25)
[1] 0.008197536
```

The next example involves replicating the factorial design because one of the responses is variability at the factor combinations[5].

Example 11.7: Elk electronics [mean and sd responses]

The epitaxial layer is a mono-crystalline film deposited on the substrate of wafers used for the manufacture of integrated circuit devices. In this experiment the deposition process takes place in a bell jar which is infused with gases. The wafer is mounted on a disc at the base of the jar, and the disc is either rotated or oscillated. The objective is to minimize the variability of the epitaxial layer about a target depth of 14.5 microns. The response is the measured depth of the epitaxial layer. Four factors are investigated, each at two levels, as given in Table 11.6.

TABLE 11.6: Depth of epitaxial layer: 4 factors at 2 levels, 6 replicates.

Label	Description	Low (−1)	High (+1)
A	Disc motion	Rotating	Oscillating
B	Nozzle position	Low	High
C	Deposition temperature	1210°C	1220°C
D	Deposition time	17 minutes	18 minutes

TABLE 11.7: ELK: design points for experiment – 6 replicates at each point.

A	B	C	D	Ybar	SD
−1	−1	−1	−1	13.860	0.070711
+1	−1	−1	−1	13.972	0.347851
−1	+1	−1	−1	14.165	0.063246
+1	+1	−1	−1	14.032	0.296648
−1	−1	+1	−1	13.880	0.031623
+1	−1	+1	−1	13.907	0.475395
−1	+1	+1	−1	14.037	0.044721
+1	+1	+1	−1	13.914	0.264575
−1	−1	−1	+1	14.821	0.054772
+1	−1	−1	+1	14.932	0.463681
−1	+1	−1	+1	14.888	0.054772
+1	+1	−1	+1	14.878	0.383406
−1	−1	+1	+1	14.757	0.054772
+1	−1	+1	+1	14.415	0.453872
−1	+1	+1	+1	14.921	0.126491
+1	+1	+1	+1	14.843	0.571839

A full factorial design is composed of 2^4 design points (first four columns of Table 11.7).

The design of the experiment consists of six replicates so that the variability can be assessed. Two responses are considered: the mean depth of the 6 measured depths at each factor combination, and the logarithm of the standard deviation of the 6 measured depths at each factor combination.

A saturated model is fitted for the mean response, and the only coefficients that are large in absolute magnitude correspond to the main effects. The model with only main effects is

```
> epitaxial.dat=read.table("Epitaxial.txt",header=TRUE)
> attach(epitaxial.dat)
> head(epitaxial.dat)
   A  B  C  D   Ybar       SD
1 -1 -1 -1 -1 13.860 0.070711
2  1 -1 -1 -1 13.972 0.347851
3 -1  1 -1 -1 14.165 0.063246
4  1  1 -1 -1 14.032 0.296648
5 -1 -1  1 -1 13.880 0.031623
```

---

[5]We overlook the fact that different variability infringes the assumption that the errors have constant variance.

```
6  1 -1  1 -1 13.907 0.475395
> meanmod-lm(Ybar~A+B+C+D)
Error: object 'meanmod' not found
> summary(meanmod)
Error in summary(meanmod) : object 'meanmod' not found
> meanmod=lm(Ybar~A+B+C+D)
> summary(meanmod)

Call:
lm(formula = Ybar ~ A + B + C + D)

Residuals:
     Min       1Q    Median       3Q      Max
 -0.23913 -0.03931   0.01500  0.04744  0.16862

Coefficients:
            Estimate Std. Error t value Pr(>|t|)
(Intercept) 14.38887    0.02734 526.317  < 2e-16 ***
A           -0.02725    0.02734  -0.997    0.340
B            0.07088    0.02734   2.592    0.025 *
C           -0.05462    0.02734  -1.998    0.071 .
D            0.41800    0.02734  15.290 9.32e-09 ***
---
Residual standard error: 0.1094 on 11 degrees of freedom
Multiple R-squared:  0.9571,    Adjusted R-squared:  0.9415
F-statistic: 61.37 on 4 and 11 DF,  p-value: 1.882e-07
```

Including interactions increases the estimated standard deviation of the errors from 0.1094 to 0.1328 with a loss of 6 degrees of freedom.

```
> summary(lm(Ybar~(A+B+C+D)^2))
Call:
lm(formula = Ybar ~ (A + B + C + D)^2)
.
.
Residual standard error: 0.1328 on 5 degrees of freedom
```

There is no evidence of any interaction effects on the mean deposition depth. The mean depth can be controlled by adjusting the deposition time. The higher nozzle position appears to give a slight increase increase in depth and there is weak evidence that the lower temperature does too, so these would be preferred values, because throughput will increase as the deposition time decreases, provided they don't have an adverse effect on variability.

But, before making recommendations we consider a model for the logarithm of the standard deviation[6]. A model that includes 2-factor interactions has a slightly lower estimated standard deviation of the errors, 0.290 on 5 degrees of freedom, than the model with only main effects, 0.328 on 11 degrees of freedom, but none of the individual interaction terms is statistically significant at even the 20% level, so we just show the simpler model.

```
> lnsdmod=lm(log(SD)~A+B+C+D)
> summary(lnsdmod)

Call:
lm(formula = log(SD) ~ A + B + C + D)

Residuals:
    Min      1Q   Median      3Q      Max
-0.44959 -0.19539 -0.07501  0.18934  0.58347

Coefficients:
            Estimate Std. Error t value Pr(>|t|)
(Intercept) -1.88580    0.08207 -22.978 1.20e-10 ***
A            0.95824    0.08207  11.676 1.54e-07 ***
B            0.02298    0.08207   0.280    0.785
C            0.01638    0.08207   0.200    0.845
D            0.15363    0.08207   1.872    0.088 .
---
Residual standard error: 0.3283 on 11 degrees of freedom
Multiple R-squared:  0.9271,    Adjusted R-squared:  0.9006
F-statistic: 34.99 on 4 and 11 DF,  p-value: 3.384e-06
```

Normal quantile-quantile plots of the residuals from the two models are shown in Figure 11.5, and there are no outlying values. The conclusion is clear, to reduce the variability set A to -1 which corresponds to rotation of the disc rather than oscillation. The coefficient of factor D, deposition time, is statistically significant at the 10% level ($p = 0.088$), which is unsurprising because an increase in standard deviation of depth, which is proportional to deposition time, is expected. The final recommendations are given in Table 11.8.

The deposition time was set by first estimating the mean depth when $D = 0$.

```
> newdat=data.frame(A=-1,B=1,C=-1,D=0)
> predict(meanmod,newdata=newdat)
       1
14.54162
```

---

[6]The reason for using the logarithm of standard deviation, rather than standard deviation as a response is that it is less influenced by outlying values. It also gives a response that is not bounded below at 0. Using the logarithm of the variance is equivalent to using twice the logarithm of standard deviation, and will lead to the same conclusions.

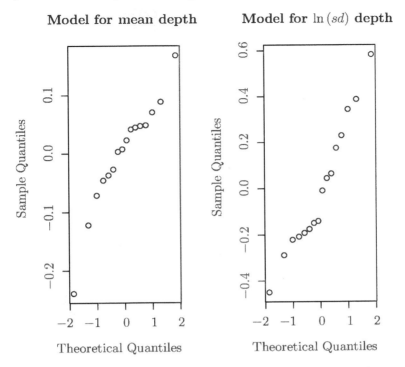

Model for mean depth **Model for $\ln(sd)$ depth**

FIGURE 11.5: Normal quantile-quantile plots of the residuals from the two models.

TABLE 11.8: Depth of epitaxial layer: recommended process settings.

Factor	Recommended value	Reason
Disc motion	Rotating	reduce variability
Nozzle position	High	slight reduction in time
Deposition temperature	1210°C	slight reduction in time and heating cost
Deposition time	17 min 27 s	achieve target depth of 14.5 microns

The target value is 14.5 so we need to lower the mean by 0.04162. The coefficient of D is 0.418 so D needs to be set to -0.10. $D = 0$ corresponds to 17 minutes and 30 seconds so $D = -0.10$ corresponds to 17 minutes and 27 seconds. We can also calculate the predicted process standard deviation at the recommended settings. The prediction for the expected value of the logarithm of the standard deviation is

```
> newdat=data.frame(A=-1,B=1,C=-1,D=-.1)
> predict(lnsdmod,newdata=newdat,interval="confidence",level=0.95)
      fit       lwr       upr
1 -2.852807 -3.214528 -2.491085
```

If we take exponential of this quantity, and assume the distribution of the logarithm of standard deviation is symmetric, the expected value of the median of the distribution of standard deviation, and the upper limit of the 95% confidence interval are

```
> exp(-2.852807)
[1] 0.05768218
> exp(-2.491)
[1] 0.0828271
```

If we also assume that the distribution of standard deviation is lognormal the adjustment factor for the mean rather than the median is

```
> fac=exp(.2898^2/2)
> fac
[1] 1.042886
```

which makes little difference. The best estimate of the standard deviation is 0.06 and we can be reasonably confident that it will be less than 0.10. Nevertheless, we should closely monitor the process under the recommended settings to confirm that the low standard deviation predicted by the experiment is achieved in routine production.

There are two limitations to full factorial experiments. The first is that the number of runs can become excessive if there are many factors. The second is that linear interpolation between the low and high value of a continuous factor may not be justified. In the case of depth of epitaxial layer there is good reason to assume a linear relationship with deposition time but in other cases there may be quadratic effects. A potential solution to the second limitation is to have three levels for each factor but this exacerbates the first. We consider better strategies in the next two sections.

11.3   Fractional factorial designs

If the number of factors, k, is large the number of runs for even a single replicate of a full factorial experiment 2^k is excessive. Moreover, the interactions that involve many factors are likely to be negligible. If we forgo estimating the interactions that involve the greatest

TABLE 11.9: Selecting half factorial design, 2^{3-1}, with generator $ABC = +1$.

A	B	C	AB	AC	BC	ABC	Design point	run
-1	-1	-1	$+1$	$+1$	$+1$	-1	1	
$+1$	-1	-1	-1	-1	$+1$	$+1$	a	✓
-1	$+1$	-1	-1	$+1$	-1	$+1$	b	✓
$+1$	$+1$	-1	$+1$	-1	-1	-1	ab	
-1	-1	$+1$	$+1$	-1	-1	$+1$	c	✓
$+1$	-1	$+1$	-1	$+1$	-1	-1	ac	
-1	$+1$	$+1$	-1	-1	$+1$	-1	bc	
$+1$	$+1$	$+1$	$+1$	$+1$	$+1$	$+1$	abc	✓

number of factors we can reduce the number of runs. We illustrate the general principle with a 2^3 design. The 3-factor interaction is estimated by the difference between the four runs for which $ABC = +1$ and the four runs for which $ABC = -1$. A half-fraction of the

2^3 design, denoted by 2^{3-1} consists of either the four runs for which $ABC = +1$ or the four runs for which $ABC = -1$. There is generally no reason to favor either the $+1$ or -1 alternative, and we'll chose the former. The **design generator** is $ABC = +1$, and we obtain the four runs checked in Table 11.9. We now look at the four selected runs in more detail in Table 11.10. The main effect of A can be estimated by

$$\frac{a + abc}{2} - \frac{b + c}{2}$$

and main effects of B and C can be estimated in a similar way. The mean is estimated as

$$\frac{a + b + c + abc}{4}$$

and this completes the fitting of the saturated model. No interactions can be fitted.

Definition 11.13: Alias

A factor, or interaction of several factors, is aliased with another factor, or interaction, if they correspond to identical design points. The effects of aliased factors or interactions cannot be distinguished because they are represented by the same variable. Aliased factors or interactions are confounded.

If you look at Table 11.10 you will see that for any run the level of A is identical to the level BC. That is, the columns of -1s and $+1$s for A and for BC are the same. We then write $A = BC$. The column BC is itself the element-by-element product[7] of the column of B and the column of C. Similarly, $B = AC$ and $C = AB$. In the 2^{3-1} design BC is an alias of A. The main effect A is confounded with the 2-factor interaction BC.

TABLE 11.10: Half factorial design, 2^{3-1}, with generator $ABC = +1$.

A	B	C	AB	AC	BC	design point
$+1$	-1	-1	-1	-1	$+1$	a
-1	$+1$	-1	-1	$+1$	-1	b
-1	-1	$+1$	$+1$	-1	-1	c
$+1$	$+1$	$+1$	$+1$	$+1$	$+1$	abc

It follows that the coefficient of A could represent half the main effect of A or half the interaction effect BC or a linear combination of the two effects. The aliases of effects can be found by simple algebra because the (element by element) product of a column with entries that are all ± 1 with itself is a column of 1s.

$$ABC = 1 \Rightarrow AABC = A1 \Rightarrow A^2BC = A \Rightarrow 1BC = A \Rightarrow BC = A.$$

The 2^{3-1} design would only be useful if it is reasonable to assume that 2-factor interactions are negligible, and it is generally better to avoid making any such assumption. Fractional factorial designs are more useful when there are at least four factors.

Definition: Design generator

A design generator is a constraint on the levels of factors that can be set to give a design

---

[7]Element-by-element, or element-wise, multiplication is $A * B$ in R and $A \cdot *B$ in MATLAB.

point. In the case of a 2^k design the design generator that the product of factor levels equals 1, or minus 1, gives a half-fraction, a 2^{k-1} design, of the 2^k design. Two design generators would lead to a quarter-fraction, a 2^{k-2} design, and m design generators would lead to a 2^{k-m} design (Exercise 11.6).

Example 11.8: Wombat Welding [2^{5-1} design]

A small engineering company Wombat Welding specializes in spot welding. The manager, and business owner, has bought a second spot welding machine and decides to run an experiment to compare it with the older machine, which was manufactured by a different company. The manager is also interested in investigating the effects of four other factors:

A: button (tip of welding rod) diameter,

B: weld time,

C: hold time,

D: electrode force, and

E: whether old machine or new machine.

She decides to test the four factors A, B, C, D at low and high levels and defines levels of factor E as low for the old machine and high for the new machine. The design is a single run of a 2^{5-1} design, with the design generator

$$ABCDE = -\mathbf{1}.$$

It follows that main effects will be aliased with 4-factor interactions because

$$A(ABCDE) \;=\; A(-\mathbf{1}) \Rightarrow BCDE \;=\; -A$$

and similarly for the other factors. Also 2-factor interactions will be aliased with 3-factor interactions because

$$AB(ABCDE) \;=\; AB(-\mathbf{1}) \Rightarrow A^2B^2CDE \;=\; -AB \Rightarrow CDE \;=\; -AB$$

and so on. The set of alias relations is known as the **alias structure** of the design. The response is the tensile strength of test pieces measured as the force (N) required to break the welded joint. The results are given in Table 11.11. The order of runs in the table corresponds to a full factorial for B, D, D, E with A set to give $ABCDE = -1$, but the runs were performed in a random order[8].

We begin the analysis by fitting the saturated model, which includes all 2-factor interactions, and plotting a normal quantile-quantile plot of the coefficients (Figure 11.6 left panel), other than the intercept.

```
> Wombat.dat=read.table("Wombat.txt",header=TRUE)
> attach(Wombat.dat)
> head(Wombat.dat)
   A  B  C  D  E strength
```

---

[8]A random order can be obtained in R, for example, using `set.seed(42)`; `sample(16)`.

TABLE 11.11: Tensile strength of test pieces.

A	B	C	D	E	run order	strength
−1	−1	−1	−1	−1	15	1 330
+1	+1	−1	−1	−1	16	1 935
+1	−1	+1	−1	−1	5	1 775
−1	+1	+1	−1	−1	11	1 275
+1	−1	−1	+1	−1	8	1 880
−1	+1	−1	+1	−1	6	1 385
−1	−1	+1	+1	−1	12	1 220
+1	+1	+1	+1	−1	2	2 155
+1	−1	−1	−1	+1	13	1 715
−1	+1	−1	−1	+1	14	1 385
−1	−1	+1	−1	+1	3	1 000
+1	+1	+1	−1	+1	4	1 990
−1	−1	−1	+1	+1	7	1 275
+1	+1	−1	+1	+1	1	1 660
+1	−1	+1	+1	+1	10	1 880
−1	+1	+1	+1	+1	9	1 275

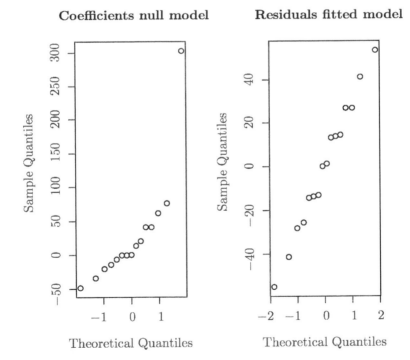

Coefficients null model   **Residuals fitted model**

FIGURE 11.6: Normal quantile-quantile plots of the residuals.

```
1 -1 -1 -1 -1 -1      1330
2  1  1 -1 -1 -1      1935
3  1 -1  1 -1 -1      1775
4 -1  1  1 -1 -1      1275
```

```
5  1 -1 -1  1 -1      1880
6 -1  1 -1  1 -1      1385
> n=length(strength)
> m0=lm(strength~(A+B+C+D+E)^2)
> par(mfrow=c(1,2))
> qqnorm(m0$coef[2:n],main="Coefficients null model")
> summary(m0)

Call:
lm(formula = strength ~ (A + B + C + D + E)^2)

Residuals:
ALL 16 residuals are 0: no residual degrees of freedom!

Coefficients:
              Estimate Std. Error t value Pr(>|t|)
(Intercept) 1570.9375         NA      NA       NA
A            302.8125         NA      NA       NA
B             61.5625         NA      NA       NA
C              0.3125         NA      NA       NA
D             20.3125         NA      NA       NA
E            -48.4375         NA      NA       NA
A:B           -0.3125         NA      NA       NA
A:C           75.9375         NA      NA       NA
A:D           -0.3125         NA      NA       NA
A:E          -14.0625         NA      NA       NA
B:C           40.9375         NA      NA       NA
B:D          -34.0625         NA      NA       NA
B:E           -6.5625         NA      NA       NA
C:D           40.9375         NA      NA       NA
C:E           13.4375         NA      NA       NA
D:E          -20.3125         NA      NA       NA

Residual standard error: NaN on 0 degrees of freedom
Multiple R-squared:      1,       Adjusted R-squared:     NaN
F-statistic:   NaN on 15 and 0 DF,  p-value: NA
```

There is one outstanding effect associated with factor A. We now fit a model including the main effects and the four interactions with coefficients that have the largest absolute values. Since fractional factorial designs are orthogonal the coefficients will be unchanged.

```
> m1=lm(strength~A+B+C+D+E+A:C+B:C+B:D+C:D)
> qqnorm(m1$res,main="Residuals fitted model")
> summary(m1)

Call:
lm(formula = strength ~ A + B + C + D + E + A:C + B:C + B:D +
    C:D)

Residuals:
    Min     1Q  Median     3Q     Max
-55.000 -17.187   0.625  17.500  53.750
```

```
Coefficients:
              Estimate Std. Error t value Pr(>|t|)
(Intercept) 1570.9375    11.7911 133.231 1.21e-11 ***
A            302.8125    11.7911  25.681 2.30e-07 ***
B             61.5625    11.7911   5.221 0.001974 **
C              0.3125    11.7911   0.027 0.979716
D             20.3125    11.7911   1.723 0.135718
E            -48.4375    11.7911  -4.108 0.006301 **
A:C           75.9375    11.7911   6.440 0.000663 ***
B:C           40.9375    11.7911   3.472 0.013273 *
B:D          -34.0625    11.7911  -2.889 0.027736 *
C:D           40.9375    11.7911   3.472 0.013273 *
---
Residual standard error: 47.16 on 6 degrees of freedom
Multiple R-squared:  0.9924,    Adjusted R-squared:  0.9809
F-statistic: 86.73 on 9 and 6 DF,  p-value: 1.166e-05
```

There is little doubt about the importance of the main effects of A, and B. The 2-factor AC interaction, between button diameter and hold time, also seems important but it is aliased with the 3-factor interaction BDE. A physical interpretation of this 3-factor interaction is that the 2-factor BD interaction between weld time and force differs between the old and new machine. The BD interaction has a non-negligible coefficient so the 3-factor interaction is quite plausible. Taking the other half of the 2^5 design, with the generator $ABCDE = +1$, for the follow up experiment would enable the independent estimation of the AC and BDE interactions, if the results from the two experiments are combined for the analysis. The statistical significance of the other effects is dependent on the choice of which interactions to include in the model, and we recommend a follow up experiment. The conclusions from this experiment are summarized in Table 11.12 The standard deviation of the strength of welds on test pieces is estimated to be 47. So the estimated effect of using the higher force, 41, is close to one standard deviation of strength and would be worthwhile if it can be confirmed with a follow up experiment.

11.4 Central composite designs

The critical limitation of 2^k designs is that quadratic terms are not included. If a factor is defined over a continuous scale, interpolation is based on an assumption of a linear response over the range of the variable. One remedy would be to have all factors at three levels and use 3^k designs, but the number of design points soon becomes excessive and one-third fractions are awkward to deal with. A better solution is to use a **central composite design**. A central composite design for k factors is given by a 2^k design, with each factor at ± 1 for low and high level, augmented with a star design. The star design has:

- one point with all the factors at 0;

- two points for each factor with that factor set very low and very high, and all the other factors set at 0.

So, such a central composite design has $2^k + 2k + 1$ design points and the cases of $k = 2$ and $k = 3$ are shown in Figure 11.7.

TABLE 11.12: Wombat Welding: conclusions for half factorial experiment.

Factor	Description	Coefficient	Effect	Recommendation
A	button diameter	303	606	use larger diameter
B	weld time	62	123	use longer weld time
C	hold time	0.3	0.6	see interaction
D	force	20	41	investigate further
E	machine	−48	−97	investigate further
A : C	AC interaction	76	152	longer hold time? caution AC alias BDE
B : C	BC interaction	41	82	(beneficial)
B : D	BD interaction	−34	−68	prefer smaller force
C : D	CD interaction	41	84	prefer larger force

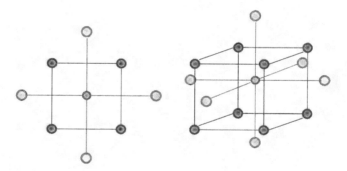

FIGURE 11.7: Central composite designs $k = 2$ (left), $k = 3$ (right).

We now consider the detail of central composite designs.

- Very low and very large are $\pm d$ where, generally, $1 \le d \le \sqrt{k}$. If $d = \sqrt{k}$ all the design points are the same distance from the origin (Pythagoras's Theorem in k dimensions), which is neat. However, in some situations it may not be safe, or practical, to allow factors to take values below −1 or above 1, in which case $d = 1$.

- For $k = 2$ the central composite design has the same number of points, 9, as a 3^k design, but for $k = 3$ the number of points are 15 and 27 respectively, and for $k = 4$ the number of points are 25 and 81 respectively. A 3^4 is a good design if you need the precision that 81 runs will provide, but it is a poor use of resources if a single replicate of the central composite design will suffice.

- Central composite designs are not generally precisely orthogonal. The coefficients of remaining quadratic terms, and the intercept, will change slightly if some of them are dropped from the model. This is not a practical limitation.

- The design point at the center is sometimes replicated to give an estimate of the standard deviation of the errors that is independent of the model fitted.

- A fractional factorial design can be used with a star design, in which case the number of runs is reduced to $2^{k-m} + 2k + 1$.

- Some of the factors may be restricted to two values rather than being defined on a continuous scale. In this case the factorial design is augmented by a star design for the factors that can be set to center, very high, and very low values.

- The correlations between estimators of coefficients in the model for the analysis of any experiment can be calculated before the experiment is run, because they do not depend on the response. The design defines the values of the predictor variables which are typically main effects, 2-factor interactions, and quadratic effects if factors are at more than two levels. Thus X is known in advance and the correlations follow from the covariance matrix which is proportional to $(X'X)^{-1}$.

Example 11.9: Electrophoresis [central composite $k = 3$ design]

[Morris et al., 1997] consider an experiment for optimizing the separation of ranitinide hydrochloride from four related compounds by electrophoresis[9]. Screening experiments had identified three factors as important:

- the pH of the buffer solution (A),
- the voltage (B), and
- the concentration of alpha-cyclodextrin (C).

The very low, low, center, high and very high levels were set at $-1.68, -1, 0, +1, +1.68$ respectively, 1.68 being just slightly less than the square root of 3. The physical values of the factors corresponding to these levels are shown in Table 11.13. The objective is to minimize the response, the logarithm of CEF. The design points in a convenient stan-

TABLE 11.13: Ratinide hydrochloride central composite design: factors and levels.

| | | \multicolumn{5}{c}{Level} | | | | |
		-1.68	-1	0	1	1.68
	pH A	2	3.42	5.5	7.58	9
Factor	voltage (kV) B	9.9	14	20	26	30.1
	alpha-cyclodextrin (mMole) C	0	2	5	8	10

dard order, the randomized run order, and the responses are shown in Table 11.14. We read in the data for the analysis and we can check the correlations between estimators in a model that includes main effects, 2-factor interactions, and quadratic terms.

```
> Pharma.dat=read.table("Pharma.txt",header=TRUE)
> attach(Pharma.dat)
> head(Pharma.dat)
    Stdord Runord  A  B  C    CEF  ln.CEF.
```

---

[9]Ratinide hydrochloride is the active ingredient in a medication to treat stomach ulcers.

```
1       1      4 -1 -1 -1     17.3    2.85
2       2     10  1 -1 -1     45.5    3.82
3       3     11 -1  1 -1     10.3    2.33
4       4      3  1  1 -1  11757.1    9.37
5       5     16 -1 -1  1     16.9    2.83
6       6      8  1 -1  1     25.4    3.23
> n=length(A)
> AA=A*A;BB=B*B;CC=C*C;AB=A*B;AC=A*C;BC=B*C
> X=matrix(cbind(rep(1,n),A,B,C,AA,BB,CC,AB,AC,BC),ncol=10)
> XTXI=solve(t(X)%*%X)
> print(round(XTXI,4))
        [,1]   [,2]  [,3]  [,4]   [,5]    [,6]    [,7]  [,8]  [,9]  [,10]
 [1,]  0.166 0.000 0.000 0.000 -0.057 -0.057 -0.057 0.000 0.000 0.000
 [2,]  0.000 0.073 0.000 0.000  0.000  0.000  0.000 0.000 0.000 0.000
 [3,]  0.000 0.000 0.073 0.000  0.000  0.000  0.000 0.000 0.000 0.000
 [4,]  0.000 0.000 0.000 0.073  0.000  0.000  0.000 0.000 0.000 0.000
 [5,] -0.057 0.000 0.000 0.000  0.070  0.007  0.007 0.000 0.000 0.000
 [6,] -0.057 0.000 0.000 0.000  0.007  0.070  0.007 0.000 0.000 0.000
 [7,] -0.057 0.000 0.000 0.000  0.007  0.007  0.070 0.000 0.000 0.000
 [8,]  0.000 0.000 0.000 0.000  0.000  0.000  0.000 0.125 0.000 0.000
 [9,]  0.000 0.000 0.000 0.000  0.000  0.000  0.000 0.000 0.125 0.000
[10,]  0.000 0.000 0.000 0.000  0.000  0.000  0.000 0.000 0.000 0.125

> corQQ=XTXI[6,7]/(sqrt(XTXI[6,6])*sqrt(XTXI[7,7]))
> print(corQQ)
[1] 0.09832097
```

TABLE 11.14: Electrophoresis responses.

Stdord	Runord	A	B	C	CEF	$\ln(CEF)$
1	4	-1	-1	-1	17.3	2.85
2	10	1	-1	-1	45.5	3.82
3	11	-1	1	-1	10.3	2.33
4	3	1	1	-1	11 757.1	9.37
5	16	-1	-1	1	16.9	2.83
6	8	1	-1	1	25.4	3.23
7	18	-1	1	1	31 697.2	10.36
8	5	1	1	1	12 039.2	9.40
9	12	-1.68	0	0	16 548.7	9.71
10	20	1.68	0	0	26 351.8	10.18
11	15	0	-1.68	0	11.1	2.41
12	2	0	1.68	0	6.7	1.90
13	17	0	0	-1.68	7.5	2.01
14	7	0	0	1.68	6.3	1.84
15	19	0	0	0	9.9	2.29
16	13	0	0	0	9.6	2.26
17	6	0	0	0	8.9	2.18
18	9	0	0	0	8.8	2.17
19	14	0	0	0	8.0	2.08
20	1	0	0	0	8.1	2.09

The design is not precisely orthogonal. The estimators of the coefficients of the quadratic terms are slightly correlated amongst themselves, 0.098 for any pair, and with the estimator of the intercept. It follows that the intercept will change as quadratic terms are added or removed, and the coefficient of AA, for example, will change slightly if BB and CC are removed from the model rather than being included. However, this is not a practical concern. The fitted model[10] is

```
> m1=lm(ln.CEF.~(A+B+C)^2+AA+BB+CC)
> summary(m1)

Call:
lm(formula = ln.CEF. ~ (A + B + C)^2 + AA + BB + CC)

Residuals:
     Min       1Q    Median       3Q       Max
-2.84588 -0.71305  0.01977  0.65922   2.31428

Coefficients:
            Estimate Std. Error t value Pr(>|t|)
(Intercept)   2.1552     0.7552   2.854 0.017139 *
A             0.6039     0.5013   1.205 0.256122
B             1.3099     0.5013   2.613 0.025915 *
C             0.5251     0.5013   1.047 0.319597
AA            2.8983     0.4886   5.932 0.000145 ***
BB            0.1382     0.4886   0.283 0.783070
CC            0.0567     0.4886   0.116 0.909910
A:B           0.5887     0.6547   0.899 0.389673
A:C          -1.0713     0.6547  -1.636 0.132848
B:C           1.0837     0.6547   1.655 0.128872
---
Residual standard error: 1.852 on 10 degrees of freedom
Multiple R-squared:  0.8363,    Adjusted R-squared:  0.6889
F-statistic: 5.675 on 9 and 10 DF,  p-value: 0.006038
> par(mfrow=c(2,2))
> plot(A,m1$res)
> plot(B,m1$res)
> plot(C,m1$res)
> qqnorm(m1$res)
```

The main recommendation is to set A to minimize

$$0.6039A + 2.8983A^2,$$

which is at $A = -0.104$. The coefficients of B and C suggest setting temperature and alpha-cyclodextrin to very low, but the coefficients have substantial standard errors so we would recommend further runs with B and C in the range $[-1.68, 0]$ before changing the standard process.

---

[10] If AA and BB are dropped from the model the coefficient of AA becomes 2.8808.

Example 11.10: Kookaburra Kilns [central composite $2^{6-1} + 2 \times 6 + 4$]

Portland cement is a closely controlled chemical combination of calcium, silicon, aluminum, iron and small amounts of other ingredients to which gypsum is added in the final grinding process. The main process is burning limestone in a large rotating kiln. The target range for the free lime (calcium oxide) in the cement is between 1% and 2%.

P1: The aim of this experiment was to fit an equation for predicting the free lime content from the values of six variables that can be set to chosen values to control the operation of the kiln and two variables that are measured on the limestone that is input to the kiln, water content and a measure of chemical composition referred to as burnability. Neither the water content nor the burnability can be controlled, but they are known to affect the free lime content and they can be monitored.

The equation was to be incorporated in an expert system[11] which would measure the water content and burnability of the limestone before it enters the kiln; and set the control factors to suitable values to maintain free lime at a target value of 1.5%, subject to maximizing profit. The profit depends on the throughput and fuel costs [Norman and Naveed, 1990].

P2: The response variable is the percentage free lime.

P3: The factors are shown in a Table 11.15.

All the factors can be set on a continuous scale and took values: $-2, -1, 0, +1, +2$ corresponding to very low, low, center, high and very high. The standard settings for the kiln were the center values. The design was a 2^{6-1} augmented by a star design with 4 center points, a required a total of $2^5 + 6 \times 2 + 4 = 48$ runs

P4: Keep everything else as constant as is possible.

P5: The concomitant variables were water content and burnability. The concomitant variables were crucial for this application.

P6: The runs were carried out in a random order.

P7: The design has 48 runs. The kiln is operated continuously, with three shifts in a 24 hour day, and the lime content is monitored for each shift. It was convenient to set the process up for a run at the beginning of a shift and maintain the settings throughout the shift. The standard deviation of line content from shifts was known to be around 0.15, so the standard error of an effect would be around:

$$\sqrt{\frac{0.15^2}{24} + \frac{0.15^2}{24}} = 0.043.$$

if the concomitant variables were not associated with the lime content, and potentially lower as the concomitant variables are expected to have an effect on the lime content. The standard errors of the coefficients of main effects in the model would be less than 0.024% and this was considered adequate for the initial model in the expert system. The expert system could update the regression equation during use.

P8: The limestone was sourced from a single quarry close to the kiln.

---

[11] A robotic system that can operate without operator interaction. The system includes the algorithm coded in a computer together with computerized sensors and activators.

P9: No confounding variables were identified.

P10: 2-factor interactions are aliased with 4-factor interactions, and the latter were considered negligible.

The design is with star points extending as far as -2 and 2 in the scaled units. Notice that $2 < \sqrt{6}$ so all the design points are not the same distance from the origin. The design is not exactly orthogonal[12] because the correlation between any two quadratic terms is 0.06. Also, the concomitant variables are likely to be correlated with the control factors[13]. The data are given on the website and the beginning of the file is listed in the R output. The free lime y is in parts per million (ppm), and the water content and burnability have been centered. The free lime is plotted against water content and against burnability in Figure 11.8. The free lime appears to increase with water content and to as lesser extent with burnability.

TABLE 11.15: Cement kiln control factors.

A	Feed rate
B	Rotation speed
C	Fuel to oxidant ratio
D	Fuel rate
E	Fan 1 speed
F	Span 2 speed

We fit a regression including all main effects and squared terms for the 6 factors and the 15 2-factor interactions and linear and quadratic effects of burnability and water content together with their interaction[14].

```
> kiln.dat=read.table("Kiln.txt",header=TRUE)
> attach(kiln.dat)
> print(head(kiln.dat))
   A  B  C  D  E  F burn water     y
1 -1  1  1 -1  1  1  -19     5 21092
2 -1 -1  1 -1 -1  1  -17    44 20073
3  1 -1 -1 -1 -1  1   -5    28 15772
4  0  0  2  0  0  0   14   -14 18411
5 -1 -1 -1 -1 -1 -1   -2    -1 18846
6 -1 -1 -1 -1  1  1    3    37 21363
> AA=A*A;BB=B*B;CC=C*C;DD=D*D;EE=E*E;FF=F*F
> m1=lm(y~(A+B+C+D+E+F)^2+AA+BB+CC+DD+EE+FF+burn+water+water:burn
+ +I(water^2)+I(burn^2))
> summary(m1)

Call:
lm(formula = y ~ (A + B + C + D + E + F)^2 + AA + BB + CC + DD +
    EE + FF + burn + water + water:burn + I(water^2) + I(burn^2))
```

---

[12]It is not exactly orthogonal even if the star points do extend to $\pm\sqrt{6}$.

[13]The largest such correlation, between D and water content, turned out to be -0.25. This is of no practical consequence since it is clear that both D and water content should be retained in the model.

[14]interactions between either burnability or water content and the factors $A, B, \ldots, F$ were negligible and have been included in the error on 15 degrees of freedom.

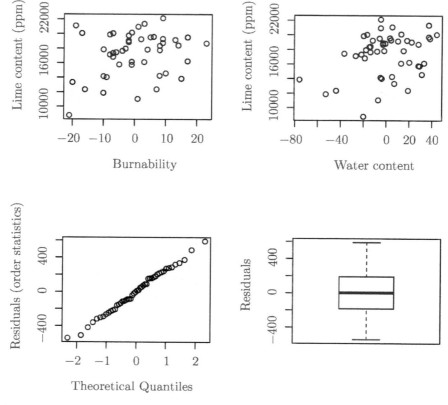

FIGURE 11.8: Kiln: scatter plots of free lime content (ppm) against burnability and water content (upper), normal score plot of residuals from regression model (lower left), box plot of residuals (lower right).

```
Residuals:
     Min      1Q  Median      3Q      Max
 -542.94 -171.41    2.99  185.28   590.43

Coefficients:
              Estimate Std. Error t value Pr(>|t|)
(Intercept) 18170.4606   229.2949  79.245  < 2e-16 ***
A            -2080.2388    94.1504 -22.095 7.42e-13 ***
B             -232.6608    73.9003  -3.148 0.006629 **
C              815.5342    75.2456  10.838 1.72e-08 ***
D            -1697.2679    89.2337 -19.020 6.52e-12 ***
E              913.4601    86.8853  10.513 2.58e-08 ***
F              446.9768    79.0159   5.657 4.56e-05 ***
AA            -237.7687   113.6283  -2.093 0.053807 .
BB             -82.2581    89.9149  -0.915 0.374746
CC            -455.4216    85.8165  -5.307 8.79e-05 ***
DD             -59.1908   109.7418  -0.539 0.597551
EE            -208.9922    89.8050  -2.327 0.034368 *
FF            -384.7774    91.0839  -4.224 0.000736 ***
```

burn	106.5220	11.5672	9.209	1.46e-07	***
water	37.6621	4.6914	8.028	8.25e-07	***
I(water^2)	0.3115	0.1429	2.180	0.045566	*
I(burn^2)	1.0278	1.0306	0.997	0.334445	
A:B	224.4928	107.0408	2.097	0.053329	.
A:C	-80.8264	92.0325	-0.878	0.393665	
A:D	-545.8210	82.4877	-6.617	8.19e-06	***
A:E	29.2251	95.1856	0.307	0.763039	
A:F	124.2015	105.4807	1.177	0.257349	
B:C	31.8874	83.6183	0.381	0.708295	
B:D	-643.2878	97.7496	-6.581	8.72e-06	***
B:E	203.3790	90.7175	2.242	0.040514	*
B:F	256.5799	91.5694	2.802	0.013405	*
C:D	71.7123	88.9363	0.806	0.432650	
C:E	128.2020	88.0971	1.455	0.166212	
C:F	104.5182	90.5747	1.154	0.266578	
D:E	126.3724	95.0270	1.330	0.203434	
D:F	1.4533	91.8456	0.016	0.987584	
E:F	-6.0782	98.2127	-0.062	0.951469	
burn:water	1.2880	0.5269	2.445	0.027327	*

```
---
Residual standard error: 452.4 on 15 degrees of freedom
Multiple R-squared:  0.9926,    Adjusted R-squared:  0.9767
F-statistic: 62.64 on 32 and 15 DF,  p-value: 2.847e-11
```

There are no outlying points in a normal quantile-quantile plot (Figure 11.8) of the residuals from this full quadratic model, and the adjusted R^2 is around 0.98. The estimated standard deviation of the errors is 452 ppm (0.045%), which is small when compared with the specified range of (1%, 2%).The model was implemented in the expert system.

11.5 Evolutionary operation (EVOP)

The yield of a pharmaceutical product is known to depend on the pressure and temperature inside the reactor. Both variables can be controlled quite precisely and the specified values have been 160 kPa and 190 degrees Celsius for as long as anyone can remember. The chemistry of the reaction is well understood and it is known that the reaction will proceed, and that it will be safe to operate the plant, with changes in temperature and pressure up to at least 10%. A recently appointed process engineer, Bianca Bradawl, suggests experimenting with small changes in temperature and pressure during routine production with the aim of finding settings that give the highest yield. The plant manager is interested but skeptical because this was how the specified values were obtained.

Bianca explains that previous experiments might have relied on a strategy of changing one variable at a time, which might not lead to the optimum settings. Her explanation was based on a possible scenario for the dependence of yield on temperature and pressure shown in Figure 11.9. The figure shows contours of yield against temperature and pressure and is an example of a **response surface**. In this scenario the pressure and temperature interact, so the effect of changing pressure depends on the temperature and vice-versa.

In practice the response surface is unknown and the aim of the experiment is to infer its shape. In Figure 11.9 the investigator sets the pressure to 150 and changes the temperature from 180 to 200 in steps of 1 degree. Random variation is low, and the investigator obtains a highest yield of 32% at a temperature of 190. The investigator now sets the temperature at 190 and changes the pressure from 140 to 170 in steps of 1 and finds that yield increases with pressure up to 160 (shown as 0 in Figure 11.9), but then declines. At this stage the conclusion would be to operate the plant at a pressure of 160 and a temperature of 190, and expect a yield of 34%. The potential optimum yield of 48% would be missed. It would be possible to find the optimum if the investigation continued by fixing the pressure at 160 and again varying the temperature, and so on. However, factorial designs offer a far more efficient experimental program. We describe a strategy in general terms for two factors A and B and a response y.

Step 1. The principle is that if we are some way from the peak of the response surface then a plane will provide a reasonable local approximation to the surface. Perform a 2^2 factorial experiment centered on the current operating conditions. Estimate the plane by fitting the regression model:

$$Y_i = \beta_0 + \beta_1 A + \beta_2 B + \varepsilon_i.$$

If β_1 and β_2 are significantly different from 0, proceed to Step 2. If β_1 and β_2 are not significantly different from 0 then replicate the 2^2 experiment and fit a plane to the combined results. If β_1 and β_2 are still not significantly different from 0 proceed to Step 3.

Step 2. The direction of steepest ascent is to change A and B in proportion to the absolute values of the estimated coefficients and in the direction of increase in y (Exercise 11.12). Move the centre of the 2^2 design to a point in the direction of steepest ascent. If the factors A and B are standardized, changing the factor with the coefficient that has the larger absolute value by 1 standardized unit will give a reasonable distance for the move.

Step 3. If the 2^2 design is centered near the peak of the surface, fitting a plane will not be helpful. The 2^2 design is augmented with a star design to give a central composite design. A quadratic surface is fitted to the results from the central composite design as it can provide a close approximation to the response surface. Use the quadratic surface to identify the location of the peak and the expected value of the response at this point. If the peak is located within 1 standardized unit of the center of the central composite design, confirm the finding with another replicate of the central composite design. If the location of the peak is more than 1 standardized unit away from the center of the central composite design move 1 standardized unit in the direction of steepest ascent and perform another central composite design.

The three steps are just guidelines, and the detail can be modified to suit the application. The same principles apply if the aim is to minimize the response, and the strategy can be used with more than two factors.

The manager was convinced by Bianca's explanation and was keen to carry out the experiment, having been assured that it would not disrupt production. The principle of experimenting by making small changes to factors during routine production is known as **evolutionary operation** (EVOP) or, more colloquially, as hill-climbing.

Example 11.11: Manufacturing medicines [EVOP]

Bianca started with a 2^2 design centered on the specified values of 160 and 190 for pressure and temperature respectively. The results are shown in Table 11.16.

TABLE 11.16: EVOP for pharmaceutical product - Experiment 1.

Pressure A (5 kPa from 160 kPa)	Temperature B ($5°C$ from $190°C$)	Yield y %
-1	-1	43.2
1	-1	44.9
-1	1	43.7
1	1	46.2

The relevant excerpt from the R analysis is

```
> A=c(-1,1,-1,1);B=c(-1,-1,1,1)
> y=c(43.2,44.9,43.7,46.2)
> m1=lm(y~A+B)
> summary(m1)
Coefficients:
            Estimate Std. Error t value Pr(>|t|)
(Intercept)   44.50      0.20    222.50  0.00286 **
A              1.05      0.20      5.25  0.11983
B              0.45      0.20      2.25  0.26625
```

Bianca decided that it would be prudent to obtain smaller standard errors for the effects before moving from the specified values and decided to augment the 2^2 design with a star design. The additional points and the associated yields are given in Table 11.17. Bianca combined the results from the two experiments to give a central composite

TABLE 11.17: EVOP for pharmaceutical product - Experiment 2.

Pressure A (5 kPa from 160 kPa)	Temperature B ($5°C$ from $190°C$)	Yield y %
-1.4	0	41.9
1.4	0	43.9
0	-1.4	42.1
0	1.4	45.1
0	0	43.8

design and fitted both a quadratic model and a linear model, a plane.

```
> A=c(A,-1.4,1.4,0,0,0);B=c(B,0,0,-1.4,1.4,0)
> y=c(y,41.9,43.9,42.1,45.1,43.8)
> m2q=lm(y~A*B+I(A^2)+I(B^2))
> summary(m2q)
```

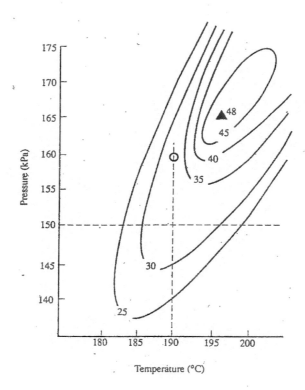

FIGURE 11.9: Contour plot of pressure and temperature.

```
Coefficients:
            Estimate Std. Error t value Pr(>|t|)
(Intercept)  43.7496    1.1689   37.428  4.2e-05 ***
A             0.8838    0.4155    2.127   0.123
B             0.7576    0.4155    1.823   0.166
I(A^2)       -0.1121    0.6936   -0.162   0.882
I(B^2)        0.2451    0.6936    0.353   0.747
A:B           0.2000    0.5847    0.342   0.755
Residual standard error: 1.169 on 3 degrees of freedom
> m21=lm(y~A+B)
> summary(m21)
Coefficients:
            Estimate Std. Error t value Pr(>|t|)
(Intercept)  43.8667    0.2971 147.631 6.52e-12 ***
A             0.8838    0.3167    2.790   0.0316 *
B             0.7576    0.3167    2.392   0.0539 .
Residual standard error: 0.8914 on 6 degrees of freedom
```

The linear model is clearly a better fit because the estimated standard deviation of the errors, 0.89, is smaller than that for the quadratic model (1.12). Bianca decided to increase A by 1 unit to 165 KPa and to increase temperature by .758/.884 of 1 unit, which was rounded to 4 degrees Celsius, to give the direction of steepest ascent.

TABLE 11.18: EVOP for pharmaceutical product - Experiment 3.

Pressure A (5 kPa from 165 kPa)	Temperature B (5°C from 194°C)	Yield y %
1	-1	46.7
-1	1	47.1
-1.4	0	45.9
1.4	0	43.7
0	-1.4	45.9
0	0	48.0
1	1	44.2
0	1.4	45.1
-1	-1	47.7

A third experiment was performed, centered on 165 kPa and 194 degrees Celsius. Bianca decided to carry out a central composite design as she considered at least nine runs were necessary to identify the direction of steepest ascent with acceptable precision. The results are given, in the random order in which the 9 runs were performed, in Table 11.18.

She was able to estimate the location of the maximum yield from these results, and you are asked to duplicate her analysis in Exercise 11.11. She follows up with a confirmatory experiment.

11.6 Summary

11.6.1 Notation

For additional notation, see Chapters 8 and 9.

$A, B, C, \ldots$	factors and variables representing factors
$1, a, b, ab, \ldots$	design points and observations of the response at design point
2^k	factorial design with k factors at 2 levels
2^{k-p}	fractional factorial design

11.6.2 Summary of main results

Factorial designs for factors at two levels: These are used to investigate the effects of k factors, each at two levels referred to as high and low and coded $+1$ and -1 respectively, including interactions. If there are k factors, then there are 2^k possible factor combinations.

Fractional factorial designs: These can be used to reduce the number of runs in the full factorial design by forgoing estimates of high order interactions. A consequence of forgoing high order interactions is that an alias structure is introduced. If there are k factors, then there are 2^{k-p} possible factor combinations.

Central composite designs: These are made up of a 2^{k-p} design with additional points: at the center; and very high and very low for each factor, with all the other factors at their center value. The number of runs needed is $2^{k-p} + 2k + 1$. The central composite design enables the investigation of quadratic effects.

Further points:

- Designs can be adapted if, for example, some factors are limited to just two levels. The corresponding variance-covariance matrix of estimators of the coefficients can be checked by evaluating $(X'X)^{-1}\sigma^2$ with some assumed value of σ^2.

- If a design is replicated the variance at each design point can be estimated, and variability can be included as a response.

- Experiments can be performed with small changes in factors during routine production. If this is combined with movement towards optimal operating conditions it is known as EVOP.

11.6.3   MATLAB and R commands

In the following A, B, C and D are column vectors containing predictor data and y is a column vector containing response data. For more information on any built in function, type `help(function)` in R or `help function` in MATLAB.

R command	MATLAB command
`AA=A^2;BB=B^2;CC=C^2`	`X=dataset(y,A,B,C,D);`
`m1=lm(y~A*B*C*D)`	`m1=fitlm(X, 'y ~ A*B*C*D')`
`m2=lm(y~(A+B+C+D)^2)`	`m2=fitlm(X, 'y ~ (A+B+C_D)\^2')`
`m3=lm(y~(A+B+C)^2+AA+BB+CC)`	`m3=fitlm(X, 'y~(A+B+C)^2+A^2+B^2+C^2)')`

11.7   Exercises

Section 11.2 Factorial designs with factors at two levels

Exercise 11.1: Half fraction

An experiment is run to investigate four factors A, B, C, D at two levels.

(a) How many runs are there in a single replicate of a 2^{4-1} design ?

(b) What are the aliases of the main effects?

(c) What are the aliases of the 2-factor interactions?

(d) How can the design be used to obtain some information about interactions?

Exercise 11.2: Emu engineering

Refer to Example 11.4. Estimate the standard deviation of the errors in the regression by averaging the variances of the 4 responses at the 4 factor combinations and taking square root. Verify that it is equal to 2.445.

Exercise 11.3: Silicon wafers

After [Vining, 1998]. This experiment was designed to investigate the uniformity of the layer formed during a tungsten deposition process for silicon wafers. Three factors were considered:

Level	Temperature (degrees C)	Pressure (Torr)	Argon flow (sccm)
Low	440	0.80	0
High	500	4.00	300

Let coded temperature be x_1, coded pressure be x_2, and coded argon flow be x_3. All factors were coded -1 for low and $+1$ for high. The coded values are sometimes referred to as design units. The response (y) is the range of thickness measurements, made over the layer, divided by the average thickness and expressed as a percentage. The objective is to minimize this measure of uniformity.

x_1	x_2	x_3	y
-1	-1	-1	4.44
1	-1	-1	6.39
-1 mean	1	-1	4.48
1	1	-1	4.44
-1	-1	1	8.37
1	-1	1	8.92
-1	1	1	7.89
1	1	1	7.83

(a) Fit a regression of y on x_1, x_2 and x_3. Write down the fitted model. What is the estimated standard deviation of the errors, and how many degrees of freedom is it estimated on?

(b) Fit a regression of y on the three main effects and all possible 2-factor interactions. Write down the fitted model. What is the estimated standard deviation of the errors, and how many degrees of freedom is it estimated on?

(c) The t-ratio for x_3 increases when the interaction terms are included in the model, but the associated p-value increases too. Explain this.

(d) All the factors can be varied on a continuum so a ninth run is carried out at the centre of the design, i.e. x_1, x_2 and x_3 equal to 0 in design units. The response is 7.53. Calculate the three squared terms. What do you notice about them?

(e) Fit a regression of y on x_1, x_2, x_3, and $(x_3)^2$. Write down the fitted model. What is the estimated standard deviation of the errors, and how many degrees of freedom is it estimated on? Calculate the fitted values and the residuals. Plot the residuals against the fitted values.

(f) Express the regression model of (e) for variables in original units.

(g) What action do you recommend the company should take?

Section 11.3 Fractional factorial designs

Exercise 11.4: Quarter fraction 1

The design generators for a 2^{6-2} design with 6 factors are:

$$ABCE \;=\; 1 \qquad BCDF \;=\; 1.$$

(a) Explain why these two generators imply a third: $ADEF = 1$.
(b) Write down the aliases of A.
(c) Write down the aliases of all the 2-factor interactions.
(d) Write down the aliases of ABC.

Exercise 11.5: Quarter fraction 2

An experiment is performed to investigate the effects of eight factors $A, \ldots, H$ on the capacitance of a design of capacitor. The design generators are: $ABCDG = 1$ and $ABEFH = 1$

(a) What are the aliases of A?
(b) What are the aliases of DE?

Exercise 11.6: Quarter fraction 3

After [Montgomery, 2004]. The results of a 2^{6-2} experiment are given in Table 11.19. The objective is to minimize shrinkage in an injection molding process. There are 6

TABLE 11.19: Three reps 2^{5-1} design.

A	B	C	D	E	F	Shrink
-1	-1	-1	-1	-1	-1	6
1	-1	-1	-1	1	-1	10
-1	1	-1	-1	1	1	32
1	1	-1	-1	-1	1	60
-1	-1	1	-1	1	1	4
1	-1	1	-1	-1	1	15
-1	1	1	-1	-1	-1	26
1	1	1	-1	1	-1	60
-1	-1	-1	1	-1	1	8
1	-1	-1	1	1	1	12
-1	1	-1	1	1	-1	34
1	1	-1	1	-1	-1	60
-1	-1	1	1	1	-1	16
1	-1	1	1	-1	-1	5
-1	1	1	1	-1	1	37
1	1	1	1	1	1	52

factors at two levels:

A mold temperature

B screw speed

C holding time

D cycle time

E gate size

F holding pressure

(a) Fit a model that includes as many 2-factor interactionsas is possible.

(b) Fit a model with three predictor variables.

(c) Consider the residuals from the model in (b)

(i) Draw a normal qq-plot of the residuals.

(ii) Plot the residuals against A,...,F and comment.

Exercise 11.7: Taguchi L8 orthogonal array design

The Taguchi L8 orthogonal array design given in Table 11.20 can be used to investigate up to 7 variables, each at two levels low (1) and high (2), which are assigned to the columns in the array. There are 8 runs, corresponding to the rows of the array, and these should be carried out in a random order.

TABLE 11.20: Taguchi L8 orthogonal array design.

x_1	x_2	x_3	x_4	x_5	x_6	x_7
1	1	1	1	1	1	1
1	1	2	1	2	2	2
1	2	1	2	1	2	2
1	2	2	2	2	1	1
2	1	1	2	2	1	2
2	1	2	2	1	2	1
2	2	1	1	2	2	1
2	2	2	1	1	1	2

(a) A 2^3 design for three factors x_1, x_2, x_3 has 8 runs. Write down the columns corresponding to x_1, x_2, x_3, the 3 two-factor and 1 3-factor interactions. Use -1 and $+1$ for high and low levels respectively.

(b) What do you notice about the part above and the L8 array if low and high are designated -1 and $+1$ respectively?

(c) Investigating 7 variables with an L8 design can be useful at the screening stage of product development, but what critical assumption is being made? What fractional factorial design is being used?

Exercise 11.8:

The effects of five factors on the tensile strength of spot welds were investigated using three replicates of a 2^{5-1} design of experiment. The results are in Table 11.21.

The factors are:

A button diameter

B weld time

TABLE 11.21: Three reps 2^{5-1} design.

A	B	C	D	E	Strength rep1	rep2	rep3
−1	−1	−1	−1	−1	1 330	1 330	1 165
+1	+1	−1	−1	−1	1 935	1 935	1 880
+1	−1	+1	−1	−1	1 775	1 770	1 770
−1	+1	+1	−1	−1	1 275	1 270	1 275
+1	−1	−1	+1	−1	1 880	1 935	1 880
−1	+1	−1	+1	−1	1 385	1 440	1 495
−1	−1	+1	+1	−1	1 220	1 165	1 440
+1	+1	+1	+1	−1	2 155	2 100	2 100
+1	−1	−1	−1	+1	1 715	1 715	1 660
−1	+1	−1	−1	+1	1 385	1 550	1 550
−1	−1	+1	−1	+1	1 000	1 165	1 495
+1	+1	+1	−1	+1	1 990	1 990	1 980
−1	−1	−1	+1	+1	1 275	1 660	1 550
+1	+1	−1	+1	+1	1 660	1 605	1 660
+1	−1	+1	+1	+1	1 880	1 935	1 935
−1	+1	+1	+1	+1	1 275	1 220	1 275

C hold time

D electrode force

E machine type

The objective is to maximize strength.

(a) State the alias structure.

(b) Fit a model with just main effects. Based on this model what recommendations would you make.

(c) Now fit a model with all possible 2-factor interactions. How would you modify the advice in (b)?

(d) What alternative design, restricted to 48 runs, might have been preferable?

(e) If there are resources for a follow up experiment, what advice would you give on the design of that experiment?

Section 11.4 Central composite designs

Exercise 11.9: Maximization

Consider the function

$$\phi = -12x_1 - 12x_2 + 2x_1x_2 - 4x_1^2 - 4x_2^2$$

(a) Find the value of x_1 and x_2 at which the function has its maximum.

(b) What is the maximum value?

(c) Now suppose x_1 and x_2 are restricted to the domain −1 up to 1.

   (i) At what point within this domain does the function take its greatest value?

    (ii) What is the maximum value?

    (iii) What is the relevance of this to process optimization using regression models?

Exercise 11.10: Design

A central composite design with two factors has the following 10 runs.

	1	2	3	4	5	6	7	8	9	10
x_1	-1	1	-1	1	-k	k	0	0	0	0
x_2	-1	-1	1	1	0	0	-k	k	0	0

Note that x_1 and x_2 are uncorrelated.

(a) For what value of k are x_1^2 and x_2^2 uncorrelated?

(b) Are x_1 and x_1^2 uncorrelated?

(c) Are x_1 and x_2^2 uncorrelated?

Section 11.5 Evolutionary operation (EVOP)

Exercise 11.11: Bianca

Refer to the results from Bianca's third experiment in her EVOP investigation. Show that a quadratic surface is a better fit then a plane. Plot contours of yield and hence estimate the values of A and B that give the maximum yield in standardized unit. Give the estimated values for pressure and temperature that give the maximum yield. Estimate the maximum yield and give a 95% confidence interval for the maximum yield.

Exercise 11.12:

Suppose we have a plane $y = a + bx_1 + cx_2$, change in x_1 by 1 unit and let the change in x_2 be θ.

(a) Express the change in y, Δy, in terms of b, c, and θ.

(b) Express the change in magnitude of the vector $(x_1, x_2)'$ as a function of θ. Call this Δx.

(c) Maximize $\dfrac{\Delta y}{\Delta x}$ with respect to θ, and hence show that $\theta = c/b$.

(d) A regression plane $y = 50 + 3x_1 - 2x_2$ has been fitted to the results of an experiment. The objective is to maximize y. If x_1 is increased by 1, what change in x_2 corresponds to moving in the direction of steepest ascent? What is then the magnitude of the change in the vector x? What is the change in y?.

(e) A regression plane $y = 50 + 3x_1 + x_2$ has been fitted to the results of an experiment. The objective is to minimize y. What change in vector x corresponds to moving in the direction of steepest descent if the change in vector x is to have magnitude 1?

(f) If you are familiar with the concept of gradient in vector calculus, how can you obtain the result in (c) more succinctly?

Miscellaneous problems

Exercise 11.13: Optical density

The response, optical density, is labeled y, and the predictor variables band frequency and film thickness are labeled $bfreq$ and $film$ respectively in Table 11.22. Calculate

TABLE 11.22: Optical density (y), band frequency and film thickness.

y	$bfreq$	$film$
0.231	740	1.1
0.107	740	0.62
0.053	740	0.31
0.129	805	1.1
0.069	805	0.62
0.030	805	0.31
1.005	980	1.1
0.559	980	0.62
0.321	980	0.31
2.948	1235	1.1
1.633	1235	0.62
0.934	1235	0.31

three new columns

$$bfsq = bfreq^2$$
$$filmsq = film^2$$
$$int = bfreq \times film$$

(a) This is a designed experiment. State the number of levels for each of the two factors.

(b) Calculate the correlations between: $bfreq$ and $film$; $bfsq$ and $filmsq$; $film$ and $filmsq$; $film$ and int. Now calculate x_1 and x_2

$$x_1 = bfreq - mean(bfreq)$$
$$x_2 = film - mean(film)$$

and then

$$x_1 x_1 = x_1^2$$
$$x_1 x_2 = x_1 \times x_2$$
$$x_2 x_2 = x_2^2$$

(c) Calculate the correlations between all pairs of predictor variables from the set of 5 predictor variables: x_1, x_2, $x_1 x_1$, $x_1 x_2$, $x_2 x_2$.

(d) Regress the optical density y on x_1 and x_2. Write down the fitted equation, the estimated standard deviation of the errors (s), R^2 and R^2 adjusted.

(e) Regress y on x_1, x_2, $x_1 x_1$, $x_1 x_2$ and $x_2 x_2$. Write down the fitted equation, the estimated standard deviation of the errors (s), R^2 and R^2 adjusted.

(f) Repeat the analysis using $\ln(y)$ as the response. Investigate the residuals from all the models.

(g) Which model would you recommend the company use to predict optical density?

12

Design of experiments and analysis of variance

We consider experiments in which the response depends on one or two categorical factors. The emphasis is on balanced designs, in which all factor combinations are replicated the same number of times, and analysis of variance is used as the first step for the analysis. Least significant differences is used as a follow up procedure.

12.1   Introduction

In this chapter we consider a response that depends on one or more factors that take categorical values. The first example (Example 12.1) is an experiment to compare tensile strengths of material from different suppliers. The factor is supplier and the categorical values are the four suppliers. The categorical values are often referred to as levels. In Example 12.6 we investigate how the pressure drop across a prosthetic heart valve depends on the four different designs of the valve, and a second factor, simulated pulse rate at six different rates. The pulse rate is a continuous variable, but we don't want to model pressure drop as a linear or quadratic function of pulse rate. We allow for each of the rates to have a different effect on pressure drop, without imposing any constraints, by treating the six rates as categorical values.

The designs will be balanced inasmuch as there is an equal number of runs at each combination of factor levels. Such designs could be analyzed with a multiple regression routine using indicator variables, but an equivalent analysis of variance approach is neater.

The analysis begins with a plot of the data followed by an **analysis of variance (ANOVA)**. In ANOVA the total sum of squares is defined as the sum of squared deviations of the response from its mean. The total sum of squares is partitioned into sums of squares that can be attributed to various sources of variation, and a sum of squared residuals. The residuals are estimates of the random errors which allow for unexplained variation.

The ANOVA facilitates tests of hypotheses that factors have no effect on the response. It is sensible to perform such tests before investigating the effects of specific factor levels.

12.2   Comparison of several means with one-way ANOVA

In the following example we compare simple random samples (SRS) of the same size from different populations that might be equivalent.

An engineer in a company that manufactures filters wanted to compare the tensile strength of several different membrane materials. The factor is membrane material with I different levels corresponding to I types of material. An equal number J of pieces of each material will be tested. The test pieces will be as close an approximation to an SRS as is practical. The order of the $I \times J$ tensile strength tests will be randomized.

12.2.1   Defining the model

It is convenient to express a linear model using double subscripts

$$Y_{ij} = \mu + \alpha_i + \varepsilon_{ij}$$

for $i = 1, 2, \ldots, I$ and $j = 1, 2, \ldots, J$ where

$\quad Y_{ij}$ is the response (tensile strength in this example)

$\quad \mu$ is the overall mean (the mean of the I population mean strengths)

$\quad \mu + \alpha_i$ is the population mean strength for material i with $\sum\limits_{i=1}^{I} \alpha_i = 0$ and

$\quad \varepsilon_{ij}$ is a random variable.

The constraint $\sum\limits_{i=1}^{I} \alpha_i = 0$ is necessary for μ to be the mean of the population means, as we now show. The mean of the population means is

$$\frac{(\mu + \alpha_1) + (\mu + \alpha_2) + \cdots + (\mu + \alpha_I)}{I} \;=\; \mu + \frac{\sum\limits_{i=1}^{I} \alpha_i}{I},$$

which equals μ if and only if $\sum\limits_{i=1}^{I} \alpha_i = 0$. Notice that given this constraint we have I parameters to be estimated (μ and $(I-1)$ of the α_i) corresponding to the I population means.

The error terms ε_{ij} are independently and identically distributed (iid) with mean 0 and variance σ_ε^2 ($\varepsilon_{ij} \sim iid(0, \sigma_\varepsilon^2)$). For the statistical tests we also assume the ε_{ij} are normally distributed ($\varepsilon_{ij} \sim N(0, \sigma_\varepsilon^2)$). So, the populations of responses at different levels are assumed to have equal variances.

In the following development, we distinguish the response as a random variable, Y_{ij}, from the observed value y_{ij}. We also use "." in place of a subscript to indicate a mean over that subscript. For example, the random variable $\overline{Y}_{3.} = \mu + \alpha_3 + \bar{\varepsilon}_{3.}$ is the mean of the sample from population 3, and $\bar{y}_{3.}$ is the observed mean of the sample from population 3. Similarly $\overline{Y}_{..} = \mu + \bar{\varepsilon}_{..}$ is the overall mean of the responses for the experiment as a random variable, and $\bar{y}_{..}$ is the overall mean of all the observations from the experiment.

12.2.2   Limitation of multiple t-tests

An overall null hypothesis of no difference between the materials is set up.

$$H_0^O : \alpha_1 = \alpha_2 = \cdots = \alpha_I = 0$$

The alternative hypothesis is that the population means are not all equal

$$H_1^O : \text{not all } \alpha_i = 0.$$

It is not satisfactory to carry out the $\binom{I}{2}$ possible two-sample t-tests and to reject H_0^O if any one is statistically significant at the α level[1] because these are multiple tests of

---

[1] The "α" for the test is unrelated to the α_i.

hypotheses that two population means are equal. Call the individual tests H_0^1 up to H_0^m, where $m = \binom{I}{2}$, and consider

$$P(\text{reject } H_0^1 \ \cup \ \text{reject } H_0^2 \ \cup \ \cdots \cup \ \text{reject } H_0^m | \ H_0^O \text{ true}).$$

The tests are not independent, but whatever the dependencies the above probability is less than

$$P(\text{reject } H_0^1 \ | \ H_0^O \text{ true}) + P(\text{reject } H_0^2 \ | \ H_0^O \text{ true}) + \cdots + P(\text{reject } H_0^m \ | \ H_0^O \text{ true}) = m\alpha.$$

This is known as the Bonferroni inequality (see Exercise 12.18), and it gives an upper bound for the p-value. It is referred to as a conservative procedure because it tends to underestimate the strength of evidence against the overall null hypothesis. A far better approach is to use an F-test, which provides a precise p-value if the errors are $N(0, \sigma_\varepsilon^2)$.

12.2.3 One-way ANOVA

Here we consider a single factor at I levels. The one-way analysis of variance (ANOVA) partitions the overall variability of the observations into two additive components. One is the variation between sample means, and the other is the variation within samples.

The F-test of H_0^O compares an estimate of the variance of the errors made within samples with an estimate based on the variance of the I sample means.

The **within samples estimator of the variance of the errors** is the mean of the I sample variances. It is independent of any differences in the I population means.

If the population means are equal (H_0^O is true), then the variance of the I sample means will be an unbiased estimator of σ_ε^2/J. This is because the I sample means are means of J independent observations with the same mean and variance σ_ε^2. It follows that the product of J with the variance of the sample means is an unbiased estimator of σ_ε^2 if H_0^O is true. We refer to this product as the **between samples estimator of the variance of the errors**.

If H_0^O is true, then we have two independent estimators of the same variance and the ratio of the between samples estimator of the variance of the errors to the within sample estimator of the errors has an F-distribution with $I-1$ and $I(J-1)$ degrees of freedom. The degrees of freedom is the number of squared deviations in the sum, less any constraints. For the between samples estimate there are I deviations $((\bar{y}_{1.} - \bar{y}_{..}), (\bar{y}_{2.} - \bar{y}_{..}), \cdots, (\bar{y}_{I.} - \bar{y}_{..}))$ and one constraint $\sum_{i=1}^{I}(\bar{y}_{i.} - \bar{y}_{..}) = 0$, so the degrees of freedom is $I - 1$. Turning to the within samples estimate there are I samples of size J, each sample leads to an estimate of σ_ε^2 on $J - 1$ degrees of freedom, so the overall degrees of freedom is $I(J - 1)$.

If H_0^O is not true, then the population means differ, and the expected value of the variance of the sample means will exceed σ_ε^2/J. In this case the ratio of the between samples estimator of variance to the within samples estimator of variance is expected to exceed 1. An F-test, with a one-sided alternative, is used to test the overall null hypothesis that the variances are equal. The calculations are usually set out in an ANOVA table (Table 12.1), which extends to balanced experiments with several factors.

The principle behind the ANOVA table is that the difference between observation j from sample i (y_{ij}) and the overall mean ($\bar{y}_{..}$) can be expressed as the sum of the deviation of the mean of sample i ($\bar{y}_{i.}$) from the overall mean ($\bar{y}_{..}$) and the deviation of the observation from the mean of sample i. That is,

$$y_{ij} - \bar{y}_{..} \ = \ (\bar{y}_{i.} - \bar{y}_{..}) + (y_{ij} - \bar{y}_{i.}),$$

TABLE 12.1: One-way ANOVA.

Source of variation	Corrected sum of squares (SS)	Degree of freedom (Df)	Mean Square (MS)	$E[MS]$
Between samples	$SSB = J \sum\limits_{i=1}^{I} (\bar{y}_{i.} - \bar{y}_{..})^2$	$I-1$	$\dfrac{SSB}{I-1}$	$\sigma_\varepsilon^2 + \dfrac{J \sum\limits_{i=1}^{I} \alpha_i^2}{I-1}$
Within samples	$SSW = \sum\limits_{i=1}^{I} \sum\limits_{j=1}^{J} (y_{ij} - \bar{y}_{i.})^2$	$I(J-1)$	$\dfrac{SSW}{I(J-1)}$	σ_ε^2
Total	$SST = \sum\limits_{i=1}^{I} \sum\limits_{j=1}^{J} (y_{ij} - \bar{y}_{..})^2$	$IJ-1$		

since the $+\bar{y}_{i.}$ and $-\bar{y}_{i.}$ on the right hand side sum to 0. The $y_{ij} - y_{i.}$ are residuals as we can see by rearranging the model which can be written as:

$$Y_{ij} - \mu = \alpha_i + \varepsilon_{ij}.$$

Estimates of μ, α_i, and ε_{ij} are $\hat{\mu} = \bar{y}_{..}$, $\hat{\alpha}_i = \bar{y}_{i.} - \bar{y}_{..}$, and the residuals, r_{ij} are defined by

$$
\begin{aligned}
r_{ij} &= y_{ij} - \hat{\mu} - \hat{\alpha}_i \\
&= y_{ij} - \bar{y}_{..} - (\bar{y}_{i.} - \bar{y}_{..}) \\
&= y_{ij} - \bar{y}_{i.}.
\end{aligned}
$$

Substituting estimates into the rearranged model gives

$$y_{ij} - \bar{y}_{..} = (\bar{y}_{i.} - \bar{y}_{..}) + (y_{ij} - \bar{y}_{i.}).$$

The same principle holds for the other ANOVA considered in this chapter. The ANOVA table is obtained as follows. The total sum of squares (SST) in decomposed as

$$SST = \sum_{i=1}^{I} \sum_{j=1}^{J} (y_{ij} - \bar{y}_{..})^2 = \sum_{i=1}^{I} \sum_{j=1}^{J} ((y_{ij} - \bar{y}_{i.}) + (\bar{y}_{i.} - \bar{y}_{..}))^2.$$

Continuing

$$SST = \sum_{i=1}^{I} \sum_{j=1}^{J} (y_{ij} - \bar{y}_{i.})^2 + \sum_{i=1}^{I} \sum_{j=1}^{J} (y_{i.} - \bar{y}_{..})^2 + 2 \sum_{i=1}^{I} \sum_{j=1}^{J} (y_{ij} - \bar{y}_{i.})(\bar{y}_{i.} - \bar{y}_{..}).$$

Now, the sum of cross-products (the third term on the right hand side) is equal to 0. This because $(\bar{y}_{i.} - \bar{y}_{..})$ is a common factor of any summation over j. Thus

$$
\begin{aligned}
2 \sum_{i=1}^{I} \sum_{j=1}^{J} (y_{ij} - \bar{y}_{i.})(\bar{y}_{i.} - \bar{y}_{..}) &= 2 \sum_{i=1}^{I} (\bar{y}_{i.} - \bar{y}_{..}) \sum_{j=1}^{J} (y_{ij} - \bar{y}_{i.}) \\
&= 2 \times 0 \times \sum_{j=1}^{J} (y_{ij} - \bar{y}_{i.}) = 0,
\end{aligned}
$$

since the sum of deviations from their mean is 0, that is $\sum\limits_{j=1}^{J}(y_{ij}-\bar{y}_{i.})=0$ for any i. Therefore

$$\sum_{i=1}^{I}\sum_{j=1}^{J}(y_{ij}-\bar{y}_{..})^2 \;=\; \sum_{i=1}^{I}\sum_{j=1}^{J}(y_{ij}-\bar{y}_{i.})^2 + J\sum_{i=1}^{I}(\bar{y}_{i.}-\bar{y}_{..})^2$$

using the fact that $(\bar{y}_{i.}-\bar{y}_{..})$ is the same for all j. More succinctly, writing SSB and SSW for the between samples and within samples corrected sum of squares respectively, we have shown that

$$SST \;=\; SSW + SSB.$$

The within samples sum of squares (SSW) is also referred to as the residuals sum of squares. There are $I(J-1)$ degrees of freedom associated with the SSW because there are I sums of J squared mean-adjusted observations. There are $I-1$ degrees of freedom associated with the SSB because it is the sum of I squared mean-adjusted observations. The mean squares are the sum of squares divided by their degrees of freedom. It remains to justify the expressions for the expected values of the mean squares.

These expected values are algebraic expressions in terms of the parameters of the model. They are used to justify the F−tests. For each material (i) we have an SRS of size J from a population with a variance σ_ε^2. It follows that for each i:

$$\mathrm{E}\left[\frac{\sum\limits_{j=1}^{J}(Y_{ij}-\overline{Y}_{i.})^2}{(J-1)}\right] \;=\; \sigma_\varepsilon^2.$$

If we now average I such estimators

$$\mathrm{E}\left[\frac{\sum\limits_{i=1}^{I}\sum\limits_{j=1}^{J}(Y_{ij}-\overline{Y}_{i.})^2}{I(J-1)}\right] \;=\; \sigma_\varepsilon^2.$$

Moreover the I estimators are independent, because they are based on distinct SRS from the I populations. We use

$$s_\varepsilon^2 \;=\; \frac{\sum\limits_{i=1}^{I}\sum\limits_{j=1}^{J}(y_{ij}-\bar{y}_{i.})^2}{I(J-1)}$$

as an estimate of σ_ε^2. The expected value of the between samples mean square follows directly from the model. For each i,j:

$$\begin{aligned}Y_{ij}-\overline{Y}_{i.} &= (\mu+\alpha_i+\varepsilon_{ij})-(\mu+\alpha_i+\bar{\varepsilon}_{i.})\\ &= \varepsilon_{ij}-\bar{\varepsilon}_{i.}\end{aligned}$$

and for each i:

$$\begin{aligned}\overline{Y}_{i.}-\overline{Y}_{..} &= (\mu+\alpha_i+\bar{\varepsilon}_{i.})-(\mu+\bar{\varepsilon}_{..})\\ &= \alpha_i+\bar{\varepsilon}_{i.}-\bar{\varepsilon}_{..}\,.\end{aligned}$$

Therefore

$$\frac{\sum_{i=1}^{I}(\overline{Y}_{i.} - \overline{Y}_{..})^2}{(I-1)} = \frac{\sum_{i=1}^{I}\alpha_i^2}{(I-1)} + \frac{\sum_{i=1}^{I}(\bar{\varepsilon}_{i.} - \bar{\varepsilon}_{..})^2}{(I-1)} + \frac{2\sum_{i=1}^{I}\alpha_i(\bar{\varepsilon}_{i.} - \bar{\varepsilon}_{..})^2}{(I-1)}.$$

The middle term is an unbiased estimator of the variance of the mean of J errors. Taking expectation gives

$$E\left[\frac{\sum_{i=1}^{I}(\overline{Y}_{i.} - \overline{Y}_{..})^2}{(I-1)}\right] = \frac{\sum_{i=1}^{I}\alpha_i^2}{(I-1)} + \frac{\sigma_\varepsilon^2}{J} + 0.$$

The 0 on the right hand side is a consequence of the errors being uncorrelated with the α_i. It follows that

$$E\left[\frac{J\sum_{i=1}^{I}(\overline{Y}_{i.} - \overline{Y}_{..})^2}{(I-1)}\right] = \sigma_\varepsilon^2 + \frac{J\sum_{i=1}^{I}\alpha_i^2}{(I-1)}.$$

12.2.4   Testing H_0^O

If H_0^O is true, then the expected value of the between samples mean square equals the expected value of the within samples mean square. These two estimators of variance are independent[2] and the equality can be tested with an $F-$test with $I-1$ and $I(J-1)$ degrees of freedom. If H_0^O is not true, then the expected values of the between samples variance is higher by the addition of

$$\frac{J\sum_{i=1}^{J}\alpha_i^2}{(I-1)}.$$

So the critical value for a test at the $\alpha \times 100\%$ level is the upper α quantile: $F_{I-1,I(J-1),\alpha}$.

12.2.5   Follow up procedure

If there is evidence to reject H_0^O, then we will want to follow up the ANOVA and state which of the means are statistically significantly different. The **least significant difference (LSD)** is commonly used and corresponds to the p-values reported in a regression analysis. Using LSD is equivalent to performing all possible t-tests, with the modification that all the samples are used to estimate the common standard deviation of the populations, and it does not allow for multiple comparisons (Section 12.2.2). But, we do now have evidence that there are some differences and multiple comparisons is less of an issue than it is at the beginning of the analysis (see Exercise 12.5). According to the LSD procedure, two sample means will be significantly different at a nominal α level if their difference exceeds

$$\left(\sqrt{\frac{s_\varepsilon^2}{J} + \frac{s_\varepsilon^2}{J}}\right)t_{I(J-1),\alpha/2}$$

---

[2]Changing the $\widehat{\alpha_i}$, where $\widehat{\alpha_i}$ is $(\bar{y}_{i.} - \bar{y}_{..})$, will have no effect on the within samples estimate of variance and multiplying $(y_{ij} - \bar{y}_{i.})$ for $j = 1, \cdots, J$ by any factor will have no effect on the between samples estimate of variance.

and this is known as the $LSD(\alpha)$. Moreover, a $100(1 - \alpha)\%$ confidence interval for the difference in two means is given by the difference in the sample means plus or minus the LSD.

A conservative alternative to LSD, that does allow for multiple comparisons during the follow up analysis, is based on the studentized range and discussed in Exercise 12.5. We demonstrate an ANOVA, and follow up procedure, with an example from a company that manufactures liquid filtration equipment.

Example 12.1: Strength of filter membranes [one-way ANOVA]

The production engineer of a company which manufactures filters for liquids, for use in the pharmaceutical and food industries, wishes to compare the burst strength of four types of membrane. The first (A) is the company's own standard membrane material, the second (B) is a new material the company has developed, and C and D are membrane materials from other manufacturers. The engineer has tested five filter cartridges from ten different batches of each material. The mean burst strengths for each set of five cartridges are given in Table 12.2 and are used as the response[3]. So the response Y_{ij} is the mean burst strength of the five cartridges made from batch j of material i.

TABLE 12.2: Burst strength of filter membranes (kPa).

Type	Burst strength									
A	95.5	103.2	93.1	89.3	90.4	92.1	93.1	91.9	95.3	84.5
B	90.5	98.1	97.8	97.0	98.0	95.2	95.3	97.1	90.5	101.3
C	86.3	84.0	86.2	80.2	83.7	93.4	77.1	86.8	83.7	84.9
D	98.5	93.4	87.5	98.4	87.9	86.2	89.9	89.5	90.0	95.6

The first step is to plot the burst strengths for each batch against material type[4]. The plot is shown in the left panel of Figure 12.1 and materials A and B appear to be stronger than C and D. However, the responses for different batches of the same material are quite variable and the appearance might be due to chance. The ANOVA will quantify this.

```
> qqnorm(m1$res,ylab="ordered residuals",main="")
> membrane.dat=read.table("filtermembrane.txt",header=TRUE)
> head(membrane.dat)
    material strength
1       A       95.5
2       A      103.2
3       A       93.1
4       A       89.3
5       A       90.4
6       A       92.1
> tail(membrane.dat)
    material strength
```

---

[3]We do not consider burst strength of individual cartridges as the response because they may not be independent. The strengths of cartridges made from the same batch are likely to vary less than strengths of cartridges from different batches.

[4]The material type is recorded by a letter in the data file so R treats material as a factor. The default for the R command plot() gives box plots when used with a factor, and we have chosen to change the default to plot individual points.

```
35      D       87.9
36      D       86.2
37      D       89.9
38      D       89.5
39      D       90.0
40      D       95.6
> attach(membrane.dat)
> material_num=as.numeric(material)
> par(mfrow=c(1,2))
> plot(material_num,strength,xaxt="n",xlab="Material")
> axis(1,at=c(1:4),labels=c("A","B","C","D"))
> m1=lm(strength~material)
> anova(m1)

Analysis of Variance Table

Response: strength
            Df   Sum Sq mean Sq   F value      Pr(>F)
material     3   709.23 236.409    15.538   1.202e-06 ***
Residuals   36   547.75  15.215
---
> qqnorm(m1$res,ylab="ordered residuals",main="")
```

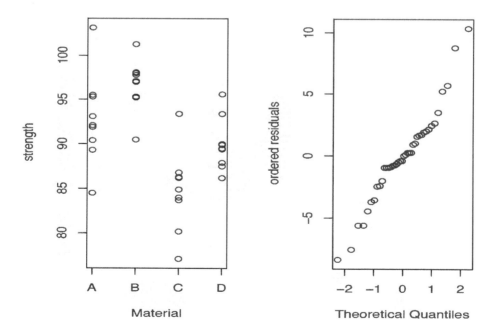

FIGURE 12.1: Scatter plot of burst strength against type of membrane material (left panel) and normal quantile-quantile plot of residuals (right panel).

Before considering the ANOVA table we check that the residuals from the model approximate a random sample from a normal distribution and that there are no extreme outlying values (right panel of Figure 12.1). In the ANOVA table the mean square (Mean Sq) for material is far higher than that of the residuals (within samples estimate of the errors) and there is very strong evidence of a difference in materials ($F = 15.5$, p-value< 0.001). We use LSD to help establish which of the materials differ, and calculate the means and the LSD(.05), LSD(.01) using R.

```
> s=summary(m1)$sigma
> LSD5=sqrt(2*s^2/10)*qt(0.975,m1$df)
> LSD1=sqrt(2*s^2/10)*qt(0.995,m1$df)
> print(c("LSD 5%",round(LSD5,2),"LSD 1%",round(LSD1,2)))
[1] "LSD(.05)" "3.54"    "LSD(.01)" "4.74"
> matmeans=tapply(strength,material,mean)
> print(sort(round(matmeans,2)))
   C     D     A     B
84.63 89.89 92.84 96.08
```

Using the LSD(0.05), which is 3.54, we can conclude that there is evidence that $\mu_C < \mu_D, \mu_A, \mu_B$ and that $\mu_D < \mu_B$, because the differences in the corresponding sample means exceed 3.54. If we use LSD(0.01) we can claim stronger evidence that $\mu_C < \mu_A, \mu_B$ and that $\mu_D < \mu_B$. So, the manufacturer can be reasonably confident that the standard material is better than C's material and that the new material is better than both C's and D's material. However, the main purpose of the experiment was to compare materials A and B. A 95% confidence interval for this difference is given by (The R command `as.numeric()` removes unhelpful labels.):

```
> dBA=matmeans[2]-matmeans[1];dBA=as.numeric(dBA
> print(round(c(dBA-LSD5,dBA+LSD5),1))
[1] -0.3  6.8
```

The estimated increase in strength by changing from material A to material B is 3.24, but as the 95% confidence interval for the difference, $[-0.3, 6.8]$, includes 0 we are not confident about achieving an increase. The company decided to check that other features of material B, such as filtration performance and ease of handling, were as good as those for material A. If B remained a viable alternative, then the company would perform a larger experiment for comparing the tensile strengths of just these two materials.

12.3   Two factors at multiple levels

In this section we consider two factors, each at two or more levels. If there is only a single run at all pairs of levels any interaction is confounded with the error. Tests of null hypotheses that levels of the factors have no effect on the response can be made from a two-way ANOVA.

It is necessary to replicate runs if interactions are to be estimated. The analysis can be performed with a three-way ANOVA and the hypothesis of no interactions should be tested before considering the main effects. If there is evidence of interactions the estimates of effects of one factor need to be made for the different levels of the other factor.

12.3.1   Two factors without replication (two-way ANOVA)

An engineer wants to investigate the effects of two factors A and B, each of which has several categories, on the response. We will assume a balanced design in which one run of the experiment is made at each factor combination. A linear model can be written as follows

$$Y_{ij} \;=\; \mu + \alpha_i + \beta_j + \varepsilon_{ij}, \quad \text{for } i = 1, \ldots, I \text{ and } j = 1, \ldots, J,$$

where

Y_{ij} is the response when factor A is at level i and factor B is at level j

μ is the overall population mean

α_i is the effect, relative to μ, of factor A being at level i with $\sum\limits_{i=1}^{I} \alpha_i = 0$

β_j is the effect, relative to μ, of factor B being a level j with $\sum\limits_{j=1}^{J} \beta_j = 0$ and

$\varepsilon_{ij} \sim idd(0, \sigma_\varepsilon^2)$.

There are now two overall null hypotheses and corresponding alternative hypotheses as follows.

$H_0^{OA} : \alpha_1 = \alpha_2 = \cdots = \alpha_I = 0$

$H_1^{OA} :$ not all $\alpha_i = 0$

$H_0^{OB} : \beta_1 = \beta_2 = \cdots = \beta_J = 0$

$H_1^{OB} :$ not all $\beta_j = 0$.

The theory follows the same principles as the one-way ANOVA of Section 12.2.3. The two-way ANOVA table is given in Table 12.3 and the details are left for Exercise 12.8. In this case the difference between an observation at level i for the first factor and level j for the second factor and the overall mean can be expressed as

$$y_{ij} - \bar{y}_{..} \;=\; (\bar{y}_{i.} - \bar{y}_{..}) + (\bar{y}_{.j} - \bar{y}_{..}) + (y_{ij} - \bar{y}_{i.} - \bar{y}_{.j} + \bar{y}_{..}).$$

The third term on the right hand side is the residual. In terms of the parameters of the model

$$\begin{aligned}
\widehat{\alpha}_i &= \bar{y}_{i.} - \bar{y}_{..} \\
\widehat{\beta}_j &= \bar{y}_{.j} - \bar{y}_{..} \\
\widehat{\mu} &= \bar{y}_{..}
\end{aligned}$$

We test H_0^{OA} by calculating the ratios of the mean square for factor A to the residual mean square and rejecting it at the $\alpha \times 100\%$ level of significance if the ratio exceeds $F(\alpha, I-1, (I-1)(J-1))$. Similarly for H_0^{OB}. These tests are based on an assumption that $\varepsilon_{ij} \sim N(0, \sigma_\varepsilon^2)$. We demonstrate such an analysis using data from a mining company.

TABLE 12.3: Two-way ANOVA.

Effect	Corrected sum of squares (SS)	Degrees of freedom (Df)	Mean Square (MS)	$E[\text{MS}]$
A	$J \sum\limits_{i=1}^{I} (\bar{y}_{i.} - \bar{y}_{..})^2$	$I - 1$	$\dfrac{SSA}{I-1}$	$\sigma_\varepsilon^2 + \dfrac{J \sum\limits_{i=1}^{I} \alpha_i^2}{I-1}$
B	$I \sum\limits_{j=1}^{J} (\bar{y}_{.j} - \bar{y}_{..})^2$	$J - 1$	$\dfrac{SSB}{J-1}$	$\sigma_\varepsilon^2 + \dfrac{I \sum\limits_{j=1}^{J} \beta_j^2}{J-1}$
Residuals	$\sum\limits_{i=1}^{I} \sum\limits_{j=1}^{J} (y_{ij} - \bar{y}_{i.} - \bar{y}_{.j} - \bar{y}_{..})^2$	$(I-1)(J-1)$	$\dfrac{SSR}{(I-1)(J-1)}$	σ_ε^2
Total	$\sum\limits_{i=1}^{I} \sum\limits_{j=1}^{J} (y_{ij} - \bar{y}_{..})^2$	$IJ - 1$		

Example 12.2: Water content of rock [two-way ANOVA]

Samples of rock were taken from five different locations within a large deposit. The locations were chosen in regions where the porosity was around 20%, 25%, 30%, 35% and 40% respectively. The porosity is the proportion of the rock by volume which is void, and the void can be filled by air, oil or water. A drill core was taken at each location and six samples were taken from each drill core at depths of 10 up to 60 in steps of 10 meters. The response was the water content, as a percentage by volume, of the rock. The two factors are porosity, with five categories, and depth with six categories. Although porosity and depth are continuous variables their values will be treated as categories, rather than assuming a linear or polynomial relationship between water content and the factors. The data are given in Table 12.4 and the aim is to investigate whether the water content depends on depth or porosity. The marginal means are given in the table and it appears that the water content increases with depth but not necessarily linearly. The water content also seems to decrease with increasing porosity but this is less marked. An analysis of variance can be used to assess the strength of the evidence for these claims.

We read the data illustrated in Table 12.4 into R, check the head and tail of the data frame, and then plot the data in Figure 12.2.

```
> rock.dat=read.table("rockwater.txt",header=TRUE)
> print(head(rock.dat))
    depth  poros  water
1      10     20   6.01
2      10     25   5.41
3      10     30   2.71
4      10     35   3.17
5      10     40   1.25
```

```
6      20    20    6.32
> print(tail(rock.dat))
     depth  poros  water
25     50    40    7.36
26     60    20   12.15
27     60    25   11.79
28     60    30   10.37
29     60    35    8.36
30     60    40    9.58
> attach(rock.dat)
> plot(depth,water,pch=poros/5-3)
> legend(10,12,legend=c(20,25,30,35,40),pch=poros/5-3)
```

TABLE 12.4: Water content (%) of rocks by porosity and depth.

Depth	Porosity					mean water content
(meter)	20	25	30	35	40	for given depth
10	6.01	5.41	2.71	3.17	1.25	3.71
20	6.32	7.03	5.61	2.91	3.82	5.14
30	8.76	9.65	7.34	5.62	4.93	7.26
40	10.42	6.71	9.35	6.07	5.10	7.53
50	11.82	9.70	8.95	9.02	7.36	9.37
60	12.15	11.79	10.37	8.36	9.58	10.45
Mean water content for given porosity	9.25	8.38	7.39	5.86	5.34	7.24

We notice that the water content increases with depth, although there is a suggestion that it might level off beyond a depth of 50 meters. There is also a tendency for the water content to decrease with porosity. The analysis of variance confirms that these effects are highly statistically significant.

```
> depth_f=factor(depth);poros_f=factor(poros)
> m1=lm(water~depth_f+poros_f)
> anova(m1)

Analysis of Variance Table

Response: water
           Df   Sum Sq  mean Sq  F value    Pr(>F)
depth_f     5  159.024   31.805   33.611  4.509e-09 ***
poros_f     4   65.226   16.307   17.233  2.896e-06 ***
Residuals  20   18.925    0.946
---
```

A normal quantile-quantile plot of the residuals (not shown) is quite compatible with an assumption of *iid* normal errors.

The effects, that is the estimates of μ, α_i, β_j can be obtained by

```
> deptheffect=tapply(water,depth,mean)-mean(water)
> poroseffect=tapply(water,poros,mean)-mean(water)
```

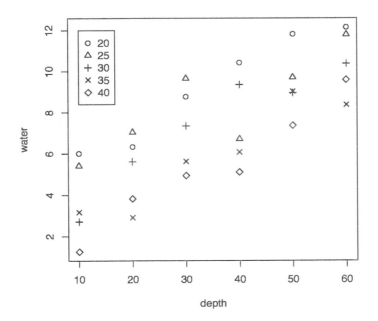

FIGURE 12.2: Scatter plot of water content (%) against depth (m) by porosity (% shown by plotting symbol).

```
> print(mean(water))
[1] 7.243
> print(deptheffect)
    10     20     30     40     50     60
-3.533 -2.105  0.017  0.287  2.127  3.207
> print(poroseffect)
       20         25         30         35         40
2.0036667  1.1386667  0.1453333 -1.3846667 -1.9030000
```

The conclusion is that the water content increases with the depth over the range 10 to 60 meters, and that it decreases with porosity over a range 20 to 40. The reason for the decrease in water content when the porosity is increased is that the pores fill with oil, leaving less space for air and water. This model does not allow for interactions, so it is assumed that the effect of increased porosity will be the same over the range of depths. If there are any interactions they will be confounded with the error. You are asked to consider an alternative regression analysis that includes an interaction term in Exercise 12.7.

12.3.2   Two factors with replication (three-way ANOVA)

We now consider two factors with I and J levels respectively, with K replicates at all combinations of levels. The model becomes

$$Y_{ijk} = \mu + \alpha_i + \beta_j + (\alpha\beta)_{ij} + \varepsilon_{ijk},$$
$$\text{for } i = 1, 2, \cdots, I, \ j = 1, 2, \cdots, J \text{ and } k = 1, 2, \cdots, K,$$

where

Y_{ijk} is the response for replicate k when factor A is at level i and factor B is at level j

μ is the overall mean

α_i is the main effect, relative to μ, of factor A being at level i with $\sum_{i=1}^{I} \alpha_i = 0$

β_j is the main effect, relative to μ, of factor B being at level j with $\sum_{j=1}^{J} \beta_j = 0$

$(\alpha\beta)_{ij}$ is the interaction effect, relative to $\mu + \alpha_i + \beta_j$, of factor A being at level

i and factor B being at level j with $\sum_{i=1}^{I} (\alpha\beta)_{ij} = 0$ and $\sum_{j=1}^{J} (\alpha\beta)_{ij} = 0$

ε_{ijk} are errors $\sim iid(0, \sigma_\varepsilon^2)$.

TABLE 12.5: Three-way ANOVA.

Effect	Corrected sum of squares (SS)	Degree of freedom (Df)	Mean Square (MS)	E[MS]
A	$JK \sum_{i=1}^{I} (\bar{y}_{i..} - \bar{y}_{...})^2$	$I-1$	$\dfrac{SSA}{I-1}$	$\sigma_\varepsilon^2 + \dfrac{JK \sum_{i=1}^{I} \alpha_i^2}{(I-1)}$
B	$IK \sum_{j=1}^{J} (\bar{y}_{.j.} - \bar{y}_{...})^2$	$J-1$	$\dfrac{SSB}{J-1}$	$\sigma_\varepsilon^2 + \dfrac{IJ \sum_{k=1}^{K} \gamma_k^2}{K-1}$
$A{:}B$	$K \sum_{i=1}^{I} \sum_{j=1}^{J} \left(y_{ij.} - \bar{y}_{i..} - \bar{y}_{.j.} - \bar{y}_{...}\right)^2$	$(I-1)(J-1)$	$\dfrac{SSA{:}B}{(I-1)(J-1)}$	$\sigma_\varepsilon^2 + \dfrac{K \sum_{i=1}^{I} \sum_{j=1}^{J} (\alpha\beta)_{ij}^2}{(I-1)(J-1)}$
Residuals	$\sum_{i=1}^{I} \sum_{j=1}^{J} \sum_{k=1}^{K} (\bar{y}_{ijk} - \bar{y}_{ij.})^2$	$(IJ)(K-1)$	$\dfrac{SSR}{(IJ)(K-1)}$	σ_ε^2

The order of the IJK runs of the experiment should be randomized. The breakdown of sum of squares in Table 12.5 again follows the same principles as in Section 12.2.3 (refer Exercise 12.9). We express a deviation from the overall mean as follows.

$$y_{ijk} - \bar{y}_{...} = (\bar{y}_{i..} - \bar{y}_{...}) + (\bar{y}_{.j.} - \bar{y}_{...}) + (\bar{y}_{ij.} - \bar{y}_{i..} - \bar{y}_{.j.} + \bar{y}_{...}) + (y_{ijk} - \bar{y}_{ij.}).$$

The fourth term on the right hand side is the residual. In terms of the parameters of the model

$$\hat{\alpha}_i = \bar{y}_{i..} - \bar{y}_{...}$$
$$\hat{\beta}_j = \bar{y}_{.j.} - \bar{y}_{...}$$
$$\widehat{\alpha\beta}_{ij} = \bar{y}_{ij.} - \bar{y}_{i..} - \bar{y}_{.j.} + \bar{y}_{...}$$
$$\hat{\mu} = y_{ijk} - \bar{y}_{ij.}$$

The expected values of the mean squares also follows the same principles as in Section 12.2.3 (refer Exercise 12.10).

Example 12.3: Load transfer of paving flags [three-way ANOVA]

An experiment [Bull, 1988] was performed to compare the load transfer of three types of paving flags (A, B and C) laid with three different gaps (3 mm, 6 mm and 10 mm). For each run 16 flags were laid in a square. There were two runs at each of the 9 factor combinations, different flags being used for each run. The order of the 18 runs was randomized.

We read the data, which are summarized in Table 12.6 into R, check the head and tail of the data frame, and then plot the data in Figure 12.3.

TABLE 12.6: Maximum load transfer of flags by type and gap and replicate.

Type	3 mm	Gap 6 mm	10 mm
A	0.257	0.171	0.172
A	0.219	0.151	0.149
B	0.492	0.368	0.242
B	0.628	0.333	0.296
C	0.279	0.226	0.236
C	0.329	0.190	0.192

The R code for this is given below:

```
> flags.dat=read.table("flags.txt",header=TRUE)
> print(head(flags.dat))
   type gap   load
1    A    3  0.251
2    A    3  0.219
3    A    6  0.171
4    A    6  0.151
5    A   10  0.172
6    A   10  0.149
> print(tail(flags.dat))
    type gap   load
13    C    3  0.279
14    C    3  0.329
```

```
15    C    6    0.226
16    C    6    0.190
17    C   10    0.236
18    C   10    0.192
> attach(flags.dat)
> type_n=as.numeric(type)
> plot(gap,load,pch=type_n)
> legend(8,0.5,legend=c("A","B","C"),pch=c(1,2,3))
```

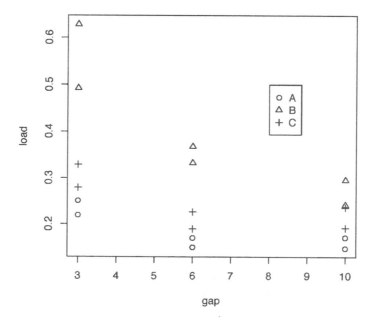

FIGURE 12.3: Scatter plot of load transfer (tonnes) by gap (mm).

We notice that load transfer is best with the narrowest gap and that this narrowest gap is particularly favorable for flag type B. Moreover, flag type B appears to be better than type C which is in turn better than type A for any gap, although the difference is slight with the widest gap. The analysis of variance confirms that the interaction is statistically significant, and that the main effects of flag type and gap are highly statistically significant.

```
> gap_f=factor(gap)
> m1=lm(load~gap_f+type+gap_f:type)
> anova(m1)

Analysis of Variance Table

Response: load
              Df      Sum Sq   mean Sq    F value       Pr(>F)
```

```
gap_f       2   0.079395   0.039697   23.5655   0.0002647 ***
type        2   0.138338   0.069169   41.0607   2.991e-05 ***
gap_f:type  4   0.029667   0.007417    4.4027   0.0302688 *
Residuals   9   0.015161   0.001685
---
```

A normal quantile-quantile plot of the residuals (not shown) is quite compatible with an assumption of *iid* normal errors.

Given the statistically significant interaction we should consider the means for all nine factor combinations.

```
> LSD5=qt(.975,m1$df)*summary(m1)$sigma/sqrt(1/2+1/2)
> print(LSD5)
[1] 0.09284648
> print(round(tapply(load,gap_f:type,mean),2))
  3:A  3:B  3:C  6:A  6:B  6:C 10:A 10:B 10:C
 0.24 0.56 0.30 0.16 0.35 0.21 0.16 0.27 0.21
```

The conclusion is that the highest load transfer, estimated as 0.56, is given by flag type B, laid with the narrowest gap (3 m).

12.4 Randomized block design

The **randomized block design** (RBD) is an extension of the matched pairs design (Section 7.8.2) to two or more treatments. A **block** is a grouping of experimental units called **plots** which are relatively similar when compared with differences between blocks[5]. The number of plots equals the number of treatments to be compared. The rationale for the design is that the effects of different blocks on the response can be removed from the comparison of treatments by randomly assigning the full set of treatments to the plots within each block.

In the first example of Section 7.8.2 the treatments are the two weigh-bridges, the five blocks are lorry loads of building materials, and the plots correspond to weighing a particular lorry load twice, once on each weigh-bridge. In the second example of that section the two treatments are oxy-propane gas cutting and oxy-natural gas cutting of steel plates. The blocks are eight steel plates of different thickness and grade. The plots are the two halves of a plate. The two gas cutting methods were randomly assigned to the two halves of each plate. The oxy-propane cut was made first on four randomly selected plates, and the oxy-natural gas cut was made first on the other four plates.

The RBD is an example of a **balanced design** because every treatment appears once in each block. The model is

$$Y_{ij} = \mu + \alpha_i + \beta_j + \varepsilon_{ij}$$

for $i = 1, 2, \cdots, I$ and $j = 1, 2, \cdots, J$ where

The block effects are considered a random sample from a hypothetical population of all possible blocks. They are described as **random effects**. In contrast the treatment effects

---

[5]The terms "blocks" and "plots" come from the agricultural applications which motivated these designs for experiments.

Y_{ij} is the response when factor A is at level i in *block* j

μ is the mean of the population means of the I treatments

$\mu + \alpha_i$ is the mean for treatment i with $\sum\limits_{i=1}^{I} \alpha_i = 0$

β_j are the block effects and are $\sim iid(0, \sigma_\beta^2)$ and

$\varepsilon_{ij} \sim iid(0, \sigma_\varepsilon^2)$.

are specific to the treatments under investigation and are described as **fixed effects**. The mean of the random effects is itself a random variable with a mean of 0 and a variance of α_β^2/J. In contrast, the mean of the treatment effects is constrained to equal to 0. The breakdown of the sum of squares is the two-way ANOVA of Section 12.2.3. However, the expected value of the blocks mean squares is different as shown in Table 12.7.

TABLE 12.7: Two-way ANOVA for RBD.

Effect	Corrected sum of squares (SS)	Degrees of freedom (df)	mean Square (MS)	E[MS]
A	$J \sum\limits_{i=1}^{I} (\bar{y}_{i.} - \bar{y}_{..})^2$	$I-1$	$\dfrac{SSA}{I-1}$	$\sigma_\varepsilon^2 + \dfrac{J \sum\limits_{i=1}^{I} \alpha_i^2}{I-1}$
blocks	$I \sum\limits_{j=1}^{J} (\bar{y}_{.j} - \bar{y}_{..})^2$	$J-1$	$\dfrac{SSB}{J-1}$	$\sigma_\varepsilon^2 + I\sigma_\beta^2$
Residuals	$\sum\limits_{i=1}^{I} \sum\limits_{j=1}^{J} (y_{ij} - \bar{y}_{i.} - \bar{y}_{.j} - \bar{y}_{..})^2$	$(I-1)(J-1)$	$\dfrac{SSR}{(I-1)(J-1)}$	σ_ε^2
Total	$\sum\limits_{i=1}^{I} \sum\limits_{j=1}^{J} (y_{ij} - \bar{y}_{..})^2$	$IJ-1$		

The proof of these expected values follows from the model by noting that averages of fixed effect are 0 whereas average of random effects are sample means. So

$$\begin{aligned}
\overline{Y}_{.j} &= \mu + 0 + \beta_j + \bar{\varepsilon}_{.j} \\
\overline{Y}_{..} &= \mu + 0 + \bar{\beta}_{.} + \bar{\varepsilon}_{..} \, .
\end{aligned}$$

$$E\left[\frac{I \sum\limits_{j=1}^{J} (\overline{Y}_{.j} - \overline{Y}_{..})^2}{J-1}\right] = E\left[\frac{I \sum\limits_{j=1}^{J} ((\beta_j - \bar{\beta}_{.}) + (\bar{\varepsilon}_{.j} - \bar{\varepsilon}_{..}))^2}{J-1}\right]$$

$$= I\sigma_\beta^2 + \frac{\sigma_\varepsilon^2}{I} = \sigma_\varepsilon^2 + I\sigma_\beta^2.$$

Example 12.4: Bioleaching of copper [RBD]

Several species of fungi can be used for extraction of metals by bioleaching. Fungi can be grown on many different substrates, such as electronic scrap, catalytic converters, and fly ash from municipal waste incineration (Wikipedia - Bioleaching). An experiment was performed to compare four fungal strains, A, B, C, and D for leaching copper (Cu) from recycled electronics and the response was percentage mobilization of Cu ions. The researcher thought that the response could vary with the type of the recycled electronic product which could be, for example: industrial; automotive; cell phones; domestic computers; or domestic appliances. The researcher decided to allow for this by sourcing material from three electronics recycle centers, which might have a different mix of products, and treating the material from each center as a block. The material of each block was divided into four parts, the plots in this application, the plots within a block being as similar as possible. Then the four fungi treatments were randomly assigned to plots within each block. The experimental arrangement is shown in Figure 12.4. The results of the experiment are given in Table 12.8.

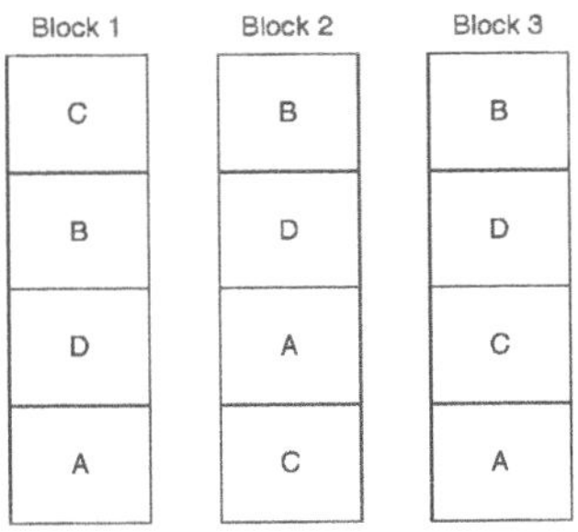

FIGURE 12.4: RBD with 3 blocks and 4 plots per block. Four treatments A, B, C, D are randomly allocated to the 4 plots within blocks.

TABLE 12.8: Percentage mobilization of Cu ions.

Fungal strain	Block		
	1	2	3
A	44	56	50
B	62	66	52
C	40	50	48
D	50	56	52

We read the data into R, check the head of the data frame, and then plot the data (Figure 12.5).

```
> bioleach.dat=read.table("bioleach.txt",header=T)
```

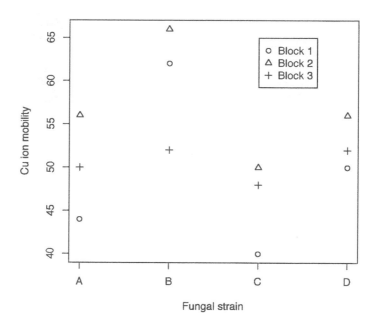

FIGURE 12.5: Scatter plot of Cu ion mobility against strain.

```
> print(head(bioleach.dat))
strain block ionmobil
1      A     1      44
2      A     2      56
3      A     3      50
4      B     1      62
5      B     2      66
6      B     3      52
> attach(bioleach.dat)
> strain_num=as.numeric(strain)
> plot(strain_num,ionmobil,pch=block,xaxt="n",xlab="Fungal strain",
+ ylab="Cu ion mobility")
> axis(1,at=c(1:4),labels=c("A","B","C","D"))
> legend(3,65,legend=c("Block 1","Block 2","Block 3"),pch=1:3)
```

There is considerable variability but fungal strain B appears to give a higher mobility. Block 2 also appears to be associated with higher mobility which suggests that there is variability amongst blocks. We continue with an ANOVA analysis.

```
> block_f=factor(block)
> m1=lm(ionmobil~strain+block_f)
> anova(m1)

Analysis of Variance Table
```

```
Response: ionmobil
          Df  Sum Sq  mean Sq  F value  Pr(>F)
strain     3  313.00  104.333  5.9057   0.03187 *
block_f    2  144.67   72.333  4.0943   0.07562 .
Residuals  6  106.00   17.667
---
```

There is evidence of a difference in fungal strains at a 5% level of significance ($p = 0.032$). We follow up by calculating the sample means for the four fungal strains, together with the LSD(0.05).

```
> strain=tapply(ionmobil,strain_num,mean)
> print(round(strain,1))
   1    2    3    4
50.0 60.0 46.0 52.7
> LSD5=qt(.975,m1$df)*summary(m1)$sigma*sqrt(1/4+1/4)
> print(round(LSD5,1))
[1] 7.3
> s2b=(anova(m1)[2,3]-anova(m1)[3,3])/4
> print(round(c(s2b,sqrt(s2b)),2))
[1] 13.67  3.70
> print(round(summary(m1)$sigma,2))
[1] 4.2
```

Using an LSD(.05) criterion, There is evidence that fungal strain B, with a sample mean of 60.0, gives a higher Cu ion mobility than A, C and D. We finish by comparing the variance, and standard deviation, of blocks with the variance and standard deviation of the errors.

```
> s2b=(anova(m1)[2,3]-anova(m1)[3,3])/4
> print(round(c(s2b,sqrt(s2b)),2))
[1] 13.67  3.70
> print(round(summary(m1)$sigma,2))
[1] 4.2
```

The estimated variance of the errors is 17.67, from the ANOVA table, and the estimated standard deviation of the errors is 4.20. The estimated standard deviation of the blocks is 3.70. So, the estimated standard deviation of blocks happens to be slightly less than the estimated standard deviation of the errors. Although the variance, and so the standard deviation of the blocks, is only statistically significantly different from 0 at a 10% level of significance we retain the blocks in the model. This is because there was reason to expect variation between blocks and blocking was incorporated in the design of the experiment.[6]

The overall conclusion is that fungal strain B gives the highest Cu ion mobility. The estimated mean mobility with B is 60. The estimated standard deviation of a single run of the process, allowing for variability within and between blocks, is $\sqrt{3.70^2 + 4.20^2}$ which is 6.0 when rounded to the nearest integer. The errors (ε_{ij}) are confounded with fungi strain interactions, but this is reasonable if blocks are random as any interaction between fungi strain and random blocks is a component of the overall error.

---

[6]Moreover, the F-test for the hypothesis of no difference in fungal strains is no longer statistically significant at the 5% level if blocks are dropped from the model.

12.5   Split plot design

A split plot design is an extension of the randomized block design (RBD). There are two factors, each at two or more levels. The first is known as the **main-plot factor**, and the second as the **sub-plot factor**. Each block is divided into **plots**, so that the number of plots equals the number of levels of the main-plot factor. The levels of the main-plot factor are randomly allocated to the plots in each block. Each plot is itself divided into **sub-plots**, so that the number of sub-plots is equal to the number of levels of the sub-plot factor. The levels of the sub-plot factor are randomly allocated to the sub-plots within each plot. The principle is shown in Figure 12.6 for two blocks, divided into four plots with three sub-plots within each plot. A split plot design is used when it is far easier to change the levels of the sub-plot factor than it is to change the levels of the main-plot factor.

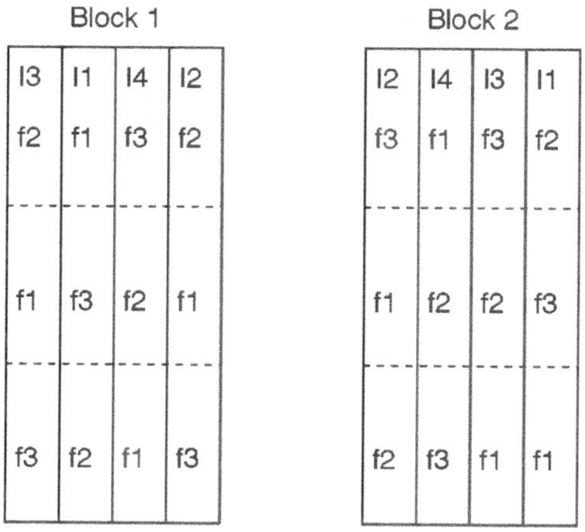

FIGURE 12.6: Split plot design with two blocks and four plots per block shown as vertical strips. Each plot is divided into three sub-plots by the broken lines. The four levels of the main-plot factor, $I1, I2, I3, I4$ are randomly allocated to the plots within blocks. The three levels of the sub-plot factor, $f1, f2, f3$ are randomly allocated to the 3 sub-plots within each plot.

The model is

$$Y_{ijk} = \mu + \alpha_i + \beta_j + (\alpha\beta)_{ij} + \gamma_k + (\alpha\gamma)_{ik} + (\beta\gamma)_{jk} + \varepsilon_{ijk}$$
$$\text{for } i = 1, 2, \ldots, I, \ j = 1, 2, \ldots, J \text{ and } k = 1, 2, \ldots, K,$$

where

Y_{ijk} is the response with main-plot factor at level i on block j with sub-plot factor at level k

μ is the overall population mean

α_i is the fixed effect, relative to μ, of main-plot factor at level i with $\sum_{i=1}^{I} \alpha_i = 0$

$\beta_j \sim iid(0, \sigma_\beta^2)$ is random effect of block j with β_j

$(\alpha\beta)_{ij} \sim iid(0, \sigma_{\alpha\beta}^2)$ is the interaction between main-plot factor level i and block j and is taken as the main-plot error because there is no replication of main-plot factor levels in blocks

TABLE 12.9: Sum of squares of split plot design.

Source of variation	degrees of freedom	Sum of squares
Main-plot factor	$I - 1$	$JK \sum_{i=1}^{I} (\bar{y}_{i..} - \bar{y}_{...})^2$
Blocks	$J - 1$	$IK \sum_{j=1}^{J} (\bar{y}_{.j.} - \bar{y}_{...})^2$
Main-plot residuals	$(I-1)(J-1)$	$K \sum_{i=1}^{I} \sum_{j=1}^{J} (y_{ij.} - \bar{y}_{i..} - \bar{y}_{.j.} + \bar{y}_{...})^2$
Sub-plot factor	$K - 1$	$IJ \sum_{k=1}^{K} (\bar{y}_{..k} - \bar{y}_{...})^2$
Main $\times$ sub- interaction	$(I-1)(K-1)$	$J \sum_{i=1}^{I} \sum_{k=1}^{K} (\bar{y}_{i.k} - \bar{y}_{i..} - \bar{y}_{..k} + \bar{y}_{...})^2$
Sub $\times$ blocks interaction	$(K-1)(J-1)$	$I \sum_{j=1}^{J} \sum_{k=1}^{K} (\bar{y}_{.jk} - \bar{y}_{.j.} - \bar{y}_{..k} + \bar{y}_{...})^2$
Sub-plot residuals	$(I-1)(J-1)(K-1)$	$\sum_{i=1}^{I} \sum_{j=1}^{J} \sum_{k=1}^{K} \Big(y_{ijk} - \bar{y}_{ij.} - \bar{y}_{i.k} - \bar{y}_{.jk} + \bar{y}_{i..} + \bar{y}_{.j.} + \bar{y}_{..k} - \bar{y}_{...} \Big)^2$

γ_k is the fixed effect, relative to μ, of sub-plot factor at level k with $\sum_{k=1}^{K} \gamma_k = 0$

$(\alpha\gamma)_{ik}$ is the fixed interaction effects, relative to μ, of the main-plot and sub-plot factor levels with $\sum_{i=1}^{I} (\alpha\gamma)_{ik} = 0$ and $\sum_{k=1}^{K} (\alpha\gamma)_{ik} = 0$

The breakdown of the total sum of squares is a three-way ANOVA and as shown in Table 12.9. The expected values of mean squares (Table 12.10) in the ANOVA all follow from considering means and noting that averages of fixed effects are 0 whereas averages of random effects are sample means and so random variables with mean 0.

Therefore a mean of main-plot factor level i over the sub-plot levels is

$$\overline{Y}_{ij.} = \mu + \alpha_i + \beta_j + (\alpha\beta)_{ij} + 0 + 0 + \overline{(\beta\gamma)}_{j.} + \bar{\varepsilon}_{ij.}$$

A mean of main-plot factor level i over blocks and the sub-plot factor levels is

$$\overline{Y}_{i..} = \mu + \alpha_i + \overline{\beta}_. + \overline{(\alpha\beta)}_{i.} + 0 + 0 + \overline{(\beta\gamma)}_{..} + \bar{\varepsilon}_{i..}$$

and the overall mean is

$$\overline{Y}_{...} = \mu + 0 + \overline{\beta}_. + \overline{(\alpha\beta)}_{..} + 0 + 0 + \overline{(\beta\gamma)}_{..} + \bar{\varepsilon}_{...}$$

Then, for example, it follows that the

$$
\mathrm{E}\left[\frac{\sum_{i=1}^{I}\sum_{j=1}^{J}\sum_{k=1}^{K}(\overline{Y}_{i..}-\overline{Y}_{...})^2}{I-1}\right] = \mathrm{E}\left[\frac{\sum_{i=1}^{I}\sum_{j=1}^{J}\sum_{k=1}^{K}(\alpha_i-+(\overline{\alpha\beta})_{i.}-(\overline{\alpha\beta})_{..}+\bar{\varepsilon}_{i..}-\bar{\varepsilon}_{...})^2}{I-1}\right]
$$

$$
= \frac{JK\sum_{i=1}^{I}\alpha_i^2}{I-1} + \frac{JK\sigma_{\alpha\beta}^2}{J} + \frac{JK\sigma_\varepsilon^2}{JK}
$$

$$
= \sigma_\varepsilon^2 + K\sigma_{\alpha\beta}^2 + \frac{JK\sum_{i=1}^{I}\alpha_i^2}{I-1},
$$

as shown in Table 12.10.

TABLE 12.10: Expected values of mean squares ($E[MS]$) of split plot design.

Source of variation	degrees of freedom	$E[MS]$
Main-plot factor	$I-1$	$\sigma_\varepsilon^2 + K\sigma_{\alpha\beta}^2 + \dfrac{JK\sum_{i=1}^{I}\alpha_i^2}{(I-1)}$
Blocks	$(J-1)$	$\sigma_\varepsilon^2 + I\sigma_{\beta\gamma}^2 + K\sigma_{\alpha\beta}^2 + IK\sigma_\beta^2$
Main-plot residuals	$(I-1)(J-1)$	$\sigma_\varepsilon^2 + K\sigma_{\alpha\beta}^2$
Sub-plot factor	$K-1$	$\sigma_\varepsilon^2 + I\sigma_{\beta\gamma}^2 + \dfrac{IJ\sum_{k=1}^{K}\gamma_k^2}{(K-1)}$
Main $\times$ sub-interaction	$(I-1)(K-1)$	$\sigma_\varepsilon^2 + \dfrac{J\sum_{i=1}^{I}\sum_{k=1}^{K}(\alpha\gamma)_{ik}^2}{(I-1)(K-1)}$
Sub $\times$ blocks interaction	$(K-1)(J-1)$	$\sigma_\varepsilon^2 + I\sigma_{\beta\gamma}^2$
Sub-plot residuals	$(I-1)(J-1)(K-1)$	σ_ε^2

Notice that the random effect $(\overline{\beta\gamma})_{..}$ is the same in both $\overline{Y}_{i..}$ and $\overline{Y}_{...}$ because they have both been averaged over j and k. The other expected values follow by applying the same principles. We use Table 12.10 to test null hypotheses that the levels of the main-plot factor, that the levels of the sub-plot factor, and that their interactions have no effect on the response. For the F-tests we assume the errors are *iid* normal.
For the main-plot factor:

$$
H_0^{main} : \alpha_1 = \alpha_2 = \cdots = \alpha_I = 0
$$
$$
H_1^{main} : \text{not all } \alpha_i = 0.
$$

If H_0^{main} is true then

$$\frac{\text{Main-plot factor } MS}{\text{Main-plot residuals}} \sim F_{I-1,(I-1)(J-1)}.$$

If H_0^{main} is not true then we expect the ratio to be larger and we can test H_0^{main} at the α-level by rejecting it if the calculated value of the ratio exceeds $F_{\alpha,I-1,(I-1)(J-1)}$. For the sub-plot factor:

$$H_0^{sub} : \gamma_1 = \gamma_2 = \cdots = \gamma_K = 0$$
$$H_1^{sub} : \text{not all } \gamma_k = 0.$$

If H_0^{sub} is true then

$$\frac{\text{Sub-plot factor } MS}{\text{Sub-plot by blocks interaction}} \sim F_{K-1,(K-1)(J-1)}.$$

The initial value for the test H_0^{sub} at the α-level is $F_{\alpha,K-1,(K-1)(J-1)}$. For the interaction between levels of the main-plot factor and the sub-plot factor:

$$H_0^{int} : (\alpha\gamma)_{11} = (\alpha\gamma)_{12} = \cdots = (\alpha\gamma)_{IK} = 0$$
$$H_1^{int} : \text{not all } (\alpha\gamma)_{ik} = 0.$$

If H_0^{int} is true then

$$\frac{\text{Main by sub interaction } MS}{\text{Sub-plot residuals } MS} \sim F_{(I-1)(K-1),(I-1)(J-1)(K-1)}.$$

The initial value for the test H_0^{int} at the α-level is $F_{\alpha,(I-1)(K-1),(I-1)(J-1)(K-1)}$. We can also use the expected values of the mean squares in Table 12.10 to produce estimators of components of variance. For example

$$\widehat{\sigma}_{\alpha\beta}^2 = \frac{\text{Main-plot residuals} - \text{Sub-plot residuals}}{K}.$$

We can use the same principle to obtain the standard error of differences in means for main-plot factor levels and sub-plot factor levels. For the main-plot factor, consider the estimation of the differences between level l and level m.

$$\overline{Y}_{l..} - \overline{Y}_{m..} = \left(\mu + \alpha_l + \overline{\beta}_{.} + \overline{(\alpha\beta)}_{l.} + 0 + 0 + \overline{(\beta\gamma)}_{..} + \overline{\varepsilon}_{l..} \right)$$
$$- \left(\mu + \alpha_m + \overline{\beta}_{.} + \overline{(\alpha\beta)}_{m.} + 0 + 0 + \overline{(\beta\gamma)}_{..} + \overline{\varepsilon}_{m..} \right)$$
$$= \alpha_l - \alpha_m + \overline{(\alpha\beta)}_{l.} - \overline{(\alpha\beta)}_{m.} + \overline{\varepsilon}_{l..} - \overline{\varepsilon}_{m..}$$

Then

$$\text{var}(\overline{Y}_{l..} - \overline{Y}_{m..}) = \left(\frac{\sigma_{\alpha\beta}^2}{J} + \frac{\sigma_\varepsilon^2}{JK} \right) + \left(\frac{\sigma_{\alpha\beta}^2}{J} + \frac{\sigma_\varepsilon^2}{JK} \right)$$

and the standard error of the difference in the main-plot factor levels is

$$\sqrt{2 \left(\frac{\sigma_{\alpha\beta}^2}{J} + \frac{\sigma_\varepsilon^2}{JK} \right)}.$$

If we refer to the expected values of the mean squares in the ANOVA table Table 12.10, we see that the main plot error divided by JK is an unbiased estimator of $\dfrac{\sigma_{\alpha\beta}^2}{J} + \dfrac{\sigma_{\varepsilon}^2}{JK}$. A similar argument (Refer to Exercise 12.16) gives the standard error of the difference in two levels l and m of the sub-plot factor as

$$\sqrt{2\left(\frac{\sigma_{\beta\gamma}^2}{K} + \frac{\sigma_{\varepsilon}^2}{IK}\right)}.$$

Referring to Table 12.10 the sub- factor level by blocks interaction divided by IK is an unbiased estimator of $\dfrac{\sigma_{\beta\gamma}^2}{K} + \dfrac{\sigma_{\varepsilon}^2}{IK}$.

We demonstrate the design and analysis of split plot experiments with two examples: the effects of irrigation type and fertilizer level on rice yields, and the effects of design type and pulse rate on the pressure drop across prosthetic heart valves.

Example 12.5: Irrigation [split plot design]

An experiment compares the effects of four irrigation methods and three levels of fertilizer on rice production in Africa [Clarke and Kempson, 1994] [7].

We read the data, summarized in Table 12.11, into R, check the head and tail of the data frame, and then plot the data in Figure 12.7. The main-plot factor was irrigation method, of which four were compared, on two blocks. The sub-plot factor was fertilizer for which there were three levels of application. A diagram of the design of this split plot experiment is shown in Figure 12.6, where $I1, I2, I3$ and $I4$ are the four irrigation methods and $f1, f2$ and $f3$ are the three fertilizer methods.

TABLE 12.11: Yields of rice by irrigation method (I1, I2, I3, I4) and block (b1, b2) with fertilizer low (L), medium (M) and high (H).

Irrigation	Block	Fertilizer level L	M	H
I1	b1	2.16	2.38	2.77
I1	b2	2.52	2.64	3.23
I2	b1	2.03	2.41	2.68
I2	b2	2.31	2.50	2.48
I3	b1	1.77	1.95	2.01
I3	b2	2.01	2.06	2.09
I4	b1	2.44	2.63	3.12
I4	b2	2.23	2.04	2.33

```
> irrig.dat=read.table('irrigation.txt',header=TRUE)
> attach(irrig.dat)
> print(head(irrig.dat))
     block  irrig  fert      y
```

---

[7][Colaizzi et al., 2004] describe a similar but larger experiment comparing subsurface drip irrigation (SDI), low-energy precision application (LEPA) and spray irrigation for grain sorghum. The blocks were locations over three years and the main-plot factor was irrigation method. One of the responses was yield of sorghum but others included soil water parameters. The sub-plot factors included volume of water used and fertilizer application [Colaizzi et al., 2004].

```
1      b1      i1      f1      2.16
2      b1      i1      f2      2.38
3      b1      i1      f3      2.77
4      b1      i2      f1      2.03
5      b1      i2      f2      2.41
6      b1      i2      f3      2.68
> x=irrig:block;x_n=as.numeric(x);fert_n=as.numeric(fert)
> plot(x_n,y,pch=fert_n,xaxt="n",xlab="Irrigation method|block",ylab=
      "Yield")
> axis(1,at=c(1:8),labels=levels(x))
> legend(5,3.1,legend=c("L","M","H"),pch=fert_n)
```

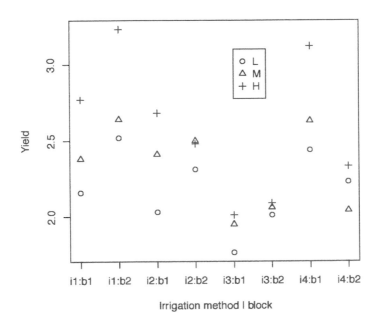

FIGURE 12.7: Scatter plot of rice yields by irrigation method | block.

The third irrigation method appears to be associated with a lower yield, and to increase with the fertilizer level. There is no suggestion of any additive block effect, and any interaction between blocks and irrigation method is confounded with the error. A standard three-way ANOVA will give the correct breakdown of the total sum of squares, but the default F-ratios and associated p-values will relate to three factors. Here we use the breakdown given by the R aov() command and complete the analysis by direct calculations[8].

```
> if.aov=aov(y~irrig*block*fert)
> ifat=summary(if.aov)[[1]]
```

---

[8]Other options in R are to use aov() with additional syntax or to use lmer(). These are considered in Exercise 12.17.

```
> print(ifat)
                  Df   Sum Sq  mean Sq
irrig             3   1.32971  0.44324
block             1   0.00034  0.00034
fert              2   0.67530  0.33765
irrig:block       3   0.65105  0.21702
irrig:fert        6   0.20110  0.03352
block:fert        2   0.08320  0.04160
irrig:block:fert  6   0.07927  0.01321
```

The F-ratios for testing the null hypothesis of no difference in irrigation types, no difference in fertilizer levels and no interaction are 0.44324/0.21702, 0.33765/0.04160, and 0.03352/0.01321 respectively. These quotients are evaluated and rounded to two decimal places in the following code.

```
> #note that if is reserved word in R
> i=ifat[1,3];b=ifat[2,3];f=ifat[3,3]
> ib=ifat[4,3];fi=ifat[5,3];bf=ifat[6,3];ibf=ifat[7,3]
> Fi=i/ib
> Ff=f/bf
> Fif=fi/ibf
> print(c("Fmain",round(Fi,2),"Fsub",round(Ff,2),"Fint",round(Fif,2)))
[1] "Fmain" "2.04"  "Fsub"  "8.12"  "Fint"  "2.54"
> pvi=1-pf(Fi,ifat[1,1],ifat[4,1])
> pvf=1-pf(Ff,ifat[3,1],ifat[6,1])
> pvif=1-pf(Fif,ifat[5,1],ifat[7,1])
> print(c(pvi,pvf,pvif))
[1] 0.2862214 0.1096902 0.1409901
```

This is a rather small experiment, with only two blocks, and there is no evidence to reject any of the null hypotheses at a 5% level of significance. However, there is weaker evidence of an interaction between irrigation method and fertilizer ($p = 0.11$). From the plot it appears that a benefit of high fertilizer level is particularly marked with irrigation method 1 and hardly apparent with irrigation method 3. This could be accounted for by irrigation method 3 tending to wash away the fertilizer before it has any effect on the crop. The lack of significance of fertilizer levels, when addition of fertilizer is generally considered beneficial in this region, can be attributed to the small size of the experiment. Also, if the interaction between fertilizer levels and blocks is considered negligible and pooled with the residual mean square, then the fertilizer levels become highly significant (Exercise 12.17). We continue to calculate means for irrigation methods and fertilizer levels and comment on these.

```
> mean(y)
[1] 2.36625
> irrigmean=tapply(y,irrig,mean);print(round(irrigmean,2))
   i1   i2   i3   i4
 2.62 2.40 1.98 2.46
> sedm=sqrt(ib/6+ib/6);print(round(sedm,2))
[1] 0.27
> fertmean=tapply(y,fert,mean);print(round(fertmean,2))
   f1   f2   f3
```

```
2.18 2.33 2.59
> seds=sqrt(bf/8+bf/8);print(round(seds,2))
[1] 0.1
```

Irrigation method 1 is estimated as increasing yield to 2.62 from a mean yield of 2.37, and such an increase would be valuable. Moreover, it exceeds the estimated yield from irrigation method 3 by 0.64. This difference of 0.64 exceeds twice the estimated standard error of the difference in means ($2 \times 0.27 = 0.54$), but it is not statistically significant, at even the 10% level, with only two blocks. Also, the estimated differences in yield at the different fertilizer levels exceed the estimated standard error of the difference in means (0.1). A tentative conclusion is that irrigation method 1 with the high level of fertilizer will give the highest yield. However, a follow up experiment is recommended.

Example 12.6: Prosthetic heart valves [split plot design]

Frank Fretsaw and Rita Rivet had been asked to run an experiment to compare four designs, type A, B, C, and D, of prosthetic heart valves in a test rig at six simulated pulse rates from 20 beats per minute to 220 beats per minute in steps of 40. The response was pressure gradient (mm Hg) across the valve, and the aim was for this pressure drop to be as low as possible. Frank suggested manufacturing one prototype heart valve of each type, treating the four different heart valves as blocks and testing each pulse rate once on each block. The blocks would be treated as fixed effects and interpreted as the differences in types.

Rita pointed out that they would have no estimate of the variation between valves of the same type. They would not know whether significant differences between blocks were due to variation in manufacturing values or differences in the types. Rita suggested manufacturing six valves of each type, so that every pulse rate would be tested with a different valve of each type. This would require 24 valves, the 6 valves of each type being randomly assigned to the 6 pulse rates. The 24 runs would be performed in a random order.

Frank accepted that they would need replicate valves of each type but he added that the research budget would not cover the manufacture of six prototype valves of each type. They agreed that two valves of each type would be a reasonable compromise. Each valve would be tested at all 6 pulse rates, and there would now be 48 runs. The analysis needs to allow for the repeated measurements on the same valves. Their results are given in Table 12.12.

TABLE 12.12: Prosthetic heart valves: pressure gradient (mm Hg) by valve type(valve number within type) and pulse rate (beats per minute).

Design(number)	20	60	100	140	180	220
$A(1)$	12	8	4	1	8	14
$A(2)$	7	5	7	5	13	20
$B(1)$	20	15	10	8	14	25
$B(2)$	14	12	7	6	18	21
$C(1)$	21	13	8	5	15	27
$C(2)$	13	14	7	9	19	23
$D(1)$	15	10	8	6	10	21
$D(2)$	14	12	7	6	18	21

The analysis follows a split-plot design, if the blocks are defined as the first and second valve of each type, and the main-plot factor is the type. The sub-plot factor is the pulse rate. We begin the analysis by reading the data and drawing a graph.

```
> phv.dat=read.table("pros_heart_valve.txt",header=TRUE)
> attach(phv.dat)
> print(head(phv.dat))
  valve type rate grad
1    1    A   20   12
2    1    A   60    8
3    1    A  100    4
4    1    A  140    1
5    1    A  180    8
6    1    A  220   14
> #treat first and second valve of each type as two blocks
> #valve number modulo 2 will coded blocks as 1 and 0
> block = valve %% 2
> #if you prefere to code blocks as 1 and 2
> block[block==0]=block[block==0]+2
> #Then valve type is a main plot factor
> #We can change the letters to integers for the plot by
> convert_a2n <- function(x) {
+     if (x == "A") { x <- 1}
+   else if (x == "B") { x <- 2}
+   else if (x == "C") {x <- 3 }
+   else if (x == "D") { x <- 4}
+   else { x <- NA}
+     return(x)
+ }
> typen=sapply(type,convert_a2n)
> #now we can plot
> plot(rate,grad,xlab="Pulse rate",ylab="Flow gradient",pch=typen)
> legend(100,25,c("A","B","C","D"),pch=c(1:4))
```

Figure 12.8 shows a clear tendency for the flow gradient to increase at the lower and higher ends of the range. It also seems that valve type A and, perhaps, valve type D have lower flow gradients. We continue with the ANOVA.

```
> #valve type is main-plot factor
> #Flow gradient is sub-plot factor
> rate_f=factor(rate)
> b_f=factor(block)
> type_f=factor(type)
> grad.aov=aov(grad~type_f*b_f*rate_f)
> gradat=summary(grad.aov)[[1]]
> print(gradat)
```

	Df	Sum Sq	Mean Sq
type_f	3	261.90	87.299
b_f	1	1.02	1.021
rate_f	5	1192.35	238.471

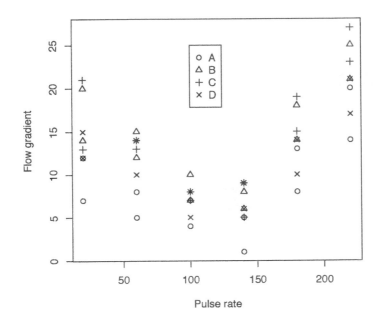

FIGURE 12.8: Scatter plot of flow gradient by pulse rate with type shown by plotting symbol.

```
type_f:b_f           3    25.06   8.354
type_f:rate_f       15    55.73   3.715
b_f:rate_f           5   112.35  22.471
type_f:b_f:rate_f   15    61.06   4.071

> t=gradat[1,3];b=gradat[2,3];p=gradat[3,3]
> tb=gradat[4,3];tp=gradat[5,3];bp=gradat[6,3];tbp=gradat[7,3]
> Ft=t/tb
> Fp=p/bp
> Ftp=tp/tbp
> pvt=1-pf(Ft,gradat[1,1],gradat[4,1])
> pvp=1-pf(Fp,gradat[3,1],gradat[6,1])
> pvtp=1-pf(Ftp,gradat[6,3],gradat[7,3])
> print(c(pvt,pvp,pvtp))
[1] 0.04265181 0.01074965 0.61662409
> typemean=tapply(grad,type,mean);print(round(typemean,2))
   A     B     C     D
8.67 14.17 14.50 11.75
> sedm=sqrt(tb/2+tb/2);print(round(sedm,2))
[1] 2.89
> pulsemean=tapply(grad,rate,mean);print(round(pulsemean,2))
   20    60   100   140   180   220
14.25 11.38  7.00  6.12 13.88 21.00
```

```
> seds=sqrt(bp/8+bp/8);print(round(seds,2))
[1] 2.37
```

There is no evidence of an interaction between the pulse rate and the valve type (p=0.62), so we can consider the main effect of valve type. There is evidence of a difference between valve types (p=0.043). The estimated variance of the difference in two types of value is

$$2 \times \left(\frac{8.354}{6 \times 2}\right) = 1.392.$$

The degrees of freedom for the t-distribution for the LSD are taken as the degrees of freedom for the main plot error, 3 in this case. So, the LSD(5%) is $t_{0.025,3} \times \sqrt{1.392} = 3.18 \times 1.18 = 3.76$. Using this criterion there is evidence that valve Type A has a lower flow gradient than valve Type C and valve Type B (the difference in sample means being 5.5). There is weaker evidence that the flow gradient of valve Type A may be less than the flow gradient of valve Type D. A 90% confidence interval for the difference in flow gradient between value type A and value type D is $-3.1 \pm t_{0.95,3} \times 1.18$ which gives $[-5.9, -0.3]$. There is no evidence of a difference in the flow gradients of valve Type D and either valve Type B or valve Type C (the differences in sample means being 2.4 and 2.75 respectively).

We can continue the analysis and estimate the standard deviation of valves of the same type $(\sigma_{\alpha\beta})$. Referring to the expected values of the mean squares in the ANOVA table, this is given by

```
> s_tb=sqrt((tb-tbp)/(gradat[3,1]+1))
> print(s_tb)
[1] 0.8449195
> s_e=sqrt(gradat[7,3])
> print(s_e)
[1] 2.017631
```

The standard deviation of the flow gradient for different valves of the same type is estimated as 0.84, which is less than half the estimated standard deviation of flow gradients measured on different runs with the same valve. The variability of flow gradient between different valves of the same type appears to be practically negligible.

12.6   Summary

12.6.1   Notation

There is duplication of notation which depends on context.

Y_{ij} response considered as a random variable for factor level i and replicate j

y_{ij} observed response for factor level i and replicate j

μ overall mean

α_i effect of factor level i relative to μ

ε_{ij} random error $iid\ (0, \sigma_\varepsilon^2)$

y_{ij} observed response for factor A at level i and factor B at level j

α_i, β_j effects of A, B relative to μ

y_{ijk} observed response for replicate k factor A at level i and factor B at level j

α_i, β_j main effects of factor levels i for A and j for B relative to μ

$(\alpha\beta_{ij})$ interaction effects of A at level i and B at level j relative to $\mu + \alpha_i + \beta_j$

y_{ij} observed response for factor level i in block j

α_i effect of factor level i relative to μ

β_j factor level block j effect $iid\ (0, \sigma_\beta^2)$

y_{ijk} observed response in split design. Main plot factor at level i, block j and sub plot factor at level k.

12.6.2 Summary of main results

The breakdowns of the total sum of squares, and the expected values of the mean squares, have been presented for the following balanced designs.

- Comparison of several means using independent samples.

- Two factors each at several levels with no replication. interactions confounded with errors.

- Two factors each at several levels with replication allowing for interactions.

- Comparison of several means over blocks.

- Split-plot designs

The ANOVA facilitates tests of overall null hypotheses. The least significant difference (LSD) is demonstrated as a follow up procedure when there is evidence to reject the null hypotheses.

12.6.3 MATLAB and R commands

In the following `mydata` is a dataset or matrix containing data with variables in columns and `formula` describes the relationships between variables in `mydata` being investigated. For more information on any built in function, type `help(function)` in R or `help function` in MATLAB.

R command	MATLAB command
`aov(formula, data=mydata)`	`anova1(mydata)` *

*There are also `anova2` and `anovan` commands for two way and n-way ANOVA.

12.7 Exercises

Section 12.2 Comparison of several means with one-way ANOVA

Exercise 12.1: Burn times

The results of an experiment to compare the burn times (s) of marine distress flares from three manufacturers (A,B, C) are reproduced in the following table.

A	B	C
122	109	127
118	57	199
131	135	136
110	94	117
39	114	183
41	157	143
103	108	228
84	140	204

1. Plot the data.

2. Calculate the means and standard deviations.

3. Does the ratio of the largest of the three sample variances to the smallest provide evidence of a difference in the corresponding population variances at the 0.10 level of significance?

4. Test the hypothesis of no difference in population means at the 0.05 level of significance.

5. Compare the three means using the LSD(0.05) criterion.

6. Compare the results of the LSD(0.05) procedure with the three two-sample t-tests and explain why the results are not identical.

7. Do the three two-sample t-tests lead to a different conclusion from LSD(0.05)?

Exercise 12.2: Limestone cores

Eighteen limestone cores from the same quarry were randomly allocated into three groups of six before testing for compressive strength. One group was tested in the natural state, another group was tested after drying in an electric oven and the third group was tested after saturation in water at room temperature for two months [Bajpai et al., 1968]. The compressive strengths in N mm^{-2} were

Natural	Dried	Saturated
60.6	75.2	72.8
65.1	64.8	82.7
96.9	82.9	59.8
57.9	64.3	61.8
85.6	70.2	57.1
82.7	54.8	91.8

1. Plot the data.

2. Calculate the means and standard deviations.

3. Test the hypothesis that the treatment of the three groups has no effect on compressive strength measurements at the 0.10 level of significance.

4. Suppose a difference of 5.0 between strength tests made on cores in their natural state and cores after drying in an oven is of practical importance. What size samples would you recommend for a follow up experiment comparing two groups, natural state with oven drying before testing, for compressive strength?

Exercise 12.3: Comparing several means with unequal sample sizes

Consider the model

$$Y_{ij} = \mu + \alpha_i + \varepsilon_{ij}$$

where $i = 1, \cdots, I$, $j = 1, \cdots, n_i$, $\sum_{i=1}^{I} \alpha_i = 0$, and ε_{ij} are iid mean 0 and variance σ_ε^2.

(a) Show that $\sum_{i=1}^{I} \sum_{j=1}^{n_i} (y_{ij} - \bar{y}_{..})^2 = \sum_{i=1}^{I} \sum_{j=1}^{n_i} (\bar{y}_{i.} - \bar{y}_{..})^2 + \sum_{i=1}^{I} \sum_{j=1}^{n_i} (\bar{y}_{ij} - \bar{y}_{i.})^2$ where

$$\bar{y}_{..} = \sum_{i=1}^{I} \sum_{j=1}^{n_i} y_{ij} / \sum_{i=1}^{I} n_i$$

(b) Explain why the residual sum of squares has $\sum_{i=1}^{I} (n_i - 1)$ degrees of freedom.

(c) Find expected value of treatments mean square.

Exercise 12.4: Studentized range distribution

The studentized range was introduced by WS Gosset (Student) in 1927 and its use as a follow up procedure to an ANOVA was described by John Tukey in 1949. The studentized range statistic is defined as the ratio of the range of an SRS of size m from a normal distribution to an independent estimate of the standard deviation of that distribution based on ν degrees of freedom. So, the distribution has two parameters m and ν.

(a) Explain why the mean and variance of the normal distribution are irrelevant.

(b) Run the following R code (or MATLAB equivalent) by replacing the "?" with some choice of m and n, and investigate the sampling distribution of the studentized range.

```
K=10000
m=?;n=?
Q=rep(0,K)
for (k in 1:K){
x=rnorm(m)
z=rnorm(n)
Q[k]=(max(x)-min(x))/sd(z)
}
histogram(Q)
```

(c) Compare the upper 5% point of your empirical distribution with the theoretical value that can be obtained from, for example, the qtukey() function in R.

```
Qs=sort(Q)
print(Qs[0.95*K])
print(qtukey(.95,m,(n-1)))
```

Exercise 12.5: Tukey's studentized range procedure

(a) In the context of the ANOVA for comparison of I means based on SRS of size J from each population explain why Q defined by

$$Q = \frac{\mathrm{range}(\{\overline{Y}_{i.}\})}{S/\sqrt{J}}$$

has a studentized range distribution with parameters I and $I(J-1)$.

(b) Any means that differ by more than $\dfrac{q(\alpha, I, I(J-1))}{\sqrt{J}}$ where $q(\alpha, I, I(J-1))$ is the upper α quantile of a studentized range distribution can be declared statistically significant with a family error rate less than or equal to α. That is the probability of declaring any of the $\binom{I}{2}$ differences statistically significant, when they are identical, is less than or equal to α for any configuration of the population means. Refer to Example 12.4 and calculate the studentized range criterion for a family error rate of 0.10.

The error rate of a multiple comparison method (overall or family-wise error rate) as the supremum of the probability of making at least one incorrect assertion is defined

$$\mathrm{error\ rate} = \sup_{\mu} \times \mathrm{P}_{\mu}(\text{at least one incorrect assertion}).$$

An assertion is a claim that a difference in population means is not equal to zero.

(i) Is the difference between the sample means for A and B statistically significant using an LSD(0.1)?

(ii) Is the difference between the sample means for A and B statistically significant using the studentized range criterion with a family error rate of 0.10?

(iii) If you use a Bonferroni correction with an LSD what value would you need to use?

(c) Refer to Example 12.4. Possible configurations of the means include:

$$\mu_A = \mu_B = \mu_C = \mu_D \qquad \mu_A = \mu_B = \mu_C \neq \mu_D \qquad \dots$$

(i) Explain why the initial F-test is of limited use in the context of multiple comparisons if one mean is substantially different from the other three which are equal.

(ii) Why does the studentized range procedure lead to an upper bound on the family error rate?

Section 12.3 Two factors at multiple levels ANOVA

Exercise 12.6: Water content of rocks

Refer to Example 12.2. The regression model with depth and porosity treated as categorical variables provides estimates of coefficients relative to a baseline depth of 10 m and baseline porosity of 20. Verify that these estimates of coefficients are equivalent to the effects relative to the overall mean.

Exercise 12.7: (Continued) Water content of rocks

Refer to Example 12.2.

(a) Fit a regression model of water content on linear effects of depth and porosity (m2). Compare this to the fitted model, with depth and porosity treated as categorical variables (m1). What is the estimated standard deviation of the errors in m2 and how does this compare with m1? Comment.

(b) Now include an interaction in m2. Is the -effect statistically significant?

(c) Scale the depth and porosity to $-5, -3, \cdots, 5$ and $-4, -2, \cdots, 4$ respectively and fit a regression of water content on linear and quadratic terms including the interaction. Is this model an improvement on m2?

Exercise 12.8: Two-way ANOVA

Refer to the model for two factors without replication.

(a) Expand $(a + b + c)^2$ as a sum of squares and two factor products.

(b) By writing

$$y_{ij} - \bar{y}_{..} = (\bar{y}_{i.} - \bar{y}_{..}) + (\bar{y}_{.j} - \bar{y}_{..}) + (y_{ij} - \bar{y}_{i.} - \bar{y}_{.j} + \bar{y}_{..})$$

obtain the breakdown of the sum of squares given in the ANOVA.

(c) Give a brief explanation for the degrees of freedom in the ANOVA table.

(d) Express:

$$\overline{Y}_{i.}, \overline{Y}_{.j} \text{ and } \overline{Y}_{..}$$

in terms of the parameters of the model and means of errors. Hence justify the expected values of the mean squares in the ANOVA table.

Exercise 12.9: Three-way ANOVA 1

(a) Expand $(a + b + c + d)^2$ as a sum of squares and two factor products.

(b) By writing

$$y_{ijk} - \bar{y}_{...} = (\bar{y}_{i..} - \bar{y}_{...}) + (\bar{y}_{.j.} - \bar{y}_{...}) + (\bar{y}_{ij.} - \bar{y}_{i..} - \bar{y}_{.j.} + \bar{y}_{...}) + (y_{ijk} - \bar{y}_{ij.})$$

obtain the breakdown of the sum of squares given in the ANOVA.

(c) Give a brief explanation for the degrees of freedom in the ANOVA table.

Exercise 12.10: Three-way ANOVA 2

(a) Refer to the model for two factors with replicates. Express:

$$\overline{Y}_{i..}, \overline{Y}_{.j.}, \overline{Y}_{ij.} \text{ and } \overline{Y}_{...}$$

in terms of the parameters of the model and means of errors.

(b) Use the results of (a) to justify the expected values of the mean squares in the ANOVA table.

Exercise 12.11: Storage battery

The maximum output voltage of a particular type of storage battery is thought to be influenced by the material used in the plates and the temperature in the location at which the battery is installed. Four replicates of a factorial experiment are run in the laboratory for three materials and three temperatures. The order in which the 36 observations are taken is randomly determined. The means for each material/temperature combination are given below.

Material type	Temperature		
	10	18	26
A	135	57	58
B	157	120	50
C	144	146	86

(a) Plot the data on a diagram. Write down a suitable model for the situation.

(b) The corrected sum of squares attributable to material types, temperature and interaction were 10 758, 39 100 and 9 840 respectively. The total corrected sum of squares was 77 647. Write down the ANOVA table including the expected values of the mean squares.

(c) Complete the analysis including an estimates of the standard error (standard deviation) of the treatment means.

Section 12.4 Randomized block design

Exercise 12.12: Bioleaching of copper

Consider the RBD for the bioleaching experiment in Example 12.4. Suppose that there was only sufficient material from the third recycling center (block 3) to test strains A, B and C. The design is no longer balanced, and the ANOVA table depends on whether strains or blocks are entered into the model first. The two tables are

```
> block=c(rep(1:3,3),1,2)
> strain=c(rep("A",3),rep("B",3),rep("C",3),rep("D",2))
> y=c(44,56,50,62,66,52,40,50,48,50,56)
> m2=aov(y~factor(strain)+factor(block))
> m3=aov(y~factor(block)+factor(strain))
> summary(m2)

Df Sum Sq Mean Sq F value Pr(>F)
factor(strain)  3   313.6    104.5   5.026 0.0571 .
factor(block)   2   146.0     73.0   3.510 0.1116
Residuals       5   104.0     20.8
---
> summary(m3)

Df Sum Sq Mean Sq F value Pr(>F)
factor(block)   2   147.6    73.82   3.549 0.1098
factor(strain)  3   312.0   104.00   5.000 0.0577 .
```

Residuals   5   104.0   20.80

(a) Explain why the ANOVA tables differ.

(b) Although the two p-values for strain only differ slightly in this application, only one is correct. Which is it?

(c) Estimate the strain effects relative to Strain A, and explain why it is not correct to calculate the differences in the means for the four strains.

Exercise 12.13: (Continued) Bioleaching of copper

Analyze the data of Example 12.4 using multiple regression with indicator variables for fungal strains and blocks. Use the following coding for the fungal strains and blocks in bioleaching experiment.

Fungal strain	x_1	x_2	x_3
A	1	0	0
B	0	1	0
C	0	0	0
D	0	0	1

Block	x_4	x_5
1	0	0
2	1	0
3	0	1

Exercise 12.14: Smoke emission

The smoke emission measurements (coded units) for three urban power stations A, B and C for 12 consecutive months were analyzed, and a part of the ANOVA table is given below.

Source of Variation	Corrected Sum of Squares	degrees of freedom
Between stations	10.98	2
Between months	140.36	11
Total	181.92	35

(a) Explain the difference between a randomized block design and two-way ANOVA.

(b) Explain the difference between considering blocks as fixed effects and considering them as random effects.

(c) Write down a suitable model for the situation above. State whether it represents a two-way ANOVA or randomized block design. If it represents a randomized block design indicate whether the blocks are fixed or random.

(d) Complete the ANOVA table including expected values of the mean squares.

(e) Test the hypothesis that there is no difference between stations against the alternative that there is a difference at the 5% level.

(f) Use the studentized range procedure to complete the analysis if the means for A, B and C are 52.2, 53.6 and 52.5 respectively. $[q(5\%, 3, 22) = 3.57]$. Compare this with the least significance difference (LSD).

Section 12.5 Split plot design

Exercise 12.15: Pulp preparation

The following data [Montgomery, 2004] are from an experiment to investigate the effects of three pulp preparation methods and four temperatures on the tensile strength of paper. The experiment was run as a split plot design with pulp preparation method being the main factor and temperature the subplot factor. Blocks are days.

Temperature	Block1 Pulp preparation method			Block2 Pulp preparation method			Block3 Pulp preparation method		
	1	2	3	1	2	3	1	2	3
200C	30	34	29	28	31	31	31	35	32
225C	35	41	26	32	36	30	37	40	34
250C	37	38	33	40	42	32	41	39	39
275C	36	42	36	41	40	40	40	44	45

(a) Plot the data.

(b) Calculate the means for the pulp preparation methods and their standard error.

(c) Calculate the means at the different temperatures and their standard error.

(d) What are your conclusions, and how would you operate the process to achieve the highest tensile strength?

Exercise 12.16: Split plot design 1

Starting from the model for a split plot design in Section 12.5, find expression for $\overline{Y}_{.l.}$ and $\overline{Y}_{.m.}$ in terms of the parameters of the model. Hence obtain the expression for the standard error of their difference.

Exercise 12.17: Split plot design2

The `aov()` function in R is quite versatile and allows random effects to be included under `Error()`.

(a) Analyze the rice yield data (Example 12.5) using the following lines of R code.

```
modela=aov(y~irrig+fert+irrig:fert + Error(block+block:irrig+block:fert))
summary(modela)
```

Verify that the results are the same as those presented in Example 12.5.

(b) Analyze the rice yield data (Example 12.5) using the following lines of R code.

```
modelb=aov(y~irrig+fert+irrig:fert + Error(block+block:irrig))
summary(modelb)
```

Explain how this analysis differs from that presented in Example 12.5. What additional assumption is being made and how would the conclusions change? Do you think the additional assumption is reasonable?

Miscellaneous problems

Exercise 12.18: Bonferroni inequality

Consider the Bonferroni inequality.

(a) Explain why $P(A \cup B) \leqslant P(A) + P(B)$ by drawing a Venn diagram. Under what circumstances is there equality? Show that the inequality is a consequence of the addition rule of probability.

(b) Explain why $P(A \cup B \cup C) \leqslant P(A) + P(B) + P(C)$ by drawing a Venn diagram.

(c) Deduce that $P(A \cup B \cup \cdots M) \leqslant P(A) + P(B) + \cdots + P(M)$.

(d) Justify the Bonferroni inequality.

Exercise 12.19: Cross over

A cross-over trial is a paired comparison experiment in which the order of treatments is allowed for. Sixty air traffic controllers were asked to rate two proposals (A, B) for a new system according to a detailed check-list. It was thought that there might be a tendency to prefer the second system so 30 controllers were randomly assigned to try A then B and the others tried B then A. Consider the model

$$
\begin{aligned}
Y_{1j1} &= \mu + S_{1j} + \varepsilon_{1j1} \\
Y_{1j2} &= \mu + \alpha + \tau + S_{1j} + \varepsilon_{1j2} \\
Y_{2j1} &= \mu + \alpha + S_{2j} + \varepsilon_{2j1} \\
Y_{2j2} &= \mu + \tau + S_{2j} + \varepsilon_{2j2}.
\end{aligned}
$$

where Y_{ijk} is the response for controller j, $j = 1, \cdots, n$, in group i, $(i = 1, 2)$, and time period j, $j = 1, 2$, μ is an overall mean and ε_{ijk} are $NID(0, \sigma^2)$. Group 1 try A before B and Group 2 try B before A. The parameter α is the effect of B relative to A and τ is the time period effect. Define

$$
\begin{aligned}
D_{1j} &= Y_{1j2} - Y_{1j1} = \alpha + \tau + \varepsilon_{1j2} - \varepsilon_{1j1} \\
D_{2j} &= Y_{2j2} - Y = -\alpha + \tau + \varepsilon_{2j2} - \varepsilon_{2j1}.
\end{aligned}
$$

(a) Explain how we can use a procedure based on the t-distribution to produce CI for α and τ.

(b) The data from the air traffic controllers is in

airtrafficcontrol_crossover.txt

Construct a 95% CI for α and τ.

Exercise 12.20: Comparison of sea planes balanced incomplete blocks

A sport aviation magazine is running a comparison of seven different designs of micro-light sea planes $A, B, \cdots, G$ that have been loaned by the manufacturers. The magazine reporter thinks that it is easier to make subjective comparisons between three items than between seven items, and has recruited seven pilots to each test three of the planes. The pilots will return total scores based on a variety of characteristics. The experimental design and the scores are shown below. The pilots tested their three planes in a random order.

Pilot 1	Pilot 2	Pilot 3	Pilot 4	Pilot 5	Pilot 6	Pilot 7
B 23	F 20	C 24	E 15	G 25	D 17	F 20
A 20	D 21	F 19	A 16	E 16	C 19	G 24
C 16	B 25	E 18	D 14	B 20	G 22	A 19

(a) (i) How many pilots test each plane?

   (ii) On how many occasions are A and B tested by the same pilot? Is this the same number for any choice of two designs?

   (iii) Fit a suitable model using linear regression and indicator variables for designs and pilots. Is there evidence of a difference in designs and, if so, which have the higher scores?

(b) In some experiments the blocks are too small to accommodate all the treatments. In a balanced incomplete block design (BIBD), let ν be the number of treatments, b be the number of blocks, r be the number of replicates of each treatment, k be the number of units in each block, and N be the number of units in the experiment. A BIBD is a design in which any pair of treatments occur together in the same block the same number of times, λ, known as the number of concurrences.

   (i) Explain why $r\nu = bk = N$.

   (ii) Explain why

$$\lambda = \frac{r(k-1)}{\nu - 1}.$$

   (iii) Is the design in Part (a) a BIBD, and if so what are the values of ν, k and λ?

   (iv) A necessary condition for a BIBD to exist is that λ is integer. Is it a sufficient condition?

Exercise 12.21: Latin square

In a Latin square design of order n there is a square divided into n rows and n columns to give n^2 plots. There is one factor with n levels customarily referred to by the Latin letters $A, B, C, ...$ up to n. Each letter appears once in each row and once in each column. A standard Latin square has the first row and first column in alphabetical order.

(a) Write down standard Latin squares of order 2, 3 and 4. Are they unique?

(b) A Martindale Abrasion tester,
    (http://www.worldoftest.com/martindale-abrasion-tester) consists of four brass plates with fine emery attached. Samples of fabric are held against the plates with weights and the plates abrade the fabric by rapid rotation in a figure of eight movement. The response is weight loss. The plates are located at four stations, four different fabrics are tested in four runs. Each run is performed by a different operator. The allocation of test pieces to run and station follows a Latin square design. The different fabrics are randomly allocated to letters. The data from such an experiment are given in the following table.

	Station 4	Station 2	Station 1	Station 3
Run 2	A (251)	B (241)	D (227)	C (229)
Run 3	D (234)	C (273)	A (274)	B (226)
Run 1	C (235)	D (236)	B (218)	A (268)
Run 4	B (195)	A (270)	C (230)	D (225)

(i) Set up indicator variables for Station and for Run. Analyze the data and state whether there is evidence of a difference in the fabrics.

(ii) Calculate the LSD(.05) for comparing means. What conclusions can you draw?

(c) The general model has the form

$$Y_{ijk} \;=\; \mu + \alpha_i + \beta_j + \gamma_k + \varepsilon_{ijk}$$

where $i, j, k = 1, \cdots, n$, $\sum_{i=1}^{n} \alpha_i = 0$, $\sum_{j=1}^{n} \beta_j = 0$, $\sum_{k=1}^{n} \gamma_k = 0$ and ε_{ijk} are *iid* mean 0 and variance σ_ε^2. Depending on the context, it may be more appropriate to take rows or columns as random effects. Then, for example, we specify $\beta_j \sim N(0, \sigma_\beta^2)$. However this does not affect the test of the null hypothesis about the treatments represented by the α_i.

(i) Explain the breakdown for the ANOVA

$$\sum_{ijk=1}^{n} (Y_{ijk} - \overline{Y}_{...})^2 \;=\; \sum_{ijk=1}^{n} ((\overline{Y}_{i..} - \overline{Y}_{..}) + (\overline{Y}_{.j.} - \overline{Y}_{...}) + (\overline{Y}_{..k} - \overline{Y}_{...}) +$$
$$(Y_{ijk} - \overline{Y}_{i.0} - \overline{Y}_{.j.} - \overline{Y}_{..k} + 2\overline{Y}_{...}))^2.$$

(ii) Show that the expected value of the treatment mean square is

$$\sigma^2 + n \sum_{i=1}^{n} \alpha_i^2 / (n-1).$$

13

Probability models

We now consider the reliability of systems given the reliability of their components. We first assume that components are not repairable, but that the system may be able to function despite some component failures because of redundancy. Then we apply Markov models to the analysis of repairable systems and a variety of other applications including queues. In some cases theoretical results can be obtained, but computer simulation is typically used for more complex models.

13.1 System reliability

The reliability of a component, indexed by i, is the probability, p_i, that it functions for some stated length of time.

Suppose a system consists of n components. The state, x_i, of the ith component is either working (1) or failed (0), for i from 1 up to n.
The state (s) of the system is either working (1) or failed (0), and depends on the states of the components.

Definition 13.1: Structure function

The structure function (φ), also called the system function, is a deterministic model for system failure and is defined by:

$$s \;=\; \varphi(x_1, \dots, x_n) \;=\; \varphi(\boldsymbol{x}),$$

where $\boldsymbol{x}$ is the $1 \times n$ array of the x_i. The domain of the function is $\{0,1\} \times \dots \times \{0,1\}$, this product set has 2^n elements, and the range is $\{0,1\}$.

In the following diagrams the system works if there is a flow from left to right. The flow passes through a component that works, but is blocked by a failed component.

13.1.1 Series system

FIGURE 13.1: Series system.

All components need to work, as this is a system where failure is 'not tolerated'. The structure function can be expressed in different ways, with two useful representations given by the following.

$$\varphi(\boldsymbol{x}) \;=\; \min\{x_1,\ldots,x_n\}$$

$$\text{and} \quad \varphi(\boldsymbol{x}) \;=\; x_1 \ldots x_n = \prod_{i=1}^{n} x_i.$$

Example 13.1: Petrol engined car

To start a petrol engined car you need all of the following:

- fuel in the tank,
- battery charged,
- fuel pump operative,
- starter motor operative,
- LT circuit to starter motor operative and
- HT circuit operative.

13.1.2 Parallel system

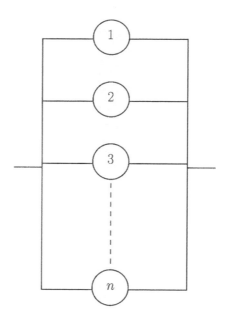

FIGURE 13.2: Parallel system.

Only one component needs to work here and again the structure function can be expressed in two ways.

$$\varphi(\boldsymbol{x}) \;=\; \max\{x_1,\ldots,x_n\}$$

$$\text{and} \quad \varphi(\boldsymbol{x}) \;=\; 1 - (1 - x_1)\ldots(1 - x_n).$$

If there is a large number of components in parallel it is useful to have a more succinct notation, and an **inverted upper case pi symbol** is used[1].

$$1 - (1 - x_1)\ldots(1 - x_n) \quad = \quad \coprod_{i=1}^{n} x_i.$$

Example 13.2: CD/DVD player

Consider a CD/DVD player that has

- a electrical lead,
- a battery pack and
- a spare battery pack.

You can play a CD/DVD provided any one of these power supplies works.

13.1.3 *k-out-of-n* system

The *k-out-of-n* system is also called a 'voting system', where there are n sensing devices and a function is performed when a set number, k, of these devices function. Notice that

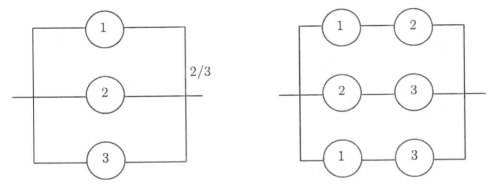

FIGURE 13.3: Alternative graphical representations for 2-out-of-3 system.

the right hand side representation in Figure 13.3 has duplicate nodes. There are at least two ways of expressing the structure function:

$$\varphi(\boldsymbol{x}) \quad = \quad 1 - (1 - x_1 x_2)(1 - x_2 x_3)(1 - x_1 x_3)$$

$$\text{and} \quad \varphi(\boldsymbol{x}) \quad = \quad int\left[\frac{\sum x_i}{2}\right],$$

where $int[\cdot]$ is the integer value of $[\cdot]$, also referred to as the floor function. A k-out-of-n system has the structure function

$$\varphi(\boldsymbol{x}) \quad = \quad int\left[\frac{\sum_{i=1}^{n} x_i}{k}\right].$$

---

[1]Look carefully to distinguish $\coprod$ from $\prod$.

Example 13.3: Boeing 747

A Boeing 747 has 4 engines, but is certified to fly on only 3 of its 4 engines.

Example 13.4: Citroën DS

The four wheel Citroën DS produced between 1955 and 1975 could run on three wheels due to its hydropneumatic suspension.

Example 13.5: Computer controls

Some computer controlled systems have three separate micro-processors and will operate if at least two provide identical output.

13.1.4 Modules

Example 13.6: A five component system

We can split a large system up into modules as shown in Figure 13.4. These modules often, but not always, form a series or parallel system.

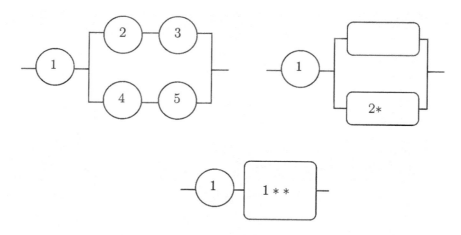

FIGURE 13.4: Decomposition into modules.

Define

$$s = x_1 z_1,$$
$$z_1 = 1 - (1 - y_1)(1 - y_2)$$
$$y_1 = x_2 x_3 \quad \text{and} \quad y_2 = x_4 x_5.$$

Variable	State of
s	system
z_1	module 1**
y_i	module i^*
x_i	component i

Successive substitution gives

$$s = x_1 \left(1 - (1 - x_2 x_3)(1 - x_4 x_5)\right).$$

Example 13.7: A bridge system

Look at the **bridge system** shown in Figure 13.5. It cannot be reduced to a set of series and parallel networks.

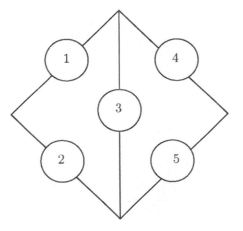

FIGURE 13.5: Bridge system.

However, for *fixed* x_3 the system can be reduced to sets of series and parallel networks and we can write

$$\varphi(\boldsymbol{x}) = x_3 z_1 + (1 - x_3)z_2,$$

where z_1 is two sets of two elements in parallel, in series and z_2 is two sets of two elements in series, in parallel. We then determine z_1 and z_2, so that $\varphi(\boldsymbol{x})$ can be written

$$x_3 \left(\left[1 - (1 - x_1)(1 - x_2)\right] \left[1 - (1 - x_4)(1 - x_5)\right] \right) + (1 - x_3)\left(1 - (1 - x_1 x_4)(1 - x_2 x_5)\right).$$

13.1.5   Duality

The dual of a system with structure function $\varphi(\boldsymbol{x})$ is defined as the system with structure function:

$$\varphi_D(\boldsymbol{x}) = 1 - \varphi(\mathbf{1} - \boldsymbol{x}).$$

Some examples are:

(i) The dual of a system with k components in parallel is a system with k components in series.

(ii) The dual of a single element is itself.

(iii) The dual of a k-out-of-n system is an $(n + 1 - k)$-out-of-n system.

Example 13.8: Duality

Consider the system with the block diagram shown in Figure 13.6.

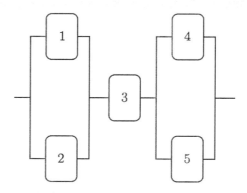

FIGURE 13.6: Original system.

It has a structure function:

$$\varphi(\boldsymbol{x}) \;=\; \Big(1 - (1 - x_1)(1 - x_2)\Big) x_3 \Big(1 - (1 - x_4)(1 - x_5)\Big).$$

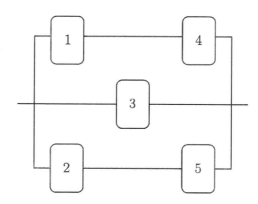

FIGURE 13.7: Dual system.

The dual system has a block diagram as shown in Figure 13.7 and has structure function:

$$\varphi_D(\boldsymbol{x}) \;=\; 1 - (1 - x_1 x_2)(1 - x_3)(1 - x_4 x_5).$$

13.1.6 Paths and cut sets

We assume from now on that the structure function is **monotone**, so that for all vectors $\boldsymbol{x}$ and $\boldsymbol{y}$

$$\boldsymbol{x} < \boldsymbol{y} \implies \varphi(\boldsymbol{x}) \leq \varphi(\boldsymbol{y}),$$

where $\boldsymbol{x} < \boldsymbol{y}$ means that all the elements of $\boldsymbol{x}$ are less than or equal to the corresponding elements of $\boldsymbol{y}$ and at least one is strictly less than the corresponding element of $\boldsymbol{y}$. In practical terms this means that repairing a single element cannot cause a working system to fail.

Definition 13.2: Minimal path vector and minimal path set

A **minimal path vector** minimal path vector(MPV) is a vector $\boldsymbol{x}$ such that

$$\varphi(\boldsymbol{x}) = 1$$

and

$$\varphi(\boldsymbol{y}) = 0, \quad \text{for all} \quad \boldsymbol{y} < \boldsymbol{x}.$$

The **minimal path set** corresponding to the MPV $\boldsymbol{x}$ is the set of indices i for which $x_i = 1$.

Definition 13.3: Minimal cut vector and minimal cut set

A **minimal cut vector** (MCV) is a vector $\boldsymbol{x}$ such that

$$\varphi(\boldsymbol{x}) = 0$$

and

$$\varphi(\boldsymbol{y}) = 1, \quad \text{for all} \quad \boldsymbol{y} > \boldsymbol{x}.$$

The **minimal cut set** corresponding to the MCV $\boldsymbol{x}$ is the set of indices i for which $x_i = 0$.

Example 13.9: Bridge system

For the bridge system shown in Figure 13.5, the minimal path sets are

$$\{1, 4\}, \{1, 3, 5\}, \{2, 5\}, \{2, 3, 4\}$$

and the minimal cut sets are

$$\{1, 2\}, \{4, 5\}, \{1, 3, 5\}, \{2, 3, 4\}.$$

The minimal path vector corresponding to the minimal path set $\{1, 4\}$ is the vector $(1, 0, 0, 1, 0)$.
The minimal cut vector corresponding to the minimal cut set $\{1, 2\}$ is the vector $(0, 0, 1, 1, 1)$.

Let $P_1, \ldots, P_m$ be the minimal path sets and $K_1, \ldots, K_k$ be the minimal cut sets of a system with structure function φ. The system works provided one or more minimal paths work and it can be represented by the minimal paths in parallel (Figure 13.8, which generally includes elements repeated several times). Then,

$$\varphi(\boldsymbol{x}) \;=\; \coprod_{i=1}^{m} \prod_{i \in P_j} x_i$$

Turning attention to the minimal cuts, a minimal cut will not itself cause the system to fail if any one of its elements is restored. If we restore at least one element in every MC the system will work (Figure 13.9), and

$$\varphi(\boldsymbol{x}) \;=\; \prod_{j=1}^{k} \coprod_{i \in K_j} x_i.$$

Try using these results to calculate the structure function for the bridge system.

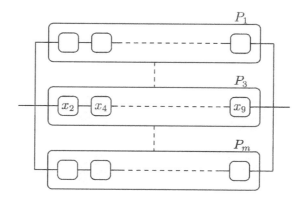

FIGURE 13.8: The system works provided at least one MP works.

13.1.7 Reliability function

Consider a system with n components and structure function φ. Let p_i be the *reliability* of the ith component, $i = 1, \ldots, n$. What is the reliability of the system?
The state, X_i, of component i is a random variable with

$$P(X_i = 1) \;=\; p_i \quad \text{and} \quad P(X_i = 0) \;=\; 1 - p_i.$$

We make extensive use of the following result

$$E[X_i] \;=\; 1 \times p_i + 0 \times (1 - p_i) \;=\; p_i.$$

Now assume that $X_1, X_2, \ldots, X_n$ are *independent*.
Let $\boldsymbol{x} = (X_1, \ldots, X_n)$. The *system reliability* is given by

$$P(\text{System works}) \;=\; P(\varphi(\boldsymbol{x}) = 1) \;=\; E[\varphi(\boldsymbol{x})].$$

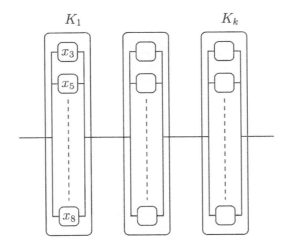

FIGURE 13.9: Every MC must fail.

The function $r : \boldsymbol{p} \rightarrow \mathrm{P}(\varphi(\boldsymbol{x}) = 1)$, where $\boldsymbol{p} = (p_1, \ldots, p_n)$, is called the **reliability function** of the system.

Some examples follow, and in each example we use the result that the expected value of a product of independent random variables is the product of their expectations.

Series system:

$$r(\boldsymbol{p}) \;=\; \mathrm{E}\left[\prod_{i=1}^{n} X_i\right] \;=\; \prod_{i=1}^{n} p_i.$$

Parallel system:

$$r(\boldsymbol{p}) \;=\; \mathrm{E}\left[1 - \prod_{i=1}^{n}(1 - X_i)\right] \;=\; 1 - \prod_{i=1}^{n}(1 - p_i).$$

k-out-of-n system, with $p_i = p$

$$r(\boldsymbol{p}) \;=\; \sum_{i=k}^{n} \binom{n}{i} p^i (1-p)^{n-i}.$$

For the *bridge system* we have

$$\varphi(\boldsymbol{x}) \;=\; 1 - (1 - X_1 X_3 X_5)(1 - X_2 X_3 X_4) \times (1 - X_2 X_5)(1 - X_1 X_4).$$

When we take the expectation of this, we may not simply replace X_i with p_i because products contain repeats of random variables. However, using the fact that $X_i = X_i^2 = X_i^3$, we may write

$$\begin{aligned}
\varphi(\boldsymbol{x}) \;=\;\; & X_1 X_3 X_5 + X_2 X_3 X_4 + X_2 X_5 + X_1 X_4 - X_1 X_2 X_3 X_5 \\
& - X_1 X_2 X_4 X_5 - X_1 X_3 X_4 X_5 - X_1 X_2 X_3 X_4 - X_2 X_3 X_4 X_5 + 2 X_1 X_2 X_3 X_4 X_5.
\end{aligned}$$

Because here all the terms are products of independent random variables we have

$$\begin{aligned}
r(\boldsymbol{p}) \;=\; \mathrm{E}[\varphi(\boldsymbol{x})] \;=\;\; & p_1 p_3 p_5 + p_2 p_3 p_4 + p_2 p_5 + p_1 p_4 - p_1 p_2 p_3 p_5 \\
& - p_1 p_2 p_4 p_5 - p_1 p_3 p_4 p_5 - p_1 p_2 p_3 p_4 - p_2 p_3 p_4 p_5 + 2 p_1 p_2 p_3 p_4 p_5.
\end{aligned}$$

13.1.8   Redundancy

Redundancy can be at the system level or at the component level. An example is shown in Figure 13.10 in which components 3 and 4 are duplicates of components 1 and 2 respectively.

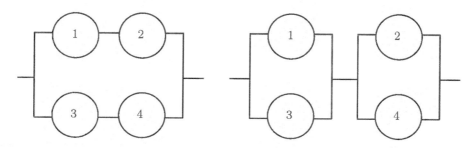

FIGURE 13.10: Redundancy at the system level (LH) and at the component level (RH). Components 1 and 3 and components 2 and 4 are duplicates.

Which system is more reliable? Suppose the probability of each component working is 0.90, and that failures are independent. What are the numerical reliabilities for the two systems?

13.1.9   Non-repairable systems

Consider a system with n components and structure function φ.
Suppose each component functions from time 0, until a random time T_i. Assume all these T_i lifetimes are independent.
Let $X_i(t)$ be the state of component i at time t an let T_i have cdf F_i:

$$F_i(t) \;=\; P(T_i \leq t) \;=\; P(X_i(t) = 0).$$

Let T be the failure time of the system. We express the cdf F of T in terms of the cdfs of the components.
We have

$$F(t) \;=\; P(T \leq t) \;=\; P(\varphi(\boldsymbol{x}(t)) = 0).$$

The reliability of the system at time t is given by

$$P(\varphi(\boldsymbol{x}(t)) = 1) \;=\; r(1 - F_1(t), \ldots, 1 - F_n(t)).$$

Hence,

$$F(t) \;=\; 1 - r(1 - F_1(t), \ldots, 1 - F_n(t)).$$

Example 13.10:   Christmas tree

A Christmas tree is illuminated by a string of 20 light bulbs. If one bulb fails the lights go out. The lifetime of each light bulb has an exponential distribution with an expectation of 1000 hours. If T is the failure time of the system, then

$$F(t) \;=\; 1 - \left(e^{-t/1000}\right)^{20}.$$

This is an exponential distribution with an expected value of 50 hours.

13.1.10   Standby systems

The system operates through 1 unless it fails, when 2 is called in by the switch. Let X_i be

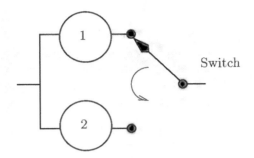

FIGURE 13.11: Switched standby system.

the state of component i, $i = 1, 2$.
If the switch is perfect, the reliability of the system is

$$r \;=\; \mathrm{P}(X_1 = 1) + \mathrm{P}(X_1 = 0, X_2 = 1).$$

If X_1 and X_2 are independent, with reliability p_1 and p_2, then

$$r \;=\; p_1 + (1 - p_1)p_2.$$

Component 2 is said to be in **hot standby** when it fails independently of component 1. An example is the tires on two wheels at the end of each axle on a semi-trailer.
Component 2 is said to be in **cold standby** when it can only fail after the switchover. An example is a spare tyre in the trunk of a car.

Example 13.11:   High definition LCD screens

High definition LCD screens, with novel design features, for a particular make of TV have lifetimes T with an exponential distribution. The mean lifetime of screens is 7.5 years.

  (i) What is the probability that the screen lasts at least 7.5 years?

 (ii) What is the probability that the screen lasts another 7.5 years once it has reached 7.5 years and is still working.

(iii) I have just bought one of these TVs and a spare screen. What is the probability I have a TV with a working screen in 15 years time if the spare screen has had the same lifetime distribution as the original screen since the purchase? Note that it is quite possible that the spare screen will have failed if I need to use it!

(iv) I have just bought one of these TVs in a special promotion with a guarantee of one replacement screen. What is the probability I have a TV with a working screen in 15 years time if the replacement screen has a lifetime with an exponential distribution with mean 7.5?

 (v) You are now told that the manufacturer put screens aside, in order to cover the guarantees at the time of the promotion, when the TVs were sold. Explain whether or not this will change your answer to (iv).

The solutions to these questions follow, where the parameter of the exponential distribution is

$$\lambda = 1/7.5$$

(i) The probability that the screen lasts at least 7.5 years is:

$$e^{-(1/7.5)(7.5)} = 0.3679$$

(ii) The probability that the screen lasts another 7.5 years once it has reached 7.5 years and is still working is the same as in (i). This follows from the following argument.

$$P(T > 15 \,|\, T > 7.5) = \frac{P(T > 15 \text{ and } T > 7.5)}{P(T > 7.5)} = \frac{P(T > 15)}{P(T > 7.5)}$$

$$= \frac{e^{-2}}{e^{-1}} = e^{-1}.$$

(iii)

$$P(\text{at least one screen works after 15 years}) = 1 - P(\text{both fail within 15 years})$$
$$= 1 - (1 - e^{-2})^2 = 0.2524$$

(iv) I have just bought one of these TVs in a special promotion with a guarantee of one replacement screen. What is the probability I have a TV with a working screen in 15 years time if the replacement screen has a lifetime with an exponential distribution with mean 7.5?

Let T be the time that the TV has a working screen. The quickest argument, and the one that easily generalizes to more than one replacement, is that T will exceed t if there are just 0 or 1 events in a Poisson process with failure rate λ equal to 1/7.5. The replacement screen, which is certain to work at the time of replacement, means that the TV can survive 0 or 1 screen failures with a constant hazard rate λ. So

$$S(t) = e^{-\lambda t} + (\lambda t)e^{-\lambda t}$$
$$F(t) = 1 - e^{-\lambda t} - (\lambda t)e^{-\lambda t}$$

Alternatively you can argue as follows. The probability the TV doesn't have a working screen after t years is equal to the sum, over years, of the probabilities that screen 1 fails in year k and screen 2 fails within $(t - k)$ years. The time increment can be refined from years to days to hours and in the limit we have the convolution:

$$F(t) = \int_0^t \lambda e^{-\lambda t}(1 - e^{-\lambda(t-\tau)})d\tau$$
$$= \left[-e^{-\lambda t} - (\lambda \tau)e^{-\lambda t}\right]_0^t = 1 - e^{-\lambda t} - (\lambda t)e^{-\lambda t}$$

We have t equal to 15 and λ equal to 1/15, so the probability the TV has a working screen in 15 years is: $e^{-2} + 2e^{-2}$.

(v) You are now told that the manufacturer put screens aside, in order to cover the guarantees at the time of the promotion, when the TVs were sold. If screens have lifetimes with an exponential distribution, the age of the replacement screen, given that it is working at the time of replacement, will not affect the probability that the TV has a working screen in 15 years time.

13.1.11 Common cause failures

Further complications are introduced when the components are not independent. This introduces the idea of **common cause failures** (CCF). Reasons for common cause failure include: identical manufacturing faults; same maintenance engineer; same software writer, or common software code; same power source; same operator.

For a 2-component parallel system with component reliability p, the system reliability is

$$r = 1 - (1 - p)^2,$$

if the components are *independent*. However, if there is a 'common cause failure' (Figure 13.12), the reliability could be as low as p.

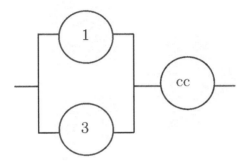

FIGURE 13.12: 2-component parallel system with common cause failure.

We can model dependencies by including an imaginary component in series with a reliability of p_{cc}. Then

$$r = p_{cc}(1 - (1 - p)^2).$$

13.1.12 Reliability bounds

Consider a monotone system with structure function φ. Let $K_1, \ldots, K_k$ be the minimal cut sets.

Let E_i be the event that all the components in K_i fail. Then, the reliability of the system is

$$r(p) = 1 - P\left(\bigcup_{i=1}^{k} E_i\right)$$

A useful general result in probability is the **Principle of Inclusion and Exclusion**, which states that the probability of the union of any sequence of events $E_1, E_2, \ldots,$ is equal to

$$\sum_i P(E_i) - \sum_{i<j} P(E_i \cap E_j) + \sum_{i<j<k} P(E_i \cap E_j \cap E_k) \ldots$$

Since successive approximations alternate around the true value, we can bound the reliability function.

Example 13.12: Electric car

A new design of a small electric car can have: a traction fault (A); an electrical fault (B); electronic system fault (C). Suppose the probabilities of these faults for a randomly selected car are 0.03, 0.02 and 0.01 respectively.

Suppose also that faults occur at random. What is the probability that a randomly selected car has at least one of these three faults? How good are the approximations?

$$
\begin{aligned}
\text{P(A or B or C)} &= 0.03 + 0.02 + 0.01 - 0.0006 - 0.0003 - 0.0002 + 0.000006 \\
&= 0.058906
\end{aligned}
$$

The approximations are 0.06 and 0.0589. Notice that they bound the precise value. The first term is the 'rare event approximation'; it holds when all p's are close to 1.

13.2   Markov chains

In this section, we consider a particular class of random process, known as a Markov process, or a Markov chain. We begin by recalling and extending some previous definitions and then go on to define a discrete time Markov chain. We often want to talk about the probability of an event A occurring conditional on the occurrence of another event B. Recall from Definition 2.13 that this is denoted $P(A \mid B)$ and is defined by

$$
P(A \mid B) = \frac{P(A \cap B)}{P(B)}, \quad \text{provided } P(B) \neq 0 .
$$

Also, if A is an event and $\{B_i\}$ is a set of disjoint and exhaustive events (that is, a set of events such that $B_i \cap B_j = \emptyset$ and $\cup_i B_i = \Omega$, or in other words $\{B_i\}$ partitions the sample space Ω), then from Theorem 2.1 we have that

$$
P(A) = \sum_i P(A \mid B_i) P(B_i) .
$$

Similarly, if A and C are events and $\{B_i\}$ a set of disjoint and exhaustive events, then

$$
P(A \mid C) = \sum_i P(A \mid B_i \cap C) \ P(B_i \mid C) .
$$

Definition 13.4: Random processes (stochastic processes)

A **random process** is a sequence of random variables X_t defined on sample space Ω, where t is a counter running over a suitable index set T (usually time).

For any fixed Ω, there are many **realizations** or **sample paths** of X: $\{X_t, t \in T\}$.

13.2.1 Discrete Markov chain

Definition 13.5: A discrete time Markov Chain

Consider a random process that is observed at each point t in time, where t is a subset of the non-negative integers, usually denoted $t \in \mathbb{Z}_+$.

The **state** of the process at time $t \in \mathbb{Z}_+$ is given by a random variable X_t, which takes values from the **sample space** Ω to a set of integers known as the state space S (in accordance with definition 4.1). $\mathrm{P}(X_t = i)$, for $i \in S$, is the probability that the state of the process at time point t is i.

Such a process is said to be a discrete time **Markov chain** if

$$\mathrm{P}(X_{t+1} = i_{t+1} \mid X_0 = i_0 \cap X_1 = i_1 \cap \ldots \cap X_t = i_t) \ = \ \mathrm{P}(X_{t+1} = i_{t+1} \mid X_t = i_t),$$

where $i_0, \ldots, i_{t+1} \in S$.

This means that the future state of the process X_{t+1} depends on its history only through its present state X_t and not on earlier history $X_0, X_1, \ldots, X_{t-1}$. In other words, Markov chains have the property of being "memoryless", which means that the next state that is visited by the Markov chain depends only on the present state, and not on any previous states.

Definition 13.6: Time homogeneous Markov chain

A Markov chain X_t is called **time-homogeneous** if we have

$$\mathrm{P}(X_{t+1} = j \mid X_t = i) \ = \ \mathrm{P}(X_1 = j \mid X_0 = i) \text{ for all } t \text{ and } i, j \in S.$$

In the case of a time homogeneous Markov chain, we can simply write

$$p_{i,j} \ = \ \mathrm{P}(X_{t+1} = j \mid X_t = i) \qquad \text{for all } t \quad \text{and } i, j \in S.$$

These conditional **transition probabilities** govern the evolution of the Markov chain. That is, the evolution of the sequence $X_1, X_2, \ldots$ from some starting state X_0, where the starting state, X_0 can be specified deterministically, or it can be chosen randomly from some distribution across the states in Ω.

Definition 13.7: Transition matrix

The **transition matrix** $\mathbb{P}$ of a discrete time-homogeneous Markov chain is the $|\Omega| \times |\Omega|$ matrix of transition probabilities

$$p_{i,j} \ = \ \mathrm{P}(X_{t+1} = j \mid X_t = i) \qquad \rightarrow \qquad \mathbb{P} \ = \ [p_{i,j}].$$

The transition probability entries of this matrix $\mathbb{P}$ satisfy the following properties

$$0 \leq p_{i,j} \leq 1, \quad \forall i, j \in S \qquad \text{and} \qquad \sum_{j \in S} p_{i,j} = 1, \quad \forall i \in S \text{ (each row sum is 1)}$$

The first property is a statement that the transition probability must lie between the same limits, 0 and 1, as any other probability. The second property is a statement that the chain must be in one state at any one time, this state can be the same as the previous state since $P_{i,i}$ is generally non-zero.

Definition 13.8: m-step transition matrix

The m-**step transition matrix** $\mathbb{P}^{(m)} = [p_{i,j}^{(m)}]$ of a time-homogeneous Markov chain is the $|\Omega| \times |\Omega|$ matrix of transition probabilities

$$p_{i,j}^{(m)} = \mathrm{P}(X_{n+m} = j \mid X_n = i).$$

The m-step transition probability $p_{i,j}^{(m)}$ is the probability that the process starting in state i at time n, finds itself in state j at time $n+m$. In general, there are multiple possible "paths" that result in this outcome and the probability $p_{i,j}^{(m)}$ is essentially the sum of probabilities of all such paths. A matrix of such probabilities provides a complete picture of the evolution of our Markov process. The following Theorem gives us access to these m-step probabilities.

Theorem 13.1 Calculating the m-step transition matrix

$$\mathbb{P}^{(m)} = \mathbb{P}^m.$$

An informal proof of Theorem 13.1 follows Example 13.14.

Example 13.13: A weather model

Daily rainfall records during the period from March of 1965 to October of 1986 for a weather station site in the West Riding of Yorkshire, can be summarized by

(i) 1316 transitions from a wet day to a wet day

(ii) 691 transitions from a wet day to a dry day

(iii) 686 transitions from a dry day to a wet day and

(iv) 1749 transitions from a dry day to a dry day

out of a total number of 4442 transitions[2]. We can use these as proportions to estimate the transition probability matrix for a Markov chain having two states, "Wet" for a wet day and "Dry" for a dry day given by

$$\mathbb{P} = \begin{array}{cc} & \begin{array}{cc} \text{Wet} & \text{Dry} \end{array} \\ \begin{array}{c} \text{Wet} \\ \text{Dry} \end{array} & \left(\begin{array}{cc} 0.6557 & 0.3443 \\ 0.2817 & 0.7183 \end{array} \right) \end{array}$$

---

[2]The West Riding has a mild humid temperate climate with warm summers and no dry season. The seasonal variation in the probability that precipitation will be observed at this location is around $\pm.05$, with November being highest and May lowest, and has been ignored in this simple model.

If it is wet on a particular day, the probability that it is raining in five days time is given by the first element of the matrix $\mathbb{P}^{(5)}$, or $\mathbb{P}^5$, which is 0.4540.

```
> P=matrix(c(0.6557,0.3443,0.2817,0.7183),nrow=2,byrow=TRUE)
> pw=matrix(c(1,0),ncol=2)
> M1=pw%*%P%*%P%*%P%*%P%*%P
> M1
          [,1]      [,2]
[1,] 0.4540246 0.5459754
```

Now suppose it is dry on a particular day. The probability of a wet day in 5 days' time is 0.4467, close to the probability if it is wet on a particular day.

```
> pd=matrix(c(0,1),ncol=2)
> M2=pd%*%P%*%P%*%P%*%P%*%P
> M2
          [,1]      [,2]
[1,] 0.4467072 0.5532928
```

Notice that although the probability that tomorrow will be wet is noticeably different for today being wet, 0.6557, and for today being dry, 0.2817, the probability of a wet day in five days' time is close to being independent of today's state.

Example 13.14: An internet router

An internet router is regularly observed every time unit[3] and the number of packets in the buffer is determined. For $0 < p < q < 1$, it is observed that the number of packets in the buffer

- increases by one, between observations, with probability p,
- decreases by one with probability q (provided it is not empty) and
- stays the same with probability $1 - p - q$.

When the buffer is empty, it increases by one with probability p and stays empty with probability $1 - p$. The sample space Ω is the number of packets in the buffer and the state space is the number. Letting $X_t \in S = \{0, 1, 2, \ldots\}$ represent the number of packets in the buffer at time $t \in \{0, 1, 2, \ldots\}$, using the above notation we can establish the following transition probabilities for $i \geq 1$,

$$p_{i,i+1} = p, \quad p_{i,i-1} = q, \quad p_{i,i} = 1 - p - q,$$
$$p_{0,1} = p \quad \text{and} \quad p_{0,0} = 1 - p.$$

Suppose that the buffer begins with X_0 packets at time 0 and that the router is operational for at least 100 time units. The probability that the buffer would contain j packets after 100 time units is given by

$$p_{X_0,j}^{(100)} \quad \text{and the probability that it is empty is} \quad p_{X_0,0}^{(100)}.$$

---

[3] A suitable time unit is a function of such things as the backplane switching speed, the averaged network packet size and the bandwidth of the connections. This could be in the range of milliseconds to microseconds.

Also of interest would be the expected buffer size after 100 time units is given by

$$\sum_{j=0}^{\infty} j\, p_{X_0,j}^{(100)}.$$

In order to find these probabilities and expected values, we set up the transition probability matrix $\mathbb{P}$

$$\mathbb{P} = \begin{pmatrix} 1-p & p & 0 & 0 & 0 & \cdots \\ q & 1-p-q & p & 0 & 0 & \cdots \\ 0 & q & 1-p-q & q & 0 & \cdots \\ \vdots & \ddots & & \ddots & \ddots & \ddots \end{pmatrix},$$

and calculate $\mathbb{P}^{(100)} = \mathbb{P}^{100}$ (see Exercise 13.3)

An infinite buffer size can be used to assess what buffer sizes are required for particular loads to avoid losing packets with some probability that is not negligibly small.

Note that that all row sums of $\mathbb{P}$ are $\mathbf{1}$, reflecting the fact that given a transition, the process must change to some state (could be the same one) with probability 1. That is, it gives a probability mass function across all the states following a transition from a particular state. The above model of the number of packets at a switch or router in the internet, is a simple queueing model that could also model, for example, the number of airplanes waiting to land at Los Angeles International Airport, the number of cars waiting for service at a drive through, or the number of container ships waiting to be unloaded at the seaport in Seattle. If a Markov chain begins in state X_0 at time 0, it will be in state k at time 1 with probability $p_{X_0,k}$, and then the probability that the chain is in a particular state j at time 2 is given by

$$p_{X_0,j}^{(2)} = \sum_{k \in S} p_{X_0,k}\, p_{k,j},$$

which is the sum of all possible intermediate path probabilities of being in some other state k at time 1 as shown in Figure 13.13. More generally we have the following result

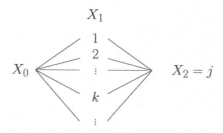

FIGURE 13.13: Interim state transitions.

(Equation (13.1)), known as the **Chapman-Kolmogorov equation**, which gives the probability of being in state j at time m, given the process starts in state X_0 at time 0.

$$p_{X_0,j}^{(m)} = \sum_k p_{X_0,k}^{(\nu)} p_{k,j}^{(m-\nu)}, \quad \text{for all } 0 < \nu \le m. \tag{13.1}$$

Let $\boldsymbol{X}(m) = \big(\mathrm{P}(X_m = 0), \mathrm{P}(X_m = 1), \ldots\big)$ be the probability mass function describing the probability that the process is in each of the states at time m.

Assuming now that the Markov chain starts with an initial probability distribution $\boldsymbol{X}(0)$, we can describe the distribution at time m by

$$\boldsymbol{X}(m) = \boldsymbol{X}(0)\mathbb{P}^m.$$

Assume you look at the queue after it has been operating for a long time and let N be the random variable denoting the state of the Markov chain that you observe. Note that N is a **discrete** random variable and let $\pi_n = \mathrm{P}(N = n)$ denote its probability mass function.

Definition 13.9: Irreducible Markov chain

A Markov chain is called an irreducible Markov chain if it is possible to go from each state to every other state (not necessarily in one move).

13.2.2 Equilibrium behavior of irreducible Markov chains

Definition 13.10: Recurrence and transience

An irreducible Markov chain is recurrent if the probability of return to a state that has just been vacated is one, otherwise it is transient. In addition:

- A recurrent Markov chain where the expected time taken to return is finite, is called **positive recurrent**.
- A recurrent Markov chain where the expected time taken to return is infinite, is called **null recurrent** (only possible for an infinite state space).

Assume that

$$p_{i,j}^{(m)} \;\to\; \pi_j > 0 \quad \text{as} \quad m \to \infty. \tag{13.2}$$

so that the probability of being in state j after many steps is independent of the initial state X_0. That is, the Markov chain has a limiting distribution. Note that this is not always true and we will consider Markov chains later where this does not occur.

Assumption (13.2) implies that if we let $\nu = 1$ and $m \to \infty$ in equation (13.1), we get

$$\lim_{m \to \infty} p_{X_0,j}^{(m)} \;=\; \lim_{m \to \infty} \sum_k p_{X_0,k}^{(m-1)} p_{k,j}, \quad \text{so that}$$

$$\pi_j \;=\; \sum_k \pi_k \, p_{k,j}, \quad \text{for all } j. \tag{13.3}$$

Equations (13.3) are known as the equilibrium equations for a Markov chain and can be written in vector/matrix form as

$$\boldsymbol{\pi} \;=\; \boldsymbol{\pi}\mathbb{P}, \tag{13.4}$$

where $\boldsymbol{\pi} = (\pi_1, \pi_2, \ldots)$ is known as the equilibrium probability distribution[4].

---

[4]We can then see that $\boldsymbol{\pi}$ is a left-eigenvector of   associated with the eigenvalue 1.

Before going on, it is worth noting that for any Markov chain, equations (13.3) may always be written down. If there exists a collection of non-negative numbers π_j satisfying both Equation (13.3) and the normalizing equation (13.5)

$$\sum_j \pi_j = 1, \tag{13.5}$$

then the collection of non-negative numbers π_j is the equilibrium probability distribution of the Markov chain, which is positive recurrent. If these equations cannot be satisfied, then the Markov chain is either transient or null recurrent and does not have an equilibrium probability distribution.

Note that all finite state space irreducible Markov chains have an equilibrium probability distribution.

Definition 13.11: Periodicity

An irreducible Markov chain is called **periodic** if there is an $m > 1$ such that $p_{j,j}^{(n)} = 0$, whenever n is not divisible by m and m is the smallest integer with this property, otherwise it is called **aperiodic**.

This definition means that a revisit to state j is impossible except in a multiple of m transitions such as $m, 2m, 3m, \ldots$

There are two important physical interpretations of the equilibrium distribution $\boldsymbol{\pi}$, if it exists:

1. Limiting: By definition $\lim_{m \to \infty} p_{i,j}^{(m)} = \pi_j$ and so π_j is the limiting probability of the process being in state j, which is only true for aperiodic chains. This means that so long as the process has been going for quite a long time, the probability that the process is in state j will be approximately π_j.

2. Stationary: In Equation (13.4) we saw that $\boldsymbol{\pi}\mathbb{P} = \boldsymbol{\pi}$, and so $\boldsymbol{\pi}$ is the equilibrium (or stationary) probability distribution of the process, which is true for both periodic and aperiodic chains. This means that once the process has the probability distribution of $\boldsymbol{\pi}$ it will persist with that distribution forever. The practical importance of this result is that $\boldsymbol{\pi}$ gives the relative proportion of time that the process spends in each state.

Example 13.13: (Continued) A weather model

The equilibrium equations for the weather model are given by

$$\begin{aligned}
\pi_{Wet} &= 0.6557\pi_{Wet} + 0.3443\pi_{Dry} \\
\pi_{Dry} &= 0.2817\pi_{Wet} + 0.7183\pi_{Dry},
\end{aligned}$$

which have solution $\boldsymbol{\pi} = (\pi_{Wet}, \pi_{Dry}) = (0.4500, 0.5500)$.

Example 13.14: (Continued) An internet router

Let's return to Example 13.14, where the equilibrium equations (13.3) become

$$\begin{aligned}
\pi_0 &= \pi_1 q + \pi_0(1 - p), \\
\pi_j &= \pi_{j+1} q + \pi_j(1 - p - q) + \pi_{j-1}p, \quad j \geq 1
\end{aligned}$$

and which have solution

$$\pi_j \;=\; \left(1 - \frac{p}{q}\right)\left(\frac{p}{q}\right)^j , \quad \text{for all } j = 0, 1, 2, \dots$$

Example 13.15: Moran Dam Model [Moran, 1959]

A country has a rainy season and a dry season each year. A reservoir is modeled as having a capacity of 4 units of water and 5 states are defined, $\{0, 1, 2, 3, 4\}$, corresponding to the reservoir being empty, and containing $1, 2, 3$ or 4 units of water respectively. The inflows to the reservoir during the rainy season are $0, 1, 2, 3$ or 4 units of water with probabilities $0.4, 0.2, 0.2, 0.1$ and 0.1 respectively. Overflow is lost down the spillway. There is no inflow to the dam during the dry season. Provided the reservoir is not

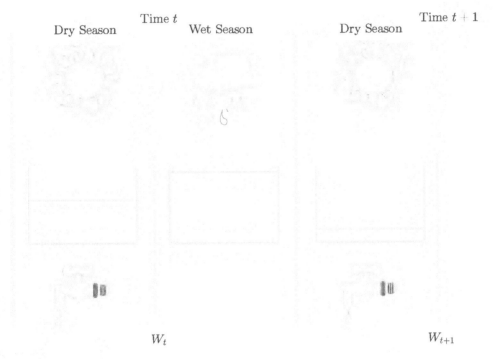

FIGURE 13.14: The state of the system at time t, W_t, is defined as the content of the reservoir at the end of the dry season and can take integer values between 0 and 3.

empty, 1 unit of water is released during the dry season. The random variable W_t is the number of units in the reservoir at the end of the dry season and it follows that its range is $\{0, 1, 2, 3\}$ (see Figure 13.14). The transition matrix is

$$\begin{pmatrix} 0.6 & 0.2 & 0.1 & 0.1 \\ 0.4 & 0.2 & 0.2 & 0.2 \\ 0.0 & 0.4 & 0.2 & 0.4 \\ 0.0 & 0.0 & 0.4 & 0.6 \end{pmatrix}$$

The reasoning for the top left element is that the reservoir will remain empty if the inflow is either 0 or 1 because in the latter case the inflow will be released during the dry season. The second element in the top row follows because if the reservoir is 0 at the end of one dry season the only way it can transition to 1 at the end of the next dry season is if the inflow during the wet season is 2 units. Similar arguments lead to the other entries. The following R code raises the transition matrix to the power[5] of 100.

```
> P=matrix(c(0.6,0.2,0.1,0.1,
+ 0.4,0.2,0.2,0.2,
+ 0.0,0.4,0.2,0.4,
+ 0.0,0.0,0.4,0.6),nrow=4,byrow=TRUE)
> matrix.power <- function(A, n) {
+     e <- eigen(A)
+     M <- e$vectors    # matrix for changing basis
+     d <- e$values     # eigen values
+     return(M %*% diag(d^n) %*% solve(M))
+ }
> matrix.power(P,100)
          [,1]      [,2]       [,3]       [,4]
[1,] 0.173913 0.173913 0.2608696 0.3913043
[2,] 0.173913 0.173913 0.2608696 0.3913043
[3,] 0.173913 0.173913 0.2608696 0.3913043
[4,] 0.173913 0.173913 0.2608696 0.3913043
```

We see that the probabilities of being in the states $0, 1, 2$ and 3 after 100 time steps is $0.174, 0.174, 0.261$ and 0.391 respectively, regardless of the initial state. Furthermore these probabilities are the solution to the equilibrium equations. In the long run the dam would be empty at the end of the dry season for 17.4% of years. Although 100 is a large number of time steps, you can check that

```
matrix.power(P,10)
```

is close to P^{100}. You are asked to investigate a policy of releasing 2 units during the dry season, if it is possible to do so in Exercise 13.4. In practical applications the number of states can be increased to give a more realistic model and stochastic dynamic programming, an extension of the decision tree, can be used (e.g. [Fisher et al., 2010, Fisher et al., 2014]).

13.2.3   Methods for solving equilibrium equations

We have seen an example of how to set up the equilibrium equations, of which there is no all-purpose automatic method of solution. However, we will now present a reasonably systematic way for finding their solution.

1. If the process has a finite number of states, N, there is always exactly one redundant equation (because the row sums are all one and so the last column can be deduced from this fact and all the other columns). However, there is also the normalizing equation (13.5) and there are N linearly independent equations in N unknowns and so the unique solution can be found by any of the usual matrix methods.

---

[5]If you are familiar with eigenvalues and eigenvectors you will recognize that the function `matrix.power()` uses the spectral decomposition. MATLAB provides matrix powers directly.

2. If the process has a countably infinite number of states and the transition probabilities $p_{j,k}$ do not depend on the actual value of j for $j \geq J$, but just $k - j$, so that we have a homogeneous Markov chain for states above J, the there are two natural methods to be used in this case.

   (a) Difference (or recurrence) equation methods if you just wish to determine the π_j.

   (b) Probability generating functions if you wish to extract summary statistics from the distribution, as well as the π_j.

3. If the process has a countably infinite number of states and the transition probabilities $p_{j,k}$ **do depend** on the actual value of j and not just $k - j$, which is the most general class of problems; many problems within this class have a set of equilibrium equations which are impossible to solve analytically. However, there is a large sub-class of problems that have a very useful property, known as **partial balance** (13.11) for which there is a method of determining the equilibrium distribution.

Example 13.14: (Continued) A weather model

Using the the first of the equilibrium equations for the weather model and the normalizing equation (13.5),

$$
\begin{aligned}
\pi_{Wet} &= 0.6557\pi_{Wet} + 0.3443\pi_{Dry} \\
1 &= \pi_{Wet} + \pi_{Dry}, \quad \text{we see that} \\
\pi_{Dry} &= 1 - \pi_{Wet}, \quad \text{and hence} \\
\pi_{Wet} &= 0.6557\pi_{Wet} + 0.3443\left(1 - \pi_{Wet}\right), \quad \text{which yields} \\
\pi_{Wet} &= 0.4500, \quad \text{and} \quad \pi_{Dry} = 0.5500
\end{aligned}
$$

Note also that in this example

$$
\mathbb{P}^m \rightarrow \begin{pmatrix} 0.4500 & 0.5500 \\ 0.4500 & 0.5500 \end{pmatrix} = \begin{pmatrix} \pi_{Wet} & \pi_{Dry} \\ \pi_{Wet} & \pi_{Dry} \end{pmatrix} \quad \text{as } m \text{ gets large.}
$$

This says that $p_{i,j}^{(m)} \rightarrow \pi_j$ as $m \rightarrow \infty$ and so $\mathbb{P}$ in this case is aperiodic[6].

Example 13.15: (Continued) An internet router using difference equations

The equilibrium equations (a set of difference equations with constant coefficients) can be rewritten as

$$
\begin{aligned}
(p+q)\pi_j &= p\pi_{j-1} + q\pi_{j+1}, \quad j \geq 1 & (13.6) \\
p\pi_0 &= q\pi_1, & (13.7)
\end{aligned}
$$

in which we try a solution of the form $\pi_j = M^j$ in (13.6), where $M \in \mathbb{R}$, to get something known as the characteristic equation

$$
(p+q)\,M^j = p\,M^{j-1} + q\,M^{j+1} \quad j \geq 1.
$$

---

[6]Note that $= \begin{pmatrix} 0 & 1 \\ 1 & 0 \end{pmatrix}$ has equilibrium probability distribution $(0.5, 0.5)$ but $\lim_{m\to\infty} \mathbb{P}^m \neq \begin{pmatrix} 0.5 & 0.5 \\ 0.5 & 0.5 \end{pmatrix}$. This Markov chain is periodic with period 2 and does not have a limiting distribution, yet it does have an equilibrium probability distribution that can be interpreted as being stationary.

Dividing by M^{j-1} and rearranging yields

$$\Rightarrow (M-1)(qM-p) = 0 \qquad \Rightarrow M = 1 \text{ or } \frac{p}{q}.$$

The general solution is then of the form $\pi_j = a\left(\frac{p}{q}\right)^j + b$, for $a,b \in \mathbb{R}$, which we substitute into equation (13.7) to get

$$p(a+b) = q\left(\frac{ap}{q}+b\right), \quad \text{or} \quad b(p-q) = 0.$$

The assumption that $p < q$ implies that $b = 0$ and in order to uniquely determine a (and hence $\pi_n, n \geq 0$) we use the normalizing equation (13.5), to get

$$\sum_{j=0}^{\infty} a\left(\frac{p}{q}\right)^j = 1, \quad \text{which because } \frac{p}{q} < 1, \text{ yields}$$

$$a = 1 - \frac{p}{q}, \quad \text{(sum of a geometric series) and hence}$$

$$\pi_j = \left(1 - \frac{p}{q}\right)\left(\frac{p}{q}\right)^j, \quad \text{for all } j = 0,1,2,\ldots$$

We note here that if $p \geq q$, the model does not have an equilibrium probability distribution as the normalizing equation cannot be satisfied and the router buffer will be unstable with the number of buffered packets inexorably increasing as time goes on. That is, any finite sized buffer will eventually see lost packets.

Definition 13.12: Probability generating function

If π_n, for $n = 0,1,2,\ldots$ is a discrete probability distribution, then its **probability generating function** is defined by

$$P(z) = \sum_{n=0}^{\infty} \pi_n z^n, \quad \text{for } z \in [0,1].$$

The probability generating function of a discrete distribution has many interesting properties (see Exercise 13.5).

(i) $P(1) = 1$.

(ii) $P(0) = \sum_{n=0}^{\infty} \pi_n 0^n = \pi_0,$

(iii) $j!\pi_j = \left.\dfrac{d^j}{dz^j}P(z)\right|_{z=0},$

(iv) $\left.\dfrac{d}{dz}P(z)\right|_{z=1} = \mathrm{E}[N]$ and

(v) $\left.\dfrac{d^2P(z)}{dz^2}\right|_{z=1} + \left.\dfrac{dP(z)}{dz}\right|_{z=1} - \left[\left.\dfrac{dP(z)}{dz}\right|_{z=1}\right]^2 = \mathrm{var}(N).$

Example 13.15: (Continued) An internet router

We can also use probability generating functions to solve the equilibrium equations. We start by multiplying equation (13.6) for π_j by z^j and summing over the legitimate range $j \geq 1$ to get

$$(p+q) \sum_{j=1}^{\infty} \pi_j z^j = p \sum_{j=1}^{\infty} \pi_{j-1} z^j + q \sum_{j=1}^{\infty} \pi_{j+1} z^j.$$

We then add equation (13.7) for π_0 to get

$$p\pi_0 + \left[(p+q) \sum_{j=1}^{\infty} \pi_j z^j \right] = \left[p \sum_{j=1}^{\infty} \pi_{j-1} z^j \right] + \left[q \sum_{j=1}^{\infty} \pi_{j+1} z^j \right] + q\pi_1.$$

To get an expression for the generating function $P(z)$, we make each entry under the summation signs look like a $\pi_j z^j$ and then change the index of summation. That is,

$$\Rightarrow \left[(p+q) \sum_{j=0}^{\infty} \pi_j z^j \right] - q\pi_0 = \left[pz \sum_{j=1}^{\infty} \pi_{j-1} z^{j-1} \right] + \left[\frac{q}{z} \sum_{j=1}^{\infty} \pi_{j+1} z^{j+1} \right] + \frac{q}{z} \pi_1 z$$

$$\Rightarrow (p+q) P(z) - q\pi_0 = \left[pz \sum_{j=0}^{\infty} \pi_j z^j \right] + \left[\frac{q}{z} \sum_{j=2}^{\infty} \pi_j z^j \right] + \frac{q}{z} \pi_1 z$$

$$\Rightarrow (p+q) P(z) - q\pi_0 = pz P(z) + \frac{q}{z} [P(z) - \pi_0]$$

$$\Rightarrow P(z) \left[p + q - pz - \frac{q}{z} \right] = q\pi_0 \left[1 - \frac{1}{z} \right].$$

We now re-arrange to get an expression for $P(z)$

$$\Rightarrow P(z) \left[q \left(1 - \frac{1}{z} \right) - pz \left(1 - \frac{1}{z} \right) \right] = q\pi_0 \left(1 - \frac{1}{z} \right)$$

$$\Rightarrow P(z) [q - pz] = q\pi_0 \quad \text{and so} \quad P(z) = \frac{q\pi_0}{q - pz}.$$

Then by the properties of the generating function,

$$P(1) = \frac{q}{q-p} \pi_0 = 1, \quad \text{and so} \quad \pi_0 = \frac{q-p}{q} = 1 - \frac{p}{q}$$

and so

$$P(z) = \frac{q \left(1 - \frac{p}{q} \right)}{q - pz} = \frac{q-p}{q-pz} = \left(1 - \frac{p}{q} \right) \frac{1}{1 - \frac{pz}{q}}$$

$$= \sum_{j=0}^{\infty} \left(1 - \frac{p}{q} \right) \left(\frac{p}{q} \right)^j z^j \quad \text{for all } z \text{ s.t. } \left| \frac{pz}{q} \right| < 1, \quad \text{i.e. } |z| < \frac{q}{p}.$$

By equating coefficients of $P(z) = \sum_{j=0}^{\infty} \pi_j z^j$, we have

$$\pi_j = \left(1 - \frac{p}{q}\right)\left(\frac{p}{q}\right)^j.$$

Note, that, as long as $\{\pi_j\}$, is a genuine distribution (i.e. sums to 1), $P(1) = 1$ and so we know that $P(z)$ converges for all $|z| \le 1$.

Consequently, it must be that $\frac{p}{q} < 1$, again establishing the requirement for the existence of the equilibrium probability distribution.

Given the probability generating function, it is straight forward to find the mean and variance (and higher moments) of the distribution of the number of packets in the buffer.

Example 13.16: Optical fibre using partial balance

A single optical fibre can transmit digital signals at much higher frequencies than a copper wire of the same diameter. Optical fibers can also carry much higher frequency ranges than copper wire and using wavelength multiplexing techniques, an optical fibre link (made up of multiple fibers) can essentially be assumed to have an almost infinite carrying capacity for telephone calls. Let's consider the state of such a link as the number of calls in progress, observed at each time a new call arrives or a call ends. The transition probabilities are

$$p_{i,i+1} = \frac{p}{p+iq}, \text{ for } i \ge 0, \quad p_{i,i-1} = \frac{iq}{p+iq}, \text{ for } i \ge 1 \text{ or } p_{i,k} = 0 \text{ otherwise.}$$

The equilibrium equations (13.3) then become

$$\pi_j = \left(\frac{p}{p+(j-1)q}\right)\pi_{j-1} + \left(\frac{(j+1)q}{p+(j+1)q}\right)\pi_{j+1}, \quad j \ge 1 \quad (13.8)$$

$$\pi_0 = \left(\frac{q}{p+q}\right)\pi_1. \quad (13.9)$$

These equations do not have constant coefficients (the $p_{j,k}$ **do depend** on the actual value of j and not just $k - j$) and to solve these equations, observe that we can rewrite (13.8) as

$$\left(\frac{p+jq}{p+jq}\right)\pi_j = \left(\frac{(j+1)q}{p+(j+1)q}\right)\pi_{j+1} + \left(\frac{p}{p+(j-1)q}\right)\pi_{j-1},$$

therefore,

$$\left(\frac{(j+1)q}{p+(j+1)q}\right)\pi_{j+1} - \left(\frac{p}{p+jq}\right)\pi_j = \left(\frac{jq}{p+jq}\right)\pi_j - \left(\frac{p}{p+(j-1)q}\right)\pi_{j-1} \quad (13.10)$$

Letting $A(j) = \left(\frac{jq}{p+jq}\right)\pi_j - \left(\frac{p}{p+(j-1)q}\right)\pi_{j-1}$, equation (13.10) says

$$A(j+1) = A(j) \text{ for all } j \Rightarrow A(j) = A(1), \text{ for all } j$$

$$\Rightarrow A(j) = \frac{q}{p+q}\pi_1 - \pi_0 = 0 \text{ for all } j \text{ by (13.9)}$$

$$A(j) = 0 \Rightarrow \pi_j = \pi_{j-1}\frac{p(p+jq)}{jq(p+(j-1)q)}$$

$$= \pi_{j-2}\frac{p(p+jq)}{jq(p+(j-1)q)}\frac{p(p+(j-1)q)}{(j-1)q(p+(j-2)q)}$$

$$= \vdots$$

$$= \pi_0\left(\frac{p}{q}\right)^j\frac{1}{j!}\left(\frac{p+jq}{p}\right),$$

where π_0 can be obtained by normalization, and the expected number of calls evaluated.

The condition $A(j) = 0$, yields the equations

$$\pi_j p_{j,j+1} = \pi_{j+1}p_{j+1,j}, \tag{13.11}$$

which are known as partial balance. If a solution to the equilibrium equations in problems of this type can be found under an assumption that the partial balance equations (13.11) hold, then the assumption is valid and the solution is the equilibrium probability distribution.

13.2.4 Absorbing Markov chains

There are many occasions where we wish to use Markov chains to model situations in which the process stops making transitions if it gets into a particular state. Prime examples of this are gambling games in which the process stops if one of the players loses all their money, or population processes which stop if a species dies out. States in which the process stops are called **absorbing states** .

In such a Markov chain, the states can always be re-ordered so that the transition matrix looks like

$$\mathbb{P} = \begin{array}{c} \text{non-absorbing} \\ \text{absorbing} \end{array}\begin{pmatrix} R & S \\ 0 & I \end{pmatrix}$$

where 0 is the zero matrix and I is the identity matrix. We write the transition probability in this way so that we can find neat expressions for

$$\mathbb{P}^n = \begin{pmatrix} R^n & \sum_{i=0}^{n-1}R^iS \\ 0 & I \end{pmatrix} \quad \text{and} \quad \lim_{n\to\infty}\mathbb{P}^n = \begin{pmatrix} 0 & (I-R)^{-1}S \\ 0 & I \end{pmatrix}.$$

This then means that for a non-absorbing state i and an absorbing state j, the $(i,j)^{\text{th}}$ entry of $\lim_{n\to\infty}\mathbb{P}^n$ contains the probability that the process is eventually absorbed into state j conditional on starting in i.

Example 13.17: Machining castings

Consider a manufacturing process where there are a series of n milling machines that perform various machining tasks on castings supplied by the company's foundry. Immediately after each machining stage, the casting is inspected both for casting defects that become apparent after machining and for the suitability of the casting proceeding to the next stage. There are three possibilities after inspection at machining stage i that will see the casting

- proceed to the next machining stage with probability p_i,
- being reprocessed at the current stage with probability q_i or
- being scrapped with probability $r_i = 1 - p_i - q_i$.

Considering each stage of machining and subsequent inspection collectively as a state i, we have a state space $S = \{0, 1, 2, \ldots, n+1\}$ over which we can model the flow of castings through this system. State 0 corresponds to the casting being scrapped and state $n+1$ corresponds to the packing and shipping of a completed casting. Both states 0 and $n+1$ correspond to absorbing states , where the casting is no longer being processed. The Markov chain which models this system is written

$$\mathbb{P} = \begin{array}{c} 0 \\ 1 \\ 2 \\ \vdots \\ n-1 \\ n \\ n+1 \end{array} \left(\begin{array}{ccccccc} 1 & 0 & 0 & 0 & 0 & \cdots & 0 \\ r_1 & q_1 & p_1 & 0 & 0 & \cdots & 0 \\ r_2 & 0 & q_2 & p_2 & 0 & \cdots & 0 \\ \vdots & & \ddots & \ddots & \ddots & & \vdots \\ r_{n-1} & 0 & \cdots & 0 & q_{n-1} & p_{n-1} & 0 \\ r_n & 0 & \cdots & 0 & 0 & q_n & p_n \\ 0 & 0 & \cdots & 0 & 0 & 0 & 1 \end{array} \right).$$

For a numerical example, let $n = 2$ and set $p_1 = 0.98$, $p_2 = 0.95$, $q_1 = 0.01$, $q_2 = 0.02$ so that the 4 state transition matrix for the Markov chain describing this system is

$$\mathbb{P} = \begin{array}{c} 0 \\ 1 \\ 2 \\ 3 \end{array} \left(\begin{array}{cccc} 1 & 0 & 0 & 0 \\ 0.01 & 0.01 & 0.98 & 0 \\ 0.03 & 0 & 0.02 & 0.95 \\ 0 & 0 & 0 & 1 \end{array} \right).$$

Then writing state 0 after state 3 we get

$$\begin{array}{c} 1 \\ 2 \\ 3 \\ 0 \end{array} \left(\begin{array}{cccc} 0.01 & 0.98 & 0 & 0.01 \\ 0 & 0.02 & 0.95 & 0.03 \\ 0 & 0 & 1 & 0 \\ 0 & 0 & 0 & 1 \end{array} \right), \quad \text{where}$$

$$R = \begin{pmatrix} 0.01 & 0.98 \\ 0 & 0.02 \end{pmatrix} \quad \text{and} \quad S = \begin{pmatrix} 0 & 0.01 \\ 0.95 & 0.03 \end{pmatrix},$$

so that

$$(I - R)^{-1}S = \begin{pmatrix} 0.9596 & 0.0404 \\ 0.9694 & 0.0306 \end{pmatrix}.$$

This tells us that a casting has probability 0.9596 of being successfully machined, packed and shipped and has probability of 0.9694 of being successfully machined, packed and shipped given that it has been successfully machined at stage 1.

Correspondingly each casting has probability 0.0404 of being scrapped and probability 0.0306 of being scrapped given that it has been successfully machined at stage 1.

Note also that any Markov chain can be modified so that a particular state is made into an absorbing state in order to find the expected time to reach that state from some other starting state.

Example 13.18: An alternative model for the internet router

Consider the internet router of example 13.14, where we now observe a finite buffer of size N at points in time where there is a change in the number of packets present. For $0 < p < 1$, it is observed that the number of packets in the buffer

- increases by one, between observations, with probability p (provided it is not full),
- decreases by one with probability $1 - p$ (provided it is not empty).

Note that when the buffer is full, any arriving packets are lost to the system. In this model[7], when the buffer is empty or full we will consider these states as absorbing states so that we can investigate the probability that the buffer will be become empty or full from some state k, where $0 < k < N$. Letting $X_n \in \Omega = \{0, 1, 2, \ldots, N\}$ be the number of packets in the buffer at time $n \in \{0, 1, 2, \ldots\}$, we can establish the following transition probability matrix $\mathbb{P}$

$$\mathbb{P} = \begin{pmatrix} 1 & 0 & 0 & 0 & 0 & \cdots & 0 \\ 1-p & 0 & p & 0 & 0 & \cdots & 0 \\ 0 & 1-p & 0 & p & 0 & \cdots & 0 \\ \vdots & \ddots & \ddots & \ddots & \ddots & & \\ 0 & \cdots & & 1-p & 0 & p & 0 \\ 0 & \cdots & & 0 & 1-p & 0 & p \\ 0 & \cdots & \cdots & 0 & 0 & 0 & 1 \end{pmatrix}.$$

In Example 13.18, we could find the probability of being in each state after n time steps, and hence absorption in less than or equal to n time steps including the probability of eventual absorption into the different absorbing states as we did in the previous example. However, if the latter is all we want, then we can evaluate them more efficiently using the following theorem. In exercise 13.2, you are asked to verify the equivalence of the two approaches for a particular numerical case.

Theorem 13.2 Absorption probabilities

Let $X_j^{(N)}$ be the probability that the process is absorbed in state N given that it starts in state j. Let state 0 be another absorbing state, otherwise $X_j^{(N)} = 1$ (a single absorbing state into which absorption is sure). Then $X_j^{(N)}$ satisfies the equation

$$X_j^{(N)} = \sum_k p_{j,k} X_k^{(N)}, \quad 1 \le j \le N - 1,$$

with boundary conditions $X_N^{(N)} = 1$ and $X_0^{(N)} = 0$.

---

[7]This model is a random walk (with boundaries) and can also be used to model such things as the path traced by a molecule as it travels in a liquid or a gas, the search path of a foraging animal, the price of a fluctuating stock or the financial status of a gambler.

The boundary conditions arise because 0 and N are the absorbing states . That is,

$$P\big(\text{absorbed in state } N \big| \text{ in state } N \text{ at time } 0\big) \;=\; 1 \quad \text{therefore } X_N^{(N)} = 1 \quad \text{and}$$
$$P\big(\text{absorbed in state } N \big| \text{ in state } 0 \text{ at time } 0\big) \;=\; 0 \quad \text{therefore } X_0^{(N)} = 0.$$

The above theorem uses the premise that if the process is not yet absorbed, then we can use a one-step analysis based on the transition probabilities and move from a non-absorbing state to another state.

Example 13.18: (Continued) An alternative model for the internet router

For the general problem, Theorem 13.2 says that

$$X_j^{(N)} \;=\; pX_{j+1}^{(N)} + (1-p)X_{j-1}^{(N)}, \quad 1 \le j \le N-1 \tag{13.12}$$
$$X_0^{(N)} \;=\; 0 \tag{13.13}$$
$$X_N^{(N)} \;=\; 1 \tag{13.14}$$

We shall solve (13.12) using the method for solving difference equations that we discussed earlier, by trying a solution of the form $X_j^{(N)} = M^j$ to get the characteristic equation

$$m^j \;=\; pm^{j+1} + (1-p)m^{j-1} \quad 1 \le j \le N-1$$
$$\Rightarrow \quad m = 1 \text{ or } \frac{1-p}{p}.$$

1. If $p \neq 1/2$, the general solution is of the form $X_j^{(N)} = a\left(\frac{1-p}{p}\right)^j + b.$

   Now from (13.13) we have that $\quad X_0^{(N)} \;=\; 0 \quad \Rightarrow b = -a$

   also from (13.14) we have that $\quad X_N^{(N)} \;=\; 1 \quad \Rightarrow a = \dfrac{p^N}{(1-p)^N - p^N}$

   so that $\quad X_j^{(N)} \;=\; \dfrac{p^N}{(1-p)^N - p^N}\left[\left(\dfrac{1-p}{p}\right)^j - 1\right].$

2. If $p = 1/2$ then $(1-p)/p = 1$ and we have a repeated root and therefore our solution will be of the form $X_j^{(N)} = aj + b$. Applying the boundary equations as before we see that

   From (13.13) we have that $\quad X_0^{(N)} \;=\; 0 \quad \Rightarrow b = 0$

   also from (13.14) we have that $\quad X_N^{(N)} \;=\; 1 \quad \Rightarrow a = \dfrac{1}{N}$

   so that $\quad X_j^{(N)} \;=\; \dfrac{j}{N}.$

Therefore putting it all together, the probability that the buffer eventually overflows before becoming empty given that it starts with j packets is given by

$$
X_j^{(N)} = \begin{cases} \dfrac{p^N}{(1-p)^N - p^N}\left[\left(\dfrac{1-p}{p}\right)^j - 1\right] & \text{if } p \neq \dfrac{1}{2} \\[3mm] \dfrac{j}{N} & \text{if } p = \dfrac{1}{2}. \end{cases}
$$

It is not always the probability of eventually reaching a given state that is of interest, but how long it actually takes to get there. That is, what is the expected time until a Markov chain reaches a particular state. A one-step analysis can be used again here, noting that one-step takes an expected time of 1, which must appear in our equations.

Theorem 13.3 Mean time to absorption

Consider a general Markov chain with state space $0, 1, 2, \ldots$ and a single absorbing state 0 into which absorption is certain. Let M_j be the mean time until absorption of the chain, conditional on starting in state j. Then M_j satisfies the equations

$$
M_j = 1 + \sum_k p_{j,k} M_k, \quad \text{for} \quad j \geq 1, \quad \text{with } M_0 = 0.
$$

Example 13.19: Another alternative model for the internet router

In the Internet router buffer model 13.18, we could calculate the expected time until either the buffer is emptied or full. However, if we are now particularly interested in the expected time until the buffer is full (packets are lost), we must modify the Markov chain model to achieve this. We no longer want state 0 to be an absorbing state ($p_{0,0} \neq 1$) and as we observe the buffer at points in time where there is a change in the number of packets present in the buffer, we can set the transition probabilities $p_{0,0} = 0$ and $p_{0,1} = 1$ (when the buffer is empty, the only change occurs when a packet arrives), so the equations of Theorem 13.3 become

$$
\begin{aligned}
M_0 &= 1 + M_1 \quad \text{and} \quad M_N = 0 \\
M_j &= 1 + pM_{j+1} + (1-p)M_{j-1}, \quad \text{for } 0 < j < N \\
\Rightarrow -1 &= pM_{j+1} - M_j + (1-p)M_{j-1}, \quad \text{for } 0 < j < N.
\end{aligned}
$$

The general equation is an inhomogeneous version of the equations that we solved to get the absorption probabilities before. There we saw that the general solution to the homogeneous equation (where the RHS is 0), before using the boundary conditions was

$$
M_j = a\left(\frac{1-p}{p}\right)^j + b.
$$

We now need to calculate a particular solution to the inhomogeneous equation. Normally, to do this we would substitute a function of the form that appears on the RHS (i.e. a constant C). However, this is one of the solutions to the homogeneous equation, so instead we substitute a function of the form $M_j = jC$, so that

$$
\begin{aligned}
jC &= 1 + p\,(j+1)C + (1-p)\,(j-1)C \\
\Rightarrow C &= \frac{1}{1-2p} \quad > 0, \quad \text{since } p < \frac{1}{2}.
\end{aligned}
$$

The general solution to the inhomogeneous equation is, therefore,

$$
M_j = a\left(\frac{1-p}{p}\right)^j + b + \frac{j}{1-2p}, \quad 0 < j < N.
$$

Substituting the first boundary condition $M_0 = 1 + M_1$ yields

$$
a + b = 1 + a\left(\frac{1-p}{p}\right) + b + \frac{1}{1-2p} \quad \Rightarrow a = -\frac{2p(1-p)}{(1-2p)^2}.
$$

and so we have

$$
M_j = -\frac{2p(1-p)}{(1-2p)^2}\left(\left(\frac{1-p}{p}\right)^j\right) + b + \frac{j}{1-2p}, \quad 0 < j < N.
$$

The other boundary condition $M_N = 0$ yields

$$
\begin{aligned}
M_N &= -\frac{2p(1-p)}{(1-2p)^2}\left(\frac{1-p}{p}\right)^N + b + \frac{N}{1-2p} = 0 \\
\Rightarrow b &= \frac{2p(1-p)}{(1-2p)^2}\left(\left(\frac{1-p}{p}\right)^N - \frac{N(1-2p)}{2p(1-p)}\right) \quad \text{and so} \\
M_j &= \frac{2p(1-p)}{(1-2p)^2}\left(\left(\frac{1-p}{p}\right)^N - \left(\frac{1-p}{p}\right)^j - \frac{N(1-2p)}{2p(1-p)}\right) + \frac{j}{1-2p}, \quad 0 \le j \le N.
\end{aligned}
$$

Questions involving the mean time to reach a given state are of particular interest in population models, where eventual extinction is certain and it is the mean time until extinction that is of interest. Extinction events also take on an added importance for much wider environmental considerations as extinction events are claimed to amplify diversity-generation by creating unpredictable evolutionary bottlenecks [Lehman and Miikkulainen, 2015a] and accelerate evolution [Lehman and Miikkulainen, 2015b].

Example 13.20: An extinction model

Consider a discrete time population process, where at time point t, the population increases by one with probability p or decreases by one with probability $(1-p)$, where $p < \frac{1}{2}$. Theorem 13.3 says that the expected time M_j until the population of initial size j becomes extinct satisfies

$$
\begin{aligned}
M_j &= 1 + pM_{j+1} + (1-p)M_{j-1}, \qquad \text{for } j \ge 2 \\
M_1 &= 1 + pM_2 \\
M_0 &= 0. \\
\Rightarrow \quad & pM_{j+1} - M_j + (1-p)M_{j-1} = -1.
\end{aligned}
$$

We now need to calculate a particular solution to the inhomogeneous equation, so we substitute a function of the form $M_j = jC$, so that

$$
\begin{aligned}
jC &= 1 + p\,(j+1)C + (1-p)\,(j-1)C \\
\Rightarrow C &= \frac{1}{1-2p} \;>\; 0, \quad \text{since } p < \frac{1}{2}.
\end{aligned}
$$

The general solution to the inhomogeneous equation is, therefore,

$$
M_j = a\left(\frac{1-p}{p}\right)^j + b + \frac{j}{1-2p}, \quad j \ge 0.
$$

Now $M_0 = 0$ implies that $b = -a$ and therefore

$$
M_j = a\left(\frac{1-p}{p}\right)^j - a + \frac{j}{1-2p}.
$$

There still is one arbitrary constant left and we need some more advanced theory which says that we are looking for the **minimal** non-negative solution in Theorem 13.3. That is, the smallest solution which remains non-negative. Re-arranging, we can write

$$
M_j = a\left[\left(\frac{1-p}{p}\right)^j - 1\right] + \frac{j}{1-2p}
$$

and since we have assumed that $p < \frac{1}{2}$, we have $\frac{1-p}{p} > 1$ and so $\left[\left(\frac{1-p}{p}\right)^j - 1\right] > 0$, for all j. So the smaller a is, the smaller the solution and in fact the minimal non-negative solution is when $a = 0$. That is, the expected time to extinction, given that the population starts with j individuals is given by

$$
M_j = \frac{j}{1-2p}.
$$

Note that if we cannot find a minimal non-negative solution to the system of equations given in Theorem 13.3, then the mean time to absorption is infinite. This is not to say that a finite absorption time cannot occur, but only that absorption is not always certain into the state of interest, such as the mean time to reach state N (the buffer is full) in Example 13.18, where state 0 (the buffer is empty) is also an absorbing state. In some infinite state space models, it is also possible that excursions away from the state of interest can take an infinitely long time.

13.2.5 Markov chains in continuous time

The Pascal distribution is the discrete time equivalent of the exponential distribution. In a similar fashion we can move from Markov chains in discrete time to what are commonly called **Markov processes** in continuous time. We rely on the memoryless (Markov) property of the exponential distribution.

Example 13.21: Machine repair

A machine can be in two states working (1) or failed (0). The time until failure has an exponential distribution with rate λ and hence mean time to failure of $1/\lambda$. The repair time also has an exponential distribution, with a mean of $1/\theta$. The corresponding repair rate is θ. Let $X(t)$ be the state of the machine at time t and define $p_1(t)$ and $p_0(t)$ as the probabilities that the machine is working and failed at time respectively. Consider a small length of time δt. If δt is small enough such that $(\delta t)^2$ and higher powers are negligible, and the machine is working, then the probability of failure in time δt is $\lambda \delta t$. The probability that it is still working in in time δt is $1 - \lambda \delta t$. Similar results apply for repairs and hence

$$p_0(t + \delta t) = p_0(t)(1 - \theta \delta t) + p_1(t)\lambda \delta t,$$

because if the machine is to be in the failed state in time δt, it is either failed now and not repaired or working now and fails[8]. Similarly

$$p_1(t + \delta t) = p_0(t)\theta \delta t + p_1(t)(1 - \lambda \delta t).$$

In matrix notation, defining

$$\boldsymbol{p}(t) = \big(p_0(t), p_(t)\big)$$

and

$$M \;=\; \begin{pmatrix} 1 - \theta \delta t & \theta \delta t \\ \lambda \delta t & 1 - \lambda \delta t \end{pmatrix}$$

$$\boldsymbol{p}(t + \delta t) \;=\; \boldsymbol{p}(t)M.$$

The crucial result follows by considering

$$\frac{\boldsymbol{p}(t + \delta t) - \boldsymbol{p}(t)}{\delta t} = \boldsymbol{p}(t)\frac{M - I}{\delta t}$$

and taking the limit as δt tends to 0.

$$\dot{\boldsymbol{p}}(t) \;=\; \boldsymbol{p}(t)\Lambda,$$

where

$$\Lambda \;=\; \frac{M - I}{\delta t} \;=\; \begin{pmatrix} -\theta & \theta \\ \lambda & -\lambda \end{pmatrix},$$

is known as the **rate matrix**. Since the rows of the transition matrix must sum to 1 the sum of the rows of the rate matrix must sum to 0. If we assume that the probabilities will tend to a fixed value $\boldsymbol{pi}$ the derivative will be 0 and we solve

$$\boldsymbol{\pi}\Lambda = \boldsymbol{0}, \qquad \sum_i \pi_i = 1.$$

---

[8] We can ignore the possibility of repair and subsequent failure because the probability of such an event is of order $(\delta t)^2$ and so negligible.

For the single machine

$$\pi = \left(\frac{\lambda}{\lambda + \theta}, \frac{\theta}{\lambda + \theta} \right).$$

The proportion of time that the machine is working is given by $\theta/(\lambda + \theta)$. This result is as expected, and division of the numerator and denominator by $\lambda\theta$ gives the ratio of the mean working time to the sum of the mean working time and mean repair time. But, the solutions to more complex applications are not so straightforward.

Example 13.22: N identical extrusion machines

Suppose there are N identical extrusion molding machines in a factory. The machines are expected to operate continuously but breakdowns occur. For each machine the working time until it breaks down has an exponential distribution with mean $1/\lambda$, so the failure rate for each machine is λ. Machine failures are independent. There is always one mechanic on duty to repair machines and repair times have an exponential distribution with mean $1/\theta$, and therefore a repair rate θ. A repair is completed before work begins on the next machine waiting for repair. If there are m machines working the probability that 1 of the m fails in time $\lambda\delta t$ is given by the binomial probability

$$m(\lambda\delta t)(1 - \lambda\delta t)^{m-1}$$

which reduces to $m(\lambda\delta t)$ when terms of order $(\delta t)^2$ and above are negligible. Similarly the probability that 2 or more machines fail is of order $(\delta t)^2$ and negligible. This approximation becomes exact in the limit. It follows that the rate matrix Λ is

$$
\begin{array}{c}
\begin{array}{ccccccc} 0 & \quad 1 & \quad 2 & \quad \cdots & \quad N-1 & \quad N \end{array} \\
\begin{array}{c} 0 \\ 1 \\ 2 \\ \vdots \\ N-1 \\ N \end{array}
\left(
\begin{array}{cccccc}
-\theta & \theta & 0 & \cdots & 0 & 0 \\
\lambda & -\lambda - \theta & \theta & & 0 & 0 \\
0 & 2\lambda & -2\lambda - \theta & & 0 & 0 \\
\vdots & & & \ddots & & \\
0 & 0 & 0 & \cdots & -(N-1)\lambda - \theta & \theta \\
0 & 0 & 0 & \cdots & N\lambda & -N\lambda
\end{array}
\right).
\end{array}
$$

The equation for the equilibrium probabilities is

$$\pi\Lambda = 0.$$

It is straightforward to obtain the solution

$$\pi_j = \pi_0 \left(\frac{\theta}{\lambda} \right)^j \frac{1}{j!} \quad \text{for all } j = 0, 1, 2, \ldots, N,$$

where

$$\pi_0 = \left[\sum_{j=0}^{N} \left(\frac{\theta}{\lambda} \right)^j \frac{1}{j!} \right]^{-1}.$$

The same principle can be used if there is more than one mechanic (Exercise 13.6). The assumption of exponential repair times, in particular, may be unrealistic and it can be relaxed by introducing **hidden states** (Exercise 13.7) and utilizing the result that the sum of exponential variables (the gamma distribution) tends towards a normal distribution [Green and Metcalfe, 2005].

Example 13.23: Erlang B (blocking) formula

The solution to the N machine problem was obtained in a different context by the Danish mathematician Agner Erlang (1878-1929) [Erlang, 1917] who worked for the Copenhagen Telephone Company. It is known as the Erlang B formula and: N represents the number of phone lines and i for 0 up to N is the number of lines in use; θ now represents the call arrival rate; and $1/\lambda$ is the mean length of a call or call holding time. Then π_N is the proportion of time that all the lines are in use and because Poisson arrivals see time averages[9], it is the blocking or loss probability for arriving calls. Hence the expression for π_N is known as the Erlang B formula or Erlang loss formula. The rate of lost calls is therefore given by

$$\theta\,\pi_N = \theta\,\frac{\left(\frac{\theta}{\lambda}\right)^N \frac{1}{N!}}{\sum_{j=0}^{N}\left(\frac{\theta}{\lambda}\right)^j \frac{1}{j!}} = \frac{\frac{\theta^{N+1}}{\lambda^N}\frac{1}{N!}}{\sum_{j=0}^{N}\left(\frac{\theta}{\lambda}\right)^j \frac{1}{j!}}.$$

13.3 Simulation of systems

An independent medical practice, the Campus Medical Center (CMC) is situated on a university campus. Students and staff of the university can register with the CMC along with any people living in the neighborhood of the university. Patients are encouraged to make appointments for regular consultations, but urgent situations can be dealt with as emergencies by arriving and waiting to be seen by a duty doctor. There are many patients using this facility and the practice manager is considering various options for reducing the waiting times of patients. The feature of this problem that makes it a typical candidate for simulation modeling is the random nature of arrival patterns, consultation times and types of patient.

Simulation is used to answer **"What if?"** questions and allows the medical practice manager in the above example to explore what the possible outcomes would be if particular changes were made to the operation of the practice. In the PC sales team Example 10.1, we used simulation to demonstrate the effect of Bill's methods of avoiding interaction with his manager Benito. This is essentially answering, "what if Bill did not disclose his sales to Benito correctly?" Simulation is particularly useful in systems that have random features, for example, in the areas related to engineering decision making such as in system design and redesign. It is also increasingly being used in on-line system control and as a training aid. By using simulation, decisions can be made quickly, with the ability to test alternatives so that the engineer can be confident that he or she is making the justifiable decisions. Some of the reasons engineers use simulation are as follows

[9]The memoryless nature of the exponential distribution implies that arriving calls see the stationary distribution of the queueing system, a condition that is generally termed Poisson arrivals; see time averages or PASTA.

- Simulation of complex systems can yield valuable insight into which variables are most important and how they interact.

- Simulation makes it possible to study the complex interactions of a given system. This may help in identifying and resolving problem areas.

- Simulation can be used to investigate and experiment with new and modified systems prior to physically implementing them such as when purchasing new equipment, replacing or deleting resources or when intending new strategies, methods, or layouts.

- Simulation can be used where it may be impossible, extremely expensive or dangerous to observe certain processes.

Simulation is often used when the observed system is so complex, that it is impractical to describe it in terms of a set of mathematical equations[10], but it can be also be used where the observed system may be described in terms of a set of Mathematical equations, but where solutions cannot be gained using straight-forward techniques, or it may inform the solutions of those equations such as what variability in solution can you reasonably expect.

13.3.1   The simulation procedure

The problem must clearly be defined and its objectives stated so that a decision can be made on a set of criteria for evaluating how well the objectives are satisfied. An abstract model must be developed and a decision made on the required complexity, validity, adequacy and therefore on how many variables are to be included.

1. **Exogenous:** independent (inputs) are assumed to be given.

2. **Endogenous:** Dependent (outputs) are generated from the interactions of the system.

During the development a computer simulation model, relevant data must be collected (inputs), which sometimes can be expensive, very time consuming or even impossible. Coding of the model can be done using a simulation package or by writing some code. In either case, the model must first be validated, which can be a difficult but absolutely necessary procedure to establish confidence in any output results [Pawlikowski et al., 2002]. This can be done by comparing with historical data to check if the simulation model is representative of the current system. For a proposed system, where there is no existing historical data, expected behavior or behavior based on similar systems must be used. After validation, model variations must be selected for testing during the simulation runs and outputs processed, analyzed and interpreted. Statistical analysis of the results are mandatory here along with any graphical or tabular display of the results. Initial simulation results may also inform a direction to pursue to gain further improvements. That is, iterations of simulation and interpretation of outputs is a useful method of approach here.

The most common form of simulation is called **discrete event simulation**, where the state of a system is modeled by discrete variables such as the number of patients in a waiting room, the number of items in stock or the number of machines that are out of service. The behavior of the system is characterized by times between events that change the system state and the clock essentially advances when these events occur. An alternative method is called **synchronous**, where the clock "ticks" at small fixed time increments and a check is made as to whether the state of the system has changed.

---

[10]Systems of equations will only give a mean value analysis in most cases, where stochastic variation may in fact be more important.

Discrete event simulation is demonstrated in the following by way of example and the synchronous method, which should be able to be understood in the light of this presentation is left to the reader. A single system is considered and modeled in two ways, initially in a very simple fashion and then by incorporating more realistic inputs in the second model. The reliability of the actual outputs and hence results from the simulations must therefore be interpreted accordingly to the level of the model's realism.

Example 13.24: Machine shop

A machine shop contains three identical machines, each machine reported to take three minutes to process a casting. Twenty castings were observed to have arrived at the machine shop at the following times (in minutes) from the start of day at time 0

$$0.10, 0.15, 0.65, 1.95, 2.50, 3.30, 5.10, 6.00, 7.12, 9.30, 9.83,$$
$$10.30, 10.69, 11.30, 12.43, 12.60, 12.90, 13.10, 14.41, 15.70$$

A casting arriving at the machine shop is processed on any one of the machines when it becomes available. Castings arriving while all the machines are busy are held in a buffer store until a machine is free. Your objective as the production engineer, is to calculate the average time spent by the castings in the buffer store and to calculate the required size of the buffer store.

Drawing a picture as in Figure 13.15 is useful to envisage the flow Whenever a simula-

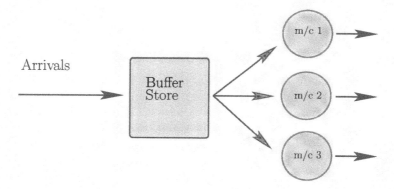

FIGURE 13.15: Workshop layout and flow.

tion model is to be used, a question that arises is whether there is enough information available or is there a need to make some assumptions. This comes about for two reasons, either there is not enough information given by an assessment of the system or there is a need to make simplifying assumptions to actually model the system. The above system is simple enough, but there are questions that still need to be addressed such as, does the machining shop start empty, which available machine is the casting sent to, what is the frequency of machine breakdowns, what is the size of the buffer store?

In this simulation the machines will start empty, available machines will be loaded in numerical order and there will be no machine breakdowns. We will also assume that the buffer store is unlimited so that any number castings may be temporarily stored and so we can see what size buffer store is required.

TABLE 13.1: Simulation record for the deterministic machine shop processing of twenty castings.

Event Time	Occupancy Machine 1	Occupancy Machine 2	Occupancy Machine 3	Occupancy Buffer Store	Cumulative Number In Buffer	Cumulative Time Spent In Buffer
0.00	0	0	0	0	0	0
0.10	1(3.10)	0	0	0	0	0
0.15	1	1(3.15)	0	0	0	0
0.65	1	1	1(3.65)	0	0	0
1.95	1	1	1	1	1	0
2.50	1	1	1	2	2	0.55
3.10	1(6.10)	1	1	1	2	1.75
3.15	1	1(6.15)	1	0	2	1.80
3.30	1	1	1	1	3	1.80
3.65	1	1	1(6.65)	0	3	2.15
5.10	1	1	1	1	4	2.15
6.00	1	1	1	2	5	3.05
6.10	1(9.10)	1	1	1	5	3.25
6.15	1	1(9.15)	1	0	5	3.30
6.65	1	1	0	0	5	3.30
7.12	1	1	1(10.12)	0	5	3.30
9.10	0	1	1	0	5	3.30
9.15	0	0	1	0	5	3.30
9.30	1(12.30)	0	1	0	5	3.30
9.83	1	1(12.83)	1	0	5	3.30
10.12	1	1	0	0	5	3.30
10.30	1	1	1(13.30)	0	5	3.30
10.69	1	1	1	1	6	3.30
11.30	1	1	1	2	7	3.91
12.30	1(15.30)	1	1	1	7	5.91
12.43	1	1	1	2	8	6.04
12.60	1	1	1	3	9	6.38
12.83	1	1(15.83)	1	2	9	7.07
12.90	1	1	1	3	10	7.21
13.10	1	1	1	4	11	7.81
13.30	1	1	1(16.3)	3	11	8.61
14.41	1	1	1	4	12	11.94
15.30	1(18.30)	1	1	3	12	15.50
15.70	1	1	1	4	13	16.70
15.83	1	1(18.83)	1	3	13	17.22
16.30	1	1	1(19.3)	2	13	18.63
18.30	1(21.3)	1	1	1	13	22.63
18.83	1	1(21.83)	1	0	13	23.16
19.30	1	1	0	0	13	23.16
21.30	0	1	0	0	13	23.16
21.83	0	0	0	0	13	23.16

Having decided on initial conditions, what information has to be collected to answer the objectives? Here the use of the times at which machines are occupied and emptied is appropriate to get the mean time spent by castings in the buffer store and the buffer store occupation.

The data given in Table 13.1 is a record of the discrete event simulation for processing 20 castings, which keeps track of the state of the system at each event point in time. The entries are occupation levels and the bracketed values show the completion time of machining for each newly arrived casting.

The simulation performed using these inputs completes at time 21.83 minutes, with 23.16 minutes of sojourn time spent by 13 castings in the buffer from a total of 20 castings actually processed. The average time in the buffer store, conditional on going there $= \frac{23.16}{13} = 1.7815$, the average time in the buffer store for all castings $= \frac{23.16}{20} = 1.158$ and the maximum number of castings in the buffer store was 4.

You are asked to recreate this simulation record in Exercise 13.8 and verify these results. This will give some necessary background to set up and complete far more complex simulation models, one of which is easily produced based on the next example of this same system with more appropriate inputs.

The times of arrival although correct for the observed twenty arrivals, do not give us a realistic picture of the variability of the real system for other arrivals. The processing times similarly would probably not be fixed at 3 minutes, but rather have some distributional form about a mean value of 3 minutes and further collection of data is required to establish the processing time distribution. It is also possible that individual machines may have different processing time distributions, but we will not consider that here as it can easily be added later.

The simulation model in Example 13.24 has no variability associated with any of the inputs and therefore of the outputs. Every simulation using those inputs will yield the exact same result, which only gives us a very rough idea of what may occur if castings arrive at times other than those prescribed or for longer periods of operation. Variation in inter-arrival times and processing times will have a significant impact on the maximum number of castings in the buffer store, as will running this system for for longer periods and processing more than just 20 castings.

Example 13.25: Machine shop alternative

Consider the same machine shop as in the previous example, except that now the times to process a casting have a mean value of 3 minutes but with the following discrete distributional form of 2.50 minutes with probability 0.20, 3.00 minutes with probability 0.60 and 3.50 minutes with probability 0.20. Similarly we approximate the inter-arrival times with the following discrete distributional form based on the aforementioned times.

Time	0.1565	0.3695	0.5825	0.7955	1.2215	1.8605	2.0735
Probability	0.2000	0.1500	0.2000	0.1000	0.2500	0.0500	0.0500

These discrete distributional forms allow us to simulate much more than just a small run of twenty castings. In practice, much more data would be collected to establish better distributional forms for both the inter-arrival times of castings and the machining times. Machine breakdown and repair times could also be included in a very straight-forward manner. To generate twenty each of the inter-arrival times (IAT) for the arrival process and machining times (MT) for the machines for this simulation, the following R code may be used

```
set.seed(8)
IAT<-sample(x=c(0.1565,0.3695,0.5825,0.7955,1.2215,1.8605,2.0735),
+ size=20, replace=TRUE, prob=c(0.20,0.15,0.20,0.10,0.25,0.05,0.05))
set.seed(21)
MT<-sample(x=c(2.5,3.0,3.5), size=20, replace=TRUE,
+ prob=c(0.20,0.60,0.20))
```

The set.seed($\cdot$) command enables the control of the generation of different sequences of random variates from each of the distributions to provide different results for multiple simulation runs (see later).

13.3.2 Drawing inference from simulation outputs

To draw inferences about an underlying stochastic process from a set of simulation output data we must make some assumptions about the underlying stochastic process. A typical assumption is that the stochastic process is **covariance stationary** (second order stationary). A discrete time stochastic process $X_1, X_2, \ldots, X_n$ is covariance stationary if

$$
\begin{aligned}
\mathrm{E}[X_i] &= \mu, &&\text{for all } i = 1, 2, \ldots, &&\text{where } -\infty < \mu < \infty, \\
\mathrm{var}(X_i) &= \sigma^2, &&\text{for all } i = 1, 2, \ldots, &&\text{where } \sigma^2 < \infty \quad \text{and} \\
\end{aligned}
$$
$$
\mathrm{cov}(X_i, X_{i+j}) \text{ is independent of } i \text{ for } j = 1, 2, \ldots
$$

This means that for a covariance stationary process the mean μ and variance σ^2 are stationary over time and the covariance between X_i and X_{i+j} depends on the **lag** j and not the actual values of i and $i + j$. Most simulations are unlikely to be covariance stationary from time zero, because there will usually be transient effects (due to the initial state of the process). It is often the case however, that after some k (the **warmup period**) the process

$$
X_{k+1}, X_{k+2}, \ldots, X_n
$$

will be approximately covariance stationary.

To estimate the common mean μ of a sequence of independent and identically distributed (IID) random variables $X_1, X_2, \ldots, X_n$, we use the fact that the sample mean

$$
\overline{X}(n) = \frac{1}{n} \sum_{i=1}^{n} X_i
$$

is an unbiased estimator of the true mean μ. That is, $\mathrm{E}\big[\overline{X}(n)\big] = \mu$. Thus, for a given sample $x_1, \ldots, x_n$ an unbiased estimate of μ is $\dfrac{1}{n} \sum_{i=1}^{n} x_i$. That this choice is a good one is supported by the Strong Law of Large Numbers, which essentially says that

$$\lim_{n\to\infty} \overline{X}(n) \;\; = \;\; \lim_{n\to\infty} \frac{1}{n}\sum_{i=1}^{n} X_i \;\; = \;\; \mu.$$

Thus as n gets large,

$$\overline{X}(n) \;\; = \;\; \frac{1}{n}\sum_{i=1}^{n} X_i$$

will (almost surely) settle down to the value μ.
The sample variance

$$S^2(n) \;\; = \;\; \frac{1}{n-1}\sum_{i=1}^{n}[X_i - \overline{X}(n)]^2$$

is an unbiased estimator of the true variance σ^2. That is, $\mathrm{E}\big[S^2(n)\big] = \sigma^2$. Thus, if we are asked to give our best estimate, (a **point estimate**) of μ and σ^2 from a particular sample $x_1, x_2, \ldots, x_n$, we would say

$$\bar{x}(n) \;\; = \;\; \frac{1}{n}\sum_{i=1}^{n} x_i, \text{ and } s^2(n) \;\; = \;\; \frac{1}{n-1}\sum_{i=1}^{n}[x_i - \bar{x}(n)]^2.$$

However, if we just do this, we have no way of assessing how close our estimates are to the real values of μ and σ^2 respectively. The usual way to assess the precision of the estimator $\overline{X}(n)$ is to construct a confidence interval. To do this, we need to look at the random variables $\overline{X}(n)$ and $S^2(n)$ more closely. First, let's calculate $\mathrm{var}\big(\overline{X}(n)\big)$ (Recall Section 6.4.3).

$$\mathrm{var}\big(\overline{X}(n)\big) \;\; = \;\; \mathrm{var}\left(\frac{1}{n}\sum_{i=1}^{n} X_i\right) \;\; = \;\; \frac{1}{n^2}\mathrm{var}\left(\sum_{i=1}^{n} X_i\right)$$

$$= \;\; \frac{1}{n^2}\sum_{i=1}^{n}\mathrm{var}(X_i) \;\; = \;\; \frac{1}{n^2}n\sigma^2 \;\; = \;\; \frac{\sigma^2}{n}.$$

This still depends on the true but unknown variance σ^2. If we replace σ^2 by $S^2(n)$ then we get a random variable

$$\overline{V}(n) \;\; = \;\; \frac{S^2(n)}{n} \;\; = \;\; \frac{\sum_{i=1}^{n}[X_i - \overline{X}(n)]^2}{n(n-1)},$$

which is an unbiased estimator of $\mathrm{var}\big(\overline{X}(n)\big) = \frac{\sigma^2}{n}$. That is, $\mathrm{E}\big[\overline{V}(n)\big] = \frac{\sigma^2}{n}$.
We now know some information about the mean and variance of $\overline{X}(n)$. The final piece of information that we need to be able to construct a confidence interval is the distribution of $\overline{X}(n)$. We get this from the Central Limit Theorem (see Section 6.4.3), probably the most important result in statistics.

In the above analysis we have assumed IID data and used the relations

$$\mathrm{var}\big(\overline{X}(n)\big) \;\; = \;\; \sigma^2/n \quad \text{and} \quad \mathrm{E}\big[S^2(n)\big] = \sigma^2$$

to justify using $S^2(n)/n$ as an estimate for $\mathrm{var}\big(\overline{X}(n)\big)$. A quote from Law and Kelton, [2000] states the following:

"It has been our experience that simulation output data are always correlated."

This indicates that we cannot use the analysis techniques from Chapter 7 directly on simulation output data. To see this, consider when we have correlated observations how this goes wrong on two counts for the expected value of the sample variance, which can be shown to be given by

$$\mathrm{E}\big[S^2(n)\big] \;=\; \sigma^2 \left[1 - 2\frac{\sum_{j=1}^{n-1}(1-j/n)r_j}{n-1} \right] \neq \sigma^2 \quad \text{in general,}$$

where r_j is the auto-correlation at lag j defined in Chapter 6. If for example $r_j > 0$, (positive correlation) then $\mathrm{E}\big[S^2(n)\big] < \sigma^2$ and $S^2(n)$ has a negative bias. Thus the variance will be underestimated. Apparently, several major simulation languages use $S^2(n)$ to estimate the variance of a set of simulation output data, which may lead to serious errors in analysis. For a covariance stationary process it can be shown that

$$\mathrm{var}\big(\overline{X}(n)\big) \;=\; \sigma^2 \frac{\left[1 + 2\sum_{j=1}^{n-1}(1-j/n)r_j \right]}{n} \quad \text{and so using } \mathrm{E}\big[S^2(n)\big] \text{ in place of } \sigma^2 \text{ gives}$$

$$\approx \;\; \mathrm{E}\big[S^2(n)/n\big] \frac{(n-1)\left(1 + 2\sum_{j=1}^{n-1}(1-j/n)r_j\right)}{n - \left(1 + 2\sum_{j=1}^{n-1}(1-j/n)r_j\right)}.$$

If $r_j > 0$, then $\mathrm{var}\big(\overline{X}(n)\big) > \mathrm{E}\big[S^2(n)/n\big]$ and the value of any test statistic constructed using $S^2(n)/n$ as the estimator for the variance, will be too large and so a null hypothesis will be rejected too often. Similarly, any confidence intervals will be too small.

[Law and Kelton, 2000] give an example (based on the simple M/M/1 queue), where

$$\mathrm{var}\big(\overline{X}(n)\big) \;=\; 294\mathrm{E}\big[S^2(n)/n\big].$$

We therefore cannot use statistics based on IID formulae directly to analyze simulation output data, as they are generally correlated. However, it is often possible to group data into new "observations" to which formulae based on IID observations can be applied. The easiest and most commonly used approach is to **replicate** simulation runs.

Let $Y_1, Y_2, \ldots, Y_m$ be an output stochastic process from a single simulation run. In general, the Y_i will not be independent and maybe not even identically distributed. However, let $y_{11}, y_{12}, \ldots, y_{1m}$ be a realization of the random variables $Y_1, Y_2, \ldots, Y_m$ resulting from a simulation run of length m observations. Now run the simulation again with a different stream of random numbers, so that we will obtain a different realization $y_{21}, y_{22}, \ldots, y_{2m}$ of the random variables $Y_1, Y_2, \ldots, Y_m$. In general, from n independent replications we will get the observations

$$
\begin{array}{cccccc}
y_{11}, & y_{12}, & \cdots & y_{1i} & \cdots & y_{1m} \\
y_{21}, & y_{22}, & \cdots & y_{2i} & \cdots & y_{2m} \\
\vdots & \vdots & \vdots & \vdots & \vdots & \vdots \\
y_{n1}, & y_{n2}, & \cdots & y_{ni} & \cdots & y_{nm}
\end{array}
$$

The observations from a particular run are not IID. However, the observations $y_{1i}, y_{2i}, \ldots, y_{ni}$ from the ith column are IID observations of the random variable Y_i.

We can then use the analysis methods for independent random variables which we have described above. Thus, for example, unbiased estimates of $E[Y_i]$ and $\text{var}(Y_i)$ are given by

$$\bar{y}_i(n) \;=\; \frac{1}{n}\sum_{j=1}^{n} y_{ji} \quad \text{and} \quad s_i^2(n) \;=\; \frac{1}{n-1}\sum_{j=1}^{n} [y_{ji} - \bar{y}_i(n)]^2,$$

respectively. Then, an approximate $100(1-\alpha)\%$ CI for μ_{Y_i} is

$$\left(\bar{y}_i(n) - z_{1-\alpha/2}\sqrt{s_i^2(n)/n}, \quad \bar{y}_i(n) + z_{1-\alpha/2}\sqrt{s_i^2(n)/n} \right).$$

We can also apply this to any other statistic derived from the different runs.

$$\bar{y}(n) \;=\; \frac{1}{n}\sum_{j=1}^{n}\left[\sum_{i=1}^{m}\frac{y_{ji}}{m}\right]$$

the mean of sample means, and

$$s^2(n) \;=\; \frac{1}{n-1}\sum_{j=1}^{n}\left[\sum_{i=1}^{m} y_{ji}/m - \bar{y}(n)\right]^2$$

the variance of $\bar{y}(n)$.

Therefore to perform an adequate simulation of the system given in Example 13.25, we should perform multiple simulation runs of much longer duration than just the processing time for twenty castings (see Exercise 13.9).

Although the two examples given have specific discrete distributional forms for the inputs, other discrete and continuous distributional forms may also be used as appropriate and deviates generated using those techniques described in Chapters 4 and 5. The discrete event simulation procedure is then performed using the generated input data in the exact same way by keeping track of state changes at event times. Replication and the method for getting approximately covariance stationary data remain the same.

13.3.3 Variance reduction

We have discussed the drawbacks associated with correlation in simulation, but there can also be beneficial results with certain forms of intentional correlation. A neat trick that can often be used in simulation that results in what is known as variance reduction is to use common random numbers (CRN) to drive two alternative configurations. For example, if an initial simulation model is created and validated as an acceptable model of an existing system, the system model can be changed to reflect some proposed changes for the real system and the same inputs may be used where appropriate to drive both simulations. This can induce a positive correlation between the output variables W_{1i} from the original model and W_{2i} from the modified model that may be constructive rather than destructive. There is no guarantee that this technique will always work. For example if the simulation model is very complex, we may have no way of determining if we will induce a positive correlation.

The rationale behind this idea is as follows. Suppose we wish to compare the outputs from the original and a changed system and we perform a matched pairs test by considering $V_i = W_{1i} - W_{2i}$, for $i = 1, \ldots, n$. The V_is are IID random variables and $\overline{V}(n)$ is an unbiased estimator of their difference and

$$\text{var}\big(\overline{V}(n)\big) \;=\; \frac{\text{var}(V_i)}{n} \;=\; \frac{Var[W_{1i}] + Var[W_{2i}] - 2Cov[W_{1i}, W_{2i}]}{n}.$$

If W_{1i} and W_{2i} are IID then $2Cov[W_{1i}, W_{2i}] = 0$, otherwise, if there exists positive correlation between W_{1i} and W_{2i}, then $2Cov[W_{1i}, W_{2i}] > 0$ and so $\text{var}(\overline{V}(n))$ will be reduced. In order to show how positive correlation can affect a result, consider the following example.

Example 13.26: Useful correlation

Suppose from a given simulation model of some system, we find that the existing system's observed average waiting times are

$$W_{1i} = (9.9365, 8.8701, 8.9179, 11.2187, 9.9205, 10.7262,$$
$$10.7262, 10.5610, 11.0695, 10.5664, 8.1030)$$

After modifying the operation of the system with the intention of reducing the average waiting time, a new set of inputs is generated and the average waiting times are observed to be

$$W_{2i} = (8.4106, 8.5554, 12.6029, 8.7324, 4.5124,$$
$$7.6202, 10.3992, 10.5449, 9.9952, 9.7833)$$

Here, the sample means and variances are

$$\overline{W}_1 = 9.9890, \quad s_1^2 = 1.0941$$
$$\overline{W}_2 = 9.1157, \quad s_2^2 = 4.6061$$

To perform a paired t-test (see Section 7.8.2) for the difference of the two means we use

```
> W1=c(9.9365, 8.8701, 8.9179, 11.2187, 9.9205, 10.7262, 10.5610,
+ 11.0695, 10.5664, 8.1030)
> W2=c(8.4106, 8.5554, 12.6029, 8.7324, 4.5124, 7.6202, 10.3992,
+ 10.5449, 9.9952, 9.7833)
> t.test(W1,W2,paired=TRUE)
Paired t-test
data:  W1 and W2
t = 1.0983, df = 9, p-value = 0.3006
alternative hypothesis: true difference in means is not equal to 0
95 percent confidence interval:
 -0.9254779  2.6721379
sample estimates:
mean of the differences
          0.87333
```

This shows that there is no statistical evidence for a difference in the two mean waiting times.

Suppose now that the same inputs (CRNs) are used to drive both simulation models and that the modified model yields the following data

$$W_{3i} = (9.0080, 6.8200, 6.9181, 11.6388, 8.9752, 10.6283,$$
$$10.2894, 11.3327, 10.3004, 5.2461)$$

with the same sample mean and variance as the W_{2i} data given by

$$\overline{W}_3 = 9.1157, \quad s_3^2 = 4.6061$$

To perform a paired t-test for the difference of the two means based on this data, we use

```
> W3=c(9.0080, 6.8200, 6.9181, 11.6388, 8.9752, 10.6283, 10.2894,
+ 11.3327, 10.3004, 5.2461)
> t.test(W1,W3,paired=TRUE)
Paired t-test
data:  W1 and W3
t = 2.5101, df = 9, p-value = 0.0333
alternative hypothesis: true difference in    means is not equal to 0
95 percent confidence interval:
 0.08626538 1.66029462
sample estimates:
mean of the differences
            0.87328
```

There now appears to be statistical evidence for a difference between the means in this test as the p-value is 0.033. This is brought about by the fact that there exists a positive correlation between each of the W_{1i} and the W_{3i} for $i = 1, 2, \ldots, 10$. Note that in both test cases the mean of differences is approximately 0.8733.

There are many more techniques available to achieve a variance reduction in simulation analysis, which can be found in for example [Ross, 2013].

13.4   Summary

13.4.1   Notation

φ	structure (system) function
r	reliability function
$\mathbb{P}$	transition matrix
$\boldsymbol{X}(m)$	distribution of Markov chain at time m
$\boldsymbol{\pi}$	equilibrium distribution of Markov chain
X_j^i	probability of absorption in state i, given it started in state j
M_j	expected time until absorption starting in state j

13.4.2   Summary of main results

System reliability: The reliability of a component, indexed by i, is the probability, p_i, that it functions for some stated length of time. If component i is working, then $x_i = 1$, otherwise $x_i = 0$. The structure function, φ is a model for system failure and is a function of x_i for all i. Depending on the structure of the component (series, parallel or k-out-of-n), the structure function will take multiple forms.

A complicated system, consisting of many different structures can be split into modules, which often form a series or parallel system.

The reliability function r represents the probability the system is functioning and is given by $r = \mathrm{E}[\varphi(\boldsymbol{x})]$.

Markov chains: A discrete time Markov process X_t is a random process with state space S observed at discrete time points, which satisfies

$$\mathrm{P}(X_{t+1} = i_{t+1} \mid X_0 = i_0 \cap X_1 = i_1 \cap \ldots \cap X_t = i_t) \;=\; \mathrm{P}(X_{t+1} = i_{t+1} \mid X_t = i_t),$$

where $i_0, \ldots, i_{t+1} \in S$. To specify a Markov chain, an initial distribution $\boldsymbol{X}(0)$ and the transition matrix $\mathbb{P} = (p_{ij})$ are required. The elements of $\mathbb{P}$ are given by $p_{i,j} = \mathrm{P}(X_{t+1} = j \mid X_t = i)$. To describe the distribution of that Markov chain at time m, we use $\boldsymbol{X}(m) = \boldsymbol{X}(0)\mathbb{P}^m$.

If the Markov chain has a limiting distribution, then the equilibrium distribution is given by the solution to $\boldsymbol{\pi} = \boldsymbol{\pi}\mathbb{P}$ and the normalizing equation $\sum_j \pi_j = 1$.

A state is called an absorbing state if the probability of leaving the state is zero. If state 0 and N are absorbing and $X_j^{(N)}$ is the probability that the process is absorbed in state N given that it starts in state j, then $X_j^{(N)}$ satisfies the equation

$$X_j^{(N)} = \sum_k p_{j,k} X_k^{(N)}, \quad 1 \le j \le N-1,$$

with boundary conditions $X_N^{(N)} = 1$ and $X_0^{(N)} = 0$. If M_j is the mean time until absorption of the chain, conditional on starting in state j, then M_j satisfies the equations

$$M_j \;=\; 1 + \sum_k p_{j,k} M_k, \quad \text{for} \quad j \ge 1,$$

subject to $M_j = 0$ for all j absorbing.

Simulation:

- Simulation of complex systems can yield valuable insight into which variables are most important and how they interact.

- Simulation makes it possible to study the complex interactions of a given system. This may help in identifying and resolving problem areas.

- Simulation can be used to investigate and experiment with new and modified systems prior to physically implementing them such as when purchasing new equipment, replacing or deleting resources or when intending new strategies, methods, or layouts.

- Simulation can be used where it may be impossible, extremely expensive or dangerous to observe certain processes.

13.5   Exercises

Section 13.1 System reliability

Exercise 13.1: System dual

(a) Find the dual of a 2-out-of-3 system.

(b) State the dual of a k-out-of-n system.

(c) Prove the result in the previous part.

Section 13.2 Markov chains

Exercise 13.2: Absorbing probabilities

In Example 13.18, consider the case when $N = 6$ and $p = 0.4$. That is when

$$
\mathbb{P} =
\begin{pmatrix}
1 & 0 & 0 & 0 & 0 & 0 & 0 \\
0.6 & 0 & 0.4 & 0 & 0 & 0 & 0 \\
0 & 0.6 & 0 & 0.4 & 0 & 0 & 0 \\
0 & 0 & 0.6 & 0 & 0.4 & 0 & 0 \\
0 & 0 & 0 & 0.6 & 0 & 0.4 & 0 \\
0 & 0 & 0 & 0 & 0.6 & 0 & 0.4 \\
0 & 0 & 0 & 0 & 0 & 0 & 1
\end{pmatrix} .
$$

Find the absorbing probabilities to states 0 and 6 from each state $1, 2, 3, 4, 5$

(a) using the technique shown in Example 13.17,

(b) using the technique shown in Theorem 13.2.

Exercise 13.3: Multistep probabilities

Consider the infinite matrix in Exercise 13.14, with $p = 0.4$ and $q = 0.5$, so that

$$
\mathbb{P} =
\begin{pmatrix}
0.6 & 0.4 & 0 & 0 & 0 & \dots \\
0.5 & 0.1 & 0.4 & 0 & 0 & \dots \\
0 & 0.5 & 0.1 & 0.4 & 0 & \dots \\
\vdots & \ddots & \ddots & \ddots & \ddots & \ddots
\end{pmatrix} .
$$

Given that the buffer starts with 10 packets at time 0, show that the probability that it still has 10 packets after 10 time units is 0.1345. Note that even though this matrix is infinite, it is sparse and so a truncated matrix multiplication can be performed to get this result.

Exercise 13.4: Moran reservoir

Refer to Example 13.15. Consider a policy of: releasing 2 units during the dry season if the dam contains 2 or more units and releasing 1 unit if the dam contains 1 unit.

(a) In the long run, for what proportion of time would the dam be empty?

(b) If the value of each unit of water during the dry season is 1 and the value associated with no water supply during the dry season is -4, would you recommend the policy of releasing 2 units over that of releasing 1 unit?

(c) Calculate the long term expected value of:

(i) releasing 2 units if possible,

(ii) releasing 1 unit if possible, and

(iii) releasing 1 unit provided the reservoir contains 2 or more units at the start of the dry season.

Exercise 13.5: Properties of probability generating functions

The probability generating function in Definition 13.12 for a discrete probability distribution $\{\pi_n\}$, for $n = 0, 1, 2, \ldots$ is

$$P(z) = \sum_{n=0}^{\infty} \pi_n z^n, \quad \text{for } z \in [0, 1].$$

Prove the following interesting properties of the probability generating function.

(a) $j!\pi_j = \dfrac{d^j}{dz^j} P(z)\bigg|_{z=0}$,

(b) $\text{E}[N] = \dfrac{d}{dz} P(z)\bigg|_{z=1}$ and

(c) $\text{var}(N) = \dfrac{d^2 P(z)}{dz^2}\bigg|_{z=1} + \dfrac{dP(z)}{dz}\bigg|_{z=1} - \left[\dfrac{dP(z)}{dz}\bigg|_{z=1}\right]^2$.

Section 13.3 Simulation of systems

Exercise 13.6: Multiple repairs

Suppose there are 4 machines and the the breakdown and repair rates are 1 per 24h-day and 2 per 24h-day respectively. Compare the policy of having two mechanics who work independently with a single repair crew of two mechanics who always work together and achieve a repair rate of 2.2 per 24h-day.

Exercise 13.7: Hidden states

Suppose there are 2 machines and breakdowns have independent exponential distributions with rate of 1 per day. There is one mechanic and the repair is in two stages: find the fault, and fix the fault. The rate for both stages is 4 per day and the times taken are independent. Compare the equilibrium probabilities with those pertaining if repair is in a single stage with rate 2 per day.

Exercise 13.8: Machine shop

Using the inter-arrival times and service times given in Example 13.24,

(a) Recreate the simulation record given in Table 13.1.

(b) Use the record of the simulation run to verify that given in Example 13.24 and

(c) For those castings that enter a buffer, verify the mean time spent by such castings in the buffer.

Exercise 13.9: Machine shop alternative

Using the given R code in Example 13.25,

(a) Generate 1 000 each of inter-arrival times and service times.

(b) Using the simulation model created in Exercise 13.8, perform a simulation and collect the output data for the buffer occupancy, leaving out the data corresponding to the first 100 castings.

(c) Give a reason for not including the data corresponding to the first 100 castings.

(d) Repeat the simulation 5 times using different seeds for each of the simulation runs.

(e) For each of the simulation runs and for those castings that enter a buffer, calculate the mean time spent by such castings in the buffer.

(f) Using the data from all 5 simulation runs, for those castings that enter a buffer, calculate a 95% confidence interval for the mean time spent by such castings in the buffer.

(g) What is your recommendation for the capacity of the buffer store and is it substantially larger than 4 as observed in Example 13.24?

14

Sampling strategies

In this chapter we consider sampling from a well defined finite population when sample is an appreciable proportion of the population. A typical objective is the estimation of the mean, or the total, of some quantity over the population. If we know something about all the items in the population then we can use this information to set up more efficient random sampling schemes than simple random sampling. A case study of an asset management plan for a water company illustrates the ideas. See example in Appendix E:

Experiment E.3 *Robot rabbit.*

14.1   Introduction

So far we have assumed that we have taken a simple random sample (SRS) from some population and that the sample size is small by comparison with the population size. In practical applications the population is often defined as infinite, for example all production if a process continues on its present settings, and some plausible approximation to an SRS scheme is implemented.

In this chapter we consider finite populations with a clearly defined size. If the sample is a substantial proportion of the population, then we can account for the increased precision. Other ways of increasing precision are: to divide the population into sub-populations; and to use regression methods if predictor variables have known values for all members of the population. We also consider multi-stage sampling schemes for large populations. We begin with some definitions.

Definition 14.1: Target population

The **target population** is the set of elements about which information is required.

Example 14.1: Light aircraft [target population]

An example of a target population is all light civilian aircraft in the U.S. The information sought could include: number of flights in the last year together with the number of passengers and weight of freight; distance flown over the past year; airports visited over the past year; and maintenance records.

Example 14.2: Ultra-light aircraft [target population]

In some countries, solo ultra-light aircraft do not have to be registered. An aviation authority intends to estimate the number of solo ultra-light aircraft in the country. The target population is the total number of all solo ultra-light aircraft in the country, and the first objective is to estimate the size of the population.

Definition 14.2: Population characteristic (parameter)

A **population characteristic**, also referred to as a **parameter**, is a numerical summary of some variable that is defined for every element in the population. Typical population characteristics are means, standard deviations, proportions, totals, and ratios.

Example 14.3: Aviation [parameters]

The mean of the distances flown by all the light civilian aircraft in the U.S. over the past year is a parameter of that population. The population size is also a parameter, but it is known from registration details.

The number of solo ultra-light aircraft in a country is a parameter of that population. If there is no registration then the number is not known, and would have to be estimated.

Definition 14.3: Sampling unit

A **sampling unit** is a unit that can be investigated if drawn in the sampling procedure. If the sampling is **single-stage** then the set of sampling units will ideally be the elements of the target population.

Definition 14.4: Survey population

The **survey population** is the set of sampling units. Ideally the survey population is the target population. In practice the survey and target populations can differ. For example, in some cases a list of members of the population may not be up to date.

Example 14.4: Light aircraft [survey population]

The target population is all civilian light aircraft in the country. The survey population is a list of all registered light aircraft in the country at the time of planning the survey. The list will exclude aircraft that are about to be registered and may include aircraft that are no longer flown. There may also be some illegally operated light aircraft that are not registered and so do not appear in the list.

Definition 14.5: Sampling frame

A **sampling frame** is a list of all the sampling units in the survey population. The list might be names in alphabetical order or it might relate to grid squares on a map.

Example 14.5: Oil wells [sampling frame]

An intends estimating the number of pumpjacks over a particular region. A map of the region is divided by a grid into 100 m squares. These squares form a sampling frame, and one SRS of squares will be surveyed by drones.

Definition 14.6: Random sampling scheme

A **random sampling scheme** is a sampling scheme in which the following two conditions hold.

(i) Every unit in the population to be sampled has a non-zero probability of selection.

(ii) Every member of the population to be sampled has a known probability of selection.

Unbiased estimators of population parameters are obtained by weighting the sample observations with weights inversely proportional to the probabilities of selection.

Example 14.6: Aircraft [random sampling scheme]

An SRS of n elements from a population of size N is a random sampling scheme. Every element in the population has known non-zero probability of selection, n/N. There are 15,300 aircraft registered in Australia in 2015. A researcher takes an SRS of 100 aircraft. In this case $N = 15,300$, $n = 100$. The sample could be obtained from the ordered list using R with the function call:

```
sample(1:15300, 100, replace=FALSE)
```

Every aircraft has equal probability of $100/15,300$ of being chosen, so it satisfies the requirements for a random sampling scheme.

An alternative random sampling scheme would be to take a uniformly distributed integer between 1 to 153 to identify a first aircraft on the list for the sample, in R sample(1:153, 1), and every 153rd aircraft on the list thereafter. This is an example of **a systematic sampling scheme**. Every aircraft has an equal probability of $100/15\,300$ of being chosen, and it satisfies the requirements for a random sampling scheme. But, the systematic sample is not an SRS. In the SRS scheme there are $\binom{15\,300}{100}$ equally likely possible samples of 100 aircraft. In the systematic sampling scheme there are 153 equally likely possible samples of 100 aircraft.

It is not necessary that all elements of the population have equal probabilities of selection for a random sampling scheme.

Example 14.7: Farm dams [random sampling scheme]

The Murray-Darling basin extends over four Australian states, from Queensland (Q) in the north through New South Wales (NSW) and Victoria (V) to the mouth of the Murray in South Australia (SA). A researcher intends estimating the total capacity of farm dams within the basin. The area of the basin is divided into 106 grid squares, each of an area of 10 000 km^2. The number of squares in the states is shown in Table 14.1. Suppose that thestates agree to undertake aerial surveys of SRSs, with sizes shown in Table 14.1, of the squares within their borders.

TABLE 14.1: Number of grid squares and sample size for each state within Murray-Darling basin.

State	Number of squares	Sample size
Q	27	6
NSW	59	8
V	12	3
SA	8	3

This is a random sampling scheme. Each square has a known non-zero probability of selection. The probabilities of selection are different in the states, 6/27, 8/59, 3/12 and 3/8 for Q, NSW, V, and SA respectively. The estimate of the total capacity is the sum of the estimated capacities in each selected grid square divided by the probability of selection (Exercise 14.4). [Nathan and Lowe, 2012] quote estimates of around 2 000 giga-liters.

14.2   Simple random sampling from a finite population

Suppose we wish to estimate a population mean and that we take an SRS, without replacement, from the population. If the population size is large relative to the sample size then the variance of the sample mean is the population variance divided by sample size. However, if a substantial proportion of the population is being sampled, the variance of the sample mean will be reduced. In the following we show that the variance of the sample mean reduces by a factor of 1 less the proportion of the population sampled.

14.2.1   Finite population correction

Consider estimation of the mean of a finite population, that consists of N elements, from a sample. An SRS scheme for a sample of size n elements from the population is a scheme in which every possible choice of n from the N elements has an equal probability of selection.

For this chapter we will define a finite population mean and variance by[1]

$$\bar{y}_U = \frac{\sum_{i=1}^{N} y_i}{N} \quad \text{and} \quad s_U^2 = \frac{\sum_{i=1}^{N}(y_i - \bar{y}_U)^2}{N-1}.$$

Then the precise result for the variance of the sample mean is

$$\text{var}(\bar{y}) = \frac{s_U^2}{n}\left(1 - \frac{n}{N}\right).$$

The factor $(1 - n/N)$ is known as the finite population correction (FPC). The FPC is 1 if the population is infinite and close to 1 if the sampling fraction n/N is small, but it represents a useful reduction in the standard error if the sampling fraction exceeds around one tenth. If the entire population is in the sample, then the FPC is 0 and the standard error of the estimator of the population mean is 0. We will justify the FPC in two ways. The first assumes only that we have an SRS from the finite population, and is based on randomization theory. The second model based analysis assumes the finite population is itself a random sample from an infinite population.

14.2.2 Randomization theory

14.2.2.1 Defining the simple random sample

The elements of a finite population of size, N, are indexed by the set

$$U = \{1, \ldots, N\}.$$

The variable for which we wish to estimate the population total (primary variable) is denoted by y_i for element i, $i = 1, \ldots, N$, and the objective is to estimate the population mean $\bar{y}_U$ and hence the population total T_U.

$$\bar{y}_U = \frac{\sum_{i=1}^{N} y_i}{N} \qquad T_U = N\bar{y}_U.$$

We have an SRS of size n from the population, and the sample is indexed by the set

$$S = \{i \mid \text{unit } i \text{ in sample}\}.$$

Now define a random variable

$$Z_i = \begin{cases} 1 & \text{if unit } i \text{ in sample} \\ 0 & \text{otherwise} \end{cases}$$

The probability that Z_i equals 1 is the ratio of the number of samples of size n that include i to the number of equally likely samples of size n. For any i, the number of samples that include it is equal to the number of choices of $n - 1$ from the remaining $N - 1$ items in the population. That is:

$$P(Z_i = 1) = \frac{\binom{N-1}{n-1}}{\binom{N}{n}} = \frac{n}{N}.$$

---

[1]The subscript U represents universe from the sampling point of view, and the reason for not using μ and σ will become clear later in the section. Moreover, the division by $N - 1$ in the definition of the finite population variance may seem odd, but it leads to simpler formulae.

The following results follow from the fact that Z_i can only take the values 0 and 1, and so equals Z_i^2. The expected value is

$$E[Z_i] \;=\; E[Z_i^2] \;=\; 0 \times \left(1 - \frac{n}{N}\right) + 1 \times \frac{n}{N} \;=\; \frac{n}{N}$$

and the variance is

$$\mathrm{var}(Z_i) \;=\; E[Z_i^2] - (E[Z_i])^2 \;=\; \frac{n}{N}\left(1 - \frac{n}{N}\right).$$

In a similar fashion

$$E[Z_i Z_j] \;=\; P(Z_i = 1 \cap Z_j = 1) \;=\; P(Z_i = 1)\,P(Z_j = 1 | Z_i = 1) \;=\; \frac{n}{N}\frac{n-1}{N-1}$$

and

$$\mathrm{cov}(Z_i Z_j) \;=\; E[Z_i Z_j] - E[Z_i]\,E[Z_j] \;=\; -\frac{1}{N-1}\left(\frac{n}{N}\right)\left(1 - \frac{n}{N}\right).$$

14.2.2.2   Mean and variance of sample mean

We define the sample mean as a random variable by introducing the Z_i.

$$\bar{y} \;=\; \sum_{i \in S} \frac{y_i}{n} \;=\; \sum_i^N Z_i \frac{y_i}{n}$$

and

$$E[\bar{y}] \;=\; \sum_i^N E[Z_i]\frac{y_i}{n} \;=\; \sum_i^N \frac{n}{N}\frac{y_i}{n} \;=\; \sum_i^N \frac{y_i}{N} \;=\; \bar{y}_U \;.$$

The variance of $\bar{y}$ is

$$\mathrm{var}(\bar{y}) \;=\; \frac{1}{n^2}\mathrm{var}\left(\sum_i^N Z_i y_i\right) \;=\; \frac{1}{n^2}\mathrm{cov}\left(\sum_i^N Z_i y_i, \sum_j^N Z_j y_j\right)$$

$$=\; \frac{1}{n^2}\left(\sum_{i=1}^N y_i^2\,\mathrm{var}(Z_i) + \sum_{1=1}^N \sum_{j \neq i}^N y_i y_j\,\mathrm{cov}(Z_i Z_j)\right).$$

We now substitute for the variances and covariances to obtain

$$\mathrm{var}(\bar{y}) \;=\; \frac{1}{n^2}\frac{n}{N}\left(1 - \frac{n}{N}\right)\left(\sum_{i=1}^N y_i^2 - \frac{1}{N-1}\sum_{1=1}^N \sum_{j \neq i}^N y_i y_j\right)$$

$$=\; \frac{1}{n^2}\left(1 - \frac{n}{N}\right)\frac{1}{N(N-1)}\left((N-1)\sum_{i=1}^N y_i^2 - \left(\sum_{i=1}^N y_i\right)^2 + \sum_{i=1}^N y_i^2\right)$$

$$=\; \frac{1}{n^2}\left(1 - \frac{n}{N}\right)\frac{1}{N(N-1)}\left(N\sum_{i=1}^N y_i^2 - \left(\sum_{i=1}^N y_i\right)^2\right) \;=\; \left(1 - \frac{n}{N}\right)\frac{s_U^2}{n}$$

In the last line we have used the definition of population variance, and its identical form

$$s_U^2 \;=\; \frac{\sum_{j=1}^N (y_j - \bar{y}_U)^2}{N-1} \;=\; \frac{1}{N-1}\left(\sum_{i=1}^N y_i^2 - \left(\frac{1}{N}\sum_{i=1}^N y_i\right)^2\right).$$

The sample variance is defined in the usual way

$$s^2 = \frac{\sum_{i=1}^{n}(y_i - \bar{y})^2}{n-1}$$

and the proof that it is unbiased for the finite population is similar to proving the corresponding result for an infinite population.

14.2.2.3 Mean and variance of estimator of population total

We denote the population total by T_U. The **number raised estimator**, T, is the product of the number of units in the population and the sample mean. That is

$$T = N\bar{y}$$

and the estimated standard error of T is

$$N\frac{s}{\sqrt{n}}\sqrt{1 - \frac{n}{N}}.$$

Example 14.8: Specialist engineering company

A light engineering company specializes in manufacturing prototype devices and small runs of components, such as engine parts for vintage cars. The company operates 11 small factories and the CEO wanted to estimate the financial loss due to scrap over the next month. The CEO selected an SRS of three factories and asked three engineering students, on an industrial placement, to monitor scrap in the three factories over a four week period, and to suggest ways in which scrap might be reduced. At the end of the four weeks the results, were $2\,300, \$1\,500$, and $4\,100$. We use R for the calculations.

```
> N=11;n=3
> y=c(2.3,1.5,4.1)
> print(c("mean",round(mean(y),2),"sd",round(sd(y),2)))
[1] "mean" "2.63" "sd"    "1.33"
> semean=sqrt(1-n/N)*sd(y)/sqrt(n)
> print(c("standard error of mean",round(semean,2)))
[1] "standard error of mean" "0.66"
> poptotal=N*mean(y)
> sepoptotal=N*semean
> print(c("estimated population total",round(poptotal,2),
+ "standard error",round(setotal,2)))
[1] "estimated population total" "28.97"
[3] "standard error"             "7.21"
> tval=qt(.90,(n-1))
> L90=N*(mean(y)-tval*semean)
> U90=N*(mean(y)+tval*semean)
> print(c("Lower 80",round(L90,1),"Upper 80",round(U90,1)))
[1] "Lower 80" "15.4"    "Upper 80" "42.6"
```

The estimate for the financial loss to the company due to scrap for this particular four week period is $29 000 with a standard error of $7 000. An 80% confidence interval is [15, 43], based on an assumption that the sampling distribution of the sample mean is reasonably approximated as normal. The interval is inevitably wide when the standard deviation based on such a small sample because the t-distribution has only two degrees of freedom. Increasing the confidence level gives uselessly wide intervals, partly because the assumption of normality somewhat overestimates the sampling variability of the sample mean.

Example 14.9: Rock Quarry (case contributed by Ward Robinson)

The client operates a quarry that produces granite rocks that are used as armor for coastal engineering projects and to provide scour protection of hydraulic structures. Rocks are transported from the quarry to a wharf, where they are weighed and stacked in bunkers that are designated for specific ranges of mass (see Figure 14.1). The total

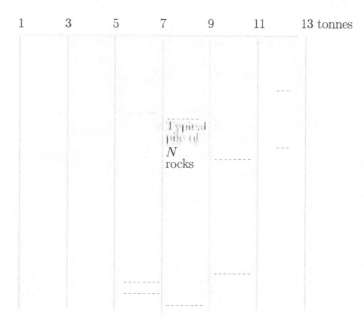

FIGURE 14.1: Piles of rocks in bunkers.

mass of rocks in each bunker is known. The rocks from bunkers are loaded onto barges, and the barge skipper takes all the rocks from allocated bunkers with the exception of the last. Reasons for not taking all the rocks from the last bunker include reaching the Plimsoll line and wanting to catch the tide. A consequence is that the precise mass of the rocks loaded from the last bunker is unknown. The number and masses of all the rocks in the last bunker are known, as is the number of rocks loaded. The client wanted to produce prediction intervals for the mass of rocks taken from the last bunker, under an assumption that the rocks in a bunker have been stacked in a haphazard way so that the more accessible rocks that have been loaded are equivalent to an SRS from the rocks in the bunker. In general, n rocks from N have been loaded and the sample mean

$$\bar{y} \sim N\left(\bar{y}_{\mathrm{U}}, \frac{s_{\mathrm{U}}^2}{n}\left(1 - \frac{n}{N}\right)\right),$$

where $\bar{y}_{\mathrm{U}}$ and s_{U} are known and the normal distribution is a reasonable approximation. It follows that there is a probability of $(1 - \alpha)$ that the mass loaded, $M = n\bar{y}$, lies within the interval

$$n\left(\bar{y}_{\mathrm{U}} \pm z_{\alpha/2}\frac{s_{\mathrm{U}}}{\sqrt{n}}\sqrt{1 - \frac{n}{N}}\right)$$

and a $(1 - \alpha)100\%$ prediction interval for the mass loaded is given by

$$n\bar{y}_{\mathrm{U}} \pm z_{\alpha/2}s_{\mathrm{U}}\sqrt{n - n^2/N}$$

For example, a skipper loads 30 from 40 rocks in the 5 to 7 tonne bunker. The mean and standard deviation of the 40 rocks were recorded as 6.12 and 0.56 respectively. Assuming rocks are loaded at random a 90% prediction interval for the total mass is:

```
> L=30*6.12-qnorm(.95)*0.56*sqrt(30-30^2/40)
> U=30*6.12+qnorm(.95)*0.56*sqrt(30-30^2/40)
> print(round(c(L,U),2))
[1] 181.08 186.12
```

14.2.3 Model based analysis

The randomization theory treats the finite population values y_i as constants that are known if the item i is in the sample. The random variables Z_i indicate whether or not an item is in the sample. In contrast, in a model based approach the finite population values are treated as a sample of size N, obtained by independent draws from a probability distribution with mean μ and standard deviation σ. The SRS justifies the assumption of independent draws. The population total is then

$$T_{\mathrm{U}} = \sum_{i=1}^{N} Y_i = \sum_{i \in S} Y_i + \sum_{i \notin S} Y_i.$$

The estimator of the population total is

$$T = \frac{N}{n}\sum_{i \in S} Y_i.$$

The estimator is unbiased for T_{U} since

$$\mathrm{E}[T - T_{\mathrm{U}}] = \mathrm{E}\left[\frac{N}{n}\sum_{i \in S} Y_i - \sum_{i=1}^{N} Y_i\right] = \frac{N}{n}n\mu - N\mu = 0.$$

The estimator has a mean squared error given by

$$\mathrm{E}[(T - T_{\mathrm{U}})^2] = \mathrm{E}\left[\left(\frac{N}{n}\sum_{i \in S} Y_i - \sum_{i=1}^{N} Y_i\right)^2\right]$$

$$= \mathrm{E}\left[\left(\left(\frac{N}{n} - 1\right)\left(\sum_{i \in S} Y_i - n\mu\right) - \left(\sum_{i \notin S} Y_i - (N - n)\mu\right)\right)^2\right]$$

and taking expectation, remembering that all the Y_i are independent:

$$\mathrm{E}\left[(T-T_\mathrm{U})^2\right] = \left(\frac{N}{n}-1\right)^2 n\sigma^2 + (N-n)\sigma^2 = N^2\frac{\sigma^2}{n}\left(1-\frac{n}{N}\right).$$

This is equivalent to the result obtained by randomization theory[2] when σ^2 is replaced by s_U^2. In practice s_U^2 is itself estimated by the sample variance s^2, and the results are identical.

14.2.4 Sample size

Suppose we wish to estimate a mean with a $(1-\alpha)100\%$ confidence interval of width 2Δ. We require

$$z_{\alpha/2}\sqrt{\frac{\sigma^2}{n}\left(1-\frac{n}{N}\right)} = \Delta.$$

Straightforward algebra leads to the expression

$$n = \frac{n_\infty}{1+n_\infty/N}, \quad \text{where} \quad n_\infty = \left(\frac{z_{\alpha/2}\sigma}{\Delta}\right)^2$$

is the sample size if the FPC is not applied.

Example 14.10: A water company

A water company has 57 water towers and wishes to estimate the total cost of refurbishment. The company would like a 90% confidence interval for the total cost to be around plus or minus 20% of that cost. An experienced engineer thinks that the coefficient of variation of cost is likely to be around 0.5 but could be as high as 1. In the latter case the ratio $\sigma/\Delta = 5$. The sample size calculations in R are

```
> z=qnorm(.95)
> sigdelratio=5
> ninf=(z*sigdelratio)^2
> print(ninf)
[1] 67.63859
> n=ninf/(1+ninf/57)
> print(n)
[1] 30.93263
```

The sample size of 31 is higher than the company is prepared to take. With a coefficient of variation of 0.5, the sample size would be 13. The company decides to take an initial sample of 10, and will review the situation when it has a better estimate of the population standard deviation based on the 10 sampled water towers.

14.3 Stratified sampling

A population can often be considered as a union of disjoint sub-populations, as in the U.S. and its states. Information may be needed for sub-populations. Also, if sub-populations are

---

[2]This gives a rationale for defining the finite population variance with the denominator $(N-1)$.

relatively homogeneous, compared to the whole, we can improve the precision of estimators for a given sample size.

Definition 14.7: Stratum

A population is stratified if it is divided into sub-populations according to criteria that are known for all elements in the population. Each sub-population is a **stratum**.

14.3.1 Principle of stratified sampling

The population is divided into sub-populations called **strata**, and samples are taken from each *stratum*. The division into strata can be based on any criteria that take account of known characteristics of elements of the population or on subjective assessment, but to be efficient the items within a stratum need to be relatively similar, in terms of the variable of interest, when compared with items from different strata. Totals can be estimated for each stratum as well as for the population as a whole.

14.3.2 Estimating the population mean and total

Suppose we have a population of N sampling units divided into H strata. The strata have sizes of N_h $h = 1, \ldots, H$ sampling units. The value of the variable for the ith unit in stratum h is denoted by

$$y_{hi} \quad \text{for} \quad i = 1, \ldots, N_h \quad \text{and} \quad h = 1, \ldots, H.$$

The stratum totals, and hence the population total, are

$$T_{\text{U},h} \;=\; \sum_{i=1}^{N_h} y_{hi} \quad \text{and} \quad T_{\text{U}} \;=\; \sum_{h=1}^{H} T_{\text{U},h},$$

It follows that the stratum means, and population mean, are

$$\bar{y}_{\text{U},h} \;=\; \frac{T_{\text{U},h}}{N_h} \quad \text{and} \quad \bar{y}_{\text{U}} \;=\; \frac{T_{\text{U}}}{N}.$$

The stratum variances are defined as

$$s_{\text{U},h}^2 \;=\; \sum_{i=1}^{N_h} \frac{(y_{hi} - \bar{y}_{\text{U},h})^2}{N_h - 1}.$$

Now suppose we take SRS of size n_h from each stratum. If we use number raised estimators the following formulae follow from applying the results for a SRS from a population to the sub-populations.

$$\bar{y}_h \;=\; \frac{\sum_{i \in S_h} y_{hi}}{n_h}, \quad T_h \;=\; N_h \bar{y}_h \quad \text{and} \quad s_h^2 \;=\; \frac{\sum (y_{hi} - \bar{y}_h)^2}{n_h - 1}.$$

Adding the means and their sampling variances leads to

$$T \;=\; \sum_{h=1}^{H} T_h \quad \text{and} \quad \widehat{\text{var}(T)} \;=\; \sum_{h=1}^{H} \left(1 - \frac{n_h}{N_h}\right) N_h^2 \frac{s_h^2}{n_h}$$

It may be that an estimate of the population total and its standard error will suffice for a report, but if a confidence interval for the total is required it is usual to ignore the fact that variances have been estimated and to use a normal approximation rather than a t-distribution. So an approximate $(1 - \alpha)100\%$ confidence interval for T_U is

$$T \;\pm\; z_{\alpha/2}\sqrt{\widehat{\mathrm{var}(T)}}.$$

Example 14.11: Stratified Sampling

[Jackson et al., 1987] compared the precision of systematic sampling with the precision of stratified sampling for estimating the average concentration of copper and lead in soil. A 10 by 10 equally spaced grid was drawn on a map of the 1 kilometer squared area, and soil samples were collected at each of the 121 (11×11) grid intersections. The mean and standard deviation of lead concentration were 127 mg kg^{-1} and 146 mg kg^{-1} respectively. The investigators then stratified the region into three strata: farmland away from roads (A); areas within 50 m of roads (B); and woodlands (C). It was expected that areas close to roads would have a higher concentration of lead from vehicle exhaust emissions, and that woodlands would have a higher concentration of lead because tree leaves capture airborne particles. The samples from each stratum were taken as the soil samples corresponding to the grid intersections within the areas of each stratum and the results are shown in Table 14.2[3].

TABLE 14.2: Lead and copper concentrations (ppm) in soil samples.

Stratum	Sample size	Lead mean	Lead sd	Copper mean	Copper sd
A	82	71	28	28	9
B	31	259	232	50	18
C	8	189	79	45	15

The mean lead concentration, and its standard error, from the systematic sample are 127 and $146/\sqrt{121} = 13.3$ respectively. The mean lead concentration and its standard error for the stratified sample are calculated in R.

```
> nh=c(82,31,8)
> stratmean=c(71,259,189)
> stratsd=c(28,232,79)
> m=sum(stratmean*nh)/sum(nh)
> v=sum((nh/sum(nh))^2*stratsd^2/nh)
> print(c("mean",round(m,1),"standard error",round(sqrt(v),1)))
[1] "mean"            "127"              "standard error" "11"
```

The mean is identical but the standard error is reduced to 11 as a consequence of the post-stratification. We also have estimates of the mean lead concentrations, and variability of lead concentrations, in the different strata.

---

[3]Stratification after the sample has been selected is known as **post-stratification**. It is not ideal because there is no control over the sample sizes in each stratum.

14.3.3 Optimal allocation of the sample over strata

For a given total cost, the minimum standard error of the number raised estimator of the population total is achieved when the size of the sample from each stratum is: proportional to the number of sampling units in the stratum, proportional to the standard deviation of the primary variable in the stratum; and inversely proportional to the square root of the cost of investigating a sampling unit in the stratum. We will use the method of Lagrange multipliers to prove this result, using a model based analysis and ignoring the FPC. Then

$$\text{var}(T) = \sum_{h=1}^{H} N_h^2 \frac{\sigma_h^2}{n_h}.$$

We aim to minimize this variance with respect to the n_h, subject to a total cost C of the survey. If the cost of investigating a sampling unit in stratum h is c_h, then the constraint is

$$c_0 + \sum_{h=1}^{H} c_h n_h = C,$$

where c_0 represents a set up cost that is incurred regardless of the number of units sampled. The function to be minimized with respect to n_h is

$$\phi = \sum_{h=1}^{H} N_h^2 \frac{\sigma_h^2}{n_h} + \lambda \left(c_0 + \sum_{h=1}^{H} c_h n_h - C \right).$$

Necessary, and as it turns out sufficient, conditions for a minimum are that the partial derivatives with respect to n_h are 0.

$$\frac{\partial \phi}{\partial n_h} = -N_h^2 \frac{\sigma_h^2}{n_h^2} + \lambda c_h = 0 \implies n_h = \frac{1}{\sqrt{\lambda}} \frac{N_h \sigma_h}{\sqrt{c_h}}.$$

Since $\frac{1}{\sqrt{\lambda}}$ is a constant of proportionality, this proves the result. The cost constraint gives

$$\frac{1}{\sqrt{\lambda}} = \frac{C}{\sum_{k=1}^{H} N_k \sigma_k \sqrt{c_k}}.$$

Example 14.12: Asset management plans

Water companies, and other utilities such as electricity supply companies, own a network of assets that need to be maintained and refurbished. Companies are typically expected to estimate the cost of the work needed to match specified standards of performance over a five year period, and to give an indication of the precision of the estimate. These estimates are known as *asset management plans* (AMPs) and companies use AMPs to justify increases in charges and convince shareholders that the business is being run in a sustainable way. Government regulators peruse AMPs to ensure that they provide unbiased estimates that have been obtained in a statistically rigorous fashion. One approach to producing an AMP for a water company is to stratify the water distribution system ([Metcalfe, 1991]). We use aspects of work done for the erstwhile Northumbrian Water Authority to illustrate aspects of sampling theory. To begin with, we assume that costs, at current prices, can be determined exactly once the scope of the work has been identified. This is unrealistic, and we describe the allowances made for uncertainty about costs in the final section.

The local water distribution network consisted of 146 zones which could reasonably be considered independent of each other inasmuch as a failure, such as a burst pipe in one zone would have a negligible effect on a neighboring zone. The number of properties in each zone was known. The population of 146 zones was partitioned into 8 strata by land use (rural, mixed but mainly rural, mixed but mainly industrial, suburban, and inner city) crossed with a low or high anticipated number of problems per property. The classification into low or high problems per property was based on the opinion of an experienced engineer, and there were two empty categories leaving the eight strata shown in Table 14.3). The water authority was prepared to carry out a detailed investigation of around 30 zones and for the purpose of the allocation of the sample over strata the sampling costs, c_h, were all considered equal to 1. The following R script gives optimal allocation.

```
> Nh=c(2,13,12,28,51,19,12,9)
> Sh=c(2,1,2,2,1,2,1,2)
> c0=0;ch=1;C=30
> conprop=C/sum(Nh*Sh*sqrt(ch))
> nh=conprop*Nh*Sh/sqrt(ch)
> print(round(nh,1))
[1] 0.6 1.8 3.3 7.8 7.1 5.3 1.7 2.5
```

TABLE 14.3: Stratification of local water distribution network into 8 strata, and stratum sizes (number of zones) and sample sizes (number of zones) are given as n_h/N_h.

Land use	Low anticipated problems/property	High anticipated problems/property
Rural		2/2
Mixed (mainly rural)	3/13	3/12
Mixed (mainly industrial)		5/28
Suburban	9/51	5/19
Inner City	3/12	3/9

The sample sizes were adjusted somewhat[4] to give those shown in Table 14.3. The detailed investigation of the selected zones included water pressure tests, water quality tests, some CCTV robotic inspection of pipes, and a review of the history of burst pipes. Schemes to be completed within the next five years were identified and costed using a unit cost data base. The results for are given in Table 14.4 The following R script estimates the total cost for the LDN and its standard error. The total cost and its variance are estimated for each stratum and then summed to give total LDN cost. The square root of the sum of the variances is the estimated standard error of the total LDN cost.

---

[4]The rationale for the changes was to increase the smallest sample sizes and to increase the size of the sample from the suburban fewer anticipated problems at the expense of the mixed (mainly industrial) as it was thought that the ratio of the corresponding standard deviations was likely to be nearer 1 than the assumed 0.5.

TABLE 14.4: Zone costs within strata.

Land use	Low anticipated problems/property					High anticipated problems/property				
Rural						1145	5281			
Mixed (mainly rural)	262	880	156			675	351	1588		
Mixed (mainly industrial)						3048	3982	2179	1684	2177
Suburban	679 625 475 1223 800 870 583 755 1841					2062	1068	744	1087	1125
Inner City	196	983	220			1261	2538	687		

```
;Nh=c(2,13,12,28,51,19,12,9);nh=c(2,3,3,5,9,5,3,3)
> m=rep(0,H);s=rep(0,H)
> Rmp=c(1145,5281);m[1]=mean(Rmp);s[1]=sd(Rmp)
> MMRfp=c(262,880,156);m[2]=mean(MMRfp);s[2]=sd(MMRfp)
> MMRmp=c(675,351,1588);m[3]=mean(MMRmp);s[3]=sd(MMRmp)
> MMImp=c(3048,3982,2179,1684,2177);m[4]=mean(MMImp);s[4]=sd(MMImp)
> Sfp=c(679,625,475,1223,800,870,583,755,1841);m[5]=mean(Sfp);s[5]=sd(Sfp)
> Smp=c(2062,1068,744,1087,1125);m[6]=mean(Smp);s[6]=sd(Smp)
> ICfp=c(196,983,220);m[7]=mean(ICfp);s[7]=sd(ICfp)
> ICmp=c(1261,2538,687);m[8]=mean(ICmp);s[8]=sd(ICmp)
> #Estimate the total and its standard error
> T=sum(Nh*m)
> V=sum((Nh^2*s^2/n)*(1-nh/Nh))
> se=sqrt(V)
> print(c("total",round(T),"se",round(se)))
[1] "total"  "182368" "se"      "14373"
```

14.4 Multi-stage sampling

In large studies it is usual to sample in stages (**multi-stage sampling**). For example, consider the estimation of the capacity of farm dams in the Murray-Darling basin (see Example 14.7). The area of the Murray-Darling basin is around 10^6 km^2. This could be divided into one hundred 100 km by 100 km squares. We could take an SRS of 5 of these large squares. We then consider more detailed maps for these 5 squares, sub-divide them into 100 squares of size 10 km by 10 km, and undertake a detailed field investigation of 5 from these 100. The final sample size would be 25 10 km^2 squares. The sampling units at the first stage are 100 km by 100 km squares, which each contain 100 second stage sampling units (10 km by 10 km squares). The 100 km by 100 km squares are examples of clusters.

Definition 14.8: Cluster

Clusters are disjoint collections of elements that together make up the population.

Definition 14.9: Multi-stage sampling

In the case that sampling proceeds by stages the sampling units at each stage will be different. Each sampling unit is a collection of elements of the population. The sampling units are disjoint. The set of all sampling units covers the entire population. That is every element in the population belongs to one, and only one, sampling unit.

Example 14.13: NOX pollution [multi-stage sampling]

Taking an SRS, or assuming that our sample was obtained by an SRS scheme, is the best strategy if the variable we are investigating is not associated with any known feature of the population. But, if there is an association between that variable and known features of the population we can do better. Suppose an engineer in the Environmental Protection Agency has been asked to estimate the mean nitrogen oxides pollution from SUVs, and has been allocated resources to test 20 SUVs from auto rental companies. The engineer knows that there are four major manufacturers of SUVs in the US and should ensure they are all represented in the sample. A colleague suggests obtaining a list of all the SUVs in all auto rental companies in the Washington area and then taking a SRS of 20 SUVs from the list. There are three drawbacks to this suggestion.

- All the selected SUVs might be from just three, or two, or even one of the four manufacturers. The we would have no information about SUVs produced by one, or more of the major manufacturers.

- If the engineer is to draw up a list of all SUVs in all auto rental companies in the Washington area then he or she will need to contact all these companies and ask for a list of the SUVs in the fleet.

- SUVs from auto rental companies in the Washington area may not be representative of SUVs in the US.

The first drawback can be can be avoided by dividing the population of auto rental companies in the Washington area into sub-populations according to the manufacturer they obtain their fleet from, which will probably be shown on the company website, and taking samples from each sub-population. This strategy is known as **stratification** and the sub-populations are known as **strata**. The second drawback can be mitigated by drawing a SRS of 5 auto rental companies from each stratum, asking these companies for a list of their SUVs and taking a SRS of one SUV from each list. The auto rental companies form **clusters** and the sampling is **two-stage**. The third drawback is that the population of SUVs held by auto rental companies in the Washington area, the **survey population**, does not correspond to SUVs in the US nominal **target population**. This is not ideal because the SUVs in the study population will be relatively new and well maintained, but the study will nevertheless provide useful information about nitrogen oxide pollution. A random sampling scheme for the whole of the US would be prohibitively expensive and take too long.

Once the sample has been drawn it may be useful to consider gasoline and diesel SUVs as separate groups and to use regression analysis to investigate how the age, or mileage, of the vehicle affects emissions. If something is known, or can be estimated, about the proportion of diesel cars or the age of vehicles in the population, then estimates of emissions in the population can be adjusted.

Stratifying will generally improve the precision of estimates of population values with a given sample, and becomes more efficient when items within strata vary less than items between strata. Clustering saves time and effort, and the reduction in precision with a given sample size is minimized if elements within the clusters are as variable as elements in the population.

Example 14.14: Overhead lockers

An engineer working for an airline has been asked to estimate the average weight of baggage in overhead lockers. Rather than take an SRS of lockers on all incoming flights in a week, it would improve the estimation to stratify flights by domestic and international. It would be more convenient to take an SRS of flights from each stratum, and then take SRS of lockers from those flights.

Example 14.15: Systematic sample

A **systematic sample** is obtained by taking every k^{th} item on the list for some choice of k, starting from a randomly selected integer between 1 and k. It is a random sampling scheme since every item has a known non-zero probability of selection $(1/k)$. But, it is not a SRS because items can only be in the same sample if they are separated by k items. It is a two stage cluster sampling scheme. At the first stage the population has been divided into k clusters and a SRS of one cluster is selected from k clusters. At the second stage all the items in the selected cluster are taken as the sample.

Systematic samples are often used as an approximation to a SRS. They work well with a list of customer names as the possible samples will contain the same proportion of, for example, Scottish names beginning with "Mac". However, a systematic sampling scheme should not be used if there are possible periodicities in the list. Sampling every 10th grid square of a 10×10 grid would correspond to taking a line of grid squares.

Example 14.16: The US EPA

The U.S. Environmental Protection Agency sampled drinking water wells to estimate the prevalence of pesticides and nitrates between 1988 and 1990. The agency adopted a multistage sampling scheme. The first stage sampling units were counties. Counties were stratified into 12 strata by perceived pesticide use (high, moderate, low, uncommon) crossed with estimated groundwater vulnerability (high, moderate, low).

14.5   Quota sampling

Random sampling provides a basis for unbiased, or approximately unbiased estimators, of population characteristics and associated standard errors. However, marketing divisions in companies are more concerned with quickly identifying customers' reactions to new products and changes in market share. They will typically use non random sampling strategies such as: phoning people who have recently purchased one of their autos, or setting up a panel of reviewers who will rank compact cars on the basis of sales literature and advertised prices. If the panel is set up with fixed proportions of, for example, men and women in a range of age groups it is said to be a **quota sample**.

14.6   Ratio estimators and regression estimators

If the known features of the population are measured on a continuous scale, rather than being categories, fitting a regression of the variable to be estimated on the known features will improve the estimation. Regression estimation is often used within strata.

14.6.1   Introduction

The estimation of the total cost for the LDN has not used the number of properties in the zones, which is known for all zones. If the number of properties in a zone is associated with the costs of schemes we can fit a regression of zone cost on the number of properties to the zones in the sample and then use this regression to predict costs for the remaining zones in the stratum.

14.6.2   Regression estimators

If values of explanatory variables are known for all the sampling units within a stratum, a regression of the response variable on the explanatory variables can be fitted to the sampling units in the sample. This regression can then be used to predict values of the response variable for the remaining sampling units. Regression can also be used instead of stratification, and may be preferable if the variable used for stratification is measured on a precise continuous scale. However, in the case of a response that is reasonably considered proportional to the predictor variable a simpler estimator known as a ratio estimator is commonly used. The ratio estimator is not equivalent to an unweighted regression through the origin (see Exercise 14.7).

14.6.3   Ratio estimator

Each sampling unit i in the population, or stratum, has a pair of observations (x_i, y_i) associated with it. The values of x_i are known for all the sampling units, and the values of y_i are found for the sampling units in the sample. If it is reasonable to assume that y is proportional to x then the constant of proportionality, **ratio**, can be estimated by

$$\widehat{B} \;=\; \frac{\sum_{i \in S} y_i}{\sum_{i \in S} x_i} \;=\; \frac{\overline{y}}{\overline{x}}.$$

An approximation to its mean squared error (MSE) is given by

$$MSE = \mathrm{E}\left[(\widehat{B}-B)^2\right] = \mathrm{E}\left[\left(\frac{\bar{y}-B\bar{x}}{\bar{x}}\right)^2\right] = \mathrm{E}\left[\left(\frac{\bar{y}-B\bar{x}}{\bar{x}_\mathrm{U}}\left(1-\frac{\bar{x}-\bar{x}_\mathrm{U}}{\bar{x}}\right)\right)^2\right]$$

$$\approx \mathrm{E}\left[\left(\frac{\bar{y}-B\bar{x}}{\bar{x}_\mathrm{U}}\right)^2\right] = \frac{1}{\bar{x}_\mathrm{U}^2}\mathrm{E}\left[(\bar{y}-B\bar{x})^2\right].$$

Now define

$$d_i = y_i - Bx_i.$$

so that $\bar{y}-B\bar{x}=\bar{d}$ and $E[(\bar{y}-B\bar{x})^2]$ can be estimated by

$$\frac{\sum_{i\in S}(y_i-\widehat{B}x_i)^2}{(n-1)n} = \frac{\widehat{sd}(d_i)}{\sqrt{n}}.$$

Finally, if we overlook the fact that $\widehat{B}$ is a slightly biased estimator of B, then the MSE is an approximation to the standard error of $\widehat{B}$. A FPC is applied if the sample exceeds around one tenth of the population.

The population mean, and total, of y_i are estimated by

$$\bar{y}_{ratio} = \widehat{B}\bar{x}_\mathrm{U}, \qquad T_y = N\bar{y}_{ratio}$$

and the approximate standard errors follow from that of $\widehat{B}$. In particular, the standard error of T_y is estimated by $N\widehat{sd}(d_i)/\sqrt{n}$.

Example 14.17: Ratio estimator

We demonstrate the use of the ratio estimator with the mixed mainly rural stratum with more identified problems. There were 12 zones in the stratum and the size, measured by the number of properties (x_i), was known for each zone. The costs y_i were estimated for a random sample of 3 zones and are given in Table 14.4. The sizes of all 12 zones, together with the costs for the three zones in the sample are given in Table 14.5. Engineers considered that zone costs within a stratum would tend to increase in proportion to the number of properties, despite considerable variation in individual zones, and the data are consistent with this assumption. In the following R code we estimate the total cost for the stratum, and its standard error, using the ratio estimator.

```
> sizepop=c(212,629,712,858,1068,1541,2011,2316,2582,2714,3138,3539)
> N=length(sizepop)
> size=c(629,858,2582)
> n=length(size)
> cost=c(675,351,1588)
> x=size;y=cost;xbarU=mean(sizepop)
> B=sum(y)/sum(x);ybar=B*xbarU;T=N*ybar
> d=y-B*x
> seT=sqrt(1-n/N)*N*sd(d)/sqrt(n)
> print(c("Est tot cost",round(T),"se",round(seT)))
[1] "Est tot cost" "13696"        "se"            "1460"
```

TABLE 14.5: Mixed mainly rural more anticipated problems: zone sizes and estimated costs for sampled zones.

Zone size (number properties) problems/property	Estimated cost problems/property
212	
629	675
712	
858	351
1068	
1541	
2011	
2316	
2582	1588
2714	
3138	
3539	

The estimate of the standard error of the cost estimate is based on only three observations and is therefore not precise. However, cost estimates from the eight strata will be added, and the overall standard error, which is the square root of the sum of the squared standard errors for each stratum, will be more reliable. Overall, the sampling error is quite well estimated and relatively small when compared with the potential error in the assessment of costs of schemes in the zones.

14.7 Calibration of the unit cost data base

14.7.1 Sources of error in an AMP

A unit cost data base was developed for the AMP, and was used to cost the schemes identified for sampled zones. There are many reasons why the realized costs of maintaining the system over the five year period will differ from the AMP estimate. These include:

- sampling error;
- uncertainty over the specification of individual schemes;
- a systematic tendency to underestimate, or overestimate, the work needed to complete schemes;
- uncertainty over the costs of work associated with individual schemes;
- a systematic error in the unit costs;
- surveys of the sampled zones missing schemes that will need to be completed during the five year period;
- changes in engineering techniques and changes in legislation;

- and inflation.

The sampling error can be estimated and quantified by its standard error. A strategy for estimating the errors in specification and cost of schemes is given in the next section. The chance of missing schemes can be reduced by making a thorough survey[5]. It is better to survey a few zones carefully than it is to take large samples and survey them in a cursory manner. Changes in engineering procedures, such as may arise from improvements in no-dig tunneling equipment, could lead to reduced costs. Inflation is more of an accounting issue, and an AMP is typically presented in terms of prices at some fixed date.

14.7.2 Calibration factor

The unit cost data base was calibrated by taking a sample of historic schemes, costing them using AMP procedures to give an estimated cost x_i, and comparing these estimated costs with the actual costs, increased by the price index to bring them to AMP price date, which we refer to as out-turn costs y_i. The ratio of out-turn cost to projected cost is used as a calibration factor for the unit cost data base.

Example 14.18: Projected out-turn costs

The projected and out-turn costs for eight schemes are given in Table 14.6.

TABLE 14.6: Estimated costs, out-turn costs adjusted to AMP price data, and ratios of out-turn to estimate, for eight historic schemes.

Estimated cost	Out-turn cost	Ratio
81	94	1.16
20	18	0.90
98	107	1.09
79	75	0.95
69	52	0.75
144	216	1.50
119	176	1.48
124	207	1.67

The R script for plotting the data, application of a ratio estimator, and an estimate of the variance of individual scheme the estimated population ratio follows[6].

```
> projected=c(81,20,98,79,69,144,119,124)
> out-turn=c(94,18,107,75,52,216,176,207)
> x=projected;y=out-turn;n=length(x)
> plot(x,y,xlim=c(0,210),ylim=c(0,210),xlab="projected cost",
+ ylab="out-turn cost")
> B=sum(y)/sum(x);seB=sd(y-B*x)/(sqrt(n)*mean(x))
> print(c("Calib factor",round(B,2),"se",round(seB,2)))
[1] "Calib factor" "1.29"        "se"          "0.12"
```

---

[5]There will inevitably be unexpected emergency situations, such as burst pipes, to deal with and there needs to be a contingency fund for these events. However, the contingency fund would usually be considered outside the scope of the AMP. An objective of the AMP is to reduce bursts by maintaining the system.

[6]In this application the values of x_i are only known for the sample so a further approximation used in the calculation of the standard error is to replace $\bar{x}_U$ by $\bar{x}$.

```
> xp=seq(1,200,0.1)
> yp=B*xp
> lines(xp,yp)
> res=y/x-B
> print(c("ratio-B",round(mean(res),2),"sd(ratio-B)",round(sd(res),2)))
[1] "ratio-B"      "-0.1"        "sd(ratio-B)" "0.33"
```

The calibration factor for the unit cost data base is 1.29, so all the projected costs estimated using the AMP procedures will be multiplied by 1.29. The standard error of this estimate is 0.12, so we are around 67% confident[7] that the hypothetical population ratio is between 1.17 and 1.41, and it can be reduced by increasing the number of schemes in the comparison of out-turn costs to projected costs. The uncertainty about an individual scheme out-turn cost to its projected cost is around 0.33[8]. So if a scheme has a projected cost of 200 this will be increased to $200 \times 1.29 = 258$, and the associated standard deviation of the out-turn cost is estimated as $258 \times 0.33 = 86$. The estimated standard deviation of out-turn costs of individual schemes may be large, but are reduced in relative terms when scheme costs are added because these errors can reasonably be considered independent. Continuing with the example, suppose that 16 schemes have been costed according to AMP procedures with the values given in the first line of the continuation of the R script.

```
> schemecost=c(76,72,41,11,306,48,154,105,91,109,31,138,398,41,51,81)
> y=schemecost;indiv=sd(res)
> T=sum(y*B)
> Tvarindiv=sum((y*B*indiv)^2)
> Tvarcalfac=T^2*seB^2
> Tvar=Tvarindiv+Tvarcalfac
> Tse=sqrt(Tvar)
> print(c("Total",round(T)))
[1] "Total" "2257"
> print(c("sd attributable indiv schemes",round(sqrt(Tvarindiv))))
[1] "sd attributable indiv schemes" "251"
> print(c("sd attributable calib factor",round(sqrt(Tvarcalfac))))
[1] "sd attributable calib factor" "261"
> print(c("overall sd",round(Tse),"as fraction",round(Tse/T,2)))
[1] "overall sd"   "362"          "as fraction" "0.16"
```

The uncertainty in this total due to variation of the 16 individual schemes quantified by the standard deviation 251 is slightly less than the uncertainty due to the imprecise calibration factor (quantified by the standard deviation 261). If more schemes are added the former standard deviation will decrease relative to the total, whereas the latter will remain at 0.12 relative to the total. It follows that there is little to be gained by sampling more schemes without reducing the uncertainty in the calibration factor.

---

[7]An approximate 67% confidence interval for a quantity is given by its estimate plus or minus one standard error, even if the sampling distribution is not well approximated by a normal distribution.

[8]This uncertainty for an individual scheme ignores the non-zero value of the sample mean (-0.1).

14.8 Summary

14.8.1 Notation

U	universe, from the sampling point of view
n	sample size
N	population size
$\bar{y}_U$	finite population mean
$\bar{y}$	sample mean
s_U^2	finite population variance
s^2	sample variance
y_i	value of variable for sampling unit i
S	indices of sampling units in sample
T_U	population total
T	number raised estimator
μ	infinite population mean
σ^2	infinite population variance
H	number of strata
N_h	number of sampling units in strata h
$T_{U,h}$	total for stratum h
$\bar{y}_{U,h}$	mean for stratum h
$s_{U,h}^2$	variance for stratum h
$\bar{y}_h$	sample mean for stratum h

14.8.2 Summary of main results

We looked at how to create a simple random sample (SRS) from a *target population* to estimate the value of a *population characteristic*.

Simple random sampling from a finite population: When estimating the mean of a finite population, the finite population correction (FPC) is required, $(1 - n/N)$, where n is the sample size and N is the population size. In such a circumstance, the sample mean, $\bar{y}$, is normally distributed with mean $\bar{y}_U$ (the population mean) and variance $\frac{s_U^2}{n}\left(1 - \frac{n}{N}\right)$, where s_U^2 is the population variance. Furthermore, a $(1 - \alpha) \times 100\%$ confidence interval of width 2Δ would require a sample of size $n_\infty/(1 + n_\infty/N)$, where $n_\infty = \left(z_{\alpha/2}\sigma/\Delta\right)^2$ is the sample size if the FPC were not applied.

Stratified sampling: The population is divided into sub-populations called *strata* based on some criteria that all subjects have and samples are taken from each *stratum*. The stratum means, and population mean, are

$$\bar{y}_{U,h} = \frac{T_{U,h}}{N_h} \quad \text{and} \quad \bar{y}_U = \frac{T_U}{N}.$$

The stratum variances are defined as

$$s_{U,h}^2 = \sum_{i=1}^{N_h} \frac{(y_{hi} - \bar{y}_{U,h})^2}{N_h - 1}.$$

Ratio estimators and regression estimators: If the values of some covariate that is highly correlated with the variable of interest are known for the population then this information can be used to adjust sample estimates using regression, or in the case of proportionality a ratio estimator.

Calibration of the unit cost data base: The calibration of the unit cost data base is crucial because errors persist however many individual schemes are assessed. There is little point in an extensive surbey of work to be completed if the cost of doing the work is highly uncertain.

14.9   Exercises

Section 14.2 Simple random sampling from a finite population

Exercise 14.1: Systematic sampling

Suppose a population is of size N where

$$N = n\,k$$

for integers n and k. A systematic sample of size n is selected.

1. How many different systematic samples are there and what is the probability that an item is selected?

2. How many different SRS are there and what is the probability that an item is selected?

3. Answer the same questions if $N = nk + 1$.

Exercise 14.2: Estimating population size

An island state has no register of ultralight aircraft, and an engineer has been asked to estimate the number (N) of ultralight aircraft in use. Suppose that ultralights are flown at weekend fairs that are held at different venues across the island.

(a) The engineer attends one such fair and tags all the m ultralights present, with a yellow adhesive band that can be removed after the investigation. A few weeks later the engineer will attend another fair and note the number of ultralights present (n) and the number of these that are tagged (Y).

(i) Suggest an estimator for N and state the assumptions on which it is based.

(ii) Suppose that Y has a binomial distribution. Use the first three terms of a Taylor expansion to obtain approximations for the bias and standard error of your estimator.

(iii) The engineer attends a fair with 28 ultralights present, all of which are tagged. One month later, the engineer attends a second fair at another location at which there are 41 microlights present. Seven of the 41 microlights are tagged. Estimate the population total and give an approximate standard error for the estimator.

(iv) Set up a simulation to investigate whether making or not an adjustment for bias reduces the mean squared error of the estimator.

(b) The engineer attends one such fair and tags all the m ultralights present. During the next month the engineer notes how many ultralights are observed until the yth tagged ultralight is seen. Define n as the total number of ultralights observed including the y tagged ultralights.

(i) Suggest an estimator for N and state the assumptions on which it is based.

(ii) Let X be the number of successes in a sequence of Bernoulli before the yth success. Then X has a negative binomial distribution. Under the assumptions of (i), $n = X + y$. Show that your estimator of N is unbiased and obtain an expression for its standard error.

(iii) The engineer attends a fair with 40 ultralights present. The number observed before observing 8 tagged ultralights is 64 (including the 8 tagged ultralights). Estimate the population total and the standard error of this estimator.

Section 14.3 Stratified sampling

Exercise 14.3: Copper concentrations

The mean and standard deviation of the 121 copper concentrations were 35 ppm and 16 ppm respectively. Refer to Table 14.2 and compare the standard error of the mean from the systematic sample with its standard error after post-stratification.

Section 14.4 Multi-stage sampling

Exercise 14.4: Murray-Darling basin

Suppose a SRS sample of n squares from the N squares in a state is taken. Let x_i for $i = 1, \ldots, n$ be the estimated capacity of farm dams in the sampled squares.

(a) What is the probability of selection (p_i) for each square in the state?

(b) Suggest an estimate of the total capacity of farm dams in the state as a function of N, X_i, and n.

(c) Express your estimate in terms of x_i and p_i.

Section 14.5 Quota sampling

Exercise 14.5: Mobile phone

Your company has developed a new mobile phone with significant novel features and requires market research.

(a) Suggest a possible quota sample scheme.

(b) What biases would you expect in your quota sample? Are there any other limitations of a quota sample when compare to an SRS?

(c) What are the advantages of a quota sample over an SRS in this case?

Section 14.6 Ratio estimators and regression estimators

Exercise 14.6: Manufacturing companies

Manufacturing companies in South Australia were classified into chemical sector, electrical sector, or mechanical sector. Simple random samples of companies were taken from each sector and asked what proportion of their sales were export. Details of the survey are summarized below.

Sector	Number of of companies	Number of companies in sample	Mean of the sample proportions
chemical	38	5	0.08
electrical	150	10	0.37
mechanical	146	10	0.24

(a) What are the probabilities of selection for companies in the three sectors?

(b) Estimate the average proportion of sales that are export for companies in South Australia.

(c) Why is your estimate in (b) not a plausible estimate of the proportion of South Australia manufacturing that is export?

Exercise 14.7: Least squares estimator

(a) Obtain the least squares estimator of the slope of a regression line fitted through n pairs (x_i, y_i) if the errors are assumed to be independent with mean 0 and constant variance σ^2.

(b) Show that this formula is not equivalent to a ratio estimator.

Section 14.7 Calibration of the unit cost data base

Exercise 14.8: Projected out-turn costs

Consider the data of the projected cost out-turn cost in Example 14.18. Alternatives to using the ratio estimator to analyze the relationship between AMP estimated cost and out-turn cost include a regression estimator through the origin (m_0) or a regression estimator with intercept (m_1).

Fit these regressions using the syntax $m_0 = lm(y\ x - 1)$, $m_1 = lm(y\ x)$

(a) Compare the estimate of the slope in m_0 with the ratio given by the ratio estimator.

(b) In general, how do the formulae for the slope of a regression through the origin and a ratio estimate differ. What explanation can you offer for the difference?

(c) Compare the fitted relationship between AMP estimated cost and out-turn cost obtained with m_1 to that obtained using m_0. Which do you think is more appropriate in this context and why?

Exercise 14.9: Alternative to the ratio estimator

An alternative to the ratio estimator $\hat{B} = \frac{\sum y_i}{\sum x_i}$ is $\tilde{B} = \sum \frac{y_i}{x_i}$.

(a) Compare $\hat{B}$ and $\tilde{B}$ for the data of the projected cost and out-turn cost in Example 14.18.

(b) Explain the difference in the formulae for $\hat{B}$ and $\tilde{B}$ by introducing a weighting for each data pair.

(c) Explain why $\hat{B}$ is generally preferred to $\tilde{B}$.

Appendix A - Notation

A.1   General

$$\sum_{i=1}^{n}$$ sum over $i = 1, \ldots, n$, abbreviates to $\sum$.

$$\prod_{i=1}^{n}$$ product over $i = 1, \ldots, n$, abbreviates to $\prod$.

$e^{(\cdot)}$ or $exp(\cdot)$ exponential function.

$\ln(\cdot)$ natural logarithm (logarithm base e).

π ratio of circumference to diameter of a circle.

$\cos(\cdot)$, $\sin(\cdot)$ circular functions with argument in radians.

$n!$ the factorial function $n \times (n-1) \cdots \times 1$.

$\Gamma(r)$ the gamma function. $\Gamma(n+1) = n!$.

A.2   Probability

Ω a sample space.

$\emptyset$ empty set.

$|E|$ cardinality (number of elements for countable set) of the set E.

$P(E)$ probability of an event E where $E \subset \Omega$. $P(\emptyset) = 0 \leq P(E) \leq P(\Omega) = 1$

$\cup$ union (or including both).

$\cap$ intersection (and).

$P_X(x)$ probability mass function for random variable X, abbreviates to $P(x)$.

$f(x)_X$, $F(x)_X$ pdf and cdf for random variable X, abbreviates to $f(x)$, $F(x)$.

P_r^n number of permutations (arrangements) of r from n.

$\binom{n}{r}$ number of combinations (choices) of r from n.

A.3   Statistics

$\{x_i\}$ $i = 1, \ldots, n$ a set of n data.

$x_{i:n}$ the ith order statistic (i^{th} smallest when sample in ascending order).

X, x random variable, X, and a particular value it takes, x, as in $\mathrm{P}(X = x)$.

In the following table, with the exception of the sample size n, the sample quantities (statistics) estimate the corresponding population quantities. The convention is either Roman for sample and Greek for population[9] or the sample estimate is the population symbol with the addition of a hat $\widehat{}$ above the symbol.

Sample	Name	Population
n	size	∞ or if finite N
$\bar{x}$	mean	μ
s^2, or s_x^2, or $var(x)$	variance	σ^2, or σ_X^2 or $var(X)$
s, or s_x, or $sd(x)$	standard deviation	σ
$\widehat{\gamma}$	shewness	γ
$\widehat{\kappa}$	kurtosis	κ
$\widehat{p}$	proportion	p
histogram		pdf
cumulative frequency polygon		cdf
$\sum \phi(x_i)/n$	average	$E[\phi(X)]$
$x_{i:n}$	order statistic, i^{th} smallest in sample i^{th} smallest|quantile	
$\widehat{cov}(x,y)$	covariance	$cov(X,Y)$
r	correlation	ρ
$\widehat{\alpha}, \widehat{\beta}$	linear regression intercept, slope	α, β
$\widehat{\beta}_j$	regression coefficient	β_j

---

[9]The lower case Greek letters μ, σ, and ρ correspond to Roman m, s, and r and are usually pronounced by mathematicians as mu, sigma, and rho respectively. The lower case Greek letters γ, κ, α and β are usually pronounced by mathematicians as gamma, kappa, alpha, and beta.

$\mathrm{sd}(\beta_j)$ standard deviation of the estimator β_j.

$\widehat{\mathrm{sd}}(\beta_j)$ estimate (or estimator) of standard deviation of the estimator β_j.

$\widehat{y}_i$ fitted value $(r_i = y_i - \widehat{y}_i)$.

$\widehat{Y}_p, \widehat{y}_p$ predictor, prediction of Y_p.

R^2 coefficient of determination

R^2_{adj} adjusted R^2.

Notes

1. $\overline{X}$ is the sample mean as a random variable which is an estimator of μ, and $\bar{x}$ is the sample mean of a particular sample which is an estimate of μ. Similarly for S and s.

2. For other statistics we generally rely on the context to distinguish estimators from estimates.

3. Other notation is specific to the chapter in which it is introduced.

A.4   Probability distributions

Name	Notation	Parameters
binomial	$\mathrm{Bin}(n,p)$	Number trials n probability success p
Poisson	$\mathrm{Poi}(\lambda t)$	rate λ length of continuum t

Z a standard normal random variable, $Z \sim N(0,1)$.

$\phi(\cdot), \Phi(\cdot)$ pdf and cdf of standard normal distribution.

z_α upper α quantile of standard normal distribution.

Appendix B - Glossary

2-factor interaction: The effect on the response of one factor depends on the level of the other factor.

3-factor interaction: The interaction effect of two factors depends on the level of a third factor.

absorbing state: A state that once entered cannot be left.

acceptance sampling: A random sampling scheme for goods delivered to a company. If the sample passes some agreed criterion, the whole consignment is accepted.

accuracy: An estimator of a population parameter is accurate if it is, on average, close to that parameter.

addition rule: The probability of one or both of two events occurring is: the sum of their individual probabilities of occurrence less the probability that they both occur.

aliases: In a designed experiment, sets of a factor and interactions between factors that are indistinguishable in terms of their effect on the response. In time series analysis, frequencies that are indistinguishable because of the sampling interval.

analysis of variance (ANOVA): The total variability in a response is attributed to factors and a residual sum of squares which accounts for the unexplained variation attributed to random errors.

AOQL: The average proportion of defective material leaving an acceptance sampling procedure, average outgoing quality (AOQ), depends on the proportion of defectives in the incoming material. The maximum value it could take is the AOQ limit (AOQL).

aperiodic: In the context of states of a Markov chain, not limited to recurring only at a fixed time interval (period).

asset management plan (AMP): A business plan for corporations, like utilities such as water, gas, and electricity, which own a large number of physical assets.

asymmetric: Without symmetry, in particular a pdf is described as asymmetric if it is not symmetric about a vertical line through its mean (or median).

asymptote: A line that is a tangent to a curve as the distance from the origin tends to infinity.

asymptotic: In statistics, an asymptotic result is a theoretical result that is proved in the limiting case of the sample size approaching infinity.

auto-correlation: The correlation between observations, spaced by a lag k, in a time series. It is a function of k.

auto-covariance: The covariance between observations, spaced by a lag k, in a time series. It is a function of k.

auto-regressive model: The current observation is modeled as the sum of a linear combination of past values and random error.

balanced: The same number of observations for each treatment or factor combination.

Bayes' theorem: This theorem enables us to update our knowledge, expressed in probabilistic terms, as we obtain new data.

between samples estimator of the variance of the errors: An estimator of population variance calculated from the variance of means of random samples from that population.

bias: A systematic difference - which will persist when averaged over imaginary replicates of the sampling procedure - between the estimator and the parameter being estimated. Formally, the difference between the mean of the sampling distribution and the parameter being estimated. If the bias is small by comparison with the standard deviation of the sampling distribution, the estimator may still be useful.

bin: A bin is an alternative name for a class interval when grouping data.

binomial distribution: The distribution of the number of successes in a fixed number of trials with a constant probability of success.

bivariate: Each element of the population has values for two variables;

block: Relatively homogeneous experimental material that is divided into plots that then have different treatments applied.

bootstrap: Re-sampling the sample to estimate the distribution of the population.

box plot: A simple graphic for a set of data. A rectangle represents the central half of the data. Lines extend to the furthest data that are not shown as separate points.

categorical variable: A variable that takes values which represent different categories.

causation: A change in one variable leads to a consequential change in another variable.

censored: The value taken by a variable lies above or below or within some range of values, but the precise value is not known.

centered: A set of numbers that has been transformed by subtraction of a constant that is typically the mid-range.

central composite design: A 2^k factorial design is augmented by each factor being set at very high and at very low with all other factors at 0 and by all factors being set at 0.

Central Limit Theorem: The distribution of the mean of a sample of independently drawn variables from a probability distribution with finite variance tends to normality as the sample size increases.

Chapman-Kolmogorov equation: The probability of moving from state i to state j in two steps is equal to the sum, over all states k, of the probabilities of going from i to k in one step and then k to j in the next step.

chi-squared distribution: A sum of m independent squared normal random variables has a chi-squared distribution with m degrees of freedom.

chi-squared test: A test of the goodness of fit of some theoretical distribution to observed data. It is based on a comparison of observed frequencies and expected frequencies equal to discrete values or within specific ranges of values.

class intervals (bins): Before drawing a histogram the data are grouped into classes which correspond to convenient divisions of the variable range. Each division is defined by its lower and upper limits, and the difference between them is the length of the class interval. Also known as bins.

cluster: A group of items in a population.

coefficient: A constant multiplier of some variable.

coefficient of determination: The proportion of the variability in a response which is attributed to values taken by predictor variables.

coefficient of variation: The ratio of the standard deviation to the mean of a variable which is restricted to non-negative values.

cold standby: When an item fails it can be replaced with a spare item. In contrast, hot standby is when an item is continuously backed up with a potential replacement.

common cause variation: Variation that is accepted as an intrinsic feature of a process when it is running under current conditions.

concomitant variable: A variable that can be monitored but cannot be set to specific values by the experimenter.

conditional distribution: A probability distribution of some variable(s) given the value of another associated variable(s).

conditional probability: The probability of an event conditional on other events having occurred or an assumption they will occur. (All probabilities are conditional on the general context of the problem.)

confidence interval: A 95% confidence interval for some parameter is an interval constructed in such a way that on average, if you imagine millions of random samples of the same size, 95% of them will include the specific value of that parameter.

consistent estimator: An estimator is consistent if its bias and standard error tend to 0 as the sample size tends to infinity.

continuity correction: The probability that a discrete variable takes an integer value is approximated by the probability that a continuous variable is within plus or minus 0.5 of that integer.

continuous: A variable is continuous if it can take values on a continuous scale.

control variable: A predictor variable that can be set to a particular value by a process operator.

correlation coefficient: A dimensionless measure of linear association between two variables that lies between -1 and 1.

correlogram: Auto-correlation as a function of lag

covariance: A measure of linear association between two variables, that equals the average value of the mean-adjusted products.

covariate: A general term for a variable that is associated with some response variable and that is therefore a potential predictor variable for that response.

cumulative distribution function: A function which gives the probability that a continuous random variable is less than any particular value. It is the population analogue of the cumulative frequency polygon. Its derivative is the pdf.

cumulative frequency polygon: Plotted for continuous data sorted into bins. A plot of the proportion, often expressed as a percentage, of data less than or equal to right hand ends of bins. The points are joined by line segments.

data, datum: Information on items, on one item, from the population.

degrees of freedom: The number of data values that could be arbitrarily assigned given the value of some statistic and the values of implicit constraints.

deseasonalized: A time series is deseasonalized (seasonally adjusted) if seasonal effects are removed.

design generator: A product of columns representing factor values that is set equal to a column of 1s.

design matrix: An experiment is set up to investigate the effect of certain factors on the response. The design matrix specifies values of the factors in the multiple regression model used for the analysis.

detrended: A time series is detrended if the estimated trend is removed.

deviance: A generalization of the sum of squared errors when a model is fitted to data.

deviate: The value taken by a random variable, typically used to denote a random number from some probability distribution.

discrete event simulation: A computer simulation that proceeds when an event occurs, rather than proceeding with a fixed time interval.

empirical distribution function (edf): The proportion of data less than or equal to each order statistic. A precise version of a cumulative frequency polygon.

endogenous, exogenous: Internal, or external, to a system.

ensemble: The hypothetical infinite population of all possible time series.

error: A deviation from the deterministic part of a model.

estimator, estimate: A statistic that is used to estimate some parameter is an estimator of that parameter when considered as a random variable. The value it takes in a particular case is an estimate.

equilibrium: The probabilistic structure of a model for a stochastic process does not depend on time.

evolutionary operation: An experiment that is confined to small changes in factors that can be accommodated during routine production. The idea is that optimum operating conditions will be found.

expected value: A mean value in the population.

explanatory variable: In a multiple regression the dependent variable, usually denoted by Y, is expressed as a linear combination of the explanatory variables, which are also commonly referred to as predictor variables. In designed experiments, the explanatory variables are subdivided into **control** variables, whose values are chosen by the experimenter, and **concomitant** variables, which can be monitored but not preset.

exponential distribution: A continuous distribution of the times until events in a Poisson process, and, as events are random and independent, the distribution of the times between events.

factorial experiment: An experiment designed to examine the effects of two or more factors. Each factor is applied at two or more levels and all combinations of these factor levels are tried in a full factorial design.

finite, infinite: A fixed number, without any upper bound on the number.

fixed effect, random effect: A fixed effect is a factor that has its effects on the response, corresponding to its different levels, defined by a set of parameters that change the mean value of the response. A random effect is a source of random variation.

frequency: In statistic usage - the number of times an event occurs. In physics usage - cycles per second (Hertz) or radians per second.

F-distribution: The distribution of the ratio of two independent chi-squared variables divided by their degrees of freedom.

gamma distribution: The distribution of the time until the kth event in a Poisson process. It is therefore the sum of k independent exponential variables.

gamma function: A generalization of the factorial function to values other than positive integers.

Gaussian distribution: An alternative name for the normal distribution.

Gauss-Markov theorem: In the context of a multiple regression model with errors that are independently distributed with mean 0 and constant variance: the ordinary least squares estimator of the coefficients is the minimum variance unbiased estimator among all estimators that are linear functions of the observations.

generalized linear model: The response in the multiple regression model (linear model) is other than a normal distribution.

geometric distribution: The distribution of the number of trials until the first success in a sequence of Bernoulli trials.

goodness of fit test: A statistical test of a hypothesis that data has been generated by some specific model.

Gumbel distribution: The asymptotic distribution of the maximum in samples of some fixed size from a distribution with unbounded tails and finite variance.

hidden states: Hypothetical states that are part of a system but cannot be directly observed.

highly accelerated lifetime testing (HALT): Testing under extreme conditions that are designed to cause failures within the testing period.

histogram: A chart consisting of rectangles drawn above class intervals with areas equal to the proportion of data in each interval. It follows that the heights of the rectangles equal the relative frequency density, and the total area equals 1.

hot standby: A potential replacement provides continuous back up - see *cold standby*.

hypothesis (null and alternative): The null hypothesis is a specific hypothesis that, if true, precisely determines the probability distribution of a statistic. The null hypothesis is set up as the basis for an argument or for a decision, and the objective of an experiment is typically to provide evidence against the null hypothesis. The alternative hypothesis is generally an imprecise statement and is commonly taken as the statement that the null hypothesis is false.

ill-conditioned: A matrix is ill-conditioned if its determinant is close to 0 and so its inverse will be subject to rounding error.

imaginary infinite population: The population sampled from is often imaginary and arbitrarily large. A sample from a production line is thought of as a sample from the population of all items that will be produced if the process continues on its present settings. An estimator is considered to be drawn from an imaginary distribution of all possible estimates, so that we can quantify its precision.

independent: Two events are independent if the probability that one occurs does not depend on whether or not the other occurs.

indicator variable: A means of incorporating categorical variables into a regression. The variable corresponding to a given category takes the value 1 if the item is in that category and 0 otherwise.

inherent variability: Variability that is a natural characteristic of the response.

intrinsically linear model: A relationship between two variables that can be transformed to a linear relationship between functions of those variables.

IQR: The difference between the upper and lower quartiles.

interaction: Two explanatory variables interact if the effect of one depends on the value of the other. Their product is then included as an explanatory variable in the regression. If their interaction effect depends on the value of some third variable a third order interaction exists, and so on.

interval estimate: A range of values for some parameter rather than a single value.

kurtosis: The fourth central moment, that is a measure of weight in the tails of a distribution. The kurtosis of a normal distribution is 3.

lag: A time difference.

Laplace distribution: Back-to-back exponential distributions.

least significant difference: The least significant difference at the 5% level, for example, is the product of the standard error of the difference in two means with the upper 0.025 quantile of a t-distribution with the appropriate degrees of freedom.

least squares estimate: An estimate made by finding values of model parameters that minimize the sum of squared deviations between model predictions and observations.

level of significance: The probability of rejecting the null hypothesis is set to some chosen value known as the level of significance.

linear model: The response is a linear function of predictor variables. The coefficients of the predictor variables are estimated.

linear regression: The response is a linear function of a single predictor variable.

linear transformation: The transformed variable is obtained from the original variable by the addition of some constant number and multiplication by another constant number.

linear trend: A model in which the mean of a variable is a linear function of time (or distance along a line).

logit: The natural logarithm of the odds (ratio of a probability to its complement).

lower confidence bound: A value that we are confident that the mean of some variable exceeds

main effect: The effect of changing a factor when other factors are at their notional mid-values.

main-plot factor: In a split-plot experiment each block is divided into plots. The different levels of the main plot factor are randomly assigned to the plots within each block (as in a randomized block design).

marginal distribution: The marginal distribution of a variable is the distribution of that variable. The 'marginal' indicates that the variable is being considered in a a multivariate context.

Markov chain: A process can be in any one of a set of states. Changes of state occur at discrete time intervals with probabilities that depend on the current state, but not on the history of the process.

Markov process: A process can be in anyone of a set of states. Changes of state occur over continuous time with rates that depend on on the current state, but not on the history of the process.

matched pairs: A pairing of experimental material so that the two items in the pair are relatively similar.

maximum likelihood: The likelihood function is the probability of observing the data treated as a function of the population parameters. Maximum likelihood find the values of the parameters that maximize the probability.

meal: A mixture of materials, that have been ground to a powder, used as raw material for a chemical process.

mean: The sum of a set of numbers divided by their number. Also known as the average.

mean-adjusted: A set of numbers that has been transformed by subtraction of their mean. The transformed set has a mean of 0.

mean-corrected: An alternative term for mean-adjusted.

mean-square error: The mean of squared errors.

measurement error: A difference between a physical value and a measurement of it.

median: The middle value if data are put into ascending order.

method of moments: Estimates made by equating population moments with sample moments.

mode: For discrete data, the most commonly occurring value. For continuous data, the value of the variable at which the pdf has its maximum.

monotone: Continually increasing or continually decreasing.

Monte-Carlo simulation: A computer simulation that relies on the generation of random numbers.

multiple regression: The response, is expressed as a linear combination of predictor (also known as explanatory variables) plus random error. The coefficients of the variables in this combination are the unknown parameters of the model and are estimated from the data.

multiplicative rule: The probability of two events both occurring is the product of the probability that one occurs with the probability that the other occurs conditional on the first occurring.

multivariate normal distribution: A bivariate normal distribution is $3D$ bell shaped. The marginal distributions are normal, each with a mean and variance. The fifth parameter is the correlation. This concept generalizes to a multivariate normal distribution which is defined by its means, variances and pair-wise correlations.

mutually exclusive: Two events are mutually exclusive if they cannot occur together.

m-step transition matrix: The matrix of probabilities of moving between states in m-steps.

non-linear least squares: Fitting a model which in non-linear in the unknown coefficients using the principle of least squares.

normal distribution: A bell-shaped pdf which is a plausible model for random variation if it can be thought of as the sum of a large number of smaller components.

normalizing factor: A factor that makes the area under a curve equal 1.

or: In probability 'A or B' is conventionally taken to include both.

order statistics: The sample values when sorted into ascending order.

orthogonal: In a designed experiment the values of the control variables are usually chosen to be uncorrelated, when possible, or nearly so. If the values of the control variables are uncorrelated they are said to be orthogonal.

orthogonal design: The product of the transpose of the design matrix with the design matrix is a diagonal matrix.

over-dispersed: Variance of the residuals is greater than a value that is consistent with the model that is being fitted.

parameter: A constant which is a characteristic of a population.

parametric bootstrap: The sampling distribution of a statistic is investigated by taking random samples from a probability distribution chosen to represent the population from which the sample has been taken. The parameters of the distribution are estimated from the sample.

parent distribution: The distribution from which the sample has been taken.

paver: A paving block. Modern ones are made from either concrete or clay in a variety of shapes and colors.

percentage point: The upper $\alpha\%$ point of a pdf is the value beyond which a proportion α of the area under the pdf lies. The lower point is defined in an analogous fashion.

periodic: Occurring, or only able to occur, at fixed time intervals.

point estimate: A single number used as an estimate of a population parameter (rather than an interval).

Poisson distribution: The number of events in some length of continuum if events occur randomly, independently, and singly.

Poisson process: Events in some continuum, often form a Poisson process if they are random, independent, and occur singly.

population: A collection of items from which a sample is taken.

power (of test): The probability of rejecting the null hypothesis if some specific alternative hypothesis is true. The power depends on the specific alternative hypothesis.

precision: The precision of an estimator is a measure of how close replicate estimates are to each other. Formally, it is the reciprocal of the variance of the sampling distribution.

prediction interval: An interval within which a random variable will fall with some specified probability.

predictor variable: A variable in the regression equation used to predict a response; also known as an explanatory variable.

priority controlled junction: A road junction which is controlled by 'Give Way' signs and road markings, rather than by lights.

probability: A measure of how likely some event is to occur on a scale ranging from 0 to 1.

probability density function: A curve such that the area under it between any two values represents the probability that a continuous variable will be between them. The population analogue of a histogram.

probability function: A formula that gives the probability that a discrete variable takes any of its possible values.

process capability index: The ratio of the difference between the upper and lower specification limits to six process standard deviations (C_p).

process performance index: The ratio of the larger of the differences between the upper/lower specification limits and the mean to three process standard deviations (C_p).

pseudo-random numbers: A sequence of numbers generated by a deterministic algorithm which appear to be random. Computer generated random numbers are actually pseudo-random.

pseudo-3D plot: A scatter plot in which the plotting symbol indicates the range within which some third variable lies.

p-value: The probability of a result as extreme, or more extreme, as that observed, if the null hypothesis is true.

quadrants: In a scatter plot, the positive and negative x-axis and y-axis divide the plane into four quadrants.

quantiles: The upper/lower α quantile is the value of the variable above/below which a proportion α of the data lie.

quantile-quantile plot: A plot of the order statistics against the expected value of the order statistic in a random sample from the hypothetical population.

quartiles: The upper (lower) quartile, UQ (LQ), is the datum above (below) which one-quarter of the data lie.

quota sample: A non-random sample taken to satisfy specific identifying criteria.

random digits: A sequence in which each one of the digits $0, 1, \ldots, 9$ is equally likely to occur next in the sequence.

random effect: A component of the error structure.

random numbers: A sequence of numbers drawn a specified probability distribution so that the proportion of random numbers in any range matches the corresponding probability calculated from the probability distribution, and such that the next number drawn is independent of the existing sequence.

random sample: A sample which has been selected so that every member of the population has a known, non-zero, probability of appearing.

range: Difference between the largest datum and the smallest datum when the data are sorted into ascending order.

rate matrix: A matrix of the rates of moving between states in a Markov process.

realization: A sequence of data that have been drawn at random from some probability distribution or stochastic process.

regression: A model for the value taken by a response as an unknown linear combination of values taken by predictor variables. The unknown coefficients are estimated from data.

regression line: A plot of the expected value of the response against a single predictor variable under an assumed linear relationship.

regression sum of squares: The mean-adjusted sum of squares of the response is split into the sum of squared residuals and the regression sum of squares.

regression towards the mean: If one variable is far from its mean, then the mean value of a correlated variable will be closer, in terms of multiples of standard deviations, to its marginal mean. In the case of a single variable, if one draw is far from the mean the next draw is likely to be closer to the mean.

relative frequency: The ratio of the frequency of occurrence of some event to the number of scenarios in which it could potentially have occurred. That is, the proportion of occasions on which it occurred.

relative frequency density: Relative frequency divided by the length of the bin (class interval).

reliability function: The complement of the cumulative distribution function of component lifetime.

repeatability: The ability to get similar results when you test under the same conditions.

replication: The use of two or more experimental units for each experimental treatment. The execution of an entire experiment more than once so as to increase precision and obtain a more precise estimate of sampling error.

reproducibility: The ability to get similar results when others test under conditions that satisfy given criteria designed to maintain comparability.

resampling: Taking a random sample, without replacement, from the sample.

residuals: Differences between observed and fitted values.

residual sum of squares: The sum of squared residuals.

response surface: The response is modeled as a quadratic function of covariates. The predictor variables are the covariates, squared covariates, and cross products between two covariates.

response variable: The variable that is being predicted as a function of predictor variables.

robust: A statistical technique that is relatively insensitive to assumptions made about the parent distribution.

run: A performance of a process at some specified set of values for the process control variables.

run-out: A measurement of deviation of a disc from its plane.

sample: A collection of items taken from a population.

sample path: A sequence of sample values from a stochastic process.

sample space: A list of all possible outcomes of some operation which involves chance.

sampling distribution: An estimate is thought of as a single value from the imaginary distribution of all possible estimates, known as the sampling distribution.

saturated model: A model in which the number of parameters to be estimated equals the number of data.

scatterplot: A graph showing data pairs as points.

seasonal term/effect/component: A component of a time series that changes in a deterministic fashion with a fixed period.

Simpson's paradox: An apparent relationship between variables that is a consequence of combining data from disparate sub-groups.

simple random sample: A sample chosen so that every possible choice of n from N has the same chance of occurring.

simulation: A computer model for some process.

skewness: A measure of asymmetry of a distribution. Positive values correspond to a tail to the right.

spurious correlation: A correlation that can be attributed to known relationships to a common third variable (often time).

standard deviation: The positive square root of the variance.

standard error: The standard deviation of some estimator.

standard normal distribution: The normal distribution scaled to have a mean of 0 and a standard deviation of 1.

standard order: A systematic list of runs for a process.

state: A set of values for the variables that define a process.

state space: The set of all possible states.

stationarity: Constant over time.

statistic: A number calculated from the sample.

statistically significant: A result that is unlikely to be equalled or exceeded if the null hypothesis is true.

strata: A sub-population.

stratification: Division of a population into relatively homogeneous sub-populations.

Student 's t-distribution: The sampling distribution of many statistics is normal and can therefore be scaled to standard normal. If the mean of the sampling distribution is the parameter of interest, and the unknown standard deviation is replaced by its sample estimate, with v degrees of freedom, the normal distribution becomes a t-distribution with v degrees of freedom. If v exceeds about 30 there is little practical difference.

stochastic process: A random process, sometimes referred to as a time series model.

survey population: The population that is to be surveyed, when it does not match the target population precisely. .

sub-plot factor: A factor which has its different levels applied over each of the main plots in a split-plot design.

symmetric distribution: A probability distribution with a pdf that is symmetric about a vertical line through its mean.

synchronous: Moving together over time.

systematic sample: A sample drawn as every k-th item from a list.

tail (heavy): A probability distribution with tails that tend towards 0 more slowly than those of a normal distribution.

target population: The population about which we require information.

test statistic: A statistic designed to distinguish between a null hypothesis and the alternative hypothesis.

time homogeneous: Parameters of the process do not change over time.

tolerance interval: A statistical tolerance interval is an interval that includes a given proportion of the population with some given level of confidence.

training data: A sub-set of the available data used to fit a model.

transition: A change of state.

transition matrix: A matrix of transition probabilities in a Markov chain.

transition probability: The probability of changing between two given states in one step of a Markov chain.

trend: A deterministic model for change over time.

t-ratio: The ratio of an estimate to an estimate of its standard deviation.

unbiased estimator (estimate): An estimator is unbiased for some parameter, if the mean of its sampling distribution is equal to that parameter.

uniform distribution: A variable has a uniform distribution between two limits if the probability that it lies within some interval between those limits is proportional to the length of that interval.

upper confidence bound: A value that will not be exceeded, as determined with the given confidence.

variable: A quantity that varies from one member of the population to the next. It can be measured on some continuous scale, be restricted to integer values (discrete), or be restricted to descriptive categories (categorical).

variance: Average of the squared deviation from the mean. The averaging is performed by dividing by the degrees of freedom.

variance-covariance matrix: A matrix of covariances between all possible pairs of variables when analyzing multi-variate data. The variances lie along the leading diagonal.

Weibull distribution: A versatile model for the lifetimes of components.

weighted mean: An average in which the data are multiplied by numbers called **weights**, summed, and then divided by the sum of the weights.

within samples estimator of the variance of the errors: A variance is calculated for each sample, and these variances are then averaged.

Appendix C - Getting started in R

We explain the use of inbuilt R functions throughout the book. The following is a review of some of the basic commands that we expect you to know beforehand. In addition to the inbuilt help and many textbooks (Crawley, [2012] is a useful reference), the internet is a good resource for helpful hints. The Short reference card is a valuable aide-memoire (`http://cran.r-project.org/doc/contrib/Short-refcard.pdf`).

    *R is case sensitive so X and x are different objects. Parentheses (), square brackets [], and braces {} are used to contain the argument of a function, the element of an array or matrix, and commands within a **for loop** or similar construction respectively.*

    The following is a selection of some of the basic facilities in R, but you will learn more as you go along. You will also find that there are often several ways of doing the same thing in R.

C.1 Installing R

Type CRAN in Google to navigate to the Comprehensive R Archive Network (`http://cran.r-project.org/`). To load the base package go to "Download and Install R": select your operating system from: Linux, Mac OS, Windows. Run the set up program which has a name like `R*.exe` on a PC, where * is the version number. To run R look under `Programs>R..` or click on the icon.

    R functions are followed by () which contain the argument. A few functions don't need an argument. For example:

```
citation()
```

tells you how to cite R.

```
quit()
```

finishes a session, although you can just close the window.

    One of the strengths of R is the multitude of packages, also referred to as libraries. We shall only use a few of these, They are obtained by opening R, selecting `Packages > Install Packages`, choose a mirror site and click on the package you need.

    To use the package during an R session you have to load it. `Packages > Load Package`

C.2 Using R as a calculator

You can use R as a calculator. The basic arithmetic operators are: `+ - * /` `^` Precedence follows the usual convention and is given to expressions in brackets, followed by exponentiation, followed by multiplication and division, followed by addition and subtraction. Run R and type after the prompt, which in R is `>`

    Assignment is either `=` or the **gets arrow** `<-` which is a composite symbol made up from `<` and `-` with no space between them. For example, to assign the values 0.3, 25 and 3 to x, y and z, each of the following R commands, collectively, achieve this result.

```
> x = 0.3
> y <- 25
> 3 -> z
```

Some of the mathematical functions that we will use are:

square root	`sqrt(x)`
absolute value of x	`abs(x)`
round to	`round(x,digits=m)`
smallest integer greater than x	`ceiling(x)`
greatest integer less than x	`floor(x)`
closest integer to x between 0 and x	`trunc(x)`
natural logarithm and inverse	`log(x) exp(x)`
common logarithm and inverse	`log10(x) 10^x`
circular functions and inverses	`cos(x) sin(x) tan(x)`
	`acos(x) asin(x) atan(x)`
hyperbolic functions and inverses	`cosh(x) sinh(x) tanh(x)`
	`acosh(x) asinh(x) atanh(x)`

Try the following examples, by typing the text following the prompt. You should then see the same result shown following the [1].

```
> y <- 543.27
> round(y,digits=-1)
[1] 540
> trunc(-5.8)
[1] -5
> 1/0
[1] Inf
> sqrt(-2)
[1] NaN Warning message: In
sqrt(-2) : NaNs produced
```

Despite the not a number (NaN) given by R for $\sqrt{-2}$, it does handle complex numbers. The syntax is real part + imaginary part preceding i with no space between them. For example:

```
> x <- 3+4i
> abs(x)
[1] 5
> sqrt(2+0i)
[1] 1.414214+0i
> sqrt(-2+0i)
[1] 0+1.414214i
```

C.3   Setting the path

As for any software you must specify the directory from which you will read data files and which you save R scripts and graphics files. From the top bar

```
File > Change directory
```

C.4   R scripts

If you need more than one line of code you should type lines into an R script. From the top bar `File > New script` will open the editor. Type in your commands save and the file (`Save`), which will automatically be given the extension '.r'. (e.g. `myscript.r`) Then run the script by selecting `Files > Source R code` You can edit the script after opening it with `File > Open script`. You can also copy from the script and paste to the console. This is not good practice for a long script but we have done this to demonstrate inbuilt functions in this appendix. If you run a script, you need to include `print()` commands to see the results, or just type the variable name.

C.5   Data entry

C.5.1   From keyboard

For a small data set you can use the concatenate function `c()`.

```
> w <- c(1.8,5.3,7.2)
> print(w)
[1] 1.8 5.3 7.2
> w
[1] 1.8 5.3 7.2
> x <- c(w,23)
> print(x)
[1]  1.8 5.3 7.2 23.0
> x
[1]  1.8 5.3 7.2 23.0
```

To obtain a sequence of integers you can specify the first and last separated by a colon and to obtain an equally spaced sequence on numbers use `seq()`.

```
> y <- 1:6
> print(y)
[1] 1 2 3 4 5 6
> z <- seq(0,1,.2)
> print(z)
[1] 0.0 0.2 0.4 0.6 0.8 1.0
```

To repeat a number, or patterns of numbers, use the `rep()` function.

```
> x <- rep(c(1:3),4)
> print(x)
[1] 1 2 3 1 2 3 1 2 3 1 2 3
```

C.5.2 From a file

C.5.2.1 Single variable

An ascii file, `sales.txt`, contains data for a single variable separated by spaces and extending over several lines. The file contents are

10 15 25 32 11

Reading into R

```
> x <- scan("sales.txt")
Read 5 items
> print(x)
[1] 10 15 25 32 11
```

In R we refer to x as a vector.

C.5.2.2 Several variables

An ascii file, `inventory.txt`, contains the number of fenders and wheels held at several stores. The file has a column for each variable and its contents are

```
store  fender  wheel
A        5       23
B        3       18
C        1       53
D        4       24
```

Read into a data frame in R and check the first three lines

```
>
inv.dat <- read.table("inventory.txt",header=T)
> head(inv.dat,3)
  store fender wheel
1     A      5    23
2     B      3    18
3     C      1    53
```

Now suppose we want the total number of wheels. We need to obtain the vector wheel from the data frame `inv.dat`, then we can use the function `sum()`. There are several ways of doing this. One is to specify that wheel is part of the data frame `inv.dat` using $, and a second is to use the function `with()`.

```
> sum(inv.dat$wheel)
[1] 118
> with(inv.dat,sum(wheel))
[1] 118
```

The second has the advantage that you can perform a sequence of operations within braces {}.

```
> with(inv.dat,{c(sum(fender),sum(wheel))})
[1]   13 118
```

Another way is to attach the data frame using **attach()**. Then R will look for the variable in that data frame

```
> attach(inv.dat)
> sum(wheel)
[1] 118
```

Although using **attach()** can be convenient, there can be issues. If we already have an object "wheel", R will use it rather than the data in **inv.dat**. For this reason, we have avoided using **attach()** in this book. If you do attach a data frame, detach it when you have completed the analysis using **detach(inv.dat)**.

C.6   R vectors

In R, the values of a variable are contained in a vector. Arithmetic is performed on elements. Elements can be obtained with the vector name and square brackets **[]**. The number of elements can be found with the function **length()**. Vectors can be bound together as columns with the function **cbind()**.

```
> x <- 0:5
> f <- c(9,11,7,5,2,1)
> length(f)
[1] 6
> y <- f*x
> x[4]
[1] 3
> W <- cbind(x,f,y)
> print(W)
      x  f  y
[1,]  0  9  0
[2,]  1 11 11
[3,]  2  7 14
[4,]  3  5 15
[5,]  4  2  8
[6,]  5  1  5
```

Another useful function is **which()** as in

```
> which(y > 12)
[1] 3 4
```

Notice that the result is subscripts of the vector. If you want values of a vector satisfying some condition you can give that condition within the **[]**.

```
> z <- x[f>3]
> print(z)
[1] 0 1 2 3
```

The **list()** function is is useful for bundling vectors together. The function **lapply()** applies a specified function to each element in the list.

```
> x <- 1:10
> y <- 5:11
> z <- 1:3
> xyz.list <- list(x,y,z)
> lapply(xyz.list,mean)
[[1]] [1] 5.5

[[2]] [1] 8

[[3]] [1] 2
```

C.7   User defined functions

The name you choose to give the function is assigned as function() of an argument, for example function(x), or arguments, as in f(x,y,z). Then the function is defined between braces { }. A function to convert degrees Celsius to degrees Fahrenheit centigrade follows.

```
C2F <- function(x){
  y <- x*180/100
  z <- y+32
  return(z)
}
> deg <- c(0,-40,20,100)
> C2F(deg)
[1]   32 -40  68 212
```

C.8   Matrices

You can define a matrix with the matrix() function.

```
> x <- 1:8
> M <- matrix(x,nrow <- 2,ncol <- 4,byrow <- TRUE)
> print(M)
      [,1] [,2] [,3] [,4]
[1,]    1    2    3    4
[2,]    5    6    7    8
```

Matrix transposition, multiplication, and inverse are given by the function t(), binary operator %*%, and function solve() respectively. You can set up a diagonal matrix with the function diag()

C.9   Loops and conditionals

If possible avoid them. For instance y[y<0]<-0 will set all the negative values in a vector to 0.

If you do need to use a loop the syntax is shown in the following R script which gives the first 10 Fibonacci numbers. The commands to be repeated are enclosed by {}. An example of a for loop follows.

```
n <- 10
fibn <- rep(0,n)
fibn[1] <- 1
fibn[2] <- 1
for (i in 3:n){
  fibn[i] <- fibn[i-1]+fibn[i-2]
  print(fibn[i])
}
```

Logical operators include:

!	NOT
&	AND
|	OR
==	EQUAL
<	less than
<=	less than or equal to
>	greater than
<=	greater than or equal to

An example of a while loop follows.

```
> i <- 1
> while(i<5){
  print(i^2)
  i <- i+1
}
[1] 1
[1] 4
[1] 9
[1] 16

median <- function(x){
  y <- sort(x)
  m <- length(x)+1
  if ( m/2  ==   floor(m/2) ){
    return(y[m/2])
  } else {
  return((y[(m-1)/2]+y[(m-1)/2+1])/2)
  }
}
```

You can use functions in R recursively

```
fact <- function(n){
  if (n==1){
    1
  } else {
    n*fact(n-1)
  }
}
> fact(5)
[1] 120
```

C.10 Basic plotting

As an example we consider the function

$$y = x^2$$

and evaluate it for 11 integer values from -5 up to 5.

```
x <- -5:5
y <- x^2
```

The command

```
plot(x,y)
```

will plot 11 points, shown as circles, in a rectangular Cartesian coordinates.If you would prefer to join points with line segments use

```
plot(x,y,type="l")
```

but note that the points are not shown explicitly. You can show the points explicitly and have lines with

```
plot(x,y,pch=4)
lines(x,y)
```

The points are shown by crosses because we changed the print character. To see what print character corresponds to the integers from 1 up to 25, try the following commands.

```
x=1:25
y=x
plot(x,y,pch=x)
```

It is often useful to print several graphs on the same figure. The function **par(mfrow=c(r,c))** splits the page into frames, r rows and c columns which are filled by row. To revert to one graph per page use **par(mfrow=c(1,1))**. The following script will give you four graphs in the same figure. It also shows you how to add a title and labels to axes.

```
x <- -5:5
y <- x^2
par(mfrow=c(1,3))
```

```
plot(x,y,)
plot(x,y,type="l")
plot(x,y,main="quadratic",xlab="argument(x)",ylab="f(x)")
lines(x,y)
```

To change the line type from full line to dashed line, add the argument `lty = 2`.

C.11 Installing packages

Sometimes, it is necessary to use add-on libraries to increase R's functionality. These add-ons are called libraries. If a library is installed, then you can load its functionality into R with the command:

```
library(<package-name>)
```

For example to use the **ggplot2** package for advanced plotting, then you enter the following command into R:

```
library(ggplot2)
```

If the package is not installed, then this can be achieved with the command:

```
install.packages("<package-name>")
```

Notice the necessity of quotations around the package name when installing packages. Again as an example consider installing **ggplot2**:

```
install.packages("ggplot2")
```

C.12 Creating time series objects

Some R functions take time series objects as their argument. The basic command to create a time series objects from a time series in R is the `ts()` command. The main arguments needed are **data, start = 1** and **frequency**. The **data** is the observed values at each time-point. R assumes that the observation are made at regular time-points, e.g. days, months, quarterly. Frequency is the number of observations within a period. Star can be set as some starting number, typically 1, or using `c( , )` as first period and observation within the period. For example, suppose **Font** contains the time series of monthly inflows to the Fontburn Reservoir from January 1980, then

```
Font.ts <- ts(Font,start=c(1980,1),frequency=12)
```

Appendix D - Getting started in MATLAB

We explain the use of inbuilt MATLAB functions throughout the book. The following is a review of some of the basic commands that we expect you to know beforehand. In addition to the inbuilt help and many textbooks, the internet is a good resource for helpful hints.

Just like R, MATLAB is also case sensitive so X and x are different objects. Parentheses (), square brackets [], and braces {} are used to contain the argument of a function, the elements of an array or matrix, and the elements of a cell structure respectively.

D.1 Installing MATLAB

Unlike R, installing MATLAB involves purchasing a license. This can be done at `https://www.mathworks.com/store`, where you can select the appropriate license. Your school or university may also have a way for students to use MATLAB on their personal computers. Contact your school to see if this is possible and follow their installation instructions.

If you have purchased a full license, MathWorks will provide installation instructions for your particular computer and operating system.

D.2 Using MATLAB as a calculator

You can use MATLAB as a calculator. The basic arithmetic operators are: `+ - * / ^`. Precedence follows the usual convention and is given to expressions in brackets, followed by exponentiation, followed by multiplication and division, followed by addition and subtraction. Open MATLAB and in the command window, which in MATLAB has `>>` to begin each line, type

```
>> 5 + 6
```

Assignment is done via `=`. For example, to assign the value 0.3 to x, you can use the following command

```
>> x = 0.3
```

Some of the mathematical functions that we will use are:

square root	`sqrt(x)`
absolute value of x	`abs(x)`
round to	`round(x,m)`
smallest integer greater than x	`ceil(x)`

greatest integer less than x	`floor(x)`
closest integer to x between 0 and x	`fix(x)`
natural logarithm and inverse	`log(x) exp(x)`
common logarithm and inverse	`log10(x) 10^x`
circular functions and inverses	`cos(x) sin(x) tan(x)` `acos(x) asin(x) atan(x)`
hyperbolic functions and inverses	`cosh(x) sinh(x) tanh(x)` `acosh(x) asinh(x) atanh(x)`

Try the following examples, by typing the text following the prompt. You should then see the same result shown following "`ans =`".

```
>> y = 543.27
y =
  543.2700
>> round(y,-1)
ans =
  540
>> fix(-5.8)
ans =
  -5
>> 1/0
ans =
  Inf
>> sqrt(-2)
ans =
  0.0000 + 1.4142i
```

D.3 Setting the path

As for any software you must specify the directory from which you will read data files and which you save MATLAB scripts (called "m-files") and graphics files. From the ribbon, select the "HOME" tab, look in the "ENVIRONMENT" section and select "Set Path." In here you can add one folder or multiple folders as well as adding folders with subfolders included. Be sure to click "Save" before closing.

D.4 MATLAB scripts (m-files)

If you need more than one line of code you should type lines into an m-file. From the top ribbon "HOME > New script" will open the editor. Type your commands into the new script and click "Save", and the file will automatically be given the extension .m (e.g. myscript.m). Then run the script by clicking "EDITOR > Run" or typing the name of the file into the command window. You can edit the script after opening it with "HOME > Open" or typing **open** followed by the name of the file into the command window. When typing any commands, in the command window or the editor, using the semicolon symbol (;) at the end of the line will suppress the output. If you would like to see the output, remove the semicolon.

D.5 Data entry

D.5.1 From keyboard

For a small data set you can use matrices to store the data.

```
>> w = [1.8,5.3,7.2];
>> x = [w,23];
```

To obtain a sequence of integers you can specify the first and last separated by a colon and to obtain an equally spaced sequence on numbers add the increment between two colons.

```
>> y = 1:6
y =
     1     2     3     4     5     6
>> z = 0:0.2:1
z =
        0    0.2000    0.4000    0.6000    0.8000    1.0000
```

To repeat a number, or patterns of numbers, use the **repmat()** function.

```
>> x = repmat(1:3,1,4)
x =
     1     2     3     1     2     3     1     2     3     1     2     3
```

D.5.2 From a file

D.5.2.1 Single variable

An ascii file, **sales.txt**, contains data for a single variable separated by spaces and extending over several lines. The file contents are

```
10 15 25 32 11
```

Reading into MATLAB

```
>> x = dlmread('sales.txt')
x =

    10    15    25    32     1
```

In MATLAB we refer to x as a vector.

D.5.2.2   Several variables

An ascii file, `inventory.txt`, contains the number of fenders and wheels held at several stores. The file has a column for each variable and its contents are

```
store  fender  wheel
A        5       23
B        3       18
C        1       53
D        4       24
```

Read into a data frame in MATLAB and check the first three lines

```
>> invdat = readtable('inventory.txt');
>> invdat(1:3,:)
ans =
  33 table

    store    fender    wheel

    -----    ------    -----

    'A'        5         23
    'B'        3         18
    'C'        1         53
```

Now suppose we want the total number of wheels. We need to obtain the vector wheel from the data frame **invdat**, then we can use the function **sum()**. To do this, we first specify that wheel is part of the data frame **inv.dat** using a period (.), and then perform the summation.

```
>> invdat.wheel
ans =
    23
    18
    53
    24
>> sum(invdat.wheel)
ans =
   118
```

When using a period (.), we have to be careful as the period is also used for structures in MATLAB, which is a similar but different data type.

D.6   MATLAB vectors

In MATLAB, the values of a variable are contained in a vector or matrix. It is important to note that MATLAB allows row or column vectors and so the method of entry is important. Within a matrix (using square brackets), the semicolon (;) creates a new row. The apostrophe (') will transpose a row vector into a column vector or vice versa. Hence, there are two ways to create a column vector. The first way creates a column vector initially.

```
>> x = [1 ; 2 ; 3];
```

The second way creates a row vector and then transposes the row vector into a column vector.

```
>> x = 0:5
x =
     0     1     2     3     4     5
>> x = (0:5)'
x =
     0
     1
     2
     3
     4
     5
```

By default, arithmetic is performed on the vector or matrix. If element-wise arithmetic is required, then a period (.) should be used before the operation. For example

```
>> [2 2]*[2 3]
Error using  *
Inner matrix dimensions must agree.
>> [2 2].*[2 3]
ans =
     4     6
```

Elements can be obtained with the vector name and parentheses (). The number of elements can be found with the function **length()**. Vectors can be bound together as columns using square brackets [], ensuring these vectors are column vectors and have the same length.

```
>> x = (0:5)';
>> f = [9;11;7;5;2;1];
>> length(f)
ans =
     6
>> y = f.*x;
>> x(4)
ans =
     3
>> W = [x,f,y]
W =
     0     9     0
     1    11    11
     2     7    14
     3     5    15
     4     2     8
     5     1     5
```

MATLAB can find which elements in a vector (or matrix) satisfy a certain condition. These conditions can also be used as indices for vectors.

```
>> y > 12
ans =
```

```
  6x1 logical array
     0
     0
     1
     1
     0
     0
>> y(y>12)
ans =
    14
    15
```

In the first case, $y > 12$ provides a logical vector showing which elements of y satisfy the condition (the third and fourth elements which equal 1) and which do not. The second case then selects those particular elements from y.

If the column vectors do not have the same length, then a cell data type should be used (which replaces the square brackets with braces {}) However, for element-wise arithmetic, column vectors need to be the same length. The function cellfun() applies a specified function to each element in the cell and the @ symbol must be placed before the function name.

```
>> x = 1:10;
>> y = 5:11;
>> z = 1:3;
>> xyz = {x,y,z};
>> cellfun(@mean,xyz)
ans =
    5.5000    8.0000    2.0000
```

Some functions, such as mean, can handle matrix input and treat the matrix like multiple column vectors. The operation is then performed on each column. This behavior is often the default but can be changed so that the operation is performed across a row (rather than down a column).

```
>> W
W =
     0     9     0
     1    11    11
     2     7    14
     3     5    15
     4     2     8
     5     1     5
>> mean(W)
ans =
    2.5000    5.8333    8.8333
>> mean(W,2)
ans =
    3.0000
    7.6667
    7.6667
    7.6667
    4.6667
    3.6667
```

D.7   User defined functions

The name you choose to give the function is assigned as function() of an argument, for example function(x), or arguments, as in f(x,y,z). A function to convert degrees Celsius to degrees Fahrenheit centigrade follows and is written in the editor. Be careful to include element-wise arithmetic (.) where required.

```
function z = C2F(x)
  y = x.*180/100
  z = y+32
```

In the command window, you would call this function via the following.

```
>> deg = [0,-40,20,100];
>> C2F(deg)
ans =
    32    -40    68    212
```

D.8   Matrices

Defining matrices is done in the same way as vectors. Square brackets begin and conclude the matrix, spaces or commas divide each row into columns and a semicolon creates a new row.

```
>> x = [1:4 ; 5, 6, 7, 8]
x =
    1    2    3    4
    5    6    7    8
```

When entering a new matrix, ensure that each row has the same number of columns.

```
>> x = [1:3 ; 5, 6, 7, 8]
Dimensions of matrices being concatenated are not consistent.
```

Matrix transposition, multiplication, and inverse are given by the apostrophe ', asterisk *, and function inv() respectively. You can set up a diagonal matrix with the function diag().

D.9   Loops and conditionals

MATLAB is great at computing vector operations. For instance, instead of looping through a vector, checking each entry and updating it if it is negative, the command y(y<0)=0 will set all the negative values in a vector to 0. Therefore, if a loop can be avoided by using a vector operation, the program will be faster.

If you do need to use a loop the syntax is shown in the following MATLAB script which gives the first 10 Fibonacci numbers. The commands to be repeated are indented. An example of a **for** loop follows.

```
n = 10;                 %number of Fibonacci numbers
fibn = zeros(10,1);     %initialize column vector fibn
fibn(1) = 1;
fibn(2) = 1;            %provide first two Fibonacci numbers
for i = 3:n
    fibn(i) = fibn(i-1) + fibn(i-2);
end
fibn                    %print all 10 numbers to command window
```

Logical operators include: An example of a while loop follows.

~=	NOT EQUAL
&&	AND
||	OR
==	EQUAL
<	less than
<=	less than or equal to
>	greater than
<=	greater than or equal to

```
>> i = 1
>> while i<5
    i^2
    i = i+1;
end
ans =
    1
ans =
    4
ans =
    9
ans =
    16
```

A user defined function using an **if** statement is given in the following.

```
function z = median(x)
  y = sort(x);
  m = length(x)+1;
  if m/2  ==    floor(m/2)
    z = y(m/2);
  else
    z = (y((m-1)/2)+y((m-1)/2+1))/2
  end
```

You can use functions in MATLAB recursively.

```
function z = fact(n)
    if n == 1
        z = 1;
    else
        z = n*fact(n-1);
    end
>> fact(5)
ans =
    120
```

D.10   Basic plotting

As an example we consider the function

$$y \;=\; x^2$$

and evaluate it for 11 integer values from -5 up to 5.

```
>> x = -5:5;
>> y = x.^2;
```

The command

```
>> plot(x,y)
```

will plot a line through these 11 points, in a rectangular Cartesian coordinates. If you would prefer to only plot the points without line segments use

```
plot(x,y,'.')
```

The character string '.' can be up to three characters long, with each character specifying a line color, a marker style and a line style. Type **help plot** for all the options. For example

```
plot(x,y,'rx:')
```

will plot the curve with a red line, x marks for each point and a dotted line.

It is often useful to print several graphs on the same figure. The function **subplot(r,c,n)** splits the page into frames, r rows and c columns which are filled by row and plots the next graph on the nth position. The following script will give you four graphs in the same figure, arranged in two rows and two columns. It also shows you how to add a title and labels to axes.

```
x = -5:5;
y = x^2;
subplot(2,2,1)
plot(x,y,'.')
subplot(2,2,2)
plot(x,y)
subplot(2,2,3)
plot(x,y)
```

```
title('quadratic')
xlabel('argument(x)')
ylabel('f(x)')
subplot(2,2,4)
plot(x,y,'.-')
```

D.11   Creating time series objects

Some MATLAB functions take time series objects as their argument. The basic command to create a time series object from a time series in MATLAB is the **timeseries()** command. The main arguments needed are **data** and **time**. The **data** is the observed values at each time-point. The **time** is the time of each observation in **data**, which can be in any time unit. If **time** are data strings, you must specify **time** as a cell of data strings. The command **datenum** will help convert times into correct formats. For example, suppose **Font** contains the time series of the first three months of inflows to the Fontburn Reservoir in 1980, then

```
>> Font = [23 21 20];
>> time = {'01/01/1980','02/01/1980','03/01/1980'}; %in form mm/dd/yyyy
>> Font_TS = timeseries(Font,time)
  timeseries

  Common Properties:
            Name: 'unnamed'
            Time: [3x1 double]
        TimeInfo: [1x1 tsdata.timemetadata]
            Data: [3x1 double]
        DataInfo: [1x1 tsdata.datametadata]
```

Appendix E - Experiments

E.1 How good is your probability assessment?

E.1.1 Objectives

To consider a scoring rule as an aid to assessing probabilities [Lindley, 1985]. To relate the scoring rule to other strategies for defining subjective probabilities. To consider some aspects of a simple designed experiment. There are two sets of questions, A and B. Select one at random, by flipping a coin.

E.1.2 Experiment

Read Me:

There are two sets of questions, A and B. Select one at random, by flipping a coin. Both sets contain 16 statements with alternative forms, one of which is true the other being false. For most questions, you are unlikely to be sure which form is true unless you look up Wikipedia or similar resource. You are asked to give your probability, p, to one decimal place, that the unbracketed form is true, without looking up the answer. For example, if the statement is

"A Galah is a parrot (cockatoo)."

and you know it is a cockatoo you would assess the probability of the unbracketed form being true as 0. If you weren't quite certain you might assess the probability as 0.1. If you are quite undecided you might assess as 0.5.

Don't read yet:

Now we will introduce a scoring rule for penalty points. If you assess that the unbracketed form is true with probability p

unbracketed statement	penalty
true	$100 \times (1-p)^2$
false	$100 \times p^2$

Notice that assessing a probability as 0.5 is likely to attract far fewer penalty points than guessing between 0 and 1. Now that you know the scoring rule, assess the probabilities for the statements in the second set.

E.1.3 Question sets

Questions A

1. There were 13 (15) spacecraft launched as part of the Pioneer program.

2. Daniel Hughes (Heinrich Herz) demonstrated a radio transmission in 1879.

3. The ALGOL programming language was developed before (after) FORTRAN.

4. The boiling point of butane (pentane) is below zero degrees Celsius.

5. The model T Ford was produced by Henry Ford for 19 (20) years.

6. Santiago is west (east) of New York City.

7. Alaska extends (does not extend) south of Moscow.

8. The state of Colorado is larger (smaller) in area than the state of Arizona.

9. The modern viola has 5 (4) strings.

10. Hoover (Spangler) patented his rotating brush design electric powered vacuum cleaner in 1908.

11. Lloyd Hopkins is a detective in a novel by Raymond Chandler (James Ellroy).

12. William Gascoigne invented the micrometer in the 16th (17th) century.

13. The artist Jackson Pollock was born in 1912 (1923).

14. The first De Haviland Comet had oval (square) windows.

15. The musician Aaron Copeland was born in 1896 (1900).

16. The Cha Cha Cha comes from Brazil (Cuba).

Questions B

1. Sojourner was the Mars rover landed during the Pathfinder (Viking) mission.

2. Rome is south(north) of Washington DC.

3. The Wright brothers 17 December 1903 flight was within a km of Kitty Hawk (Kill Devil Hills).

4. The model T Ford was produced by Henry Ford for 19 (20) years.

5. The siemens is the unit of electrical conductance that was previously known as the (mho) moh.

6. The Grand canyon is in the state of Arizona (Colorado).

7. The term Googol, which refers to the value 10^{100} was coined by Milton Sirotta (Princeton Engineering Anomalies Research Lab).

8. The Sator Square, one of the worlds oldest examples of a palindrome "Sator Arepo Tenet Opera Rotas" dates back to before 79 BC (AD).

9. Refrigerating rubber bands extends (shortens) their lifetimes.

10. Coal fired electricity generating plants carry less (more) radiation into the environment per kWh than nuclear plants.

11. John Calvin Coolidge was the 31st (30th) president of the United States.

12. A trumpet has three (four) valves.

13. The IBM PC was first announced on the 12th of August 1981 (1980).

14. Boeing's first jet airliner, the 707, which first went into production in 1958 only ceased production in 1975 (1979).

15. The Elephant of Celebes is a painting by Jacob Epstein (Max Ernst).

16. The musician Charles Ives was born in 1874 (1878).

E.1.4 Discussion

Q1. Did you do better once you were told the rule?

Q2. Did the class as a whole do better once they were told the rule?

Q3. What proportion of the class answered set A first?

Q4. Did the class as a whole tend to do better with Table A or Table B? Can this be reasonably attributed to chance or were the statements less obscure on one of the tables?

E.1.5 Follow up questions

1. The minimum penalty score is achieved if an event with probability p of occurring is assigned a probability p. Show this by completing the following argument. Suppose you are given a large set of alternate form questions and you have no idea about the truth of the unbracketed form of any of them. However, you are told that a proportion p of the unbracketed forms is true. Write down an expression for your penalty score if you assign a probability θ to the unbracketed forms. Now minimize your penalty with respect to θ.

2. Suppose that the probability that the unbracketed form is true is p. Suppose you set up a bet such that you receive $b - (1 - x)^2$ if the statement is true and $b - x^2$ if it is false. Show that if $x = p$ and $b = p - p^2$ then the bet is fair.

E.2 Buffon's needle

E.2.1 Objectives

- To verify that relative frequencies tend towards a probability as the sample size increases.

- To estimate π from a sampling experiment using a relationship based on geometric probability.

- To compare estimators for π.

E.2.2 Experiment

The apparatus is a cocktail stick and an A3 sheet of paper with parallel lines drawn on it. The spacing of the parallel lines is equal to the length of the cocktail stick. Work as a pair.

(a) One person should rotate the A3 sheet through an arbitrary angle, before the other haphazardly drops the cocktail stick onto the sheet. Note whether or not the stick crosses a line. Repeat 10 times in all. Record the number of occasions out of 10 that the stick crosses a line (x_1). You can now calculate the proportion $(x_1/10)$ of occasions that the stick crossed the line.

(b) Now change roles and drop the stick another 10 times. How many times did it cross the line (x_2)? Combine the results from (a) and (b) and calculate the proportion $((x_1 + x_2)/20)$ of times the stick has crossed the line.

(c) Repeat (b) eight times to obtain proportions out of $30, 40, \ldots, 100$ drops. Call the proportion of times it crossed the line in 100 drops $\widehat{p}$.

(d) Sketch a plot of the overall proportion of times the stick crossed the line against the number of times it was dropped (from 10 to 100 in steps of 10).

(e) Estimate π, as $2/\widehat{p}$ and call this estimate $\widehat{\pi}$.

(f) Record your $\widehat{\pi}$ and $\widehat{\pi}$ on the whiteboard, as leaves in the stem-and leaf plots. Leaf units should be 0.01 for $\widehat{p}$ and 0.1 for $\widehat{\pi}$ respectively. Stem values of: 0.4, 0.4, 0.5, 0.5, 0.6, 0.6, 0.7, 0.7 for $\widehat{p}$, and 2, 2, 2, 2, 2, 3, 3, 3, 3, 3, 4, 4, 4, 4, 4 for $\widehat{\pi}$ should suffice.

(g) Calculate the mean and median values of the $\widehat{\pi}$.

(h) Calculate the mean of all the $\widehat{p}$ and call this $\bar{p}$. Now estimate π by $2/\bar{p}$.

E.2.3 Questions

1. In general, which of the estimators of that we have considered in (g) and (h) would you expect to be the more reliable?

2. Probability relationship: Let the stick have length $2d$. Let x be the distance from the mid-point of the stick to the nearest line, measured along a perpendicular to that line. Let θ be the acute angle the stick makes with a parallel to the line. Then x must lie between 0 and d, and θ must lie between 0 and $\pi/2$. Assume x and θ are uniformly distributed over their domains. The continuous sample space can be represented by a rectangle of base $\pi/2$ and height d. Then θ varies along the base of the rectangle and x varies up its height. What area within this rectangle corresponds to the stick crossing the line? Hence prove the result used in (e).

E.2.4 Computer simulation

Try these sites.

    `http://www.mste.uiuc.edu/reese/buffon/bufjava.html`

    `http://www.angelfire.com/wa/hurben/buff.html`

If you type Buffon's needle in a search engine you will find many more.

E.2.5 Historical note

Buffon's needle is one of the oldest problems in the field of geometric probability.

The problem was first stated, and solved, in 1777 by Georges-Louis Leclerc (1707-88), Comte de Buffon, in his work Essai d'Arithmetique Morale. The original problem was posed in terms of needles and planked floors, but otherwise related to the same experiment as this practical.

E.3 Robot rabbit

E.3.1 Objectives

- To have experience of using a stream of random numbers.

- To understand the concept of the sampling distribution of estimators.

- To anticipate a theoretical result for the standard deviation of a sample mean.

- To understand the benefits of stratification.

E.3.2 Experiment

An engineer in a large civil engineering company has been asked to assess the possible benefits of the RobotRabbit system, an electronically controlled tunneling machine, for laying water pipes without digging trenches. Benefits will depend on the length of pipe laid. The company is responsible for water pipes in 64 zones. The engineer has resources to make a detailed survey of the lengths of pipe that are likely to need replacing within a six year period, using a robotic CCTV pipe inspection system, in a sample of 5 zones.

Read about all three sampling strategies (A, B, C), before you are assigned one of them.

A Take a simple random sample (SRS) of size 5 from the 64 zones. Sample without replacement, i.e. continue until you have numbers corresponding to 5 different zones. Identify the 5 zones. The bracketed figures are the length of pipes needing replacement in the zones. Pretend you can only see these for the 5 selected zones. Now calculate (i) the mean, and (ii) the median of your sample.

B Take an SRS of 4 from the 48 small zones, and an SRS of 1 from the large zones. Sample without replacement, i.e. continue until you have numbers corresponding to 4 different small zones. Identify the 5 zones. The bracketed figures are the length of pipes needing replacement in the zones. Pretend you can only see these for the 5 selected zones. Estimate the population mean by:

$$\frac{48 \times (\text{mean length for 4 small zones}) + 16 \times (\text{length for 1 large zone})}{64}.$$

C Take an SRS of 3 from the 48 small zones, and an SRS of 2 from the large zones. Sample without replacement, i.e. continue until you have numbers corresponding to 3 different small zones, and 2 different large zones. Identify the 5 zones. The bracketed figures are the length of pipes needing replacement in the zones. Pretend you can only see these for the 5 selected zones. Estimate the population mean by:

$$\frac{48 \times (\text{mean length for 3 small zones}) + 16 \times (\text{mean length for 2 large zones})}{64}.$$

You are given a sheet representing these 64 zones, 48 are classed as small and 16 are classed as large, and an excerpt from a long stream of computer generated pseudo-random digits. The stream of random numbers is long enough for everyone to have different pieces from a copy of it. Therefore, your results will be independent of others in the class. Sample without replacement when drawing a sample of 5 zones, but notice that each sample of 5 is a separate simulation, and you may have the same zones occurring in separate simulations. It is only within a sample of 5 that all zones must be different

1. **Strategy A.** Use your piece of the random number table to draw a random sample of 5, without replacement, according to your given strategy.
   $01 - 64$ correspond to zones $01 - 64$ (Ignore numbers $65 - 00$)
   Strategies B,C.
   When you sample from the stratum of small zones, let:

01 − 48 correspond to zones 01 − 48, and

51 − 98 correspond to zones 01 − 48

Then you would only have to ignore numbers 49, 50, 99, and 00. This is quicker than waiting for numbers in the range 01 − 48.

When you sample from the stratum of large zones, let:

09 − 24 correspond to zones 49 − 64, and

29 − 44 correspond to zones 49 − 64, and

49 − 64 correspond to zones 49 − 64, and

69 − 84 correspond to zones 49 − 64.

Then you would only have to ignore numbers 00, . . . , 08, 25, . . . , 85, . . . , 99. This is quicker than waiting for numbers in the range 49−64, only, and doesn?t involve much arithmetic.

Calculate and keep a record of the mean and median (A) or weighted means (B,C).

2. Repeat 1. as many times as you can within about 15 minutes Record your statistics in the appropriate bins on the blackboard using tally marks (e.g. ⊬ ||| for a frequency of 8).

3. Use the bin mid-points and frequencies to calculate the approximate mean and standard deviation of the column of estimates.

E.3.3   Data

There are 64 zones as shown in Table 14.8. Those numbered 1–48 are classed as small and those numbered 49–64 are classed as large. The numbers in brackets are the lengths of pipes in the zones. A detailed survey will uncover the bracketed lengths, so each simulated sample of 5 will lead to 5 such lengths.

The data in Table 14.7 will help you answer questions 1,2 and 3. They are not known to the engineer.

TABLE 14.7: Descriptive stats for Robot Rabbit.

Zones	Number	Mean	Standard deviation
small	48	372.6	93.3
large	16	715.1	204.4
all	64	458.2	196.9

E.3.4   Discussion

Simulation studies don't provide proofs, but they often provide a useful indication of the best policy. It is possible to repeat the sampling strategy many thousands of times within a few minutes on a PC.

Q1. If the engineer is restricted to one sample of size 5, would you recommend estimating the total length of pipe to be replaced by, 64 times: the mean in strategy A, the median in strategy A, the weighted mean in strategy B, or the weighted mean in strategy C? If you use strategy A, would you use the mean or median?

Q2. If you selected zones for strategy A by throwing a hypothetical random dart (equally likely to land anywhere on the sheet) at the data sheet, would you obtain a SRS? If not, why not?

TABLE 14.8: Robot Rabbit.

01 (442)	02 (199)	03 (512)	04 (387)	05 (373)	06 (341)	07 (417)	08 (396)
09 (330)	10 (218)	11 (396)	12 (551)	13 (458)	14 (336)	15 (395)	16 (299)
17 (324)	18 (392)	19 (446)	20 (496)	21 (374)	22 (303)	23 (400)	24 (342)
25 (280)	26 (294)	27 (281)	28 (384)	29 (320)	30 (442)	31 (319)	32 (495)
33 (321)	34 (427)	35 (308)	36 (435)	37 (393)	38 (623)	39 (430)	40 (299)
41 (69)	42 (398)	43 (344)	44 (491)	45 (311)	46 (405)	47 (352)	48 (335)
49 (792)	50 (447)	51 (1014)	52 (552)	53 (511)	54 (595)	55 (909)	56 (702)
57 (698)	58 (638)	59 (896)	60 (305)	61 (848)	62 (1030)	63 (830)	64 (674)

Q3. For strategy A, calculate the ratio: (variance of means)/(variance of individuals in the population) and comment.

Q4. Think of some reasons why simulation studies don't provide proofs.

E.3.5   Follow up question

What is the optimal allocation of a sample of n over two strata of known sizes and known standard deviations if the objective is to estimate the population total?

E.4   Use your braking brains

E.4.1   Objectives

- To use a reaction ruler to compare your reaction times with a standard.

- To use a reaction ruler to compare your reaction times with your neighbor's reaction times.

- To have experience of using a running a paired comparison experiment.

- To construct and interpret confidence intervals.

E.4.2   Experiment

1. Measuring reaction times.

   You need to work in pairs. To obtain a measurement of your reaction time, ask your neighbor to hold the reaction ruler so that 0 is between the thumb and forefinger of the hand you write with. Your thumb and forefinger should be about 50 mm apart. Your neighbor should then, without warning, let go of the ruler. You catch the ruler between your thumb and forefinger and read off your reaction time in 0.01 s.

   Practice the measurement technique a few times.

2. Paired comparison experiment.

   Compare your reaction times with those of your neighbor over 10 trials. People may improve over the trials so we analyze the differences between your reaction time and your neighbor's time for each trial. In this way any systematic tendency for people to improve with practice should be eliminated. Also, going second might confer an advantage because you know the reaction time you have to beat. Therefore, flip a coin before each trial, to see who goes first, subject to a restriction that you each go first for 5 out of 10 trials. This restricted randomization balances the possible advantage of being second.

3. Graph the data.

   Plot your reaction times and your neighbor's times, using different symbols, against trial number.

E.4.3   Discussion

Q1. The Royal Society for the Prevention of Accidents (RoSPA) in the UK states that 0.18 seconds is good. Let $\bar{x}$ and s be the mean and standard deviation calculated from your sample of n, which is 10, reaction times. Construct a 90% confidence interval for the mean reaction time in the corresponding population of all possible reaction times (μ) as:

$$\bar{x} \quad \pm \quad 1.833 \, \frac{s}{\sqrt{n}}.$$

Do you have evidence that your mean reaction time is better or worse than 0.18? (1.833 is the upper 5% quantile of Student's t-distribution with $(10-1)$ degrees of freedom.)

Q2. Assume your reaction times are normally distributed with a mean and standard deviation equal to the mean and standard deviation calculated from your sample. Hence estimate the probability that an individual reaction time exceeds 0.18.

Q3. Calculate the difference between your time and your neighbor's (your time minus neighbor's time) for each trial. This will give you 10 differences $\{d_i\}$. Calculate the mean $(\bar{d})$ and standard deviation (s_d) of the 10 differences from your sample. Construct a 90% confidence interval for the mean in the corresponding population of all differences (μ_d) as:

$$\bar{d} \quad \pm \quad 1.833 \, \frac{s_d}{\sqrt{n}}.$$

Is there any evidence that you, or your neighbor, has the faster mean reaction times?[10]

E.5 Predicting descent time from payload

E.5.1 Objectives

- To use a regression analysis to predict the descent time of a helicopter from the number of paper clips representing the payload.

- To carry out a regression analysis using the basic formulae.

- To compare the rough prediction interval with an accurate one.

- To understand the R^2 statistic.

- To compare the regression model with a simple differential equation model for the helicopter's descent.

E.5.2 Experiment

1. Measuring descent times.

   You need to work in small groups, of at least two, depending on the number of stop watches. Make a paper helicopter, using the diagram provided (website) or your own design if you prefer. The aim is to relate the time taken to reach the ground to the number of paper clips, in the range 1–5. Have a few practice flights and standardize the procedure. Then time 3 descents with 1 up to 5 paper clips

   This will give you 15 pairs, (n_i, y_i), where n_i is the number of paper clips and y_i is the corresponding descent time.

---

[10]If you don't have a reaction ruler, you can use an ordinary ruler. The reaction time in 0.01s is $\sqrt{2d}$ where d is the distance along the ruler in mm.

2. Analysis

The descent time is y. At least one person in each group should carry out an analysis with x taken as n, and at least one person should carry out an analysis with x equal to $\frac{1}{n_0+n}$, where n_0 is the weight of the helicopter in paper clips ($n_0 = 1$ is reasonable).

- Plot y (vertical) against x (horizontal). There should be at least 15 points. Label axes.

- Fit the regression line:

$$y = \widehat{\alpha} + \widehat{\beta}x.$$

E.5.3 Discussion

Which of the two regression models fits the data better? Ideally all the class results can be displayed using the Excel spreadsheet (website).

E.5.4 Follow up question

Take the equation of motion of your helicopter as:

$$m\ddot{y} = -mg - k\dot{y},$$

where y is vertical displacement, with away from the center of the earth taken as positive, m is the mass of the helicopter plus payload, g is gravitational acceleration on earth, and k is a constant.

i. Explain the physical significance of this equation.

ii. What is the velocity in the steady state. That is, when $\ddot{y} = 0$? How would the time taken for a long descent depend on m?

iii. Do you think this is relevant to the analysis of your experiment?

iv. Can you estimate k? (A paper clip has a mass of approximately 0.5 gram and g is approximately 9.8 ms^{-2}.)

E.6   Company efficiency, resources and teamwork

E.6.1   Objectives

- To investigate the association between resources, teamwork, and efficiency in companies.

- To compare single predictor and two predictor regressions.

E.6.2   Experiment

(groups of 5 or more)

Answer the following questions, for a company in which you have worked, or some other project team, if you haven't had experience of working in a company. Use the scale 0, 1, 2, 3, 4, 5 corresponding to whether

never, rarely, sometimes, usually, often, always,

applies.

The questions are in three categories: Efficiency (E); Resource (R); and Teamwork (T); as detailed below:

E1) Work completed within budget.

R1) Allocation of resources meets needs of departments.

E2) High quality work is achieved.

E3) Work completed on time.

T1) Departments work well together.

R2) Appropriate technology used.

R3) Available resources are well used

E4) Company has a good reputation.

T2) Groups get together and work on common problems.

T3) Meetings are effective.

T4) Good communication within the organization.

R4) Marketing is effective.

E5) company has an impressive record for innovation..

T5) Mistakes are corrected without searching for someone to blame.

R5) Staff training provided.

Add your marks for the five questions in each of the categories, and call these totals R, T and y respectively.

Collect together the results from the n people in your group. You should now have n data triples:

$$(R_i, T_i, y_i), \quad \text{for} \quad i = 1, \ldots, n.$$

Define:

$$x_{1i} = R_r - \bar{R}, \quad \text{where} \quad \bar{R} = \sum \frac{R_i}{n},$$

$$x_{2i} = T_i - \bar{T}, \quad \text{where} \quad \bar{T} = \sum \frac{T_i}{n}.$$

(a) Calculate the correlations between x_1 and x_2.

(b) Fit the regression of y on x_1.

(c) Fit the regression of y on x_2.

(d) Fit the regression of y on x_1 and x_2. Are the coefficients of x_1 and x_2 in (c) the same as in (a) and (b)?

Note

The formula for the regression coefficients is

$$\hat{B} = (X'X)^{-1}X'Y.$$

When you mean correct the predictor variables (i.e. subtract their mean so that the mean of the scaled variable is 0) the arithmetic is relatively easy. Even in the multiple regression case the matrix inversion is straightforward, because it partitions so that you only need remember the result for inverting a 2×2 matrix.

E.6.3 Discussion

Interpret the results of the regression analyses.

E.7 Factorial experiment – reaction times by distraction, dexterity and distinctness

E.7.1 Aim

To investigate the effect of left/right hand, distraction, and visibility on reaction times.

E.7.2 Experiment

Three factors A, B, C defined as:

A left hand $(x_1 = -1)$ right hand $(x_1 = +1)$;

B both eyes open: no(close one eye) $(x_2 = -1)$ yes(both eyes open) $(x_2 = +1)$; and

C standing on both legs: no (stand on one leg) $(x_3 = -1)$ yes (stand on both legs) $(x_3 = +1)$.

Alternate who catches the ruler. Each person should catch the ruler on 16 occasions according to two replicates of a 2^3 factorial experiment. The 8 runs in each replicate should be in random order.

Standard order	A x_1	B x_2	C x_3	Design point	Random order Rep 1	Random order Rep 2	Reaction time Rep 1 (0.01 s)	Reaction time Rep 2 (0.01 s)
1	−1	−1	−1	1				
2	+1	−1	−1	a				
3	−1	+1	−1	b				
4	+1	+1	−1	ab				
5	−1	−1	+1	c				
7	+1	−1	+1	ac				
8	−1	+1	+1	bc				
9	+1	+1	+1	abc				

E.7.3 Analysis

Define an additional variable x_{4i} as −1 for the first replicate and +1 for the second replicate. Let y_i be the reaction time. Fit the model:

$$Y_i = \beta_0 + \beta_1 x_{1i} + \beta_2 x_{2i} + \beta_3 x_{3i} + \beta_4 x_{1i}x_{2i} + \beta_5 x_{1i}x_{3i} + \beta_6 x_{2i}x_{3i} + \beta_7 x_{4i} + \varepsilon_i,$$

where we assume ε_i are independently distributed with a mean of 0 and a constant variance σ^2. To estimate the coefficients from the 16 y_i all you need calculate are:

$\widehat{\beta}_0$ = overall mean

$\widehat{\beta}_1$ = [(mean of $8y$ with $x_1 = +1$)−(mean of $8y$ with $x_1 = -1$)]/2

$\widehat{\beta}_2$ and $\widehat{\beta}_3$ similarly

$\widehat{\beta}_4$ = [(mean of $8y$ with $x_1x_2 = +1$)−(mean of $8y$ with $x_1x_2 = -1$)]/2

$\widehat{\beta}_5$ and $\widehat{\beta}_6$ similarly

$\widehat{\beta}_7$ = [(mean replicate 1)] − [(mean replicate 2)]/2

The fitted values are calculated as:

$$\widehat{y}_i = \widehat{\beta}_0 + \widehat{\beta}_1 x_{1i} + \widehat{\beta}_2 x_{2i} + \widehat{\beta}_3 x_{3i} + \widehat{\beta}_4 x_{1i}x_{2i} + \widehat{\beta}_5 x_{1i}x_{3i} + \widehat{\beta}_6 x_{2i}x_{3i} + \widehat{\beta}_7 x_{4i}$$

for $i = 1, 2, \cdots, 16$.

The residuals are $r_i = y_i - \widehat{y}_i$, and the estimated standard deviation of the errors is given by:

$$s = \sqrt{\frac{\sum r_i^2}{16 - 8}}.$$

90% confidence levels for β_j are constructed from:

$$\beta_j \pm t_{8,0.05}\, s\, \sqrt{\left(\frac{1}{8} + \frac{1}{8}\right)/2^2}, \quad \text{where } t_{8,0.05} = 1.860.$$

E.7.4 Discussion

A reaction time of 0.18 s would be reasonable for driving, 0.16 s would be excellent whereas 0.20 s would be too slow (after Royal Society for Prevention of Accidents).

If a confidence interval is all positive, or all negative, you can be fairly confident that the corresponding factor, or interaction between factors, has an effect on your reaction times (today, at least).

You can combine your results with those of the other person. You could redefine x_4 to distinguish persons. The experiment would be two replicates of 2^4.

E.7.5 Follow up questions

The general equation for estimating the coefficients in a multiple regression $(X'X)^{-1}X'Y$ simplifies and that simplification is a consequence of the balanced design. The covariance between any pair from the 7 predictor variables are all 0. Also the covariance between the 3-factor interaction $x_1x_2x_3$ and any one of the 7 predictor variables used in the model is also 0.

Q1 The 3-factor interaction could be included in the model. How would its coefficient be estimated?

Q2 Consider a single replicate of a 2^2 design in the shorthand notation. Show that when fitting the regression, $Y = \beta_0 + \beta_1 A + \beta_2 B + \beta_3 AB$, $(X'X)^{-1}X'Y$ reduces to:

$$\widehat{\beta}_0 = \frac{1+a+b+ab}{4}, \ \widehat{\beta}_1 = \frac{1}{2}\left(\frac{a+ab}{2} - \frac{1+b}{2}\right),$$
$$\widehat{\beta}_2 = \frac{1}{2}\left(\frac{b+ab}{2} - \frac{1+a}{2}\right), \ \widehat{\beta}_3 = \frac{1}{2}\left(\frac{1+ab}{2} - \frac{a+b}{2}\right)$$

E.8   Weibull analysis of cycles to failure

E.8.1   Aim

To use a plotting method to assess the suitability of a Weibull distribution for modeling lifetime data.
To estimate the parameters of the Weibull distribution from the graph.

E.8.2   Experiment

Bend a paper clip through ninety degrees, and then bend it back to its original flat state. Repeat, noting how many times you have bent it, until it breaks.

Continue with others in your group until you have results for 20 paper clips.

E.8.3   Weibull plot

Sort the data into ascending order: $x_{i:n}$. Record the results.

i	$x_{i:n}$	$\ln(x_{i:n})$	$p_i = \dfrac{i}{n+1}$	$\ln\left(-\ln(1-p_i)\right)$
1				
2				
3				
4				
$\vdots$	$\vdots$			

If several of you recorded the same number of bends until the paper clip breaks, there will be several rows with the samevalue of $(x_{i:n})$.

Plot $\ln(x_{i:n})$ against $\ln\left(-\ln(1 - p_i)\right)$. Draw a plausible line through the data and hence estimate the parameters of the Weibull distribution.

Now estimate the upper and lower 1% quantiles of the assumed distribution of cycles to failure of paper clips. Also estimate the probability that the clip fails on or before the 3rd cycle. Since the number of cycles is a discrete variable, and the argument of the Weibull distribution is continuous, we should apply a continuity correction. In this case you should calculate $F(3.5)$. *(In terms of the pdf, the probability of exactly 3 is approximated by the area under the curve between 2.5 and 3.5, and the probability of less than or equal to 3 is given by the area up to 3.5.)*

E.8.4 Discussion

You have n data, which are cycles to failure for a random sample of paper clips. Sort the data into ascending order, and denote this sequence by:

$$x_{i:n} \quad \text{for} \quad i = 1, \ldots, n,$$

where $x_{i:n}$ is the ith smallest out of n. Plot $\ln(x_{i:n})$ against $\ln(-\ln(1-p_i))$, where $p_i = \frac{i}{n+1}$. If the data appear to be scattered about a straight line, rather than some curve, the Weibull distribution may be a suitable model for the lifetimes. But, if the sample size is small, we can't be very sure about this.

The Weibull cdf is:

$$F(x) \;=\; 1 - \mathrm{e}^{-\left(\frac{x}{\beta}\right)^\alpha} \quad \text{for} \quad x \geq 0.$$

If you draw a line through the data, the slope of the line estimates $\left(\frac{1}{\alpha}\right)$ and the intercept estimates $\ln(\beta)$. It follows that:

$$\widehat{\alpha} \;=\; \frac{1}{\text{slope}} \quad \text{and} \quad \widehat{\beta} \;=\; \mathrm{e}^{\text{intercept}}.$$

You are just expected to draw any plausible line through the points. When you do so, try and place less emphasis on the largest values. Alternatively you can fit a regression line.

(Although it may be convenient to use a regression routine, it does not give an ideal solution because var$(i:n)$ *is not constant, and becomes much larger when i approaches n. Ideally, we would allow for this by using generalized least squares, which includes a weighting matrix. However this is rarely done because maximum likelihood estimators have greater precision.)*

E.9 Control or tamper?

Preliminary reading is the section of Chapter 11 "Experiment with Red Beads to Show Total Fault in the System" page 346 onwards in *Out of the Crisis* by W. Edwards Deming, Massachusetts Institute of Technology [Deming, 1986].

1. A manufacturer of a gold solution dispenses it into bottles. The target fill is 1000 ml. The observed fill is O_i and this is the sum of the underlying mean μ_i and an error ϵ_i or

$$O_i \;=\; \mu_i + \epsilon_i.$$

Here we assume the errors are independent from $N(0, 5^2)$.

Define y as the deviation of the observed fill from target.

$$y_i \;=\; O_i - 1\,000.$$

The underlying mean fill can be adjusted by turning a handle over a dial which is accurately graduated in ml from a center point which is marked as $1\,000$, although the operators are skeptical about the accuracy of this. Turning the wheel to the right increases the mean fill and turning it to the left decreases the mean fill.

You are given a sheet of random numbers from $N(0, 5^2)$. Take an arbitrary starting point, **the same start for everyone in your group**, and assume the numbers are the errors ϵ_i. So, the observation will be process mean plus the random number.

Within your group, simulate the strategies of 4 process operators A, B, C, D for about 20 bottles. Note that within your group you will all have the **same sequence** of random numbers. In the following y is deviation from the target of 1 000 ml.

Finally, assume the process does start on target.

A. Take no control action.

B. Adjust the process mean by $-y$ from its present position.

C. Adjust the process mean by $-y$ from the target.

D. Adjust the process mean to the observed value.

For example, suppose the random numbers are $+8, +11, -3, \ldots$

Operator A bottle #(t)	Process mean at $t-$	Random number	Observed fill	y	adjustment	Process mean at $t+$
1	1000	8	1008		0	1000
2	1000	11	1011		0	1000
3	1000	-3	997		0	1000

Operator B bottle #(t)	Process mean at $t-$	Random number	Observed fill	y	adjustment	Process mean at $t+$
1	1000	8	1008	8	-8	992
2	992	11	1003	3	-3	989
3	989	-3	986	-14	14	1003

Operator C bottle #(t)	Process mean at $t-$	Random number	Observed fill	y	adjustment	Process mean at $t+$
1	1000	8	1008	8	1000-8	992
2	992	11	1003	3	1000-3	997
3	997	-3	994	-6	1000+6	1006

Operator D bottle #(t)	Process mean at $t-$	Random number	Observed fill	y	adjustment	Process mean at $t+$
1	1000	8	1008		observed	1008
2	1008	11	1019		observed	1019
3	1019	-3	1016		observed	1016

Calculate the mean, standard deviation, and root mean squared error (RMSE) of the observed fills and comment. Why might the operators B, C, and D think their actions helpful?

2. A manufacturer has fitted a laser displacement probe to measure the depth of a solder layer on each PC processor board as it passes. Table 14.9 gives the deviations (microns) from the target of τ, for 49 boards when the process mean was fixed at τ.

The mean, standard deviation and RMSE of these data are: 19.8, 12.5 and 23.3.

Now suppose the process mean could have been adjusted, and compare the effects of applying the strategies of Operator A and Operator B.

TABLE 14.9: Deviations - read along rows for the time order.

15	11	-5	0	6	-4	12	24	22	24
24	20	27	18	39	33	31	38	34	29
18	20	6	15	31	21	28	38	18	3
11	8	16	24	29	39	41	48	29	26
22	5	9	0	17	9	15	17	8	

Operator A's strategy is to leave the process mean fixed at τ, so the mean, standard deviation and RMSE of the boards remain at: $19.8, 12.5$ and 23.3.

Remember that the deviations in Table 14.9 are not only deviations from target but also the deviations from the process mean which equaled the target. To perform the simulation for Operator B, you need to add the deviations in Table 14.9 to the adjusted process mean. This will give the observed depth. The simulation starts as follows.

Operator B board #(t)	Process mean at $t-$	Deviation from mean at $t-$	Observation	Deviation from target (y)	adjustment	Process mean at $t+$
1	τ	15	$\tau+15$	15	-15	τ-15
2	τ-15	11	τ-4	-4	4	τ-11
3	τ-11	-5	τ-16	-16	16	$\tau+5$
4	$\tau+5$	0	$\tau+5$	5	-5	$\tau+0$

Continue the simulation of B's strategy and calculate the mean, standard deviation and RMSE of the observed depths. You might also plot the observations, relative to target, using different symbols for Operator A and Operator B.

If there are several of you, you could vary the experiment by adjusting by $-\theta y$, rather than $-y$, where θ is between 0 and 1 e.g. $0.3, 0.5, 0.7$.

E.10   Where is the summit?

The objective is to find the combination of temperature and pressure that maximizes the yield of a chemical process.

Work in groups of about 4.

One person (Person C) takes a square piece of graph paper with axes ranging from -10 to $+10$ representing temperature and pressure in coded units. Draw imaginary contours of yield - the only restrictions are that there is a summit yield of 60% and that the lowest contour represents 10%. The contours don't need to be elliptical in shape and you can have more than one maximum and minimum provided there is a unique point corresponding to 60% and no minimum below 10%.

Other group members do not see the contours, but they are told that the permissible values for temperature and pressure are between -10 and $+10$ in coded units. They each ask C about the yield at one particular point, in order. Members of the group know the point that each other member has chosen.

Then C estimates the yield at these points from the contour plot, adds a random deviate from a normal distribution with mean 0 and standard deviation 4%, rounded to 0.1%, (a list of random deviates can be produced in advance) and returns the yields to the group members individually on a slip of paper.

This continues for about 20 rounds. Who is closest to the point that corresponds to the summit?

Note: Twenty rounds allows for three 2 by 2 factorial designs and one central composite design, or 3 by 3 factorial. Rough assessments of the direction of steepest ascent will suffice, though participants could use laptop computers. If the group size is five or more, then C might be two people. See the article "Blindfold Climbers" by Tony Greenfield [Greenfield, 1979] for a variation on this experiment.

References

[Adamson et al., 1999] Adamson, P., Metcalfe, A., and Parmentier, B. (1999). Bivariate extreme value distributions: an application of the gibbs sampler to the analysis of floods. *Water Resources Research*, 35(9):2825–2832.

[Aljanahi et al., 1999] Aljanahi, A., Rhodes, A., and Metcalfe, A. (1999). Speed, speed limits and road traffic accidents under free flow conditions. *Accident Analysis & Prevention*, 31(1):161–168.

[Bajpai et al., 1968] Bajpai, A., Calus, I., and Fairley, J. (1968). *Statistical Methods for Engineers and Scientists*. Wiley.

[Bass, 1969] Bass, F. M. (1969). A new product growth for model consumer durables. *Management Science*, 15(5):215–227.

[Bayes, 1763] Bayes, T. (1763). An essay towards solving a problem in the doctrine of chances. *Phil. Trans. of the Royal Soc. of London*, 53:370–418.

[Bull, 1988] Bull, J. W. (1988). The design of footway paving flags. *The Transportation Research Board, The National Academies of Sciences, Engineering, and Medicine*, 56(1936):44–45.

[Calisal et al., 1997] Calisal, S. M., Howard, D., and Mikkelsen, J. (1997). A seakeeping study of the UBC series. *Marine Technology and SNAME News*, 34(1):10.

[Caulcutt, 1999] Caulcutt, R. (1999). From measurement to action. In *Proceedings of the 2nd International Conference on the Control of Industrial Processes*, pages 149–155, University of Newcastle upon Tyne, UK.

[Cayla et al., 2011] Cayla, J., Maizi, N., and Marchand, C. (2011). The role of income in energy consumption behaviour: Evidence from french households data. *Energy Policy*, 39(12):7874–7883.

[Clarke and Kempson, 1994] Clarke, G. and Kempson, R. (1994). *Introduction to the Design and Analysis of Experiments*. Wiley.

[Colaizzi et al., 2004] Colaizzi, P., Schneider, A., Evett, S., and Howell, T. (2004). Comparison of SDI, LEPA, and spray irrigation performance for grain sorghum. *Transactions of the ASAE*, 47(5):1477.

[Dalal et al., 1989] Dalal, S. R., Fowlkes, E. B., and Hoadley, B. (1989). Risk analysis of the space shuttle: Pre-challenger prediction of failure. *Journal of the American Statistical Association*, 84(408):945–957.

[David, 1955] David, F. N. (1955). Studies in the History of Probability and Statistics I. Dicing and Gaming (A Note on the History of Probability). *Biometrika*, 42(1/2):pp. 1–15.

[Deming, 1986] Deming, W. E. (1986). *Out of the Crisis*. Cambridge University Press, illustrated, revised edition.

[Deming, 2000] Deming, W. E. (2000). *Out of the Crisis*. MIT Press. Paperback. Originally published by MIT-CAES in 1982.

[Dunn, 2013] Dunn, P. K. (2013). Comparing the lifetimes of two brands of batteries. *Journal of Statistics Education*, 21(1).

[Erlang, 1917] Erlang, A. (1917). Solution of some problems in the theory of probabilities of significance in automatic telephone exchanges. *Elektrotkeknikeren*, 13:5–13.

[Fisher et al., 2010] Fisher, A., Green, D., and Metcalfe, A. (2010). Managing river flow in arid regions with matrix analytic methods. *Journal of Hydrology*, 382:128–137.

[Fisher et al., 2014] Fisher, A., Green, D., Metcalfe, A., and Akande, K. (2014). First-passage time criteria for the operation of reservoirs. *Journal of Hydrology*, 519 (Part B):1836–1847.

[Fisher, 1921] Fisher, R. A. (1921). On the 'probable error' of a coefficient of correlation deduced from a small sample. *Metron*, 1:3–32.

[Forster, 2003] Forster, M. (2003). *Percolation: An Easy Example of Renormalization*. http://philosophy.wisc.edu/forster/Percolation.pdf.

[Frankel, 1998] Frankel, E. (1998). China's maritime developments. *Maritime Policy and Management*, 25(3):235–249.

[Green and Metcalfe, 2005] Green, D. and Metcalfe, A. (2005). Reliability of supply between production lines. *Stochastic Models*, 21:449–464.

[Greenfield, 1979] Greenfield, T. (1979). Blindfold climbers. *Teaching Statistics, An International Journal for Teachers*, 1(1):15–19.

[Hampson and Walker, 1960] Hampson, R. F. and Walker, R. F. (1960). Vapor pressures of platinum, iridium, and rhodium. *Journal of Research of the National Bureau of Standards*, 65A.

[Hampson Jr and Walker, 1961] Hampson Jr, R. and Walker, R. (1961). Vapor pressures of platinum, iridium, and rhodium. *Journal of Research of the National Bureau of Standards–A. Physics and Chemistry*, 65(4).

[Hayter, 2012] Hayter, A. (2012). *Probability and Statistics for Engineers and Scientists (4e)*. Brooks/Cole.

[Hesterberg, 2015] Hesterberg, T. C. (2015). What teachers should know about the bootstrap: Resampling in the undergraduate statistics curriculum. *The American Statistician*, 69(4):371–386.

[Hiebeler, 2010] Hiebeler, D. (2010). MATLAB/R Reference. `http://cran.r-project.org/doc/contrib/Hiebeler-matlabR.pdf`.

[Hsiue et al., 1995] Hsiue, L.-T., Ma, C.-C. M., and Tsai, H.-B. (1995). Preparation and characterizations of thermotropic copolyesters of p-hydroxybenzoic acid, sebacic acid, and hydroquinone. *Journal of applied polymer science*, 56(4):471–476.

[ISO, 2015a] ISO (2015a). Quality Management series. `http://www.iso.org/iso/home/standards/management-standards/iso_9000.htm`.

[ISO, 2015b] ISO (2015b). Quality Management series. `http://www.iso.org/iso/home/store/catalogue_tc/catalogue_detail.htm?csnumber=62405`.

[ISO, 2015c] ISO (2015c). Quality Management series. `http://www.iso.org/iso/iso14000`.

[Jackson et al., 1987] Jackson, K. W., Eastwood, I. W., and Wild, M. S. (1987). Stratified sampling protocol for monitoring trace metal concentrations in soil. *Soil Science*, 143(6):436–443.

[Kalkanis and Rosso, 1989] Kalkanis, G. and Rosso, E. (1989). The inverse power law model for the lifetime of a mylar-polyurethane laminated dc hv insulating structure. *Nuclear Instruments and Methods in Physics Research Section A: Accelerators, Spectrometers, Detectors and Associated Equipment*, 281(3):489–496.

[Kattan, 1993] Kattan, M. (1993). Statistical process control in ship production. *Quality Forum*, 19(2):88–92.

[Knuth, 1968] Knuth, D. E. (1968). *The Art of Computer Programming*. Addison-Wesley.

[Kroese et al., 2011] Kroese, D. P., Taimre, T., and Botev, Z. I. (2011). *Handbook of Monte Carlo Methods*. Wiley.

[Law and Kelton, 2000] Law, A. M. and Kelton, W. D. (2000). *Simulation Modeling and Analysis*. McGraw-Hill series in industrial engineering and management science. McGraw-Hill, New York.

[Lee, 2012] Lee, P. M. (2012). *Bayesian Statistics: A Introduction, 4th Edition*. WILEY.

[Lehman and Miikkulainen, 2015a] Lehman, J. and Miikkulainen, R. (2015a). Enhancing divergent search through extinction events. In *Proceedings of the Genetic and Evolutionary Computation Conference (GECCO 2015)*, Madrid, Spain.

[Lehman and Miikkulainen, 2015b] Lehman, J. and Miikkulainen, R. (2015b). Extinction Events Can Accelerate Evolution. *PLoS ONE*, 10(8).

[Lieblein and Zelen, 1956] Lieblein, J. and Zelen, M. (1956). Statistical investigation of the fatigue life of deep-groove ball bearings. *Journal of Research of the National Bureau of Standards*, 57(5):273–316.

[Lindley, 1985] Lindley, D. V. (1985). *Making Decisions*. Wiley, 2nd edition.

[Matsumoto and Nishimura, 1998] Matsumoto, M. and Nishimura, T. (1998). Mersenne Twister: a 623-Dimensionally Equidistributed Uniform Pseudo-Random Number Generator. *ACM Transactions on Modeling and Computer Simulation*, 8(1):3–30.

[McHale, 1977] McHale, E. T. (1977). Safety evaluation of distress flares and smokes. Unknown binding, National Technical Information Service: 29 pages.

[Meeker and Escobar, 2014] Meeker, W. Q. and Escobar, L. (2014). *Statistical Methods for Reliability Data*. Wiley & Sons.

[Metcalfe, 1991] Metcalfe, A. (1991). Probabilistic modelling in the water industry. *Journal of the Institute of Water and Environmental Management*, 4(5):439–449.

[Mocé-Llivina et al., 2005] Mocé-Llivina, L., Lucena, F., and Jofre, J. (2005). Enteroviruses and bacteriophages in bathing waters. *Applied and Environmental Microbiology*, 71(11):6838–6844.

[Montgomery, 2004] Montgomery, D. (2004). *Design and Analysis of Experiments*. Wiley & Sons.

[Moore, 1972] Moore, P. G. (1972). *Risk in Business Decision*. Longman.

[Moran, 1959] Moran, P. (1959). *The Theory of Storage*. Monographs on Applied Probability and Statistics. Methuen and Co. Ltd., London.

[Morris et al., 1997] Morris, V.-M., Hargreaves, C., Overall, K., Marriott, P.-J., and Hughes, J.-G. (1997). Optimisation of the capillary electrophoresis separation of ranitidine and related compounds. *Journal of Chromatography A*, 766:245–254.

[Myers et al., 2010] Myers, R. H., Montgomery, D. C., Vining, G. G., and Robinson, T. J. (2010). *Generalized Linear Models: with Applications in Engineering and the Sciences*. Wiley, 2nd edition.

[Nathan and Lowe, 2012] Nathan, R. and Lowe, L. (2012). The hydrologic impacts of farm dams. *Australasian Journal of Water Resources*, 16(1):75–83.

[N.D. Singpurwalla, 1975] N.D. Singpurwalla, R.E. Barlow, J. F., editor (1975). *Reliability and Fault Tree Analysis*. Society for Industrial and Applied Mathematics.

[Nemati, 2014] Nemati, M. (2014). An appraisal of aftershocks behavior for large earthquakes in persia. *Journal of Asian Earth Sciences*, 79:432–440.

[NIST website, 2017] NIST website (2017). Extreme Wind Speeds Software: Excel *Jacksonville, FL*. http://www.itl.nist.gov/div898/winds/excel.htm.

[Norman and Naveed, 1990] Norman, P. and Naveed, S. (1990). A comparison of expert system and human operator performance for cement kiln operation. *Journal of the Operational Research Society*, 41(11):1007 1019.

[Nuttli, 1973] Nuttli, O. W. (1973). Seismic wave attenuation and magnitude relations for eastern north america. *Journal of Geophysical Research*, 78(5):876–885.

[Ostertagová, 2012] Ostertagová, E. (2012). Modelling using polynomial regression. *Procedia Engineering*, 48:500–506.

[Pan and Chi, 1999] Pan, Y. and Chi, P. S. (1999). Financial performance and survival of multinational corporations in china. *Strategic Management Journal*, 20(4):359–374.

[Pawlikowski et al., 2002] Pawlikowski, K., Jeong, H.-D., and Lee, J.-S. R. (2002). On credibility of simulation studies of telecommunication networks. *IEEE Communications Magazine*, pages 132–139.

[Plackett, 1972] Plackett, R. L. (1972). Studies in the History of Probability and Statistics. XXIX The discovery of the method of least squares. *Biometrika*, 59(59/2):pp. 239–251.

[Ross, 2013] Ross, S. M. (2013). *Simulation*. Elsevier Academic Press, San Diego, CA, 5 edition.

[Sambridge et al., 2011] Sambridge, M., Tkalčić, H., and Arroucau, P. (2011). Benford's law of first digits: from mathematical curiosity to change detector. *Asia Pacific Mathematics Newsletter*, 1(4):1–6.

[Shewhart, 1939] Shewhart, W. A. (1939). *Statistical Method from the Viewpoint of Quality Control (1939)*. Dover.

[Short, 2004] Short, T. (2004). R Reference Card. `http://cran.r-project.org/doc/contrib/Short-refcard.pdf`.

[Stahl and Gagnon, 1995] Stahl, F. L. and Gagnon, C. P. (1995). *Cable Corrosion in Bridges and Other Structures: Causes and solutions*. American Society of Civil Engineers.

[StandardsUK.com, 2014] StandardsUK.com (2014). British Standards Online. In *British Standards Online*. `http://www.standardsuk.com/`.

[Stewardson and Coleman, 2001] Stewardson, D. and Coleman, S. (2001). Using the summed rank cusum for monitoring environmental data from industrial processes. *Journal of Applied Statistics*, 8(3–4):469–484.

[The Enterprise Initiative, 2015a] The Enterprise Initiative (2015a). Introducing Harold Slater - Part 1 . `https://www.youtube.com/watch?v=3ObUaQvxYw0`.

[The Enterprise Initiative, 2015b] The Enterprise Initiative (2015b). Introducing Harold Slater - Part 2 . `https://www.youtube.com/watch?v=JWXUMxF8xvY`.

[The Times, 2017] The Times (2017). Voyager Documentary is out of this World by Jennifer O'Brien. `https://www.thetimes.co.uk/article/voyager-documentary-is-out-of-this-world-nf9v28tft`.

[Vardeman, 1992] Vardeman, S. B. (1992). What about the other intervals? *The American Statistician*, 46(3):193–197.

[Vining, 1998] Vining, G. (1998). *Statistical Methods for Engineers*. Duxbury Press.

[Winderbaum et al., 2012] Winderbaum, L., Ciobanu, C. L., Cook, N. J., Paul, M., Metcalfe, A., and Gilbert, S. (2012). Multivariate Analysis of an LA-ICP-MS Trace Element Dataset for Pyrite. *Mathematical Geosciences*.

Index

9 780367 570620